Lecture Notes in Computer Science   2170

Edited by G. Goos, J. Hartmanis, and J. van Leeuwen

T0180340

Springer

Berlin
Heidelberg
New York
Barcelona
Hong Kong
London
Milan
Paris
Tokyo

Sergio Palazzo (Ed.)

Evolutionary Trends of the Internet

2001 Tyrrhenian International Workshop on
Digital Communications, IWDC 2001
Taormina, Italy, September 17-20, 2001
Proceedings

 Springer

Series Editors

Gerhard Goos, Karlsruhe University, Germany
Juris Hartmanis, Cornell University, NY, USA
Jan van Leeuwen, Utrecht University, The Netherlands

Volume Editor

Sergio Palazzo
Università di Catania, Dip. di Ingegneria Informatica e delle Telecomunicazioni
Viale A. Doria 6, 95125 Catania, Italy
E-mail: palazzo@diit.unict.it

Cataloging-in-Publication Data applied for

Die Deutsche Bibliothek - CIP-Einheitsaufnahme

Evolutionary trends of the internet : proceedings / 2001 Tyrrhenian
International Workshop on Digital Communications, IWDC 2001, Taormina,
Italy, September 17 - 20, 2001. Sergio Palazzo (ed.). - Berlin ; Heidelberg ;
New York ; Barcelona ; Hong Kong ; London ; Milan ; Paris ; Tokyo :
Springer, 2001
  (Lecture notes in computer science ; Vol. 2170)
  ISBN 3-540-42592-6

CR Subject Classification (1998): C.2, H.4, H.5.1, H.3

ISSN 0302-9743
ISBN 3-540-42592-6 Springer-Verlag Berlin Heidelberg New York

Springer-Verlag Berlin Heidelberg New York
a member of BertelsmannSpringer Science+Business Media GmbH

http://www.springer.de

© Springer-Verlag Berlin Heidelberg 2001
Printed in Germany

Typesetting: Camera-ready by author, data conversion by Steingräber Satztechnik GmbH, Heidelberg
Printed on acid-free paper   SPIN: 10840397   06/3142   5 4 3 2 1 0

Preface

In recent years the Internet has seen a tremendous growth in terms of amount of traffic and number of users. At present, the primary technical objective is to provide advanced IP networking services to support an evolving set of new applications like, for example, IP-telephony, video teleconferencing, Web TV, multimedia retrieval, remote access to and control of laboratory equipment and instruments. To achieve this objective, the Internet must move from a best-effort paradigm without service differentiation to one where specific Quality of Service (QoS) requirements can be met for different service classes in terms of bandwidth, delay, and delay jitter. Another crucial goal is the ubiquitous deployment of advanced services over the global network resulting from the effective integration of wireless and satellite segments into the future Internet.

Although many researchers and engineers around the world have been working on these challenging issues, several problems remain open.

The *Tyrrhenian International Workshop on Digital Communications 2001*, which focused on the *Evolutionary Trends of the Internet*, was conceived as a highly selective forum aimed at covering diverse aspects of the next generation of IP networks.

The members of the Technical Program Committee of the workshop concentrated their efforts on identifying a set of topics that, although far from being exhaustive, provide a sufficiently wide coverage of the current research challenges in the field.

Eight major areas were envisioned, namely *WDM Technologies for the Next Generation Internet, Mobile and Wireless Internet Access, QoS in the Next Generation Internet, Multicast and Routing in IP Networks, Multimedia Services over the Internet, Performance Modeling and Measurement of Internet Protocols, Dynamic Service Management, Source Encoding and Internet Applications.*

With the invaluable help of the session organizers, 46 papers, partly invited and partly selected on an open-call basis, were collected for presentation at the workshop and publication in this book. We believe that the contributions contained in these proceedings represent a timely and high-quality outlook on the state of the art of research in the field of multiservice IP networks, and we hope they may be of use for further investigation in this challenging area.

September 2001   Sergio Palazzo

Organization

General Chairman

S. Palazzo

Technical Session Organizers

A. Acampora
I. Chlamtac
M. Gerla

M. El Zarki
J. Kurose
J. Liebeherr

Technical Program Committee

M. Ajmone Marsan
G. Albertengo
G. Bianchi
N. Blefari Melazzi
A. Campbell
G. Corazza
F. Davoli
C. Diot
S. Fdida
L. Fratta
S. Giordano
D. Hutchison
G. Karlsson
E. Knightly
L. Lenzini

M. Listanti
A. Lombardo
S. Low
R. Melen
G. Morabito
F. Neri
G. Pacifici
A. Pattavina
G. Polyzos
C. Rosenberg
K. Ross
G. Paolo Rossi
A. Roveri
G. Schembra
T. Todd

Publications

G. Schembra

Acknowledgements

The editor is much indebted and wish to express his sincere thanks to all the components of the Technical Program Committee of the 2001 edition of the International Thyrrenian Workshop on Digital Communications, and especially to the Organizers of the Technical Sessions, namely *Anthony Acampora* from the University of California at San Diego, USA, *Imrich Chlamtac* from the University of Texas at Dallas, USA, *Mario Gerla* from the University of California Los Angeles (UCLA), USA, *Magda El Zarki* from the University of California at Irvine, USA, *Jim Kurose* from the University of Massachusetts at Amherst, USA, and *Jorg Liebeherr* from the University of Virginia, USA, whose precious cooperation was essential to the organization of the Workshop.

The editor also expresses his sincere appreciation to all the authors for contributing to the Workshop with their high quality papers.

The preparation of this volume has benefited from the hard work of many people. The editor would like to acknowledge the LNCS staff at Springer, *Alfred Hofmann* in particular, for making so many efforts to release the volume on schedule. Finally, a special thank goes to the Publication Chair, *Giovanni Schembra* from the University of Catania, whose help in collecting, processing and editing all the manuscripts was invaluable.

The Workshop has been technically co-sponsored by the
IEEE Communication Society

The Workshop would not have come into being without the support of the Italian National Consortium for Telecommunications (CNIT), without the patronage of the University of Catania and of the Municipality of Taormina, and without the sponsorship of the following companies, which are gratefully acknowledged.

Table of Contents

WDM Technologies for the Next Generation Internet

Mobile and Wireless Internet Access

QoS in the Next Generation Internet

Multicast and Routing in IP Networks

Multimedia Services over the Internet

Performance Modeling and Measurement of Internet Protocols

Dynamic Service Management

Source Encoding and Internet Applications

Progressive Introduction of Optical Packet Switching Techniques in WDM Networks

Marc Vandenhoute [1], Francesco Masetti [1], Amaury Jourdan [2], and Dominique Chiaroni [2],

(1) Alcatel USA, 1201 E. Campbell Road, Richardson, Texas 75081-1936, USA
(2) Alcatel CIT, Route de Nozay, F-91460 Marcoussis, France

Abstract. The transport network, mainly based on optical infrastructure, see a traffic increase, which introduces new requirements and challenges. This paper provides a summary of the trends that will bring bandwidth optimisation in WDM core networks, and will thus require the progressive introduction of optical packet switching techniques.

1 Introduction

The transport network is currently experiencing a doubling of the demand every 8-15 months. In the view of most operators, precedence of data traffic over voice is already a reality, and expected to represent 90% within three years. Optimising the network at all levels (core and metropolitan, figure 1) for data applications, and the underlying IP support protocol is definitely the trend in the next generation of routing and switching products prepared by all vendors.

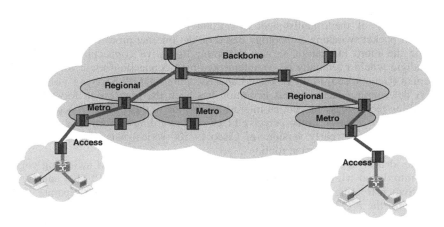

Fig. 1. Transport network schematics

S. Palazzo (Ed.): IWDC 2001, LNCS 2170, pp. 1–9, 2001.

If WDM transmission is clearly expected to meet these capacity requirements, the situation is clearly not settled for "layer-1", "layer-2" and "layer-3" routing and switching technologies. Hence the efforts to develop a new generation of networking products, putting forward scalability and flexibility as most critical specifications and trying to avoid complex protocol stacks featuring replication of functionality between different layers.

This paper will discuss how "fast" optical switching, adopting packet-based techniques, can improve network organisation and bandwidth efficiency in WDM networks, thanks to unique scalability and modularity features, and the recent maturity of the technology. In this discussion, we focus on the evolution of the core network; metropolitan network architectures will typically be subject to partially different requirements, as discussed in [8].

2 Evolution Trends

Put in an historical perspective, we can identify three major trends in optical internetworking which will clarify this functional split.

2.1 Switching Paradigm

A systematic pattern can be identified over the last 30 years of networking where the dominant switching paradigm (in high-speed networks) has been cycling back and forth between circuit switching and packet/cell switching technologies.

An earlier manifestation of this pattern was the emergence of Asynchronous Transfer Mode (ATM) cell switching as a more flexible and efficient alternative to Time Division Multiplexing (TDM). While ATM did bring a more flexible switching granularity, it (initially [1]) did retain a strict connection-oriented forwarding paradigm, in the form of the establishment of Virtual Paths and Connections. Eventually, another packet switching technology based on the Internet protocol (IP) would prove to be the more ubiquitous service platform. Its native connection-less switching paradigm would eventually be complemented by a connection-oriented mode, in the form of (Generalised) Multi-Protocol Label Switching (GMPLS).

A determining factor in this evolution has been the progress in forwarding and switching technology. Indeed, an interesting episode in this evolution occurred when an innovative solution to the perceived lack of forwarding performance of software-based routers (e.g. early versions of Cisco's 7500 series) was introduced with the concept of "IP/ATM shortcut routing".

The idea of applying high performance hardware-based ATM switching engines was introduced with Ipsilon's Flow Management Protocol (FMP) [12] and Cisco's Tag Switch Routing (TSR) [13]. A slightly different approach, with separate control planes for the ATM and IP layers, was proposed by the ATM Forum in the form of Multi-Protocol Over ATM (MPOA) [14]. The common idea behind all of them was the establishment of edge-to-edge transparent circuits (in casu, ATM VP/VCs) under the control of a native IP control layer.

---

[1] a connection-less variant of ATM was proposed, but never deployed

Eventually, of course, the emergence of native high performance IP-oriented network processors would obviate the need for ATM cell switching and allow the building of the gigabit-class IP/MPLS routers we find on the market today.

Interestingly, we can identify a lot of parallels with the current IP/optical network evolution:

- GMPLS is the native IP incarnation of a generic UNI/NNI signalling mechanism which we use today to establish edge-to-edge (wavelength) paths in WDM optical networks. This protocol evolved from a connection-oriented path establishment mechanism targeted at the IP client layer, to a generic signalling mechanism for the widest possible variety of connection-oriented transport technologies, including SDH/SONET, G.709-based and, more generally, DWDM networks [9][11].

- The current OWS models call for the signalled establishment of shortcut lightpaths between edge routers. This form of optical wavelength switching looks like the only alternative today to achieve switching capacities beyond the capabilities of current routers. Optical cross-connects (OXC) with capacities in the tens of terabits are achievable today, well beyond the capabilities of the current generation of routers

- Top-end network processors (10 Gbps arriving now) seem to lag one generation behind top-end wavelength switching subsystems (40 Gbps starting to be deployed). However, market pressure make it very unlikely that deployment of 40 (and later 80 or 160) Gbps transmission systems will be artificially slowed down to match the port capacity of high-end routers. This essentially mandates, for the foreseeable future, a network architecture where electronic processing nodes remain positioned at the edge of an optically switched/routed core network and some form of optical switching is mandated for the core nodes

- In the design of switching fabrics for individual routers, hybrid approaches are emerging with the introduction of optical (frame) switching elements to alleviate the scalability problems of current electronic switching fabrics. These hybrid network elements are a clear foreboding of a more fundamental shift to end-to-end optical switching, just as ATM HW switching elements were the foreboding of the current generation of network processors.

- The fundamentally different transfer modes between the IP and underlying transport layer (ATM cells in the former, optical frames in the latter) introduce some level of opaqueness, resulting in sub-optimal solutions wrt. multiplexing, framing, QoS control, … In the case of IP/ATM, the segmentation to which IP packets were subjected resulted in awkward AAL5-frame aware cell processing in ATM switches [16]. In the all-optical domain, the absence of any processing capabilities on optical headers makes it more difficult to perform basic processes such as congestion control (selective frame dropping) loop prevention (e.g. through header Time-To-Live decrementing), QoS diffserv-like marking/remarking, …

- The differences between the overlay, augmented and peer models currently investigated in the IETF IPO group [9] are not dissimilar to the differences between the MPOA (overlay) and FMP/TSR (integrated/peer) approaches in IP/ATM shortcut routing. They relate to the different degrees of integration between the routing layers in the underlying transport and IP client layer.

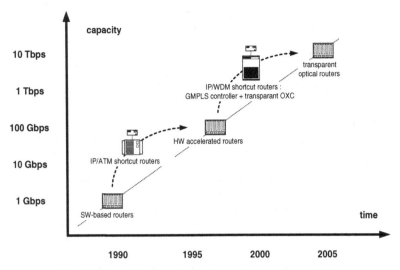

Fig. 2. Role of "shortcutting"in the router capacity evolution

Analysing the analogy further, we can also however identify key differences:

- IP over ATM shortcutting involved the mapping of 2 levels of statistical multiplexing network layers (IP and ATM), which were themselves layered on top of a deterministic multiplexing layer (SDH/SONET). Some simplification in the protocol stacks did occur since with more direct IP/WDM mapping [8]. However, this does not imply the absence of any conflicts between control mechanisms at different layers, as exemplified by the continued debate around the interworking of survivability mechanisms (protection, restoration, ...) at the IP, SDH/SONET and wavelength layer.
- The (virtual) label space in IP/ATMN shortcutting was quite large (up to 64K channels). However, in the case of a wavelength switched transport network, the GMPLS label space will have to be mapped to physical wavelengths and, hence, is unlikely to exceed a few hundred for an individual link in the foreseeable future.
- GMPLS-controlled lightpaths (and also SDH/SONET trails) are physical channels with a fixed capacity, whereas ATM VP/VCs could be virtual connections with adjustable capacity. The notion of flow and/or burst switching can be seen as the link between the two worlds, introducing also the need to identify the appropriate switching granularity according to network requirements.

Summarizing, from our experience with IP/ATM shortcutting we can draw some insights on our current IP/WDM deployment efforts :

- IP/ATM deployment scenarios typically resulted in a de facto establishment of a full mesh of channels between edges. Signalling latencies, limited switching capacity of edge routers (relative to the core OXCs) and the topological diversity of traffic seen in core networks makes it unlikely that this will be any different in optical wavelength switched networks: a full mesh of lightpaths for default connectivity will likely be complemented by additional on-demand lightpaths. As a consequence, the control scalability problems resulting from the N^2 full-mesh connectivity encountered in IP/ATM overlays will also occur in IP/WDM overlay networks. These problems related mainly to the overhead of routing peer adjacencies over fully meshed networks, as documented in [15].

- Furthermore, of particular concern for IP/optical networks, may become the (physical) label space which will be a few orders of magnitude smaller than in earlier IP/ATM deployment scenarios. Taking into account additional requirements on the label/wavelength space for features such as survivability, bandwidth-on-demand, QoS schemes [2], traffic isolation (e.g. for VPNs), … it becomes increasingly clear that the number of nodes in a fully meshed optically switched network may quickly become limited by the label space necessary to interconnect them, making it unlikely that large combined metro/core networks will be build using a flat transparent topology.
- This problem will be compounded by the fact that the capacity of the lightpaths between edge nodes will closely follow optical technology trends. In essence, in an expanding OWS network, both the large number of lightpaths from an edge node (driven by a need for full mesh connectivity, determined by the increasing number of other edge nodes) AND the fixed capacity of those lightpaths (10 Gbps -> 40 Gbps -> 80 Gbps -> 160 Gbps -> …) will increase significantly. This "high lightpath count / fat fixed pipes" combination may result in very poor efficiency overall, due to lack of bandwidth flexibility and capacity manageability.

These considerations lead us to the conclusion that, in order to tackle some of the control and scalability issues associated with OWS, a new network architecture should emerge in time with following features :

- Support for both connection-less and connection-oriented forwarding paradigms, reflecting the complementarity currently available in the IP client layer. Connection-oriented forwarding should be based on virtual, rather than physical path establishment [5].
- Low granularity optical switching, at the level of individual flows and/or (time) slots (i.e. fractions of a wavelength), rather than complete wavelengths, allowing for the flexible bandwidth management necessary in very high capacity networks.
- Hybrid switching in the optical and electronic domain : complex forwarding (incl. multicast & VPN), statistical multiplexing w/ QoS control, … decisions on control packets will continue to be performed in the electronic domain, but associated (large) payloads should remain in the optical domain in order to benefit from the bandwidth scalability associated w/ optical switching. This aspect will be further investigated in a later chapter.

A number of variants of this novel switching paradigm are currently being discussed, with Optical Burst Switching as its most obvious exponent [1][2][3][4][6].

2.2 Optical Network Transparency

An important enabler of this network evolution is the increased interest in building transparent optical networks.

---

[2] which e.g. might mandate the assignment of a separate edge-to-edge lightpath for individual (a.o. diffserv) classes of service

Optical transparency could bring a number of obvious advantages :

- Maintain the integrity of the light in the optical path to reduce the cost : O/E/O conversions in general and high-speed electronic and opto-electronic components and subsystems constitute a very important part of the cost of current network elements

- Provide some bit-rate scalability : optical switching and transmission elements are often bit-rate agnostic, hence improving the prospect for future upgradability of an optical infrastructure. Of course, genuine transparency will largely remain an elusive target since edge systems will never be truly bit-rate agnostic, neither will be monitoring subsystems within the optical network. However, it can be expected that these networks elements will not remain an impediment to build a transparent data path within the optical backbone network. Edge systems could be incrementally upgraded at the periphery with several co-existing generations of edge systems exchanging data at various bit-rates within the optical transport network. Monitoring systems targeted at specific bit-rates could coexist and would operate in a transparent (sampling) manner (i.e. attached via splitters to the optical channel, hence not obstructing the data path).

- Create a new framing transparent to the client protocol (Framing agnosticism) : the major benefit of optical network transparency would eventually be its framing agnosticism, i.e. the capability to have different kinds of client layers PHY (and higher layer) framing co-existing on a single optical infrastructure, even transparent to the switching elements on the optical path [3]. This could go as far as allowing forms of asynchronous optical switching to coexist with more traditional synchronous forms of transmission, e.g. G.709 framing with asynchronous burst framing.

Of course, transparent optical networking will remain controversial for the foreseeable future, as it essentially implies a return to analogue network engineering rules [4], and a departure from the generally heralded and clearly demarcated network model imposed by the Synchronous Digital Hierarchy (SDH/SONET) suite of specifications.

The future will show whether the potential functional advantages of network transparency will weigh up against the obvious operational disadvantages. However, we definitely foresee a future where increasing levels of transparency will be introduced in optical wavelength switched (OWS) networks. For optical packet switching, the most challenging level of transparency remains the bit-rate transparency since the existing fast technology imposes today pulse degradations in the time domain that forces the use of 3R regeneraotrs. Without regeneration, the switching throughput of a switch is limited and becomes in direct concurrence with existing electronic technologies. A reasonable transparency is probably in the adoption of a new optical framing (burst), and in the exploitation of all-optical packet regenerators to replace advantageously costly O/E/O regenerators.

---

[3] one could imagine part of an optical frame switch, featuring transparent optical switching elements, operating in a static cross-connect-mode for the establishment of long-duration OWS lightpaths.

[4] necessitating a.o. the introduction of routing rules based on optical channel impairments

2.3 Separation of Control & Data

A third enabler in optical networking is the evolution towards separation between the control plane and the data plane, already alluded to in a previous chapter.

This evolution can be documented in three steps :

- classic data network architectures where control information is exchanged in-band with normal data [5]. This can clearly be observed in current IP networks where general control (ICMP), routing (OSPF, BGP, …), signalling (MPLS, IGMP, …) and resource management (RSVP, …) information is exchanged on the same carrier as normal data, albeit, sometimes, subject to relative prioritisation [6].
- OWS network architectures where GMPLS signalling and routing control data can be exchanged out of band wrt the data-carrying lightpaths [10]
- Optical packet and burst switching solutions where not only control information, but also individual packet/burst header information is transported apart from the data payloads on separate wavelengths.

This separation is inevitable due to the complementarity between the complex processing capabilities of electronic control systems and the bit-rate agnostic switching capability of emerging optical components. It is unlikely to disappear in the foreseeable future, given the state of current research in optical computing.

The separation of control poses some unique challenges, which are only now being investigated. Issues abound, as exemplified here:

- The correlation between information elements on separate carriers (e.g. a signalling message and the actual data on its associated lightpath) are not trivial to observe, since monitoring capabilities on the optical lightpath are limited, especially if some form of non-intrusive [7] monitoring is implemented
- Time-synchronisation between information elements on separate carriers will not only be impacted by optical characteristics (e.g. differential fibre propagation speeds), but will require particular discipline in the design of control electronics (requiring latency control at the ns granularity level) and network layout (compensating for optical paths with varying lengths).

3 Optical Burst Switching

Optical burst switching has the potential to reconcile the optimisation of resource utilisation and service differentiation from packet switching, with application of both connection-less and (GMPLS-based) connection-oriented forwarding [5] techniques, and the scalability of wavelength cross-connects.

IP routers are interconnected to a layer-1/2 optical system performing both traffic aggregation and core switching in a much more flexible way than a cross-connect, but at a lower cost than an IP router. In this network approach, depicted in figure 3, IP

---

[5] we are making abstraction here of management plane information which, in current transport networks, is already often transported out-of-band

[6] e.g. through diffserv DSCP marking

[7] i.e. not involving the O/E/O conversion of the complete data stream

packets are concatenated into optical "bursts", of larger dimension, at edge nodes, and are then routed as a single entity through the network, within core optical packet routers [2].

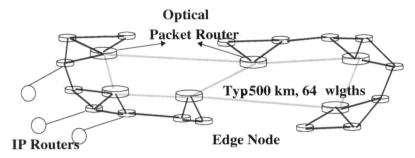

Fig. 3. Optical burst-switching network

The advantage of this aggregation is to relax drastically the forwarding speed required from core routers, and to scale up their forwarding capability by at least one order of magnitude, well within the multi-Tbps range. In addition, with this approach, it becomes possible to consider groups of wavelengths (WDM ports) as a single resource (of typ. 300-600 Gbps/s of capacity), and therefore to improve the logical performance and/or decrease the memory requirements with respect to IP routers with single wavelength processing capabilities. In a first step, this may happen within the switching fabric, where optics could be used to reach larger throughputs (i.e. 10s of Tbps) and where the control and scheduling could become the bottleneck of the system. Later, it can be applied to networking concepts.

In fact, this burst-switching principle is in addition very much optics-friendly, since fast optical switching technology requires some specific framing to avoid the loss of any payload data. Optical switching also offers the perspective of large scalability (current research realisation have demonstrated up to 10 Tbps) in a single stage configuration, avoiding complex interconnection and improving foot-print. It also is less sensitive to an increase of the line-rate, which is very important in view of the on-coming evolution towards 40 Gbps/s transmission.

In the long run, the ultimate target could be to reach an all-optical implementation of such a network, and additional key functions, such as buffering or regeneration that have been demonstrated in the laboratory using optical devices (optical fibre delay lines and non linear optical elements, respectively). However depending on cost/performance trade-offs to reduce the time to market an intermediate opto-electronic solution could be introduced first.

A more detailed analysis of the current state of research and enabling optical technologies for Optical Burst Switching can be found in [2] and [7].

4   Conclusions

Optical wavelength switched networks are today's answer to the capacity growth requirements of Internet backbones.

However, from an historical perspective, if these IP/WDM shortcut solutions do offer some relief in the short and medium term, we can only expect that they will also be subject to some of the scalability issues experienced by IP/ATM-based shortcut solutions.

Moreover, the gradual introduction of two novel concepts in OWS transport networks, optical transparency [8] and separation of data and control, will be essential milestones. Successful mastery of these new network architectural elements will be a condition sine qua non for the introduction of transparent end-to-end optical packet/burst switching.

Optical packet/burst switching technologies are a natural answer to a number of concerns with OWS networks, offering desirable benefits in term of both network efficiency and control scalability.

References

[1] J.S. Turner, "Terabit burst switching," Journal of High-Speed Networks, Vol. 8, 1999, pp. 3-16.

[2] D. Chiaroni, A. Jourdan et al. : "Data, Voice and multimedia convergence over WDM: the case for optical routers"; Alcatel Telecom Review, 3rd quarter 99.

[3] C. Qiao, M. Yoo, "Optical bust switching (OBS) – A new paradigm for an optical Internet," Journal of High-Speed Networks, Vol. 8, pp. 69-84, 1999.

[4] J.S. Turner, "WDM burst switching for petabit data networks," Technical Digest OFC 2000, paper WD2-1, pp. 47-49, Feb. 2000.

[5] C. Qiao.: "Labeled optical burst switching for IP-over-WDM integration". IEEE Comm. Mag. pp 104-114, vol. 38, Sept. 2000.

[6] S. Verma, H. Chaskar, R. Ravikanth, "Optical burst switching: A viable solution for terabit IP backbone," IEEE Network, Vol. 14, No. 6, pp. 48-53., Nov./Dec. 2000.

[7] D. Chiaroni et al: "Towards 10 Tbit/s optical packet routers for the backbone", Proceedings ECOC'2000, paper 10.4.7., September 2000.

[8] A. Jourdan, D. Chiaroni, E. Dotaro, G. Eilenberger, F. Masetti, N. Le Sauze, M. Renaud, "Perspective of optical packet switching in IP-dominant backbone and metropolitan networks" IEEE Communication Magazine, pp 136-141, vol.39, n°3, March 2001.

[9] Generalized Multi-Protocol Label Switching (GMPLS) Architecture, draft-many-gmpls-architecture-00.txt

[10] IP over Optical Networks: A Framework, draft-many-ip-optical-framework-03.txt.

[11] Generalized MPLS Control Plane Architecture for Automatic Switched Transport Network, draft-xu-mpls-ipo-gmpls-arch-00.txt .

[12] Ipsilon Flow Management Protocol Specification for IPv4, RFC1953.

[13] Cisco Systems' Tag Switching Architecture Overview, RFC2105.

[14] http://www.cis.ohio-state.edu/~jain/atm/atm_mpoa.htm.

[15] http://www.arl.wustl.edu/arl/workshops/atmip/session7/Yakov.ps.Z.

[16] http://www.aciri.org/floyd/epd.html.

---

[8] and its non-trivial OAM implications

Optical Packet Switching
for IP-over-WDM Transport Networks

Stefano Bregni, Giacomo Guerra, and Achille Pattavina

Department of Electronics and Information, Politecnico di Milano
Piazza Leonardo da Vinci 32, 20133 Milan, Italy
bregni@elet.polimi.it, guerra@cerbero.elet.polimi.it,
pattavin@elet.polimi.it

Abstract. As new bandwidth-hungry IP services are demanding more
and more capacity, transport networks are evolving to provide a reconfigurable optical layer in order to allow fast dynamic allocation of WDM
channels. To achieve this goal, optical packet-switched systems seem to
be strong candidates as they allow a high degree of statistical resource
sharing, which leads to an efficient bandwidth utilization. In this work, we
propose an architecture for optical packet-switched transport networks,
together with an innovative switching node structure based on the concept of per-packet wavelength routing. Some simulations results of node
operation are also presented. In these simulations, the node performance
was tested under three different traffic patterns.

1 Introduction

Telecomunication networks are currently experiencing a dramatic increase in demand for capacity, driven by new bandwidth-hungry IP services. This will lead
to an explosion of the number of wavelengths per fiber, that can't be easily
handled with conventional electronic switches. To face this challenge, networks
are evolving to provide a reconfigurable optical layer, which can help to relieve
potential capacity bottlenecks of electronic-switched networks, and to efficiently
manage the huge bandwidth made available by the deployment of dense wavelength division multiplexing (DWDM) systems.

As current applications of WDM focus on a relatively static usage of single
wavelength channels, many works have been carried out in order to study how
to achieve switching of signals directly in the optical domain, in a way that
allows fast dynamic allocation of WDM channels, to improve transport network
performance. Two main alternative strategies have been proposed to reach this
purpose: optical packet switching [1], [2] and optical burst switching [3], [4].

In this article, we first introduce optical packet and burst switching approaches. Then an architecture for packet-switched WDM transport networks
and a novel optical switching node are proposed. Some simulation results of
node operation, under different traffic patterns, are also presented.

S. Palazzo (Ed.): IWDC 2001, LNCS 2170, pp. 10–25, 2001.

2   Optical Packet and Burst Switching

Optical packet switching allows to exploit single wavelength channels as shared resources, with the use of statistical multiplexing of traffic flows, helping to efficiently manage the huge bandwidth of WDM systems. Several approaches have been proposed to this aim [5], [6].

Most proposed systems carry out header processing and routing functions electronically, while the switching of optical packet payloads takes place directly in the optical domain. This eliminates the need for many optical-electrical-optical conversions, which call for the deployment of expensive opto-electronic components, even though most of the optical components, needed to achieve optical packet switching, still remain too crude for commercial availment.

Optical burst switching aims at overcoming these technological limitations. The basic units of data transmitted are bursts, made up of multiple packets, which are sent after control packets, carrying routing information, whose task is to reserve the necessary resources on the intermediate nodes of the transport network (see Fig. 1). This results in a lower average processing and synchroniza-

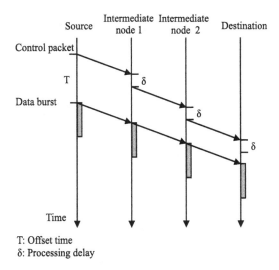

Fig. 1. The use of an offset time in optical burst switching

tion overhead than optical packet switching, since packet-by-packet operation is not required. However packet switching has a higher degree of statistical resource sharing, which leads to a more efficient bandwidth utilization in a bursty, IP-like, traffic environment.

Since optical packet-switching systems still face some technological hurdles, the existing transport networks will probably evolve through the intermediate step of burst-switching systems, which represent a balance between circuit and packet switching, making of the latter alternative a longer term strategy for network evolution.

In this work, we have focused our attention on optical packet switching, since it offers greater flexibility than the other relatively coarse-grained WDM techniques, aiming at efficient system bandwidth management.

3   Optical Transport Network Architecture

The architecture of the optical transport network we propose consists of $M = 2^m$ *optical packet-switching nodes*, each denoted by an optical address made of $m = \log_2 M$ bits, which are linked together in a mesh-like topology. A number of *edge systems* (ES) interfaces the optical transport network with IP legacy (electronic) networks (see Fig. 2).

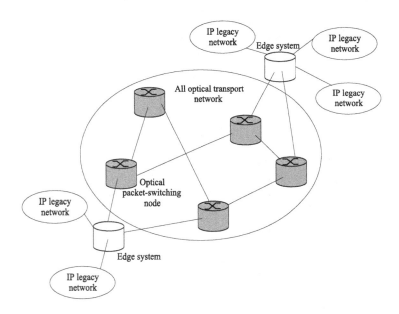

Fig. 2. The optical transport network architecture

An ES receives packets from different electronic networks and performs traffic aggregation in order to build *optical packets*. The optical packet is composed of a simple optical header, which comprises the m-bits long destination address, and of an optical payload made of a single IP packet, or, alternatively, of an aggregate of IP packets.

The optical packets are then buffered and routed through the optical transport network to reach their destination ES, which delivers the traffic it receives to its destination electronic networks.

At each intermediate node, in the transport network, packet headers are received and electronically processed, in order to provide routing information to the control electronics, which will properly configure the node's resources to switch packet payloads directly in the optical domain.

The transport network operation is asynchronous; that is, packets can be received by nodes at any instant, with no time alignment. The internal operation of the optical nodes, on the other hand, is synchronous (slotted). In the model we propose, the time slot duration, T, is equal to the amount of time needed to transmit an optical packet, with a 40-bytes long payload, from an input WDM channel to an output WDM channel.

The operation of the optical nodes is slotted since the behavior of packets, in an unslotted node, is less regulated and more unpredictable, resulting in a larger contention probability.

A contention occurs every time that two or more packets are trying to leave a switch from the same output port. How contentions are resolved has a great influence on network performance. Three main schemes are generally used to resolve contention: wavelength conversion, optical buffering and deflection routing.

In a switch node applying *wavelength conversion*, two packets trying to leave the switch from the same output port are both transmitted at the same time but on different wavelengths. Thus, if necessary, one of them is wavelength converted to avoid collision. In the *optical buffering* approach, one or more conteding packets are sent to fixed-length fiber delay lines, in order to reach the desired output port only after a fixed amount of time, when no contention will occur. Finally, in the *deflection routing* approach, contention is resolved by routing only one of the contending packets along the desired link, while the other ones are forwarded on paths which may lead to longer than minimum-distance routing paths.

Implementing optical buffering gives good network performance, but involves a great amount of hardware and electronic control. On the other hand, deflection routing is easier to implement than optical buffering, but network performance is reduced since a portion of network capacity is taken up by deflected packets.

In the all-optical network proposed, in order to reduce complexity while aiming at attaining good network performance, the problem of contention is resolved combining a small amount of optical buffering with wavelength conversion and deflection routing. Our policy can be summarized as follows:

1. When a contention occurs, the system first tries to transmit the conflicting packets on different wavelengths.
2. If all of the wavelengths of the correct output link are busy at the time the contention occurs, some packets are scheduled for transmission in a second time, and are forwarded to the fiber delay lines.
3. Finally, if no suitable delay line is available, at the time the contention occurs, for transmission on the correct output port, a conflicting packet can be deflected to a different output port than the correct one.

4   Node Architecture

The general architecture of a network node is shown in Fig. 3. It consists of
N incoming fibers with W wavelengths per fiber. The incoming fiber signals

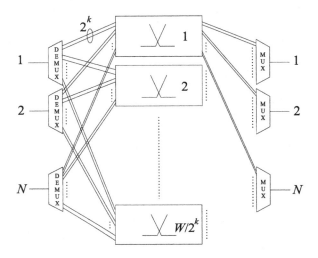

Fig. 3. Optical packet-switching node architecture

are demultiplexed and 2^k wavelengths, from each input fiber, are then fed into
one of the $W/2^k$ switching planes, which constitute the switching fabric's core.
Once signals have been switched in one of the second stage parallel planes,
packets can reach every output port on one of the 2^k wavelengths that are
directed to each output fiber. This allows the use of wavelength conversion for
contention resolution, since 2^k packets can be contemporarely be transmitted,
by each second-stage plane, on the same output link.

The detailed structure of one of the $W/2^k$ parallel switching planes is shown
in Fig. 4. Each incoming link carries a single wavelength and the switching plain
consists of three main blocks: an input *synchronization unit*, as the node is
slotted and incoming packets need to be aligned, a *fiber delay lines unit*, used to
store packets for contention resolution, and a *switching matrix unit*, to achieve
the switching of signals.

These three blocks are all managed by an *electronic control unit* which carries
out the following tasks:

- optical packet header recovery and processing;
- managing the synchronization unit in order to properly set the correct path
  through the synchronizer for each incoming packet;
- managing the tunable wavelength converters in order to properly delay and
  route incoming packets.

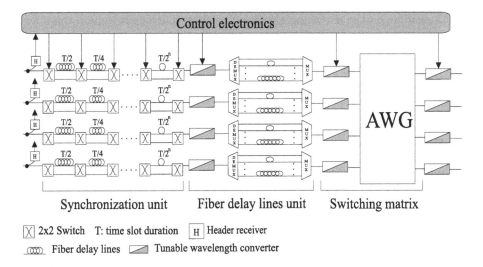

Fig. 4. Detailed structure of one of the $W/2^k$ parallel switching planes

We will now describe the second-stage switching planes mentioned above, detailing their implementation.

4.1 Synchronization Unit

This unit consists of a series of 2×2 optical switches interconnected by fiber delay lines of different lengths. These are arranged in a way that, depending on the particular path set through the switches, the packet can be delayed of a variable amount of time, ranging between $\triangle t_{min} = 0$ and $\triangle t_{max} = (1 - (1/2)^n) \times T$, with a resolution of $T/2^n$, where T is the time slot duration and n the number of delay lines.

The synchronization is achieved as follows: once the packet header has been recognized and packet delineation has been carried out, the packet start time is identified and the control electronics can calculate the necessary delay and configure the correct path of the packet through the synchronizer.

Due to the fast reconfiguration speed needed, fast 2×2 switching devices, such as 2×2 semiconductor optical amplifier (SOA) switches [7], which have a switching time in the nanosecond range, must be used.

4.2 Fiber Delay Lines Unit

After packet alignment has been carried out, the routing information carried by the packet header allows the control electronics to properly configure a set of tunable wavelength converters, in order to deliver each packet to the correct delay line to resolve contentions. To achieve wavelength conversion several devices are available [8], [9], [10].

Depending on the managing algorithm used by control electronics, the fiber delay lines stage can be used as an *optical scheduler* or as an *optical first-in-first-out* (FIFO) *buffer*.

- *Optical scheduling*: this policy uses the delay lines in order to schedule the transmission of the maximum number of packets onto the correct output link. This implies that an optical packet P_1, entering the node at time t_1 from the i-th WDM input channel, can be transmitted after an optical packet P_2, entering the node on the same input channel at time t_2, being $t_2 > t_1$. For example, suppose that packet P_1, of duration $l_1 T$, must be delayed of d_1 time slots, in order to be transmitted onto the correct output port. This packet will then leave the optical scheduler at time t_{1+d_1}. So, if packet P_2, of duration $l_2 T$, has to be delayed for d_2 slots, it can be transmitted before P_1 if $t_{2+d_2+l_2} < t_{1+d_1}$ since no collision will occur at the scheduler output.
- *Optical FIFO buffering*: in the optical FIFO buffer the order of the packets entering the fiber delay lines stage must be maintained. This leads to a simpler managing algorithm than the one used for the optical scheduling policy, yielding, however, a sub-optimal output channel utilization. In fact, suppose that optical packet P_1, entering the FIFO buffer at time t_1, must be delayed for d_1 time slots. This implies that packet P_2, behind packet P_1, must be delayed of, at least, d_1 time slots, in order to maintain the order of incoming packets. Due to this rule, if packet P_2 has to be delayed for $d_2 < d_1$ slots, in order to avoid conflict, its destination output port is idle for $d_1 - d_2$ time slots, while there would be a packet to transmit.

4.3   Switching Matrix Unit

Once packets have crossed the fiber delay lines unit, they enter the switching matrix stage in order to be routed to the desired output port. This is achieved using a set of tunable wavelength converters combined with an arrayed waveguide grating (AWG) wavelength router [11].

This device consits of two slab star couplers, interconnected by an array of waveguides. Each grating waveguide has a precise path difference with respect to its neighbours, $\triangle X$, and is characterized by a refractive index of value n_w.

Once a signal enters the AWG from an incoming fiber, the input star coupler divides the power among all waveguides in the grating array. As a consequence of the difference of the guides lengths, light travelling through each waveguide emerges with a different phase delay given by:

$$\triangle\Phi = 2\pi n_w \times \frac{\triangle X}{\lambda} \tag{1}$$

being λ the incoming signal central wavelength. As all the beams emerge from the grating array they interfere constructively onto the focal point in the output star coupler, in a way that allows to couple an interference maximum with a particular output fiber , dependig only on the input signal central wavelength.

Figure 5 shows the mechanism described above. Two signals of wavelength λ_0 and λ_3 entering an 8×8 AWG, from input fibers number 6 and number 1

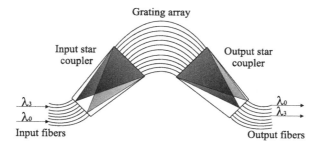

Fig. 5. Arrayed waveguide grating

respectively, are correctly switched onto the output fibers number 0 and number 3, being the wavelength of signals the only routing information needed to achieve the required permutation.

The AWG is used as it gives better performance than a normal space switch interconnection network, as far as insertion losses are concerned. This is due to the high insertion losses of all the high-speed all-optical switching fabrics available at the moment, that could be used to build a space switch interconnection network. Moreover AWG routers are strictly non-blocking and offer high wavelength selectivity.

After crossing the three stages previously described, packets undergo a final wavelength conversion, to avoid collisions at the output multiplexers, where W WDM channels are multiplexed on each output link.

5   Simulation Results

In this section, we present some simulation results of the operation of one among the $W/2^k$ parallel switching planes, which structure has been shown in Fig. 4.

These results have been obtained assuming that the node receives its input traffic directly from N edge systems. The edge systems buffers capacity is supposed to be large enough to make packet loss negligible. Each WDM channel is supposed to have a dedicated buffer in the edge system.

The packet arrival process has been modeled as a Poisson process, with packet interarrival times having a negative exponential distribution. As the node operation is slotted, the packets duration was always assumed to be multiple of the time slot duration T, which is equal to the amount of time needed to transmit an optical packet, with a 40-bytes long payload, from an input WDM channel to an output WDM channel.

As far as packet length is concerned, the following probability distributions were considered:

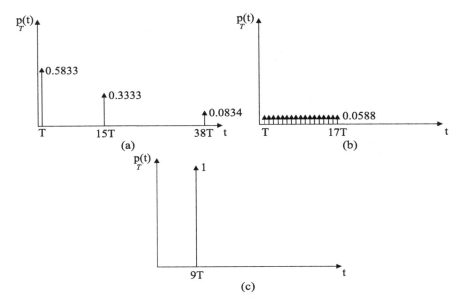

Fig. 6. Packet duration probability distributions: empirical distribution (a), uniform distribution (b), constant duration (c)

1. *Empirical distribution.* Based on real measurements on IP traffic [12], [13], we assumed the following probability distribution for the packet length L:

$$\begin{cases} p_0 = P(L = 40 \text{ bytes}) = 0.5833 \\ p_1 = P(L = 576 \text{ bytes}) = 0.3333 \\ p_2 = P(L = 1500 \text{ bytes}) = 0.0834 \end{cases} \tag{2}$$

   In this model, packets have average length equal to 341 bytes. Since a 40-bytes long packet is transmitted in one time slot of duration T, the average duration of an optical packet is approximatively $9T$. Moreover, p_0, p_1 and p_2 represent the probability that the packet duration is T, $15T$ and $38T$ respectively (see Fig. 6 (a)).

2. *Uniform distribution.* To show a comparison with the empirical model described above, we have modeled the optical packet length as a stochastic variable, uniformly distributed between 40 bytes (duration T) and 680 bytes (duration $17T$). Also in this model, packets have average duration of $9T$ (see Fig. 6 (b)).

3. *Constant length.* We have also investigated the behaviour of the system when packets have a constant duration of value $9T$ (see Fig. 6 (c)).

   These simulations were carried out assuming that no deflection routing algorithm is implemented. Under this assumption, a packet is supposed to be lost if it can't be delayed of a suitable amount of time, in order to transmit it onto the correct output port. Figures 7 through 12 show the packet loss probability at different traffic loads per wavelength, for different values of the maximum delay attainable by the fiber delay lines unit, $D = iT$ ($i = 0, 1, 2, \cdots$).

Figures 7, 9 and 11 report the simulation results for the optical FIFO buffering (OFB) policy, in the fiber delay lines unit, while Figs. 8, 10 and 12 report the results for the optical scheduling (OS) policy.

It can be seen that, regardless of packet length distribution, the OS policy yields a better performance than the OFB policy, with an increasing improvement as D grows.

Figures 13, 14 and 15 show the values of the ratio

$$\eta = \frac{\Pi_{OS}}{\Pi_{OFB}} \tag{3}$$

for different values of the maximum delay achievable, D, at different traffic loads per wavelength, where Π_{OS} and Π_{OFB} are the packet loss probability for the optical scheduling and optical FIFO buffering policy, respectively. It can be pointed out that no significant improvement is experienced as D value is 0, 1, 2 or 4.

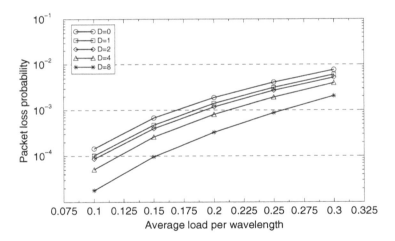

Fig. 7. Packet loss probability for the empirical distribution, with FIFO policy, at different loads per wavelength

It can also be seen that this increase is more evident for the uniform distribution, and even more for the constant length packets. This happens because the system performance is not only influenced by the maximum delay achievable, D, but also by the maximum optical packet length.

In fact, what really influences the system performance is the L_M/D rate, being L_M the maximum packet duration and D the maximum delay attainable.

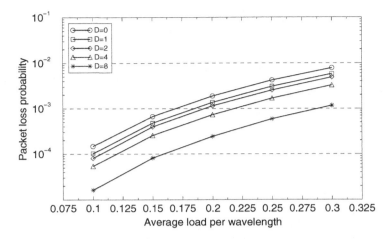

Fig. 8. Packet loss probability for the empirical distribution, with scheduling policy, at different loads per wavelength

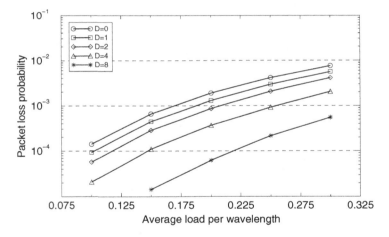

Fig. 9. Packet loss probability for the uniform distribution, with FIFO policy, at different loads per wavelength

So as the value of D is much smaller than L_M, the influence of the fiber delay lines managing discipline is negligible. When D becomes much larger than L_M, on the other hand, the OS policy efficiency improvement becomes more and more evident.

It is now interesting to show the efficiency improvement variation, yielded by the OS policy, depending on the optical packet length. Figure 16 plots this variation at different loads per wavelength, for three values of the packet length,

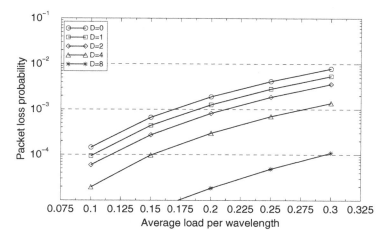

Fig. 10. Packet loss probability for the uniform distribution, with scheduling policy, at different loads per wavelength

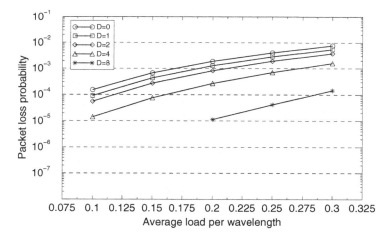

Fig. 11. Packet loss probability for constant packet length, with FIFO policy, at different loads per wavelength

for constant length packets, when the maximum delay attainable is $D = 16$. It can be seen that, for $L_M = 8$, that is $L_M = D/2$, a significative efficiency improvement is experienced, while for $L_M = 32$, that is $L_M = 2D$, the optical scheduling and optical FIFO buffering policies almost give the same performance.

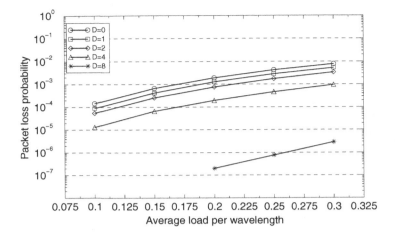

Fig. 12. Packet loss probability for constant packet length, with scheduling policy, at different loads per wavelength

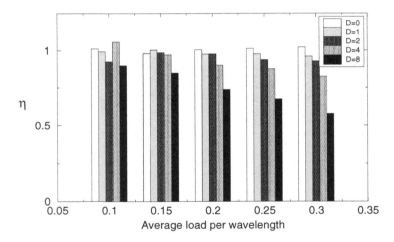

Fig. 13. Empirical distribution: values of η at different loads per wavelength and for different values of the maximum delay attainable D

6   Conclusions and Topics for Further Research

In this work, we proposed an architecture for optical packet-switched transport networks. The structure of the optical switching nodes was detailed and the basic building blocks were described. Some simulation results were also presented, showing a comparison between two different managing policies for the fiber delay lines stage: *optical scheduling* and *optical FIFO buffering*.

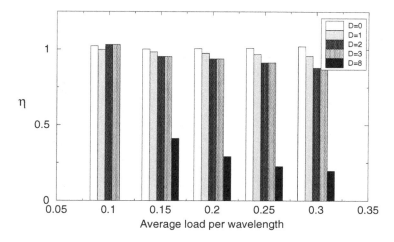

Fig. 14. Uniform distribution: values of η at different loads per wavelength and for different values of the maximum delay attainable D

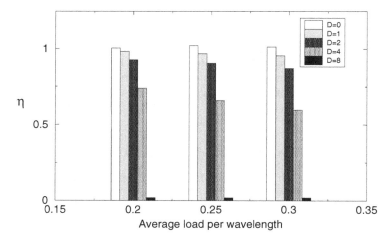

Fig. 15. Constant packet length: values of η at different loads per wavelength and for different values of the maximum delay attainable D

It was shown that, for $D \ll L_M$, OS and OFB almost give the same performance. For $D \gg L_M$, on the other hand, the optical scheduling policy yields a better performance than the optical FIFO buffering policy, because the output links are more efficiently exploited.

Many issues will have to be addressed in the future, such as the detailed study of the improvement attainable with the optical scheduling policy depending on the optical packet length. Moreover, the behaviour of an optical transport

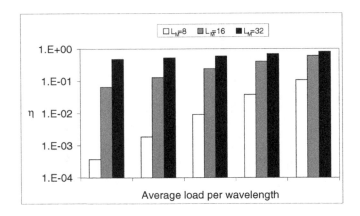

Fig. 16. Constant packet length: values of η at different loads per wavelength, for $D = 16$, $L_M = 8$, $L_M = 16$ and $L_M = 32$

network, as a whole, will have to be investigated, since a single node operation was simulated for this work. Another interesting issue is the implementation of a suitable deflection routing algorithm in order to improve network performance, varying the optical network topology.

References

1. Hunter, D.K., Andonovic, I.: Approaches to Optical Internet Packet Switching. IEEE Commun. Mag. (Sep. 2000) 116-122
2. Yao, S., Mukherjee, B., Dixit, S.: Advances in Photonic Packet Switching: An Overview. IEEE Commun. Mag. (Feb. 2000) 84-94
3. Yoo, M., Qiao, C.: Just-Enough-Time(JET): A High Speed Protocol for Bursty Traffic in Optical Networks. Proc. IEEE/LEOS Tech. for a Global Info. Infrastructure (Aug. 1997) 26-27
4. Qiao, C.: Labeled Optical Burst Switching for IP-over-WDM Integration. IEEE Commun. Mag. (Sep. 2000) 104-114
5. Hunter, D.K. et al.: WASPNET: A Wavelength Switched Packet Network. IEEE Commun. Mag. (Mar. 1999) 120-129
6. Renaud, M., Masetti, F., Guillemot, C., Bostica, B.: Network and System Concepts for Optical Packet Switching. IEEE Commun. Mag. (Apr. 1997) 96-102
7. Dorgeuille, F., Mersali, B., Feuillade, M., Sainson, S., Slempkès, S., Foucher, M.: Novel Approach for Simple Fabrication of High-Performance InP-Switch Matrix Based on Laser-Amplifier Gates. IEEE Photon. Technol. Lett., Vol. 8. (1996) 1178-1180
8. Stephens, M.F.C. et al.: Low Input Power Wavelength Conversion at 10 Gb/s Using an Integrated Amplifier/DFB Laser and Subsequent Transmission over 375 km of Fibre. IEEE Photon. Technol. Lett., Vol. 10. (1998) 878-880

9. Owen, M. et al.: All-Optical 1x4 Network Switching and Simultaneous Wavelength Conversion Using an Integrated Multi-Wavelength Laser. Proc. ECOC '98, Madrid, Spain, (1998)
10. Tzanakaki, A. et al.: Penalty-Free Wavelength Conversion Using Cross-Gain Modulation in Semiconductor Laser Amplifiers with no Output Filter. Elec. Lett., Vol. 33. (1997) 1554-1556
11. Parker, C., Walker, S.D.: Design of Arrayed-Waveguide Gratings Using Hybrid Fourier-Fresnel Transform Techniques. IEEE J. Selec. Topics Quant. Electron., Vol. 5. (1999) 1379-1384
12. Thompson, K., Miller, G.J., Wilder, R.: Wide-area Internet Traffic Patterns and Characteristics. IEEE Network, Vol 11. (1997) 10-23
13. Generating the Internet Traffic Mix Using a Multi-Modal Length Generator. Spirent Communications white paper. http://www.netcomsystems.com

MPLS over Optical Packet Switching

F. Callegati, W. Cerroni, G. Corazza, and C. Raffaelli

D.E.I.S. - University of Bologna, Viale Risorgimento 2 - 40136 Bologna – Italy
(fcallegati,gcorazza,wcerroni,craffaelli)@deis.unibo.it

Abstract. This paper deals with the problem of connection to wavelength assignment in an MPLS optical packet switched network with DWDM links. The need to adopt dynamic allocation of connections to wavelengths is outlined to avoid congestion and dynamic wavelength assignment algorithms are proposed. The main results show the effectiveness of dynamic assignment with respect to static one and show that the connection configuration can be exploited for performance enhancement

1. Introduction

The explosive growth of the Internet demands for larger and larger bandwidth, in particular in the core part of the network. The recent introduction and rapid growth of the Dense Wavelength Division Multiplexing (DWDM) technology provides a platform to exploit the huge capacity of optical fibers. At the same time the introduction of the MPLS paradigm in TCP-IP based network promises for effective network management and traffic engineering in the future.

The integration of MPLS with the all-optical networks is a widely discussed issue, and proposals such as, for instance, MPlS are emerging. Assuming the availability of all-optical packet switching technology, this paper focuses on the problem of integration of MPLS and DWDM.

Optical packet switching has been the subject of several international research projects in the last decade (see for instance [1][2][3]) and its feasibility has been proved. Most of this work is related to the case of fixed length synchronous packets, that provides easier switching matrix design but do not match easily with IP. For this reason more recently the case of asynchronous variable length has also been studied [4][5], showing that several new problems arise but also that acceptable performance may be achieved. The work here presented considers the case of MPLS traffic and therefore assumes variable length packets.

In the case of all optical packet switching congestion resolution may be achieved in the time domain by means of queuing and in the wavelength domain by means of suitable wavelength multiplexing.

Queuing is achieved by delay lines (coils of fibers) that are used to delay packets as if they were placed for some time into a memory and then retrieved for transmission. In particular buffering variable length packets with delay lines is a very critical task and the performance depends strongly on the so called buffer time scale [6]. This is the unit of delay introduced by the delay lines that, depending on their length, delay a packet of a multiple of such unit. In particular the ratio between delay

S. Palazzo (Ed.): IWDC 2001, LNCS 2170, pp. 26–37, 2001.
© Springer-Verlag Berlin Heidelberg 2001

unit and average packet length is a key parameter and for this reason most of the results presented in the following are plotted against this quantity.

Wavelength multiplexing depends on the wavelength allocation strategy in the network architecture. The possible alternatives are two [7][8]:

- Wavelength Circuit (WC) in which the elementary path within the network for a given packet flow is designated by the wavelength, therefore packets belonging to the same flow will all be transmitted on the same wavelength;
- Wavelength Packet (WP) in which the wavelengths can be used as a shared resource, the traffic load is spread over the whole wavelength set on an availability basis and packets belonging to the same flow can be spread over more wavelengths.

In the case of a purely connectionless network such as an IP network it has been proven [9][10] that the WP approach is by far superior and may largely improve the overall performance.

This works extends these ideas in the case of a connection oriented network, such as an MPLS network and provides some insight in the problem related to the application of dynamic wavelength allocation to a connection oriented environment.

The paper is structured as follows. In section 2 a brief overview of the problem of integrating MPLS with optical packet switching is presented with particular focus on the issues of integrating management of LSPs and wavelength dimension. In section 3 a formal definition of the problem of LSP grouping on the input/output wavelengths is presented providing formulas to quantitatively evaluate such grouping. In section 4 performance results are presented for a simple wavelength allocation algorithm and in section 5 improved wavelength management algorithms are presented that take into account the grouping factor. Some conclusions are presented in section 6.

2. Multiplexing MPLS over DWDM

MPLS is a connection-oriented protocol (in contrast with regular IP) setting up unidirectional Label Switched Paths (LSPs) identified by an additional label added to the IP datagrams [11]. With MPLS the network layer functions are partitioned into two basic components: *control and forwarding*. The control component uses standard routing protocols to exchange information with other routers, and based on routing algorithms, it builds up and maintains a forwarding table. The forwarding component has the task to process incoming packets, examine the headers and make forwarding decisions, based on the forwarding table. The entire set of packets that a router can forward is split into a finite number of subsets, called Forwarding Equivalence Classes (FECs). Packets belonging to the same FEC are, from a forwarding point of view, indistinguishable and are forwarded in a connection-oriented fashion from source to destination along an LSP. The MPLS approach has important consequences with respect to e.g. traffic engineering in the IP-layer: one can set up explicit routes through the network to optimize the usage of available network resources, or to create distinct paths for different classes of quality of service.

In present proposals for optical networking, such as MPλS, the LSPs are mapped into wavelengths, in order to realize end-to-end wavelength switched paths. These may not be the most efficient solutions for multiplexing, since the incoming flow of information is bursty by nature. To achieve maximum flexibility in terms of bandwidth allocation and sharing of capacity the challenge is to combine the advantages of DWDM with emerging all-optical packet switching capabilities to yield optical routers able to support label switching of packets.

Optical routers require a further partitioning of the forwarding component into forwarding algorithm, that is routing tables look up to determine the datagram next hop destination and switching function that is the physical action of transferring a datagram to the output interface properly chosen by the forwarding algorithm. The main goal is to limit the electro-optical conversion to the minimum to achieve better interfacing with optical WDM transmission systems:

- the header is converted from optical to electrical and the execution of the forwarding algorithm is performed in electronics;
- the datagram payload is optically switched without conversion to the electrical domain.

This approach suites very well with the MPLS paradigm and exploits the best of both electronic and optical technology. Electronics is used in the routing component and forwarding algorithm, while optics is used in switching and transmission where high date rates are required.

This paper focuses on this scenario, assuming the availability of an all-optical switching matrix able to switch variable length packets, for instance like the one described in [5]. The problem of congestion due to temporary overloads is addressed by means of queuing in a fiber delay line buffer and by means of wavelength multiplexing in the assumption of a Wavelength Packet use of the wavelength resource. A very limited number of fiber delay lines are used to solve congestion in the time domain. The fibers are shared among all input/output pairs but, due to the need for delay allocation before feeding a packet into the buffer, the buffering architecture is equivalent to pure output queuing.

Should a packet be forwarded to a wavelength that is busy, it may be queued or re-routed towards a less congested wavelength on the same fiber, thus keeping the same network path (same fiber) and exploiting the WP multiplexing. This scheme has proved effective in the case of a connectionless network. In particular it has been shown that, by means of intelligent wavelength selection algorithms aiming at minimizing queue occupancy a reduction in packet loss probability of several order of magnitudes can be achieved [9][10]. These results are obtained in the assumption of a complete freedom in output wavelength assignment to incoming packets.

Unfortunately in a connection-oriented network scenario, such as MPLS, the wavelength hopping of packets belonging to the same LSP will cause out of order arrivals and updates of the forwarding table. The former issue means more complex interfaces at the edge of the optical network for re-sequencing and the latter a possible overload for the control function of the optical packet switch. Therefore a trade-off has to be found between wavelength hopping for congestion resolution and forwarding of packets of the same LSP on the same wavelength as long as possible.

The reference configuration is an optical packet switch with $N{\times}N$ input/output fibers, W wavelengths per fiber. We use the following notation:

- n: input fiber index ($n=1...N$);
- w: wavelengths index on a fiber ($w=1...W$);

Therefore a wavelength can be identified with the couple of indexes (n,w). We assume that wavelength (n,w) carries $M_{n,w}$ LSPs and that we can use the index

- m: LSP index for a given wavelength and fiber ($m=1 \ldots M_{n,w}$).

Each LSP can be identified by the triplet (n,w,m), where m is the LSP index on wavelength w of fiber n.

In the assumption of uniform traffic distribution each LSP will provide on average the same traffic load and therefore the same number of M LSPs will be carried by input and output wavelengths. Nevertheless, *congestion is not uniformly distributed* and depends on the forwarding set up of LSPs. This concept can be understood intuitively as follows: let us assume that the M LSPs incoming on a given wavelength are routed to the same output fiber n and all together forwarded to the same wavelength w. In the assumption of uniform traffic wavelength w will not carry other LSPs. Well, it happens that wavelength w will never experience any congestion and its buffer will be totally unused. Obviously the packets arriving cannot overlap since incoming from the same input wavelength and no queuing will arise. Therefore the buffering resource devoted to wavelength w is useless in this forwarding configuration.

By means of this trivial example we understand that the congestion phenomena are related to the forwarding table of the LSPs and in particular to the per wavelength grouping of LSPs routed to the same fiber. The previous example for instance suggests that, if possible, it is worthwhile to group LSPs incoming from the same wavelength and also that the unused buffering resource could be temporarily used to relieve the congestion undergoing at some other wavelength (queuing may still arise because of random fluctuations in the packet arrival process when no grouping is possible).

The basic idea is to modify the MPLS forwarding table and shift temporarily one or more LSPs from the overloaded wavelength to another where some buffering resource is available. The choice of the new wavelength among those available is a matter of optimization and has to take into account the grouping of LSPs in some form. The first step towards the definition of an algorithm to perform this task is to define a quantitative measure of the grouping of LSPs. A proposal is presented in the following section, followed by numerical examples and by an algorithm that aims at minimizing packet loss by exploiting the connection oriented nature of the input traffic.

3. The Grouping Index

In this section we define an index that is used to take into account and compare the different grouping configurations of the LSPs on the input/output wavelengths. Let us refer to a *group* of LSPs as the set of LSPs incoming on the generic wavelength (n,w) that are addressed to the same output fiber n'. For the generic LSP (n,w,m) it is possible to define a parameter $g_{n,w,m}$ that measures whether the LSP belong to a group and how numerous is that group, defined as:

$$g_{n,w,m} = \frac{k_{n,w,m}}{M_{n,w}} \tag{1}$$

Here $k_{n,w,m}$ is the cardinality (that is the number of LSPs belonging to the group) of the group which the LSP belongs to. In the case the LSP does not belong to a group because there are no other LSPs on the same wavelength addressed to the same output, $k_{n,w,m} = 0$ and consequently $g_{n,w,m} = 0$. The parameter $g_{n,w,m}$ assumes values in the range between 0 and 1, being 1 when the all LSPs on the wavelength belong to the same group with $k_{n,w,m} = M_{n,w}$ for all $m : 1 \le m \le M_{n,w}$.

Let us focus on the input ports of the switch and define the input grouping index G. On the basis of the given definitions the grouping index $G_{n,w}$ for input wavelength (n,w) is defined as:

$$G_{n,w} = \frac{1}{M_{n,w}} \sum_{m=1}^{M_{n,w}} g_{n,w,m} \tag{2}$$

and the average input grouping index G for the switch is:

$$G = \frac{1}{WN} \sum_{n=1}^{N} \sum_{w=1}^{W} G_{n,w} \tag{3}$$

In particular G is equal to 1 only when all the LSPs on (n,w) form a single group and $G=0$ when no LSPs belongs to any group. In all other cases $0 < G < 1$.

The same definition can be applied to the output ports of the switch. Let us use the index (i,j), i=1...N', j=1...W' for the output wavelengths, meaning that i is the output fiber and j is the wavelength on that fiber. In this case the output grouping index G' is defined in a similar way as follows. We first focus on the single wavelength (i,j):

$$G'_{i,j} = \frac{1}{M'_{i,j}} \sum_{m=1}^{M'_{i,j}} g'_{i,j,m} \tag{4}$$

where $M'_{i,j}$ is the number of LSPs on the wavelength and $g'_{i,j,m}$ is the corresponding of the index per LSP defined in equation (1) but referred to the output groups. An *output group* is the set of LSPs assigned to the same output wavelength of the same output fiber coming from the same input wavelength of the same input fiber. Consequently output group definition takes into account the wavelength assignment performed by the switch.

The average output grouping index G' is then given by

$$G' = \frac{1}{W'N'} \sum_{i=1}^{N'} \sum_{j=1}^{W'} G'_{i,j} \tag{5}$$

While G always depends only on LSP routing, G' does it only in case of static assignment, because otherwise, when dynamic wavelength assignment is performed, it depends also on the wavelenth assignment algorithm. The upper bound of G' is in any case G, and this represents the optimal wavelength assignment because it keeps the same grouping as the input on the output.

4. Dynamic Wavelength Assignment

In the normal switch operation it is said that an output wavelength is congested when its queue is full. To avoid unbalanced usage of wavelengths and maximum exploitation of wavelength multiplexing, the following algorithm, called Round-Robin Wavelength Selection (RRWS), is defined:

- an output wavelength is randomly assigned to each LSP at call set up;
- any time a packet arrives at queue (i,j) and finds it congested another wavelength is searched in a round-robin fashion starting from $(i,j+1)$;
- when a not congested wavelength is found, the LSP is assigned to that wavelength by updating the forwarding table so that the packets of the LSP will be forwarded to the new wavelength.

This algorithm is very simple and requires a very limited amount of processing. To prove its effectiveness we have compared the packet loss probability for the RRWS algorithm with:

1. a purely connectionless environment, like the one considered in [9], where the wavelength is changed packet by packet, packets being considered as IP datagrams not associated to specific LSPs;
2. an MPLS environment without dynamic wavelength assignment multiplexing (static case), meaning that the output wavelength per LSP is chosen at call set up and never changed regardless of the congestion state of the assigned wavelength.

The simulation set up is an optical switching matrix with $N = 4$ fibers each carrying $W = 16$ wavelengths, simulated by means of an ad-hoc event driven simulator. The number of MPLS connections per wavelength is $M = 3$ and a fiber delay buffer with $B = 16$ delay lines is placed in front of any output wavelength. The input traffic is random, with average packet length of 500 bytes and load equal to 0.8 per wavelength, and is uniformly distributed.

In this simulation for 1 set out of 16 the LSPs are forwarded to the same output fiber and grouped on the same wavelength, for 9 sets out of 16 there are 2 LSPs out of 3 that are forwarded to the same fiber and are grouped on the same wavelength and, finally, 6 sets out of 16 are such that the 3 LSPs are forwarded to different output fibers and can not be grouped. Such a configuration results from the evaluation of the average assignment of the LSPs to the outputs. The assumption of uniform traffic pattern leads to $1/N$ for the probability that a given LSP is addressed to a given output fiber. The probability P_1 that all the 3 LSPs on the same input wavelength are directed to the same output is also the probability that, given the output of the first LSP, the same value has to be extracted twice more, which leads to $P_1 = 1/N^2$. On the other hand when the 3 LSPs are forwarded to 3 different outputs, given the first value the other ones can be chosen between the remaining $N-1$ and $N-2$ respectively. This leads to a probability $P_3 = (N-1)(N-2)/N^2$. The last case of only two LSPs directed to the same output has probability $P_2 = 1 - P_1 - P_3$. With $N = 4$ the probabilities are $P_1 = 1/16$, $P_2 = 9/16$ and $P_3 = 6/16$. Assuming the previous routing configuration allows us to provide average results with a single simulation.

In this average case the input grouping index G can be evaluated as follows. For each triplet (n,w,m) the corresponding LSP may have parameter

$$g_{n,w,m} = \begin{cases} 0 & \text{if the LSP does not belong to any group} \\ 2/3 & \text{if the LSP belongs to a couple} \\ 1 & \text{if the LSP belongs to a triplet} \end{cases}$$

Because each input wavelength (n,w) carries at most one group, for $N = 4$ and $W = 16$ it follows

$$G_{n,w} = \begin{cases} 0 & \text{if } (n,w) \text{ carries only single LSPs, i.e. for 24 wavelengths out of 64} \\ 4/9 & \text{if } (n,w) \text{ carries a couple and a single LSP, i.e. for 36 wavelengths out of 64} \\ 1 & \text{if } (n,w) \text{ carries a triplet, i.e. for 4 wavelength out of 64} \end{cases}$$

The resulting input grouping index is

$$G = \frac{1}{64}\left(4 \times 1 + 36 \times \frac{4}{9} + 24 \times 0\right) = 0.3125.$$

In figure 1 the packet loss probability is plotted as a function of D that is the buffer time unit normalized to the average packet length. As already outlined in the introduction this is a critical parameter in the case of variable length packets, because there is an optimal choice of D due to the trade off between a small buffer (D small) and an inefficient utilization of the buffer itself (D large) [6].

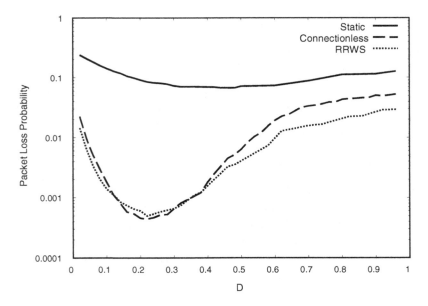

Fig. 1. Comparison of the RRWS with the static and connectionless case for N=4, W=16, M=3, G=0.3125, B = 16, average load per wavelength 0.8.

The figure shows that the RRWS algorithm significantly improves the performance with respect to the static case, being very close to the connectionless case.The reason of this behavior is motivated by the connection-oriented nature of the simulated traffic. Because of the grouping some wavelengths experience less congestion than the purely connectionless case or, in 1 case out of 16, even no congestion at all. The RRWS algorithm exploits such situations and, when congestion arises on one wavelength, it shifts the overload traffic on a wavelength not congested. Because of the grouping it is very likely that a non-congested wavelength is available to relieve a temporary overload.

To support this intuitive statement in table 1 the packet loss probability is plotted for different configurations of the LSPs forwarding table. In particular we have varied the number of triplets and couples of LSPs that can be grouped on the outputs. The table shows that the more the triplets the lower the overall packet loss probability. It also shows that the packet loss probability decreases when the grouping factor G increases, that is the quantitative description of the intuitive concepts presented

Table 1. Packet loss probability with varying numbers of LSPs following the same path at the fiber level.

# of triplets of LSPs grouped	# of couples of LSPs grouped	G	Packet loss probability
0	0	0	8.395e-04
1	9	0.3125	4.837e-04
6	5	0.5139	3.395e-04
10	3	0.7083	1.482e-04
14	2	0.9306	2.801e-05
16	0	1	0

To further support these ideas in figure 2 the output grouping index G' and the packet loss probability is plotted as a function of the simulation time. It is shown that the two curves have a complementary behavior meaning that the packet loss probability increases when the grouping index decreases and consequently it exhibits a local minimum (or a maximum) when the grouping index is maximum (or minimum).

5. Algorithm for Grouping Exploitation

The previous results show that the RRWS algorithm can significantly improve the switch performance. The cost to pay is an update of the LSPs forwarding tables any time a congestion event occurs and a wavelength hopping is performed. As outlined in section 2 this may results in a significant increase in the complexity of the network. A possible way around this problem could be to try to exploit as much as possible the grouping of LSPs that is not taken into account by the RRWS algorithm.

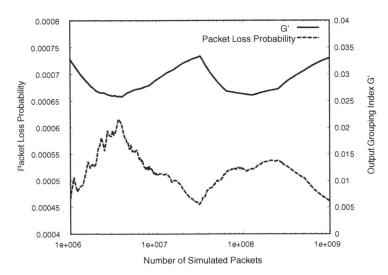

Fig. 2. Time diagram of packet loss probability and output grouping index G' for N=4, W=16, M=3, G=0.3125, B = 16, average load per wavelength 0.8.

Here we propose an extension of the RRWS algorithm that takes into account the grouping configuration inside the switch. The new algorithm is called Grouping Wavelength Selection (GWS). It is based on the assumption that connection assignment to wavelengths is performed trying to maintain the connections of a given group together. The aim is to reduce the number of connection re-assignment to wavelengths.

The algorithm performs as follows:

- a wavelength is assigned at each logical connection at call set up: a choice that optimizes the output grouping index is performed.
- connections that are grouped with gouping index equal to 1 are never moved
- any time a packet arrives at queue (i,j) and finds it congested another wavelength is searched among the whole set with the minimum number of connections already assigned and the minimum value of the grouping index. This avoids to move connections to wavelengths with high grouping index causing it to decrease (in particular no connections are added when group with grouping index equal to 1 are present)
- once a not congested wavelength l is found, the LSP is assigned to (i,l) by updating the forwarding table so that the packets of the LSP use the new wavelength l; the number of connections and the grouping index for wavelength l are updated.

Figure 3 shows the percentage of reassignment of LSPs to a new wavelength because of congestion for the RRWS and GWS algorithms. The percentage is calculated in terms of number of packets, that is the figure plots the percentage of packets that hop from a wavelength to another. This percentage is around 30% for the RRWS, that is a fairly big number. GWS significantly reduces the number of re-assignment, up to almost the half in the case of optimal values of D. This is the result of the wavelength assignment performed for groups on output link that maximizes output grouping index and of the wavelength re-assignment policy that tends to maintain the grouping index high thus reducing the average rate of congestion events.

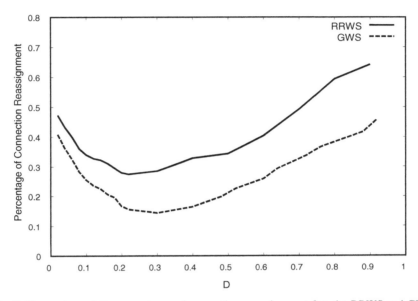

Fig. 3. Comparison of the percentages of connection re-assignment for the RRWS and GWS for N=4, W=16, M=3, G=0.3125, B = 16, average load per wavelength 0.8.

Figure 4 compares the packet loss probability for the RRWS and GWS, showing that the saving outlined in figure 3 is obtained while also maintaining practically the same packet loss performance.

6. Conclusions

In this paper we have shown that, in the presence of MPLS connections (LSPs) over a DWDM optical packet switched network, the wavelength domain can be used to improve performance. In particular dynamic re-assignment of connections to wavelengths has been considered, which exploits connection topology properties. The results showed that with the application of limited contention algorithms, dynamic re-assignment is particularly effective in relation to static mapping, being very close to a connectionless environment. An algorithm that exploits the grouping properties has also been defined to reduce the number of connections re-assigned thus simplifying switching functions.

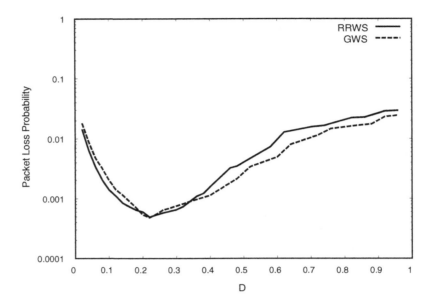

Fig. 4. Comparison of packet loss probability for the RRWS and GWS for N=4, W=16, M=3, G=0.3125, B = 16, average load per wavelength 0.8.

Acknowledgements

Work partially funded by the Italian Minister of Scientific Research, project "IP Optical Packet Networks - IPPO" and by the Commission of the European Community, Project IST-1999-11742 "DAVID - Data And Voice Integration over DWDM". The authors wish to thank Mr. Paolo Zaffoni for his help in developing the simulation programs.

References

[1] F. Masetti et al., "High speed, high capacity ATM optical switches for future telecommunication transport networks", *IEEE Journal on Selected Areas in Communications*, Vol. 14, No. 5, pp. 979-999, 1996.

[2] P. Gambini et al., "Transparent optical packet switching: network architecture and demonstrators in the KEOPS project", *IEEE Journal on Selected Areas in Communications*, Invited paper, Vol. 16, No. 7, pp. 1245-1259, 1998.

[3] L. Chlamtac et. al., "CORD: contention resolution by delay lines", *IEEE Journal on Selected Areas in Communications*, Vol. 14, No. 5, pp. 1014-1029, 1996.

[4] L. Tancevski, S. Yegnanarayanan, G. Castanon, L. Tamil, F. Masetti, T. McDermott, "Optical routing of asynchronous, variable length packets", *IEEE Journal on Selected Areas in Communications*, Vol. 18, No 10 , pp. 2084 –2093, 2000.

[5]   F. Callegati, G. Corazza, C. Raffaelli, "Design of a WDM Optical Packet Switch for IP traffic", Proc. of IEEE Globecom 2000, San Francisco, USA, November 2000.

[6]   F. Callegati, "Optical Buffers for Variable Length Packets", *IEEE Communications Letters*, Vol. 4, N.9, pp. 292-294, 2000.

[7]   C. Guillemot et al., "Transparent Optical Packet Switching: the European ACTS KEOPS project approach", IEEE/OSA Journal of Lightwave Technology, Vol. 16, No. 12, pp. 2117-2134, 1998.

[8]   D. K. Hunter et al., "WASPNET: A Wavelength Switched Packet Network", *IEEE Communication Magazine*, Vol. 37, No. 3, pp. 120-129, 1999.

[9]   F. Callegati, W. Cerroni, "Wavelength Selection Algorithms in Optical Buffers", Proc. of IEEE ICC 2001, Helsinki, June 2001.

[10] L. Tancvski, A. Gee, G. Castanon, L.S. Tamil, "A New Scheduling Algorithm for Asynchronous, Variable Length IP Traffic Incorporating Void Filling", Proc. of Optical Fiber Communications Conference - OFC'99, paper ThM7, San Diego, Feb. 1999.

[11] E. Rosen, A. Viswanathan, R. Callon, "Multiprotocol Label Switching Architecture", IETF RFC 3031, January 2001.

Access Control Protocols for Interconnected WDM Rings in the DAVID Metro Network

A. Bianco[1], G. Galante[1], E. Leonardi[1], F. Neri[1], and M. Rundo[1]

[1] Dipartimento di Elettronica, Politecnico di Torino,
Corso Duca degli Abruzzi, 24 – 10129 Torino – Italy
{bianco, galante, leonardi, neri}@polito.it

Abstract. DAVID (Data And Voice Integration over D-WDM) is a research project sponsored by the European Community aimed at the design of an optical packet-switched network for the transport of IP traffic. The network has a two level hierarchical structure, with a backbone of optical packet routers interconnected in a mesh, and metropolitan areas served by sets of optical rings connected to the backbone through devices called Hubs. The paper focuses on nodes and Hubs architecture, and on the operations of the media access protocol to be used in the DAVID metropolitan area network. A simple access protocol for datagram (not-guaranteed) traffic is defined and its performance are examined by simulation.

1 Introduction

The DAVID (Data And Voice Integration over D-WDM) project is part of the IST (Information Society Technology) Program sponsored by the European Community. Its aim is the design of an optical packet-switched network for the transport of IP traffic over metropolitan, national and international distances.

The DAVID network is designed to offer an optical transport format independent of the traffic type; the clients of the DAVID network are mainly IP routers and/or switches that collect traffic from legacy networks. The network is based on a hierarchical architecture consisting of several metropolitan area networks, named DAVID Metro networks, interconnected by a wide area optical backbone. We focus on the DAVID Metro network in this paper.

The DAVID Metro Network consists of several uni-directional slotted optical physical rings interconnected in a star topology by a Hub. No optical buffering is required in the Metro; all the buffering is done in electronics at access nodes. The Hub functionality is ring interconnection; since the Hub is buffer-less, it behaves basically as a space switch. Ring interconnections are dynamically modified at the Hub following a scheduling algorithm. The aim of the scheduling algorithm is to provide an amount of bandwidth to ring pairs close to instantaneous (short-term) bandwidth requirements. The scheduling is based both on measurements at the Hub and on congestion signals issued by nodes. A WDMA/TDMA based MAC (Medium

S. Palazzo (Ed.): IWDC 2001, LNCS 2170, pp. 38–55, 2001.

Access Control) protocol is defined to regulate access to shared network resources. A fairness protocol is proposed to guarantee throughput fairness among nodes on each ring.

The remainder of the paper is organized as follows. In Section 2 we give an overview of the DAVID network architecture. In Section 3 we focus on the metropolitan network describing both the node and the Hub architecture. In Section 4 the MAC protocol and the scheduling algorithm at the Hub are described. In Section 5 we present some preliminary simulation results to assess the performance of the proposed scheme. We conclude the paper in Section 6, where we describe future research directions.

2   Network Architecture

An overview of the two-level DAVID network architecture is shown in Fig. 1: several Metro networks are interconnected by a wide area network (WAN) backbone. Both network parts operate in packet switched mode. The backbone network consists of optical packet routers interconnected by a mesh network, while each Metro network comprises one or more rings interconnected through a Hub. Each ring collects traffic from several nodes and each Hub is connected to an optical packet router in the WAN. Access points to the network are provided both in the Metro network and in the WAN, and the traffic is collected by IP routers and switches connected to local area networks (LANs).

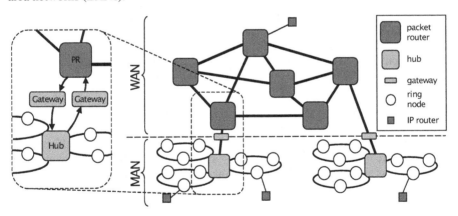

Fig. 1. General overview of the DAVID network

The network uses a mixed WDMA/TDMA access protocol: each fiber carries up to 32 wavelength channels at 2.5 or 10 Gbit/s and time is divided into fixed size slots, each carrying an optical packet which consists of a header and a payload.

In packet switched networks buffering inside routers is needed to solve contentions arising among packets arriving in a given node and headed to the same output port. In the DAVID WAN, optical packet routers provide buffering in the optical domain by means of fibre delay lines.

No packet buffering in the optical domain is instead performed for packets flowing among ring nodes in the same Metro network. In a similar way, optical buffering is completely avoided along the node-to-Hub path for traffic exchanged among Metro nodes and nodes outside the Metro. Indeed, packets are buffered in ring nodes in the electrical domain, and are sent on the Metro network only when there are enough free resources on the Metro to travel from source to destination without being stored at any intermediate node. Thus, buffers are pushed towards the edge of the Metro network and sharing of rings resources among nodes must be regulated by a properly designed MAC protocol.

The interfaces between WAN and Metro network are critical points where contentions involving traffic flowing between the backbone and the Metro networks might arise. This is worsened by the fact that optical packets could need either format or bit-rate translation (or, eventually, both) while travelling up and down the network hierarchy. Therefore, buffering and translation functions are implemented in Gateways placed between optical packet routers and Hubs (see Fig.1).

In DAVID, the Hub has connection points to the Metro rings and towards a WAN Optical Packet Router through a Gateway. A certain number of Hub ports (wavelengths) are devoted to connections towards the Gateway. The remaining Hub ports connect the Hub to the optical packet rings of the MAN. Since the Hub is buffer-less, as described later, it performs space switching and wavelength conversion only. Optical/electrical memories are present in the Gateway to solve contentions in the time domain for optical packets going from Metro network to WAN and vice versa. Moreover, the Gateway will participate in the MAC protocol, such that, from a logical point of view, the connections from and to the Gateway appear to the Hub as additional Metro ring connections.

We will focus on the DAVID Metro network in the remainder of the paper.

3 Metro Network

In general, a DAVID Metro Network consists of several uni-directional optical physical rings interconnected in a star topology by a Hub. On each fibre, a fixed number of wavelengths is available by WDM partitioning. Logical rings can either be physically disjoint (i.e., run on different fibres), or be obtained by partitioning the optical bandwidth of one fibre into disjoint portions. Nodes belonging to the same logical ring access the same set of shared resources. Recall that one logical ring may represent the WAN/MAN gateway functionality. In the remainder of the paper we use the term ring to identify a logical ring; any reference to physical rings will be explicit. The number of rings in a Metro network is denoted by N_{ring}

While the number of wavelengths on each ring can be in general different, we assume that it is a multiple of the same number. In DAVID demonstrators this is dictated by technological constraints, since SOA arrays are used at each ring node to select the wavelengths from/to which packets are received/transmitted. Up to 32 wavelengths are available on each physical ring (fibre), and all wavelengths run at either 2.5 or 10 Gbit/s. We also assume that all the nodes of a ring can transmit and receive on any wavelength used in that ring. The latter is a rather essential assumption, since the access scheme would be much more complex if nodes could have a limited tunability on the wavelengths of the ring they belong to. In particular, in this paper we assume for simplicity that the same number of wavelengths (N_{chan}=4 wavelengths) is available on any ring.

Ring resources are shared by the nodes of the Metro network using a statistical time/wavelength/space division scheme. Indeed,

- each wavelength is time slotted (TDM) and the slot duration is about 500 ns,
- several slots are simultaneously transmitted through wavelength division (WDM),
- rings can be disjoint in space (SDM).

Thus, resource sharing is based on a WDMA/TDMA scheme, i.e. a combination of Wavelength Division Multiple Access and Time Division Multiple Access.

Time slots are aligned on all wavelengths of the same ring, so that a multi-slot (a slot in each wavelength) is available to each node in each time slot. Slot alignment among different Metro rings is dealt with at the Hub; we assume for simplicity that the propagation delay on each ring is an integer multiple of the slot size. One of the wavelengths (hence a slot in each multi-slot) is devoted to management and network control purposes. We assume that this control slot can be read and written by all nodes independently of their data transmissions and receptions in other slots of the multi-slot. The control information contained in a multi-slot refers to data slot in the same multi-slot; thus, a delay is added in each node to process information contained in the control slot. Wavelengths are (dynamically) assigned to ring-to-ring communications by the Hub on a time-slot basis: all the wavelengths in the multi-slot are devoted to transmissions to a given destination ring, identified with a label in the control slot. Any wavelength in the multi-slot may be used by a ring node to reach any node in the destination ring.

Metro ring nodes are subject to collisions and receiver contentions. By collision, we mean multiple transmissions in the same time slot, the same wavelength and the same physical ring. By receiver contention we mean having in the same multi-slot and the same ring a number of packets (in different wavelengths) to be received by a given node larger than the number of receivers available at that node.

Both collisions and contentions are avoided at each source node thanks to the MAC protocol, by monitoring the state of the incoming multi-slot, and giving priority to in-transit traffic. To avoid collisions, no new packet can be transmitted on a busy channel; to avoid contentions, if the number of packets in the current multi-slot for a given destination exceeds its capacity (i.e. number of receivers), no new packet can be transmitted to that destination.

It is important to observe that contentions may arise also at the Hub; contentions are avoided by defining the Hub as a space switch and by running a proper slot scheduling algorithm.

3.1 Ring Node Architecture

We assume that the number K of transceivers at each Metro ring node is smaller than the number of WDM channels; this means that a node can only transmit and receive on at most K channels at the same time, i.e. in each multi-slot. We typically consider the case $K=1$; thus, each node has a single tunable transceiver: tuning actions are executed before transmitting and receiving independently at the transmitter and the receiver. We also assume that all the nodes of a ring can transmit and receive on any WDM channel used in the ring they belong to.

The board of a ring node is basically composed of two parts: an optical part and an electronic one. For the optical part, the ring node can drop, add and erase any packet on any wavelength at each time slot; switching is forbidden for in-transit traffic, in the sense that no operation is allowed on data not addressed to the node. Data are taken off the ring when they arrive at their destination.

The electronic part is composed of the following portions: a segmentation (reassembly) stage to create fixed size data units from variable size packets (viceversa), a queuing stage, in which packets are grouped and stored per destination ring to avoid HoL (Head of the Line) blocking [1], and a load balancing stage, to distribute the packets evenly over the available wavelengths. The HoL blocking is typical of FIFO queues: a packet at the head of the FIFO queue that cannot be transmitted to avoid collisions or contentions on the ring may prevent a successful transmission of another packet following in the FIFO order. Note that this queue architecture is very similar to the VOQ (Virtual Output Queue) architecture used in IQ (Input Queued) switches [2], where, at each input port, packets are stored in separate queues on the basis of the destination port they should reach.

Since resources (multi-slots and wavelengths) in DAVID are allocated to ring-to-ring communication, queues are organized per ring destination, i.e., at each node a FIFO queue is available to store packets directed to all the nodes belonging to a given ring. This avoids HoL blocking due to collision avoidance (since multi-slots are associated with destination rings), but does not solve HoL blocking due to receiver contentions, which would require a per-destination-node queuing scheme. The considered per-destination-ring queuing is however simpler to implement and to control, and scales much better to large network configurations.

The electronic interface is used also to solve the contention problem by running the MAC protocol and to drive the packet insertion on the ring in a free slot.

3.2 Hub Architecture

The role of the Hub is to switch packets between Metro rings, and from Metro rings towards the WAN (and vice-versa). Being all-optical, the Hub includes only a space switching stage, a wavelength conversion stage, and a WDM synchronisation stage; 3R regeneration may be added if necessary. Note that the target switching capacity in DAVID, given that in a typical Metro network $N_{ring}=4$ rings running 32 wavelengths at 10 Gbit/s are envisioned, is 1.28 Tbit/s.

In every time slot, the Hub operates a permutation from input rings to output rings, as depicted in Fig. 2 for the case of four rings. This permutation is the same for all wavelengths of each ring and is known for each time slot in each ring: we can assume that each multi-slot is labeled by the Hub with the identity of the ring to which packets transmitted in the multi-slot will be forwarded by the Hub.

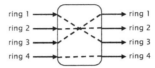

Fig. 2. A ring-to-ring permutation at the Hub

Since we are assuming that the number of wavelengths in each ring is the same, no congestion occurs at the Hub: each incoming multi-slot can be forwarded to Hub outputs. The Hub must act as a non-blocking switch that is re-configured in every time slot. It does not have to operate in the time domain, but it may have to perform wavelength conversion when the wavelengths used in the input ring are different from those used in the output ring (this always happens when the two rings are obtained in wavelength division on the same fibre).

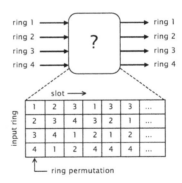

Fig. 3. Scheduling at the Hub

The computation of the sequence of permutations operated by the Hub is a scheduling problem [3, 4], as shown in Fig. 3. Several approaches can be envisaged to solve this problem, ranging from complex optimisations to simple heuristics, and are based onto an estimation of the ring-to-ring traffic pattern (note that the complexity of the scheduling problem depends on the number of rings, not on the number of nodes: this allows good scalability features). The scheduling algorithm is described in Section 4.4.

Given this Hub behaviour, each multi-slot traverses a sequence of rings, e.g. as illustrated in Fig. 4, where roman number indicate successive positions of the multi-slot, the upper slot is the control slot where the multi-slot destination ring is written, and numbers within the multi-slot represent node destinations. Nodes of ring x transmit data to be received by nodes of ring y (Steps II to IV). Ring x can be viewed as the "upstream" ring, where transmissions occur, while ring y can be viewed as the "downstream" ring, where receptions occur. Note however that when the considered multi-slot traverses the downstream ring y (Steps VI to VIII), it gathers transmissions for the next ring, say ring z, so that the traversal of a ring can be viewed as a downstream path for transmissions done in the previous ring, and as an upstream path for receptions in the following ring.

Space reuse of slots is possible in the DAVID Metro: a node receiving a packet leaves free the corresponding slot, which can be reused in the same ring, possibly by the same receiving node, for another transmission (see the transmission from node 1 to node 3 in Step I of Fig. 4). This also means that, in the example above, transmissions on upstream ring x can also be directed to other nodes of ring x (in addition to transmissions to nodes of downstream ring y). Note that transmissions to destinations belonging to the same ring of the source node must go through the Hub when the destination precedes the source in the ring, hence Hub permutations in which the input and the output ring are the same are possible and required.

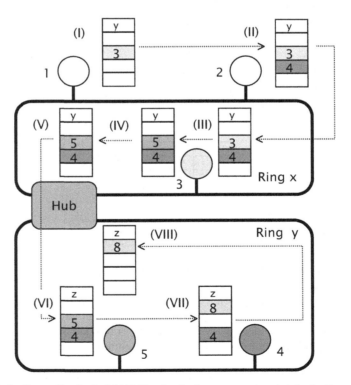

Fig. 4. Multi-slot forwarding in the MAN. Number in slots represent packet destinations

We inhibit these slot reuse capabilities in our simulation experiments, and force all traffic to pass through the Hub before being removed from the ring.

4 MAC Protocol and Scheduling at the Hub

In this section, we first describe the contention and collision problem in a DAVID Metro. Then, a simple access control scheme is proposed, and a fairness control is introduced to overcome the unfair behaviour of ring architectures. Finally, the scheduling algorithm at the Hub is discussed in detail.

4.1 Contention and Collision Resolution

Receiver contentions are not recoverable (packets would be lost), unless very complex receiver architectures are used. The proposed approach to solve contentions and collisions avoids packet losses in the path from the source node to the destination node, and is presented in the sequel. It is mainly achieved by the nodes, so that the operations and the implementation of the Hub are drastically simplified. In particular, no packet buffering, nor packet switching in the time domain, is required at the Hub.

4.2 The Access Control Scheme

In the description of the access control scheme, we assume for simplicity that the number of wavelengths supported on each ring is the same.

The choice of a ring for the DAVID Metro network significantly impacts the underlying framework in which the MAC protocol operates. Although the generic solutions befitting switches with VOQ architecture can be adapted to the ring topology, the nature of the ring, where the signal has to pass through all nodes taking a round trip time for the collection of reservations, makes token based solutions more advantageous for this environment.

The status of each slot of the multi-slot is reflected in suitable fields of the control slot. Each node that has packets to send must monitor the control wavelength seeking an empty slot in any λ of a multi-slot that will be forwarded by the Hub to the corresponding destination ring. The node grabs the slot by setting the corresponding slot status field also adding the destination address in the relevant field. The node must check before grabbing the slot that the intended destination does not already appear in as many other λs as the number of available tunable receivers ($K=1$ in this paper), in which case it refrains from getting this slot and waits for the next opportunity.

Ring nodes also monitor the control wavelength looking for any instance of their address, in which case they tune to the indicated λ to receive the data contained in the corresponding slot. Again we assume at each node a delay in processing multi-slots larger than the tuning time required to set the receiver to the proper λ.

In summary, receiver contentions are solved assuming that the source node knows how many receivers are available at the destination node: transmission of a packet is forbidden if the number of packets sent by upstream nodes in the current multi-slot to the destination exceeds the reception capacity. To avoid collisions, an empty-slot protocol is used: incoming slots are inspected, and transmission is permitted only if the slot in some wavelength is free, i.e., no upstream node transmitted in that slot and that wavelength. Note that this gives some advantage to upstream nodes, i.e., to nodes preceding others along the signal propagation direction: a given node can be completely starved by continuous transmissions of upstream nodes. This raises fairness issues, so that a protocol that provides fairness control is needed.

4.3 Fairness Control

As noted above, the proposed empty-slot operation can exhibit fairness problems under unbalanced traffic; this is particularly true in the ring topology, in which upstream nodes have generally better access chances than downstream nodes.
Credit-based schemes, such as the Multi-MetaRing [5] previously studied in the context of single ring can enforce throughput fairness. MetaRing [6] was proposed by Y. Ofek for ring-based, electronic metropolitan area networks. It is basically a generalisation of the token-ring technique: a control signal or message, called SAT, is circulated in store-and-forward mode from node to node along the ring. A node forwarding the SAT is granted a transmission quota: the node can transmit up to Q packets before the next SAT reception. When a node receives the SAT, it immediately forwards the SAT to the next node on the ring if it is satisfied (hence the name SAT), i.e. if

- no packets are waiting for transmission on the ring, or
- Q packets were transmitted since the previous SAT reception.

If the node is not satisfied, the SAT is kept at the node until one of the two conditions above are met. Thus, SAT are delayed by nodes suffering throughput limitations, and SAT rotation times increase whit the network load. To be able to provide the full bandwidth to a single node, the quota Q must be at least equal to the number of data slots contained in the ring, i.e., proportional to the ring latency (propagation delay) measured in slot times. In overload, each node sends exactly Q packets per SAT rotation time.

In the case of the DAVID MAN, several rings exist, and multi-slots traverse pairs of rings. We therefore need a SAT for each ring pair (upstream ring, downstream ring). SAT signals can be carried in the multi-slot control wavelength.

The Hub must be able to store N_{ring}^2 SATs, where N_{ring} is the number of rings attached to the Hub. Since SATs do not carry any information, N_{ring}^2 boolean variables **SAT**$_{i,j}$ do the job; **SAT**$_{i,j}$ is TRUE when the SAT regulating transmissions from ring i to ring j is at the Hub. When the Hub issues on ring i a multi-slot that will be switched, upon return to the Hub, to ring j, if the SAT $i{\rightarrow}j$ is currently at the Hub (i.e., if **SAT**$_{i,j}$=TRUE), the SAT is loaded in the control slot of the multi-slot, by setting a suitable bit, and by setting **SAT**$_{i,j}$ to FALSE.

Each node inspects the control slot of incoming multi-slots, and operates on SATs as described above for the single ring case. Recall that each queue is regulated by a different SAT and transmission opportunities are regulated by a MetaRing quota Q that may be different for each queue; however, the quota Q must be greater or equal to the ring latency to allow a single node to grab all the available bandwidth; thus, since in this paper we assume that all ring latencies are equal, we use the same value of the quota Q for all queues.

SAT are also used to trigger congestion notification signals from ring nodes to the Hub. This information is used by the Hub to determine the scheduling in successive frames as described later.

4.4 A Simple Scheduling Algorithm

We describe the approach followed to compute the scheduling at the Hub; the algorithm is run in a centralised fashion at the Hub. Multi-slots are labelled at the Hub according to the outcome of the scheduling algorithm, using the control slot to identify the ring to which the multi-slot will be forwarded upon return to the Hub. Only unicast transmissions are considered, i.e. multicast transmission are considered as multiple unicast transmissions.

The Hub scheduler is driven by an $N_{ring} \times N_{ring}$ request matrix $\mathbf{R}$. Each element $\mathbf{R}_{i,o}$ in $\mathbf{R}$ contains the number of multi-slots that must be transmitted from input ring i to output ring o, i.e., the number of multi-slots labelled with o in the control channel that the Hub must send on ring i, and, upon arrival at the Hub, switch to ring o. This request matrix is obtained by mixing periodic measurements and congestion signals issued by nodes as described in Section 4.4.1.

According to combinatorial theory [7], $\mathbf{R}$ can be scheduled in at most F time slots, where the frame length F is equal to:

$$F = \max_{i,o} \left\{ \sum_i \mathbf{R}_{i,o}, \sum_o \mathbf{R}_{i,o} \right\}$$

by using a sequence of F switching matrices $\mathbf{P}(i)$, $i \in \{1, 2, ..., F\}$, of size $N_{ring} \times N_{ring}$. A switching matrix is a binary matrix whose element $\mathbf{P}_{i,o}$ is 1 when input ring i is connected to output ring o, and 0 otherwise. The resulting scheduling in F is then repeated an integer number of times, until a new value for $\mathbf{R}$ becomes available and a new matrix decomposition can be computed. Traffic flows from ring i to ring o are served with a rate proportional to $\mathbf{R}_{i,o}/F$.

Since each input ring can be connected to at most one output ring and each output ring can be connected to at most one input ring in each time slot, a switching matrix always contains at most one non-null element in each row and in each column. Thus, the sum of each row and column in $\mathbf{P}$ is either equal to 0 or 1. Each switching matrix represents the Hub switching configuration in a given time slot; recall that we need to obtain a set of F ring permutations as the outcome of the Hub scheduling algorithm. Thus, in each time slot one and only one element from each row and one and only one element from each column must be equal to 1 in $\mathbf{P}$. In other words, we are interested in doubly stochastic switching matrixes, i.e. matrices $\mathbf{P}$ such that

$$\sum_i \mathbf{P}_{i,o} = 1, \quad \forall o \qquad \sum_o \mathbf{P}_{i,o} = 1, \quad \forall i$$

The outcome of the scheduling algorithm is a sequence of F doubly stochastic switching matrices; this scheduling satisfies a matrix $\mathbf{R}^F$ where each row and each column sum to F, a condition that in general does not hold for $\mathbf{R}$. We artificially add integer quantities, representing ring to ring multi-slot requests, to some elements in the original matrix $\mathbf{R}$, to obtain the matrix $\mathbf{R}^F$ to be scheduled. Any algorithm can be used to obtain a matrix $\mathbf{R}^F$ satisfying this condition; see e.g. [8].

The matrix $\mathbf{R}$ may be associated with a bipartite graph G having $2\,N_{ring}$ nodes. Each node represents either one input or one output of the switch, and input node i is connected to output node j by one edge only if $\mathbf{R}_{i,o} \neq 0$. A *matching* on G is a subset E of the edges in G such that, each node in G is incident to at most one edge in E. The number of edges in E is the *size* of the matching, and a matching is said to be *maximum* when it has maximum size.

We may apply a maximum size algorithm [9] on $\mathbf{R}^F$ to obtain the Hub scheduling, i.e., a sequence of doubly stochastic $\mathbf{P}(i)$, $i \in \{1, 2,..., F\}$.

Another possible algorithm that may be used is a critical maximum matching on $\mathbf{R}$. Any input i for which

$$\sum_o \mathbf{R}_{i,o} = F$$

and any output o for which

$$\sum_i \mathbf{R}_{i,o} = F$$

is said to be *critical*, since it must be served in every time slot if $\mathbf{R}$ must be scheduled in F slots. The request matrix $\mathbf{R}$ is decomposed into F switching matrices through iterated application of the critical maximum matching algorithm [4]. A *critical maximum matching* is a maximum matching which covers all the critical input and output nodes.

At step i, the decomposition algorithm computes the switching matrix $\mathbf{P}(i)$ as a critical maximum matching on $\mathbf{R}$. When the matching has size lower than N_{ring}, matrix $\mathbf{P}(i)$ is completed so that all input rings are always connected to all output rings. Finally, $\mathbf{P}(i)$ is subtracted from $\mathbf{R}$, and a new iteration is started.

At the end, the matrices $\mathbf{P}(i)$ are randomly shuffled to uniformly distribute ring to ring pairs on the F time slots of the frame, to reduce traffic burstiness.

4.4.1 Traffic Measurement

The request matrix $\mathbf{R}$ used by the scheduling algorithm is estimated on the basis of traffic measurements performed at the Hub during consecutive observation windows

(OW); the duration of each OW is fixed and roughly equal to 10 ring propagation times.

The key idea of the algorithm is that, as long as the network is not overloaded, the throughput is a good estimator of the offered load. When one or more traffic relations among different ring pairs become overloaded, congestion control mechanisms are introduced to modify the bandwidth allocation in the network. Note that overloading conditions depend on the scheduling at the Hub. If the scheduling determined at the Hub is not matched to the traffic distribution, some ring experience overloading conditions until the scheduling is not modified, since the scheduling determines bandwidth allocation among ring-to-ring pairs.

The matrix $\mathbf{R}$ that must be scheduled is computed, at the end of each OW, as the sum of 3 contributions $N_{ring} \times N_{ring}$ matrices):

$$\mathbf{R} = \lceil \mathbf{SM} + \beta\, \mathbf{IC} + \gamma\, \mathbf{EC} \rceil$$

with β and γ positive constants where:

- **SM** (smoothed measure) is a measure of the (long term) average number of multi-slots transmitted among ring pairs, where each element is a real number ranging between 0 and OW; this is an absolute throughput measure

- **IC** (implicit congestion) is the percentage of filled slots, where each element is represented as a real number between 0 and 1; this is a relative throughput measure

- **EC** (explicit congestion) takes into account explicit congestion signals sent by ring nodes, where each element is either 0 or 1.

The Hub stores in each element $\mathbf{M}_{i,o}$ of matrix $\mathbf{M}$ (measured) the number of packets flowing from ring i to ring o during each OW. The matrix $\mathbf{M}$ is then passed through an exponential filter to smooth out measurement errors, obtaining matrix $\mathbf{SM}$. Thus, a new value for $\mathbf{SM}$ is computed at the end of each OW as a function of the last measured matrix $\mathbf{M}$ and of the values assumed by $\mathbf{SM}$ at the end of the previous OW:

$$\mathbf{SM}_{new} = \alpha\, \mathbf{SM}_{old} + (1-\alpha)\, \mathbf{M}/N_{chan}$$

where $\alpha \in [0,1]$ is a constant, N_{chan} is the number of wavelengths channels available on a logical ring, which we assume to be equal for all Metro rings; matrix $\mathbf{M}$ is divided by N_{chan} to convert number of packets in number of multi-slots. Therefore, element $\mathbf{SM}_{i,o}$ of $\mathbf{SM}$ is the average number of multi-slots transmitted from ring i to ring o during one OW, roughly averaged over the last $1/\alpha$ observation windows.

Matrix $\mathbf{IC}$ gives the ring to ring connections throughput measured at the Hub, i.e., the occupation of scheduled slots. Each element $\mathbf{IC}_{i,o}$ is the ratio of the number of packets sent from ring i to ring o over the number of slots available for transmission on the same traffic relation in one OW. If $\mathbf{IC}_{i,o}$ is close to 1, this is a signal of potential congestion between i and o.

Matrix $\mathbf{EC}$ is a binary matrix which provides information on the ring congestion level on the basis of nodes queue length. Congestion signals are triggered at nodes by SAT transmissions. Each node on ring i, when releasing $\mathbf{SAT}_{i,o}$, checks the length of the queue toward ring o; if the queue exceed a given threshold, i.e. it contains more than $L_{thr} \geq Q$ packets (where Q is the MetaRing quota), the node sends, on the control channel, a congestion signal to the Hub. Note that we use the control channel to send congestion signal to the Hub instead of SAT messages, since SAT messages may be delayed by downstream nodes experiencing difficulties in channel access. Each element $EC_{i,o}$ in EC is set to 1 at the Hub, if the Hub has received at least one congestion signal toward ring o from a node on ring i during the last OW. The value of L_{thr} is related (equal in our simulation experiments) to the MetaRing quota Q; the rationale is that the quota represents, for each node, transmission opportunities toward a given ring between two consecutive node SAT reception. We assume congestion if the number of packets already in the queue when releasing the SAT is greater than the MetaRing quota, since the node will not be able to transmit all the packets in the queue in the following SAT rotation time.

Note that the two congestion signals operate on two different time scales: the first indication, stored in IC, is related to the observation window, which is fixed; the second indication, stored in EC is triggered by SAT arrivals, and depends on the SAT rotation time, which in turn depends on the number of nodes in the network. Moreover, the implicit congestion signal can be used as an early congestion signal indication, to trigger an increase in slot allocation to a given ring pair without waiting until the queue size in a node exceeds the threshold.

The presented algorithm has some important properties that we want to highlight. Suppose that the network is not overloaded, since the scheduling algorithm at the Hub provides enough slots (bandwidth) to each ring-to-ring traffic relation. This means that the scheduling determines a slot allocation "matched" to the offered traffic, i.e. a slot allocation that satisfies all traffic relations, which are never congested. This is the solution we would like to obtain with our algorithm under stationary traffic conditions. Congestion signals are never issued, since nodes do not experience congestion. Thus, the frame length is determined by the scheduling on matrix $R=SM$; the measured average slot occupation is proportional to OW via the network load ρ. All the simulation results show that if the network is not overloaded, the frame length is close to this value. This feature is obtained because the measurement interval is fixed. If we had a variable measurement interval proportional to the frame length, we would have obtained a shrinking frame length, since each measurement would create a matrix R where each element is on average reduced by a factor ρ with respect to the value assumed in the previous interval. On the other hand, a fixed measurement interval raises the problem of deciding a value for such interval, which indirectly decides also the granularity in bandwidth allocation and control. Recall that we chose the measurement interval to be equal to 10 ring latencies in our simulation experiments.

Finally, we must ensure that the scheduling provides at least a multi-slot for each ring to ring pair, i.e. at least a set of covering permutations must be scheduled in the frame, so that at least one multi-slot is available in each ring to send packets to any other ring. Otherwise, if no traffic exist on a given ring-to-ring pair, it is not possible to measure any slot occupation, the SAT cannot be sent and explicit congestion signals cannot be raised by nodes and no implicit congestion signal may be measured at the node. We enforce the scheduling to provide this set of N_{ring} covering permutations in each frame.

5 Simulation Results

We present some simulation results to assess the performance of the proposed access scheme. We do not exploit the space reuse capability described at the end of Section 3.2: if a multi-slot on ring x is labelled with destination ring z, it is used only to send traffic to nodes in ring z. Moreover, we force each multi-slot on ring x labelled with destination ring x (inter-ring traffic) to pass through the Hub; this is required to allow the Hub to perform traffic measurement for all ring pairs.

In our simulation experiments the Metro network comprise N_{ring}=4 rings, with 10 nodes on each ring. For each ring-to-ring communication N_{chan}=4 data channels are available; thus, each multi-slot comprise 5 slots, 4 for data traffic and 1 for control and management. Each nodes store packets in 4 queues, one for each destination ring. Each queue is 1000 packets long, and the packet size is matched to the slot size.

The values used in our simulation experiments for the parameters defined in the measurement algorithm are the following: β=1, γ=3, α=0.9. The ring round trip time is assumed equal to 44 time slots, the MetaRing quota is Q=44 and the threshold L_{thr}=Q. The observation window is OW=440 time slots.

We consider two traffic patterns: a uniform traffic pattern and an unbalanced traffic pattern. Define the weight matrix $\mathbf{W}$, of size $N_{ring} \times N_{ring}$, where the value assumed by each element $\mathbf{W}_{i,o}$ is a real number ranging between 0 and 1 representing the percentage of traffic generated on ring i toward ring o with respect to the total network load ρ. Clearly,

$$\sum_i \mathbf{W}_{i,o} \leq 1, \quad \forall o \qquad \sum_o \mathbf{W}_{i,o} \leq 1, \quad \forall i$$

In the uniform traffic pattern $\mathbf{W}_{i,o}$=1/N_{ring} $\forall i,o$. For the unbalanced traffic pattern $\mathbf{W}_{i,o}$=0.7 when i=o, and $\mathbf{W}_{i,o}$=0.1 otherwise; in other words, the ratio among intra-ring traffic and inter-ring traffic is 7.

Packets are generated at ring nodes according to a Bernoulli distribution whose average is derived from the weight matrix described above.

We first plot the throughput (ratio between used and allocated slots) for each destination ring on ring 0; this is a steady-state value obtained using statistically significant measures by simulation. Note that, although we plot the throughput for a single ring, the same behaviour holds for all other rings due to ring symmetries. Nodes on the same ring do not exhibit throughput unfairness thanks to the MetaRing algorithm.

In Fig. 5 we report the throughput for each destination ring on ring 0, and the overall network throughput (black square markers) as a function of offered load under uniform traffic. Each destination ring is treated fairly and the total network utilisation is close to 0.95. Note that we significantly overload the network, since ρ ranges from 0.1 to 3, but the algorithm behaves well even under this extreme condition.

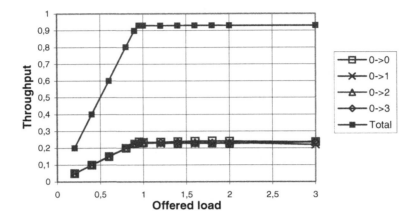

Fig. 5. Throughput under uniform traffic

The 5% utilisation loss is small, given the complexity of the system, and it can be shown to be mainly due to receiver contentions, which can be analytically evaluated with a combinatorial analysis.

Fig. 6 shows the frame length as a function of time. We start with a uniform scheduling with a frame equal to N_{ring} slots. As expected, the system converges to a frame length roughly equal to $OW \times \rho$; once this value is reached, the frame length changes slowly following traffic fluctuations. The convergence speed is determined by the value of the parameter α.

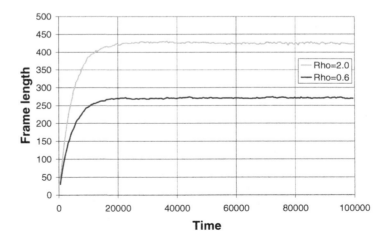

Fig. 6. Frame length as a function of time under uniform traffic

In Fig. 7 we report the throughput for each destination ring on ring 0, and the overall ring throughput (black square markers) as a function of offered load under unbalanced traffic. For values of ρ ranging from 0.1 to 1, the throughput is proportional to the weight matrix defined for the unbalanced traffic scenario. As soon as the offered load ρ increases to values that create congestion, the scheduling algorithm treats all ring-to-ring connections fairly according to a max-min like fairness criteria [10]; the intra-ring throughput decreases steadily until it reaches the same throughput obtained by inter-ring connections. Also in this scenario each destination ring is treated fairly and the total network utilisation is close to 0.95.

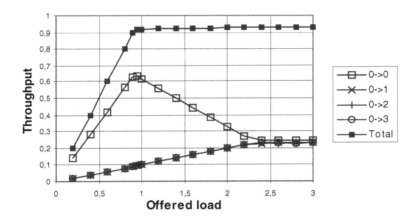

Fig. 7. Throughput under unbalanced traffic

We examine in Fig. 8 the bandwidth allocation determined by the scheduling algorithm for ring 0 under unbalanced traffic for $\rho=0.6$ (similar curves are observed for other values of ρ). The allocation is sampled at interval lasting OW, the observation window. The ideal scheduling algorithm would allocate steadily bandwidth equal to 0.7 for the connection from ring 0 to ring 0, and 0.1 to all other inter-ring connections. In our experiment, the initial scheduling algorithm is matched to a uniform traffic pattern, which is clearly not optimal for unbalanced traffic. We can observe a transient behaviour of less than 2000 slot times (roughly 4 observation windows); this value depends on the choice made for the parameters defined in the measurement algorithm. Then, the allocation is close to the optimal one, with some small variations of few % around the ideal value; these differences are due to traffic fluctuations, to which the scheduler tries to adapt the bandwidth allocation, and to inaccuracies in the traffic measurement process. The choice of the parameters should be optimised to control these fluctuations under all traffic conditions. We observed that the algorithm does not exhibit any drift from the optimal values also under heavily loaded conditions.

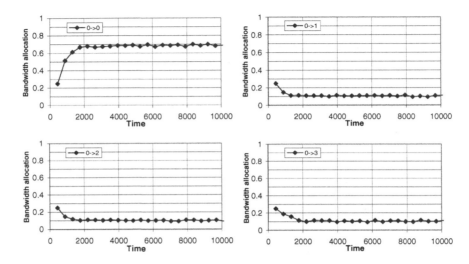

Fig. 8. Bandwidth allocation for unbalanced traffic with $\rho=0.6$

Finally, in Fig. 9 we show the queue occupancy (in packets) in overload ($\rho=2.0$) for a given node on ring 0 (all other nodes show similar queue length behaviours), sampled every 100 slot times. Whereas the queue length for intra-ring traffic saturates since this connection is overloaded, all other queues show oscillating behaviours, since each inter-ring connection becomes congested only when the scheduling does not allocate enough slots to this connection. Remarkably, although the algorithm aims only at fair bandwidth allocation, the queue occupancy level is fairly well controlled, at values smaller that 100 packets, a value not far from the ring propagation time, the time constant under which any bandwidth control cannot be achieved in this network.

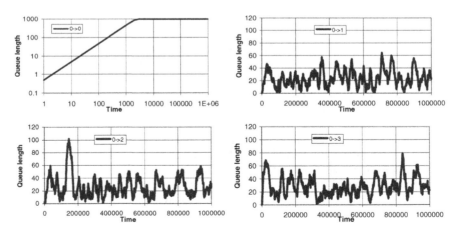

Fig. 9. Queue length for unbalanced traffic with $\rho=2.0$

6 Conclusions and Future Work

Although the presented simulation results are encouraging and the algorithm shows good performance, several issue remain to be addressed.

First, other traffic scenarios should be studied to prove the algorithm robustness to different environments. Different traffic pattern should be examined, and traffic generation should be extended from Bernoulli to on-off and/or heavy-tailed traffic models.

Then, an accurate analysis of the effect of the parameter setting must be provided, to obtain a set of values that provides good performance under different conditions. Transient behaviours must be carefully analysed to test the ability of the algorithm to follow short-term traffic fluctuations.

Finally, we want to extend the proposal to deal with multiple classes of traffic, to provide QoS guarantees similar to those of the DiffServ environment defined by the IETF for Internet.

References

1. M. Karol, M. Hluchyj, S. Morgan, *"Input Versus Output Queuing on a Space Division Switch"*, IEEE Transactions on Communications, Vol.35, No.12, December 1987, pp.1347-1356
2. N. McKeown, A. Mekkittikul, V. Anantharam, J. Walrand, *"Achieving 100% throughput in an input-queued switch"*, IEEE/ACM Transactions on Communications, Vol. 47, No. 8, August 1999
3. T. Inukai, "An efficient SS/TDMA time slot assignment algorithm", IEEE Transactions on Communications, Vol. 27, pp. 1449-1455, October 1979
4. B. Hajek, T. Weller, *"Scheduling Non-Uniform Traffic in a Packet-Switching System with Small Propagation Delay"*, IEEE/ACM Transactions on Networking, Vol.5, No.6, December 1997, pp. 813-823
5. M. Ajmone Marsan, A. Bianco, E. Leonardi, A. Morabito, F. Neri, *"All-Optical WDM Multi-Rings with Differentiated QoS"*, IEEE Communications Magazine, Feature topic on Optical Networks, Communication Systems and Devices, M. Atiquzzaman, M. Karim (eds.), Vol. 37, No.2, pp.58-66, February 1999
6. I. Cidon, Y. Ofek, *"MetaRing - a Full-Duplex Ring with Fairness and Spatial Reuse"*, IEEE Transactions on Communications, Vol.41, No.1, January 1993, pp.110-120
7. M. Hall, Jr., Combinatorial Theory, Waltham, MA, Blaisdell, 1969
8. C.S. Chang, W.J. Chen, H.Y. Huang, *"Birkhoff-von Neumann Input Buffered Crossbar Switches"*, IEEE Conference on Computer Communications (INFOCOM 2000), Tel Aviv, Israel, pp. 1614-1623, March 2000
9. R.E. Tarjan, *Data Structures and Network Algorithms*, Society for Industrial and Applied Mathematics, Pennsylvania, November 1983
10. D. Bertsekas, R. Gallager, *Data networks*, Prentice-Hall, 1987

Reactive Search for Traffic Grooming in WDM Networks

Roberto Battiti and Mauro Brunato

Università di Trento, Dipartimento di Matematica
via Sommarive 14, I-38040 Pantè di Povo (TN), Italy
{battiti|brunato}@science.unitn.it

Abstract. In this paper the *Reactive Local Search* (RLS) heuristic is proposed for the problem of minimizing the number of expensive Add-Drop multiplexers in a SONET or SDH optical network ring, while respecting the constraints given by the overall number of fibers and the number of wavelengths that can carry separate information on a fiber. RLS complements local search with a history-sensitive feedback loop that tunes the prohibition parameter to the local characteristics of a specific instance.
Simulation results are reported to highlight the improvement in ADM cost when RLS is compared with greedy algorithms used in the recent literature.

1 Introduction

During the last few years, dramatic advances in optical communications have led to the creation of large capacity optical WANs, usually in the form of hierarchies of rings. As shown in Fig. 1, these rings are joined together via hubs in hierarchies at various levels.

In the framework of Wavelength Division Multiplexing (WDM) the overall bandwidth of the fiber is divided into densely packed adjacent frequency bands. In this manner, every fiber can carry a large number of high-capacity channels. Dividing these channels into lower-bandwidth virtual connections between couples of nodes gives rise to technical difficulties: wavelengths are at most a few hundreds, therefore they need to be time-multiplexed to be shared among many couples of communicating nodes. The key components to achieve time multiplexing, the Add-Drop Multiplexers (ADM), are need edeach time a wavelength has to be processed electronically at a node in order to add or drop packets. Therefore they represent a significant fraction of the total cost of the infrastructure.

We consider the problem of dividing the available bandwidth of a single ring into many channels to establish as many node-to-node connections as requested. In the following, the term *virtual connection* shall be used to refer to an elementary node-to-node communication channel.

In Fig. 2 we describe the structure of a node in the ring. Optical bandwidth is shared among communication channels in two ways (both of which are implemented in state-of-the-art optical networks): the incoming signal is initially

S. Palazzo (Ed.): IWDC 2001, LNCS 2170, pp. 56–66, 2001.
© Springer-Verlag Berlin Heidelberg 2001

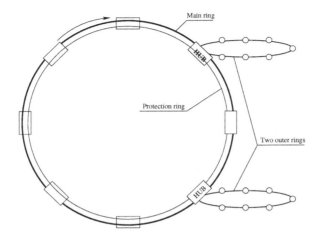

Fig. 1. A SONET/SDH hierarchy of rings

split into its component wavelengths by means of a Wavelength Demultiplexer (WDEMUX), then some wavelengths are directly relayed to the next node, while some others are further split into packets via a time division multiplexing mechanism operated by Add-Drop Multiplexers (ADM). ADMs are critical components (they need to exploit fast packet conversion, header examinations, and must be tightly synchronised), therefore they are expensive.

In this paper we shall consider unidirectional WDM rings where each wavelength can be time-shared among g virtual connections. In other words, the bandwidth of a wavelength is assumed to be g (also called *grooming factor*), while single virtual connections, carried by a wavelength in a single time slot, have a unit bandwidth.

The paper is organized as follows. Section 2 describes the Reactive Local Search heuristic and summarizes the reasons leading to its choice for the WDM Traffic Grooming problem. Section 3 defines the problem, and summarizes approaches that have already been used for its solution. Section 4 is devoted to the application of RLS to the WDM traffic grooming problem, with a description of the design choices and of the data structures involved in the process. Section 5 analyzes simulation results, giving an experimental comparison among previous heuristics and the proposed technique.

2   Reactive Local Search

Let us consider a discrete optimization problem, where the system configuration x is given by parameters varying in a discrete set C, for example a binary string ($C = \{0,1\}^n$) or an integer vector ($C = \mathbb{N}^n$), and the optimality of a given configuration is evaluated by a cost function $f : C \to \mathbb{R}$.

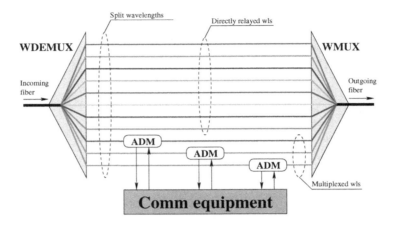

Fig. 2. Structure of a node in a WDM SONET/SDH ring

Many interesting optimization problems are known to be computationally unaffordable, and the WDM minimization problem has been proved NP-hard [10]. Local Search techniques are a family of heuristics aimed at finding near-optimal solutions to hard problems, optimizing the value of the cost function by local modifications of the system configuration.

A simple example is given in Fig. 3, where a repeated *local search* technique is applied to the problem of minimizing a function of a binary string. Here "local" means that the step from a solution to the next is performed by flipping a single bit of the string. In other words, the *neighborhood* of a solution is given by all solutions having Hamming distance 1 from it.

The basic steps of the algorithm are lines 4–12. Here a random configuration is generated (line 4), then local steps are repeatedly performed by evaluating the objective function on all neighbors and finding the best improvement (lines 6–12). If f is bounded from below, this leads to a local minimum of the cost function[1], and by repeating the whole procedure with new random starting points (lines 3–16) the optimum will eventually be found. Once a local minimum is found the algorithm will update the best configuration (lines 13–15) and jump to a completely new starting point.

A substantial improvement to the previous scheme is obtained when one considers that problem instances are not completely unstructured. Usually, local minima tend to cluster, and once one is found it is advisable to continue exploring its vicinities rather than starting immediately from a new random point. The basic scheme cannot do that, because it is forced to explore only towards improving solutions. A possible modifications of this approach is Simulated Annealing, which accepts - with a given probability - solutions that lead to increases of the

---

[1] Unless a neighborhood with equal cost is reached (a "plateau"): modifications of the algorithm in Fig. 3 would be necessary in this case

```
 1.   var x, best_x: binary string; n, cost, best_cost: integer;
 2.   best_cost ← +∞
 3.   repeat
 4.    ⌈  generate a random solution into x
 5.    │  cost ← f(x)
 6.    │  repeat
 7.    │   ⌈  let n be one of the bits of x whose flipping most decreases
 8.    │   │      the value of f(x)
 9.    │   │  if a decrease is possible then
10.    │   │   ⌈  x[n] = not x[n]
11.    │   └   └  cost ← f(x)
12.    │  while an improvement is possible
13.    │  if cost < best_cost then
14.    │   ⌈  best_cost ← cost
15.    └   └  best_x ← x
16.   until some termination condition is satisfied
```

Fig. 3. The basic local search algorithm on a binary string

cost function; notice that Simulated Annealing is a Markovian heuristic, where the next configuration produced is not influenced by past history, but only by the current configuration.

Another branch of local search heuristics consists of non-Markovian techniques, such as prohibition-based heuristics that forbid some moves according to criteria that take into account past moves. Prohibition-based schemes date back to the sixties [8,9], and have been proposed for a growing number of problems in the eighties with the term *Tabu Search* (TS) [6] or *Steepest Descent-Mildest Ascent* [7]. Fixed TS (see the classificatin proposed in [1]) implements the Local Search technique with two major modifications:

1. it does not terminate a run when the local minimum is found and
2. once a move (for example a bit flipping) is made, the reverse move (i.e. flipping that bit again) is prohibited for a given number of iterations T (called *prohibition period*).

Although various TS schemes have been shown to be effective for many problems, some of them are complex and contain many possible choices and parameters, whose appropriate setting is a problem shared by many heuristic techniques In some cases the parameters are tuned through a trial-and-error feedback loop that includes the user as a crucial *learning* component: depending on preliminary tests, some values are changed and different options are tested until acceptable results are obtained. The quality of results is not automatically transferred to different instances and the feedback loop can require a lengthy process.

On the other hand, *reactive schemes* aim at obtaining algorithms with an internal feedback (*learning*) loop, so that the tuning is automated. Reactive schemes are therefore based on *memory*: information about past events is collected and used in the future part of the search algorithm. The TS-based Reactive Local Search (RLS) adopts a reactive strategy that is appropriate for the neighborhood structure of the problem: the feedback acts on a single parameter (the *prohibition period*) that regulates the search diversification and an explicit memory-influenced restart is activated periodically. RLS has been successfully applied to a growing number of problems, from Maximum Clique [3] to Graph Partitioning [2].

In this paper, RLS is adapted for the ADM minimization problem in WDM traffic grooming.

3   The WDM Traffic Grooming Problem

Let N be the number of nodes in the ring, and let M be the number of available wavelengths, computed as the number of fibers times the number of wavelengths per fiber. The problem is not affected by the values of the two factors, since the only important number is the overall number of wavelengths.

Every physical link from a node to its neighbor is capable of carrying M wavelengths; at every node, some of these wavelengths will be simply relayed to the outgoing link by an internal fiber without electronic conversion, while some others will be routed through an ADM, which is therefore necessary only if some of the traffic contained in that wavelegth is directed to, or originated from, that node.

Let g be the *grooming factor*, i.e. the number of low-bandwidth channels that can be packed in a single wavelength on a fiber. For example, if the wavelengths are carrying an OC-48 channel each, a grooming factor $g = 16$ means that traffic is quantized into 16 OC-3 time-multiplexed channels, while if $g = 4$ only 4 OC-12 time-multiplexed channels will be carried.

Another fundamental parameter is the *traffic pattern*, an $N \times N$ matrix T whose entry t_{ij} is the number of time-multiplexed low-bandwidth unidirectional virtual connections required from node i to node j. Note that, being the ring unidirectional, there is no reason to consider the matrix as symmetric, and that the diagonal elements must be null: $t_{ii} = 0$.

Let P be the overall number of virtual connections along the ring: $P = \sum_{ij} t_{ij}$. A solution to the problem can be given by an $N \times N$ matrix W whose entry w_{ij} is an integer array of t_{ij} elements (thus empty if $i = j$), one for each virtual connection from node i to node j. Thus, the wavelength assigned to the k-th virtual connection from i to j ($1 \leq i, j \leq N$, $1 \leq k \leq t_{ij}$) is w_{ijk} ($1 \leq w_{ijk} \leq M$).

The number of ADMs required at node i ($1 \leq i \leq N$) is the cardinality of the set of wavelengths assigned to virtual connections that originate from, or go to, node i:

$$CH_i(W) = \left\{ w_{ijk} : 1 \leq j \leq N \wedge 1 \leq k \leq t_{ij} \right\} \bigcup$$

$$\bigcup \left\{ w_{jik} : 1 \leq j \leq N \wedge 1 \leq k \leq t_{ji} \right\}.$$

Because $CH_i(W)$ is a set, multiple occurrences of the same wavelength are counted once (indeed, only one ADM is required to deal with them). The overall number of ADMs needed for a given wavelength assignment W is

$$f(W) = \sum_{i=1}^{N} \#CH_i(W).$$ (1)

Wavelength assignment matrix W is subject to the constraint that no wavelength should be overloaded in any fiber of the ring. The fiber segment exiting from node n (and going to node $(n \mod N) + 1$) is traversed by all virtual connections (i, j) where $i \le n < j$ or $n < j < i$ or $j < i \le n$. The load of wavelength l on the outgoing fiber of node n is given by the cardinality of the set

$$WL_{nl} = \Big\{ (i, j, k) :$$

$$(1 \le i \le n < j \le N \vee 1 \le n < j < i \le N \vee 1 \le j < i \le n \le N)$$

$$\wedge \quad 1 \le k \le t_{ij} \quad \wedge \quad w_{ijk} = l \Big\}.$$

The load constraint will assert that

$$\forall n, l \quad (1 \le n \le N \wedge 1 \le l \le M) \Rightarrow \#WL_{nl} \le g.$$ (2)

The WDM traffic grooming problem can be stated as follows.

WDM GROOMING PROBLEM:
Given integers $N > 0$, $M > 0$, $g > 0$ and the $N \times N$ traffic matrix $T = (t_{ij})$
Find a wavelength asignment

$$W = (w_{ijk}) \quad (1 \le i, j \le N, \quad 1 \le k \le t_{ij}, \quad 1 \le w_{ijk} \le M)$$

that minimizes the objective function (1) subject to the load constraint (2).

Most papers on the traffic grooming problem propose combinatorial greedy algorithms [4,5,10]. For example, [5] suggests some techniques for different kinds of traffic matrices, notably the egress-node case where all traffic is directed towards a single hub node, the uniform all-to-all case and a more general distance dependent traffic. Two types of algorithms are presented. Some algorithms attempt to maximize the number of nodes requiring just one ADM, then those requiring two and so forth. Others try to efficiently pack the wavelengths by dividing nodes into groups and assigning to different wavelengths intra-group traffic.

4   Reactive Local Search for the Grooming Problem

To implement RLS in an efficient way, the problem needs to be formulated in terms of an integer array optimization. To transform the wavelength assignment matrix into an integer array we simply associate an array entry to each virtual connection.

First of all, all indices start from 0, so in the following sections nodes indices will vary from 0 to $N - 1$ and channel numbers from 0 to $M - 1$. All virtual connections between nodes are enumerated consecutively, so that the k-th virtual connection from node i to node j is assigned index

$$p_{ijk} = \sum_{i'=0}^{i-1} \sum_{j'=0}^{N-1} t_{i'j'} + \sum_{j'=0}^{j-1} t_{ij'} + k,$$

so that any node couple (i, j) is assigned a consecutive group of t_{ij} paths. The overall number of paths is, of course,

$$P = \sum_{i=0}^{N-1} \sum_{j=0}^{N-1} t_{ij}.$$

The wavelength assignment matrix W is stored into an P-entries integer array $S = (s_i)$, where the wavelength assigned to the k-th virtual connection from node i to node j is stored into $s_{p_{ijk}}$.

The objective function (1) and the load constraint (2) are given in Sect. 3.

To implement a local search heuristic we need to define the neighborhood of a given configuration S. The basic move we chose is equivalent to changing the wavelength assignment of a given virtual connection, i.e. to changing a single entry of the current configuration array S.

Last, we modified the objective function to take into account load constraint violations by adding to it a penalty term given by the number of violations multiplied by a very large constant (larger than NM, i.e. the maximum number of ADMs in the system). For this reason we needn't explicitly check for a non-violating string: while minimizing the objective function the number of violations is automatically reduced.

We show in Fig. 4 an outline of the Reactive Local Search algorithm used for the WDM grooming problem. The function accepts three parameters (line 1): the maximum number of moves *max_moves*, the integer array *best_x* which is going to contain the best found solution and the cost function *ADMs*.

In the first section (lines 4–7) variables are declared. The array x contains the current problem configuration, T contains the current prohibition period (time after which a given move can be repeated), while the array *LastExecuted* is indexed by moves and contains the last iteration a move was performed ($-\infty$ if it was never executed). By *move* we mean the local step from a configuration to the next. In our tests a move is represented by a couple of integers (i, n) meaning that wavelength n is assigned to the i-th virtual connection.

The initialization section (lines 8–15) generates a random configuration, initializes the prohibition period T and sets all entries in *LastExecuted* to minus infinity (they have never been performed).

The solution improvement cycle (lines 16–28) is made of two distinct parts. At first a suitable move is found: the move must be the "best" in the sense that, after

```
1.   function RLS_for_Grooming (max_moves: integer;
2.          by_ref best_x: array of integer;
3.          function ADMs (array of integer): integer);
4.   var
5.   ⌈  x: array of integer;
6.   │  n, cost, best_cost, T, mv, step: integer;
7.   ⌊  LastExecuted: array of moves;
8.   Initialization:
9.   ⌈  generate a random assignment into x
10.  │  best_x ← x;
11.  │  cost ← ADMs(x);
12.  │  best_cost ← cost;
13.  │  T ← 1;
14.  │  for each mv
15.  ⌊      LastExecuted[mv] ← −∞;
16.  Improvement:
17.  ⌈  for step in 1..max_moves do
18.  │  ⌈  find the best move mv such that LastExecuted[mv] < step - T
19.  │  │  modify x[i] according to mv
20.  │  │  cost ← nADM(x);
21.  │  │  LastExecuted[mv] ← step
22.  │  │  if cost < best_cost then
23.  │  │  ⌈  best_cost ← cost
24.  │  │  ⌊  best_x ← x
25.  │  │  if x has been visited too often then
26.  │  │      increase T
27.  │  │  else if T hasn't been increased for some time
28.  ⌊  ⌊      decrease T
```

Fig. 4. The RLS algorithm for the ADM minimization problem. The details about the reactive prohibition scheme determining T (lines 25–28) are given in the text.

some nonprohibited moves have been checked (as it may be impractical to check all possible moves, we chose to stop after checking 1000 possible candidates), a move is chosen that most decreases the *ADMs* function or, if it is impossible, it increases *ADMs* as little as possible. After the move has been performed, the configuration string is updated and the new value of the cost function is computed, compared with the best value and eventually stored.

After every move, a *reaction* step is performed (lines 25–28): a configuration dictionary is checked for the current configuration. If it has already been visited, then the prohibition period T is increased by some amount (10% in our test), while if no configuration has been repeated for some time the value of T is reduced (again, by 10%).

The implementation of the Reactive Local Search algorithm has been done in C++ language. The program operates on the P-entry array, while a 64-bit (`long long int`) hash fingerprint of each visited configuration is used to index a LEDA `dictionary` structure containing relevant data such as the iteration number and the number of times that configuration has been visited.

The initial prohibition period is 1 (in this case a move cannot be undone in the next step). If the number of configurations that have been visited more than three times exceeds 3, the prohibition time is increased by a 10% amount and rounded to the next integer. If prohibition time has not been raised for a certain number of steps, then it is decreased by the same amount (a high prohibition time facilitates escaping from local minima, but prevents a large fraction of neighboring configurations from being explored).

5 Simulation Results

We tested the algorithm on the all-to-all uniform traffic case, where the traffic requirement is equal to 1 for all couples of nodes:

$$t_{ij} = \begin{cases} 1 & i \neq j \\ 0 & i = j \end{cases} \quad \Rightarrow \quad T = \begin{pmatrix} 0 & 1 & \dots & 1 \\ 1 & 0 & \dots & 1 \\ \vdots & \vdots & \ddots & \vdots \\ 1 & 1 & \dots & 0 \end{pmatrix}.$$

Figure 5 shows a comparison between the RLS results (best value after 10 10^5-step runs) and Modiano's algorithm for all-to-all uniform traffic with one channel request per couple of nodes. We considered cases $g = 4$ and $g = 16$, as they are analitically studied in [5]. In both cases RLS results in a considerable reduction of the number of ADMs in the ring, up to 28% for $g = 4$ and up to 31% for $g = 16$.

In Fig. 6 we compare the best solution found after 10 RLS runs (already reported in Fig. 5) with the average value for the same set of runs of the algorithm. In order to distinguish the two lines we were forced to restrict the graph to the higher portion of tested values (12 to 16 nodes) and to a grooming factor equal to 16. In fact, most runs return the best found value, and only one or two in the total end with one or two more ADMs.

The RLS algorithm took about 8 minutes of CPU time per run on a 500MHz PentiumIII computer with 64 MB RAM running the Linux operating system. The size of the problem (the number of nodes) did not affect the execution time because we fixed to 1000 the number of neighboring configurations to check.

6 Conclusions

Experimental results obtained by simulations of the all-to-all uniform traffic case show that the proposed RLS technique is competitive with other greedy techniques used in the literature. Of course, local search heuristics need to explore a large set of solutions to find good local minima; this is acceptable, because circuit planning is an off-line operation, where a computation taking a few minutes is perfectly tolerable, in particular when it cuts down hardware costs in a significant way.

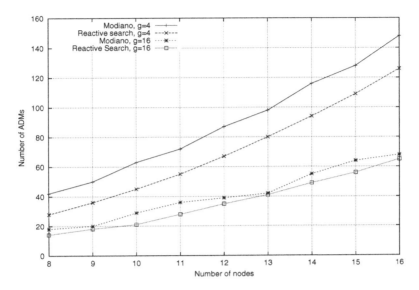

Fig. 5. Comparison between RLS and Modiano

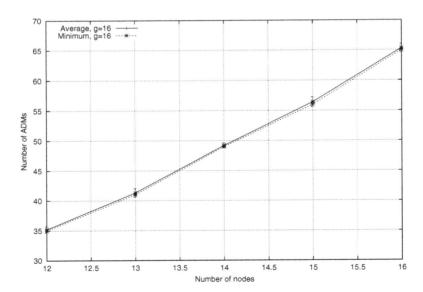

Fig. 6. Minimum, average and confidence interval for RLS, $g = 16$

References

1. Battiti, R.: Reactive search: Toward self-tuning heuristics. In V. J. Rayward-Smith, editor, Modern Heuristic Search Methods, chapter 4, pages 61–83. John Wiley and Sons Ltd, 1996.
2. Battiti, R., Bertossi, A.: Greedy, prohibition, and reactive heuristics for graph partitioning. IEEE Transactions on Computers **48** (1999) 361–385
3. Battiti, R., Protasi, M.: Reactive local search for the maximum clique problem. Algorithmica **29** (2001) 610–637
4. Berry, R., Modiano, E.: Reducing electronic multiplexing costs in SONET/WDM rings with dynamically changing traffic. IEEE Journal on selected areas in communications **18** (2000) 1961–1971
5. Chiu, A., Modiano, E.: Traffic grooming algorithms for reducing electronic multiplexing costs in WDM ring networks. IEEE Journal of Lightwave Technology **18** (2000) 2–12
6. Glover, F.: Tabu search — part I. ORSA Journal on Computing **1** (1989) 190–206
7. Hansen, P., Jaumard, B.: Algorithms for the maximum satisfiability problem. Computing **44** (1990) 279–303
8. Kernighan, B., Lin, S.: An efficient heuristic procedure for partitioning graphs. Bell Systems Technical Journal **49** (1970) 291–307
9. Steiglitz, K, Weiner, P.: Some improved algorithms for computer solution of the traveling salesman problem. Proceedings of the Sixth Allerton Conference on Circuit and System Theory (Urbana, IL, 1968) 814–821
10. Wan, P.-J., Călinescu, G., Liu, L., Frieder, O.: Grooming of arbitrary traffic in SONET/WDM BLSRs. IEEE Journal on Selected Areas in Communications, **18** (2000) 1995–2003

Micromobility Strategies
for IP Based Cellular Networks

Joseph Soma-Reddy and Anthony Acampora

University of California, San Diego
9500 Gilman Drive, La Jolla, CA 92093 USA
soma@cwc.ucsd.edu, acampora@ece.ucsd.edu

Abstract. With the increasing popularity of IP based services on mobile devices, future cellular networks will be IP based and designed more as an extension to the Internet rather than the telephone network. Managing mobile hosts within a IP based network is a challenge especially when fast handoff are required. We propose a micromobility scheme that achieves very fast handoffs with no wireless overhead. We compare our protocol with other proposed schemes.

1 Introduction

The evolution of cellular networks into the third generation and beyond will drastically change the nature of mobile applications and services. While mobile telephony has been the dominant application so far, the high data rates available in future will cause IP based services like Internet access, email etc. to become more prevalent. The nature of the mobile devices will also change from simple cellphones to IP based hosts like PDA's, handheld computers, laptops etc.

Eventually, the hybrid voice and data 3rd generation networks will evolve into All-IP cellular networks which will be just an extension to the Internet. Base stations would simply be routers with wireless links and will be connected directly to the Internet. Telephony will be offered as an application using VoIP(voice over IP). Such a network will be truly integrated with the Internet and will enable easy creation and deployment of new mobile services.

One of the main challenges facing the development of such networks is mobility management. The Internet protocols have been designed for fixed hosts. The IP address of a host identifies not a particular host but rather its point of attachment. Thus a mobility management protocol is needed to handle mobile users and to ensure that they receive service as they move within the All-IP cellular network.

The Mobile IP protocol[1] is the Internet standard for managing mobile hosts. While it is a scalable protocol, it has been designed for nomadic hosts and its high latency and overhead make it unsuitable for cellular networks that require fast and frequent handoffs. Instead a hierarchical approach is needed with two levels of mobility. A micromobility protocol that handles fast mobility within a small region and a macromobility protocol that handles mobility between these regions.

S. Palazzo (Ed.): IWDC 2001, LNCS 2170, pp. 67–75, 2001.

Various micromobility protocols have been proposed. However, all these protocols require exchange of signaling packets between the mobile and the cellular network during a handoff. Since cellular links have high delay, these protocols suffer from significant handoff latency. When VoIP is used, these protocols introduce noticeable speech disruption during a handoff.

In this paper, we propose a new micromobility protocol that does not require any exchange of signaling packets over the air during a handoff. When the mobile moves to a new base station, it is able to acquire and register a new IP address without any *over the air* signaling. This ensures that the handoff can be completed quickly and with no overhead on the wireless link. There is no disruption of speech during a VoIP call. The protocol does not impose any more infrastructure requirements or scalability constraints compared to other proposed schemes. We conclude that the protocol compares favorably with other proposed schemes and can be used along with a macromobility protocol like Mobile IP to achieve fast, scalable handoffs.

In section 2 we describe the Mobile IP protocol and the hierarchical mobility management architecture. Micromobility schemes proposed in the literature are described in section 3. We introduce our micromobility protocol in section 4 and compare it with other schemes in section 5.

2   Mobile IP

The Mobile IP protocol[1] has been standardized by the IETF to provide mobility to Internet hosts. This protocol allows a mobile host to maintain its permanent IP address while changing its point of attachment. The operation of the protocol is illustrated in Fig. 1. The mobile host has a permanent address(called the Home Address) assigned to it. The mobile is identified by this address and other hosts that wish to communicate with the mobile use this address. In addition, the mobile acquires a temporary Foreign Agent address(also called the *care of* address) at its current location. The mobile registers the *care of* address with its Home Agent node, which is located on the mobile's home network. When the mobile is at a foreign location, the Home Agent intercepts packets addressed to the mobile and encapsulates them in new packets addressed to the mobile's *care of* address. The Foreign Agent receives these packets and, after decapsulation, delivers to the mobile. The encapsulation and decapsulation is transparent to the corresponding host and to the transport and higher layers in the mobile which see only the permanent address of the mobile. A slight variation of this protocol, called the *colocated care of address option*, has no Foreign Agent and the decapsulation function is performed by the mobile.

The Mobile IP protocol is a very scalable protocol. It only requires a Home Agent and Foreign Agent(not necessary with *colocated care of address* option) and the rest of the network infrastructure can be left unchanged. Since the mobile always has a topologically correct address at its current location, packet routing is done on the basis of network prefixes, which is a very scalable method. Hence the protocol can operate well in a network of global scale, like the Internet.

However, when it is applied in situations that demand fast mobility, like cellular networks, the high overhead and latency of the protocol make it inefficient. At each handoff, the mobile node has to acquire a *care of* address and register it with the Home Agent(which may be in a distant location). This process can take a long time(up to one second) and requires the exchange of several packets between the mobile, foreign network and Home Agent.

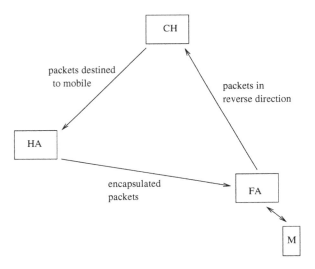

Fig. 1. Mobile IP

Scalability and low latency are two important requirements for a mobility management protocol for the global cellular network. However, these two requirements are contradictory and it is difficult to design a protocol that is both fast and scalable. Hence, a hierarchical mobility management architecture with two levels of mobility is used. This is illustrated in Fig. 2. The global cellular network is divided into domains, each of which spans a small geographic area. A domain consists of a Gateway router, several intermediate routers and a collection of base stations all of which are under the administrative control of a single authority. A micromobility protocol would handle mobility between base stations within a single domain and a macromobility protocol would handle mobility between domains. The micromobility protocol must be able to achieve fast handoffs but need not be scalable since it is used only within a domain which is of limited size. The macromobility needs to be scalable but not necessarily fast since it only handles handoffs between domains, which is not a frequent occurrence. It needs to be a standard protocol, since it has to operate between different domains which could be under different administrative authorities. Mobile IP, being an IETF standard and a scalable protocol, is a natural choice for this role. Thus we need a micromobility protocol that can execute fast handoffs in order to achieve global mobility management in a fast, efficient and scalable manner.

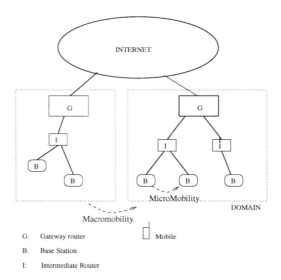

Fig. 2. Hierarchical Mobility Management

3   Other Micromobility Protocols

3.1   Description

Several micromobility protocols have been proposed in the literature[2,3,4]. In[2], Mobile IP itself has been proposed to be used as a micromobility protocol. Since the Home Agent will be located in the Gateway router, close to the mobile(because a domain will consist of a limited geographical area), the latency due to mobile registration with the Home Agent will be small. Other optimizations like using link layer triggers to detect a handoff immediately(rather than waiting for Foreign Agent or Router advertisements) can help reduce handoff latency. Work is progressing in this direction in the Mobile IP Working Group.

   In[3], a new micromobility protocol called Cellular IP has been proposed. This protocol is different in that the mobile uses the same IP address at different locations. Instead, the routing tables are modified at each handoff so that packets addressed to the mobile are routed to its new location. When the mobile moves to a new base station, it sends a route update packet to the Gateway router. As the route update packet travels to the Gateway router, the routers along the way change their routing entry for the mobile in the reverse direction. Thus symmetric routing paths are created for the mobile and updated at each handoff.

3.2   Performance

A common feature of all the proposed micromobility protocols is the exchange of packets between the mobile and a node within the fixed network(Home Agent or Gateway Router) at every handoff. This exchange of packets causes both high latency and overhead in the mobility protocol. The cellular channel is unreliable and time varying due to effects like path loss, shadowing and fading. In order to counter these effects, particularly fading, diversity techniques like coding, interleaving etc. are used. The effect of these techniques is to increase the delay on the cellular link. For example, the cdma2000 air interface specifies a 20ms interleaver. Since the complete frame must be received before it can be de-interleaved, the interleaver alone introduces a delay of between 20 and 40ms. Various other factors like coding, processing and queuing delays also contribute to the overall delay. The typical cellular link has been known to suffer a delay of between 60-100ms. Delays for packet data systems could be much large r(due to higher queuing delays, ARQ schemes etc.). GPRS networks have a link delay of between 80-200ms and Ricochet networks(a cellular packet data network available in some cities in US) have link delays of the order of 230-500ms.

Clearly, a handoff protocol that relies on the exchange of packets over cellular links would suffer significant latency. For example, Mobile IP(with colocated care of address) would require the exchange of 4 packets(two packets to acquire an IP address and two to register the address with the Home Agent). This alone would cause a handoff latency of 240-400ms. Assuming a 20ms delay for the radio level handoff and a 10ms delay on the wired network, the total handoff latency would be 280-440ms. Cellular IP, on the other hand, requires exchanging 2 packets(route update packet and acknowledgment). The handoff latency would be 160-240ms.

In a delay tolerant data connection, a handoff latency of a few hundred milliseconds will lead to either delayed or lost data packets and will not substantially affect the quality of the connection. However, for a delay intolerant service like a VoIP call, handoff latency leads to lost packets. Thus a VoIP call made over a IP based cellular network using a Mobile IP based mobility protocol would suffer a speech loss of about 240-400ms(12-20 consecutive frames at 20ms a frame) at each handoff and a speech loss of 160-240ms(8-12 consecutive frames) at each handoff if Cellular IP were the mobility protocol. The speech loss could be worse if header compression techniques are used(and they must, otherwise the header overhead would be too much for VoIP), since such techniques usually propagate packet loss when several consecutive packets are lost. Thus, it is not possible to execute handoffs without audible glitches during a VoIP call. And this problem will be exacerbated in future cellular networks which will achieve higher system capacity through smaller cell sizes. This leads to increased rate of handoffs which will worsen the above problem.

4   Our Micromobility Protocol

4.1   Protocol Description

The high handoff latency in the proposed schemes is caused largely by the high
delay on the cellular links. Thus, to achieve low latency handoffs, we eliminate
any signaling over the cellular link. This not only achieves low latency handoffs,
but also reduces signaling overhead on the cellular link.

When the mobile moves to a new base station, it needs to acquire a new IP
address that is topologically consistent with its new location. Since we propose
to not do any signaling over the cellular link during a handoff, the new IP
address must either be assigned to the mobile ahead of time or the mobile must
be somehow capable of determining the new IP address. While it is possible
to pre-assign to the mobile, IP addresses at every base station in the domain,
a better alternative would be for the mobile to determine the IP address at
each base station based on some unique radio level parameters that have been
already exchanged as part of the radio level handoff. The radio level identifiers
that can be used to determine the IP address of the mobile will depend on the
particular cellular network that is in operation. Thus, when the mobile moves
to the new base station, it(and the network) can determine its new IP address
without exchanging any IP level packets over the air.

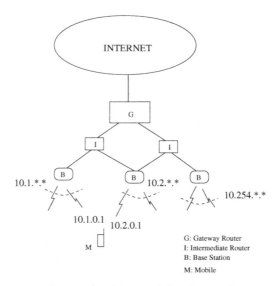

Fig. 3. Our Micromobility Protocol

For example, consider the domain shown in Fig 3. In this domain, the temporary IP address of a mobile connected to a base station is derived as follows:

a mobile identifier m_i of length 16 bits is derived from some unique radio level identifier of the mobile, say its IMSI(international mobile subscriber identifier). Similarly, a base station identifier b_i of length 8 bits is derived from a radio level identifier of the base station. The two identifiers are then concatenated and prefixed with the network id to form an IP address. Thus, mobile m_i when connected to base station b_i has a temporary IP address $10.b_i.m_i$(where 10.*.*.* is the network id of this domain). Thus mobile 1 when connected to base station 1 has temporary IP address of 10.1.0.1 and when connected to base station 2 has temporary IP address of 10.2.0.1. When the mobile moves to a new base station, the identifiers m_i and b_i are derived from the radio level identifiers exchanged during the radio level handoff and the new temporary IP address is constructed by both the mobile and the new base station without exchanging any packets.

A link layer trigger is used to inform the network layer(in the mobile and base station) of the handoff so that there is no delay in handoff detection. Finally, the change of IP address of the mobile is kept transparent to the higher layer through encapsulation.

4.2   Protocol Operation

When a mobile firsts connects to a base station within a domain, it is assigned a public IP address that will be used by the mobile as long as it is within the domain. In addition, the mobile will determine its temporary IP address at the base station based on the radio level parameters exchanged during the radio level handoff. The base station informs the gateway of the mobile's temporary IP address.

Packets addressed to the public IP address of the mobile are intercepted by the Gateway router and encapsulated in a new IP packet that is addressed to the temporary IP address of the mobile. Since this is a topologically consistent address, the packets will reach the mobile. At the mobile, the packets are de-encapsulated and sent to the higher layers which are unaware of the temporary address.

The mobile continuously measures the signal strengths of the pilot transmissions of nearby base stations. When the mobile decides to handoff to another base station(because the received pilot signal strength from it is stronger than the current base station), it simply registers with the new base station at the radio level. A link layer trigger informs the network layer in the mobile and the new base station of the handoff. The mobile and the base station independently determine the new temporary IP address of the mobile at that base station without exchanging any packets. The base station informs the Gateway router of the mobile's new temporary IP address. The Gateway now begins to encapsulate packets addressed to the mobile with the new temporary IP address and they reach the mobile at the new location.

Since no packets are exchanged over the air, the latency with this mobility management protocol is independent of the delay on the cellular link and is determined solely by the latency of the radio level handoff and the delay on the wired network(between the base station and the Gateway router). Using the numbers from section 3, the handoff latency is 40ms. This corresponds to a loss of 2 speech frames in a VoIP call, which is unnoticeable by the user.

5   Comparison

Latency: With our micromobility protocol, the handoff latency is only 40ms while it is about 280-440ms with Mobile IP as the micromobility protocol and about 160-240ms with Cellular IP as the micromobility protocol. While this difference may not be noticeable in a data connection, a VoIP call would experience noticeable speech disruption at every handoff with both Mobile IP and Cellular IP protocols, while with our protocol there would be no noticeable speech disruption.

Signaling Overhead: In our protocol, there is no *over the air* signaling and thus no wastage of wireless bandwidth. In contrast, with Mobile IP as the micromobility protocol, 4 packets are sent over the air at every handoff and with Cellular IP as the micromobility protocol, 2 packets are sent over the air at every handoff. The amount of wireless bandwidth consumed by these signaling messages is not significant at the current rate of handoffs, but with increasing rate of handoffs in the future(because of shrinking cell sizes), this could become important.

Scalability: Since the mobile always uses topologically consistent addresses, only one routing table entry per base station is required in the intermediate routers. This is true for Mobile IP also. However, with Cellular IP as the micromobility protocol, the intermediate routers need to have one routing table entry per mobile(because the mobile uses the same address at different base stations). As the size of the domain increases and the number of mobiles per domain increases, our protocol and Mobile IP will be able to scale better.

Infrastructure Requirements: The Gateway router and the base stations need to be modified to operate our protocol. The rest of the cellular infrastructure does not need any changes and standard IP equipment can be used. This is also true for Mobile IP. However, with Cellular IP as the micromobility protocol, the intermediate routers also need to be modified to support the protocol.

6   Conclusion

In this paper, we presented a micromobility scheme for an IP based cellular network. It achieves fast handoffs with no overhead no the wireless links. We compared the performance of our protocol with other proposed schemes and determined it is better. In conjunction with Mobile IP, it can be used for mobility management over a global IP based cellular network.

References

1. Charles Perkins, editor: IP Mobility Support. Internet RFC 2002, October 1996
2. Claude Castelluccia and Ludovic Bellier: A Hierarchical Mobility Management Framework for the Internet. IEEE Intl. Workshop on Mobile Multimedia Communications, Nov. 1999
3. A. G. Valko: Cellular IP - A New Approach to Internet Host Mobility. ACM Computer Communication Review, January 1999
4. Ramachandran Ramjee et. al.: HAWAII: A Domain-based Approach for Supporting Mobility in Wide area Wireless Networks. International Conference on Network Protocols, ICNP'99

Satellite Systems Performance
with TCP-IP Applications

P. Loreti[1], M. Luglio[1], R. Kapoor[2], J. Stepanek[2], M. Gerla[2],
F. Vatalaro[1], and M. A. Vázquez-Castro[3]

[1] Dipartimento di Ingegneria Elettronica, Università di Roma Tor Vergata
[2] Computer Science Department, University of California Los Angeles
[3] Dept. de Tecnologías de las Comunicaciones, Universidad Carlos III de Madrid

Abstract. Satellites are well suited for mobile Internet applications because of their capability to enhance coverage and support long-range mobility. Satellites are an attractive alternative for providing mobile access to the Internet in sparsely populated areas where high bandwidth UMTS cells cannot be economically deployed or in impervious regions where deployment of terrestrial facilities is not practical. In this paper we analyze various mobile Internet applications for both GEO and LEO satellite configurations (Iridium-like and Globalstar-like.) For the simulations, we use 'ns2' (Network Simulator 2), enhanced to support LEO and GEO satellites and mobile terminals. As part of the ns-2 upgrade, we developed a channel propagation model that includes shadowing data from surrounding building skylines. We compute via simulations the performance of FTP applications when users are mobile, traveling along "urban canyons". The results show that throughput and delay performance is strongly affected by skyline shadowing and that shadowing degradation can be compensated by satellite diversity, such as provided by Globalstar.

1. Introduction

The use of satellites for Internet traffic is a very attractive proposition since the wired network is highly congested. On the average, an Internet request needs to travel through 17 to 20 nodes, and hence may go across multiple bottlenecks. On the other hand, with satellites, just one hop is sufficient to connect two very distant sites.

Several systems, operating or in development, have been planned to support the satellite traffic. In particular, it is worth mentioning Spaceway (Hughes) [1], Astrolink (Lockheed Martin) [2], Cyberstar (Loral) [3], SES-Astra [4], Eutelsat [5], SkyBridge (Alcatel) [6] and Euroskyway (Alenia Spazio) [7].

UMTS/IMT2000 (Universal Mobile Telecommunication System/International Mobile Telecommunications 2000) aims to be the platform for supporting new multimedia services. In this new system, the satellite component is supposed to be integrated with the terrestrial component [8]. Satellites, thus, assume a particularly important role, and not only complementary, especially if aiming at real global coverage and ensuring access to maritime, aeronautical and remote users.

S. Palazzo (Ed.): IWDC 2001, LNCS 2170, pp. 76–90, 2001.
© Springer-Verlag Berlin Heidelberg 2001

The main goal of this paper is to evaluate the performance of typical mobile Internet applications, using different TCP schemes. We analyze some representative satellite scenarios, two existing LEO systems (Iridium [9] and Globalstar [10]) and the classical GEO configuration. We develop a channel propagation model based on shadowing from surrounding building skylines. The model parameters are based on actual data in a built-up area.

The paper is organized as follows. Section 2 reviews the role of satellites. In Section 3, we highlight the characteristics of satellites likely to have an effect on TCP and other Internet applications. Section 4 introduces the "urban canyon" model for the study of the satellite channel in urban environments. Section 5 presents the simulation platform (NS-2) used in our experiments and describes the extensions required for LEO satellite support. In Section 6, we describe the experiments and present the experimental results in Section 7. Section 8 presents the conclusions.

2. The Role of Satellites

In the future, satellites will play an important role in providing wireless access to mobile users, for enhancing coverage of existing or developing land mobile systems and to ensure long-range mobility and large bandwidth. The presence of satellite links provides extra capacity and an alternate and less congested route to existing wired and wireless systems, thus offering unique opportunities for improving efficiency and reliability. The broadcast nature of the satellite channel also makes satellites suitable for broadcasting/multicasting.

The satellite may also be integrated with short-range wireless networks (Bluetooth) in order to provide Internet services to mobile vehicles. In this scenario, the satellite will connect one vehicular terminal to the Internet while the Bluetooth system will be able to interconnect equipment inside the vehicle (car, bus, train, plane or ship).

The above-mentioned systems [1-7], providing large bandwidth directly to users, are based on a geosynchronous orbital configuration. A single GEO satellite is able to cover a very wide area with multiple narrow spot beams using multibeam antennas. In this way, very reliable links can be provided even with small terminals. Such GEO systems provide (or are designed to provide) high availability high data rate links (up to 2 Mbit/s and more) and huge total capacity (of the order of 10 Gbit/s). Many of them are expected to provide regional limited coverage in a few years and global coverage (excluding poles) finally.

As an alternative to GEO satellites, LEO constellations may also be considered. These have low latency and thus are more suitable for real time applications but they may need more than one hop to reach remote destinations. They may provide global or limited coverage depending on the inclination of orbits. An important issue with LEO satellites is that satellites move and therefore, connections have to be handed off. The only available LEO system (Globalstar [10]) provides low bit rate voice and data services; no broadband system is foreseen for the next few years. One LEO system, called Teledesic is being developed for the provision of broadband services [11].

Previous work [12-16] has addressed the performance of TCP/IP over geostationary satellites. This has mainly focused on evaluating performance of different TCP schemes (Tahoe, Reno, New Reno, SACK), where the connection between two fixed stations goes across a GEO satellite.

In this paper, we also address other TCP schemes (Westwood [16]) and evaluate TCP performance when terminals are mobile. Such an environment creates a very realistic scenario for the evaluation of performance in a satellite environment.

3. Satellite Features Impacting Performance of Internet Applications

First, when considering the performance of real time (e.g. voice and video) applications the large propagation delays represent a critical issue. The problem is particularly acute in TCP applications over links with large bandwidth-delay products, where a large TCP window is required for efficient use of the satellite link. A lossy satellite link can cause frequent "slow start" events, with significant impact on throughput. In the case of LEO networks, a delay variation is introduced due to the fast movement of satellites, which causes frequent handovers. The key satellite network features that need to be considered in order to evaluate the impact of satellite on internet application performance are: propagation delay, Delay-Bandwidth Product (DBP), frequent handover, signal-to-noise ratio (SNR), satellite diversity, routing strategy.

Satellite systems are changing their role from the "bent-pipe" (transparent) channel paradigm to the on-board routing (regenerative) paradigm associated with packet transmission and switching. In this process, which involves both GEO satellites and non-GEO constellations, each satellite is an element of a network, and new design problems are involved both at the network and the transport layers.

While considering the performance of standard Internet protocols over paths including satellite links, the large propagation delays have to be taken into account. The problem is particularly relevant to TCP due to its delay-sensitive nature. For voice and other real time applications too, the delay may be critical. The delay problem can be further accentuated in the case of LEO networks, where the delay may also be very variable and the satellite or gateway handover may occur frequently.

Some of the main aspects that need to be considered in order to evaluate the impact of system architecture on the performance of network protocols are:

Propagation Delay

Due to the large distance that packets need to travel in satellite networks, they can experience a significant transmission delay (depending on the type of constellation and the routing strategy adopted). In addition, this delay may also show a lot of variability when a LEO architecture is used.

The transmission delay in a GEO architecture depends on the user-gateway distance (or user-user if direct connection is allowed) when a single satellite configuration with no ISLs (inter-satellite links) is considered, or on the connection strategy if ISLs are used. This delay can be considered constant in the case of fixed users while it is variable in the case of mobile users.

In the case of LEO constellations, the delay variability is much greater than in the GEO case. In fact, the time variant geometry of the constellation induces a fast variability of the user-satellite-gateway distance both for fixed and for mobile users. This phenomenon is even more critical if ISLs are implemented because the routing strategy may also play a very important role. In [20], a simplified analysis, considering the delay as constant, is presented.

The delay variability is particularly relevant to TCP, since it causes the round-trip-time to be variable and may affect the estimate of Round Trip Time (RTT) by TCP. Whenever a change in RTT occurs, it takes some time for TCP to adapt to the change. If the RTT keeps changing (as can happen in case of LEO constellations), TCP may not be able to update its estimate of RTT quickly enough. This may cause premature timeouts/retransmissions, reducing the overall bandwidth efficiency.

Delay-Bandwidth Product (DBP)

The performance of TCP over satellite links is also influenced by the Delay-Bandwidth-Product (DBP). This has been evaluated through simulations and experiments [13, 14, 15]. This performance can be enhanced by adopting techniques such as selective acknowledge and window dimensioning [13][15]. Also, techniques such as TCP spoofing and cascading may be used [15].

Frequent Handover

In a connection-oriented service with LEO satellites, each time a handoff occurs, a sizeable number of TCP packets can get lost, especially if the signaling exchange is not performed carefully. Also, after the handoff is complete (and packets are lost), TCP has to scale down its congestion window to 1.

Satellite Diversity

Satellite diversity is a very efficient technique for improving link availability or SNR by utilizing more than one user-satellite link to establish communication. In a connectionless system, the satellite diversity feature has the advantage of increasing the probability of packet delivery to/from the satellite network from/to Earth. However, this increases the number of busy channels, which causes a reduction in capacity. In the experiments in this paper, we see how the satellite diversity in Globalstar helps to improve performance over Iridium.

Signal-to Noise Ratio (SNR)

In GEO constellations, the SNR (or equivalently Bit Error Rate, BER) is characterized by a great variability due to free space losses and tropospheric propagation (including rain, over 10 GHz) for fixed communications and also due to shadowing in case of mobile communications. Both shadowing and deep rain fading can cause packet loss.

In LEO systems, the BER variability may be caused, in addition to the previously mentioned phenomena, by the radio link dynamics due to the continuously changing link geometry (antenna gain and free space losses).

Poor SNR is an extensively studied issue for GEO satellites. Since TCP has been designed mainly for wired links, it assumes that the BER is very low (of the order of 10^{-10}, or less) so that when a packet is lost, TCP attributes it to network congestion. This causes TCP to employ its congestion control algorithms, which reduces the overall throughput. This is detrimental in satellite links, and causes an unnecessary reduction in throughput. The effects due to poor SNR extend to LEO and hybrid LEO/GEO networks.

4. Mobile Satellite Channel Model

It is well accepted that signal shadowing is the dominant critical issue influencing land mobile satellite (LMS) systems availability and performance. While multipath fading can be compensated through fade margins and advanced transmission techniques, blockage effects are not easy to mitigate, resulting in high bit error rates and temporary unavailability. The solution to reduce such shadowing effects is path or satellite diversity [19].

Shadowing is mainly due to blockage and diffraction effects of buildings in urban areas and mountains or hills in rural areas, although absorption losses through vegetation may also significantly impair the radio channel in a number of environments including urban. Non-GEO orbit satellite networks have dynamic, yet deterministic topologies and the use of multiple visible satellites and satellite diversity techniques is needed to improve link availability.

Constellations for positioning satellite systems are designed so that even if them some of the satellites are shadowed, the number of satellites is sufficient to get the data needed for position calculation. In the case of communication satellite systems too, constellations must take into account the shadowing occurrence. Also, since voice and data applications are affected by shadowing differently, the impact of satellite diversity needs to be studied separately for them.

Models based on physical or physical-statistical approaches make use of detailed information of the actual environment or a synthetic environment. Consequently, these models are appropriate to study the use of satellite diversity since geometry information is inherently included in such models. This is true for other modeling approaches like purely statistical or empirical models.

In this paper we are using a physical-statistical land mobile satellite channel model. The model proposed here is based on computing the geometrical projection of buildings surrounding the mobile, described through the statistical distributions of their height and width [19]. The existence or absence of the direct ray defines the line-of-sight (LOS) or shadowed state, respectively. The model can be divided into two parts:

Deterministic or Statistical Parameterization of Urban Environment

The physical-statistical approach used here proposes a canonical geometry for the environment traversed by a mobile receiver, typically a street as shown in Figure 1. The canyon street, composed of buildings on both sides, may block the satellite link along the mobile route depending on the satellite elevation.

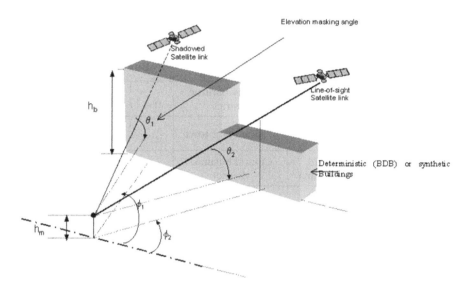

Fig. 1. Shadowed and Line-of-sight satellite links. Buildings can be obtained either from a BDB or through generating synthetic environment (h_b: building height, h_m: mobile height).

In the case of deterministic characterization of the urban environment, a Building Data Base (BDB) is used to obtain the canyon street data. Then a receiver is placed at a given position (right or left lane or sidewalk) and the skyline (building profile in the sky direction) as seen by the receiver terminal is computed. For fixed users, the skyline obtained is fixed; only the satellites will be moving according to the constellation dynamics. In the case of a user moving along a given street, the skyline seen by the mobile as well as the satellites will be time varying. Note that for satellite systems using Gateways (GW), the signal goes from transmitter to receiver through the GW. However, the GW links must be considered free of shadowing effects due to the environment since the GW antennas are sufficiently elevated and directive and keep tracking the satellites in view.

In order to also address a statistical approach, which means computing synthetic canyon streets, we investigated urban canyon street geometry to parameterize real street canyons. Statistical approach is clearly of interest towards general results provided that we use real data to generate the canyon streets. In addition, statistical approaches are generally less time consuming and BDB are not always available.

Heights of urban environments from two European countries were studied and statistically parameterized enabling the generation of realistic streets. High and medium built-up density areas from England and Spain were investigated. Heights of buildings in London and Guildford were found to be log-normally distributed while three different sectors of Madrid were found to adhere to a truncated Normal distribution. These results are summarized in Table 1.

For the simulations performed in this work, we used the Madrid Street that has an average masking angle of 30°.

Table 1. Fitted distributions of building heights.

Country	Location	General Description	Building Heights
England	London	Densely built-up district	Log-normal $\mu = 17.6$ m, $\sigma = 0.31$ m
	Guildford	Medium-size town	Log-normal $\mu = 7.2$ m, $\sigma = 0.26$ m
Spain	Madrid, Castellana	Central business district	Normal $\mu = 21.5$ m, $\sigma = 8.9$ m
	Madrid, Chamberí	Residential area	Normal $\mu = 12.6$ m, $\sigma = 3.8$ m

Calculation of the Elevation Angle to the Skyline (Masking Angles)

Once the canyon shaped street is available, either by extracting data form DBDs or by computing it, the elevation angle to the skyline can be computed. At the user position, a scan of 360 degrees is performed to compute the elevation angle to the skyline, i.e., the masking angle is computed for every azimuth angle around the user terminal. Figure 2 shows an example of the skyline surrounding the user. Buildings are synthetic and are generated with parameters corresponding to Madrid-Castellana. Figure 3 shows an example of computed masking angles for four streets in Madrid.

To determine link conditions, we use these computed elevation masking angles for different values of the azimuth angle under which the satellite is seen. If the satellite elevation angle is larger than the masking angle, it is assumed to be in line-of-sight, otherwise it is blocked. The procedure is repeated for all satellites in view. In case more than one visible satellite is found the one with the highest elevation angle is chosen, and the packet is made available to this satellite. If no satellites are found, the packet is dropped.

Our canyon street geometry includes modeling of crossroads by setting buildings to zero height. What is not considered in the canyon shaped street model is the eventual presence of second-row buildings, but their effect can be considered negligible for the purposes of this paper.

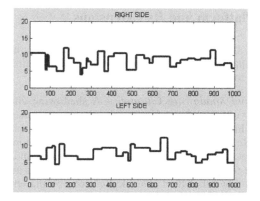

Fig. 2. Example (1000 samples) of skyline generated with parameters of London

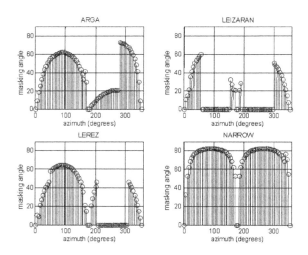

Fig. 3. Masking angles computed for 4 streets of Madrid.

On the other hand, the time the mobile needs to move along a given canyon street may be too short (depending on the mobile speed) to obtain statistically significant TCP simulation times. In order to obtain long simulation runs very long canyon streets were generated to simulate the statistical variation of heights and the occurrence of crossroads with the movement. In addition, longer routes are also easily and realistically obtained by changing the azimuth reference of the masking angle series.

5. Simulation Platform and Its Extensions

The simulations have been performed using ns-2 (Network Simulator) [17], enhanced to provide better support for satellite environments. The following were the key enhancements added to ns-2:

Terminal Mobility and Shadowing Channel – A shadowing channel was added to simulate the behavior of a mobile terminal in an urban environment. The channel was derived from the skyline of a street in London. The terminal is considered shadowed if the elevation angle of the satellite is less than the elevation of the skyline, as explained in Section 4. The channel has an ON-OFF behavior and the link is assumed to be down when the terminal is shadowed. Also, mobility was added to the terminal by moving it up and down the street. The skyline seen by the terminal changes as it moves and this combined with the current position of the satellite network determines the shadowing state of the terminal.

Gateway – The concept of a "Gateway" node was added to the simulator. This node was introduced to model the "Gateways" present in satellite networks like Iridium and Globalstar. The Gateway can be used as an interface between the satellite constellation and a terrestrial network and this feature can be used to model hybrid satellite-terrestrial networks. An important feature of the Gateway node is that it maintains links to all satellites that are visible to it. Also, these links typically belong to different orbits in a non-polar constellation. This "multiple links" property is used to enhance the Globalstar constellation to a "full constellation" in which inter-orbit switches are performed by going through the Gateway node.

Diversity – Since the Gateway node can maintain links to more than one satellite, it supports "selection diversity" of satellites. This means it periodically (every 0.4 s) evaluates the elevation angle of all the visible satellites to select the best and forwards packets to it.

Mobility Modeling and Handoff - In the simulations, mobility is modeled by moving the terminal continuously up and down the street over a straight path of about 10 km. The position of the terminal at any time is determined by its speed and the time elapsed and the skyline seen by the terminal at that position gives the minimum elevation angle below which a satellite is shadowed. While modeling handoffs, we assume that the handoff execution time is negligible. The handoff procedure is invoked every 0.4 s and a handoff takes place when either the current satellite goes below the horizon or the current satellite is shadowed by the skyline. While performing the handoff, we look for the "unshadowed" satellite with the highest elevation angle. Note that the skyline gives us the minimum elevation angle above which a satellite is visible for a certain value of azimuth of the satellite. Thus, the azimuth of a satellite together with the information provided by the skyline determines whether a satellite is shadowed or not.

6. Description of the Experiments

To measure TCP performance for LEO satellite networks (both Iridium and Globalstar constellations) we ran a set of 24 half-hour FTP sessions in various topologies. For all the experiments, we ran FTP transfers using different TCP versions (Reno, SACK, and Westwood).

In the first topology, both the satellite gateway and a terminal were co-located in Europe (lat: 47°N, long: 7°E), and connected with a single LEO satellite (the LEO constellation has no ISLs). In this topology, we measured TCP throughput using a range of uniformly distributed packet error rates between 10% and 40%. We also performed one experiment for a low packet error rate of 10^{-4} in order to compare with higher packet error rates. In addition to the packet error rate, we varied link capacity between 16 Kbit/s and 144 Kbit/s. The results for this topology are presented in Figures 4 and 5.

In the second topology, the terminal was located as in the earlier topology, but the gateway was relocated to Los Angeles. In this case, the LEO constellation also has inter-satellite links. Again, we performed TCP throughput experiments using packet error rates between 10% and 40%. The link capacity was also varied as earlier. Figure 6 shows the results for this topology.

The third topology used a GEO satellite to connect two terminals, one located in Rome and the other in Washington D.C. respectively, with the satellite positioned over the Atlantic Ocean. TCP throughput experiments were performed using packet error rates between 10^{-3} and 10^{-1}. Link capacities were 1.024 Mbit/s and 2.048 Mbit/s. In this case, only six half-hour FTP sessions were used for the GEO simulations. The LEO case required more FTP sessions in order to appropriately capture changes due to the time-varying constellation topology. The results for this experiment are shown in Figure 7.

In the fourth set of experiments, the two terminals are again located at Rome and Los Angeles. These are connected using two GEO satellites, connected by an ISL. The parameters (link capacity, packet error rate) are the same as in the previous experiment; the results are shown in Figure 8.

Finally, we perform shadowing and mobility experiments both for GEO and LEO configuration. We use the mobile satellite channel model explained in Section 4. In the first shadowing experiment, a mobile terminal, located in Madrid and a Gateway, located in Los Angeles are connected using a LEO satellite. The terminal and the Gateway run an FTP transfer between them. We do this experiment for the 3 TCP versions (Reno, SACK and Westwood) for both Iridium and Globalstar. The link capacity is 2.048 Mbit/s and the terminal speed is varied. The results are shown in Figure 9.

7. Simulation Results

Figure 4 shows performance of FTP as a function of packet error rate for single hop LEO configuration. In this case performance is acceptable up to an error rate of 10% and there is no significant difference in performance among different TCP schemes; the choice of constellation (Iridium or Globalstar) also does not cause much difference in performance.

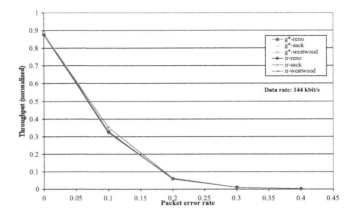

Fig. 4. Performance of FTP as a function of packet error rate for single hop LEO configuration

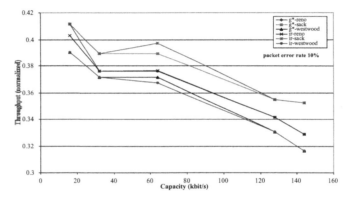

Fig. 5. Performance of FTP as a function of capacity for single hop LEO configuration

Figures 5 shows performance of FTP as a function of capacity for a single hop LEO configuration. The graph shows the normalized throughput (the utilization) as a function of link capacity.

In Figure 6, we show performance of FTP as a function of packet error rate for a full LEO configuration.

Figure 7 shows performance of FTP as a function of capacity and packet error rate for a single hop GEO configuration, with users located in Rome and Washington D.C. It can be seen that the faster recovery algorithm employed by TCP Westwood, which sets the congestion window equal to the estimated link capacity * RTT, shows a better performance compared to TCP Reno. It may be noted that TCP Reno halves its window each time a packet is lost due to corruption. This new window size will be less than the estimated window size by TCP Westwood except when the pipe is full (in which case both are equal).

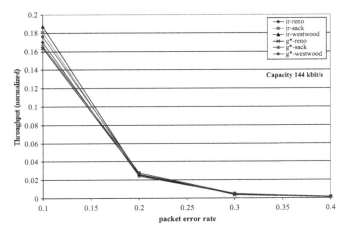

Fig. 6. Performance of FTP as a function of packet error rate for full LEO configuration

Figure 8 shows the results for configuration containing two GEO satellites, (connected by an ISL) with the users located in Rome and Los Angeles respectively. TCP Westwood shows better performance in this configuration too.

All TCP experiments demonstrate a high sensitivity of TCP performance to random errors. The impact of errors is particularly egregious for the GEO case as a result of increased Bandwidth Delay Product (BDP). TCP Westwood performs significantly better in the presence of such random errors.

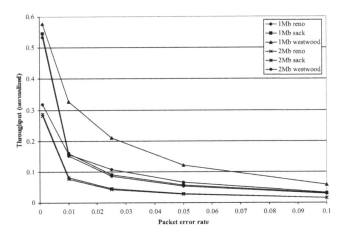

Fig. 7. Performance of FTP as a function of capacity and packet error rate for single hop GEO configuration

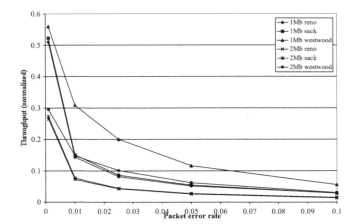

Fig. 8. Performance of FTP as a function of capacity and PER for double GEO (with ISL) configuration

Figure 9 shows that, when using a LEO constellation, there is not much difference between the TCP schemes, but Globalstar outperforms Iridium due to its "diversity capability". In the Globalstar constellation, most areas are covered by more than one satellite at any point. Iridium, on the other hand, provides only one satellite to cover an area. When one satellite is shadowed, Globalstar is able to select an alternate satellite to establish a connection. This enables to provide better service compared to Iridium in the presence of shadowing. The performance of TCP Westwood is also seen to be slightly better than that of the other TCP versions.

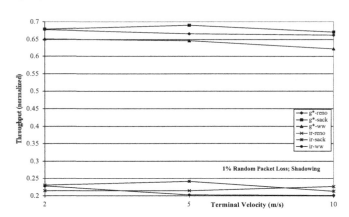

Fig. 9. Performance of FTP vs. terminal speed for a full LEO configuration in a shadowed environment

8. Conclusions

The access to the Internet in the presence of wide range mobility represents one of the key issues for future telecommunication systems. In such a scenario, the satellite assumes a very important role.

In this paper, we investigated the performance of various TCP schemes in satellite environments characterized by variable propagation conditions. We evaluated different architectures (LEO, GEO, single hop, full) in representative fixed/mobile terminal scenarios.

The simulations show that TCP Westwood is able to outperform TCP Reno and SACK in the presence of random errors or shadowing. The faster recovery algorithm used by TCP Westwood helps it to recover quickly from packet error. Also, the satellite diversity provided by Globalstar may be used to provide better connectivity in a shadowed environment.

References

1) http://www.spaceway.com/spaceway/index.htm
2) http://www.astrolink.com
3) http://www.cyberstar.com/
4) http://www.astra.lu/
5) http://www.eutelsat.com/
6) http://www.skybridgesatellite.com/
7) http://www.alespazio.it/program/tlc/eurosk/eurosk.htm
8) D. O'Mahony, *UMTS: the fusion of fixed and mobile networking*, IEEE Internet Computing, vol. 21, Jan.-Feb. 1998, pp. 49-56.
9) R. J. Leopold, *Low-Earth Orbit Global Cellular Communications Network*, Mobile Sat. Comm. Conf., Adelaide, Australia, 23 Aug. 1990.
10) R. A. Wiedeman, A. J. Viterbi, *The Globalstar Mobile Satellite System for worldwide Personal Communications*, Proc. of 3rd Int. Mobile Satellite Conference (IMSC '93), June 1993, pp. 285-290.
11) http://www.teledesic.com
12) W. D. Ivancic, D. Brooks, B. Frantz, D. Hoder, D. Shell, D. Beering, *NASA's broadband satellite networking research*, IEEE Communications Magazine, Volume: 37, n. 7, July 1999, pp. 40–47.
13) C. P. Charalambous, V. S. Frost, J. B. Evans, *Performance evaluation of TCP extensions on ATM over high bandwidth delay product networks*, IEEE Comm. Magazine, Vol. 37, n. 7, July 1999, pp. 57–63.
14) H. Kruse, S. Ostermann, M. Allman, *On the Performance of TCP-based Data Transfers on a Faded Ka-Band Satellite Link*, Proceedings of the 6th Ka-Band Utilization Conference, June 2000.
15) M. Allman, H. Kruse, S. Ostermann, *A History of the Improvement of Internet Protocols Over Satellites Using ACTS*, Invited paper for ACTS Conference 2000, May 2000.
16) C. Casetti, M. Gerla, S. S. Lee, S. Mascolo, M. Sanadidi, *TCP with Faster Recovery*, UCLA CS-Technical Report #200017, 2000
17) "Network Simulator (NS-2)", www.isi.edu/nsnam/ns/.
18) *INTELSAT Internet Technical Handbook*, April 20th, 1998.

19) M. A. Vázquez-Castro, S. Saunders, C. Tzaras, F. Pérez-Fontán, *Shadowing correlation for mobile satellite diversity*, AP2000 Millennium Conference on Antennas & Propagation, April 2000.
20) S. Maloo, *TCP/IP Issues and Performance over LEO Satellite Networks*, International Conference on Telecommunications, Acapulco, Mexico 22-25 May, 2000, pp. 369-373.

Performance of TCP
on a Wideband CDMA Air Interface

Michele Zorzi, Michele Rossi, and Gianluca Mazzini

University of Ferrara, 44100 Ferrara, Italy
{mzorzi, mrossi, gmazzini}@ing.unife.it
http://www.ing.unife.it

Abstract. We present a study on the performance of TCP, in terms of both throughput and energy consumption by considering a Wideband CDMA radio interface. The results show that the relationship between throughput and average error rate is largely independent on the network load, making it possible to introduce a universal throughput curve, which gives throughput predictions for each value of the user error probability. Furthermore, the possibility to select an optimal power control threshold to maximize the tradeoff between throughput and energy, is discussed.

1 Introduction

Based on the evolutionary trend of the telecommunications market, a significant step has been taken by the wireless industry, by developing a new generation of systems whose capabilities are intended to considerably exceed the very limited data rates and packet handling mechanisms provided by current second generation systems.

The goal of third generation systems is to provide multimedia services to the wireless terminals and to offer transport capabilities based on an IP backbone and on the Internet paradigm calls for the extension of widespread Internet protocols to the wireless domain.

In particular, extension of the Transmission Control Protocol (TCP) [4] has received considerable attention in recent years, and many studies have been published in the open literature, which address the possible performance problems that TCP has when operated over a connection comprising wireless links, and propose solutions to those problems (see for example [5,6,7,8,9,10]).

When TCP is run in a wireless environment, two major considerations must be made regarding its performance characterization: wireless links exhibit much poorer performance than their wireline counterparts and the effects of this behavior are erroneously interpreted by TCP, which reacts to network congestion every time it detects packet loss, even though the loss itself may have occurred for other reasons.

Furthermore, it has been shown that the statistical behavior of packet errors has a significant effect on the overall throughput performance of TCP, and that different higher-order statistical properties of the packet error process may lead

S. Palazzo (Ed.): IWDC 2001, LNCS 2170, pp. 91–107, 2001.
© Springer-Verlag Berlin Heidelberg 2001

to vastly different performance even for the same *average* packet error rates [11]. It is therefore important to be able to accurately characterize the actual error process as arising in the specific environment under study, as simplistic error models may just not work.

Another critical factor to be considered when wireless devices are used is the scarce amount of energy available, which leads to issues such as battery life and battery recharge, and which affects the capabilities of the terminal as well as its size, weight and cost. Since dramatic improvements on the front of battery technology do not seem likely, an approach which has gained popularity in the past few years consists in using the available energy in the best possible way, trying to avoid wasting power and to tune protocols and their parameters in such a way as to optimize the energy use [12].

It should be noted that this way of thinking may lead to completely different design objectives or to scenarios in which the performance metrics which have traditionally been used when evaluating communications schemes become less important than energy-related metrics. This energy-centric approach therefore gives a different spin to performance evaluation and protocol design, and calls for new results to shed some light on the energy performance of protocols.

Studies on the energy efficiency of TCP have been very limited so far [13,14], [15,16]. In addition, they do not address specifically the Wideband CDMA environment typical of third generation systems, and therefore do not necessarily provide the correct insight for our scenario.

The purpose of this paper is to provide a detailed study of the performance of TCP, in terms of both throughput and energy, when a Wideband CDMA radio interface is used. In particular, the parameters of the UMTS physical layer will be used in the study. As a first step of the study, we present results in the absence of link-layer retransmissions. This is done in order to more clearly understand the interactions between TCP's energy behavior and the details of the radio interface. Results on the throughput performance of TCP in the presence of link-layer retransmissions can be found, e.g., in [9,10,17], where on the other hand energy performance is not studied.

2   System Model

2.1   Simulation at the System Level

In order to carry out the proposed research, a basic simulation tool has been developed. It simulates the operation of a multicellular wideband CDMA environment, where user signals are subject to propagation effects and transmitted powers are adjusted according to the power control algorithm detailed in the specifications.

We consider a hexagonal cell layout with 25 cells total. This structure is wrapped onto itself to avoid border effects. Each simulation is a snapshot in time, so that no explicit mobility is considered while running the simulation.

The fact that users may be mobile is taken into account in the specification of the Doppler frequency, which characterizes the speed of variation of the

Rayleigh fading process. For the same reason, long-term propagation effects, namely path loss and shadowing, are kept constant throughout each simulation. Path loss relates the average received power to the distance between the transmitter and the receiver, according to the general inverse-power propagation law $P_r(r) = Ar^{-\beta}$, where in our results we chose $\beta = 3.5$. Shadowing is modeled by a multiplicative log-normal random variable with dB-spread $\sigma = 4$ dB, i.e., a random variable whose value expressed in dB has Gaussian distribution with zero mean and standard deviation equal to 4 [18].

At the beginning of each simulation, all user locations are randomly drawn with uniform distribution within the service area, and the radio path gains towards all base stations are computed for each. Users are then assigned to the base station which provides the best received power. Such assignment does not change throughout the simulation. Also, note that each user is assigned to a single base station, i.e., soft handover is not explicitly considered here. Extension of the program to include soft handover is currently in progress, although qualitatively similar results are expected.

During a simulation run, the fading process for each user is dynamically changed, according to the simulator proposed by Jakes [18] and to the selected value of the Doppler frequency. Note that the instantaneous value of the Rayleigh fading does not affect the base station assignment. In order to take into account the wideband character of the transmitted signals, a frequency-selective fading model is used, with five rays whose relative strengths are taken from [19]. Maximal ratio combining through RAKE receiver is assumed at the base station. Only the uplink is considered in this paper, although similar results have been obtained for the downlink as well.

Each connection runs its own power control algorithm as detailed in the specifications. The resulting transmitter power levels of all user, along with the instantaneous propagation conditions, determine the received signal level at each receiver, and therefore the level of interference suffered by each signal. We assume here perfect knowledge of the Signal-to-Interference Ratio (SIR) which is used to make the decision about whether the transmitted power should be increased or decreased by the amount Δ. A finite dynamic range is assumed for the power control algorithm, so that under no circumstances can the transmitted power be above a maximum or below a minimum value. The delay incurred in this update is assumed to be one time unit (given by the power control frequency of update), i.e., the transmitted power is updated according to the SIR resulting from the previous update. The effect of late updates on the overall performance has been studied in [25].

Table 1 summarizes the various parameters used in the simulation.

2.2 Block Error Probability

The output of the system level simulations is a log of the values of the SIR, transmitted power and fading attenuation for all users, which allows us to gain some understanding on the time evolution of these quantities, as well as on the

Table 1. System-level simulation parameters

PARAMETER	VALUE
Cell Side	200 m
β (path loss model)	3.5
A (path loss model)	-30 dB
Max. TX Power	-16 dBW
Power Range	80 dB
σ (shadowing)	4 dB
Time Unit	0.667 ms
number of Oscillators (Jakes)	8
n_rays (Selective Channel)	5
Chip Rate	3.84 Mcps
Data Rate	240 kbps
SF (Spreading Factor)	16
Δ (Power Control Step)	0.5 dB
Noise	-132 dBW

behavior of the network at the system level. A post-processing package translates the SIR traces into sequences of block error probabilities (BEP).

This is done while taking into account how the radio frames are deinterleaved and decoded. The interleaving schemes have been taken from the specifications, and a convolutional code with rate 1/2 and constraint length 8 with Viterbi decoding has been considered. An analytical approximation has been used to relate a string of SIR values (one per time unit) to the probability that the corresponding block (transmitted within one or more radio frames) is in error.

The resulting trace of the block error probabilities can then be used in the simulation of higher-layer protocols, as is done in this paper, or to perform some statistical analysis of block errors. The latter approach is explored in [26], where the burstiness of the error process is looked at in some detail.

2.3   TCP Simulation

For each simulation run (which corresponds to a given set of parameters), the SIR traces of all users are produced, which are then mapped into BEP traces as just explained. The latter are then used to randomly generate a block error sequence (BES).

The BES is generated from the BEP traces by just flipping a coin with the appropriate probability in each Time Transmission Interval (TTI), which is the time used to transmit a block. The BES obtained is then fed to the TCP simulator that uses it to specify the channel status in each TTI. By doing so, the SIR traces generated for all users in a simulation run are used by the TCP simulator to compute the throughput.

The average throughput is defined as the fraction of the channel transmission rate (considered at the IP level output) which provides correct information bits at the receiver, i.e., not counting erroneous transmissions, spurious retransmissions, idle time and overhead.

In addition to the throughput value, the TCP simulator computes other parameters of interest, i.e.: the average block error probability, P_e (obtained for every user simply by averaging all the values of its BEP sequence); the average energy spent in transmitting data (obtained from the transmit power traces).

These traces have been generated assuming continuous channel transmission and by updating the transmitted power according to the power control algorithm. However, when TCP is considered, transmission is bursty, due to the window adaptation mechanism implemented by the TCP algorithms, i.e., idle times occur when the window is full or the system is waiting for a timeout.

To account for idle times, we have considered the *actual* transmitted power, equal to the one obtained from power traces when TCP transmits, and to zero otherwise. The average transmitted power is then computed by summing the actual transmitted power throughout the simulation (all slots) and dividing it by the total simulation time (in number of slots). Moreover, to obtain the average consumed energy per correctly received bit, we simply divide the average transmit power by the correct information bit delivery rate.

In the following, we report the throughput and average consumed energy expressions:

$$S = \frac{CIB}{CBR}$$

$$ACE = \frac{ATP}{CIB} = \frac{ATP}{S \cdot CBR}$$

where S=average throughput, CIB=correct information bits per second, CBR= channel bit rate at the IP level output, ACE=average consumed energy per bit, and ATP=average transmitted power. We remark that, with our definitions, ACE is the average consumed energy per *correctly received bit*, i.e., the energy cost of delivering a single bit to the destination. Notice that this quantity is equal to the inverse of the *energy efficiency* of the protocol as defined in [27].

The TCP simulator implements fragmentation of TCP segments, window adaptation and error recovery. It simulates a simple unidirectional ftp session, where the direct link packet generation is assumed continuous as in a long ftp transfer. The TCP algorithm considered is *New Reno* [21]. Data flow is unidirectional, i.e., data packets flow only from sender to receiver, while ACKs flow in the reverse direction. Receiver generates non-delayed ACKs, i.e., one ACK is sent for each packet received. The TCP/IP stack is version 4, with a total of 40 bytes (including both TCP and IP overhead) for each header (compression is not taken into account in the results presented) and MTU size of 512 bytes. We have not considered RLC and MAC levels, which are assumed here to operate in transparent mode. RLC/MAC level implementation and characterization are currently under study.

To compute the bit rate at the output of the IP level, we have to account for overhead added by the physical layer as well as possibly due to multiplexing of other channels. For the purpose of discussion, we assume the following figures:

a transport block is 1050 bits, including 16 bits of overhead due to transparent
RLC/MAC operation; a CRC and tail bits for code termination are added to
this block, and the result is convolutionally encoded at rate 1/2. The resulting
encoded block is then brought to 2400 encoded symbols by the rate matching
algorithm, so that the raw physical layer symbol rate is 240 kbps. The application
of a spreading factor $SF = 16$ makes it 3.84 Mcps, which is the standard channel
transmission rate. At the IP level output, we then have a block of 1034 bits of
data every 10 ms, thereby yielding a net bit rate of 103.4Kbps, which is the bit
rate used in the TCP simulator.

The use of TCP New Reno algorithm has been motivated by its implementation of fast recovery and fast retransmit algorithms [22], as recommended in
[23], especially for wireless environments. This is an optimization over previous
TCP versions, and is currently at the Proposed Standard level.

3   TCP Throughput Performance

In the graphs presented we will indicate with N_u the number of users in the
simulation, with TTI the number of radio frames over which interleaving is
performed, $SIR_{th} = t + \Delta$ dB indicates that the threshold used in the power
control algorithm is t, while Δ is its increment as described in Section 2. Finally,
with the term f_d we refer to the Doppler frequency used in the Rayleigh fading
simulator. In the following graphs, the results will be represented as average
TCP throughput, S, vs. average block error probability, P_e, thereby assigning a
single point in the graph to each user.

3.1   Doppler Frequency Effect

Figure 1 shows the TCP throughput performance for different values of the
Doppler frequency. The graph is plotted by reporting throughput vs. P_e for each
of the 90 users involved in the simulation; each user is identified by a marker.
The case of independent errors is also reported for comparison purpose (here the
markers are used only to identify the curve and are not related to the users).

Note that, for a given value of the Doppler frequency the points representing
the various users of a simulation appear to lie along a fairly well-defined curve.
It is worth stressing that this was not obvious a priori since different users are
placed in different locations and are subject to different propagation conditions,
both in terms of slow impairments (log-normal shadowing) and in terms of fast
fading.

This allows us to introduce the concept of "universal throughput curve" for a
given situation, in the sense that users which suffer similar values of P_e will enjoy
about the same throughput. Again, this is not obvious since different users in a
simulation may see different statistical behaviors of the errors, which could in
principle lead to different performance even in the presence of the same average
error rate [11].

An explanation can be drawn from the results in [26], where it was found
that for a given value of f_d there is a strong correlation between P_e and the

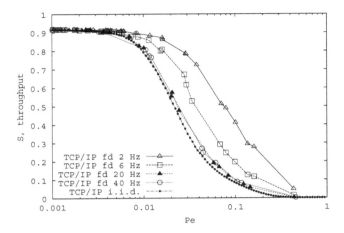

Fig. 1. TCP/IP Sensitivity to Doppler,($N_u = 90$, $TTI = 1$, $SIR_{th} = 5.5 + 0.5$dB, $f_d = 2, 6, 20, 40$Hz).

error burstiness. In this situation, P_e essentially determines the full second-order characterization of the error process, which in turn almost fully specifies the value of the TCP throughput. On the other hand, for different values of f_d we observe different curves. In fact, even in the presence of the same P_e the different extent of the channel memory results in different performance.

Another interesting observation from Figure 1 is that as the Doppler frequency increases the performance degrades, i.e., slower channels correspond to better performance, as already observed in [8]. As expected, for sufficiently high values of the Doppler frequency, the behavior of the system is close to the iid case.

Finally, we note that the shape of the curves appears to be fairly regular, with a smooth transition from highest throughput values (essentially limited only by the percentage of overhead in the TCP packets, far left of the graph), to essentially zero throughput when errors are very likely (right end of the graph). This shape, which has been observed by other authors, lends itself nicely to numerical fitting, as detailed later.

3.2 Interleaving Depth Effect

In UMTS, besides the so-called *intra-frame* interleaving, which is always used to scramble the bits within a radio frame (10 ms) before encoding, it is also possible to use a second, *interframe*, interleaving, which mixes bits *across* frames. By doing so, the performance of the decoder is of course improved also in the presence of burst errors, but a larger interleaving delay is introduced.

Therefore, another interesting sensitivity analysis regards the interleaving span allowed by the application. While keeping in mind the price to be paid in

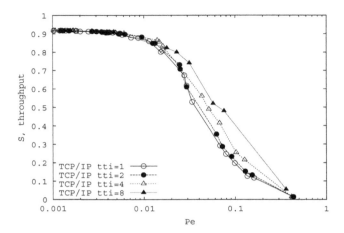

Fig. 2. Average throughput vs TTI,($N_u = 90$, $TTI = 1, 2, 4, 8$, $SIR_{th} = 5.5 + 0.5$dB, $f_d = 6$Hz).

terms of delay (an increase of the TTI corresponds to a larger delay), we can note from Figure 2 how a deeper interleaving gives a beneficial effect. Notice also that the shape of the curves is similar to what observed in the previous case.

3.3   Network Load Effect

The effect of the network load is shown in Figures 3 and 4. In Figure 3, results from three simulations are shown, with 80, 90 and 100 users in the network, respectively. For each simulation, all users are assigned a point with the same abscissa (which is given by the number of users in the system) and with vertical coordinate given by their average TCP throughput.

We can see how for increasing load the presence of disadvantaged users becomes more noticeable, as expected, and the system is more and more unfair. The same results are represented in Figure 4 by reporting the throughput as a function of P_e where the curve along which the various points are aligned is relatively insensitive to the system load.

In Figure 1, it was observed that, in a given scenario, two users whose average block error probability is the same will have essentially the same throughput, regardless of the specific situation of each. What we see here is that this behavior still holds across simulations in which different levels of network load are considered (but for the same Doppler frequency). For higher network load, each user will certainly see worse performance due to the increased interference, but the relationship between average throughput and average error rate is essentially unaffected.

This highlights the power of the concept of "universal curve" which can be used to study typical cases and to infer TCP throughput performance based

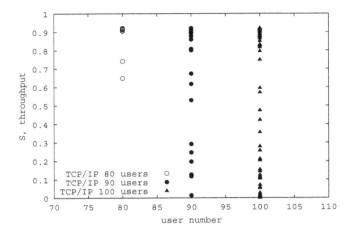

Fig. 3. TCP/IP average throughput *vs* network load, ($N_u = 80, 90, 100$, $TTI = 1$, $SIR_{th} = 5.5 + 0.5$dB, $f_d = 6$Hz).

only on easily measurable physical layer parameters. In Figure 4 a possible form for the universal curve is given for comparison with the simulator output points; more details about this fitting function will be presented below. Similar behavior has been observed for other values of the Doppler frequency.

3.4 Analytical Throughput Prediction

The observed shape of the throughput curves, which tend to a constant equal to one minus the percentage of overhead for $P_e \to 0$ and to zero for $P_e \to 1$, suggests a numerical fit involving the logarithm of P_e. Also, the shape of the transition is seen to depend on the value of the Doppler frequency.

In Figure 5 we show a proposal for the modeling of the TCP throughput behavior through a heuristic function $f(x)$, independent of the network load and parameterized only by the Doppler frequency, as suggested by the curves of Figure 4 as well as by other results not shown in this paper. The proposed expression for $f(x)$ is as follows:

$$f(x) = S(0) \cdot \frac{10^{\alpha \cdot \ln(\frac{1}{\tilde{x} - x_s} - 1)}}{10^{\alpha \cdot \ln(\frac{1}{\tilde{x} - x_s} - 1)} + 1}$$

with $\tilde{x} = 1 + \frac{\log_{10}(x)}{3}$, $\alpha = 1.3$, $x_s = \frac{1}{Af_d + B} - k$ and where $A = 1.39$, $B = 2.78$, $k = 0.03$, and $S(0)$ is the average throughput for $P_e = 0$ and f_d is the Doppler frequency in Hz.

The accuracy of the proposed fit has been tested for various values of the parameters involved. Examples of these tests are given in Figures 4 and 6, in which the fitting expression is compared against the simulation results for two values

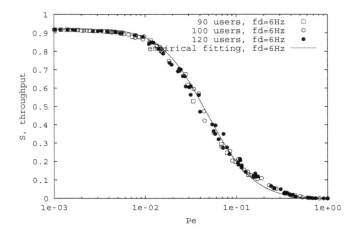

Fig. 4. Throughput curve, independence from network load with Doppler 6Hz, ($N_u =$ 80, 90, 100, $TTI = 1$, $SIR_{th} = 5.5 + 0.5$dB, $f_d = 6$Hz).

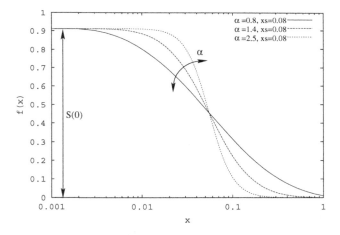

Fig. 5. Throughput heuristic function.

of the Doppler frequency. These graphs show that the analytical expression is reasonably close to the actual points obtained by simulation.

4   TCP Energy Performance

All previous results were obtained for a given value of the power control threshold, SIR_{th}, which is used to drive the transmit power dynamics at each user and

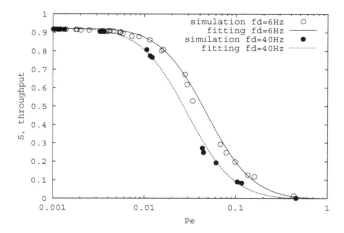

Fig. 6. Throughput heuristic function: approximation of simulator outputs, ($N_u = 90$, $TTI = 1$, $SIR_{th} = 5.5 + 0.5$, $f_d = 6, 40$Hz).

which directly affects the error performance. In fact, choosing a higher value of this threshold has the double effect of forcing the users to transmit more power in order to achieve a higher SIR (thereby consuming more energy) but also of causing the SIR experienced by the typical user to be higher (thereby improving the error rates and therefore the TCP throughput).

It is therefore of interest to study how varying the power control SIR threshold makes it possible to cut a tradeoff between QoS and energy consumption.

4.1 Throughput and Consumed Energy

Figures 7 and 8 show the trade-off between TCP throughput and consumed energy. Each curve corresponds to a given user for a given set of parameters, whereas different points on the same curve refer to different values of the power control threshold. Figure 7 shows the effect of using different threshold values in terms of S as a function of ATP. In this figure only the behavior of some selected users has been reported. These users can be seen as representative of all users in the network, in the sense that they illustrate typical behaviors as they arise in the system.

In Figure 8 the same results are shown, by considering ACE instead of ATP; as expected, for low throughput the obtained curves are shifted to the right, since ACE is obtained by dividing ATP by the throughput S (except for an inessential constant scaling factor. This is intuitively explained by the fact that, for a given average consumed power, the energy per correct information bit is greater for low throughput values, i.e., when it is hard to deliver bits correctly, the cost associated to each one of them is higher. Notice that TCP already does the right thing by stopping transmission when the channel is very bad (the

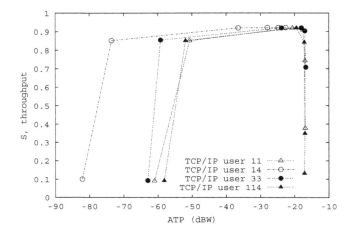

Fig. 7. Throughput *vs* average transmitted power for different threshold values, (N_u = 120, $TTI = 1$, $SIR_{th} = \{1.5, 2.5, 3.5, 4, 4.5, 5\} + 0.5$dB, $f_d = 6$Hz).

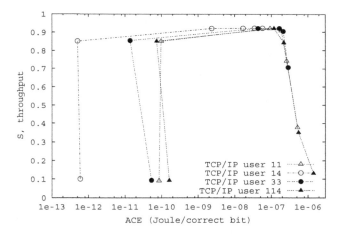

Fig. 8. Throughput *vs* average consumed energy for different threshold values, (N_u = 120, $TTI = 1$, $SIR_{th} = \{1.5, 2.5, 3.5, 4, 4.5, 5\} + 0.5$dB, $f_d = 6$Hz).

timeout event), whereas it tries to recover from errors whenever possible through retransmissions, which may waste some power.

In general, increasing the threshold should lead to better performance for many users, since the SIR experienced is expected to be higher; this is not necessarily true, however, since, in order to achieve a higher SIR threshold, many users will transmit more power, thereby causing more interference in the system. If the SIR objective is not achievable for all users in the system, some users will actually see degraded performance for higher values of the threshold since, although the threshold value would correspond to better performance, they cannot achieve its value.

From the obtained results we have noted different users behavior. For some users an increasing power threshold always corresponds to a greater throughput: for these users the throughput as a function of the power threshold is a monotonic curve. For others, the throughput vs. consumed power curve increases up to a breakpoint, after which an increment of the power threshold (and thereby of the transmitted power) actually leads to worse performance, due to greater interference as discussed above.

In our results, user 14 is the one showing monotonic behavior, since it experiences favorable propagation conditions, and therefore is not significantly affected by the increased interference level in the system. For users 11, 33 and 114, a different situation can be observed. In particular, user 11 is the one with the worst behavior as the power threshold increases.

In any event, from these results we can conclude that increasing the target value of the SIR in the system does not necessarily translate into improved quality, but there exists an optimal value of the threshold, beyond which some users will experience negligible throughput improvements, whereas others will even see degraded performance. In the cases studied in this paper, this optimal value is seen to be close to 3.5 dB.

4.2   Energy and Threshold

Another important remark regards the numerical values shown in Figures 7 and 8. It can be clearly seen that unlike for throughput, which except for badly chosen values of the threshold exhibits relatively small variations, the range spanned by the energy performance extends over multiple orders of magnitude. This indicates that the choice of the proper power control threshold, while certainly important for error and throughput considerations, becomes critical when energy performance is considered.

Figure 9 shows the average consumed power as a function of the power threshold. As expected, a greater threshold value always corresponds to a higher consumed power, i.e., all curves have a strictly monotonic behavior. This is due to the fact that for higher threshold values, more power is necessary in order to obtain the required SIR target.

A different behavior is observed in Figure 10 in which the consumed energy per bit is reported instead. In particular, notice that in the far left of the graph, a decrease of the threshold, although corresponding to smaller average power (see Figure 9), results in a higher energy cost per bit. This behavior corresponds to users suffering from low throughput performance, where the consumed energy per correct information bit grows, as explained before. As a last observation, we note from Figure 9 that users 11, 33 and 114, i.e., the ones that suffer from system interference as the threshold grows, show a transmitted power which is essentially constant for values of the threshold beyond 4 dB, This is due to the fact that these users, in trying to achieve the required SIR and to make up

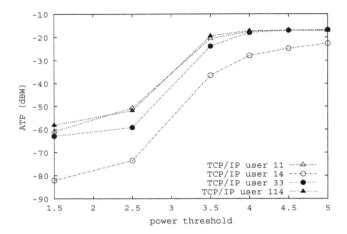

Fig. 9. Average consumed power *vs* power threshold, ($N_u = 120$, $TTI = 1$, $SIR_{th} = \{1.5, 2.5, 3.5, 4, 4.5, 5\} + 0.5$dB, $f_d = 6$Hz).

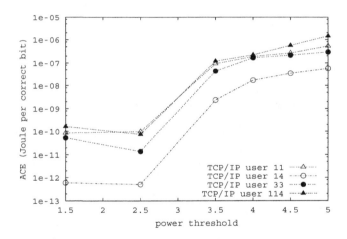

Fig. 10. Average consumed energy *vs* power threshold, ($N_u = 120$, $TTI = 1$, $SIR_{th} = \{1.5, 2.5, 3.5, 4, 4.5, 5\} + 0.5$dB, $f_d = 6$Hz).

for the increased interference, have reached the maximum allowed value for the transmit power, and therefore their power can not be increased any more.

4.3 Error Probability and Consumed Energy

A similar QoS-energy trade-off relates to the error rate performance instead of the TCP throughput. In Figure 11 the block error probability P_e has been

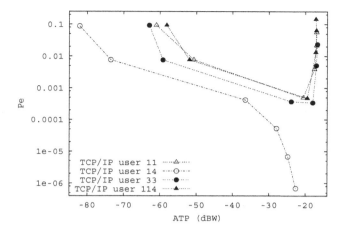

Fig. 11. P_e *vs* average transmitted power for different threshold values, ($N_u = 120$, $TTI = 1$, $SIR_{th} = \{1.5, 2.5, 3.5, 4, 4.5, 5\} + 0.5$dB, $f_d = 6$Hz).

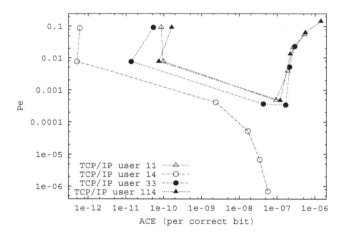

Fig. 12. P_e *vs* average consumed energy for different threshold values, ($N_u = 120$, $TTI = 1$, $SIR_{th} = \{1.5, 2.5, 3.5, 4, 4.5, 5\} + 0.5$dB, $f_d = 6$Hz).

reported against the average transmitted power ATP by using different values of the threshold. From the graph, we note that P_e decreases as the threshold grows until a minimum is reached. After this point P_e starts to grow, again due to the increased interference in the system. The points in which P_e has a minimum are the same on which the throughput of the system is maximized (see Figure 7).

As before, user 14 is the only user considered for which P_e never grows, i.e., increasing the threshold always leads to better performance. In Figure 12 ACE

is reported instead of ATP and, as in the previous cases, some points are shifted to the right due to poor throughput performance.

From the above results, we may conclude that the selection of the power control SIR threshold is critical in cutting the right trade-offs between QoS and energy performance. In particular, we observe that the potential for energy gains is very significant compared to similar effects on throughput, being measured over multiple orders of magnitude. Therefore, it seems that more attention should be given to energy consumption issues at the Radio Resource Management level, which is responsible for the power control parameter selection.

5   Conclusions

In this paper, some results on the behavior of TCP over a Wideband CDMA air interface have been reported. In particular, TCP throughput curves have been obtained, and their dependence on various parameters, such as the number of users, the interleaving depth, and the Doppler frequency, has been investigated.

From this study, we have found that the relationship between average TCP throughput and average block error rate is largely independent of the number of users in the system. For this reason, it is possible to empirically characterize such a curve with a matching function that only depends on the Doppler frequency. A study of the energy consumption has also been performed, showing that a trade-off between the throughput and the power control threshold exists. It is therefore possible to trade-off QoS of the data transfer for increased energy efficiency. For many users, it has been shown that an optimal value of the threshold exists, potentially leading to very significant energy savings in return for very small throughput degradation. In the considered case, values of the power control threshold close to 3.5-4 dB cut the best tradeoff.

In order to focus exclusively on the interactions of TCP with the WCDMA radio technology and as a preliminary step in characterizing TCP's energy performance in this scenario, no link-layer retransmissions have been considered in this study. Future work includes extension of the study to the presence of a radio link layer which improves the wireless link performance through block retransmission. Similar performance studies for TCP versions other than New-Reno also seem worth pursuing.

References

1. G. Patel, S. Dennett. The 3GPP and 3GPP2 movements toward an all-IP mobile network. *IEEE Personal Communications*, vol. 7, pp. 62-4, Aug. 2000.
2. B. Sarikaya. Packet mode in wireless networks: overview of transition to third generation. *IEEE Communications Magazine*, vol. 38, pp. 164-72, Sept. 2000.
3. Third Generation Pasrtnership Project website: www.3gpp.org
4. W. R. Stevens. *TCP/IP Illustrated, Volume 1*. Addison Wesley, 1994.
5. H. Balakrishnan, V. N. Padmanabhan, S. Seshan, and R. H. Katz. A comparison of mechanisms for improving TCP performance over wireless links. *ACM/IEEE Trans. on Networking*, Dec. 1997.

6. R. Caceres, and L. Iftode. Improving the performance of reliable transport protocols in mobile computing environments. *IEEE Jl. Sel. Areas Commun.*, vol. 13, no. 5, pp. 850-857, June 1995.

7. A. Kumar. Comparative performance analysis of versions of TCP in a local network with a lossy link. *ACM/IEEE Trans. on Networking*, Aug. 1998.

8. M. Zorzi, A. Chockalingam, R.R. Rao. Throughput Analysis of TCP on Channels with Memory. *IEEE J. Selected Areas Comm.*, vol. 18, pp. 1289–1300, Jul. 2000.

9. A. Chockalingam, Gang Bao. Performance of TCP/RLP protocol stack on correlated fading DS-CDMA wireless links. *IEEE Trans. on Vehicular Technology*, vol. 49, pp. 28-33, Jan. 2000.

10. N.M. Chaskar, T.V. Lakshman, U. Madhow. TCP over wireless with link level error control: analysis and design methodology. *IEEE/ACM Trans. on Networking*, vol. 7, pp. 605-15, Oct. 1999.

11. M. Zorzi, R.R. Rao. Perspectives on the Impact of Error Statistics on Protocols for Wireless Networks. *IEEE Personal Communications*, vol. 6, Oct. 1999.

12. *IEEE Personal Communications Magazine*. special issue on "Energy Management in Personal Communications and Mobile Computing," vol. 5, June 1998.

13. M. Zorzi, R.R. Rao. Is TCP energy efficient?. in *Proc. IEEE MoMuC*, Nov. 1999.

14. V. Tsaoussidis, H. Badr, X. Ge, K. Pentikousis. Energy/throughput tradeoffs of TCP error control strategies. In *Proceedings ISCC 2000* pp. 106-12, Jul. 2000.

15. V. Tsaoussidis, A. Llahanns, H. Badr. Wave & wait protocol (WWP): high throughput and low energy for mobile IP-devices. In *Proceedings IEEE ICON 2000*, pp. 469-73, Sept. 2000.

16. M. Zorzi, R.R. Rao. Energy Efficiency of TCP. To appear in *Mobile Networks and Applications*, 2001.

17. F. Khan, S. Kumar, K. Medepalli, S. Nanda. TCP performance over cdma2000 RLP. In *Proc. VTC2000-Spring*, pp. 41-5, May 2000.

18. W. C. Jakes, Jr. *Microwave mobile communications*. New York: John Wiley & Sons, 1974.

19. ETSI SMG2. Evaluation report for ETSI UMTS Terrestrial Radio Access (UTRA) ITU-R RTT candidate. Sep. 1998 (http://www.itu.int/imt/2-radio-dev/reports/etsi/eva_utra.pdf)

20. Harri Holma, Anttiq Toskala. *WCDMA for UMTS: Radio Access For Third Generation Mobile Communications*. New York: John Wiley & Sons, 2000.

21. S. Floyd, T. Henderson. The New Reno Modification to TCP's Fast Recovery Algorithm. Request For Comment 2582, April 1999.

22. M. Allman, V. Paxson, W. Stevens. TCP congestion control. Request For Comment 2581, April 1999.

23. S. Dawkins, G. Montenegro, M. Kojo, V. Magret, N. Vaidya. End-to-end Performance Implications of Links with Errors. Internet Draft, November 17, 2000.

24. Anna Calveras, Josep Paradells Aspas. TCP/IP over wireless links: Performance Evaluation. VTC'98. 48th IEEE. Volume: 3, 1998, pages: 1755-1759.

25. A. Giovanardi, G. Mazzini, V. Tralli, M. Zorzi. Some results on power control in Wideband CDMA cellular networks. In *Proc. WCNC2000*, Chicago (IL), Sep. 2000.

26. M. Zorzi, G. Mazzini, V. Tralli, A. Giovanardi. Some results on the error statistics over a WCDMA air interface. In *Proc. MMT2000*, Florida (USA), Dec. 2000.

27. M. Zorzi, R.R. Rao. Energy constrained error control for wireless channels. *IEEE Personal Communications Magazine*, vol. 4, pp. 27–33, Dec. 1997.

Differentiated Services
in the GPRS Wireless Access Environment

Sergios Soursos, Costas Courcoubetis, and George C. Polyzos

Department of Informatics
Athens University of Economics and Business
Athens 10434, Greece
{sns, courcou, polyzos}@aueb.gr

Abstract. The General Packet Radio Service extends the existing GSM mobile communications technology by providing packet switching and higher data rates in order to efficiently access IP-based services in the Internet. Even though Quality-of-Service notions and parameters are included in the GPRS specification, no realization path for Quality-of-Service support has been proposed. In this paper we adapt the Differentiated Services framework and apply it over the GPRS air interface in order to provide various levels of service differentiation and a true end-to-end application of the Internet Differentiated Services architecture.

1 Introduction

The convergence of mobile technologies with the technologies of the Internet was of great importance this last decade. One step towards this direction was made by the introduction of the General Packet Radio Service (GPRS) over the Global System for Mobile communications (GSM). GPRS is a packet-switched service offered as an extension of GSM. In contrast to the classic circuit-switched service provided by GSM, GPRS offers the efficiency of packet-switching desirable for bursty traffic, higher transfer speeds than the ones available today to a single end-terminal (theoretically up to 115 kbps) and instantaneous connectivity with any IP-based external packet network.

An important issue in this context is the Quality-of-Service (QoS) provided by GPRS. Even though GPRS specifications define QoS parameters and profiles, we are unaware of specific implementation plans and strategies in order to support specific QoS models, particularly over the wireless access network. Recent proposals in the area of GPRS QoS focus on providing QoS support in the core GPRS network (which is typically non-wireless and IP based) [11] using the standard Internet QoS frameworks (i.e., Integrated Services or Differentiated Services).

On the other hand, we believe that the critical part for the support of QoS to the applications and the end users is the access network where, because of the scarcity of the radio spectrum, greater congestion problems can result. Therefore, we have developed an architecture that provides QoS in the form of support for Differentiated Services over the radio link and integration with the Internet DiffServ architecture, thus providing end-to-end QoS "guarantees" [12]. As described later in this paper,

S. Palazzo (Ed.): IWDC 2001, LNCS 2170, pp. 108–119, 2001.

GPRS operators can easily implement this proposal, with no need for radical changes to their existing GPRS network architecture.

The structure of the remainder of this paper is as follows. First we provide a short overview of the GPRS technology and architecture. We then review briefly the Internet Differentiated Services architecture and we focus particularly on the description of the two-bit DiffServ scheme. In the following section we adapt the two-bit DiffServ scheme in the GPRS environment, describing all the new tasks that are required to be performed by the GPRS Serving Nodes (GSNs), the key new elements in the GSM architecture introduced to support GPRS. Finally, we discuss some open issues and present our conclusions.

2 The GPRS Environment

GPRS [2] is a new service offered by the GSM network. In order for the operators to be able to offer such services two new types of nodes must be added to the existing GSM architecture. These two nodes are the serving GPRS support node (SGSN) and the gateway GPRS support node (GGSN), as shown in the Fig. 1.

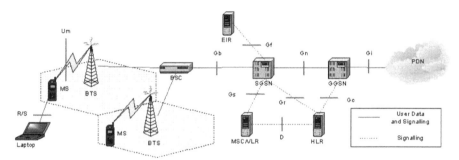

Fig. 1. The GPRS network

The SGSN keeps track of the location of mobile users, along with other information concerning the subscriber and its mobile equipment. This information is used to accomplish the tasks of the SGSN, such as packet routing and switching, session management, logical link management, mobility management, ciphering, authentication and charging functions. The GGSN, on the other hand, connects the GPRS core network to one or more external Packet Data Networks (PDNs). Among its tasks, is to convert the incoming packets to the appropriate protocol in order to forward them to the PDN. Also, the GGSN is responsible for the GPRS session management and the correct assignment of a SGSN to a Mobile Station (MS), depending on the MS's location. The GGSN also contributes to the gathering of useful information for the GPRS charging subsystem.

The core GPRS network is IP based. Among the various GSNs (SGSN and GGSN) the GPRS Tunnel Protocol (GTP) protocol is used. The GTP constructs tunnels between two GSNs that want to communicate [1]. GTP is based on IP. At the radio link, the existing GSM structure is used, making it easier for operators to offer GRPS services. The uplink and downlink bands are divided through FDMA into 124 frequency carriers each. Each frequency is further divided through TDMA into eight timeslots, which form a TDMA frame. Each timeslot lasts 0.5769 ms and is able to transfer 156.25 bits (both data and control). The recurrence of one particular timeslot defines a Packet Data CHannel (PDCH). Depending on the type of data transferred, a variety of logical channels are defined, which carry either data traffic or traffic for channel control, transmission control or other signaling purposes.

The major difference between GPRS and GSM concerning the radio interface is the way radio resources are allocated. In GSM, when a call is established, a channel is permanently allocated for the entire period. In other words, one timeslot is reserved for the whole duration of the call, even when there is no activity on the channel. This results in a significant waste of radio resources in the case of bursty traffic. In GPRS the radio channels, i.e. the timeslots, are allocated on a demand basis. This means that when a MS is not using a timeslot that has been allocated to it in the past, this timeslot can be re-allocated to another MS. The minimum allocation unit is a radio block, i.e. four timeslots in four consecutive TDMA frames. One RLC/MAC packet can be transferred in a radio block.

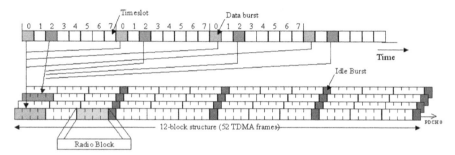

Fig. 2. Radio Channels

One or more (multi-slot capability) timeslots per TDMA frame may be assigned to a MS for the transfer of its data. During the transfer, the Base Station Subsystem (BSS) may decrease (or increase in some cases) the number of timeslots assigned to that particular MS, depending on the current demand for timeslots. This is accomplished by the use of flags (Uplink State Flag) and counters (Countdown Value) in the headers of the packets transferred on the radio link.

In order to make an exchange of data with external networks, a session must be established between the MS and the appropriate GGSN. This session is called Packet Data Protocol (PDP) context [4]. During the activation of such a context, an address (compatible with the external network, i.e. IP or X.25) is assigned to the MS and is mapped to its IMSI and a path from the MS to the GGSN is built. The MS is now visible from the external network and is ready to send or receive packets. The PDP context concerns the end-to-end path in the GPRS environment (MS ↔ GGSN).

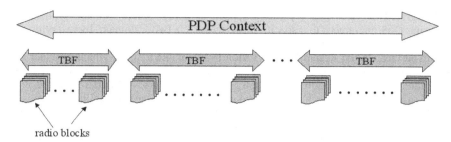

Fig. 3. PDP context and TBF

At the (lower) radio link level, when the MS starts receiving/sending data, a Temporary Block Flow (TBF) is created [5]. During this flow a MS can receive and send radio blocks uninterrupted. For a TBF establishment, the MS requests radio resources and the network replies indicating the timeslots available to the MS for data transfer. A TBF may be terminated even if the session has not ended yet. The termination of a TBF depends on the demand for radio resources and the congestion of the link. After the termination, the MS must re-establish a new TBF to continue its data transfer.

ETSI has also specified a set of QoS parameters and the corresponding profiles that a user can choose. These parameters are precedence, reliability, delay, and peak and mean throughput [3]. Precedence (priority) defines three classes (high, medium and low). Three classes are also defined for reliability. Four classes for delay, nine classes for peak throughput and thirty-one classes for mean throughput (including best-effort). A user's profile may require that the level of all (or some) parameters is defined. This profile is stored in the HLR and upon activation of a PDP context the mobile station is responsible for the required uplink traffic shaping. On the downlink, the GGSN is responsible to perform traffic shaping. It is obvious that such an implementation will not guarantee that a user will conform to the agreed profile. Also, the QoS profiles are not taken into consideration by the resource allocation procedures. Thus, it is up to the GPRS operator to use techniques that provide QoS "guarantees" and to police user traffic.

A first step in this direction is to use only the precedence parameter to define QoS classes and link allocation techniques. Precedence was chosen because of its simplicity and effectiveness and because it can be directly implemented in the GPRS architecture, as we will see in the following sections. Also, precedence can introduce very easily the idea of Differentiates Services, which is the preferred (realistic) approach for QoS in the Internet, gaining wide acceptance.

3   Differentiated Services

The Internet is experiencing increased publicity lately and great success. Multimedia and business applications have increased the volume of data traveling across the Internet, causing congestion and degradation of service quality. An important issue of practical and theoretical value is the efficient provision of appropriate QoS support.

Integrated Services [6], [7] was proposed as a first solution to the problem of ensuring QoS guarantees to a specific flow across a network domain, by reserving the needed resources at all the nodes from which the specific flow goes through. This is achieved through the Resource Reservation Protocol (RSVP) [6], which provides the necessary signaling in order to reserve network resources at each node. Although the Integrated Services solution works well in small networks, attempts to expand it to wider (inter-)networks, such as the Internet, has revealed many scalability problems.

An alternative architecture, Differentiated Services [7], was designed to address these scalability problems by providing QoS support on aggregate flows. In a domain where Differentiated Services are used, i.e. a DS domain, the user keeps a contract with the service provider. This contract, the Service Level Agreement (SLA) will characterize the user's flow passing through this DS domain, so as to include it in an aggregate of flows. The SLA also defines behavior of the domain's nodes to the specific type of flow, i.e. the Per-Hop Behavior (PHB). SLAs are also arranged between adjacent DS domains, so as to specify how flows directed from one domain to another will be treated.

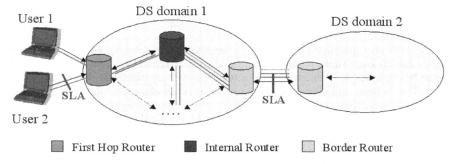

Fig. 4. The Differentiated Services Architecture

The DS field in an IP packet defines the PHB that each packet of a particular flow type shall have. This field uses reserved bits in the IP header-the "Type Of Service" field in IPv4 and the "Traffic Class" field in IPv6. In Fig. 4 we depict the DS architecture. The first-hop router is the only DS node that handles individual flows. It has the task to check whether a flow originated from a user conforms to the contract that this user has signed and to shape it, if found to be out of bounds. This is achieved by using traffic conditioners. The internal routers handle aggregates of flows and treat them according to the PHB that characterizes them. The border router checks whether the incoming (or outgoing) flows conform to the contract that has been agreed to between the neighbor DS domains. All the traffic that exceeds the conditions of the contract is (typically) discarded.

Currently, there are no standardized PHBs, but two of the basic PHBs are widely accepted. These are the Premium (or Expedited) Service [9] and the Assured Services [8]. In Premium Service, the key idea is that the user negotiates with the ISP a minimum bandwidth that will be available to the user no matter what the load of the link will be. Also, the ISP sets a maximum bandwidth allowed for this type of flow, so as to prevent the starvation of other flows. In most cases these two limits are equal, making Premium Service to act like a virtual leased line or, better, like the CBR service of ATM. The exceeding packets are discarded while the remaining ones are forwarded to the next node.

The Assured Service does not provide any strict guarantees to the users. It defines four independent classes. Within each class, packets are tagged with one of three different levels of drop precedence. So, whether a packet will be forwarded or not depends on the resources assigned to the class it belongs, the congestion level of that class and the drop precedence with which it is tagged. In other words, Assured Service provides a high probability that the ISP will transfer the high-priority-tagged packets reliably. Exceeding packets are not discarded, but they are transmitted with a lower priority (higher drop precedence).

It has been realized that there are many benefits from the deployment of both Premium and Assured services in a single DS domain. Premium service is thought of as a conservative assignment, while Assured service gives a user the opportunity to transmit additional traffic without penalty. Nowadays, Differentiated Services are known as the combination of these two services. This new architecture uses a two-bit field to distinguish the various types of services and is called Two-bit Differentiated Service [10].

Each packet is tagged with the appropriate bit (A-bit and P-bit, with null for best-effort). The ISP has previously defined the constant rate that Premium Service should guarantee. Also, exceeding packets that belong to a Premium flow are dropped or delayed, while exceeding packets of Assured Service are forwarded as best effort. In Fig. 5 we depict the tasks accomplished by the first hop router of the two-bit DiffServ architecture.

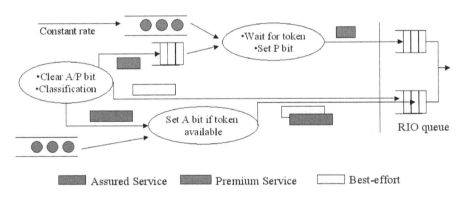

Fig. 5. First Hop Router

In the first hop router, packets that are tagged by users are checked for their conformity with the agreed SLA. In the case of Premium Service, all packets tagged with the P-bit wait in the first queue until there is a token available in the token pool. When a token becomes available, the packets are forwarded to the output queue. In the case of Assured Service, the packets for which there is no token available are forwarded to the output queue as best-effort packets, with a null tag. The queue that is used by both Assured Service and best-effort packets is a RIO (RED with In and Out) queue. RIO queues are RED (Random Early Detection) type queues with two thresholds instead of one, one for in-profile packets and one for out-of-profile packets. In this case, in-profile are the packets marked with the A-bit, while the rest (best-effort packets) are assumed to be out-of-profile. The threshold for in-profile packets is higher than the threshold for the out-of-profile packets, so that the later are discarded more often than the former. With this technique, a "better than best-effort" service is given to the packets using Assured Service.

Note that in the above figure, only the architecture concerning flows from one user is depicted. This is because the first hop router is the first, and only, router that controls and shapes individual flows. Therefore, we can assume that for each user there are two pools of tokens and a queue. The output queues are the same for all users and their characteristics depend on the outbound transfer rate of the router. The output queues can be served either by a simple priority scheme or by a more complex algorithm, such as the Weighted Fair Queuing (WFQ) algorithm.

At the border router the same basic tasks are performed, with a small variation. Since the border router manages and controls flow aggregates, it cannot buffer the packets that exceed the agreements. Thus, the packets tagged with the P-bit are not queued, as in the first hop router, but they are discarded, as shown in Fig. 6.

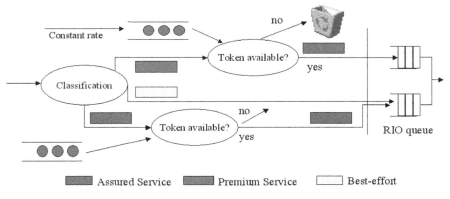

Fig. 6. Border Router

4 Differentiated Services over the GPRS Air Interface

In this section, we apply the Differentiated Services framework to the existing GPRS architecture. Specifically, we will see how the two-bit DiffServ architecture fits in GPRS, what changes must be made, and how it will be implemented.

We will give a simple example in order to make clear the reasons why we want to apply the Differentiated Services framework in the GPRS environment. Let us suppose that the GPRS network is attached to an external IP data network that uses Differentiated Services to provide QoS. The MS sends its IP packets to the GGSN, over the air interface where they are fragmented into RLC/MAC packets (frames). When these packets arrive at the GGSN, they are reassembled to IP packets and they are forwarded to the external network. Each IP packet is tagged according to the service that the user wants to receive. Thus, the GGSN acts like the first hop router in the Internet context, since there is only one IP hop from the MS to the GGSN, and checks whether the user flow conforms to the existing SLA. The next task of the GGSN is to forward the packets to the external network, where its nodes behave towards the packets as specified by the tag. We can easily conclude that any mobile user can use the Differentiated Services, as long as the external PDN supports them, in order to specify the way these packets will be treated in the external network. However, it is obvious that with the present techniques, the mobile user cannot control the way these packets are treated within the GPRS network. Our purpose is to design such a mechanism.

Before we proceed to the application of the Two-bit DiffServ architecture in the GPRS environment, we must make some assumptions. First, we assume that the core GPRS network has sufficient resources for all traffic. In other words, the point of congestion is not the GPRS backbone, but the radio link, i.e. the access link that connects the MS with the appropriate BSS. This is an important but reasonable assumption given that the scarce resource in the GPRS network is the radio spectrum. Also, we assume that the size of the frames transferred over the radio link is fixed and equal to the size of a GPRS RLC/MAC packet (frame).

As described in the previous section, the two-bit DiffServ architecture involves two types of nodes in a DS domain: the first hop and the border router. In the case of our design for GPRS, we decided to have the GPRS network act as an independent DS domain. As far as the border router is concerned, it is obvious that the GGSN is the most appropriate node for this task. It is the node that connects two DS domains. The GGSN monitors the incoming and outgoing flow aggregates in order to check their consistency with the SLAs between the two DS domains. Non-conforming traffic should be either discarded or degraded, as depicted in Fig. 6. No special changes need to be made to the GGSN in order for it to act as a border router since it communicates via the IP protocol with both sides (both the SGSN and the border router of the neighbor domain).

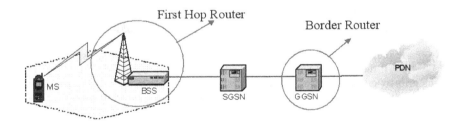

Fig. 7. Two-bit Differentiation Architecture in GPRS

When a PDP context is activated, the user can request a specific QoS level using the quality parameters mentioned earlier. In this case, the user sets the precedence parameter equal to one of the three available values. The highest priority makes use of the Premium Service, the medium priority of the Assured Service and the lowest priority of the best-effort service. This parameter is used to specify the behavior that the flow should receive in the GPRS core network, in the external network, if the later one uses Differentiated Services, and also the default radio priority used over the radio link.

As for the first hop router, this should be the BSS. Although its tasks will be the same with the ones described in Section 3, its structure will be totally different from the one depicted in Fig. 5. This happens because of some differences in the architecture between an IP network and a GPRS network. Taking into account that the MSs send their data only when the BSS instructs them to and that they use the timeslot(s) defined by the USF field, we can assume that the traffic conditioner does not reside on the BSS, but it is distributed. The queues are realized in the MS (or in the notebook connected to the MS) and the tokens come from the BSS. Actually, the USF values are the tokens transferred over the radio link.

Another important difference in having the BSS as a first hop router is that within the BSS there is just an emulation of the system depicted in Fig. 5, as described later in this section. Therefore, the BSS only needs a software upgrade in order to act as a first hop router, which makes it easier for implementation. No complex data structures are required. For queue implementation, linked lists can be used. Timers, counters and constants are all that is needed to realize the constant fill rate of the token pools and the thresholds of the RIO queues.

In the system described above, no packets do actually circulate, just requests for transfer. To be more precise, for each packet that the MS wants to transfer over the air, a pair (MS identity, service class) enters the above system. When the request exits the system then the BSS instructs the corresponding MS to transfer its packet by transmitting in a specified timeslot. The service class that a MS desires is declared with the use of the radio priority field at the TBF establishment request message. This field is two bits long, resulting into four values. We decided to have the following encoding: "1" for Premium Service, "2" for Assured Service and "3" for best-effort service. "0" specifies that the priority chosen at the PDP context activation will be used. The default value of the radio priority field is zero.

When a pair is inserted into the system, three possible actions may occur:

- the pair is forwarded to the appropriate output queue, if the counters of the Premium or Assured Service's pools are bigger than zero, or if the priority chosen is equal to "3"
- the pair is inserted into the waiting queue of Premium Service, if the corresponding counter is equal to zero, or
- the pair is forwarded to the corresponding output queue with its priority set to "3", if the Assured Service's pool counter is equal to zero.

If the priority chosen is zero, then the corresponding value in the pair inserted into the system will not be zero. Instead, the real value from the default PDP context is used.

After the transmission of a packet (i.e., after four TDMA frames, since the packet is a radio block) the MS must make a new request to the BSS to transfer another packet. This makes clear that a TBF lasts for the transmission of only one radio block, after which the TBF is terminated and another one must be established to continue the transfer.

The architecture described above provides good results in both directions of the radio link. On the downlink, when data enter the GPRS network in order to reach a mobile user, the traffic is either characterized with, or translated to, one of the available service classes (Premium, Assured, best-effort). This is done at the GGSN. If the neighbor PDN does not support Differentiated Services, then the GGSN tags the incoming packets according to the profile of the user they are directed to. If, on the other hand, the neighbor PDN supports Differentiated Services, then the GGSN translates the incoming tags according to the SLA between the two DS domains.

On the uplink, the mobile user is able to tag his IP packets, activate a service class during PDP context activation or request a service class during the TBF establishment phase. The decision of which method to use depends on the user and on the network and is discussed in the next section.

5 Discussion

In this section we discuss some issues concerning the proposals we made in this paper. One first issue concerns the transfer rate offered by the Premium Service. It is obvious that if the GPRS operator defines the Premium Service's constant rate, then he can calculate how many simultaneous users a BSS can handle, taking into consideration the number of channels that the BSS serves, the number of timeslots in each frequency carrier assigned to GPRS traffic, the size of radio blocks and, for statistical decisions, user profiles. Thus, the operator will be able to perform Call Admission Control on Premium Service requests, which is required since this type of service is the only that offers strict guarantees.

A second issue is the length of a TBF, in the case of adapting Differentiated Services to the GPRS environment. As described in Section 4, the length of a TBF is set equal to the time to transmit one radio block. This happens because it is necessary for the BSS to receive a request for every packet that must be transferred on the uplink. Furthermore, the BSS must know the radio priority of each packet. Since the radio priority is defined only during the establishment of the TBF, when the MS requests permission to transfer its data, the result is to limit the duration of a TBF to the transmission of one radio block. This makes the emulation system easier to implement and keeps the computational load to the BSS very low. However, it also results in an unnecessary use of extra TBFs (and TFIs) for the transfer of packets from the same MS. On the downlink things are simpler since the BSS is the one that does all the scheduling and buffering.

Another important issue is which service class should be assigned to the IP packets that are reassembled at the GGSN and forwarded to the external network, in the case where Differentiated Services are also supported by the external PDN. There are many possibilities. The user's application may use the "Type of Service" or the "Traffic Class" field of the IP packet to define what service should be used to the external network. Another solution is to use the default priority class defined at the PDP Context activation phase. The first solution gives the user the ability to have his packets treated differently inside and outside the GPRS network. The second solution allows the user to have his packets treated uniformly in both networks. It is desirable that the user should be able to make the final choice, so the GPRS network should probably implement both solutions.

One last issue is the charging and pricing of such services. Although it is outside the scope of this paper, we should mention that the architecture described here enables charging using congestion pricing techniques. A first step in this direction is described in [12], where the existing congestion pricing theory is extended to the DiffServ environment described here.

6 Conclusions

We have presented a way to apply the Differentiated Services framework to the GPRS wireless access environment. Our purpose was to enhance the GPRS network with QoS support that will be taken into consideration by the radio resource allocation procedures. For this purpose, the precedence QoS parameter and the radio-priority field were used, in combination with an adapted Two-bit Differentiated Services architecture. Note that the wireless access part is expected to be the most congested part of the GPRS network because of the scarcity of the wireless spectrum and therefore the part of the system where QoS support is most critical. At the same time, dynamic charging techniques can be combined with the service differentiation in order to make the resource allocation decisions efficient.

With the proposed architecture, GPRS operators will be able to provide end-to-end service differentiation fully compatible with the rest of the Internet and in cooperation with content providers. Mobile users will be able to select what service they want to be used for the transfer of their data and they will be charged accordingly. Even if the external networks do not provide service differentiation, GPRS operators will manage to offer a first level of differentiation to the wireless access network that they own.

Acknowledgments

This research was supported by the European Union's Fifth Framework Project M3I (Market-Managed Multiservice Internet - RTD No IST-1999-11429).

References

1. R. Kalden, I. Meirick and M. Meyer, "Wireless Internet Access Based on GPRS," IEEE Personal Communications, vol. 7, no. 2, pp. 8-18, April 2000.
2. C. Bettstetter, H.-J. Vogel, and J. Eberspacher, "GSM Phase 2+, General Packet Radio Service GPRS: Architecture, Protocols and Air Interface," IEEE Communications Surveys, vol. 2, no. 3, 1999 (http://www.comsoc.org/pubs/surveys/).
3. GSM 02.60: "Digital cellular telecommunications system (Phase 2+); General Packet Radio Service (GPRS); Service Description; Stage 1"
4. GSM 03.60: "Digital cellular telecommunications system (Phase 2+); General Packet Radio Service (GPRS); Service Description; Stage 2"
5. GSM 04.60: "Digital cellular telecommunications system (Phase 2+); General Packet Radio Service (GPRS); Mobile Station (MS) – Base Station (BSS) Interface; Radio Link Control/Medium Access Control (RLC/MAC) protocol."
6. P.F. Chimento, "Tutorial on QoS support for IP," CTIT Technical Report 23, 1998.
7. F. Baumgartner, T. Braun, P. Habegger, "Differentiated Services: A new approach for Quality of Service in the Internet," Proceedings of Eighth International Conference on High Performance Networking, Vienna, Austria, 21-25 Sept. 1998. Edited by: Van As, H.R., Norwell, MA, USA: Kluwer Academic Publishers, 1998. p. 255-73.
8. J. Heinane, F. Baker, W. Weiss, J. Wroclawski, "Assured Forwarding PHB Group," RFC 2597, June 1999.
9. V. Jacobson, K. Nichols, K. Poduri, "An Expedited Forwarding PHB," RFC 2598, February 1999.

10. K. Nichols, V. Jacobson, L. Zhang, "A Two-bit Differentiated Services Architecture for the Internet," RFC 2638, July 1999.
11. G. Priggouris, S. Hadjiefthymiades, L. Merakos, "Supporting IP QoS in the General Packet Radio Service," IEEE Network, Sept.-Oct. 2000, vol.14, (no.5), p. 8-17.
12. S. Soursos, "Enhancing the GPRS Environment with Differentiated Services and Applying Congestion Pricing," M.Sc. thesis, Dept. of Informatics, Athens University of Economics and Business, February 2001.

Wireless Access to Internet via IEEE 802.11: An Optimal Control Strategy for Minimizing the Energy Consumption

R. Bruno, M. Conti, and E. Gregori

Consiglio Nazionale delle Ricerche
Istituto CNUCE
Via G. Moruzzi, 1, 56124 Pisa - Italy
Tel: (050) 315 3062, Fax: (050) 3138091,
{Marco.Conti,Enrico.Gregori}@cnuce.cnr.it
Raffaele.Bruno@guest.cnuce.cnr.it

Abstract. The IEEE 802.11 standard is the most mature technology to provide wireless connectivity for fixed, portable and moving stations within a local area. Wireless communications and the mobile nature of devices involved in constructing WLANs generate new research issues compared with wired networks: dynamic topologies, limited bandwidth, energy-constrained operations, noisy channel. In this paper, we deal with the issue of minimizing the energy consumed by each station to perform a successful transmission. Specifically, by exploiting analytical formulas for the energy consumption, we derive the theoretical lower bound for the energy consumed to successfully transmit a message. This knowledge allows us to define a novel transmission control strategy based on simple and low-cost energy consumption estimates that permits each station to optimize at run-time its power utilization. Our strategy is completely distributed and it does not require any information on the number of stations in the network. Simulation results prove the effectiveness of our transmission control strategy. Specifically, the IEEE 802.11 extended with our algorithm approaches the theoretical lower bound for the energy consumption in all the configurations analyzed.

Keywords: power saving, Wireless LAN (WLAN), MAC protocol, IEEE 802.11, analytical modeling, performance analysis

1. Introduction

In the near future we will witness to a rapid growth in the need to have a mobile and ubiquitous connection to the Internet information services. Hence, the wireless technologies will become more an more utilized as means to access the Internet . In this paper we focus our attention on the Wireless Local Area Networks (WLANs) technologies, which are designed to provide wireless connectivity for fixed, portable and moving stations within a local area. A key success factor of WLANs is connected to

This work was carried out in the framework of the NATO Collaborative Linkage Grant project "Wireless Access to Internet exploiting the IEEE 802.11 technology" (PST.CLG.977405).

S. Palazzo (Ed.): IWDC 2001, LNCS 2170, pp. 120–138, 2001.

the availability of global standards to develop networking products that can provide wireless network access at competitive price. In this sense, the most mature technology is the one defined by the IEEE 802.11 standard [8], which follows the *Carrier Sensing Multiple Access with Collision Avoidance* (CSMA/CA) paradigm.

Two different approaches can be followed in the implementation of a WLAN: an *infrastructure-based* approach, or an *ad hoc networking* one [10]. An infrastructure-based architecture imposes the existence of a centralized controller for each cell, often referred to as *Access Point*. The Access Point is normally connected to the wired network thus providing the Internet access to mobile devices. In contrast, an ad hoc network is a peer-to-peer network formed by a set of stations within the range of each other that dynamically configure themselves to set up a temporary network. The IEEE 802.11 can be utilized to implement both wireless *infrastructure networks* and wireless *ad hoc networks*. The IEEE 802.11 WLAN is a single-hop ad hoc network. However, it is emerging also as one of the most promising technologies for constructing multi-hop mobile ad hoc networks [6].

Wireless communications and the mobile nature of devices involved in constructing WLANs generate new research issues compared with wired networks: dynamic topologies, limited bandwidth, energy-constrained operations, noisy channel. In WLANs, the medium access control (MAC) protocol is the main element that determines the efficiency of the resource utilization, since it performs the coordination of transmissions of the network stations and manages the congestion situations that may occur inside the network. The congestion level in the network negatively affects both the link utilization, i.e., the fraction of channel bandwidth used from successfully transmitted messages, and the energy consumed to successfully transmit a message. Specifically, each collision removes a considerable amount of channel bandwidth from that available for successful transmissions. At the same way, each collision represents significant energy wastage, since transmitting data is one of the most power consuming activities to perform. To reduce the collision probability, the IEEE 802.11 protocol uses a set of slotted windows for the backoff, whose size doubles after each collision. However, the time spreading of the accesses that the standard backoff procedure accomplishes can have a negative impact on both the link utilization and the energy consumption. Specifically, the time spreading of the accesses can introduce large delays in the message transmissions and additional energy wastage due to the carrier sensing. Furthermore, the time spreading is obtained at the cost of second collisions. Previous works have shown that an appropriate tuning of the IEEE 802.11 backoff algorithm can significantly increase the protocol capacity [4], [3]. Specifically, in [3], by exploiting exact analytical formulas for the link utilization, we define a backoff tuning algorithm based on simple and low-cost load estimates (as they are obtained from the information provided by the carrier sensing mechanism, i.e., by observing idle slots, collisions and successful transmissions) that enables each station to estimate at run-time the average size of the backoff window that permits to achieve the theoretical upper bound for the link utilization.

In this paper, we deal with the issue of minimizing the energy consumed by each station to perform a successful transmission. Specifically, this work is based on the approach followed in [3], but we will extend it by adding power-saving features to our backoff tuning algorithm. The impact of network technologies and interfaces on the energy consumption has been investigated in depth in [1], [7]. The power saving features of the emerging standards for WLANs have been analyzed in [11], [5]. Distributed (i.e., independently executed by each station, hence fitting to the ad hoc networking paradigm) strategies for power saving have been proposed and investigated in [2], [9]. Specifically, in [9] the authors propose a power controlled wireless MAC protocol based on a fine-tuning of network interface transmitting power. In [2] the authors propose a mechanism that dynamically adapts, by estimating the slot utilization and the average message length, the time spreading of accesses to asymptotically approach the minimum energy consumption for a large network population. As in [2], we exploit exact analytical formulas for the energy consumption to derive the theoretical lower bound for the energy consumed to successfully transmit a message. The knowledge of the station behavior that minimizes the energy consumption allow us to define a novel transmission control strategy based on simple and low-cost energy consumption estimates that permits each station to optimize at run-time its power utilization. Our strategy does not require any information about the network population, but it is adaptive to the number of stations in the network. Simulation results prove the effectiveness of our transmission control strategy to approach the theoretical lower bound for the energy consumption in an IEEE 802.11 network in all the configurations analyzed.

The rest of the paper is organized as follows. Section 2 describes the protocol details of the wireless MAC protocol considered. Section 3 discusses the analytical model for the energy consumption required to successfully transmit a MAC frame. Section 4 defines the Power Saving Simple Dynamic 802.11 Protocol (PS-SDP). In Section 5, PS-SDP 802.11 is compared to standard IEEE 802.11 protocol. Finally, Section 6 summarizes key results.

2.   Description of the *p-persistent* IEEE 802 .11 Protocol

The IEEE 802.11 MAC protocol provides asynchronous, time-bounded and contention free access control on a variety of physical layers. The basic access method in the IEEE 802.11 MAC protocol is the *Distributed Coordination Function* (DCF) which is a *Carrier Sense Multiple Access with Collision Avoidance* (CSMA/CA) MAC protocol. An exhaustive description of the DCF MAC protocol features can be found in [8]. In [4] it has been shown that the performances of the standard protocol can be derived by analyzing the corresponding *p-persistent* IEEE 802.11 protocol. The *p-persistent* IEEE 802.11 protocol differs from the standard protocol only in the selection of the backoff interval: at the beginning of an empty slot a station transmits (in that slot) with a probability p, while the transmission differs with a probability $1 - p$, and then repeats the procedure at the next empty slot. Hence, in this protocol the average backoff time is completely identified by the p value. It is worth remembering that choosing a p value is equivalent to identify, in the standard protocol, the average backoff window size [4]. This means that the procedure analyzed in this paper to tune the *p-persistent* IEEE 802.11 protocol by observing the network status, can be exploited in an IEEE 802.11 network to select, for a given congestion level, the appropriate size of the contention window. Due to the equivalence between the standard IEEE 802.11 protocol and the *p-persistent* IEEE 802.11 protocol, we will provide an analytical model of the *p-persistent* protocol.

In the rest of this section we detail the *p-persistent* IEEE 802.11 protocol behavior. Before a station initiates a transmission, it senses the channel to determine whether another station is transmitting. If the medium is found to be idle for an interval that exceeds the *Distributed InterFrame Space* (DIFS), the station continues with its transmission. On the other hand (i.e., the medium is busy), the transmission is deferred until the end of the ongoing transmission. The idle time immediately following an idle DIFS after the transmission completion is slotted, and a station is allowed to transmit only at the beginning of each *slot time*. Hereafter, we will refer to the slot time duration as t_{slot}. The t_{slot} is equal to the time needed at any station to detect the transmission of a packet from any other station. The decision to begin a transmission or to defer the transmission to the next empty slot is accomplished according to the transmission probability p. Immediate positive acknowledgements are employed to ascertain the successful reception of each packet transmission[1]. This is accomplished by the receiver (immediately following the reception of the data frame), which initiates the transmission of an acknowledgement frame (ACK) after a time interval, *Short InterFrame Space* (SIFS), which is less than the DIFS. If an acknowledgement is not received, the data frame is presumed to have been lost and a retransmission is scheduled.

The model used in this paper to evaluate the performance figures does not depend on the technology adopted at the physical layer (e.g., infrared and spread spectrum). However, the physical layer technology determines some network parameter values, e.g., SIFS, DIFS and t_{slot}. In Table 1 we report the parameter setting we will adopt in all the numerical evaluation and simulation runs performed in this paper. The choice of the values for the technology-dependent parameter is compliant to the frequency-hopping-spread-spectrum technology at a 2 Mbit/s transmission rate [8].

Table 1. Parameter setting for the *p-persistent* IEEE 802.11 protocol.

t_{slot}	Propagation Delay (τ)	DIFS	SIFS	ACK	Bit Rate
50 μsec	≤1 μsec	2.56 t_{slot}	0.56 t_{slot}	112 bits	2 Mbps

3. Power Consumption Analysis

The analysis of the energy consumption in a *p-persistent* IEEE 802.11 network has been performed in [2]. Specifically, in that paper, the authors focus their attention on a tagged station and observe the system at the end of each successful transmission attempt of the tagged station. By assuming that the message lengths are random variables i.i.d., and considering the *p-persistent* protocol behavior, it follows that all the processes that characterize the occupancy pattern of the channel (i.e., idle periods, collisions and successful transmissions) are regenerative with respect to the time instants corresponding to the completion of the tagged-station successful transmissions[2]. Therefore, the energy consumption analysis can be performed by studying the system

---

[1] Let us remeind that CSMA/CA does not rely on the capability of the stations to detect a collision by hearing their own transmission

[2] Hereafter, we will assume that successive transmission attempts of a station have independent lengths.

behavior in a generic renewal period, also referred to as *virtual transmission time*. By assuming that *PTX* and *PRX* are the power consumptions (expressed in mW) of the network interface during the transmitting and receiving phase, respectively, then the average energy (in mJ) required to a station to perform a successful transmission can be expressed as follows [2]:

$$E[Energy] = E[Nc + 1] \cdot E[N_{not\_used\_slot}] \cdot E[Energy_{not\_used\_slot}] + \tag{1}$$

$$+ E[Nc] \cdot E[Energy_{tagged\_collision}] + E[Energy_{tagged\_success}]$$

where, $E[Nc]$ is the average number of collisions experienced by the tagged station during two successful transmissions, $E[N_{not\_used\_slot}]$ is the average number of *not_used* slots (from the tagged station standpoint) before the transmission attempt of the tagged station, $E[Energy_{not\_used\_slot}]$ is the average energy consumption during *not_used* slots, $E[Energy_{tagged\_collision}]$ is the average energy consumption during collisions, conditioned to observe a tagged-station collision, and $E[Energy_{tagged\_success}]$ is the average energy consumption during tagged-station successful transmissions.

To correctly evaluate the energy consumed during both a collision and a successful transmission, we have to take in account the protocol overheads introduced. Let us to denote with *Collision* the average length of a collision (not involving the tagged station), and with S the average length of a successful transmission, including their overheads. Hence, according to the protocol behavior described in Section 2 and considering a geometric distribution with parameter q for the message length (expressed as number of t_{slot}), it follows (see [4] for the proofs):

$$S \leq 2\tau + SIFS + ACK + DIFS + t_{slot}\left(1/(1-q)\right) \tag{2.a}$$

$$Collision \leq \tau + DIFS + \frac{t_{slot}}{1 - \left[(1-p)^{M-1} + (M-1)p(1-p)^{M-2}\right]} \cdot \tag{2.b}$$

$$\left[\sum_{h=1}^{\infty}\left\{h\left[\left(1-pq^{h}\right)^{M-1} - \left(1-pq^{h-1}\right)^{M-1}\right]\right\} - \frac{(M-1)p(1-p)^{M-2}}{1-q}\right]$$

The unknown quantities in Equation (1) are given by the following Lemma (see [2] for the proofs).

Lemma 1. In a network with M stations, by assuming that each station is operating in asymptotic conditions (i.e., stations have always at least a packet waiting to be transmitted), by denoting with p the transmission probability adopted by each station, and with q the parameter of the geometric distribution that defines the message length (expressed as number of t_{slot}), it follows:

$$E[Nc] = \frac{1 - (1-p)^{M-1}}{(1-p)^{M-1}} \quad , \quad E[N_{not\_used\_slot}] = \frac{1-p}{p} \tag{3}$$

$$E[Energy_{not\_used\_slot}] = PRX \cdot [t_{slot}(1-p)^{M-1} + S \cdot (M-1)p(1-p)^{M-2} + \quad\quad (4)$$
$$Collision \cdot (1 - (1-p)^{M-1} - (M-1)p(1-p)^{M-2})]$$

$$E[Energy_{tagged\_collision}] = PTX \cdot t_{slot} \cdot (1/(1-q)) + PRX \cdot t_{slot} \cdot \sum_{x=1}^{\infty} q^{x-1}(1-q) \cdot \quad\quad (5)$$
$$\left[\sum_{y=1}^{\infty} y \cdot \frac{\left(1 - pq^{y+x}\right)^{M-1} - \left(1 - pq^{y+x-1}\right)^{M-1}}{1 - (1-p)^{M-1}} + DIFS + \tau \right]$$

$$E[Energy_{tagged\_success}] = PTX \cdot (1/(1-q)) + PRX \cdot (2\tau + SIFS + ACK + DIFS) \quad\quad (6)$$

By assuming that all the times are expressed in t_{slot}, and that also PTX and PRX are expressed in mW/t_{slot}, then $E[Energy]$ (in mJ) is a $f(p,q,M,PTX,PRX)$. Hereafter, we refer to p_{opt} as the p value that minimizes the energy consumption, fixed M, q and the PTX and PRX system parameters. We also assume that a network is in its *optimal operating state*, if each station adopts the p_{opt} value as its transmission probability. The p_{opt} values for several settings of the proposed parameters can be computed by numerically minimizing the Equation (1).

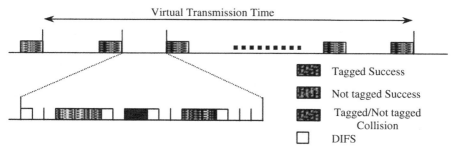

Fig. 1. Channel structure during to successful transmission of the tagged station.

The optimization of the station's power utilization requires the knowledge of the p_{opt} value. However, it is computationally expensive to afford at run time the minimization of Equation (1). Furthermore, the minimization of Equation (1) requires a robust estimation of the M parameter, i.e., the number of stations in the network, but it is extremely complex and unreliable to retrieve this information from the channel status in a *CSMA*-like protocol. Hence, it would be convenient to find out a simpler relationship to provide an approximation of the p_{opt} value. To this end, in the following we will investigate the role of the various terms of Equation (1) in determining the overall energy consumption. Specifically, we will separate the energy consumptions that are increasing function of the p value from the energy consumptions that are decreasing function of the p value.

In Figure 1 we have plotted the sequence of idle periods, collisions and successful transmissions that occur on the channel between two consecutive tagged-station successful transmissions. During the virtual transmission time, we observe on the channel an average number of not tagged-station successful transmissions equal to $M-1$ due to the symmetry of the system. Specifically, none station can be privileged during the access to the channel, hence all the stations have the same probability to experience a successful transmission. Therefore, the successful transmissions, from an energy consumption standpoint, have a cost that is invariant with the p value, so we cannot minimize the energy consumed by a station during successful transmissions, but only the energy consumed by a station during idle periods and collisions. Specifically, when the channel is not occupied by successful transmissions (i.e., the white fraction of the channel, as shown in Figure 1) we have a sequence of tagged-station and not tagged-station collisions that follow idle periods. Hereafter, we refer to the average energy consumed by a station during two consecutive transmission attempts as $E[Energy_{transmission\_attempt}]$. Since all the processes that characterize the occupancy pattern on the channel are still regenerative between two transmission attempts, we derive the closed formula for $E[Energy_{transmission\_attempt}]$ by using the regenerative property (see Appendix A for the closed formulas):

$$E[Energy_{transmission\_attempt}] = E[Energy_{Idle\_p}] + E[Energy_{success}] \cdot \Pr\{success \mid N_{tr} \geq 1\} + \qquad (7)$$

$$E[Energy_{tagged\_collision}] \cdot \Pr\{tagged\_collision \mid N_{tr} \geq 1\} +$$

$$E[Energy_{not\_tagged\_collision}] \cdot \Pr\{not\_tagged\_collision \mid N_{tr} \geq 1\}$$

where $E[Energy_{Idle\_p}]$ is the average energy consumed by a generic station listening the channel, the second term in the right-hand side is the average energy consumed during a generic success conditioned to the observation of a transmission attempt, let it say $E[Energy_{success|trasm}]$, the third term in the right-hand side is the average energy consumed during tagged-station collisions conditioned to the observation of a transmission attempt[3], let it say $E[Energy_{tagged\_collision|transm}]$, and the fourth term in the right-hand side is the average energy consumed during not tagged-station collisions, conditioned to the observation of a transmission attempt, let it say $E[Energy_{not\_tagged\_collision|transm}]$.

It is straightforward to derive that $E[Energy_{Idle\_p}]$ is a decreasing function of the p value, whereas $E[Energy_{tagged\_collision|transm}]$ and $E[Energy_{not\_tagged\_collision|transm}]$ are increasing functions of the p value. In [3] and [4], we have shown that the p value that maximizes the throughput is well approximated by the p value that permits to have the average time spent by sensing the channel equal to the average time the channel is occupied by collisions Following the same approach, we suggest that the minimum energy consumption is achieved when each station adopts a transmission probability that permits to have the average energy spent by sensing the channel equal to the

---

[3] Since we will not take in account the energy consumed during the successful transmissions, in Appendix A we avoid to report the formula related to $E[Energy_{success|trasm}]$

average energy wasted during collisions. We can express this condition with the following relationship:

$$E\left[Energy_{Idle\_p}\right] = E\left[Energy_{tagged\_collision|trasm}\right] + E\left[Energy_{not\_tagged\_collision|trasm}\right] \tag{8}$$

The right-hand side in Equation (8) represents the average energy consumption during a generic collision, given a transmission attempt, i.e., $E[Energy_{Collision}]$. To easier compute $E[Energy_{Collision}]$ we further expand the $\Pr\{tagged\_collision \mid N_{tr} \geq 1\}$ and $\Pr\{not\_tagged\_collision \mid N_{tr} \geq 1\}$ expressions by conditioning on the tagged-station transmission, it say tag_tr, and the not tagged-station transmission, say not_tag_tr, respectively, Equation (8) can be written as (see Appendix A for the closed formulas):

$$E\left[Energy_{Idle\_p}\right] = E\left[Energy_{tagged\_collision}\right] \cdot \Pr\{tagged\_collision \mid tag\_tr\} \cdot \tag{9}$$

$$\Pr\{tag\_tr \mid N_{tr} \geq 1\} + E\left[Energy_{not\_tagged\_collision}\right] \cdot$$

$$\Pr\{not\_tagged\_collision \mid not\_tag\_tr\} \cdot \Pr\{not\_tag\_tr \mid N_{tr} \geq 1\}$$

As previously said the approach followed to derive Equation (9) is similar to the one followed in [4], [3] to maximize the throughput. Therefore, afterwards we will provide further considerations on the relationship between the energy-consumption minimization problem and the throughput maximization problem.

Table 2. Energy consumption analysis ($PTX = 2$, $PRX = 1$).

m (t_{slot})	p_{opt} value				Minimum Energy Consumption			
	Exact Value		Approximated Value		Exact Value		Approximated Value	
	M=10	M=100	M=10	M=100	M=10	M=100	M=10	M=100
2	5.08e-2	5.10e-3	5.43e-2	5.54e-3	99.4944	987.412	99.6043	988.999
5	3.84e-2	3.84e-3	4.06e-2	4.14e-3	144.895	1411.79	144.988	1413.14
10	2.98e-2	3.01e-3	3.11e-2	3.17e-3	214.851	2063.92	214.928	2065.03
20	2.24e-2	2.26e-3	2.32e-2	2.36e-3	346.766	3290.21	346.828	3291.11
50	1.49e-2	1.51e-3	1.53e-2	1.55e-3	721.198	6759.72	721.244	6760.38
100	1.08e-2	1.09e-3	1.10e-2	1.12e-3	1321.73	12310.1	1321.77	12310.6

In Tables 2 and 3 we compare the minimum energy consumption, calculated by minimizing Equation (1), and the energy consumption measured when each stations adopts the p value that satisfies Equation (9). In Tables 2 and 3 we also compare the p value that satisfies Equation (9) with the p_{opt} value. The numerical results are obtained by assuming $PRX = 1$ and considering either $PTX = 2$ or $PTX = 10$ [7]. To fix $PRX = 1$ is done to avoid useless details. Specifically, our energy units is the energy consumption in a t_{slot} when the network interface is in the receiving state.

Table 3. Energy consumption analysis ($PTX = 10$, $PRX = 1$).

m (t_{slot})	p_{opt} value				Minimum Energy Consumption			
	Exact Value		Approximated Value		Exact Value		Approximated Value	
	M=10	M=100	M=10	M=100	M=10	M=100	M=10	M=100
2	4.15e-2	4.96e-3	4.42e-2	5.38e-3	123.874	1013.78	123.987	1015.39
5	2.99e-2	3.74e-3	3.14e-2	3.99e-3	199.335	1470.15	199.425	1471.51
10	2.25e-2	2.88e-3	2.34e-2	3.04e-3	315.967	2171.08	316.039	2172.20
20	1.66e-2	2.16e-3	1.71e-2	2.26e-3	537.148	3489.52	537.204	2490.41
50	1.09e-2	1.44e-3	1.11e-2	1.48e-3	1169.70	7222.64	1169.74	7223.30
100	7.88e-3	1.04e-3	8.00e-3	1.06e-3	2190.53	13199.3	2190.56	13199.9

The results reported in the Tables 2 and 3 show that Equation (8) gives a good approximation of the minimum energy consumption for all the network parameter settings we have considered. It is worth pointing out that the numerical results show that a precise approximation of the minimum energy consumption does not require the same accuracy level for the p_{opt} approximation.

From the discussion performed so far, it results that the energy consumption minimization can be achieved not only by computing the p value that minimizes Equation (1), but it is sufficient to identify the p value that permits to have $E[Energy_{Idle\_p}]$ equal to $E[Energy_{Collision}]$. By exploiting the latter, we are able to define an efficient and simple transmission control strategy, see Section 4. In the remaining of this section, we will give an insight in some of the most relevant characteristics of the network behavior when each station operates in such a way to minimize the energy consumption. This analysis is significant not only to investigate the network characteristics, which our transmission control strategy would be based on, but also to point out how the energy-consumption minimization issue is related to (and differs from) the throughput maximization issue.

First of all, we give some approximations for the complex expressions for $E[Energy_{tagged\_collision}]$ and $E[Energy_{not\_tagged\_collision}]$ when the network is in its optimal operating state. Specifically, we can assume that the collision probability is low when the network is in its optimal operating state. Hence, the approximation for the collision probability is obtained by assuming that no more than two stations collide. According to this assumption, for a geometric message-length distribution, it follows:

$$E[Energy_{not\_tagged\_collision}] \approx PRX \cdot \left[\frac{1 + 2q - q^2 - 2q^3}{(1-q)^2(1+q)^2} + DIFS + \tau \right] = \overline{E}_{CNT} \quad \textbf{(10.a)}$$

$$E[Energy_{tagged\_collision}] \approx \frac{1}{1-q} \left[PTX + PRX \cdot \frac{q}{1+q} \right] + PRX \cdot (DIFS + \tau) = \overline{E}_{CT} \quad \textbf{(10.b)}$$

In Tables 4 and 5 we compare the exact values of the average energy consumed during tagged-station collisions, given by relationship (5), and during not tagged-station collisions, given by relationship (A.2), with the approximations provided by (10.a) and (10.b), respectively, when all the stations adopt the p_{opt} value as their transmission probability.

Table 4. $E[Energy_{tagged\_collision}]$.

m (t_{slot})	$PTX/PRX = 2$			$PTX/PRX = 10$		
	Exact Value		**Approx.**	**Exact Value**		**Approx.**
	M=10	**M=100**	**Value**	**M=10**	**M=100**	**Value**
2	7.34877	7.37080	7.246667	23.32946	23.36726	23.24667
10	27.7045	27.7934	27.31684	107.6080	107.7734	107.3164
100	253.757	254.098	252.3287	1053.358	1054.013	1052.328

The numerical results reported in Tables 4 and 5, confirm that expressions (10.a) and (10.b) provide precise approximations of $E[Energy_{tagged\_collision}]$ and $E[Energy_{not\_tagged\_collision}]$. Furthermore, the numerical results show that both $E[Energy_{tagged\_collision}]$ and $E[Energy_{not\_tagged\_collision}]$ do not depend significantly on the M value, but only on the average packet length, and on the PTX and PRX values. This characteristic will be used during the definition of our transmission control strategy in Section 4.

Table 5. $E[Energy_{not\_tagged\_collision}]$.

m (t_{slot})	Exact Value $PTX/PRX = 2$		Exact Value $PTX/PRX = 10$		Approx. Value
	M=10	**M=100**	**M=10**	**M=100**	
2	5.30675	5.32942	5.29517	5.32702	5.246667
10	17.5174	17.6106	17.4525	17.5966	17.31684
100	152.015	152.556	151.367	152.456	152.3287

Let us now focus on the average number of stations that try to access the channel at the same time to achieve the minimum energy consumption. It is straightforward to derive that this average number is given by the Mp_{opt} product.

Figures 2.a and 2.b show the Mp_{opt} product for an average message length m equal to 2 and 100 time slots, respectively. In each figure we plot the Mp_{opt} value related to different PTX/PRX values. The curves labeled with $PTX/PRX = 1$ correspond to the average number of transmitting stations that maximizes the throughput. Specifically, with $PTX/PRX = 1$, Equation (1) reduces to the average length of a virtual transmission time. The minimization of Equation (1) corresponds to minimize the time necessary to complete a successful transmission for the tagged station, hence to maximize the throughput.

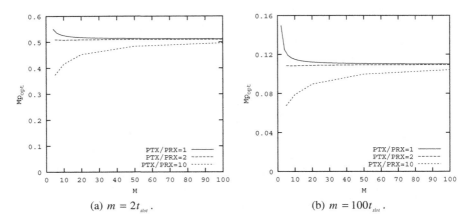

(a) $m = 2t_{slot}$.   (b) $m = 100t_{slot}$.

Fig. 2. The Mp_{opt} product.

From the plotted curves we derive that the Mp_{opt} product exhibit a dependency on the M value for small network-size population. This effect is more marked when $PTX/PRX = 10$. Specifically, for $PTX/PRX = 2$, the Mp_{opt} product can be considered a constant function independent of the M value. In this case, the energy-consumption minimization problem is very similar to the throughput maximization problem, due to the not significant difference between the Mp_{opt} values. In other words, to adopt the p_{opt} value as transmission probability, beyond providing power saving, contributes to obtain quasi-optimal channel utilization. Instead, for $PTX/PRX = 10$, the Mp_{opt} product shows a significant dependency on the M value. This dependency reduces only for large M value (e.g., $M \geq 50$). This behavior is explained by observing that the energy consumed during not tagged collisions depends only on the PRX parameter, as $E[Energy_{Idle\_p}]$, whereas the PTX parameter has a significant impact in the energy consumed during tagged collisions. However (in a large network) the probability to have a not tagged collision is much higher than the probability to have a tagged collision, hence in this case the impact of the PTX parameter is significantly reduced.

From Figures 2.a and 2.b it follows that the transmission probability that minimizes the energy consumption is always lower than the transmission probability that permits to maximize the throughput. To better explain this behavior, we refer to Figure 3 where we have plotted $E[Energy_{Idle\_p}]$ and $E[Energy_{Collision}]$ versus the p value, for various PTX/PRX values. $E[Energy_{Idle\_p}]$ is equal to $E[Idle\_p]$ due to the assumption of $PRX = 1$. At the same way, the $E[Energy_{Collision}]$ related to $PTX/PRX = 1$ is equal to the average length of a collision given a transmission attempt, also referred to as $E[Coll]$. The p value that corresponds to the intersection point of the $E[Energy_{Idle\_p}]$ and $E[Energy_{Collision}]$ curves is the approximation of the p_{opt} value. It is worth remembering that for $PTX/PRX = 1$ the p value that corresponds to the intersection point of the $E[Idle\_p]$ with $E[Coll]$ provides a good approximation of the p value that maximizes the throughput (see [4]). We note that by increasing the PTX value also $E[Energy_{Collision}]$ grows due to the rise in the energy consumed during tagged collisions. However, $E[Energy_{Idle\_p}]$ does not depend on the PTX value, hence, only a decrease in the p_{opt} value can balance the increase in $E[Energy_{Collision}]$.

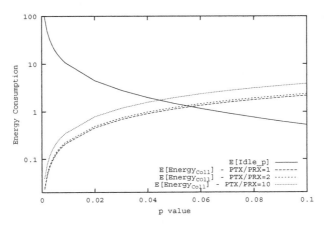

Fig. 3. p_{opt} approximation with $M = 10$ and $m = 2t_{slot}$.

4. The *Power Saving – Simple Dynamic* IEEE 802.11 Protocol (*PS-SDP*)

In this section, we will define the *Power Saving – Simple Dynamic* IEEE 802.11 protocol (PS-SDP), that is the standard *p-persistent* IEEE 802.11 protocol enhanced by a transmission control strategy based on the run-time estimate of the p_{opt} value. Le us remind that in [3] we have defined a Simple Dynamic IEEE 802.11 protocol to achieve the throughput maximization. The PS-SDP expends SDP with new power-saving features. To define a transmission control strategy that would not be cumbersome, we have to rewrite Equation (9) in a simpler, even though approximated, way. To this end we introduce the following polynomial approximations:

$$F_M(p) = (1-p)^M \approx 1 - Mp + O(p^2) \tag{11.a}$$

$$G_M(p) = Mp(1-p)^{M-1} \approx Mp - M(M-1)p^2 + O(p^3) \tag{11.b}$$

By exploiting relationships (11.a) and (11.b), it is straightforward to derive:

$$\Pr\{tagged\_collision \mid tag\_tr\} = 1 - F_{M-1}(p) \approx (M-1)p \tag{12.a}$$

$$\Pr\{not\_tagged\_collision\mid not\_tag\_tr\} = \frac{1 - [F_{M-1}(p) + G_{M-1}(p)]}{1 - F_{M-1}(p)} \approx (M-2)p \qquad \textbf{(12.b)}$$

The approximations provided by relationships (12.a) and (12.b) are not intended to be the optimal approximation for the $\Pr\{tagged\_collision \mid N_{tr} \geq 1\}$ and $\Pr\{not\_tagged\_collision \mid N_{tr} \geq 1\}$, but they well show the dominant role of the Mp product in determining these probabilities. By exploiting relationships (11.a) and (11.b), and by assuming that $\Pr\{tag\_tr \mid N_{tr} \geq 1\} \approx 1/M$ and $\Pr\{not\_tag\_tr \mid N_{tr} \geq 1\} \approx (M-1)/M^4$, it follows:

$$E[Energy_{Idle\_p}] \approx PRX \cdot \frac{1 - Mp}{Mp} \qquad \textbf{(13.a)}$$

$$E[Energy_{tagged\_collision\,transm}] \approx \overline{E}_{CT} \cdot (M-1)p \cdot \frac{1}{M} \qquad \textbf{(13.b)}$$

$$E[Energy_{not\_tagged\_collision\,transm}] \approx \overline{E}_{CNT} \cdot (M-2)p \cdot \frac{M-1}{M} \qquad \textbf{(13.c)}$$

We cannot directly exploit relationships (13.a)-(13.c) to find out the p_{opt} value, due to the presence of the unknown M parameter. However, by exploiting relationships (13.a)-(13.c) we can rewrite Equation (9) in such a way that the p_{opt} estimation does not require any information about the M value, or its distribution.

Let us consider the n-th transmission attempt, and say p_n the transmission probability adopted by all the stations during the n-th transmission attempt. From relationships (13.a)-(13.c), it follows that at the end of the n-th transmission attempt:

$$E[Energy_{Idle\_p}]\Big|_{p=p_n} \approx \tilde{E}[Energy_{Idle\_p}]_n = PRX \cdot \frac{1 - Mp_n}{Mp_n} \qquad \textbf{(14.a)}$$

$$E[Energy_{Coll}]\Big|_{p=p_n} \approx \tilde{E}[Energy_{Coll}]_n = \overline{E}_{CT} \cdot (M-1)p_n \cdot \frac{1}{M} + \overline{E}_{CNT} \cdot (M-2)p_n \cdot \frac{M-1}{M} \qquad \textbf{(14.b)}$$

If $p_n \neq p_{opt}$, then $\tilde{E}[Energy_{Idle\_p}]_n \neq \tilde{E}[Energy_{Coll}]_n$ and Equation (9) does not hold. For the $(n+1)$-th transmission attempt our algorithm searches a new transmission probability p_{n+1} such as to balance in the future the energy consumed during idle periods and collisions, namely to have $\tilde{E}[Energy_{Idle\_p}]_{n+1} = \tilde{E}[Energy_{Coll}]_{n+1}$. to this end, we first express p_{n+1} as a function of an unknown quantity x, such that $p_{n+1} = p_n(1+x)$. Then, by assuming $\tilde{E}[Energy_{Idle\_p}]_{n+1} = \tilde{E}[Energy_{Coll}]_{n+1}$, from (14.a) and (14.b), after some algebraic manipulations, we obtain:

---

[4] As lower is the p value as more correct as these assumptions

$$(1 + x) = \tag{15}$$

$$\frac{\sqrt{1 + 4PRX(1 + E[\ Idle\_p])\left(\dfrac{E[Energy_{tagged\ collision|trasm}]}{M} + E[Energy_{not\_tagged\_collision\ trasm}]\right)} - 1}{2PRX(1 + E[\ Idle\_p])\left(\dfrac{E[Energy_{tagged\ collision|trasm}]}{M} + E[Energy_{not\_tagged\_collision|trasm}]\right)}$$

The relationship (15) is valid for $M \gg 1$. Hence, to completely eliminate the algorithm dependency on the M value, we will do the conservative assumption that $M = 10$. It means that the ten percent of $E[Energy_{tagged\_collision\ trasm}]$ always contributes to the $(1 + x)$ evaluation. This is a conservative assumption because the percentage of $E[Energy_{tagged\_collision\ trasm}]$ that impacts the $(1 + x)$ evaluation decreases below the ten percent for M values grater than ten. However, we believe that this conservative behavior still guarantees a significant improvement of protocol efficiency, and our assessment will be extensively validated through the performance analysis executed in the following section.

5. Performance Analysis in Steady-State Conditions

In this section we investigate, via simulation, the performances achieved by the PS-SDP when the network operates in steady state conditions. In our simulation we assumed an ideal channel with no transmission errors and no hidden terminals, i.e., all the stations can always hear all the others. The effectiveness of PS-SDP has been analyzed with different network configurations. Specifically, we run simulation experiments for several network populations (i.e., $M \in [5...100]$) and message lengths ($m = 2t_{slot}$ and $m = 100t_{slot}$).

As we have previously explained, PS-SDP operates at the completion of each transmission attempt with the target to adjust the p value so that Equation (9) holds. The updating rule provided by relationship (15), requires the knowledge of the average length of idle periods, i.e., $E[Idle\_p]$, the average energy consumed during tagged-station collisions given a transmission attempt, i.e., $E[Energy_{tagged\_collision\ transm}]$, and the average energy consumed during not tagged-station collisions given a transmission attempt, i.e., $E[Energy_{not\_tagged\_collision\ transm}]$. Each station, by using the carrier sensing mechanism can observe the channel status and measure the length of idle periods and busy periods. In the latter case, we assume that it can distinguish successful transmission from collisions by observing the ACK. Among collisions, each station obviously knows the collisions in which it is involved or not. From these values, $E[Idle\_p]$, $E[Energy_{tagged\_collision\ transm}]$ $E[Energy_{not\_tagged\_collision\ transm}]$ can be approximated by exploiting a moving average window:

$$E[Idle\_p]_n = \alpha \cdot E[Idle\_p]_{n-1} + (1 - \alpha) \cdot Idle\_p_n \tag{16.a}$$

$$E\left[Energy_{tagged\_collision|trasm}\right]_n = \alpha \cdot E\left[Energy_{tagged\_collision|trasm}\right]_{n-1} + (1-\alpha) \cdot [PTX \cdot \tag{16.b}$$

$$Coll\_tagged_n + PRX \cdot \max\left(0, Coll_n - Coll\_tagged_n\right)]$$

$$E\left[Energy_{not\_tagged\_collision|trasm}\right]_n = \alpha \cdot E\left[Energy_{not\_tagged\_collision|trasm}\right]_{n-1} + \tag{16.c}$$

$$(1-\alpha) \cdot PRX \cdot Coll\_not\_tagged_n$$

where:

- $E[Idle\_p]_n$, $E[Energy_{tagged\_collision|trasm}]_n$ and $E[Energy_{not\_tagged\_collision|trasm}]_n$ are the approximations at the end of the n-th transmission attempt;
- $Idle\_p_n$ is the length of the n-th idle period;
- $Coll_n$ is zero if either the n-th transmission attempt is successful or a not tagged-station collision, otherwise it is the collision length;
- $Coll\_tagged_n$ is zero if either the n-th transmission attempt is successful or a not tagged-station collision, otherwise it is length of the tagged-station transmission;
- $Coll\_not\_tagged_n$ is zero if either the n-th transmission attempt is successful or a tagged-station collision, otherwise it is the collision length;
- $\alpha \in [0,1]$ is a smoothing factor.

The use of a smoothing factor, α, is widespread in the network protocols to obtain reliable estimates from the network estimates by avoiding harmful fluctuations. In the following we summarize the steps performed independently by each station to compute the p_{n+1} value for the current network and load conditions, given the p_n value.

```
begin

    step 1: measure of the n-th idle period Idle_p_n;

    step 2: measure of Coll_n, Coll_tagged_n and Coll_not_tagged_n;

    step 3: update of E[Idle_p]_n, E[Energy_tagged_collision|trasm]_n and
    E[Energy_not_tagged_collision|trasm]_n [by exploiting (16.a)-(16.c)];

    step 4: calculate
```

$$p_{new} = p_n \cdot$$

$$\frac{\sqrt{1 + 4PRX(1 + E[Idle\_p]_n)\left(\dfrac{E\left[Energy_{tagged\_collision|trasm}\right]_n}{10} + E\left[Energy_{not\_tagged\_collision|trasm}\right]_n\right)} - 1}{2PRX(1 + E[Idle\_p]_n)\left(\dfrac{E\left[Energy_{tagged\_collision|trasm}\right]_n}{10} + E\left[Energy_{not\_tagged\_collision|trasm}\right]_n\right)}$$

```
    step 5: p_{n+1} = α · p_n + (1-α) · p_new

end
```

In Figures 4.a and 4.b we compare (for an average message length equal to $2t_{slot}$) the energy consumption, measured in the standard IEEE 802.11 protocol (STD 802.11) and the PS-SDP against the theoretical lower bound (OPT 802.11). In figures 4.a and 4.b we vary the network configuration (i.e., $M \in [5...100]$) and the network interface characteristics (i.e., $PTX/PRX = 2$ and $PTX/PRX = 10$). The logarithmic scale for the y-axis has been chosen to better highlight the energy consumption behavior both for the low values measured within small networks and for the high values measured within large networks.

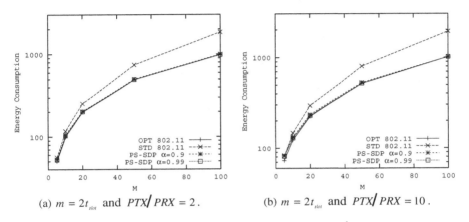

(a) $m = 2t_{slot}$ and $PTX/PRX = 2$.   (b) $m = 2t_{slot}$ and $PTX/PRX = 10$.

Fig. 4. Energy Consumption for $m = 2t_{slot}$ and different PTX/PRX values.

The numerical results clearly show that the effectiveness of PS-SDP does not depend on the number of stations in the network. These results validate the effectiveness of the conservative assumption of assigning to $E[Energy_{tagged\_collision\,trasm}]$ a fixed weight of 0.1 as we have done in step 4 of our transmission control strategy. Furthermore, by comparing the curves obtained with $PTX/PRX = 2$ and $PTX/PRX = 10$ it is confirmed that the PS-SDP effectiveness is independent of the network interface behavior. For the PS-SDP we have considered different smoothing factors to investigate the impact of the memory-estimate length on the protocol performances. The numerical results show that as the memory-estimate length increases (i.e., increasing the α values), as closer PS-SDP approaches the theoretical lower bound for the energy consumption. However, the differences between the energy consumptions measured are not meaningful and they cannot be appreciated from the figures.

In Figures 5.a and 5.b we compare (for an average message length equal to $100t_{slot}$) the energy consumption, measured in the standard IEEE 802.11 protocol (STD 802.11) and the PS-SDP against the theoretical lower bound (OPT 802.11). In figures 4.a and 4.b we vary the network configuration (i.e., $M \in [5...100]$) and the network interface characteristics (i.e., $PTX/PRX = 2$ and $PTX/PRX = 10$). The discussion performed in the case of short messages ($m = 2t_{slot}$) can be repeated in the case of long messages ($m = 100t_{slot}$), shown in Figures 5.a and 5.b. Hence, we have confirmed that PS-SDP is adaptive to both the number of stations in the network and the traffic characteristics.

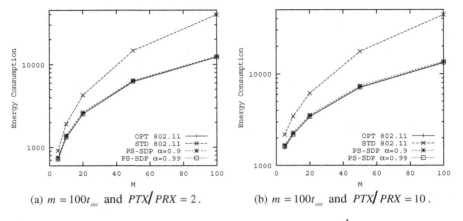

(a) $m = 100t_{slot}$ and $PTX/PRX = 2$.   (b) $m = 100t_{slot}$ and $PTX/PRX = 10$.

Fig. 5. Energy Consumption for $m = 100t_{slot}$ and different PTX/PRX values.

6. Conclusions

In this paper, by considering the *p-persistent* IEEE 802.11 protocol we have investigated the station behavior that guarantees to minimize the energy consumption required to successfully transmit a message. By exploiting this analysis we have found out that the minimum energy consumption is closely approximated when each station adopts a transmission probability that permits to have the average energy spent by sensing the channel equal to the average energy wasted during collisions. This property has been used to propose a Power Saving – Simple Dynamic IEEE 802.11 Protocol that allows each station to estimate at run-time the p_{opt} value. Ps-SDP is based on simple and low-cost energy-consumption estimates. Our strategy does not require any information about the network population, but it is adaptive to the number of stations in the network. Furthermore, it is completely distributed so as to fit well the ad hoc networking paradigm. We have demonstrated through simulative results that the PS-SDP is effective to approach the minimum energy consumption in all the configurations analyzed. Further research involves the investigation of PS-SDP behavior in dynamic conditions related to either bursty arrivals of transmission requirements or network topology changes.

References

[1] N. Bambos, "Toward power-sensitive network architectures in wireless communications: Concepts, issues and design aspects", *IEEE Personal Comm*, 1998, pp. 50-59.

[2] L. Bononi, M. Conti, L. Donatiello, "A Distributed Mechanism for Power Saving in IEEE 802.11 Wireless LANs", *ACM/Kluwer Mobile Networks and Applic. Journal*, Vol 6., N.3 (2001), pp. 211-222.

[3] R. Bruno, M. conti, E. Gregori, "A simple protocol for the dynamic tuning of the backoff mechanism in IEEE 802.11 networks", *Computer Networks*, to be published.

[4] F. Calì, M. Conti, E. Gregori, "Dynamic Tuning of the IEEE 802.11 Protocol to Achieve a Theoretical Throughput Limit", *IEEE Transactions on Networking*, Volume 8, No. 6 (Dec. 2000), pp. 785 - 799.

[5] J. Chen, K. Sivalingam, P. Agrawal, S. Kishore, "A Comparison of MAC Protocols for Wireless Local Networks Based on Battery Power consumption", *Proc. IEEE Infocom'98*, San Francisco, U.S.A. (1998).

[6] M. S. Corson, J.P. Maker, J.H. Cerincione, "Internet-based Mobile Ad Hoc Networking", *Internet Computing*, July-August 1999, pp. 63-70.

[7] R.H. Katz, M. Stemm, "Measuring and reducing energy consumption of the network interfaces in hand-held devices", *Proc. 3rd International Workshop on Mobile Multimedia Communications (MoMuC-3)*, Princeton, NJ, 1996.

[8] IEEE standard for Wireless LAN - Medium Access Control and Physical Layer Specification, P802.11, November 1997.

[9] J.P. Monks, V. Bharghavan, W.W. Hwu, "A Power Controlled Multiple Access Protocol for Wireless Packet Networks", *Proc Infocom'01*, Anchorage, Alaska (2001)

[10] W. Stallings, "Local & Metropolitan Area Networks", *Prentice Hall,* 1996

[11] H. Woesner, J.P. Ebert, M. Schlager, A. Wolisz, "Power-saving mechanisms in emerging standards for wireless LANs: The MAC level perspective", *IEEE Personal Comm*, 1998, pp. 40-48.

Appendix A.

From the analytical results presented in [4], it follows:

$$E\left[Energy_{Idle\_p}\right] = PRX \cdot E[Idle\_p] = PRX \cdot \frac{(1-p)^M}{1-(1-p)^M} \qquad (A.1)$$

$$E\left[Energy_{not\_tagged\_collision}\right] = PRX \cdot Collision \qquad (A.2)$$

$$\Pr\{tagged\_collision \mid tag\_tr\} = \frac{E[Nc]}{E[Nc]+1} = 1 - (1-p)^{M-1} \qquad (A.3)$$

$$\Pr\{not\_tagged\_collision \mid not\_tag\_tr\} = \frac{1 - \left[(1-p)^{M-1} + (M-1)p(1-p)^{M-2}\right]}{1-(1-p)^{M-1}} \qquad (A.4)$$

$$\text{Pr}\left\{tag\_tr \mid N_{tr} \geq 1\right\} = \sum_{k=1}^{M} \frac{k}{M} \cdot \frac{\dbinom{M}{k} p^k (1-p)^{M-k}}{1-(1-p)^M} \tag{A.5}$$

$$\text{Pr}\left\{not\_tag\_tr \mid N_{tr} \geq 1\right\} = \sum_{k=1}^{M} \frac{M-k}{M} \cdot \frac{\dbinom{M}{k} p^k (1-p)^{M-k}}{1-(1-p)^M} \tag{A.6}$$

A Study on QoS Provision
for IP-Based Radio Access Networks[1]

Alberto López[1], Jukka Manner[2], Andrej Mihailovic[3], Hector Velayos[4],
Eleanor Hepworth[5], and Youssef Khouaja[6]

[1] Department of Communications and Information Engineering, University of Murcia,
Facultad de Informática, Campus Universitario de Espinardo - 30071 Murcia, Spain
alberto@dif.um.es
[2] Department of Computer Science, University of Helsinki,
P.O. Box 26, FIN-00014, Helsinki, Finland
jmanner@cs.Helsinki.FI
[3] Center for Telecommunications Research,
King's College London, Strand, London, WC2R 2LS, UK
andrej.mihailovic@kcl.ac.uk
[4] Agora Systems S.A., c/ Aravaca 12 3°B - 28040 Madrid, Spain
hvelayos@agora-systems.com
[5] Siemens/Roke Manor Research, Roke Manor, Romsey, SO51 0ZN, UK
eleanor.hepworth@roke.co.uk
[5] France Télécom R&D/DMR/DDH, 4 rue du Clos Courtel - BP 59
35512 Cesson Sévigné, France
youssef.khouaja@rd.francetelecom.fr

Abstract. The fast adoption of IP-based communications for hand-held devices
equipped with wireless interfaces is creating new challenges for the Internet
evolution. Users expect flexible access to Internet based services, including not
only traditional data services but also multimedia applications. This generates a
new challenge for QoS provision, as it will have to deal with fast mobility of
terminals being independent of the technology of the access network. Various
QoS architectures have been defined, but none provides full support for guaranteed service levels for mobile hosts. This paper discusses the problems related
to providing QoS to mobile hosts and identifies the existing solutions and future work needed.

1 Introduction

The emerging wireless access networks and third generation cellular systems constitute the enabling technology for "always-on" personal devices. IP protocols, tradi-

---

[1] This work has been performed in the framework of the IST project IST-1999-10050 BRAIN, which is
partly funded by the European Union. The authors would like to acknowledge the contributions of their
colleagues from Siemens AG, British Telecommunications PLC, Agora Systems S.A., Ericsson Radio Systems
AB, France Télécom - CNET, INRIA, King's College London, Nokia Corporation, NTT DoCoMo, Sony
International (Europe) GmbH, and T-Nova Deutsche Telekom Innovations-gesellschaft mbH.

S. Palazzo (Ed.): IWDC 2001, LNCS 2170, pp. 139–157, 2001.
© Springer-Verlag Berlin Heidelberg 2001

tionally developed by the Internet Engineering Task Force (IETF), have mainly been designed for fixed networks. Their behaviour and performance are often affected when deployed over wireless networks.

The telcom world has created various systems for enabling wireless access to the Internet. Systems such as the General Packet Radio Service (GPRS), Enhanced Data Rate for GSM Evolution (EDGE), Universal Mobile Telecommunications System (UMTS) and International Mobile Telecommunications (IMT-2000) are able to carry IP packets using a packet switching network parallel to the voice network. These architectures use proprietary protocols for traffic management, routing, authorisation or accounting, to enumerate some, and are governed by licenses and expensive system costs.

From the QoS point of view, the problems with mobility in a wireless access network and mobility-related routing schemes are related to providing the requested service even if the mobile node changes its point of attachment to the network. Handovers between access points, change of IP-addresses, and mechanisms for the intra-domain micro mobility mechanisms may create situations where the service assured to the mobile node cannot be provided, and a violation of the assured QoS may occur. A QoS violation may result from excess delays during handovers, packet losses, or even total denial of service. In the case where the user only requested differentiation according to a relative priority to flows, a short QoS violation may fit within acceptable limits. If the flows were allocated explicit resources, the new network access point and route from the domain edge should provide the same resources.

Several research projects within the academic community, e.g. INSIGNIA [Lee00], and in the industrial community, e.g. ITSUMO [Chen00], have sought to combine mobility with guaranteed QoS. In the BRAIN project [BRAI00], we are envisioning an all IP network, where seamless access to Internet based services is provided to users. By using IETF protocols, we are designing a system that would be able to deliver high-bandwidth real-time multimedia independent of the wireless access network or the wireless technology used to connect the user to Internet. This implies the need for IP mobility support and also end-to-end QoS enabled transport. The provision of QoS guarantees over heterogeneous wireless networks is a challenging issue; especially because over-provisioning is not always possible and the performance of the wireless link is highly variable. We focus our architecture on wireless LAN networks, since these provide high bandwidths but may also create frequent handoffs due to fast moving users - this type of architecture is most demanding in view of mobility management and QoS.

2   QoS and Mobility Background

This sections presents QoS and mobility architectures relevant to the further discussion. We have not covered all existing architectures in or study but at least those considered most important or promising in order to understand completely all the issues concerning QoS and mobility interactions.

In the following discussion, the term mobile node (MN) is used to refer to a mobile host or mobile router. If mobile host (MH) is used, the term mobile router does not apply, and vice versa.

Regarding *QoS* we have considered IETF-presented architectures for providing different levels of services to IP flows, although much work has been done within the academic community and the telcom industry; for example INSIGNIA and ITSUMO are mature proposals for providing QoS to data flows. INSIGNIA has its own in-band signalling mechanism and ITSUMO is based on the DiffServ framework.

The IETF architectures can be classified into three types according to their fundamental operation; the Integrated Services framework [Wroc97] and the Resource Reservation Protocol (RSVP [BZB+97]) provides explicit reservations end-to-end; the Differentiated Services architecture (DiffServ, [BBC+98], [BBGS01]) offers hop-by-hop differentiated treatment of packets. There are a number of 'work in progress' efforts, which are directed towards these aggregated control models. These include aggregation of RSVP [BILD00], the RSVP DCLASS Object [Be00] to allow DSCPs to be carried in RSVP message objects, and the operation of Integrated Services over Differentiated Services networks ([Bern00], [WC00]) proposed by the Integrated Services over Specific Link Layer (ISSLL) Working group. On the application level the Real-Time Transport Protocol (RTP, [SCFJ96]) provides mechanisms for flow adaptation and control above the transport layer.

For *Mobility Management* we have based or study on an analytical method we call the Evaluation Framework [EMS00], which has been adopted for facilitating detailed analysis and comparative evaluation of mobility protocols. This framework facilitates the selection of the most promising candidates for mobility management and introduce a categorisation for distinguishing protocols and their associated purposes. This analysis is closely related to QoS development, since both mobility and QoS protocols are expected to have awareness of certain, if not all, of their functionality.

For the interaction study we have considered several mobility architectures present today. On the macro-mobility side Mobile IP [Perk00] is the current standard for supporting macroscopic mobility in IP networks and its Ipv6 counterpart, Mobile IP support in IPv6 [JP00], based on the experiences gained from the development of Mobile IP support in IPv4, and the opportunities provided by the new features of the IP version 6 protocol.

For the support of regional mobility we identified two major categories: *Proxy-Agent Architectures (PAA)* which extend the idea of Mobile IP into a hierarchy of Mobility Agents and *Localized Enhanced-Routing Schemes (LERS)* which introduce a new, dynamic Layer 3 routing protocol in a 'localised' area.

In the first group (PAA) examples include the initial Hierarchical Mobile IP [Perk97] and its alternatives, which place and interconnect Mobility Agents more efficiently: Mobile IP Regional Registration [GJP01], Transparent Hierarchical Mobility Agents (THEMA) [MHW+99] and Fast Handoff in Mobile IPv4 [El01]. The new Mobile IP version 6 [JP00] has had some optional extensions by applying a hierarchical model where a border router acts as a proxy Home Agent for the Mobile Nodes. They include "Hierarchical MIPv6 mobility management" [SCEB01] and "Mobile IPv6 Regional Registrations [MP01].

In the second group (LERS) there are several distinctive approaches: *Per host forwarding schemes* where soft-state host-specific forwarding entries are installed for each MN (HAWAII [RLT+99], Cellular IP [CGK+00], Cellular Ipv6 [SGCW00]); *Multicast-based schemes* which make use of multicast protocols for supporting point-to-multipoint connections (dense mode multicast-based [SBK95][MB97][TPL99] and the recent sparse-mode multicast-based [MSA00]); and *MANET-based schemes* adapted for mobile ad-hoc networks (MER-TORA [OTC00] [OT01]).

Figure 1 shows some of the many IP mobility protocols, which category they fall into and very roughly how they relate to each other.

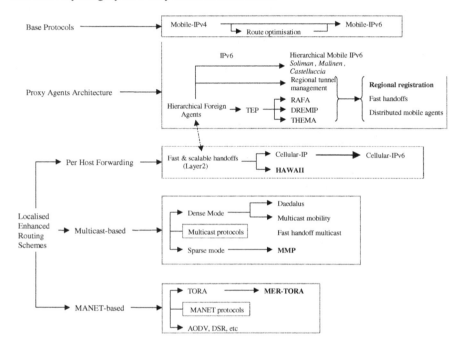

Fig. 1. Classification of mobility protocols

We will pay special attention to Handover Management, as it is considered one of the most important features of the mobility protocols when considering the interaction with QoS protocol because of the likely re-negotiation of QoS parameters. Handover refers in general to support for terminal mobility wherever the mobile node changes its point of attachment to the network.

We can identify several handover types: A **Layer-2 handover** happens if the network layer is not involved in the handover, **intra-access network[2] handover** when the new point of attachment is in the same access network, **inter-access network**

---

[2] Access Network (AN): An IP network, which includes one or more ARs and gateways.

handover when the new access router is in a different access network. **Horizontal** or **vertical** handover are said to happen if the old and the new access router[3] use the same or different wireless interface (technology) respectively.

We can also distinguish three different phases in a handover: the **Initiation Phase**, when the need for a handover (and its initiation) is recognized , the **Decision Phase**, when the best target access router is identified and the corresponding handover is triggered, based on measurements on neighbouring radio transmitters and eventual network policy information, and the **Execution Phase**, when the mobile node has been detached from the old access router and attached to the new one.

In a **planned** handover, contrary to an **unplanned** handover, some signalling messages can be sent before the mobile node is connected to the new access router, e.g. building a temporary tunnel from the old access router to the new access router.

Specific actions may be performed depending on the handover phase. For example, the events may initiate upstream buffering or advance registration procedures at the mobile node. These mechanisms characterize furthermore the handover type: **smooth** handover is a handover with minimum packet loss, **fast** handover allows minimum packet delays and **seamless** handover that is a smooth and fast handover.

3   Interaction of Mobility and QoS

This section discusses the problems related to guaranteeing service levels to mobile nodes. We classify the problem areas into three groups, namely topology related problems (3.1), and macro (3.2) and micro mobility (3.3) related issues. Solutions to these problems are presented in Section 4.

3.1 Depth of Handovers

We can identify several types of handover situations, which create different amounts of control signalling between different entities; handovers within the same Access Router (AR), between ARs and between access networks. The same physical handover can create different logical handover situations to different MN flows if the flows use different network gateways. Figure 2 shows a sample network topology to illustrate the levels of handovers while a MN moves within and between two networks.

The different levels of handovers create variable load of signalling in the access network. Also, if the QoS architecture has a signalling mechanism, such as RSVP, it adds to the need to signal in certain handover situations.

If the AR node does not change during a handover, the handover control only needs to handle radio resources since the routing paths do not change.

If the AR changes but the gateway stays the same due to similar routing, the handover affects the radio resource availability and the access network resources. In addi-

---

[3] Access Router (AR): An IP router between an Access Network and one or more access links.

tion, the new AR may need to check for admission control at the same time. All RSVP-reservations need to be refreshed.

If the gateway changes, either within the same access network or when the MN changes networks, flows may experience a drop in their QoS until the QoS signalling has updated the nodes on the paths. The time interval during which the MN is not receiving the subscribed QoS needs to be minimized.

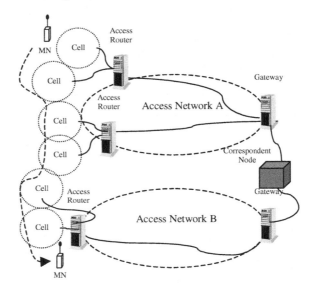

Fig. 2. Example network topology regarding different handover scenarios

3.2 Macro Mobility Issues

The first macro-mobility problem arises from the *triangular routing* phenomenon. Packets from the MN usually follow a direct path to the CNs, packets from the CNs are re-routed via the MN's home network to its point of attachment in a foreign network, from where they are forwarded to the MN's current location. Several QoS architectures operate best when packets follow the same route in the forward and reverse direction. Triangular routing can affect the service level guarantees of these schemes.

It is possible to tunnel the upstream flow to follow the downstream using *Reverse Tunnelling* [Mont01]. However, routers in the tunnel may not be able to recognize some encapsulated parameters of the QoS protocols apart from IP addresses. For example, if RSVP packets use the *Router Alert option* to indicate to routers on the path that they require special handling, when RSVP messages are encapsulated with an outer IP header, the Router Alert becomes invisible. Although solutions to this have been proposed e.g. RSVP extensions to mobile hosts [AA97], they still add complexity to the operation of QoS protocols on mobile environments.

Other main concern for QoS when the host is moving is the time needed to re-establish the routes, and hence, the time needed to re-configure resource management required to provided QoS in the new location. Even in Route Optimisation, transmission of binding updates directly to CNs result in a large update latency and disruption during handover. This effect is greatly increased if MN and HA or CN are separated by many hops in a wide area network. Data in transit may be lost until the handover completes and a new route to the MN is fully established. Route Optimisation (as a protocol specification) however includes Smooth Handoff support using Previous Foreign Agent Notification extension, which can be used to avoid the described disruption.

There are other problems related to signalling load and address management. Highly mobile MNs create frequent notifications to the home agent, which can consume a significant portion of wireless link resources. Since the current Mobile IP standard requires the mobile to change the care-of address (either FA or co-located) at every subnet transition, it is more complex to reserve network resources on an end-to-end path between the CN and the mobile. For example, if RSVP is used, new reservations over the entire data path must be set up whenever the care-of address changes. The impact on the latency for re-establishment of the new routes is critical for QoS assurances.

Mobile IPv6

Mobile IPv6 makes use of the new features provided by IPv6 protocol. They help to solve most of the problems discussed above which arise with the use of Mobile IP in IPv4 networks. For example *Route Optimisation* is included in the protocol, and there are mechanisms for movement detection that allow a better performance during handover. The *Routing Header* avoids the use of encapsulation, reducing overhead and facilitating, for example, QoS provision.

Although the Mobile IPv6 solution meets the goals of operational transparency and handover support, it is not optimised for managing seamless mobility in large cellular networks. Large numbers of location update messages are very likely to occur, and the latency involved in communicating these update messages to remote nodes make it unsuitable for supporting real-time applications on the Internet. These problems indicate the need for a new, more scalable architecture with support for uninterrupted operation of real-time applications.

3.3 Micro mobility Issues

The domain internal micro mobility schemes may use different tunnelling mechanisms, multicast or adaptable routing algorithms. The domain internal movement of MNs affects different QoS architectures in different way. IntServ stores a state in each router; thus a moving mobile triggers local repair of routing and resource reservation within the network. DiffServ on the other hand has no signalling mechanism, which means that no state needs to be updated within the network, but the offered service level may vary. At least the following design decisions of a micro mobility protocol need to be considered when combining mobility and QoS architectures within a network:

- the use of tunnelling hides the original packet information and hinders Multi-Field classification,
- changing the MN care-of-address during the lifetime of a connection,
- multicasting packets to several access routers consumes resources,
- having a fixed route to the outer network (always through the same gateway) is less scalable,
- adaptability and techniques (speed and reliability) to changing routing paths,
- having an optimal routing path from the gateway to the access router and
- support for QoS routing.

Multicast approaches can have ill effects on the resource availability, for example, because the multicast group can vary very dynamically. The required resources for assured packet forwarding might change rapidly inside the domain, triggering different QoS-related control signalling and resource reservations.

The use of tunnelling can affect the forwarding of QoS-sensitive flows since the original IP-packet is encapsulated within another IP-packet. However, as long as the tunnel end-points are capable of provisioning resources for the tunnelled traffic flows, the agreed QoS level need not be violated. Tunnelling has the advantage that multiple traffic flows can be aggregated onto a single reservation, and there is inherent support for QoS routing. Micro-mobility schemes that rely on explicit per-host forwarding information do not have such simple support for QoS routing, because there is only one possible route per host. Both IntServ and DiffServ have been extended to cope with tunnelling ([TKWZ00], [Bla00]) and the changes to the IP-address ([MH00]). Some coupling of the macro and micro mobility protocols and the QoS architecture may still be needed to ensure an effective total architecture.

4   Solutions

This section identifies various schemes for providing parts of an all-inclusive support of QoS-aware mobility. A full support of mobile terminals with QoS requirements can be accomplished by a combination of these schemes.

4.1 Strict Shaping at Network Edges

Network operators already intercept each packet arriving from an external network and decide whether the packet can be allowed into the core network. This admission control is performed by a node called the firewall and is based on IP addresses and port numbers e.g. identifying applications. Firewalls are typically deployed for security reasons and usually scan both incoming and outgoing packets.

The firewall operation can be modified by using different rules for performing the admission control. Instead of just preventing known security problems, the edge nodes would use defined bandwidth and QoS policies on a per-flow basis for controlling the traffic admitted into the network. Both the access routers and the gateways perform the admission control, the former for flows originating from mobile nodes and the latter for flows emerging from external networks.

When a previously unknown packet arrives, the edge node will check for the Service Level Agreement (SLA) and policies stored for the particular MN being contacted. A central bandwidth broker is in charge of the policy management, and once it receives a request from an edge node, it checks its databases for the proper forwarding rules and returns them to the edge node. Adjusting the load created by best-effort traffic is vital.

This method can be used to adjust the load admitted into each service class, if the network is operating with aggregate service classes, and not per-flow, as with RSVP. This can decrease the network load and thus allow for smoother handovers, especially if the traffic belonging to the best-effort class is not consuming all leftover capacity. Therefore, there is enough bandwidth left to support moving terminals.

The access routers should not need to make the primary policing decisions when the arriving load exceeds the capacity of the forward link. If we allow downlink traffic to flood the access network, mobility management schemes are affected. A bandwidth broker could be used to co-ordinate the access network resources and configure the gateways to drop excess traffic.

4.2 Coupling of Micro-mobility and QoS

In order to improve the behaviour of reservation-based QoS, as defined in the Integrated Services architecture [BCS94], in the dynamic micro-mobile environment, the QoS and micro-mobility mechanisms can be coupled to ensure that reservations are installed as soon as possible after a mobility event such as handover. Reservations are installed using a QoS signalling protocol, the most widely adopted of which is RSVP, which will be used in the following discussions as an example of an out-of-band soft state mechanism. In this study we present three levels of coupling over three different micro-mobility architectures: *proxy agent architectures* [CB00][GJP01][MS00b][MP01], *MANET-based schemes* [OTC00] and *per-host forwarding schemes* [SGCW00][RLT+99][KMTV00]. The three scales of coupling presented for consideration are described on the following sections.

4.2.1 De-coupled

In the de-coupled option, the QoS and micro-mobility mechanisms operate independently of each other and the QoS implementation is not dependent on a particular mobility mechanism. Changes in network topology are handled by the soft-state nature of the reservations.

After a mobility event, the QoS for the traffic stream will be disrupted while until a new reservation is installed via refresh messages between the node where the old route and new route intersect, known as the crossover router (figure 3), to the new access router (NAR). The reservation between the crossover router and the old access router (OAR) cannot be explicitly removed, and must be left to timeout, which is not the most efficient use of network resources. This will occur every time the MN moves AR, which may be many times during one RSVP session, and can lead to poor overall QoS for an application.

These problems are common to all micro-mobility schemes.

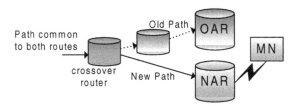

Fig. 3. Concept of a crossover router

4.2.2 Loosely Coupled

The loosely coupled approach uses mobility events to trigger the generation of RSVP messages, which distribute the QoS information along new paths across the network. The RSVP messages can be triggered as soon as the new routing information has been installed in the network. This mechanism is the Local Path Repair option, and is outlined in the RSVP specification [BZB+97] and has the effect of minimising the disruption to the application's traffic streams because there is a potentially shorter delay between handover and reservation set up. It also avoids the problem of trying to install a reservation across the network before the routing update information has been propagated. The latency for installing the reservation can also be reduced by localising the installation to the area of the network affected by the change in topology, i.e. between the crossover router and the NAR. The areas of the network affected by the topology change can have reservations installed across them almost immediately, instead of having to wait for the update to travel end-to-end, or for the correspondent node to generate a refresh message for reservations to the MN. In the case where the QoS must be re-negotiated, however, end-to-end signalling is required. The old reservation should be explicitly removed, freeing up unused resources immediately.

However, the loosely coupled approach requires additional complexity within the inter-mediate network nodes to support the interception and generation of RSVP messages when the router is acting as the crossover node. Another disadvantage is that bursts of RSVP signalling messages are generated after handover to install multiple reservations. This does not happen in the de-coupled case, because the reservation signalling messages are generated when refresh timers expire, not by the same triggering event.

In the **proxy agent architectures** the loosely coupled approach ensures that the reservation is not installed until the registration information generated by the MN has propagated across the network. In **MANET based schemes** and the **per-host forwarding schemes,** the loosely coupled ensures that the new routing information has been distributed into the network before attempting to install the reservation. The reservation is installed in the network as soon as the route to the MN is stable without having to wait until the next timeout to send QoS messages.

4.2.3 Closely Coupled

The closely coupled uses the same signalling mechanism to propagate the mobility and QoS information, either as an extension to the QoS/MM signalling protocol or via a unique QoS-routing protocol. This approach minimizes the disruption to traffic streams after handover by ensuring that the reservation is in place as soon as possible after handover by installing routing and QoS information simultaneously in a localised area. It also provides a means to install multiple reservations using one signalling message. The reservation along the old path can also be explicitly removed.

In the **proxy agent architectures**, support for the opaque transport of QoS information in the registration messages is provided, and is interpreted by the mobility agents. This allows the MN to choose a mobility agent based on the available resources and provides a degree of traffic engineering within the network. In the **MANET-based** and **per-host forwarding schemes**, the messages that install the host-specific routing information in the network also transparently carry opaque QoS information. The reservations are installed at the same time as the routing information, minimizing the disruption to the traffic flows.

4.2.4 Comparison of Approaches

Coupling reservations with micro-mobility mechanisms allow reservation set up delays to be minimised and packet loss reduced. Reservations along the new path can be installed faster because QoS messages can be generated as soon as the new route is established, reducing the disruption to the data flows. Also scalability and overhead are improved because a minor number of update messages are sent or they are localised to only the affected areas of the network. Moreover, it ensures that the request for a QoS reservation only occurs when there are valid routes to the MN in the network.

The closely coupled approach requires support from particular micro-mobility mechanisms so that the opaque QoS information can be conveyed across the network. This has the consequence that the QoS implementation will be specific to a particular micro-mobility mechanism, and extensions to the micro-mobility protocol may be needed to support the required functionality. However, the closely coupled approach maintains consistency between the reservation and the routing information within the network, and can reduce the amount of signalling required to set-up multiple reservations.

The choice between whether to use the loosely coupled approach or the closely coupled approach is a trade-off between a QoS solution that is tied to a micro-mobility protocol and the performance advantage close coupling provides. The closely coupled approach potentially provides improvements in performance and efficiency, but at the expense of additional complexity and loss of independence from the underlying micro-mobility mechanism.

4.3 Advance Reservations

The mobile host may experience wide variations of quality of service due to mobility. When a mobile host performs a handover, the AR in the new cell must take responsibility for allocating sufficient resources in the cell to maintain the QoS requested (if any) by the node. If sufficient resources are not allocated, the QoS needs may not be met, which in turn may result in premature termination of connections.

It is clear that when a node requests some QoS it is requesting it for the entire connection time, regardless of whether it is suffering handoffs or not. The currently proposed reservation protocol in the Internet, RSVP, implements so-called *immediate reservations*, which are requested and granted just when the resources are actually needed. This method is not adequate to make guaranteed reservations for mobile hosts. To obtain mobility independent service guarantees a mobile host needs to make *advance resource reservations* at the multiple locations it may possibly visit during the lifetime of the connection.

There are a number of proposals for advanced reservations in the Internet Community that can be classified into two groups, depending on the techniques they use:

- *Admission control priority*
- *Explicit advanced reservation signalling*

Those groups are not necessarily distinct, as both approaches could be used together. Admission control strategies are transparent to the mechanism using explicit advanced reservations, other than when a request is rejected.

4.3.1 Admission Control Priority

It is widely accepted that a wireless network must give higher priority to a handover connection request than to new connection requests. Terminating an established connection from a node that has just arrived to the cell is less desirable than rejecting a new connection request. *Admission control priority based mechanisms* rely on this topic to provide priorities on the admission control to handover requests without significantly affecting new connection requests.

The basic idea of these admission control strategies is to reserve resources in each cell to deal with future handover requests. The key here is to effectively calculate the amount of bandwidth to be reserved based on the *effective bandwidth* [EM93] of all active connections in a cell and the effective bandwidth of a new connection request. There are a number of different strategies to do this:

- *Fixed strategy*: One simple strategy is to reserve a fixed percentage of the AR's capacity for handover connections. If this percentage is high, adequate capacity will most likely be available to maintain the QoS needs of handover connections, but at the expense of rejecting new connections.
- *Static Strategy*: the threshold values are based on the effective bandwidths of the connection requests. There is a fraction of bandwidth reserved for each of the possibly traffic class. This fraction may be calculated from historic traffic information available to the AR.
- *Dynamic Strategy*: each AR dynamically adapts the capacity reserved for dealing with handover requests based on connections in the neighbouring cells. This will enable the AR to approximately reserve the actual amount of resources needed for handover requests and thereby accept more new connection requests as compared to in a fixed scheme. Such dynamic strategies are proposed and evaluated in [NS96] and [YL97].
- *Advanced Dynamic Strategy*: this strategy assumes an analytical model where handover requests may differ in the amount of resources they need to meet their QoS requirements, and therefore it is more suitable for multimedia applications. A proposal for this strategy is described in [RSAK99].

This kind of admission control strategy can be used on statistically access control as the one performed on non hard guaranteed QoS provision, such as some DiffServ PHBs or Controlled Load on IntServ model. It is not enough for hard guarantees in all paths followed by a mobile node.

4.3.2 Explicit Advanced Signalling

Admission Control strategies are not enough to accommodate both mobile hosts that can tolerate variations in QoS and also those that want mobility independent service guarantees in the same network. To obtain good service guarantees in a mobile environment, the mobile host makes resource reservations at all the locations it may visit during the lifetime of the connection. These are known as *advanced reservations*.

There are a number of different approaches for advanced reservation in the literature. We present here two of the most relevant for supporting Integrated Services (MRSVP [TBA98]) and other for supporting Differentiated Services (ITSUMO approach [Chen00]).

MRSVP
Mobile RSVP introduces three service classes to which a mobile user may subscribe: *Mobility Independent Guarantees* (MIG) in which a mobile user will receive guaran-

teed service, *Mobility Independent Predictive* (MIP) in which the service received is predictive and *Mobility Dependent Predictive* (MDP) in which the service is predictive with high probability.

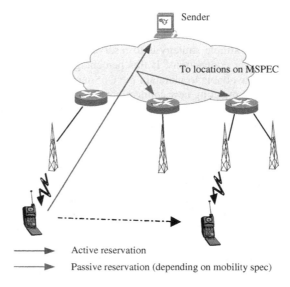

Fig. 4. MRSVP advanced reservations.

MRSVP allows the mobile node to make advance resource reservation along the data flow paths to and from the locations it may visit during the lifetime of the connection. These are specified in the Mobility Specification (MSPEC) as shown in figure 4. The advance determination of the set of locations to be visited by a mobile node is an important research problem, although several mechanisms have been proposed to approximately determine them by the network.

Two types of reservations are supported in MRSVP: active and passive. A mobile sender makes an *active* reservation from its current location and it makes *passive* reservations from the other locations in its MSPEC. To improve the utilization of the links, bandwidth of passive reservations of a flow can be used by other flows requiring weaker QoS guarantees or best effort service. However, when a passive reservation becomes active (i.e. when the flow of the mobile node who made the passive reservation moves into that link), these flows may be affected.

ITSUMO Approach
The ITSUMO approach has a different philosophy on advanced reservations. Although the mobile node itself has to explicitly request a reservation and specify a mobility profile, the advanced reservation is 'made' by Global QoS Server (GQS) on its behalf. Based on the local information and the mobility pattern maybe negotiated in the SLS, the QGS envisions how much bandwidth should be reserved in each QLN (QoS Local Node). The QGS then updates periodically the QLNs likely to be visited

by MN. Rather than actively reserving resources in each of the access points, this scheme it is likely that either a passive reservation (utilized for best effort traffic) or an "handover guard band" could be used.

The clear difference with the previous approach is that advanced reservation in MRSVP has to be signalled by the mobile node explicitly to every station according to its mobility pattern. This mobility pattern is known and processed by it. In the ITSUMO approach this information is updated periodically by the QGS, according to the mobility pattern informed by the MN but processed on the QGS. So it could be said that MN relies the explicit advanced reservation in the QGS (figure 5).

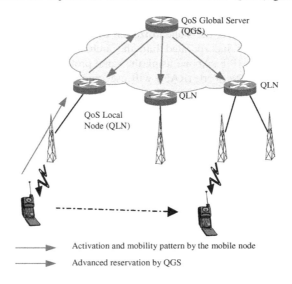

Fig. 5. ITSUMO advanced reservations

4.4 Pre-handover Negotiations

Pre-handover negotiations associate the change to a new cell to the actual resource availability in the new cell, as opposed to advance reservation schemes. When the network or the mobile node deems that a handover should occur, the access router can request some indication of resource availability from neighbouring access routers.

This needs support from the convergence layer between the IP-layer and the link layer. The link layer would need to communicate the overall resource availability of an access point in order to let the IP-layer to make a decision about a possible handover. Also an indication of a forthcoming handover is needed.

Initially, context transfer would enhance handovers between access routers, allowing access routers to communicate directly or through the MN, the QoS and other contexts of a moving MN. A further refinement to the scheme would allow both access router and gateways to communicate the mobile's context during a handover. This would allow to reduce the time during which the mobile has no specific resources allocated to it

4.5 Solutions in Third Generation Mobile Communication Systems

The currently evolving design of the third generation mobile communication systems (3G systems) aims to provide real-time multimedia services in wide area cellular networks [Walk99]. These systems will include a packet switched backbone (PSB) to carry data traffic in the form of IP datagrams, in addition to the traditional circuit switching for voice calls. As the standardization of 3G systems evolves, more and more IETF protocols are incorporated into the architecture. UMTS Release 2000 considers the PSB as an IP backbone using the same protocols as IP fixed networks, while the Radio Access Network (RAN) will use proprietary protocols. For the IP-based data transmission, this RAN is seen as a link layer.

Mobility management and the provision of QoS in 3G systems are still different from IP based fixed networks. Three types of mobility are considered in 3G systems: terminal, personal and service mobility. Service mobility provides the same set of services regardless of the current point of attachment to the 3G network. Personal mobility allows users to receive their personalized service independent of their location in the network. Terminal mobility across different operators is a key requirement in 3G systems. To this end, the support of Mobile IP is being considered with some proposed extensions [Das00]. In essence, the Internet Gateway Serving Node (IGSN) will act as Foreign Agent supporting macro mobility, while the movements of the terminal inside the Universal Terrestrial Radio Access (UTRA) are not visible outside the 3G network. The provision of QoS in 3G systems will incorporate two new features with respect to 2G systems and their evolutions: support for user/application negotiation of UMTS bearer characteristics and standardized mapping from UMTS bearer services to core network QoS mechanisms.

5 Conclusion

In this paper we discussed problems related to mobility and QoS. We deduced that the main problem in this field is following the movement of the mobile host fast enough to minimize the disruption caused to the QoS received by the application traffic flows. Also the depth of the handover signalling and the related QoS control affect the service outcome.

In Section 4 we studied solutions for the interoperability of mobility and QoS. We presented several schemes that provide parts of a total solution to mobile QoS. We discussed performing strict flow shaping at the network edge, coupling of micro-mobility and QoS protocols, advanced reservations, pre-handover negotiations and context transfer, and the 3G approaches.

It has become apparent that even though there exists several good partial solutions, we still need adaptive applications. Handovers, for example, still cause some disturbance to data streams. RTP can provide to this adaptability. The whole notion of end-to-end QoS still seems very distant. It is possible to provide adequate service to mobile hosts in a private access network, but when the corresponding node is behind some wider public network, keeping the promised QoS becomes harder.

A new IETF Working Group, Seamoby, is aiming to provide seamless mobility across access routers and even domains. The work of this group will hopefully lead to better mobility support, especially for the problematic multimedia streams. Part of the work done is on context transfer issues.

References

[AA97] Awduche, D. O., Agu, E., "Mobile Extensions to RSVP", Proceedings of ICCCN'97, Las Vegas, NV, Sep. 1997, p.132-6.

[BBC+98] Blake, S., Black, D., Carlson, M., Davies, E., Wang,, Z., Weiss, W., "An Architecture for Differentiated Services". Internet Engineering Task Force, Request for Comments (RFC) RFC 2475, Dec. 1998.

[BBGS01] Bernet, Y., Blake, S., Grossman, D., Smith, A., "An Informal Management Model for DiffServ Routers". Internet Draft (work in progress), February 2001 (draft-ietf-diffserv-model-06.txt).

[BCS94] Braden, R., Clark, D. Shenker, S., "Integrated Services in the Internet Architecture: an Overview ". Internet Engineering Task Force, Request for Comments (RFC) 1633, June1994.

[Be00] Bernet, Y., "Format of the RSVP DCLASS Object". Internet Engineering Task Force, Request for Comments (RFC) 2996, November 2000.

[Bern00] Bernet, Y. et al, "A Framework For Integrated Services Operation Over Diffserv Networks". Internet Engineering Task Force, Request for Comments (RFC) 2998, November 2000.

[BILD01] Baker, F., Ituralde, C., Le Faucheur, F., Davie, B., "RSVP Reservation Aggregation". Internet Draft (work in progress), April 2001 (draft-ietf-issll-rsvp-aggr-04.txt).

[Bla00] Black, D. "Differentiated Services and Tunnels". Internet Engineering Task Force, Request for Comments (RFC) 2983, October 2000.

[BRAI00] Broadband Radio Access for IP-based Networks, IST-1999-10050, http://www.ist-brain.org.

[BZB+97] Braden, R., Zhang, L., Berson, S., Herzog, S., Jamin, S., "Resource ReSerVation Protocol (RSVP) -- Version 1, Functional Specification". Internet Engineering Task Force, Request for Comments (RFC) 2205, September 1997.

[CB00] Castelluccia, C., Bellier, L., "Hierarchical Mobile IPv6", Internet Draft (work in progress), July 2000 (draft-castelluccia-mobileip-hmipv6-00.txt).

[Chen00] J.C. Chen, A. McAuley, A. Caro, S. Baba, Y. Ohba, P. Ramanathan. A QoS Architecture for Future Wireless IP Networks. In Twelfth IASTED International Conference on Parallel and Distributed Computing and Systems (PDCS 2000), Las Vegas, NV, November 2000)

[CGK+00] Campbell, A., Gomez, J., Kim, S., Valko, A., Chieh-Yih Wan, Turanyi, Z., "Design, implementation, and evaluation of cellular IP". IEEE Personal Communications, Vol. 7, August 2000, pp. 42-49.

[Das00] Das, S., Misra, A., Agrawal P., Das S.K., "TeleMIP: Telecommunications-Enhanced Mobile IP Architecture for Fast Intradomain Mobility", IEEE Personnal Communications, August 2000

[El01] El-Malki, K., et.al. "Low Latency Handoff in Mobile IPv4". Internet Draft (work in progress), May 2001 (draft-ietf-mobileip-lowlatency-handoffs-v4-01.txt).

[EM93] Elwalid, A. I., and Mitra, D. "Effective bandwidth of general markovian traffic sources and admisssion control of high-speed networks," IEEE/ACM Transactions on Networking, vol. 1, no. 3, pp. 329-343, June 1993.

[EMS00] Eardley, P., Mihailovic, A., Suihko, T., "A Framework for the Evaluation of IP Mobility Protocols", In Pproceedings of PIMRC 2000, London, UK, September 2000.

[GJP01] Gustafsson, E., Jonsson, A., Perkins, C,. "Mobile IP Regional Registration", Internet Draft (work in progress), March 2001 (draft-ietf-mobileip-reg-tunnel-04).

[JP00] Johnson, D., Perkins, C., "Mobility Support in IPv6", Internet Draft (work in progress), November 2000 (draft-ietf-mobileip-ipv6-13.txt).

[KMTV00] Keszei, C., Manner, J., Turányi, Z., Valko, A., "Mobility Management and QoS in BRAIN Access Networks". BRAIN International Workshop, King's College, London, November 2000.

[Lee00] Lee et al., "INSIGNIA: An IP-based quality of service framework for mobile adhoc networks". Journal of parallel and distributed computing, Vol 60 no 4, pp 374-406, April 2000.

[MB97] Mysore, J., Bharghavan, V., "A New Multicasting-based Architecture for Internet Host Mobility", Proceeding of ACM Mobicom, September 1997.

[MH00] Metzler, J., Hauth, S., "An end-to-end usage of the IPv6 flow label". Internet Draft (work in progress), November 2000 (draft-metzler-ipv6-flowlabel-00.txt).

[MHW+99] McCann, P., Hiller, T., Wang, J., Casati, A., Perkins, C., Calhoun, P., "Transparent Hierarchical Mobility Agents (THEMA)", Internet Draft (work in progress), March 1999 (draft-mccann-thema-00.txt).

[Mont01] Montenegro, G., "Reverse Tunnelling for Mobile IP". Internet Engineering Task Force, Request for Comments (RFC) 3024, January 2001.

[MP01] Malinen, J., Perkins, C., "Mobile IPv6 Regional Registrations", Internet Draft (work in progress), March 2001 (draft-malinen-mobileip-regreg6-01.txt).

[MS00] El Malki, K., Soliman, H., "Fast Handoffs in Mobile IPv4", Internet Draft (work in progress), September 2000 (draft-elmalki-mobileip-fast-handoffs-03.txt).

[MS00b] El Malki, K., Soliman, H., "Hierarchical Mobile IPv4/v6 and Fast Handoffs". Internet Draft (work in progress), March 2000, (draft-elmalki-soliman-hmipv4v6-00.txt)

[MSA00] Mihailovic, A., Shabeer, M., Aghvami, A.H., "Multicast for Mobility Protocol (MMP) for emerging Internet networks", In Proceedings of PIMRC2000, London, UK, September 2000.

[NS96] Naghshineh, M. and Schwartz, M. "Distributed call admission control in mobile/wireless networks," IEEE Journal on Selected Areas in Communications, vol. 14, pp. 711-717, May 1996.

[OT01] O'neil, A., Tsirtsis, G., "Edge mobility architecture - routeing and hand-off". British Telecom Technology Journal, Vol. 19, no 1. January 2001. (Available at: http://www.bt.com/ bttj/vol19no1/oneill/oneill.pdf)

[OTC00] O'Neill, G., Corson, S., "Homogeneous Edge Mobility for Heterogeneous Access Technologies", Proceedings of the IPCN 2000, Paris, France, May 2000.

[Per97] Perkins, C., "Mobile IP", IEEE Communication Magazine, May, 1997.

[Perk00] Perkins, C., "IP Mobility Support for IPv4, revised", Internet Engineering Task Force, Request for Comments (RFC) 2002, September 2000.

[RLT+99] Ramjee, R., La Porta, T., Thuel, S., Varadhan, K., Wang, S.Y., "HAWAII: a domain-based approach for supporting mobility in wide-area wireless networks". Proceedings of the Seventh International Conference on Network Protocols (ICNP) 1999, pp. 283 – 292.

[RSAK99] Ramanathan, P., Sivalingam, K. M., Agrawal, P., Kishore, S., "Resource allocation during handoff through dynamic schemes for mobile multimedia wireless networks". Proceedings of INFOCOM, March 1999

[SBK95] Seshan, S., Balakrishnan H., Katz, R. H., "Handoffs in Cellular Wireless Networks: The Daedalus Implementation and Experience", ACM/Baltzer Journal on Wireless Networks, 1995.

[SCEB01] Soliman, H., Castelluccia, C., El-Malki, K., Bellier, L., "Hierarchical MIPv6 mobility management". Internet Draft (work in progress), February 2001 (draft-ietf-mobileip-hmipv6-03.txt).

[SCFJ96] Schulzrinne, H., Casner, S., Frederick, R., Jacobson, V., "RTP: A Transport Protocol for Real-Time Applications". Internet Engineering Task Force, Request for Comments (RFC) 1889, January 1996.

[SGCW00] Shelby, Z.D, Gatzounas, D., Campbell, A. Wan, C., "Cellular IPv6". Internet Draft, (work in progress), November 2000 (draft-shelby-seamoby-cellularipv6-00).

[TBA98] Talukdar, A., Badrinath,, B. Acharya, A., "MRSVP: A Resource Reservation Protocol for an Integrated Services Packet Network with Mobile Hosts", In Proceedings of ACTS Mobile Summit'98, June 1998.

[TKWZ00] Terzis, A., Krawczyck, J., Wroclawski, J., Zhang, L., "RSVP Operation Over IP Tunnels". Internet Engineering Task Force, Request for Comments (RFC) 2746, January 2000.

[TPL99] Tan, C., Pink, S., Lye, K., "A Fast Handoff Scheme for Wireless Networks", In Proceedings of the Second ACM International Workshop on Wireless Mobile Multimedia, ACM, August 1999.

[Walk99] Bernhard H. Walke, "Mobile Radio Networks, Networking and protocols", chapter 5, "Third Generation Cellular", Ed. JohnWiley , England, Feb. 2000.

[WC00] Wroclawski, J. Charny, A., "Integrated Service Mappings for Differentiated Services Networks". Internet Draft (work in progress), February 2001 (draft-ietf-issll-ds-map-01.txt). (expired but rewritten as new draft)

[Wroc97] Wroclawski, J., " The Use of RSVP with IETF Integrated Services". Internet Engineering Task Force, Request for Comments (RFC) 2210, September 1997.

[YL97] Yu, O. T. W., and Leung, V. C. M., "Adaptive resource allocation for prioritised call admission over an ATM-based wireless PCN," IEEE Journal on Selected Areas in Communications, vol. 15, pp. 1208-1225, Sept. 19

Measurement Based Modeling
of Quality of Service in the Internet:
A Methodological Approach

Kavé Salamatian and Serge Fdida

LIP6-CNRS
Université Pierre et Marie Curie (UPMC)
Paris, France

Abstract. This paper introduces a new methodology for analyzing and interpreting QoS values collected by active measurement and associated with an *a priori* constructive model. The originality of this solution is that we start with observed performance (or QoS) measures and derive inputs that have lead to these observations.
This process is illustrated in the context of the modeling of the loss observed in an Internet path.
It provides a powerful solution to address the complex problem of QoS estimation and network modeling.

1 Introduction

Much research effort has been spent during the last decade on modeling and analyzing the performance of IP networks. They have contributed to the design of mechanisms aiming to enforce a level of Quality of Service. However, work focusing on the measurement and prediction of the QoS as offered in real Internet is quite new. Modeling the QoS in real Internet is a complex activity as QoS is strongly related to the traffic offered to the network. As traffic is highly fluctuating because of stochastic variations in the number of users and their demands, the relation between QoS and traffic is not straightforward, and stochastic modeling should be applied. Two main classes of approach have been used to address this problem : the constructive approach and the descriptive approach.

The constructive approach has been widely used since many years to model systems in general. It is based on the derivation of a model that ideally produces the same output than the system for an outside observer. These models provide a description of network elements that is as close as possible to the real network. The network is described as a combination of queues and routers *etc.*, and a scenario defining the parameters of the system in term of arrival process, capacity, buffer space, *etc.* The modeling phase is followed by the resolution phase that can either relies on a simulation or an analytical analysis to derive the QoS metrics for a given set of parameters. This approach is widely adopted in performance analysis and queueing theory. The generalization of the use of ns [18] as a simulation tool for complex networks had made possible very

S. Palazzo (Ed.): IWDC 2001, LNCS 2170, pp. 158–174, 2001.
© Springer-Verlag Berlin Heidelberg 2001

precise and detailed modeling of network elements and their analysis with this approach. Constructive approach has the nice property of relating directly the QoS as perceived by end users to operational traffic engineering parameters that can be controlled by network operators. It can also answer "what if" questions, arising when one want to evaluate the impact of changes in network parameters or architecture in the performance of the system under study. Nevertheless, this approach suffers from a main drawback, the assumption put on the structure of the network and on the scenario are so strong that it is very unlikely to generalize results of this approach to the real Internet. This comment restrict greatly the field of application of constructive approach for modeling and evaluating the QoS in the Internet.

On the other hand, the descriptive approach is based on measurements of QoS in operational networks. It models the QoS as seen in real world by describing it by some statistical parameters such as moment of different order (mean, variance, autocorrelation, Hurst parameter, *etc.*). In this approach the network is seen as an opaque black box without any access to its internal structure. The descriptive approach only describe the QoS measured without trying to explain the mechanism generating the observations. This process mainly aims toward predicting the QoS experienced by applications under some reproducibility or stationarity assumptions. However, as these models do not integrate the mechanism generating the observation, they cannot help on predicting what will happen if these stationarity assumptions do not hold anymore, because of change in network architecture or more simply because of change in traffic parameters. This means that this approach cannot be used for network dimensioning, capacity planning or predicting the QoS improvement consecutive to changes in network parameters. This also means that it is not possible to interpret active measurements results by means of traffic engineering parameters. These remarks clearly narrow the application of this approach to situations where the stationarity assumptions are valid.

In this paper, we try to conciliate these two classes of approaches. We propose a methodology for modeling QoS as measured in real networks, based on an *a priori* model structure. In this new approach we first choose an *a priori* model structure based on a constructive approach and we suppose that the parameters of this model are unknown. Then, descriptive methods are used to calibrate the unknown parameters to QoS as measured on a real network. We will show that this methodology can help alleviating some of the concerns expressed before about constructive and descriptive approaches.

In section 2, active and passive measurements are introduced and discussed. Then, in section 3 the new modeling framework is described. Hidden parameters estimation and the Expectation-Maximization (EM) method are presented and finally an application of the methodology for modeling packet losses is provided. At last we will conclude.

2   Measurement of QoS over Internet

Surveys campaign over Internet have been widely carried during the past years [12,9,21]. Globally two general classes of measurement were applied : active and passive measurement. In this section we will describe the specifics of each classes and introduce measurements needed for QoS estimation and modeling.

2.1   Passive Measurements

Passive measurements are by essence non-intrusive. In this class of measurements, traffic parameters are monitored in a particular point of the network such as a router or a Point of Presence (POP). This monitoring can be done at the microscopic level, by analyzing the traffic at packet level [5], as well as at the macroscopic level, where aggregate metrics as traffic per flow, or throughput, are measured. Nevertheless, passive measurement are hardly applicable for end-to-end QoS prediction as they remain local to the point of measurement.

Microscopic passive measurement proceed by storing the header of each observed packet in the monitored access point. This class of measurement generate huge volume of data and should be processed off-line. Packet header are used for reconstructing all flows crossing the monitored point at each instant of time. These measurements lead to very valuable information about the arrival distribution and the dynamic behavior of applicative flows in the network. These results are essential for constructive analysis as they provide realistic scenarii for it.

Modeling is important in the field of microscopic passive measurement as it can be seen as an "Occam's razor", describing in a compressed and concise manner a huge measurement trace.

Macroscopic passive measurements have been standardized by the RTFM group of IETF [6], and are frequently used in network management. These measurements usually monitor the overall traffic over an aggregation time scale or the traffic per flow crossing the monitoring point. However, it is clear that interpreting the measurements need *a priori* about the measured phenomena, that is provided by modeling. More specifically, the interpretation of macroscopic passive measurement is closely related to the scale of aggregation. This problem is illustrated by the different definitions provided for a flow in the literature [15]. This means that the macroscopic passive measurement can be considered by the methodology developed in this paper for QoS modeling and interpretation.

2.2   Active Measurements

Active measurement are more intrusive, as they inject traffic into the Internet. The rationale behind active measurement is that estimating the end to end QoS as sensed by real application can only be done by putting oneself in place of the real application. In this approach a probe sending process injects probe packets into the network. At the other end of the network a measurement agent records some metrics on each received probe packet. The collected metrics are used to infer about the QoS that will be seen by other packet flow crossing the network. The IPPM group of the IETF has defined different end-to-end performance metrics [11] to be collected on probe packets. Three main type of information are extracted from the received probe packet flow: packet size, packet loss process and packet delay process. they are used to derive more complex metrics as goodput, loss rate, loss run length, jitter, *etc*. The probe packets are usually sent using ICMP (in ping surveys) or UDP. ICMP probing is more difficult now because of the issue related to the Ping of the Death attack. A lot of active measurement surveys has been produced during the recent years [9,2,4,2,20,21,8] and some measurement infrastructures have been deployed [10,19].

Due to non-synchronized clocks between receivers and senders, the reliable measurement of packet delay is difficult. Strict synchronization of two entities connected by a varying delay link, can prove to be impossible without access to an external universal time reference as provided by a GPS (*Global Positioning System*) time reference [3]. In [9], complex mechanisms that converge asymptotically to the synchronization of two clocks are developed. However, GPS acquisition cards are now more and more used making delay with a resolution around 1 μsec feasible.

Active measurements are the source of some interesting and challenging problems. The first problem is related to the effect of measurement probe traffic on the network state. Probe traffic itself load the network and alter the QoS offered by the network. Precise QoS measurements using active probing needs compensation for the effect of measurement traffic. Derivation of a compensation methodology need *a priori* models describing the interaction between the measurement traffic and the background traffic. In actual practice the volume of active measurement traffic is chosen so low that its effect on network traffic can be neglected. This approach is not always viable as attaining a specific confidence level in the estimation of a QoS parameter may need higher volume of probing traffic. We will describe this problem in section 3. Another crucial problem is the estimation procedure. This issue will address in section 3. The solution to these crucial questions should motivate an Active QoS measurement methodology that will contain the measurement procedure (measurement traffic inter-arrival pdf, probe traffic rate, *etc*.) as well as the estimation procedure and confidence level (QoS estimator, compensation procedure, *etc*).

We will not deal with this important and hard problem in this paper. Our less ambitious objective is to present a framework for analysis and modeling the QoS as obtained by active measurement techniques. This framework is also applicable to the calibration of active measurement as well as to a wide spectrum of modeling problem in networking.

3 Estimation and Prediction of QoS Based on Active Measurement

In the sequel we will suppose that measurement probe packets are not fragmented during their journey in the network, and we will also assume that packets are either dropped somewhere in the network or received at the other end of the network errorless after a transit delay. These assumptions are reasonable if probe packet size are not too large and if we remain on the wired Internet. One can formalize the effect of network traffic on every packet of an application in the following general framework : the network effect can be modelled as a valve followed by a random delay element (Fig. 1). The valve is controlled by an on/off time continuous stochastic process $S(t)$ representing the fate (being dropped or passing) of a packet sent à time t and its traversal delay is represented by the time continuous stochastic process $D(t)$. Now suppose that an application generates packet i at time T_i. This specific packet will experience the valve at state $S(T_i)$ and a delay $D(T_i)$. These two processes are representative of the quality of service offered by the network to any application.

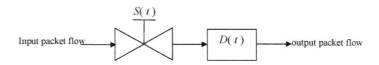

Fig. 1. Formal model of the network.

This formal model is quite general as the two processes $S(t)$ and $D(t)$ are not supposed to have any specific properties. modeling the Quality Of Service in the network will consist of specifying models and properties for these processes.

In the context of QoS evaluation based on active measurement, each active probe packet provide a sample $S(T_i)$ of the on/off process (the sampled loss process), and packets that succeed in crossing the network provide also a sample of the delay process $D(T_i)$. The descriptive approach to QoS evaluation attempts to estimate statistical characteristics of the continuous time processes $S(t)$ and $D(t)$ based on the discrete time sampled process $S(T_i)$ and $D(T_i)$. This framework is the general framework of measurement based modeling of Quality Of Service in the Internet. It is not possible to go further in the analysis without specifying some assumptions on these processes. However, to be able to interpret the observed QoS, we need to relate the two processes $S(t)$ and $D(t)$ to the traffic inside the network. Our proposed methodological approach consist in defining *a priori* classes of models relating the traffic into the network to these processes. We will first provide some details about the descriptive approach. In the sequel, we will focus on the descriptive analysis of the on/off process $S(t)$.

In the descriptive approach, we want to estimate statistical characteristics of the process $S(t)$ based on the discrete time sampled process $S(T_i)$ and $D(T_i)$. As a first analysis, we state the assumption that the processes $S(t)$ and $D(t)$ have reached a stable and stationary state. In most cases, strict sense stationarity is not needed and finite order wide sense stationarity is sufficient, meaning that k'th order statistical moment of the processes $S(t)$ and $D(t)$ and unchanged under translation. The stationary moments are obviously function of the competing Internet traffic as well as the measurement probe traffic.

We assume that active measurement has resulted in a loss trace containing K samples $\{S(T_i)\}$, $i = 1, \ldots, K$, where the samples time T_i, $i = 1, \ldots, K$ are chosen following an *iid* renewal process with lifetime distribution $F(\tau) = Prob\{T_{k+1} - T_k < \tau\}$. Under ergodic and stationary assumptions for $S(t)$, it is possible to estimate the statistical characteristics of $S(t)$, based on the samples. For example, the temporal mean $\bar{S} = \frac{1}{K} \sum_{i=1}^{K} S(T_i)$ is an unbiased estimator for the mean of $S(t)$ ($\mu = E\{S(T)\}$) which is also the loss probability. The variance of this estimator will be $\text{var}\{\bar{S}\} = \int_0^\infty R(\tau) dF(\tau)$, where $R(\tau) = E\{(S(T+\tau) - \mu)(S(T) - \mu)\}$ is the autocorrelation of $S(t)$.

If delay samples are also available, one can extend the above model to take it into account. The idea is to use the delay information to reduce the estimation variance on the loss probability. However, recent empirical studies have shown that delay and loss rates are statistically independent [9,2]. This is mainly due to the fact that losses and delays do not occur at the same location in the network. Based on this empirical observation, the joint estimation problem of statisctical parameters of $S(t)$ and $D(t)$ can be slipt into two independent estimation problems, one for the loss process and the other for the delay.

Estimations done the the probing traffic can be extended to other competing flows. This can be seen by the previous analysis : all competing flows are governed by the same open/close process, therefore under the stationary assumption for $S(t)$, the temporal mean estimator of all flows will indeed converge to the same value. However the variance of this estimator will largely depend on the autocorrelation function of the open/close process and on the dynamics of the particular flow. For example, TCP flows that send a bulk of packets on window opening will undergo higher estimate variance than competing UDP flows that send packets more regularly. This comment does not mean that a UDP flow will see lower loss rates than a TCP flow, it only says that if a UDP and a TCP flow are competing, the TCP flow may see a larger fluctuation of its loss rate than the more regularly spaced UDP flow.

This conclusion might be non-intuitive. It should be stressed on the assumption made for deriving the results. We have assumed that the processes $S(t)$ and $D(t)$ have reached a stable and stationary state. This state will depend on the background traffic as much as on the application (or the measurement probing) traffic. The reached stationary state might be different if a TCP traffic is sent instead of a UDP one (and *vice versa*). This reinstate the intuition that a reactive TCP flow should see a lower loss rate than a non reactive UDP one.

This comment emphasizes the importance of using a model relating the traffic into the network to the stationary state of $S(t)$ and $D(t)$. In the following we will deal with this problem. As stated before, it is not possible to go further than a simple descriptive analysis without any *a priori* on the process generating the on/off and the delay processes. This is the place where constructive analysis should be introduced into the analysis.

The relationship between the traffic flowing into the network and the two process $S(t)$ and $D(t)$ can be described by a constructive model defining precisely (or even roughly) the interactions between different flows resulting in the QoS as observed at the output of the network. The estimation goal is to choose some undetermined parameter of the constructive model such that an optimisation criteria (Maximum Likelihood or Least Mean Squared of error) is verified. By this way, the best set of parameters describing the measured QoS under the assumption of the *a priori* constructing model is derived.

It is clear that all choice of the *a priori* constructive model are not equivalent. Too simplistic *a priori* model, will not be able to faithfully describe the measured QoS. At the other extreme too sophisticated *a priori* models are untractable or too complex to calibrate. Nevertheless, as we are in a stochastic context no model will be perfect. To address this issue, we use a paradigm which is borrowed from the field of data compression. In data compression, a dataset is represented by a predictive model and a sequence of error terms. Each term in the dataset is divided in two parts, one obtained by the predictive model that can be discarded and an error term which is stored as is. A stronger model leads to smaller error terms and less bit needed to encode them. We can use a similar paradigm for our *a priori* Measurement based modeling of Quality Of Service. The measured QoS is a dataset that is to be represented by the predictive *a priori* model and a sequence of error terms that will be represented by a strictly descriptive model. This paradigm can be extended in a hierarchical way if error terms are themselves represented by a new level of *a priori model* followed by a simpler and smaller error term descriptive model. With this paradigm a good *a priori* model is a model that leads to a tractable parameters calibration and at the same time to a small and simple descriptive model for the error term.

The proposed methodological approach resumes to the following steps in modeling : first an *a priori* constructive model is chosen with some input parameters remaining undetermined. Second, measured QoS derived by active measurement is applied and the unknown parameters of the constructive model are estimated by some criteria optimization procedure. We will see a concrete application of this approach in section 5.

The main characteristic and difficulty of this approach is related to the hidden variable. The unknown parameters of the constructive model are not directly observed at the output of the network. The estimation procedure tries to estimate the most faithful value of the unknown (and unobservable) parameters. This estimation procedure is described in the next section (Sec. 4).

Hidden variable approach is not so non-intuitive. Reference to the network state is not so uncommon in the context of QoS evaluation in networking. However, network state is not a concrete and well defined notion. In fact, network state is an *abstract* variable, representative of the effect of all concurrent flows on one application flow. As applications have no direct access to information on router loads and characteristics, the network state is a hidden variable, that can be perceived by an application only through its effects on its data flow.

4   Hidden Variable Estimation and the Expectation-Maximisation Algorithm

The general statistical framework of the proposed methodology can be easily described. Lets $\mathbf{X} = \{X_i\}_{i=1}^{T}$ a sequence of samples of the two processes $S(t)$ and/or $D(t)$ at time $\{T_i\}$. Now let $\mathcal{M}_\theta$ represents the *a priori* model with unknown parameters θ. This *a priori* constructive model will relate the observed QoS $\mathbf{X}$ to some unobserved parameters (for example input traffic or network parameters), represented by θ, by the way of a random function $(\mathbf{X} = \mathcal{F}(\mathcal{M}(\theta)))$. Now, the objective is to find the set of parameters $\hat{\theta}$, such that an optimization criterion is satisfied. One of the most powerful and frequently used criterion is the Maximum Likelihood. It chooses the parameters $\hat{\theta}$ that maximize the probability of seeing the observed QoS samples $\mathbf{x}$ given that the *a priori* model $\mathcal{M}$ follows the parameters θ.

$$\hat{\theta} = \mathrm{Arg}\max_{\theta}\mathrm{Prob}_\theta\{\mathcal{F}(\mathcal{M}(\theta)) = \mathbf{x}\} \qquad (1)$$

The estimation procedure then reduces to an optimization problem with a cost function.

The presented framework is very general and may be difficult to manage by analytic methods. The *a priori* model can contain traffic model as well as router and link models. For a tractable analysis, we assume that we can divide the complex *a priori* model to a set of more simple models, with the same model structure $\mathcal{M}$ but with different values of parameter θ. More formally, we assume that the *a priori* model can be described by a stochastic Finite State Machine (FSM) with each state $i \in \{1,\dots,K\}$, of the FSM characterized by a set of parameter value θ_i. The observation done when the FSM is in state i is determined by a stochastic function $\mathcal{F}_j(\mathcal{M}(\theta_j))$. The sequence of state of the *a priori* FSM $(\mathbf{Y} = \{Y_j\}_{j=1}^{T})$ is unobservable and should be estimated as a side result of the model calibration. In a stochastic modeling wording, the previous assumption means that observations $\mathbf{x}$ follow a Hidden Markov Model : the unobservable sequence of state $\mathbf{Y}$ is a Markov chain and the observations at time j are a stochastic function of the state Y_j. The assumption of an *a priori* FSM is not too restrictive and a large spectrum of models are compatible with it.

4.1   The Expectation Maximization Algorithm

In this context, the Expectation Maximization (EM) algorithm [13] is a valuable approach for maximum likelihood parameter estimation. It has a number of advantages, especially its stability : each iteration of the EM algorithm increases the likelihood of the model; this ensures the convergence of the algorithm to a local, but not necessarily global extremum of the likelihood. Another major advantage of the EM algorithm is its numerical complexity. A direct computation of the likelihood would require K^T terms, where K is the number of different states for the *a priori* FSM models , and T is the number of observations. On the other hand, the numerical complexity of the EM algorithm is of the order of K^2T.

The EM algorithm involves maximizing iteratively with respect to θ a function

$$Q(\theta, \hat{\theta}_k) = E\{L(\mathbf{X}, \mathbf{Y}; \theta) \mid \mathbf{X} = \mathbf{x}; \hat{\theta}_k\} \tag{2}$$

where $\mathbf{Y}$ is the unobserved Markov state sequence, $\mathbf{X}$ is the vector of observations -a probabilistic function of $\mathbf{Y}$- and $L(\mathbf{X}, \mathbf{Y}; \theta)$ denotes the log-likelihood of the 'complete data' $\mathbf{Z} = (\mathbf{X}, \mathbf{Y})$ when θ is the parameter of the model. Expectation involved in the computation of $Q(\theta, \hat{\theta}_k)$ is the expectation given that $\mathbf{X} = \mathbf{x}$ and given that $\hat{\theta}_k$ is the parameter of the model.

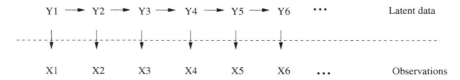

Fig. 2. Dependence structure of the Hidden Markov Model.

The dependence structure of the HMM (see Fig.2) is such that the complete log-likelihood $L(\mathbf{X}, \mathbf{Y}; \theta)$ can be split into two terms $L(\mathbf{X}, \mathbf{Y}; \theta) = L(\mathbf{Y}; \theta) + L(\mathbf{X} \mid \mathbf{Y}; \theta)$ so that the optimization can be performed independently for the parameters of the unobserved state process, and for the parameters of the observations; these optimizations are performed in spaces of a smaller dimension. In many cases there is even an analytic expression for the maximizer in θ of the function $Q(\theta, \hat{\theta}_k)$ so that the maximization step is very fast.

Each iteration of the EM algorithm can consequently be decomposed into two steps :

1. Step E (Expectation):
   Compute $Q(\theta, \hat{\theta}_k) = E\{L(\mathbf{X}, \mathbf{Y}; \theta) \mid \mathbf{X} = \mathbf{x}, \hat{\theta}_k\}$.
2. Step M (Maximization) :
   Maximize $Q(\theta, \hat{\theta}_k)$ with respect to θ :

$$\hat{\theta}_{k+1} = \operatorname*{Arg\,max}_{\theta} Q(\theta, \hat{\theta}_k).$$

The maximization involved in the M step is analytical and does not require intensive computation; the integration involved in the E step requires the computation of a non linear filter; this computation is based on the Forward Backward (or Baum-Welches) algorithm [13].

4.2   State Sequence Estimation

Application of the EM algorithm make possible to calibrate the parameters of the *a priori* model. The next step is to estimate the sequence of states $\hat{\mathbf{Y}}$.

This sequence will help on interpreting the observed vector $\mathbf{x}$ with the *a priori* constructive model. Two approaches can be used to reach this goal : the first one, the Marginal Posterior Mode - MPM, attempts to estimate the most probable state at step t ($\hat{\mathbf{Y}}_t$) based on the observations process up to step t ($\mathbf{x}_0^t$). The second approach, based on the Viterbi algorithm, estimates the most probable state sequence $\hat{\mathbf{Y}}_0^T$ knowing the overall observations $\mathbf{x}_0^T$.

The Marginal Posterior Mode. The Forward-Backward algorithm used in the EM algorithm produces as a byproduct the *a posteriori* Marginal distribution $\gamma_t(i) = Prob\{Y_t = i \mid \mathbf{X}\}$ which leads to the MPM estimate which is the most probable state at time t, given the observed sequence $\mathbf{X}$. This estimate is the maximizer in i of $\gamma_t(i)$:

$$\hat{s}_t^{MPM} = \text{Arg} \max_{1 \leq i \leq K} \gamma_t(i). \tag{3}$$

The Viterbi Algorithm. Another approach to state sequence estimation is based on the Viterbi algorithm. This algorithm produces a sequence

$$\hat{\mathbf{Y}} = (\hat{Y}_1, \hat{Y}_2, \cdots, \hat{Y}_T)$$

that is globally the most likely given the observation vector $\mathbf{X} = \mathbf{x}_0^T$. This estimate produced by the Viterbi algorithm will not coincide, in general, with the Marginal Posterior Mode at any time index t. In general, the sequence produced by the Viterbi algorithm presents longer homogeneous intervals than the sequence produced by the Maximum Posterior Mode criterion.

The *a posteriori* log-likelihood $L(\mathbf{Y} \mid \mathbf{X} = \mathbf{x}; \theta)$ is equal to the complete log-likelihood up to an additive constant, $L(\mathbf{Y} \mid \mathbf{X} = x; \theta) = L(\mathbf{Y}, \mathbf{X} = \mathbf{x}; \theta) - L(\mathbf{X} = \mathbf{x}; \theta)$ so that the maximizer of $L(\mathbf{Y} \mid \mathbf{X} = \mathbf{x}; \theta)$ is the maximizer of $L(\mathbf{Y}, \mathbf{X} = \mathbf{x}; \theta)$. The dependence structure of HMM (Fig. 2) leads to an additive expression for the complete log-likelihood:

$$L(\mathbf{Y}, \mathbf{X} = \mathbf{x}; \theta) = \sum_{t=1}^{T} \left(\log Prob\{Y_{t+1} \mid Y_t\} \right.$$
$$\left. + \log Prob\{x_{t+1} \mid Y_{t+1}\} \right).$$

This sum can be represented graphically as the length of a path in a lattice. The complete log-likelihood $L(\mathbf{Y}, \mathbf{x}; \theta)$ is the total length of the path $\mathbf{Y}$ in the lattice. The Viterbi algorithm retrieves the longest path in this lattice.

The additive form of the criterion makes it possible to construct a dynamic programming algorithm to solve this optimization problem [14].

In the previous sections we have developed the proposed modeling methodology based on *a priori* constructive model. In the following section we will show an application of this approach to the modeling of losses observed on an Internet Path.

5   Case Study: Modeling of Losses Observed on an Internet Path

In this section we apply concepts developed in previous sections. For this purpose an observed loss trace collected from the Internet following the IPPM Metrics recommendation [11] is used. We sent over the Internet a sequence of regularly spaced packets of equal size from Paris to different addresses in the United States and Europe. The packets were regularly spaced with a delay between packets of $\Delta = 50$ msec. The loss trace $\mathbf{X} = (X_t)_{t=1}^{T}$ is defined as $X_t = 0$, if the tth packet reaches its destination and $X_t = 1$, if the packet is lost. The methodology described in this paper was applied to a sequence of 10000 packets (corresponding to a period of 500 seconds with a constant inter packet time of 50 msec).

In a recent paper [17], the previously described EM algorithm is applied to derive an Hidden Markov Model (HMM) for the channel connecting a source to a destination over an Internet Path. This HMM 'switches' between K different states following an homogeneous and ergodic Markov chain. In each of these K states, the channel is uniformly blocking or passing. This defines a probability that a packet is lost at a time when the channel is in state i ($1 \le i \le K$). It is shown in [17] on 36 losses traces collected over the Internet, that not more than 4 states are needed for modeling losses by the HMM.

However, the model developped in [17] remains strictly descriptive as the states are not related to any constructive model. In this section, we use the methodology described in this paper to extend this descriptive model by an *a priori* constructive model. We will not go here through the details of the derivation steps and the evaluation of the model. This task is devoted to a companion paper [16].

Based on the descriptive HMM model develloped in [17] and on the approach proposed in [1], we assume an *a priori* constructive model (fig. 3) for the network consisting of a single bottleneck model with transmission capacity μ and buffer size M. This bottleneck is fed by a traffic following a Markov Modulated Poisson Process (MMPP). The MMPP traffic model describes the traffic entering the bottleneck, which will be the sum of the background Internet Traffic and measurement probe traffic. This model assumes that the traffic switches between K different states following a continuous time homogeneous Markov chain with infinitesimal transition matrix Q. Each state represents an homogeneous poisson process with parameter $\lambda_i + \gamma$, $i \in \{1, \ldots, K\}$, where γ is the measurement probe traffic and λ_i represent the Internet background traffic.

We suppose that the capacity μ is known and constant. This assumption is practically sound as it is possible to estimate faithfully this parameter by packet pair (or packet train) procedure that is well described in the litterature [9]. This *a priori* model can be easily fitted into the framework of FSM described in section 4.1. Each state i of the FSM is related to one state of the MMPP traffic process with a parameter set $\theta_i = (Q, \lambda_i)$. The observed packet loss trace is related to the parameter set θ_i by a stochastic function $\mathcal{F}(\mathcal{M}(\theta_i))$. In the sequel, this function is intuitively derived. A more precise derivation is presented in a companion paper [16].

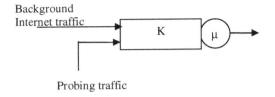

Probing traffic

Fig. 3. *A priori* constructive model of the network

The simple theory of M/M/1/K queue shows that a flow following a Poisson arrival process with parameter λ and feeding a queue with finite buffer size M and processing capacity μ will see a loss rate P calculated by [7] :

$$P = \frac{(1 - \rho)\rho^M}{1 - \rho^{M+1}} \tag{4}$$

where $\rho = \frac{\lambda}{\mu}$ is the load factor.

Nevertheless, if M is sufficiently large (more than 10) and ρ is not too small (which is the case in real network conditions) this relation simplify to :

$$P \approx 1 - \frac{1}{\rho} \tag{5}$$

which shows a simple relation between loss rate and load factor, independently of the buffer size. This relation can be used even if $\rho > 1$.

The relation given in equation 5, is the basis of the derivation of the needed function $\mathcal{F}$. If the bottleneck queue is fed by an MMPP, the behavior can be different from the simple poisson process. However, it can be shown that the overall behavior can be easily derived using the above described analysis of the M/M/1/K queue. We will assume that the mean sojourn time in each state i of the MMPP computed by $\frac{1}{\Gamma_{ii}}$ is much larger than the time scale of the queue defined by $\tau = \frac{M}{\mu}$ meaning that the queue has enough time to reach its stationary distribution in each state of the MMPP before a transition occur. The global behavior of the queue can be described as a mixture of the stationary behavior in each state. More specifically, the observed loss rate at the output of the queue when the input traffic is in state i can be derived easily by applying equation 5 to the input load factor in this state $\rho_i = \frac{\lambda_i + \gamma}{\mu}$.

In the context of our studied loss trace, measurement probes are each separated by 50 msec. This time is much larger than the time scale of the queue as defined above. This means that conditioned on the state of the input MMPP process, the losses can be assumed as independent, as the queue has enough time to be completely renewed between the arrival of two probe packets.

This means that the losses observed at the output of the *a priori* constructive model FSM for a measurement traffic is related to the state i of the FSM by the way of the following stochastic function : a probe packet is lost with probability

$$P_i = 1 - \frac{1}{\rho_i}. \tag{6}$$

This characterized the stochastic observation function $\mathcal{F}(\mathcal{M}(\theta_i))$, with $\theta_i = (Q, \lambda_i)$. With this last information the *a priori* constructive model is completely defined as a single bottleneck queue fed by an MMPP traffic (with some assumptions) and the measurement probe traffic. The observation are related to the constructive model by the above described stochastic function $\mathcal{F}(\mathcal{M}(\theta_i))$.

Now we must address the calibration task. We need to find the set of values ρ_i (or equivalently P_i) and the infinitesimal transition matrix Q that best fit an observed loss trace.

We described previously the descriptive HMM that was developped earlier in [17] for describing network channels. This descriptive model calibration follows the EM algorithm to find a set of value loss rate p_i and a state transition matrice Γ satisfying the maximum likelihood criterion. The result of this analysis can be used to calibrate our *a priori* constructive model. We have clearly $P_i = p_i$. It remains to calibrate the infinitesimal matrix Q'. The descriptive HMM provides a transition matrix Γ representing the transition of the MMPP as seen by the probe packets. If the probe packet are separated by Δ unit of time we have the following relationship between Γ and Q' :

$$\Gamma = e^{Q'\Delta}. \tag{7}$$

This equation has clearly an infinite number of solutions. Each solution will describe the observation. One of these solutions, that we use as the estimate of Q' ($\hat{Q}$), is calculated by diagonalizing the estimated transition matrice for the descriptive HMM.

$$\Gamma = VDV^{-1}$$
$$\hat{Q}' = \frac{1}{\Delta}V^{-1}\log(D)V$$

where $\log(D)$ is the diagonal matrix containing on its diagonal the logarithm of the diagonal of the matrix D. This solution has the nice property of having the same mean sojourn time in each state as the estimated HMM.

Now, let see the application of the above procedure to a real loss trace (Tab. 1). This trace was measured between France and USA and the bottleneck capacity at the time of measurement was 2 Mbps. The loss rate measured over windows of 5 sec on this loss trace are depicted in Fig 4.

Table 1. Basic parameters of the observed loss traces

Interval (msec)	Mean loss rate
50	18.58%

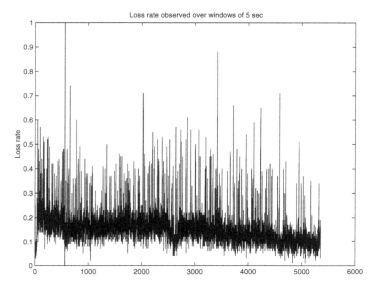

Fig. 4. Loss rate observed over windows of 5 secs.

The application of the HMM estimation process generates the following results (Tab. 2) : The stationary state distribution is also shown as π.

Table 2. HMM parameters calibrated on the observed loss process

$$p = (0.95, 0.206, 0.07)$$

$$\Gamma = \begin{bmatrix} 0.9370 & 0.0623 & 0.0006 \\ 0.0026 & 0.9973 & 0.0002 \\ 0.0000 & 0.0004 & 0.9996 \end{bmatrix}$$

$$\pi = (0.0267, 0.6581, 0.3152)$$

Following the previously described procedure, this parameter can be transformed to parameters of the *a priori* constructive model. This transformation results in the values shown in table 3. The stationary distribution of the MMPP is the same as the HMM.

This analysis shows that the observed loss process can be interpreted in the context of an *a priori* model described before by this set of input parameters. This interpretation is very helpful. It makes possible to simulate loss traces similar to the real trace by feeding the calibrated parameter to a queueing system simulator. It also help to better understand what happen inside the network. For example the stationary distribution shows us that the load factor of the

Table 3. Calibrated parameters for the *a priori* model

$$\rho = (20, 1.2594, 1.07)$$

$$Q' = \begin{bmatrix} -0.0651 & 0.0645 & 0.0006 \\ 0.0026 & -0.0028 & 0.0002 \\ 0.0001 & 0.0003 & -0.0004 \end{bmatrix}$$

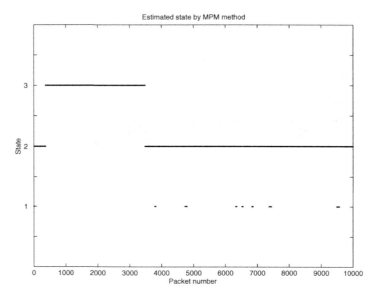

Fig. 5. Estimated state transition for the measured trace.

network was as high as 20 for at least 2.6% of the time and that 65% of the time the load factor was around 1.24. The estimated state transitions are shown in Fig. 5 for 10,000 packets (500 secs). A clear regime transition around packet 338 between state 2 and 3 can be seen. It means that the load factor went from 1.259 to 1.07. The inverse transition occurs around packet 483. Some spurious transition between state 1 and 3 representing surging traffic load that has completely congested the network (with a load factor around 20) can be observed.

6   Conclusion and Perspectives

In this paper, we developed a new modeling methodology for analyzing and interpreting QoS values collected by active measurement and associated with an *a priori* constructive model. This approach is the opposite of the classical constructive modeling approach where we start by making assumptions on the inputs and then find the performance metrics. Here, we start with observed performance (or QoS) measures and derive inputs that have lead to these observations.

This approach needs the introduction of the hidden variable statistical framework. We have described this framework and given some guidelines for the EM algorithm. It has helped in formalizing the approach in the context of the well documented Hidden Markov Model.

Finally, we have illustrated the approach in the context of the modeling of the loss observed in an Internet path. This example shows that the proposed solution is valuable in the context of the interpretation of QoS indices collected by active measurements.

Although it is more complex, this approach provides a powerful solution to address the complex problem of QoS estimation and network modeling. This methodology as well as practical case studies will be developed in the future.

References

1. Sara Alouf, Philippe Nain, and Don Towsley. Inferring network characteristics via moment-based estimators. In *infocom*, Anchorage, Alaska, April 2001.
2. J. Andren, M. Hilding, and D. Veitch. Understanding end-to-end internet traffic dynamics. In *Proceedings of SIGCOMM' 98*, 1998.
3. Hagit Attiya and Jennifer Welch. *Distributed Computing: Fundamentals, Simulations, and Advanced Topics*. McGraw-Hill Publishing Company, May 1998.
4. J.C. Bolot, S. Fosse-Parisis, and D. Towsley. Adaptive fec-based error control for internet telephony. In *Proceedings of IEEE INFOCOM*, pages 1453–1460, NY, March 1999.
5. Chuck Fraleigh, Christophe Diot, Brian Lyles, Sue Moon, Philippe Owezarski, Dina Papagiannaki, and Fouad Tobagi. Design and deployment of a passive monitoring infrastructure. In *Proceedings of PAM2001*, April 2001.
6. S. Handelman, S. Stibler, N. Brownlee, and G. Ruth. RTFM: New attributes for traffic flow measurement, 1999.
7. Leonard Kleinrock. *Queueing Systems — Theory*, volume 1. Wiley-Interscience, New York, New York, 1975.
8. S. Moon, P. Skelly, and D. Towsley. Estimation and removal of clock skew from network delay measurements. In *IEEE INFOCOM*, pages 227–234, 1999.
9. V. Paxson. *Measurements and Analysis of End-to-End Internet Traffic*. PhD thesis, UC Berkeley, February 1997.
10. V. Paxson, A. Adams, and M. Mathis. Experiences with nimi. In *Proceedings of Passive and Active Measurement: PAM-2000*, 2000.
11. V. Paxson, G. Almes, J. Mahdavi, and M. Mathis. Framework for IP performance metrics. *RFC-2330*, May 1998.
12. V. Paxson and S. Floyd. Wide-area traffic: The failure of poisson modeling. *IEEE/ACM Trans. on Networking*, 3(3):226–244, June 1995.
13. L. R. Rabiner. A tutorial on hidden Markov models and selected applications in speech recognition. *Proceedings of the IEEE*, 77(2):257–285, February 1989.
14. Sheldon M. Ross. *Introduction to Stochastic Dynamic Programming*. Academic Press, 1983.
15. Bo Ryu, David Cheney, and Hans-Werner Braun. Internet flow characterization - adaptive timeout and statistical modeling. In *Proceedings of Passive and Active Measurement: PAM2001*, April 2001.
16. K. Salamatian, T. Bugnazet, and B. Baynat. Modelling of losses on internet paths with an *priori* model, submitted to infocom 2002, 2001.

17. K. Salamatian and Sandrine Vaton. Hidden markov modelling for network channels. In *SIGMETRICS*, Cambridge, MA, June 2001.
18. VINT   project.   Ucb/lbnl/vint   network   simulator   -   ns   (version   2), http://www.isi..edu/nsnam.
19. H. Uijterwaal and R. Wilhem. Ripe ncc : Test traffic project homepage available from http://www.ripe.net/test-traffic, 1999.
20. M. Yajnik, J. Kurose, and D. Towsley. Packet loss correlation in the mbone multicast network. In *IEEE Global Internet Conf.,London,UK*, 1996.
21. Maya Yajnik, Sue Moon, Jim Kurose, and Don Towsley. Measurement and modelling of the temporal dependence in packet loss. In *infocom*, New York, March 1999.

Performance Evaluation of a Measurement-Based Algorithm for Distributed Admission Control in a DiffServ Framework

G. Bianchi[1], N. Blefari-Melazzi[2], M. Femminella[2], and F. Pugini[3]

[1]University of Palermo, Italy
bianchi@elet.polimi.it
[2]University of Perugia, Italy
{blefari,femminella}@diei.unipg.it
[3]University of Roma, Italy
pugini@infocom.uniroma1.it

Abstract. Distributed Admission Control in IP DiffServ environments is an emerging and promising research area. Distributed admission control solutions share the idea that no coordination among network routers (i.e. explicit signaling) is necessary, when the decision whether to admit or reject a new offered flow is pushed to the edge of the IP network. Proposed solutions differ in the degree of complexity required in internal network routers, and result in a different robustness and effectiveness in controlling the accepted traffic. This paper builds on a recently proposed distributed admission control solution, called GRIP (Gauge&Gate Reservation with Independent Probing), designed to integrate the flexibility and scalability advantages of a fully distributed operation with the performance effectiveness of admission control mechanisms based on traffic measurements. We show that, in the assumption that traffic sources are Dual-Leaky-Bucket shaped, GRIP allows providing deterministic performance guarantees (i.e., number of accepted flows per node never greater than a predetermined threshold). Tight QoS performance are made possible even in impulsive load conditions (i.e., sudden activation of several flows), thanks to the introduction of a "stack" mechanism in each network node. A thorough performance evaluation of the conservative effects of the stack show that the throughput reduction brought about by this mechanism is tolerable, and limited to about 15%.

1 Introduction

It is well known that the IntServ approach, while allowing hard QoS guarantees, suffers of scalability problems in the core network. This has motivated a large research effort to develop a stateless QoS provisioning approach, i.e. the DiffServ paradigm. The idea that per-flow admission control needs to be introduced in IP DiffServ networks, in order to control traffic load and therefore to provide quantifiable service enhancements, is gaining consensus in the Internet research arena. As suggested in the recent RFC [1], in a DiffServ framework, it appears necessary to define an "admission control function, which can determine whether to admit a service differentiated flow along the nominated network path". In fact, an

S. Palazzo (Ed.): IWDC 2001, LNCS 2170, pp. 175–194, 2001.
© Springer-Verlag Berlin Heidelberg 2001

apparent limit of the DiffServ framework stays in the fact that this approach lacks a standardized admission control scheme, and does not intrinsically solve the problem of controlling congestion in the Internet. Upon overload in a given service class, all flows in that class suffer a potentially harsh degradation of service.

Several Distributed Admission Control algorithms recently appeared in the literature. These proposals share the idea that each network node should accept new flows according to end-to-end congestion measurements and decision criteria. No coordination among network routers (i.e. explicit signaling) is necessary and the final decision whether to admit or reject a new offered flow is pushed to the edge of the IP network.

A critical issue is the reliability and effectiveness of end-to-end mechanisms. Therefore a novel approach, which integrates internal measurements in each node with an end-to-end lightweight probing phase, has been proposed in [2, 3], under the name GRIP (Gauge&Gate Reservation with Independent Probing). In a GRIP compliant IP domain, each network node is endowed with a measurement module, which continuously monitors the QoS traffic queues and properly decides, by means of proprietary decision criteria, to switch from an Accept state, where new admission requests can be accepted, to a Reject state, where no further QoS flows can be admitted. The internal state of each router is advertised by properly handling probe packets. These packets are emitted by end-systems before data transfer in order to test the resource availability within the relevant domain, that is to get permission to start QoS sessions. Probe packets are forwarded by nodes in Accept state and discarded by those in Reject state. GRIP is DiffServ compliant since best effort, probe, and QoS packets are distinguished and properly treated accordingly to their DS Code Point only. In addition, in [4] we have shown that the GRIP operation is semantically compatible with the AF PHB [5].

Each GRIP node is independent from others and each GRIP domain can adopt proprietary strategies to provide QoS. No information is needed to be exchanged among nodes to set internal states, thus integrating flexibility with the performance effectiveness of internal traffic measurements. The probes have the task of conveying to the end-nodes the status of the network. The end-nodes can exploit such information to take per flow accept/reject decisions. This idea is close to what TCP congestion control technique does, but it is used in the novel context of admission control. GRIP is related to the family of distributed schemes recently proposed in the literature under the denomination Endpoint Admission Control (EAC) (e.g., [6] and references therein contained). However, in our opinion, the GRIP solution overcomes some problems of early EAC schemes (see [2, 3 and 4] for more details).

In addition, we stress that GRIP i) does not use a per flow signaling protocol within the routers, for obvious reasons, that is to avoid the pitfalls of IntServ; ii) does not introduce packet marking schemes within routers (as in [7]) or any other substantial modifications of the basic IP way of operation; iii) does not force routers to parse and capture higher layer information (e.g. TCP SYN or SYN/ACK, as in [8]), which would mean modifying the basic router operation. What GRIP does is to *implicitly* convey the status of core routers to the end points, so that the latter devices can take *learned* admission control decisions, but *without violating the DiffServ paradigm*.

The problem we face in this paper is to provide strict QoS guarantees in each operational condition. To this purpose, information about QoS traffic flows characteristics is needed in order to correctly decide the internal state. Thus, we assume that traffic sources are regulated at network edges by standard Dual Leaky

Buckets, as in the IntServ framework. In addition, network nodes have to be aware of the adopted DLB parameters. In the homogeneous traffic case (that is when all flows are regulated by DLBs with the same parameters) this information is fixed once and for all. In the heterogeneous traffic case, different alternatives can be envisaged (see Section 5). In any case, GRIP does not require to signal explicit information about the *traffic mix composition*, that is how many active flows per each regulator type are present in each node. This duty is assigned to the measuring and estimator module in each node, which operates on traffic aggregates only. We stress that the above assumptions are necessary to guarantee performance. In other words, removing some key assumptions means giving up providing assured QoS levels. Finally, we note that the measurement procedure proposed in this paper could be applied also in frameworks different than GRIP.

As for the organization of the paper, in Section 2 we will describe the GRIP operation in more details. Section 3 provides a description of the adopted measurement scheme in a homogeneous traffic scenario. Section 4 is concerned with the relevant performance evaluation. Section 5 deals with a heterogeneous traffic scenario. Section 6 is dedicated to our conclusions.

2   GRIP: Gauge&Gate Reservation with Independent Probing

We envision GRIP as a mechanism composed of two components: (i) GRIP end-nodes operation, (ii) GRIP internal router operation. For clarity of presentation, in what follows, we identify the source and destination user terminals as the network end-nodes, taking admission control decisions. However, for obvious security reasons, such end-nodes should be the ingress and egress nodes, under the control of the network operator(s) (more discussion on security issues can be found in [4]).

2.1   GRIP End-Nodes Operation

GRIP's end nodes operation is extremely simple. **Fig. 1**a illustrates the setup of a monodirectional flow (from source to destination). When a user terminal requests a connection with a destination terminal, the Source Node starts a Probing Phase, by injecting in the network just one Probe Packet. Meanwhile, it activates a probing phase timeout, lasting for a reasonably low time. If no response is received from the destination node before the timeout expiration, the source node enforces rejection of the connection setup attempt. Otherwise, if a Feedback packet is received in time, the connection is accepted, the probing phase is terminated, and control is given back to the user application, which starts a Data Phase, simply consisting in the transmission of information packets. The role of the Destination Node simply consists in monitoring the incoming IP packets, intercepting the ones labeled as Probes, reading their source address, and, for each incoming probe packet, just relaying with the transmission of a feedback packet, if the destination is willing to accept the set-up request. The only mandatory requirement is that Probes and Information packets are labeled with different values of the DS codepoint field in the IP packet header. This enables destination nodes to distinguish between Probes and Information packets. Probing packets do not carry information describing the characteristics of the associated data traffic (e.g., peak bandwidth). This information is implicitly conveyed by means of the DSCP tag (i.e., a given kind of data traffic is associated with a given DSCP tag).

Note that the described GRIP operation can be trivially extended to provide setup for bidirectional connections.

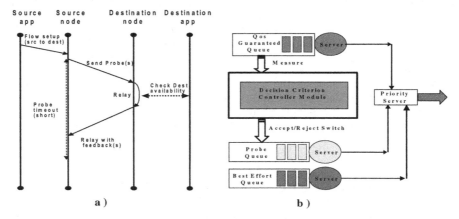

Fig. 1. End point and router operation

2.2 GRIP Router Operation

The GRIP-aware router operation is illustrated in **Fig. 1b**. For convenience of presentation, we assume that the router handles only homogeneous GRIP controlled traffic. Other traffic classes (e.g., best-effort traffic) can be handled by means of additional queues, eventually with lower priority. At each router output port, GRIP implements two distinct kind of queues, one for data packets, i.e., belonging to flows that have already passed an admission control test, and one for probing traffic. Packets are dispatched to the respective buffers according to the probe/data DSCP tag. This enables routers to provide different forwarding methods for Probes and Information packets (e.g. granting service priority to Information packets). As a consequence, the performance of the accepted traffic is not affected by congestion occurring in the probing buffer.

The GRIP router measures the overall *aggregate* accepted traffic. On the basis of running traffic measurements, the router enforces a Decision Criterion, which continuously drives the router to switch between two states: ACCEPT and REJECT. When in the ACCEPT state, the Probing queue accommodates Probe packets, and serves them according to the described priority mechanism. Conversely, when the router switches to the REJECT state, it discards all the Probing packets contained in the Probing queue, and blocks all new Probing packets arriving.

In other words, the router acts as a gate for the probing flow, where the gate is opened or closed on the basis of the traffic estimates (hence the Gauge&Gate in the acronym GRIP). The Decision Criterion may also be based on different procedures than traffic measurements, which can be signaling mechanisms used by underlying layers (e.g. ATM) or simpler approaches (e.g., limiting accepted probe packets via probe buffer limitations) or other, tunable proprietary schemes. This mechanism provides an implicit signaling pipe to the end points of which the network remains unaware, leaving each router in charge of deciding whether it can admit new flows, or it is congested. Since the distributed admission control decision is related to the successful reception at the destination of the Probing packets, locally blocking probing packets implies aborting all concurrent setup attempts of connections whose path crosses the considered router. Conversely, a connection is successfully setup when all the routers crossed by a probing packet are found in the ACCEPT state. As regards performance, it is easy to conclude that the level of QoS support provided depends on the degree of effectiveness of the Decision Criterion implementation. In particular, in this paper we propose a simple and effective criterion based on traffic measurements and suitable assumptions on the offered traffic.

3 Decision Criterion for Homogeneous Traffic Scenario

In this paper we focus on the performance evaluation of GRIP in a single DS domain. Specifically, in this section, we deal with homogeneous traffic sources.

3.1 Traffic Source Model

Traffic sources offered to the considered domain are regulated at the boundary of the domain, by standard Dual Leaky Buckets (DLBs), as in the IntServ framework. This implies that, within the domain, each source is fully characterized by its Traffic Descriptors, given in terms of three DLB parameters, namely Peak rate, P_s, Sustainable Rate, r_s, and Token Bucket Size, B_{TS}.

The Peak Rate P_s (bytes/s) represents the maximum emission rate of the source. In what follows, we neglect the Peak Tolerance, i.e. a parameter sometimes specified in DLB characterization, which accounts for the peak rate variations within a given time frame. The Sustainable Rate r_s (bytes/s), in conjunction with the Token Bucket Size B_{TS} allows to jointly characterize the average emission rate of the source as well as its variability. In addition, we specifically require the DLB to enforce that traffic does not underflow the sustainable rate specification. This is accomplished by the emission of "dummy" packets (e.g. empty packets), in order not to waste token. In other words, the traffic coming out from the DLB fully uses all the opportunities allowed by the regulating device (this does not mean that sources emit always at peak rate!).

Note that this assumption is not an unrealistic one: if a user requests a QoS service and pays for it on the basis of the selected DLB parameters, it is likely that emission opportunities will not be wasted (greedy sources). In addition, without this assumption, as in any measurement-based scheme, it is not possible to correctly estimate the admitted flows and thus guaranteeing performance. The consequence is that the number of bytes, $b(T)$, emitted by a source during an arbitrary time window of size T (seconds) is upper and lower bounded as follows:

$$max(\ r_S T - B_{TS}, 0\ ) \le b(\ T\ ) \le r_S T + B_{TS} \tag{1}$$

The upper bound results from the standard DLB operation, while the lower bound is a consequence of our assumption of greedy sources. Finally, in this Section, we assume that the sources are homogeneous, that is they are regulated by means of the same DLB parameters. No additional assumptions are done on the traffic emission pattern, i.e., $b(T)$ is a random variable with general distribution in the above range.

3.2 Decision Criterion

The localized Decision Criterion running on each router's output link is based on the runtime estimation of the number of the active sources. Our proposed Decision Criterion is founded on a strong theoretical result, proven in [9, 10], which states that, in the presence of DLB regulated sources, target performance (e.g., loss/delay) levels are hard-guaranteed whenever the maximum number of admitted sources does not overflow a suitable threshold K. For our scopes, we consider K as a "tuning knob", which allow the domain operator to set target performance levels [11]. The detailed relationship between K and the guaranteed performance levels results from an off-line computation, following the approach presented in the above references. In other words, we assume that the operator chooses target performance levels; the latter are mapped in a value of K by means of the DLB characterization and of a specific algorithm (e.g., the one of [10]) and GRIP enforces such value. GRIP is completely independent by the specific algorithm chosen to evaluate K (and thus whatever other algorithm could be selected, including evaluating K by means of trials in a specific network scenario).

We now provide a measurement technique targeted to estimate the number of admitted traffic sources on the basis of the traffic measured over the considered link during a fixed size sliding window of size T (seconds). To define the length of such window, we use the period of the worst case DLB output [10], characterized by an activity (On) period with emission at the Peak rate and a silent (Off) period, both with deterministic length. The length of this period is: $T_{min} = (B_{TS} P_s)/(r_s(P_s - r_s))$. The window size is T, with $T \ge T_{min}$, in order to catch at least the minimum periodicity of the source, associated to the worst case DLB output.

To go further, first, let us note that, in a static scenario (i.e. no arriving and departing flows), being N the *fixed* number of offered flows during the time window T, Eq. (1) yields the following inequality on the number of bytes, $A(T)$, measured in the time window T:

$$A_{MIN}^N = N(\ r_S T - B_{TS}\ ) \le A(\ T\ ) \le A_{MAX}^N = N(\ r_S T + B_{TS}\ ) \tag{2}$$

In alternative, we can read Eq. (2) as follows: given a window T and a number $A(T)$ of bytes measured within the window, the number of offered flows is a random variable in the range:

$$N_{MIN} = \left\lceil A(\ T\ )/(\ r_S T + B_{TS}\ )\right\rceil \le N \le N_{MAX} = \left\lfloor A(\ T\ )/(\ r_S T - B_{TS}\ )\right\rfloor \tag{3}$$

If no conjecture is made on the statistical properties of the emission process for each source (we do not rely on how the traffic source fills the DLB), the distribution of N between these two extremes remains unknown. To provide a conservative estimate, the admission control scheme estimates the number of allocated flows as $N_{est}=N_{MAX}$, and a new flow is accepted if $N_{est} \leq K-1$ (i.e. if there is space for at least one more flow). The router is in the ACCEPT state as long as

$$N_{est=}\lfloor A(T)/(r_S T - B_{TS})\rfloor \leq K-1 \qquad (4)$$

Finally, we note that in this static scenario (i.e. no arriving and departing flows) Eq. (3) shows that, as the window size increases, the bounds N_{MAX} and N_{MIN} become closer, until they become equal to each other, thus allowing to evaluate exactly the number of offered flows. This happens for $T \geq (2N-1)B_{TS}/r_S$. Such window size (that we can denote as exact-window) can last a significant amount of time. We stress that our measurement procedure operates in background and does not influence the flow set-up time, as in some other schemes. However, even if we are not constrained to adopt a very short measurement time, we want to avoid using always an exact-window, since its size can reach values in the order of several minutes, depending on the DLB parameters. This is because a too large window has some cons, in real non-static scenarios, which will be discussed in the sequel. The solution is to trade off the window size with the accuracy in the estimation of N. In any case, the choice of T influences the network efficiency but not the performance perceived by users, which are always guaranteed.

3.3 Stack Protection Mechanism

A limit of the above analysis is that it does not account for transient phenomena. Consider, in fact, a router in the ACCEPT state, i.e. for which condition (4) is verified. If a large number of new flows are offered in a short time frame, the measurement mechanism is not sufficient to protect the system from over-allocation, i.e. all the newly incoming flows may be accepted. In fact, when a router "accepts" the Probe packet of a given flow (i.e., the probe finds the gate opened), the relevant Data packets are not yet emitted by the source. In other words, it exists a transient time during which the router is loaded with a new flow, but $A(T)$ in Eq. (4) only partially accounts for the new traffic contribution. Such transient time is upper bounded by T, if we assume that the round trip delay is smaller than T (which is a safe assumption, since the measurement window lasts typically several seconds, see below).

To overcome the above described problem, and to provide strict guarantees in any operational condition, we introduce a protection mechanism based on a "stack" variable, which keeps memory of the amount of "transient" flows. Whenever a probe packet is accepted, the stack is incremented by one. A timer equal to the duration of the measurement window T is then started, and the stack is linearly decremented at a rate $1/T$ until the timer expires. Thus, at time t, if a single flow has been admitted at time t_1, neglecting the Round Trip Delay, the stack is equal to $1-(t-t_1)/T$ and thus it compensates the lack of packets emitted by the source in the time interval $(t-T,t_1)$. To account for the Stack variable, it is simply necessary to modify the decision criterion defined in condition (4) as follows, i.e. the router remains in the ACCEPT state as long as:

$$N_{est+STACK} = \lfloor A(T)/(r_S T - B_{TS}) + STACK \rfloor \leq K - 1 \tag{5}$$

where the STACK variable accounts for the sum of the contributes of each setting up flow. As a side note, we remark that the stack is a simple aggregate variable and hence it does not require to process probing packets and extract or maintain state information. While the stack mechanism is mandatory to protect the system from QoS impairments caused by burst arrivals of newly offered flows, in steady state conditions its effect is to reduce the system throughput, for two reasons.

1. In steady state conditions, a new flow replaces, in average, a departing flow. This means that the last bytes emitted by the departing flows compensate, in average, the lack of bytes emitted by the newly incoming flow during the measurement time T. However, since these bytes are accounted in the stack variable, condition (5) results overly conservative. The performance evaluation of this phenomenon is carried out in the next Section.

2. The stack variable is incremented for each new incoming probing packet, regardless of the fact that the locally accepted probing packet will results in a new accepted connection, or it will be blocked by a subsequent router along the path (in which case, the stack will provide transient reservation of system resources for a non incoming flow). Results that show the effect of probing packet losses in subsequent routers along the path are giving below, and show that the throughput impairment is marginal.

To discuss the effect of the stack when not all probing packets yield to a flow setup, it is in principle necessary to simulate a multi node topology. To simplify the analysis, we have considered a simulation model where only a single node is considered but where, with a given probability, a probe packet is assumed blocked in later stages of the network. This simulated scenario allows us to evaluate the performance drawbacks induced in GRIP by the effect 2) discussed above. In fact, in GRIP, each node independently decides whether to accept or reject a new flow, but there is no explicit mean to determine whether a locally accepted source has been blocked by later stages of the network.

The linear stack mechanism apparently should provide overly conservative results, as each accepted flow is accounted, during the measurement time T, as an incoming one. However, our numerical results show that the performance degradation induced by the stack implementation is almost negligible. **Fig. 2** shows the number of admitted flows in a generic router's output link versus the simulation time, for different values of the probing packets blocking probability in later stages of the network (offered load equal to 240 Erl). We set a target number of admissible flows, K, equal to 100. Ideally, the number of admitted flows should be close to K, and respect the strict constraint of not overflowing such value. The curve relevant to a probe loss equal to 0% in **Fig. 2**, after a transient period, stabilizes around a value of about 80, instead of 100. Such loss of efficiency is due to effect 1) discussed above (that is to the transient effect of the stack mechanism) and to the conservative evaluation of the number of active flows, evaluated according to Eq. (4). The curves relevant to a probe loss equal to 25% and 50% account also for the effect 2) discussed above. However, the throughput performance is very close to that obtained in the optimal case of no probing packets blocking. The most notable effect is a faster transient in the case of no probing packet block.

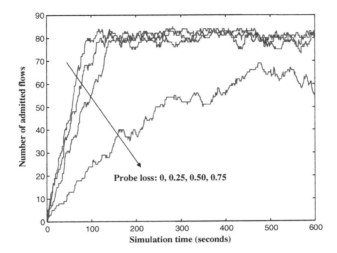

Fig. 2. Number of admitted flows (CBR) versus the simulation time: T=20 seconds, blocking probability of probing packets in subsequent routers equal to 0, 25%, 50% and 75%

In the case of 75% probing packets blocking probability, the number of allocated flows reduces simply because the "valid" offered load gets lower than the considered node capacity, which is equal to 100 Erl (while in this case we are offering 60 Erl, i.e. 25% of 240 Erl). These results hint that the effect of probe loss on the stack mechanism is very limited. As a consequence, in the following, we will focus on the evaluation of the other two effects mentioned above (i.e., transient effect of the stack mechanism and conservative evaluation of the number of active flows).

4 Performance Evaluation of the Homogeneous Traffic Scenario

A detailed performance evaluation of the effect of the stack variable requires considering a dynamic scenario, accounting for flow departures and arrivals. In this Section, we derive the utilization coefficient of a generic router's output link and an upper bound and a lower bound of the same quantity.

4.1 Evaluation of the Utilization Coefficient

Assume that the duration of offered flows is exponentially distributed, with mean value $1/\mu$. To analyze high load conditions (which are the most critical ones), we assume an impulsive load model (see e.g. [12]), in which a new flow is accepted to the system whenever condition (5) is verified, i.e. the router switches from REJECT to ACCEPT state. In other words, users continuously submit new call requests as soon as the system leaves the full occupancy status. Due to the DLB regulation, the average emission rate of each traffic source is assumed equal to the DLB sustainable rate, r_s.

Now, assume that the window size T is small with respect to the mean flow duration $1/\mu$. In these conditions, the probability that a new flow activates and deactivates within the time T is negligible. This approximation has the effect of over-estimating the utilization coefficient, since the STACK mechanism introduces inefficiency. Obviously, this effect increases when the difference $(1/\mu\text{-}T)$ decreases. In any case, this effect can be accounted for, even if, for simplicity, in this paper we neglect it (given its relatively small contribution to overall performance, see the numerical results).

Let us now define the quantity $T_{react,}$ as the measurement scheme "reaction time" i.e., the time elapsing between the instant of time a flow departs from the system, and the instant of time a state transition REJECT to ACCEPT occurs in the router (and thus, thanks to the impulsive load assumption, a new flow is admitted). Further, we define as $T_{react,ave}$, the average value of $T_{react.}$. In other words, the system can not realize immediately that a flow has switched off because, since the traffic sources are VBR, a temporary decrease of the measured bit rate could be due to source activity variations. Thus the measurement procedure needs a time $T_{react,}$ to distinguish between real flow de-activations and statistical fluctuations of active source bit rates. Considering that, owing to the impulsive load assumption, each departing flow will be replaced in average after a time $T_{react,ave}$ with a new incoming one, the average number of active flows during the time window T is given by:

$$\overline{N} = N_0 + N_d \frac{T - T_{react,ave}}{T} \qquad (6)$$

being N_0 the number of flows that remain active during the whole measurement window time T, and being N_d the number of flows departed during T, and replaced by a newly incoming flow. In turns, the average number of departing flows during the time interval T is given by

$$N_d = \overline{N}\mu T \qquad (7)$$

which combined with (6) yields:

$$N_0 = \overline{N}\left(1 - \mu\left(T - T_{react,ave}\right)\right) \qquad (8)$$

We are now able to write condition (5), replacing $A(T)$ with the sum of the average contribute of the N_0 flows that remain active during T, and of the N_d departing/arriving flows:

$$\frac{N_0 r_S T + N_d r_S (T - T_{react,ave})}{r_S T - B_{TS}} + STACK = K \qquad (9)$$

Noting that flow arrivals are uniformly distributed in the time window T, the average stack value is $STACK = N_d/2$. Substituting this value in (9), and owing to (7), (8), we can finally write a single equation which yields the average system throughput as:

$$\rho = \frac{\overline{N}r_S}{C} = \frac{K(1 - T_{OFF}/T)}{\left(1 + \mu/2(T - T_{OFF})\right)}\frac{r_S}{C} \qquad (10)$$

where $T_{OFF} = B_{TS}/r_S$ is the maximum silence period allowed by the DLB. Note that to obtain this result, we do not have to explicitly provide a value for $T_{react,ave}$ (which, in any case, is trivially shown to be given by $T_{react,ave} = (T - T_{OFF})/2$), since in the evaluation of the utilization coefficient, we have both a positive and a negative contribution of this quantity, which balance themselves.

From Eq. 10, it is straightforward to calculate the optimal value T_{opt} of the measurement window T that maximizes ρ:

$$T_{opt} = T_{OFF} + \sqrt{2T_{OFF}/\mu} \tag{11}$$

This is an important result, since it allows dimensioning the measurement window.

Equations (10) and (11) suggest also the following considerations. If the mean flow duration increases, T_{opt} will increase too (for $\mu \to 0 \Rightarrow T_{opt} \to \infty$). As regards the utilization coefficient, when $\mu \to 0$, the STACK contribution decreases and ρ increases with the measurement window size T:

$$\lim_{\mu \to 0} \rho = \frac{K(r_S - B_{TS}/T)}{C} \xrightarrow[T \to \infty]{} \frac{Kr_S}{C} \tag{12}$$

In other words, as discussed at the end of Section 3.2, in these conditions ($\mu \to 0, T \to \infty$) the number of flows is evaluated exactly and the utilization coefficient is equal to the ideal one, (i.e. $N=K$).

To give a physical meaning to the above equations, we note that the effect of the STACK protection is taken into account in (10), with respect to the static formula (i.e., when $\mu \to 0$), by means of the term in the denominator $(T - T_{OFF})\mu/2$; such term is equal to the mean reaction time multiplied by the mean rate of departures/arrivals (μ).

4.2 Evaluation of an Upper Bound of the Utilization Coefficient

In order to evaluate an upper bound of the utilization coefficient, denoted as ρ_{upp}, we make the following assumptions:
- each flow emits according to its minimum emission profile, i.e. it emits $r_S(T-T_{OFF})$ bits during the measurement window;
- we consider the minimum value for the reaction time, i.e. $T_{react,min}=0$;
- the per-flow STACK contribution is equal to zero (instead of 1/2).

Consequently, following the same approach used in the preceding section, we have:

$$N_{upp} = K, \ \rho_{upp} = (Kr_S)/C \tag{13}$$

This is a rather obvious result, reported only to show that the particularization of our derivation gives correct results. In other words, the upper bound of the utilization coefficient is equal to the ideal one, when K flows are admitted in the system.

4.3 Evaluation of a Lower Bound of the Utilization Coefficient

We first derive the measurement scheme "maximum reaction time" $T_{react,max}$, i.e., the maximum possible value of T_{react}. It is easy to deduce that the value of $T_{react,max}$ is the solution of the following equation:

$$\frac{Nr_S T}{r_S T - B_{TS}} - \frac{Nr_S(T - T_{react,max}) + (N-1)r_S T_{react,max}}{r_S T - B_{TS}} = 1 \Rightarrow T_{react,max} = T - \frac{B_{TS}}{r_S} = T - T_{OFF} \quad (14)$$

This equation is obtained by imposing that the difference between the left value of (5), computed assuming that each flow emits at its average rate and no flows depart, and the left value of (5) computed in the assumption that the departure of one flow at time $(T - T_{react,max})$ is detected after $T_{react,max}$ seconds.

In order to evaluate a lower bound to ρ, , denoted as ρ_{low}, we make the following assumptions:

- each flow emits according to its maximum emission profile, i.e. it emits $r_S(T+T_{OFF})$ bits during the measurement window;
- we consider the maximum value for the reaction time, i.e. $T_{react,max} = T - T_{OFF}$;
- the per-flow STACK contribution is equal to one (instead of 1/2).

Following the same approach of Section 4.1, we obtain:

$$N_{low} = \frac{K(T - T_{OFF})}{\mu T^2 + T + T_{OFF}(1 - \mu T_{OFF})}, \rho_{low} = \frac{N_{low} r_S}{C} \quad (15)$$

4.4 Numerical Results

In this Section, we present a simulation analysis to evaluate the effectiveness of the proposed scheme and of the analytical bounds. As discussed above, we make no assumptions on the traffic source behavior, beyond the worst case parameters supplied by the DLB characterization. However, in order to generate traffic, we have to load the DLBs with specific sources. The choice of parameters and performance figures has, here, only case study significance.

We have considered three kind of sources, loading the DLBs: constant rate sources (labeled as CBR) emitting at rate 1.7 Kbytes/s, on-off exponential voice sources (labeled as EXPO) and MPEG sources. As regards the EXPO sources, during the On state (talkspurt) the source emits packets periodically. The On and Off state durations are exponentially distributed, with average values of 352 ms and 650 ms respectively. The bit rate during the On period is equal to 4 Kbytes/s. Both sources are regulated by DLBs with parameters: $P_S = 4$ Kbytes/s; $r_S = 1.7$ Kbytes/s; $B_{TS} = 5300$ bytes. The MPEG sources will be described in the sequel. We consider a generic router's output link with parameters: link rate, $C = 2.048$ Mbps; buffer size, $B=53000$ bytes. We set, as target performance figure, a packet loss probability, P_{loss}, equal to 10^{-5}. According to the acceptance rule provided in [10], the corresponding maximum number of acceptable flows is $K=100$. The call arrival rate, modeled as a Poisson arrival process, has been set to 1 call/s, and call duration has been drawn from an exponential distribution with mean value 4 minutes. This implies that 240 Erlangs are offered to the link, i.e., more than twice the maximum number of calls that can be, in principle, simultaneously admitted $(K=100)$.

In **Fig.** 3a and **Fig.** 3b we show the utilization coefficient, denoted as ρ, as a function of the measurement window size T, for the EXPO and the CBR case, respectively. In the two figures, several curves are reported: i) the utilization coefficient in the ideal condition of $K=100$ (labeled "Reference System"); this is an horizontal line since it is independent by the measurement process (equal in this case to 0.68 and equal to the upper bound evaluated in Section 4.2); ii) the simulation results; iii) the value of ρ evaluated in Section 4.1; iv) the lower bound of ρ evaluated in Sections 4.3.

The utilization coefficient is lower than the ideal one, due to the stack mechanism and to the conservative evaluation of the number of active flows, evaluated according to Eq. (5). However, the maximum number of admitted sources is guaranteed to remain below the value $K=100$, representing the maximum number of flows that can be admitted without violating QoS requirements. This is an important advantage of our (conservative) admission control criterion, and confirms that: (i) hard guarantees can be obtained, thanks to the exploitation of the knowledge of the traffic scenario (i.e., of the DLB parameters); (ii) our scheme appears to provide an explicit performance calibration parameter, i.e., the value K.

We recall that the decision criterion is not aware of the effective number of the admitted sources, but uses a run time estimate. This has the advantage of avoiding explicit signaling or using other non DiffServ compliant schemes. In other words, by using signaling we could allocate exactly K flows. Since we want to adopt scalable procedures, while guaranteeing performance, we are forced to under-estimate K. The prices to pay are: (i) a system under utilization with respect to the "ideal" value of K (about 15% less), and (ii) the "a priori" knowledge of the traffic regulator parameters adopted at the edge of the network.

In fact, the ultimate target of the GRIP operation is to achieve a link utilization as close as possible to the maximum (i.e., K sources), but without ever exceeding this value. In order to satisfy the latter requirement, we must accept a throughput penalization. We stress also that the performance of GRIP must be evaluated in its ability to enforce a specific value of K. The overall throughput efficiency is a direct consequence of the selected value of the "tunable knob" K, and can be thus arbitrarily adjusted by the network operator. Finally, we verified that, without the stack protection, the number of admitted flows can exceed K.

Another consideration regarding **Fig.** 3a and **Fig.** 3b is that, due the presence of the DLB regulators, the performance of the two different traffic types are very similar. For both traffic types the simulation results are close to the analytical estimation of ρ, thus confirming that Eq. 11 can be used to dimension the optimal value of the measurement window size. The value of such optimal window is dependent on two parameters only: the maximum silence period allowed by the DLB (T_{OFF}) and the call departure rate μ. As shown in **Fig.** 3c, the utilization coefficient and the accuracy of the estimation of ρ (Eq. 10) increases with the mean flow duration (since T is smaller with respect to $1/\mu$, see Section 4.1). This figure reports results for EXPO sources, with the same parameter used for **Fig.** 3a, but with a greater value of $1/\mu$, equal to 720 seconds.

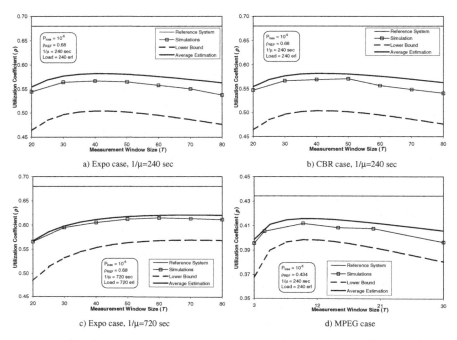

Fig. 3. Utilization Coefficient versus the Measurement Window Size.

To verify the robustness of our approach with respect to different traffic classes, we loaded the DLBs with MPEG sources. The link capacity is now set to $C=100$ Mbps and the link buffer size is $B=96000$ bytes. The values of the DLB parameters are: $P_S = 579.6$ Kbytes/s; $r_s= 54.28025$ Kbytes/s; $B_{TS}= 12480$ bytes. We set, as target performance figure, a packet loss probability, P_{loss}, equal to 10^{-5}. According to the acceptance rule provided in [9, 10], the corresponding maximum number of acceptable flows is $K=100$. The call arrival rate, modeled as a Poisson arrival process, has been set to 1 call/s, and call duration has been drawn from an exponential distribution with mean value 4 minutes (the load is of 240 Erls). The results are presented in **Fig. 3**d. As in the previous cases, the simulation results and the theoretical estimation of ρ are close to each other. The throughput penalization due to our approach with respect to the ideal value of K is smaller than in the previous cases.

Finally, in **Fig. 4** we report ρ as a function of measurement window for different values of the mean flow duration $1/\mu$ (100, 200, 400, 800 seconds); also shown is the limit curve obtained when $1/\mu$ tends to infinity, labeled "static case" (see Eq. 12). As mentioned in Section 4.1, if the mean flow duration increases, the utilization coefficient increases too and its behavior depends only on the measurement window size T. In the static case the utilization coefficient tends to the ideal one (Reference system) as T increases (see again Eq. 12).

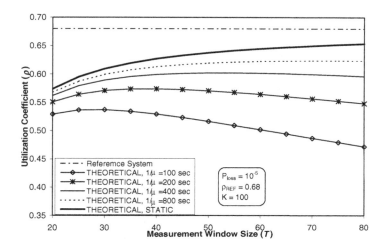

Fig. 4. Utilization Coefficient versus the Measurement Window Size T for different values of the mean flow duration $1/\mu$ (100, 200, 400, 800 seconds)

5   Heterogeneous Traffic Scenario

To extend GRIP to a heterogeneous traffic scenario (i.e. to a mix of traffic sources regulated with different DLB parameters) we need some introductory considerations. Let us assume that the sources are divided in I traffic classes, each comprising independent and homogeneous sources (i.e., with the same DLB parameters). To go further, we note that, *in the homogeneous case*, the evaluation of the number of admissible flows K is equivalent to the evaluation of the so-called equivalent bandwidth of each flow, denoted as e. The equivalent bandwidth concept is well known in the literature and represents the amount of bandwidth that must be assigned to each flow, in a statistical multiplexing framework, so as to reach some performance levels. The equivalent bandwidth is typically comprised between the mean and the peak bit rate of the traffic source. In our case, $e=C/K$, where C is the link capacity (and where K is evaluated as in [9, 10], once the buffer size B and the performance parameters of interest, e.g. $P_{loss}=10^{-5}$ have been fixed). The ideal admission control function "accept new setup requests as long as the number of admitted flows N is less than K", can then be rewritten as "accept new setup request as long as the sum of the equivalent bandwidth of accepted flows $(N*e)$ is less than C". In GRIP, due to the lack of signaling, we do not know N, so we estimate it; then we make use of the equivalent bandwidth concept by comparing the estimated N to K. Finally, we add the stack mechanism. In other words GRIP exploits the DLB characterization two times. The first time to estimate the number of admitted flows and the second time to decide if a new setup request can be accepted. Note that, to this end, each router must be implicitly aware of the DLB parameters.

The equivalent bandwidth concept can be in principle easily extended to the heterogeneous case. In [9, 10] the Authors propose an efficient algorithm to evaluate this quantity, which in the assumption of DLB regulated sources is additive even in the heterogeneous case. They evaluate the equivalent bandwidth of the i-th traffic class e_i, $\{1 \leq i \leq I\}$ as $e_i = C/K_i$, where K_i is evaluated for each class in isolation, i.e., in a homogeneous system. In other words, the limit value K_i is the number of flows such that a link with capacity C and buffer B, loaded only with traffic belonging to the i-th class, offers pre-defined performance levels. With this approach (which is shown to be conservative, even if it leads to a loss of efficiency), the ideal admission rule remains a simple sum: new setup requests are accepted as long as

$$\sum_{i=1}^{I} N_i e_i \leq C \tag{16}$$

where N_i is the number of admitted flows belonging to class i.

To extend GRIP to the heterogeneous case, we have to estimate the number of admitted flows of each class and then apply the above admission rule. To this end, we can identify three architectural alternatives. The first (trivial) alternative is shown **Fig. 5a**. Here we assume that each class is handled in a separated way and recognized by routers by assigning different *pairs* of DS codepoints to different traffic classes (i.e., each class as a DSCP for Probing packets and another one for Information packets). Thus, we need $2*I$ different DSCPs and $2*I$ different logical queues. Packets belonging to class i are classified on the basis of their DSCP tag and dispatched to the relevant queue (Probing or Information). However, we do not require to signal explicit information about the *traffic mix composition*, that is how many active flows per each class are present in each node.

The architecture is complex but the extension of GRIP is straightforward, since the heterogeneous case is reduced to a combination of homogeneous ones. The admission rules becomes:

$$\left\lfloor \frac{A_i(T)}{r_{S,i}T - B_{TS,i}} + STACK_i \right\rfloor \leq K_i - 1, \ with \ \sum_{i=1}^{I} K_i e_i \leq C \tag{17}$$

where $STACK_i = (1-(t-t_1)/T)$. In (17), $A_i(T)$ is the number of bytes emitted during the window T by traffic sources belonging to the i-th class (i.e., at the i-th Information queue) and K_i is evaluated off-line as discussed above.

Note that this alternative implies that each node is aware of the DLB parameters of all possible classes. This architectural alternative is somehow acceptable only in presence of a small number of traffic classes (e.g., a class could be IP telephony), even if it is still compliant with the DiffServ approach. An advantage of this architecture is that it allows implementing procedures to fairly divide the overall capacity among the traffic classes.

In the second alternative (see **Fig. 5b**), we assume that Probing packets belonging to different classes are handled in a separated way, with multiple probe queues. These packets are recognized by routers by assigning them different DSCPs, while the Information packets of all the I traffic classes are multiplexed together in a common queue. Each class has a DSCP for Probing packets, while the Information packets of all classes share the same DSCP. Thus, we need $I+1$ different DSCPs and $I+1$ different logical queues. As above, we do not require to signal explicit information about the traffic mix composition. This duty is assigned to the measuring module, which operates on traffic aggregates only. However, the Decision Criterion has to be suitable modified with respect to the homogeneous case, in order to take into account the presence of different traffic profiles, while still guaranteeing performance.

In this alternative, it is not possible anymore to estimate the admitted flows of each class, N_i, as done above; thus if we want still to guarantee performance, we are left with a worst case approach. Our measurement procedure evaluates now the number $A(T)$ of bytes emitted within a window of size T by *all* traffic classes. To define a GRIP allocation rule we have to evaluate the traffic mix that maximizes the overall equivalent bandwidth

$$e_{TOT} = \sum_{i=1}^{I} N_i e_i \qquad (18)$$

under the constraints

$$\begin{cases} \sum_{i=1}^{I} N_i a_i^{min} \leq A(T) & with \quad a_i^{min} = r_{S,i} T - B_{TS,i} \\ 0 \leq N_i \leq K_i \end{cases} \qquad (19)$$

In other words, we have to find the values of N_i $\{1 \leq i \leq I\}$ under the constraint that the admitted flows must be such that they emit $A(T)$ bytes, when using a minimum DLB emission profile (this is the meaning of the apex "*min*" to quantity "*a*").

The main difficulty in resolving the problem in (18) is the need to find an integer solution, since we are dealing with numbers of flows. We start by resolving the associated continuous problem (i.e. by assuming that N_i is a continuous variable) and then we apply a floor operator to the solution (which is a slightly conservative operation), obtaining:

$$\begin{cases} N_x = \left\lfloor \dfrac{A}{a_x^{min}} \right\rfloor \\ N_i = 0 & for \quad i \neq x \end{cases} \qquad (20)$$

where x is the index of the class with the greatest value of the ratio e_i / a_i^{min} $\{1 \leq i \leq I\}$.

Because of the lack of any information about the composition of the traffic mix, the measurement procedure reduces the heterogeneous case to an homogenous one, in which, for estimation purposes, only the traffic class endowed with the worst estimation of resource utilization is considered as present in the mix. Such class is labeled with the index "x". The worst estimation results from assigning the greatest equivalent bandwidth to a number of bits obtained by assuming the minimum DLB emission profile. The GRIP allocation rule relevant to the i-th probing class becomes:

$$\left\lfloor \dfrac{A}{r_{S,x} T - B_{TS,x}} + \sum_{i=1}^{I} STACK_i + \dfrac{e_i}{e_x} \right\rfloor \leq K_x \quad with \quad STACK_i = \dfrac{e_i}{e_x} \left(1 - \dfrac{t - t_I}{T} \right) \qquad (21)$$

and the ratio e_i / e_x accounts for the new incoming flow. Note that this architecture still allows implementing procedures to fairly divide the overall capacity among the traffic classes. As regards performance, this architecture is simpler than the previous one, but the price to pay is a potential smaller system efficiency.

As a third and last alternative, we propose the architecture shown in **Fig. 5c**. Here we assume that the Probing packets of all the I traffic classes are multiplexed together in the same queue and that the Information packets of all the I traffic classes are multiplexed together in another common queue. All the traffic classes share the same pair of DSCP tag: one for Probing packets and one for Information packets. Thus we need only two DSCPs. This time we have to adopt a worst case approach not only for the measurement procedure but also for the admission rule. Recall that we use the DLB characterization two times. The first one to estimate the number of admitted flows and the second time to decide if a new setup request can be accepted. Since in this alternative we can not distinguish between probes belonging to different traffic classes, we are forced to interpret each setup request as belonging to the class with the greatest equivalent bandwidth. Let us define e_{max} the maximum value of e_i $\{ 1 \leq i \leq I \}$.

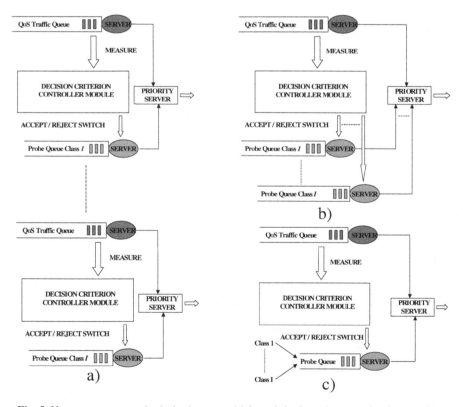

Fig. 5. Heterogeneous case logical scheme, multiple and single probe queue implementations

The GRIP allocation rule relevant to the i-th probing class becomes:

$$\left| \frac{A(T)}{r_{S,x}T - B_{TS,x}} + STACK + \frac{e_{max}}{e_x} \right| \le K_x \;\; with \;\; STACK = \frac{e_{max}}{e_x}\left(1 - \frac{t - t_1}{T} \right) \qquad (22)$$

The ratio e_{max}/e_x accounts for the new incoming flow and $A(T)$ is the number of bytes emitted within a window of size T by *all* traffic classes. The stack variable is incremented by using a scaling factor based on the greatest equivalent bandwidth, to take into account setting up flows in a conservative way. This last alternative is the simplest one. As in the previous ones, we do not require to signal explicit information about the *traffic mix composition*. In addition, this alternative implies that each node has to know the DLB parameters of only one traffic class (the worst one) and the value e_{max} (assuming that the considered traffic classes do not vary).

The prices to pay are: i) a potential smaller system efficiency with respect to the previous cases; ii) the impossibility of implementing procedures to fairly divide the overall capacity among the traffic classes. We carried out a performance evaluation of this scenario similar to the one presented in Section 4. For space limitations we report only the conclusions (for details see [3]). In the first alternative, obviously, we have no loss of efficiency with respect to the homogeneous case. The second alternative presents a loss of efficiency with respect to the homogeneous case, which depends on the "distance" between the traffic classes. This effect is even more evident in the third alternative. The conclusion, as it could be expected, is that it is convenient to jointly handle different classes when they are not too "different" from each other, i.e., when their DLB parameters are not too distant from each other.

6 Conclusions

In this paper we have presented a scalable Admission Control scheme, called GRIP. This is a novel reservation paradigm that allows an evolution from the actual best-effort Internet to a future QoS capable infrastructure. In conformance to the DiffServ principles, GRIP does not rely on explicit signaling protocols to provide an admission control function. Such a function is achieved by imposing each router to be capable of distinguish probe packets from data packets and properly enforce a suitable scheduling discipline. More specifically, we have proposed procedures that allow a tight control on the QoS experienced by admitted flows. A stack mechanism has been proposed to avoid temporary overallocation in the presence of impulsive loads. Simulation results and analytical performance evaluation have been provided in order to provide a thorough dimensioning of the system.

References

[1] G. Huston, "Next Steps for the IP QoS Architecture", RFC2990, November 2000.
[2] G. Bianchi, N. Blefari-Melazzi: "A Migration Path for the Internet: from Best-Effort to a QoS Capable Infrastructure by means of Localized Admission Control", Lecture Notes on Computer Science, Springer-Verlag, volume 1989, January 2001.

[3] G. Bianchi, N. Blefari-Melazzi, M. Femminella, F. Pugini: "GRIP: Technical report, work in progress"(http://drake.diei.unipg.it/netweb/GRIP_tech_rep.pdf)

[4] G. Bianchi, N. Blefari-Melazzi: " Per Flow Admission Control over AF PHB Classes", Internet Draft, draft-bianchi-blefari-admcontr-over-af-phb-00.txt, work in progress, http://www.ietf.org/ID.html

[5] J. Heinanen, F. Baker, W. Weiss, J. Wroclavski "Assured Forwarding PHB Group", RFC 2597, June 1999.

[6] L. Breslau, E. W. Knightly, S. Schenker, I. Stoica, H. Zhang: "Endpoint Admission Control: Architectural Issues and Performance", ACM SIGCOMM 2000, Stockholm, Sweden, August 2000.

[7] F. P. Kelly, P. B. Key, S. Zachary: "Distributed Admission Control", IEEE JSAC, Vol. 18, No. 12, December 2000.

[8] R. Mortier, I. Pratt, C. Clark, S. Crosby: "Implicit Admission Control", IEEE JSAC, Vol. 18, No. 12, December 2000.

[9] A. Elwalid, D. Mitra, R.H. Wentworth: "A New Approach for Allocating Buffers and Bandwidth to Heterogeneous, Regulated Traffic in an ATM Node", IEEE J.S.A.C. Vol. 13, N. 9, August 1995, pp. 1115-1127.

[10] A. Elwalid, D. Mitra: "Traffic shaping at a network node: theory, optimum design, admission control", IEEE Infocom 97, pp. 445-455.

[11] L. Breslau, S. Jamin, S. Schenker: "Comments on the performance of measurement-based admission control algorithms", IEEE Infocom 2000, Tel-Aviv, March 2000.

[12] M. Grossglauser, D. N. C. Tse: "A Time-Scale Decomposition Approach to Measurement-Based Admission Control", Proc. of IEEE Infocom 1999, New York, USA, March 1999.

Resource Stealing in Endpoint Controlled Multi-class Networks *

Susana Sargento[1], Rui Valadas[1], and Edward Knightly[2]

[1] Institute of Telecommunications, University of Aveiro, Portugal
[2] ECE Department, Rice University, USA

Abstract. Endpoint admission control is a mechanism for achieving scalable services by pushing quality-of-service functionality to end hosts. In particular, hosts probe the network for available service and are admitted or rejected by the host itself according to the performance of the probes. While particular algorithms have been successfully developed to provide a single service, a fundamental *resource stealing* problem is encountered in multi-class systems. In particular, if the core network provides even rudimentary differentiation in packet forwarding (such as multiple priority levels in a strict priority scheduler), probing flows may infer that the quality-of-service in their own priority level is satisfactory, but may inadvertently and adversely affect the performance of other classes, stealing resources and forcing them into quality-of-service violations. This issue is closely linked to the network scheduler as the performance isolation property provided by multi-class schedulers also introduces limits on *observability*, or a flow's ability to assess its impact on other traffic classes. In this paper, we study the problem of resource stealing in multi-class networks with end-point probing. For this scalable architecture, we describe the challenge of simultaneously achieving multiple service levels, high utilization, and a strong service model without stealing. We propose a probing algorithm termed ε-probing which enables observation of other traffic classes' performance with minimal additional overhead. We next develop a simple but illustrative Markov model to characterize the behavior of a number of schedulers and network elements, including flow-based fair queueing, class-based weighted fair queueing and rate limiters. Finally, we perform an extensive set of simulation experiments to study the performance tradeoffs of such architectures, and to evaluate the effectiveness of ε-probing.

1 Introduction

The Integrated Services (IntServ) architecture of the IETF provides a mechanism for supporting quality-of-service for real-time flows. Two important components of this architecture are admission control [3,8] and signaling [14]: the former ensures that sufficient network resources are available for each new flow, and the latter communicates such resource demands to each router along the flow's path. However, the demand for high-speed core routers to process per-flow reservation requests introduces scalability and deployability limitations of this architecture without further enhancements.

* Susana Sargento is supported by Ph.D. scholarship PRAXIS XXI/BD/15678/98. Edward Knightly is supported by NSF CAREER Award ANI-9733610, NSF Grants ANI-9730104 and ANI-0085842, a Sloan Fellowship, and Texas Instruments.

S. Palazzo (Ed.): IWDC 2001, LNCS 2170, pp. 195–211, 2001.
© Springer-Verlag Berlin Heidelberg 2001

In contrast, the Differentiated Services (DiffServ) architecture [2,9] achieves scalability by limiting quality-of-service functionalities to class-based priority mechanisms together with service level agreements. However, without per-flow admission control, such an approach necessarily weakens the service model as compared to IntServ, namely individual flows are not assured of a bandwidth or loss guarantee.

A key challenge addressed in recent research is how to simultaneously achieve the scalability of DiffServ and the strong service model of IntServ. Towards this end, several novel architectures and algorithms have been proposed. For example, architectures for scalable deterministic services were developed in [12,15]. In [12], a technique termed Dynamic Packet State is developed in which inserting state information into packet headers overcomes the need for per-flow signaling and state management. In [15], a bandwidth broker is employed to manage deterministic services without explicit co-ordination among core nodes. A scheme that provides scalable *statistical* services is developed in [5], whereby only a flow's egress node performs admission control via continuous passive monitoring of the available service on a path.

While such approaches are able to achieve scalability and strong service models, they do so while requiring specific functionality to be employed at edge and/or core nodes. For example, [5] requires packet time stamping and egress nodes to process signaling messages; [12] requires rate monitoring and state packet insertion at ingress points and special schedulers at core nodes. Thus, despite that such edge/core router modifications may indeed be feasible, an alternate and equally compelling problem is to ask whether the same goals can be achieved without *any* changes to core or edge routers, or at most with routers providing simplistic prioritized forwarding as envisioned by DiffServ extensions such as class based queueing or prioritized dropping policies.

This design constraint is quite severe: it precludes use of a signaling protocol as well as any special packet processing within core nodes. Such a constraint naturally leads to probing schemes in which end hosts perform admission control by assessing the state of the network by transmitting a sequence of probe packets and measuring the corresponding performance. If the performance (e.g., loss rate) of the probes is acceptable, the flow is admitted, otherwise it is rejected. Design and analysis of several such schemes can be found in [1,6,7]. Such approaches achieve scalability by pushing quality-of-service functionalities to the end system and indeed removing the need for a signaling protocol or any special-purpose edge or core router functions. Moreover, [4] found that such an architecture is indeed able to provide a single controlled-load like service as defined in [13].

However, can host-controlled probing schemes be generalized to support *multiple* service classes as achieved by both IntServ and DiffServ? In particular, DiffServ supports multiple service classes differentiated by simple aggregate scheduling policies (per-hop behaviors); DiffServ's Service Level Agreements (SLAs) provide aggregate bandwidth guarantees to traffic classes; IntServ provides mechanisms to associate different quality-of-service parameters (e.g., loss rate, bandwidth, and delay) with different traffic classes. Can such multi-class service models co-exist with the host-controlled architecture?

Unfortunately, a resource *stealing* problem, first described in [4], can occur in multi-class systems. In particular, the problem occurs when a user probes within its desired class and, upon obtaining no loss (or loss below the class' threshold), infers that sufficient capacity is available, which indeed it may be within the class. However, in some cases, admission of the new probing flow would force *other* classes into a situation of quality-of-service violations, unbeknownst to the probing flow. Such resource stealing, described in detail in Section 2, arises from a fundamental observability issue in a multi-class system: the performance isolation property provided by multi-class networks also inhibits flows from assessing their performance impact on other classes.

The goal of this paper is to investigate host probing in multi-class networks. Addressing the problem of resource stealing, our contribution is threefold. First, we study architectural issues and show how service disciplines and the work conservation property have important roles in the performance of probing systems. For example, while a non-work-conserving service discipline can prohibit resource borrowing across classes and remove the stealing problem, such rigid partitioning of system resources limits resource utilization. Second, we develop a probing algorithm which simultaneously achieves high utilization and a strong service model without stealing. The algorithm, termed ε-probing, provides a minimally invasive mechanism to enable flows to assess their impact on *other* traffic classes in Class-Based Fair Queueing (CBQ) and strict priority systems. Finally, we introduce a simple but illustrative analytical model based on Markov Chains. Using the model, we precisely identify stealing states, comparatively analyze several probing architectures, and quantify the aforementioned tradeoffs.

In all cases, we use an extensive set of simulations to evaluate different probing schemes and architectures under a wide range of scenarios and traffic types. The experimental results indicate that ε-probing can achieve utilizations close to the limits obtained by fair queueing, while *eliminating* resource stealing. Consequently, if core networks provide minimal differentiation on the forwarding path, ε-probing provides a scalable mechanism to control multiple service classes, achieve high utilization, and provide a strong service model without resource stealing.

The remainder of this paper is organized as follows. In Section 2, we formulate the stealing problem in multi-class networks and describe the role of the packet scheduler. Next, in Section 3, we propose a simple probing algorithm, termed ε-probing, that overcomes the observability limitations introduced by multi-class schedulers. In Section 4 we develop an analytical model to study the performance issues and tradeoffs in achieving high utilization, multiple service classes, and a strong service model without stealing. Finally, in Section 5, we describe an extensive set of simulation experiments used to investigate the design space under more realistic scenarios.

2   Resource Stealing

The stealing problem arises in multi-class systems in which resources are remotely controlled by observation. This is in contrast to systems in which resources are controlled with explicit knowledge of their load, such as in IntServ-like architectures. In this section, we describe the origins of multi-class stealing and the corresponding design and performance issues. Throughout, we consider a general definition of "class" that can be based on application types, service level agreements, etc., and with quality-of-service parameters such as loss rate and delay associated with each class.

2.1 Origins and Illustration

Probing schemes, such as those studied in [1,4,6,7], can be described with an example using the network depicted in Figure 1. To establish a real-time flow between hosts H and H', host H transmits a sequence of probes into the network at the desired rate (or peak rate for variable rate flows). If the loss rate of the probes is below a pre-established threshold for the traffic class, then the flow is admitted, and otherwise it is rejected. Scalability is achieved in such a framework by pushing all quality-of-service functionality to end-hosts, indeed removing the need for *any* signaling or storage of per-flow state.

The stealing problem can be illustrated as follows. Consider the following simple scenario with two flows sharing a single router with link capacity C and a flow-based fair queueing scheduler[1] (or similarly core stateless fair queueing [11] to achieve scalability on the data path). Suppose that the first flow requires a bandwidth of $\frac{3}{4}C$ and is admitted to an initially idle system. Further suppose that the second flow has a bandwidth requirement $\frac{1}{2}C$. Upon probing for the available service in the fair queueing system, the flow will discover that it can indeed achieve a loss-free service with throughput $\frac{1}{2}C$, and admit itself. Unfortunately, while $\frac{1}{2}C$ is indeed the fair rate for each flow, the goal here is not to achieve packet-level fairness, but rather to achieve flow-level quality-of-service objectives. Thus, in this example, abruptly reducing the first flow's capacity is a clear violation of the flow's service.

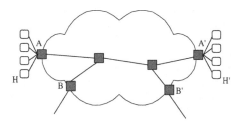

Fig. 1. Illustration of Probing and Multi-Class Stealing

This simple example illustrates an important point. The ability of fair queueing to provide performance isolation can be exploited for both flow-control (to quickly and accurately assess a flow's fair rate) and quality of service (to provide a minimum guaranteed bandwidth to a flow or group of flows). However, it is precisely this performance isolation which introduces the "stealing" problem for scalable services: since the probing flow is isolated from the established flows, it cannot assess the potentially significant performance impact that it has on them. Consequently, while a new flow can determine whether or not its own quality-of-service objectives will be satisfied, it cannot determine

---

[1] We will discuss both flow- and class-based fair queueing. In class-based, the scheduling discipline inside each class is FCFS (First Come First Served), and between classes the discipline is fair queueing. In flow based each flow is considered as a class.

its impact on others. Thus, if admitted, the new flow can unknowingly steal resources from previously admitted flows and result in service violations.

This problem is not limited to fair queueing nor to per-flow schedulers. Consider a class-based strict priority scheduler in which a new flow wishes to probe for the available service at a mid-level priority. Ideally, the flow could indeed assess the capacity remaining from higher priority flows by probing at the desired service level. However, it would not be able to assess its impact on lower priority levels without also probing at lower levels.

Thus, the stealing problem arises from a lack of *observability* of multi-class networks, namely, that assessing one's own performance does not necessarily ensure that other flows are not adversely affected.

2.2 Problem Formulation

Within a framework of scalable services based on host probing, the key challenge is to simultaneously achieve (1) multiple traffic classes (differentiated services), (2) high utilization, and (3) a strong service model without stealing. To illustrate this challenge, consider the network of Figure 1 in which each link has capacity C. Further suppose the system supports two traffic classes $\mathcal{A}$ and $\mathcal{B}$ with different traffic characteristics and QoS requirements.

A key design axis which affects these design goals is whether or not the system allows resource sharing across classes. This in turn is controlled by the scheduler and whether or not it is work conserving.

Rigid Partitioning without Work Conservation One way to ensure both classes achieve their desired QoS constraints is via hard partitioning of system resources with no bandwidth borrowing across classes allowed. Such a system can be implemented with rate limiters, i.e., policing elements with the peak rate of class i limited to $\phi_i C$ with $\phi_1 + \phi_2 \leq 1$.

Observe that a hard partitioning system can support multiple traffic classes and does not incur stealing, thereby achieving the first and third goal above. However, notice that the system is non-work-conserving in the sense that it will reject flows even if sufficient capacity is available and consequently can under utilize system resources. For example, suppose path A-A' of Figure 1 has a large class-$\mathcal{A}$ demand and no class-$\mathcal{B}$ and vice versa on path B-B'. In this case, the system would be under-utilized as only half of the flows which the system could support would be admitted. In general, whenever the current bandwidth demands are not in line with the weights ϕ_i, the system will suffer from low utilization.

Inter-class Sharing with Work Conservation In contrast to the scenario above, consider a work-conserving system which allows one class to use excess capacity from other classes. In particular, consider a two-class fair queueing system (without rate limiters) with weights ϕ_1 and ϕ_2. With the same demand as in the example above, both A-A' and B-B' flows can fully utilize the capacity due to the soft partitioning of resources in the work conserving system. Thus, the first and second goals are achieved. However, as described in Section ??, such a system suffers from the stealing problem, as a new class-$\mathcal{B}$ flow on A-A' or a new class-$\mathcal{A}$ flow on B-B' will steal bandwidth from established flows.

Targeted Behavior The targeted behavior that we strive to achieve is to combine the advantages of the hard and soft partitioning systems and allow borrowing across classes to achieve high utilization, while eliminating resource stealing to provide a strong service model. Thus, in the example, if A-A' is fully utilized by class-$\mathcal{A}$ flows, class-$\mathcal{B}$ flows (and class-$\mathcal{A}$ flows) should be blocked until class-$\mathcal{A}$ flows depart. Below, we develop new probing schemes which seek to simultaneously achieve the above three design goals and achieve this targeted behavior. This service model is a greedy one, in which all flows which can be admitted are, provided that their and all other service requirements can be satisfied. This strategy does not incorporate blocking probability as a QoS parameter. It is possible to have targeting blocking probabilities, but it is beyond the scope of this paper. Throughout the paper we will only consider the admission controlled traffic. Best-effort would have a lower priority level so it would not interfere with the admission controlled one. Also, guaranteed-like service with strict QoS assurances would have a reserved bandwidth and a higher priority.

3   Epsilon Probing

In this section, we develop probing algorithms which overcome the stealing problem in fair queueing multi-class servers. The key technique is to infer the "state" of other classes with minimal overhead in terms of probing traffic or probing duration. Throughout, we consider a simplified bufferless fluid model as in [4], in which flows and probes transmit at constant rate and probing is "perfect" in the sense that probes correctly infer their loss rate as determined by the scheduling discipline (which defines how loss is distributed among classes) and the workload (which defines the extent of the loss in the system).

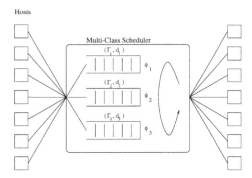

Fig. 2. Illustration of ε-Probing

Consider a class-based weighted fair queueing server with K classes, where class k has weight ϕ_k and target QoS parameters of loss rate Γ_k and delay bound d_k (for simplicity, we restrict the discussion to loss). According to the definition of WFQ, the bandwidth utilized by class k when all classes are backlogged is given by $U_k = \frac{\phi_k}{\sum_{i \in \mathcal{S}} \phi_i} C$ where $\mathcal{S}$ is the set of backlogged classes. Let the demanded bandwidth of class k be denoted by B_k. If $B_1 + \cdots + B_K > C$, then loss occurs in the system, and the loss rate of class k at that particular instant in time is given by

$$\gamma_k = (B_k - U_k)^+ / B_k. \tag{1}$$

With probing in a single level, a new class-k flow requesting bandwidth b_k is admitted if its measured loss rate is less than the class requirement, i.e., if $\gamma_k \leq \Gamma_k$. However, observe that even under congestion and arbitrarily high loss rates in other classes, the new flow would be admitted as long as

$$B_k + b_k \leq C\phi_k / (1 - \Gamma_k). \tag{2}$$

Thus, stealing across classes can occur as the probing flow fails to observe whether or not other classes' loss requirements are also satisfied. While simultaneously probing in all classes may seemingly solve the problem, it is not only unnecessary, but significantly damages the performance of the system: namely increased probing traffic forces the system to more quickly enter a thrashing regime in which excessive probing traffic causes flows to be mistakenly rejected, and in the limit, causes system collapse [4].

We propose ε-probing as a probing scheme designed to eliminate stealing in a minimally invasive way. With ε-probing, a new flow requesting bandwidth b_k *simultaneously* transmits a small bandwidth ε_i to each other class i. The motivating design principle is that the impact of the new flow on *all* classes must be observed, so that the new flow is only admitted if $\gamma_i \leq \Gamma_i$ is satisfied for all $i = 1, \cdots, K$. The admissible loss rate in each ε-probe (γ_i) is the same for all classes and globally agreed upon. In particular, addition of the new class-k flows can affect U_k for each class: the ε-probes ensure that the new U_k is sufficiently large to meet the required loss rate.

In the fluid model, ε_i can be arbitrarily small, whereas in the packet system, it must be sufficiently large to detect loss in the class. In the simulation experiments of Section 5, we consider $\varepsilon_i = 64$ kb/sec for a 45 Mb/sec link with flows transmitting at rates between 512 kb/sec and 2 Mb/sec.

Finally, we note that despite the utilization advantages of a work-conserving system, a network may still contain non-work conserving elements to achieve other objectives (e.g., to ensure that a minimum bandwidth is always available in each class, even if there is no current demand, cf. Figure 7). The goal of ε-probing is to enable inter-class resource sharing to the maximal extent allowed by the system architecture.

ε-probing is applicable to both class-based fair queueing and strict priority schedulers. In the latter type of scheduler, the ε-probes are required only in the priority levels lower than the level of the class for the flow that is requesting admission. In higher levels stealing cannot occur and no ε-probe is required. Therefore, the overhead due to ε-probes is lower in strict priority than in class-based fair queueing schedulers.

4 Theoretical Model

In this section we develop an analytical model based on continuous time Markov chains to study the problem of resource stealing in multi-class networks.

4.1 Preliminaries

In the model, each state identifies the number of currently admitted flows in each class such that with K classes, the Markov chain has K dimensions. The link capacity is C resource units and each flow of class k occupies b_k resource units.[2] We assume that new flows arrive to the system as a Poisson process with mean inter-arrival time $\frac{1}{\lambda_k}$, and that flow lifetimes are also exponentially distributed with mean $\frac{1}{\mu_k}$. Probing is considered instantaneous so that we do not consider the "thrashing" phenomenon (due to simultaneously probing flows) described in [4].

We define n_k to be the number of class k flows in the system such that the total amount of resources occupied by all flows in the system is given by $(\mathbf{b} \cdot \mathbf{n})$, where $\mathbf{b} = (b_1, \cdots, b_K)$, $\mathbf{n} = (n_1, \cdots, n_K)$, and $(\mathbf{b} \cdot \mathbf{n}) = \sum_{k=1}^{K} b_k n_k$. All classes require 0 loss probability so that we restrict our focus to multi-class stealing and do not address QoS differentiation with this model.

In the discussion below, we consider an example consisting of two traffic classes so that, for example, the transition from state $(0, 1)$ to $(1, 1)$ signifies admission of the first class 1 flow. The link capacity C is 6 resource units and the flow bandwidths are 1 and 2, i.e., $b_1 = 1$ and $b_2 = 2$.

4.2 Markov Models

Below, we model the different schedulers and probing algorithms which we compare in Section 4.3.

FIFO Here, we consider FIFO as a baseline scenario. In our working example with two classes, each flow will probe the network and will be admitted only if $b_1 n_1 + b_2 n_2 \leq C$, including the probing flow. Figure 3 depicts the corresponding state transition diagram.

In the general case, the state space is $S = \{\mathbf{n} \in I^K : (\mathbf{b} \cdot \mathbf{n}) \leq C\}$ where I is the set of non-negative integers and I^K is the set of all K-tuples of non-negative integers. The link utilization is given by $u = \frac{1}{C} \sum_{n \in S} (\mathbf{b} \cdot \mathbf{n}) \pi(\mathbf{n})$ where $\pi(\mathbf{n})$ is the probability of being in state $\mathbf{n}$, which can be computed using standard techniques [10]. Notice that the probability of stealing is zero, since probing flows are only admitted if there is available bandwidth in the link.

Flow-Based Fair Queueing As described in Section 2, larger bandwidth flows can have bandwidth stolen in flow-based fair queueing systems. Figure 4 depicts the system's state transition diagram for our working example. As shown, the state space includes all states in which the number of flows multiplied by the lowest bandwidth flow is lower or equal to the link capacity. For example, a transition from state $(4,1)$ to $(5,1)$ is possible because 1 bandwidth unit is guaranteed for each flow, and this is sufficient for class 1 flows. Alternatively, a transition from state $(5,0)$ to $(5,1)$ is not possible because class 2 flows require 2 bandwidth units. Thus, the stealing states represent admissions of

---

[2] In general the rates of flows that belong to the same class may be different; however this assumption is required for the Markov chain formulation.

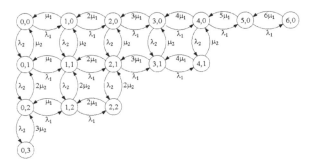

Fig. 3. FIFO Transition Diagram

low bandwidth flows when the system is at full capacity and high bandwidth flows are forced into a loss state. In the state transition diagrams, stealing states are represented by a crossed-out state.

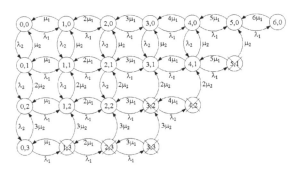

Fig. 4. Flow-Based Fair Queueing Transition Diagram

Suppose $\{n_1, n_2, ..., n_k, ..., n_K\}$ is the current admitted set of flows. Then a new class-k flow is admissible if $\{b_1 n_1 + b_2 n_2 + ... + b_k([n_k + 1] + n_{k+1} + ... + n_K) \leq C\}$ with $\{b_1 < b_2 < ... < b_k < ... < b_K\}$. There are two cases: if the total demand satisfies $\{b_1 n_1 + b_2 n_2 + ... + b_k(n_k + 1) + b_{k+1} n_{k+1} + ... + b_K n_K \leq C\}$, then all flows are correctly admissible; however if $\{b_1 n_1 + b_2 n_2 + ... + b_k(n_k + 1) + b_{k+1} n_{k+1} + ... + b_K n_K > C\}$, then stealing occurs. Since the scheduler fairly allocates bandwidth to all flows, flows with bandwidth higher than the bandwidth of the flow requesting admission are forced into a loss state.

For the case of two types of flows, the state space including stealing states is given by

$$S_{FQ} = \{\mathbf{n} \in I^2 : ((b_1 n_1 + b_1 n_2 \leq C) \wedge (b_2 n_2 \leq C) \wedge (n_1 \neq 0))$$
$$\bigvee ((b_2 n_2 \leq C) \wedge (n_1 = 0))\}$$

For example, in our transition diagram, state (2,3) (with $n_1 = 2 \neq 0$) is a possible state (a stealing state) because $b_1 n_1 + b_1 n_2 = 5 \leq C$ and $b_2 n_2 = 6 \leq C$. State (2,4) is not a possible state since $b_1 n_1 + b_1 n_2 = 6 \leq C$ but $b_2 n_2 = 8 > C$. This example explains the need of the second inequality ($b_2 n_2 \leq C$). With $n_1 = 0$, the state with the maximum number of flows is (0,3) because $b_2 n_2 = 6 \leq C$.

To generalize the state space to K types of flows, let k^* be the smallest k such that $n_{k^*} \neq 0$, then

$$S_{FQ} = \{\mathbf{n} \in I^K : (b_{k^*} n_{k^*} + b_{k^*} n_{k^*+1} + ... + b_{k^*} n_K \leq C)$$
$$\wedge (b_{k^*+1} n_{k^*+1} + ... + b_{k^*+1} n_K \leq C) \wedge ... \wedge (b_K n_K \leq C)\}$$

The mean utilization is then

$$u = \frac{1}{C} \sum_{\mathbf{n} \in S_{FQ}} (\mathbf{b} \circ \mathbf{n}) \pi(\mathbf{n}) \tag{3}$$

where

$$(\mathbf{b} \circ \mathbf{n}) = \begin{cases} (\mathbf{b} \cdot \mathbf{n}) & \text{if } (\mathbf{b} \cdot \mathbf{n}) \leq C \\ C & \text{otherwise} \end{cases} \tag{4}$$

The probability of stealing is

$$p_{st}^{FQ} = 1 - \frac{\sum_{\mathbf{n} \in S}(\mathbf{b} \cdot \mathbf{n}) \pi(\mathbf{n})}{\sum_{\mathbf{n} \in S_{FQ}}(\mathbf{b} \circ \mathbf{n}) \pi(\mathbf{n})} \tag{5}$$

computed as the percentage of bandwidth guaranteed to flows which is stolen by other flows.

Rate Limiters With rate limiters class k flows are only allowed to use a maximum of C_k bandwidth units. Here, we consider rate limiters of $C_1 = 2$ units and $C_2 = 4$ units respectively. Figure 5 depicts the corresponding state transition diagram. Given the functionality of the rate limiters, the state space is reduced to

$$S_{RL} = \{\mathbf{n} \in I^K : b_k n_k \leq C_k, k = 1, 2, ...K\}. \tag{6}$$

With this elimination of various high-utilization states, the overall system utilization in the general case of K classes, given by $u = \frac{1}{C} \sum_{\mathbf{n} \in S_{RL}} (\mathbf{b} \cdot \mathbf{n}) \pi(\mathbf{n})$, is then also reduced as compared to work-conserving systems. Clearly, the extent of this utilization reduction is a function of the system load, b_k and λ_k, and C_k. If they are properly tuned, the penalty will be minimal, whereas if they become unbalanced due to load fluctuations, the system performance will suffer.

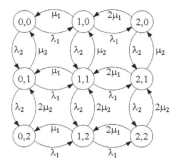

Fig. 5. Rate Limiters Transition Diagram

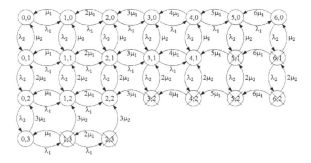

Fig. 6. Class-Based Fair Queueing Transition Diagram

Class-Based Systems In class-based fair queueing without rate limiters, resource borrowing across classes is allowed. However, as described in Section 2, stealing occurs as new flows in classes with reserved rate less than $\phi_i C$ request admission. Thus, with 1-level probing, a class k flow with bandwidth b_k will be admitted if, including the probing flow, one of two conditions occurs: $b_k n_k \leq C$ when $(\mathbf{b} \cdot \mathbf{n}) \leq C$, or $b_k n_k \leq C\phi_k$ when $(\mathbf{b} \cdot \mathbf{n}) > C$. In the state transition diagram of Figure 6, $\phi_1 = 1/3$ and $\phi_2 = 2/3$. As an example, consider the transition from (5,0) to (5,1). In the state space, the first set of inequalities is not satisfied because $b_2 n_2 = 2 \leq C$, but $b_1 n_1 + b_2 n_2 = 7 > C$. However the second set of inequalities is satisfied since $b_2 n_2 = 2 \leq C\phi_2$ and $b_1 n_1 + b_2 n_2 = 7 > C$. Therefore this transition is possible. Similarly, the transition from (4,1) to (5,1) is not allowed because neither set of inequalities are satisfied.

The state space has two parts. The first one is equal to the one of FIFO and allows borrowing between classes as long as $(\mathbf{b} \cdot \mathbf{n}) \leq C$. Thus S_{CBQ-1} includes $\{\mathbf{n} \in I^K : (\mathbf{b} \cdot \mathbf{n}) \leq C\}$. Suppose we are in one of the edge FIFO states. Due to the borrowing between classes, for some classes, $b_k n_k \leq C\phi_k$, which we will call the underload classes (UL), and for others, $b_k n_k > C\phi_k$, which we will call the overload ones (OL). Suppose there is currently no stealing in the system. New probing flows from UL classes can be admitted until $b_{ul}n_{ul} = C\phi_{ul}$, with $ul \in UL$, irrespective of the value $(\mathbf{b} \cdot \mathbf{n})$. Thus departing from an edge FIFO state, new states can be created such that $\{\mathbf{n} \in I^K, ul \in UL : b_{ul}n_{ul} \leq C\phi_{ul}\}$. The overall state space S_{CBQ-1} is then the union of the FIFO state space and the one constructed with these new states.

The utilization is $u = \frac{1}{C}\sum_{n \in S_{CBQ-1}}(\mathbf{b} \circ \mathbf{n})\pi(\mathbf{n})$ and the probability of stealing is $p_{st}^{CBQ-1} = 1 - \dfrac{\sum_{n \in S}(\mathbf{b} \cdot \mathbf{n})\pi(\mathbf{n})}{\sum_{n \in S_{CBQ-1}}(\mathbf{b} \circ \mathbf{n})\pi(\mathbf{n})}$. Note that $(\mathbf{b} \circ \mathbf{n})$ has the same definition as in flow-based fair queueing. Comparing the class-based and flow-based fair queueing, observe that the class-based system has a larger number of stealing states than the flow-based system. For example, transition from $(6,0)$ to stealing state $(6,1)$ is possible in the class-based system whereas state $(6,1)$ does not exist in flow-based fair queueing. The reason for this is that the flow based fair queueing system blocks this 7th flow as it forces the system into loss. However, the class based system admits this flow since the requested rate of 2 bandwidth units is indeed available in class 2, even though it forces class 1 into a stealing situation. Regardless, even though there are more stealing states in the class-based system, the overall stealing probability is lower (as indicated by numerical examples and simulations below) because the fraction of time spent in such stealing states is lower in the class based system, so the bandwidth stolen will also become lower.

In contrast to the above 1-level probing, with ε-probing, all classes are probed to ensure that no stealing occurs. Here, the admissible states in this scheduler are the same as in FIFO, so the state space is the same, as well as the utilization. (We note that the utilization in the real system with nonzero probe durations is not the same however.)

4.3 Numerical Examples

Here we numerically solve the Markov models for each system described above. With the solution to the state probabilities, we compute the utilization and probability of stealing using the expressions derived above. We consider the scenario of previous sections with a link capacity of 6 bandwidth units. The weights of classes 1 and 2 are 1/3 and 2/3, respectively. The bandwidths of class-1 and class-2 flows are 1 unit and 2 units, respectively. Class 1's mean flow arrival rate is 8 requests per second while class 2's is 5. The mean life time of class 1 flows is 2/3 time units while class 2's is 1/4 time units.

Table 1. Utilization and Stealing Probability

Probing Scheme	Utilization	Stealing
ε-probing/FIFO	0.789	0
Flow-FQ	0.792	0.140
Rate Limiters	0.702	0
Class-FQ (1-level)	0.789	$8.28 \cdot 10^{-4}$

We make two observations about numerical examples presented in Table 1. First, notice that ε-probing and rate limiters both have the effect of *eliminating* resource stealing.

However, ε-probing does so at higher utilizations. For example, ε-probing achieves 79% utilization as compared to 70% under rate limiters. Moreover, the difference between these two utilizations is determined by the relative class demands, which in this context are the relative flow arrival rates.

Second, note the stealing probabilities for flow- and class-based fair queueing (without ε-probing). Here, the stolen bandwidth is 0.140 for flow-based and $8.28 \cdot 10^{-4}$ for class-based. As evident from the model, CBQ incurs far less stealing than flow-based fair queueing. In simulation experiments, this relative difference still exists, however the probability of stealing for CBQ is far greater than it is in these numerical examples. The reason for this is even evident from the Markov model. In the CBQ system, stealing occurs as classes first demand bandwidths below and then later above $\phi_i C$ as defined in the state space of CBQ-1 level. It is precisely such system dynamics (changing resource demands) which are well captured by simulations but less via the Markov model. Thus, while the Markov model is useful to explore the origins and structure of multi-class resource stealing, we now turn to simulation experiments to quantitatively explore stealing under more realistic scenarios.

5   Experimental Studies

In this section, we present a set of simulation experiments with the goal of exploring the architectural design space as outlined in Section 2, evaluating ε-Probing presented in Section 3, and validating the conclusions of the analytical model of Section 4 in a more general setting.

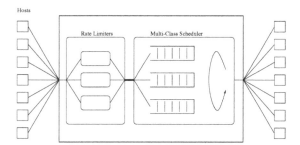

Fig. 7. Simulation Scenario

The basic scenario is illustrated in Figure 7. It consists of a large number of hosts interconnected via a 45 Mb/sec multi-class router. For some experiments, the router contains rate limiters which drop all of a class' packets exceeding the pre-specified rate. We consider several multi-class schedulers including CBQ, flow-based fair queueing, and rate limiters. We also consider FIFO for baseline comparisons. New flows arrive to the system with independent and exponential inter-arrival times through a Poisson process and probe for a constant time of 2 seconds. Flows send probes at their desired admission rate except for ε-probes, which are transmitted at 64 kb/sec. New flows are admitted if the loss rate of the probes is below the class' threshold.

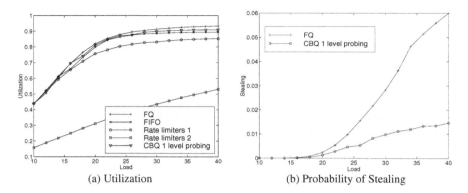

(a) Utilization   (b) Probability of Stealing

Fig. 8. Utilization and Stealing vs. Load for Various Node Architectures

Utilization and Stealing In the first set of experiments depicted in Figure 8, we investigate the challenge of simultaneously achieving high utilization and a strong service model without stealing. In this scenario, there are three traffic classes with bandwidth requirements of 512 kb/sec, 1 Mb/sec, and 2 Mb/sec respectively. We consider three variants of the system depicted in Figure 7. The flow-based fair queueing curve, (labeled "FQ") represents the case in which the scheduler allocates bandwidth fairly among *flows*, i.e., the Nth probing flow measures no loss if its rate is less than C/N. In contrast, the curves labeled "Rate Limiters 1", "Rate Limiters 2" and "CBQ 1 level probing" represent class-based scheduling. In the former case, each class is rate limited to $C/3$ so that all loss occurs in the rate limiters and none in the scheduler (cf. Figure 7). In the latter case, the classes are *not* rate limited and the scheduler performs CBQ with each class' weight set to 1/3. In all cases, probes are transmitted at the flow's desired rate and ε-probing of Section 3 is *not* performed. The x-axis, labeled load, refers to the resource demand given by $\frac{\lambda}{\mu}$.

We make the following observations about the figure. First, comparing the results with rate limiters and CBQ, Figure 8(a), indicates that CBQ achieves higher utilization than rate limiters due to the latter's non-work-conserving nature. That is, the rate limiters prevent flows from being admitted in a particular class whenever the class' total reserved rate is $C/3$, even if capacity is available in other classes. However, from Figure 8(b), it is clear that the higher utilization of CBQ is achieved at a significant cost: namely, CBQ incurs stealing in which up to 1.5% of the bandwidth (in the range shown) guaranteed to flows is stolen by flows in other classes. Hence the experiments illustrate that neither technique simultaneously achieves high resource utilization and a strong service model. Moreover, as the resources demanded by a class become mismatched with the pre-allocated weights, the performance penalty of rate limiters is further increased. That is, if the demanded bandwidth were temporarily 80/10/10 rather than 33/33/33, as is the case for the curve labeled "Rate Limiters 2" at a load of 40, then the rate limiters would restrict the system utilization to at most 53% representing a 33/10/10 allocation.

Second, observe the effects of flow *aggregation* on system performance. In particular, flow-based fair queueing achieves higher utilization and has higher stealing than CBQ. With no aggregation and flow-based queueing, smaller bandwidth flows can always steal bandwidth from higher bandwidth flows resulting in both higher utilization since more flows are admitted (in particular low bandwidth flows) as well as more flows having bandwidth stolen. In contrast, with class based fair queueing, stealing only occurs when a *class* exceeds its 1/3 allocation (rather than a flow exceeding its $1/N$ allocation) and a flow from another class requests admission, an event that occurs with less frequency.

ε-Probing Figure 9(a) depicts utilization vs. load for three cases: CBQ with one-level probing, CBQ with ε-probing, and rate limiters. Observe that compared to one-level probing, ε-probing incurs a utilization penalty. There are two contributing factors. First, the ε-probes themselves cause an additional traffic load on the system despite their small bandwidth requirement. Second, by blocking flows which will result in stealing, there are fewer flows in the system on average with ε-probing than with one class probing. Regardless, this moderate reduction in utilization has the advantage of eliminating stealing completely. Moreover, the utilization penalty of *rate limiters* can be arbitrarily high depending on the mismatch between the demanded resources and the established limits. In contrast, the performance of ε-probing does not rely on proper tuning of rate limiters, but rather the overhead of ε-probing simply increases linearly with the number of classes.

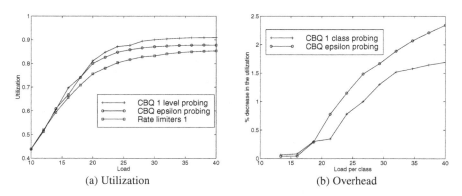

(a) Utilization   (b) Overhead

Fig. 9. Utilization and Overhead of ε-Probing

The utilization reduction solely due to probing is further illustrated in Figure 9(b). Observe that the overhead incurred in ε-probing is necessarily higher larger than that incurred by probing in only one class, as ε-probing must also ensure that other traffic classes are not in overload. However, due to the limited bandwidth required to probe in other classes, ε-probing incurs moderate utilization reductions typically below 2.5%.

Therefore, ε-probing is able to simultaneously eliminate stealing, provide multiple service levels, and enable full statistical sharing across classes.

6   Conclusions

Placing admission control functions at the network's endpoints has been proposed as a mechanism for achieving per-flow quality-of-service in a scalable way. However, if routers perform class differentiation such as multiple priority queues, the system becomes less observable to probing flows, precisely because of the performance isolation provided by the service discipline. In this paper, we have studied the resource stealing problem that arises in such multi-class networks and developed a simple probing scheme termed ε-probing which attains the high utilization of work-conserving systems while preventing stealing as in non-work-conserving systems with hard class-based rate limits. We introduced a Markov model that illustrated the design space of key network differentiation mechanisms, such as class- and flow-based weighted fair queueing and rate limiters. The model showed the different ways that stealing is manifested in the different configurations and provided a tool for formal comparison of diverse systems. Finally, our simulation experiments explored the design space under a broader set of scenarios. We quantified the severity of bandwidth stealing and found that ε-probing eliminates stealing with a modest utilization penalty required to observe the impact of a new flow on other traffic classes.

References

1. G. Bianchi, A. Capone, and C. Petrioli. Throughput analysis of end-to-end measurement-based admission control in IP. In *Proceedings of IEEE INFOCOM 2000*, Tel Aviv, Israel, March 2000.
2. S. Blake et al. An architecture for differentiated services, 1998. Internet RFC 2475.
3. L. Breslau, S. Jamin, and S. Shenker. Comments on the performance of measurement-based admission control algorithms. In *Proceedings of IEEE INFOCOM 2000*, Tel Aviv, Israel, March 2000.
4. L. Breslau, E. Knightly, S. Shenker, I. Stoica, and H. Zhang. Endpoint admission control: Architectural issues and performance. In *Proceedings of ACM SIGCOMM 2000*, Stockholm, Sweden, August 2000.
5. C. Cetinkaya and E. Knightly. Scalable services via egress admission control. In *Proceedings of IEEE INFOCOM 2000*, Tel Aviv, Israel, March 2000.
6. V. Elek, G. Karlsson, and R. Ronngren. Admission control based on end-to-end measurements. In *Proceedings of IEEE INFOCOM 2000*, Tel Aviv, Israel, March 2000.
7. R. Gibbens and F. Kelly. Distributed connection acceptance control for a connectionless network. In *Proceedings of ITC '99*, Edinburgh, UK, June 1999.
8. E. Knightly and N. Shroff. Admission control for statistical QoS: Theory and practice. *IEEE Network*, 13(2):20–29, March 1999.
9. K. Nichols, V. Jacobson, and L. Zhang. Two-bit differentiated services architecture for the Internet, 1999. Internet RFC 2638.
10. K. Ross. *Introduction to Probability Models*. Academic Press, 1997.
11. I. Stoica, S. Shenker, and H. Zhang. Core-Stateless Fair Queueing: A scalable architecture to approximate fair bandwidth allocations in high speed networks. In *Proceedings of ACM SIGCOMM '98*, Vancouver, British Columbia, September 1998.

12. I. Stoica and H. Zhang. Providing guaranteed services without per flow management. In *Proceedings of ACM SIGCOMM '99*, Cambridge, MA, August 1999.
13. J. Wroclawski. Specification of the controlled-load network element service, 1997. Internet RFC 2211.
14. L. Zhang, S. Deering, D. Estrin, S. Shenker, and D. Zappala. RSVP: A New Resource ReSerVation Protocol. *IEEE Network*, 7(5):8–18, September 1993.
15. Z. Zhang, Z. Duan, L. Gao, and Y. Hou. Decoupling QoS control from core routers: a novel bandwidth broker architecture for scalable support of guaranteed services. In *Proceedings of ACM SIGCOMM 2000*, Stockholm, Sweden, August 2000.

Pricing-Based Control of Large Networks

Xiaojun Lin and Ness B. Shroff

School of Electrical and Computer Engineering
Purdue University, West Lafayette, IN 47906, USA
{linx, shroff}@ecn.purdue.edu

Abstract. In this work we study pricing as a mechanism to control large networks. Our model is based on revenue maximization in a general loss network with Poisson arrivals and arbitrary holding time distributions. In dynamic pricing schemes, the network provider can charge different prices to the user according to the current utilization level of the network. We show that, when the system becomes large, the performance of an appropriately chosen static pricing scheme, whose price is independent of the current network utilization, will approach that of the optimal dynamic pricing scheme. Further, we show that under certain conditions, this static price is independent of the route that the connections take. This result has the important implication that distance-independent pricing that is prevalent in current domestic telephone networks in the U.S. may in fact be appropriate not only from a simplicity, but also a performance point of view. We also show that in large systems prior knowledge of the connection holding time is not important from a network revenue point of view.

1 Introduction

In the last few years there has been significant interest in using pricing as a mechanism to control communication networks. The network uses the current price of a resource as a feedback signal to coerce the users into modifying their actions (e.g. changing the rate or route). Price provides a good control signal because it carries monetary incentives.

Past works in the literature differ in their schemes to compute the price. Some of the proposed schemes use different forms of auctions, e.g., "Smart Market" [9] or "Progressive Second Price Auction" [7] to reach the right price. Other schemes use differential equations at the resource and at the users to converge to such prices [4]. A common theme behind this category of research is to have the resource calculate price based on the *instantaneous* load and available capacity. If the current condition changes, a new price is calculated. The users are expected to act on up-to-date price information to achieve the desired design objectives. Feedback delays to the user may introduce undesirable effects on the performance of the system, such as instability or slow speed of convergence [4]. In practice, this feedback delay may be difficult to control because calculating up-to-date price information continuously consumes a large amount of processing power at

S. Palazzo (Ed.): IWDC 2001, LNCS 2170, pp. 212–231, 2001.

the resource/node. Also, there is a substantial communication overhead when users try to obtain these up-to-date prices.

Pricing over multiple resources also poses additional problems. Usually the price is calculated over a given route by summing over the price of all the resources along that route [4, 8, 14]. The user has to act on the sum. This kind of *route-specific* or *distance-specific* price may introduce substantial signaling overhead in real networks.

Interestingly, in our daily lives, we observe quite the opposite phenomenon. For example, we do not see the prices of products we purchase everyday fluctuate substantially, but in fact they change very slowly. Another example of *distance-neutral* pricing is the domestic long distance telephone service in the US. Most long distance companies offer flat rate pricing to users calling anywhere within the continental US.

These observations have motivated a number of recent works. In [12], after a critique on the *optimality-pricing* approach, the authors have proposed two approximations of congestion pricing: the first approximation was to replace the instantaneous congestion conditions[1] by the *expected* or *average* congestion conditions, similar to time-of-day pricing. The second approximation was to replace the cost of the actual path the flow would traverse through the network with the cost of the *expected* path; where the charge depends only on the source and destination(s) of the flow and not on the particular route taken by the flow. However, in this work, there is no analysis of the performance implication of such approximations.

In [11], the authors investigate this problem in the case of a single resource with Poisson arrivals and exponential service times. In particular, the authors study the expected utility and revenue under both dynamic pricing and static pricing schemes in a *dynamic network*. By a dynamic network, we mean that in their system, call arrivals and departures are taken into account. This is different from the work in [4, 6, 8, 14] where the authors view congestion control as a distributed asynchronous computation to maximize the aggregate source utility. In these works, the optimization is done over a snapshot in time, i.e., it does not take into account the dynamic nature of the network. In contrast, in [11], the authors model the dynamic nature of the system by assuming that calls arrive according to a Poisson process and stay in the system for an exponentially distributed time. The authors study the expected utility and revenue under both dynamic pricing and static pricing schemes. It is shown that when the capacity is large, using a static pricing scheme suffices, in the sense that the utility (or revenue) under an appropriately chosen static price converges asymptotically to that of the optimal dynamic scheme. The implication of this work is that we can now base pricing decisions only on the *average* load rather than on the *instantaneous* load, thus reducing both processing and communication overhead.

This paper extends the results in [11] in two directions:

---

[1] Note that the congestion conditions in [12] can be obtained from the instantaneous load and available capacity in the system. Hence, these terms can be viewed interchangeably.

1) We show that the same type of invariance results hold for general loss networks (i.e., the performance of the optimal static pricing scheme asymptotically approaches the performance of the dynamic scheme). The results hold even if the service time distribution is general. We show that the right static price depends on the service time distribution only by its mean. These extensions have two implications: Firstly, while the assumption of Poisson arrivals for calls or flows in the network is usually considered reasonable, the assumption of exponential holding time distribution is not. For example, much of the traffic generated on the Internet is expected to occur from large file transfers which do not conform to exponential modeling. By weakening the exponential service time assumption we can extend our results to more realistic systems. Secondly, in the general network case, we can show that in our revenue-maximization problem, under certain conditions, the static price only depends on the price elasticity of the user, and not on the specific route or distance. This indicates that the flat pricing scheme used in the domestic long distance service in the US may, in many cases, be good enough.

2) Thus far, when we refer to dynamic pricing schemes, we mean schemes that take into account the current congestion information. We now consider an even broader set of dynamic pricing schemes and show that the performance of the static pricing scheme suffices for large systems. In this broader set of dynamic pricing schemes, the network has prior knowledge of the individual service time of the connections when they arrive. The question is whether we can gain additional advantage by pricing the resource based on this additional information. Our analysis shows that when the system grows large, this additional information does not result in significant gain. We show that the individual service time is inessential, and a static price based on the average service time suffices.

We believe that these results are important in understanding how to use pricing as a mechanism for controlling real networks. The possibility of using static and/or route-independent prices will give rise to more efficient and realistic algorithms. Also, "pricing" here can be interpreted as a general signal which is only loosely related to actual pricing. For example, pricing models have been used as a mechanism to understand resource allocation and congestion control. Therefore the results we present here will help us in better understanding how to control large networks.

1.1   Related Work

The ideas of upper bounding the performance of the optimal dynamic policy in a general loss network and showing that fixed/static policy approaches the upper bound asymptotically have been reported in the past, for example, [10, 5], and the reference therein. In their work, the objective to be optimized is a linear function of the number of users in the system, i.e., the utility value. This objective function maximizes the *utilization* of the network, but does not consider the revenue generated, hence there is no notion of the price of a resource. Maximizing the utility in a system corresponds to a linear optimization problem. However, revenue maximization is more difficult to evaluate because it is a non-linear optimization problem. Moreover, users can easily take advantage of the system by providing an incorrect utility value when price is not an issue. Pricing is a mechanism to coerce users into using the amount of resources that they need (or can afford!). Recently, in [11], Paschalidis and Tsitsiklis have investigated the problem of revenue maximization. Our work extends the result of [11] to general loss networks and general service time distributions. Further, our proof of the stationarity and ergodicity of such types of system with feedback control is new, and of value in its own right. Our treatment of pricing schemes based on the call duration sheds insight on the optimal dynamic policy.

The rest of the paper consists of three parts. In Sect. 2, we investigate the case of a general loss network with Poisson call arrivals and general service time distribution. In Sect. 3, we study the implication of individual service time on the price and in Sect. 4, we conclude.

2   General Loss Network, General Service Time

2.1   Model

Consider an abstract network providing service to multiple classes of users. There are L links in the network. Each link $l = \{1, ..., L\}$ has capacity R^l. There are I classes of users. For users of each class i, there is a route through the network. The routes are characterized by a matrix $\{C_i^l, i = 1, ..., I, l = 1, ..., L\}$, where $C_i^l = 1$ if the route of class i traverses link l, $C_i^l = 0$ otherwise. Each class i connection consumes bandwidth r_i. A call is rejected and lost if any of the links it traverses does not have enough capacity to accommodate it, otherwise it may be admitted to the network depending on the network policy.

Calls of class i arrive to the network according to a Poisson process with rate $\lambda_i(u_i)$, which is a function of the price u_i charged to users of class i. Here u_i is defined as the price per unit time of connection. We assume that $\lambda_i(u_i)$ is a non-increasing function of u_i. Therefore $\lambda_i(u_i)$ represents the *price-elasticity* of class i. Once admitted, the call will hold the resources on all links it traverses until it finishes service. The service time distribution is general with mean $1/\mu_i$.

The network provider can charge different prices for different classes of users. A *dynamic* price is one where charges are based on the current utilization of the resource, for example, how many calls of each class are already in service, and how long they have been served, etc. On the other hand, a *static* price is one which *only* depends on the class of the user and is indifferent to the current utilization of the resource.

2.2   Stability of the Model

Our first result is regarding the stability of such a system. Note that our model is very similar to a network of M/G/N/N queues. However, here we also have "feedback" introduced by the price u_i, which makes the model somewhat more complex. The arrival rate is changing over time. It is not intuitive to even define stationarity and ergodicity for the input! However, as we will see next the system will be stationary and ergodic under very general conditions.

Proposition 1. *Assume that the arrival rates $\lambda_i(u)$ are bounded above by some constant λ_0 for all classes i, the service times are i.i.d. with finite mean and independent of the arrivals. If the price is only dependent on the current state of the system, then any stochastic process that is only a function of the system state is asymptotically stationary and the stationary version is ergodic.*

Proof. To show this, let us first look at the case of a single resource with users coming from a single class i. To develop the result, we need to take a different but equivalent view of the original model. In the original system, the arrival rate is a function of the current price. In the new but equivalent system, the arrival rate is constant but is thinned by a probability as a function of the price. Specifically, in the new model, the arrivals are Poisson with constant rate λ_0. Each arrival now carries a value v that is independently distributed, with distribution function $P\{v \geq u\} = \lambda_i(u)/\lambda_0$. The value v of each arrival is independent of the arrival process and service times. If $v < u$, where u is the current price at the time of arrival, the call will not enter the system. We can see that with this construction, at each time instant, the arrivals are bifurcated with probability of success equal to $P\{v \geq u\} = \lambda_i(u)/\lambda_0$. Therefore the resultant arrivals (after thinning) in the new model are also Poisson with rate $\lambda_0 P\{v \geq u\} = \lambda_i(u)$. Thus the model is equivalent to the original model.

To show stationarity and ergodicity, we need to construct a so called "regenerative event" [1]. Let $\{\tau_n^e, \tau_n^s, v_n\}$, $-\infty < n < \infty$, be the n-th arrival's interarrival time, service time, and value, respectively (note this is the arrival of the Poisson process before bifurcation.) Let

$$Q_n = 1_{\{\tau_{n-1}^s < \tau_n^e\}} + 1_{\{\tau_{n-2}^s < \tau_n^e + \tau_{n-1}^e\}} + \ldots + 1_{\{\tau_{n-k}^s < \sum_{j=0}^{k-1} \tau_{n-j}^e\}} + \ldots$$

Also let $A_n = \{Q_n = 0\}$. Then A_n can be interpreted as the event that "all potential arrivals before time n have left the system." The event A_n is regenerative, that is, if event A_n occurs, then after time n, the system will evolve independently from the past (this is true because we assume that the price is only dependent on the current state of the network). The events A_n are stationary, i.e., if we define T as the shift operator, $T\{\{\tau_{n_i}^e, \tau_{n_i}^n, v_{n_i}\} \in B_i, i = 1\ldots k\} = \{\{\tau_{n_i+1}^e, \tau_{n_i+1}^n, v_{n_i+1}\} \in B_i, i = 1\ldots k\}$, then $A_n = T^n A_0$, and $P\{A_n\} = P\{A_0\}$. Note here we have eliminated the dependence on both v_n and the price u. Now to proceed with the proof, we need the following lemma.

Lemma 1. *Let the sequence of service times τ^s be i.i.d., and $E[\tau^s] < \infty$, then $P\{A_0\} > 0$.*

Proof. See Appendix.

Now by Borovkov's Ergodic Theorem, [1], the distribution of the state of the system converges as $n \to \infty$ to the distribution of the stationary process. Ergodicity follows from the lemma below.

Lemma 2. *The regenerative event A_n is positive recurrent, i.e., let $T_1 = \inf\{X_n \in A_n\}$. Then $E\{T_1|X_0 \in A_0\} < \infty$, where X_n is the state of the system at time n.*

Proof. See Appendix.

Since the regenerative event is positive recurrent, the state of the system is both stationary and ergodic, i.e., any random process that depends only on the state of the system is both stationary and ergodic [13].

For the case of multiple classes and multiple links, assume there are I classes. We can construct the equivalent system in the following way: we first construct Poisson arrivals with rate $I\lambda_0$. Each of these arrivals is assigned to class i with probability $1/I$, and each of these assignments is independent of each other. The service time is then generated according to the service time distribution of class i. Each class i arrival carries a value v that is independently distributed, with distribution function $P_i\{v \geq u_i\} = \lambda_i(u_i)/\lambda_0$. The value v of each arrival is independent of the arrival process and service times. If $v < u_i$ where u_i is the current price for class i at the time of the arrival, the call will not enter the system. Following the same idea as in the first paragraph of the proof, it is easy to show that such a constructed system is equivalent to the original system.

The initial Poisson arrivals with rate $I\lambda_0$ can be interpreted as "all potential arrivals from all classes." Let $\{\tau_n^e, \tau_n^s\}$ be the n-th arrival's interarrival time and service time respectively. It then follows that the sequence of service times τ_n^s is again i.i.d. with finite mean, and it is independent of the arrivals. Hence, we can construct the event A_0 as before, which is now the event that "all potential arrivals from all classes have left the system before time n." Again this event is the "regenerative event" for the system, and we can show that $P\{A_0\} > 0$, and A_0 is positive recurrent. Therefore, the system is asymptotically stationary and the stationary version is ergodic.   □

2.3   Dynamic Price, Static Price, and Upper Bound

In this section, we consider dynamic pricing schemes that are based on the current occupancy of the network resources. Let $n = \{n_i, i = 1, ..., I\}$ represent the state of the network, where n_i is the number of users from class i that are being served in the network. Let Ω denote the set of possible states, then $\Omega = \{n : \sum_i n_i r_i C_i^l \leq R^l$ for all $l\}$. Let $n(t)$ denote the state at time t, then the dynamic price for class i can be written as $u_i(t) = g_i(n(t))$, where g_i is a function from Ω to the set of real numbers R. Let $g = \{g_i, i = 1, ..., I\}$.

The expected revenue achieved by any dynamic pricing scheme is given by

$$\lim_{T \to \infty} \sum_{i=1}^{I} \frac{1}{T} E\left[\int_0^T \lambda_i(u_i(t))u_i(t)\frac{1}{\mu_i}dt\right].$$

From stationarity and ergodicity established in the last section, we have

$$\lim_{T \to \infty} \sum_{i=1}^{I} \frac{1}{T} E\left[\int_0^T \lambda_i(u_i(t))u_i(t)\frac{1}{\mu_i}dt\right] = \sum_{i=1}^{I} E\left[\lambda_i(u_i(t))u_i(t)\frac{1}{\mu_i}\right],$$

where the right hand side is independent of t (because of stationarity).

Therefore, the optimal dynamic policy is

$$J^* = \max_g \sum_{i=1}^{I} E\left[\lambda_i(u_i(t))u_i(t)\frac{1}{\mu_i}\right].$$

On the other hand, for static pricing schemes, the expected revenue per unit time is:

$$J_0 = \sum_{i=1}^{I} \lambda_i(u_i)u_i\frac{1}{\mu_i}(1 - P_{loss,i}[u]),$$

where $P_{loss,i}[u]$ is the probability of loss for class i when the price vector is $u = [u_1, ..., u_I]$. Therefore the optimal static policy is

$$J_s = \max_u \sum_{i=1}^{I} \lambda_i(u_i)u_i\frac{1}{\mu_i}(1 - P_{loss,i}[u]).$$

By definition $J_s \le J^*$.

Here we show that the upper bound in [11] is also an upper bound for our case. For convenience, we write u_i as a function of λ_i. Let $F_i(\lambda_i) = \lambda_i u_i(\lambda_i)\frac{1}{\mu_i}$. Let J_{ub} be the optimal value of the following nonlinear programming problem:

$$\max_{\lambda_i} \quad \sum_i F_i(\lambda_i) \tag{1}$$

$$\text{subject to} \quad \lambda_i = \mu_i n_i \quad \text{for all } i \tag{2}$$

$$\sum_i n_i r_i C_i^l \le R^l \quad \text{for all } l.$$

Proposition 2. *If the functions F_i are concave, then $J^* \le J_{ub}$.*

Proof. We follow the same method as in the proof of Theorem 6 in [11]. Consider an optimal dynamic pricing policy. Let $n_i(t)$ be the number of calls of class i in the system at time t. View $\lambda_i(t)$ and $n_i(t)$ as random variables. From Little's Law, we have

$$E[n_i(t)] = E[\lambda_i(t)]\frac{1}{\mu_i}.$$

At any time t, $\sum_i n_i(t)r_i C_i^l \le R^l$ for all l. Therefore

$$\sum_i E[n_i(t)r_i C_i^l] \le E\left[\sum_i n_i(t)r_i C_i^l\right] \le R^l \quad \text{for all } l. \tag{3}$$

Therefore $\lambda_i = E[\lambda_i(t)]$ and $n_i = E[n_i(t)]$ satisfy the constraint of J_{ub}.

Using the concavity of F and Jensen's inequality, we have

$$J_{ub} \ge \sum_i F_i(E[\lambda_i(t)]) \ge \sum_i E[F_i(\lambda_i(t))] = J^*.$$

$\square$

2.4   The Many Sources Regime

Consider the regime of many small users. Let $c \geq 1$ be a scaling factor. We consider a series of systems scaled by c. The scaled system has capacity $R^c = cR$, and the arrivals of each class i has rate $\lambda_i^c(u) = c\lambda_i(u)$. Let $J^{*,c}$, J_s^c and J_{ub}^c be the dynamic revenue, static revenue, and upper bound, respectively, for the c-scaled system.

Proposition 3. *If the functions F_i are concave, then*

$$\lim_{c \to \infty} \frac{1}{c} J_s^c = \lim_{c \to \infty} \frac{1}{c} J^{*,c} = \lim_{c \to \infty} \frac{1}{c} J_{ub}^c.$$

Proof. Firstly, J_{ub}^c is obtained by maximizing $\sum_i c\lambda_i u_i(\lambda_i)/\mu_i$, subject to the constraint $\sum_i c\lambda_i r_i C_i^l/\mu_i \leq cR^l$ for all l. Therefore the optimal price is independent of c, and $J_{ub}^c = cJ_{ub}$.

Now consider J_s^c, for every static price u_i^c falling into the constraint of J_{ub}, i.e.,

$$\sum_i \frac{c\lambda_i(u_i^c) r_i C_i^l}{\mu_i} \leq cR^l \quad \text{for all } l, \tag{4}$$

let J_0^c denote the revenue under this static price. We will show that as $c \to \infty$,

$$\lim_{c \to \infty} \frac{J_0^c}{c} = \sum_i \lambda_i(u_i^c) u_i^c \frac{1}{\mu_i}. \tag{5}$$

If we take the optimal price of the upper bound as our static price, then the right hand side of (5) is exactly the upper bound. Therefore,

$$\lim_{c \to \infty} \frac{J_s^c}{c} \geq \lim_{c \to \infty} \frac{J_0^c}{c} \geq J_{ub}.$$

On the other hand, $J_s^c \leq J^{*,c} \leq cJ_{ub}$, and the result follows.

Now we show (5). The key idea is to use an insensitivity result from [3]. In [3], Burman et. al. investigate a blocking network model, where a call instantaneously seizes channels along a route between the originating and terminating node, holds the channels for a randomly distributed length of time, and frees them instantaneously at the end of the call. If no channels are available, the call is blocked. When the arrivals are Poisson and the holding time distributions are general, the authors in [3] show that the blocking probabilities are still in product form, and are insensitive to the call holding-time distributions. This means that they depend on the call duration only through its mean.

Our system is a special case of [3]. Let $n = \{n_j, j = 1, ..., I\}$ be the vector denoting the state of the system, and let λ_j be the arrival rate of class j under the static price u_j^c, i.e., $\lambda_j = c\lambda_j(u_j^c)$. Let $\rho_j = \lambda_j/\mu_j$. From [3], we have the blocking probability of calls of class i as:

$$P_{loss,i}^c = \frac{\sum\limits_{n \in \Gamma'} \prod\limits_j \rho_j^{n_j}/n_j!}{\sum\limits_{n \in \Gamma_0} \prod\limits_j \rho_j^{n_j}/n_j!}, \tag{6}$$

where

$$\Gamma' = \left\{ \sum_j n_j r_j C_j^l \leq cR^l \text{ for all } l \text{ and there exists an } l \right.$$

$$\left. \text{such that } C_i^l = 1 \text{ and } \sum_j n_j r_j C_j^l > cR^l - r_i \right\}$$

$$\Gamma_0 = \left\{ \sum_j n_j r_j C_j^l \leq cR^l \text{ for all } l \right\}.$$

Since from (6) we can see that the blocking probability is exactly the same as in the case of exponential service times, we now only need to look at the case of exponential service times.

Consider an infinite channel system with the same arrival rate and holding-time distribution. Let $n_{j,\infty}$ be the number of flows of class j in the infinite channel system. Further let $n_\infty = \{n_{j,\infty}, j = 1, ..., I\}$ be the vector denoting the state of the infinite channel system. We can then rewrite $P_{loss,i}^c$ as

$$P_{loss,i}^c = P_{\Gamma'}^{c,\infty} / P_{\Gamma_0}^{c,\infty},$$

where

$$P_{\Gamma_0}^{c,\infty} = \frac{\sum\limits_{n_\infty \in \Gamma_0} \prod\limits_j \rho_j^{n_{j,\infty}} / n_{j,\infty}!}{e^{\sum_j \rho_j}} \quad \text{and} \quad P_{\Gamma'}^{c,\infty} = \frac{\sum\limits_{n_\infty \in \Gamma'} \prod\limits_j \rho_j^{n_{j,\infty}} / n_{j,\infty}!}{e^{\sum_j \rho_j}}$$

are the probabilities that $\{n_\infty \in \Gamma_0\}$ and $\{n_\infty \in \Gamma'\}$, respectively in the infinite channel system.

We will use the estimate of $P_{\Gamma_0}^{c,\infty}$ and $P_{\Gamma'}^{c,\infty}$ to bound $P_{loss,i}^c$. In the infinite channel system, there is no constraint. Therefore the number of flows $n_{j,\infty}$ in class j is Poisson (from well known M/M/∞ result) and independent of the number of flows in other classes. We can view each $n_{j,\infty}$ as a sum of c independent random variables.

First we calculate the first and second order statistics of $n_{j,\infty}$.

$$E[n_{j,\infty}] = c\frac{\lambda_j}{\mu_j}, \quad \sigma^2[n_{j,\infty}] = c\frac{\lambda_j}{\mu_j}.$$

Now by invoking the Central Limit Theorem, as $c \to \infty$, we have

$$\frac{n_{j,\infty} - c\frac{\lambda_j}{\mu_j}}{\sqrt{c}} \to N(0, \frac{\lambda_j}{\mu_j}) \quad \text{in distribution.} \tag{7}$$

Let $x_\infty^{c,l} \triangleq \sum_j n_{j,\infty} r_j C_j^l$ be defined as the amount of resource consumed at link l in the infinite channel system. We have

$$E[x_\infty^{c,l}] = c\sum_j \frac{\lambda_j}{\mu_j} r_j C_j^l, \quad \sigma^2[x_\infty^{c,l}] = c\sum_j \frac{\lambda_j}{\mu_j} r_j^2 C_j^l.$$

Therefore

$$\left\{\frac{\frac{x_\infty^{c,l}}{c} - \sum_j \frac{\lambda_j}{\mu_j} r_j C_j^l}{\sqrt{\frac{1}{c}}}\right\} \to N(0, Q) \quad \text{in distribution,} \tag{8}$$

$$\text{where } Q = \{Q_{mn}\}, \quad Q_{mn} = \sum_j \frac{\lambda_j}{\mu_j} r_j^2 C_j^m C_j^n.$$

Now since $\sum_j c \frac{\lambda_j}{\mu_j} r_j C_j^l \le c R^l$ for all l (4),

$$\liminf_{c \to \infty} P_{\Gamma_0}^{c,\infty} = \liminf_{c \to \infty} P\left\{\frac{x_\infty^{c,l}}{c} \le R^l, \text{for all } l\right\}$$

$$\ge \liminf_{c \to \infty} P\left\{\frac{x_\infty^{c,l}}{c} \le \sum_j \frac{\lambda_j}{\mu_j} r_j C_j^l, \text{for all } l\right\} \quad \text{(by (4))}$$

$$\ge \liminf_{c \to \infty} P\left\{n_{j,\infty} \le c\frac{\lambda_j}{\mu_j}, \text{for all } j\right\} \quad \text{(by definition of } x_\infty^{c,l})$$

$$= \liminf_{c \to \infty} \prod_j P\left\{n_{j,\infty} \le c\frac{\lambda_j}{\mu_j}\right\} \ge 0.5^I \quad \text{(by (7))},$$

$$\limsup_{c \to \infty} P_{\Gamma'}^{c,\infty} = \limsup_{c \to \infty} P\left\{\frac{x_\infty^{c,l}}{c} \le R^l, \text{for all } l, \text{and there exists } l\right.$$

$$\left. \text{such that } C_i^l = 1 \text{ and } \frac{x_\infty^{c,l}}{c} > R^l - \frac{r_i}{c}\right\}$$

$$\le \limsup_{c \to \infty} \sum_l P\left\{\frac{x_\infty^{c,m}}{c} \le R^m, \text{for all } m, \text{and } \frac{x_\infty^{c,l}}{c} > R^l - \frac{r_i}{c}\right\}$$

$$\le \limsup_{c \to \infty} \sum_l P\left\{R^l - \frac{r_i}{c} < \frac{x_\infty^{c,l}}{c} \le R^l\right\}$$

$$\le \limsup_{c \to \infty} \sum_l \frac{1}{\sqrt{2\pi}} \frac{\frac{r_i}{c}}{\sqrt{\frac{1}{c} \sum_j \frac{\lambda_j}{\mu_j} r_j^2 C_j^l}} \quad \text{(by (8))} \tag{9}$$

$$= \sum_l 0 = 0.$$

Therefore

$$\lim_{c \to \infty} P_{loss,i}^c = \lim_{c \to \infty} P_{\Gamma'}^{c,\infty} / P_{\Gamma_0}^{c,\infty} = 0,$$

and

$$\lim_{c \to \infty} \frac{J_s^c}{c} \ge \lim_{c \to \infty} \frac{J_0^c}{c} = \lim_{c \to \infty} \sum_i \lambda_i(u_i^c) u_i^c \frac{1}{\mu_i} (1 - P_{loss,i}^c) = \sum_i \lambda_i(u_i^c) u_i^c \frac{1}{\mu_i}.$$

Thus the result follows.   □

The above proof not only shows that the limit converges, but also shows that the speed of convergence is at least $\frac{1}{\sqrt{c}}$. To see this, we go back to (9). The convergence is the slowest when (4) is satisfied with equality, in which case

$$\sum_l \frac{1}{\sqrt{2\pi}} \frac{\frac{r_i}{c}}{\sqrt{\frac{1}{c}\sum_j \frac{\lambda_j}{\mu_j} r_j^2 C_j^l}} \leq \sum_l \frac{1}{\sqrt{2\pi}} \frac{1}{\sqrt{c}} \frac{r_i}{\sqrt{\sum_j \frac{\lambda_j}{\mu_j} r_j C_j^l \min_i r_i}}$$

In fact, we can show that if the maximizing price of the upper bound falls inside the constraint (4), (instead of at the border), then the speed of convergence is exponential.

Here we report a few numerical results. Consider the network in Fig. 1. There are 4 classes of flows. Their routes are shown in the figure. Their arrivals are Poisson. The function $\lambda_i(u)$ for each class i is of the form

$$\lambda_i(u) = \left[\lambda_{max,i}\left(1 - \frac{u}{u_{max,i}}\right)\right]^+,$$

i.e., $\lambda_i(0) = \lambda_{max,i}$ and $\lambda_i(u_{max,i}) = 0$ for some constants $\lambda_{max,i}$ and $u_{max,i}$. The price elasticity is then

$$-\frac{\lambda_i'(u)}{\lambda_i(u)} = \frac{1/u_{max,i}}{1 - u/u_{max,i}} \text{ , for } 0 < u < u_{max,i}.$$

The mean holding time is $1/\mu_i$. The arrival rates, price elasticity, service rates μ_i, and bandwidth requirement are shown in Table 1.

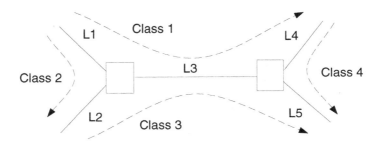

Fig. 1. The network topology

First, let us consider a base system where the 5 links have capacity 10, 10, 5, 15, and 15 respectively. The solution of the upper bound (1) is shown in Table 2. The upper bound is $J_{ub} = 127.5$. We then use simulations to verify how tight this upper bound is and how close the performance of the static pricing policy can approach this upper bound when the system is large. We use the price induced

Table 1. Traffic and price parameters of 4 classes

	Class 1	Class 2	Class 3	Class 4
$\lambda_{max,i}$	0.01	0.01	0.02	0.01
$u_{max,i}$	10	10	20	20
Service Rate	0.002	0.001	0.002	0.001
Bandwidth	2	1	1	2

Table 2. Upper bound in the constrained case, $J_{ub} = 127.5$

	Class 1	Class 2	Class 3	Class 4
u_i	9.00	5.00	12.00	10.00
λ_i	0.00100	0.00500	0.00800	0.00500
λ_i/μ_i	0.500	5.00	4.00	5.00

by the upper bound calculated above as our static price. We first simulate the case when the holding time distributions are exponential. We simulate c-scaled versions of the base network where c ranges from 1 to 1000. For each scaled system, we simulate the static pricing scheme, and report the revenue generated. In Fig. 2 we show the normalized revenue J_0/c as a function of c.

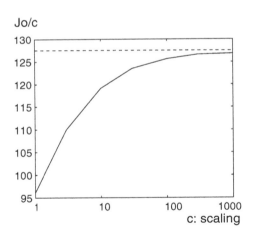

Fig. 2. The static pricing policy compared with the upper bound: the constrained case. The dotted line is the upper bound.

As we can see, as the system grows large, the performance gap between the static pricing scheme and the upper bound gets smaller and smaller. Although we do not know what the optimal dynamic policy is, its normalized revenue J^*/c must lie somewhere between that of the static policy and the upper bound. Therefore the performance gap between the static pricing scheme and the optimal dynamic scheme also gets very small. For example, when $c = 10$, which corresponds to the case when the link capacity can accommodate around 100 flows, the performance gap between the static policy and the upper bound is less than 7%. Further the gap decreases as $1/\sqrt{c}$.

Now, we change the capacity of link 3 from 5 to 15. The solution of the upper bound is shown in Table 3.

Table 3. Upper bound in the unconstrained case, $J_{ub} = 137.5$

	Class 1	Class 2	Class 3	Class 4
u_i	5.00	5.00	10.00	10.00
λ_i	0.00500	0.00500	0.0100	0.00500
λ_i/μ_i	2.50	5.00	5.00	5.00

The upper bound is $J^* = 137.5$. The simulation result (Fig. 3) confirms again that the performance of the static policy approaches the upper bound when the system is large. At $c = 10$, the performance gap between the static policy and the upper bound is less than 10%. Also note that when the system is in an unconstrained state, the price in our static scheme is the same for users with the same price-elasticity even if they traverse different routes. For example, classes 1 & 2 and classes 3 & 4 have the same price (and price-elasticity) but have different routes. In general, if there is no significant constraint of resources, the maximizing price structure will be independent of the route of the connection. To see this, we go back to the formulation of the upper bound (1). If the *unconstrained*

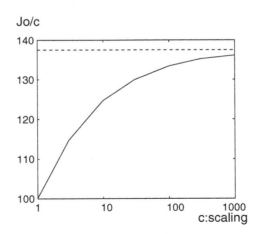

Fig. 3. The static pricing policy compared with the upper bound: the unconstrained case. The dotted line is the upper bound.

maximizer of $\sum_i F_i(\lambda_i)$ satisfies the constraint, then it is also the maximizer of the *constrained* problem. In this case the price only depends on the function $u_i(\lambda_i)$, which represents the price elasticity of the users. Readers can verify that in our second example when the capacity of link 3 is 15, if we lift the constraints in (2), and solve the upper bound again, we will get the same result. Therefore in our example, the optimal price will only depend on the price elasticity of each class and not on the specific route. Since class 1 has the same price elasticity as class 2, its price is also the same as that of class 2, even though it traverses a longer route through the network. That also explains why in some scenarios, for example, state-to-state long distance in US, prices are flat.

We also simulate the case when the holding time distribution is deterministic. The result is the same as that of exponential holding time distribution. The case with heavy tail holding time distribution is more complicated. Here is a result when the holding time distribution is Pareto, i.e., the cumulative distribution function is $1 - 1/x^a$, with $a = 1.5$. This distribution has finite mean but infinite variance. We use the same set of parameters as the constrained case above, and let the Pareto distribution have the same mean as that of the exponential distribution.

In Fig. 4, we can see that even with the Pareto distribution the performance of the static policy still follows the same trend as the case of exponential holding time distribution. It demonstrates that our result is indeed invariant of the holding time distribution. However, we also note that if the holding time distribution is Pareto with infinite variance, the sample path convergence becomes very slow. The problem is even worse when the system is large. We will briefly discuss this in the conclusion.

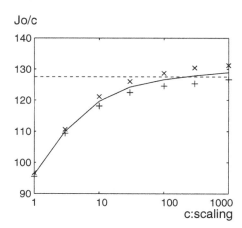

Fig. 4. The static pricing policy compared with the upper bound: the constrained case with Pareto distribution. The dotted line is the upper bound. 'x' and '+' are 0.99 confidence interval.

3   Pricing on the Incoming Traffic Parameters

Our previous result shows that dynamic pricing based on the instantaneous utilization does not provide us with significant improvement over static pricing, when the system is large. The question to then ask is: If we base our dynamic price on some other factor, can we outperform the static pricing scheme? In this section, we investigate pricing based on another factor, namely the duration of a connection. Thus, here we look at the case when the dynamic pricing scheme, in addition to the instantaneous load, also has prior knowledge of the call duration.

For convenience we restrict ourselves to the single resource and single class model. The case of multiple resources and multiple classes can be treated analogously. Assume flows consume unit bandwidth. The only difference from the system studied in Sect. 2 is that now the provider tries to price the incoming call according to additional information, i.e., the duration of the call, (here we assume that each connection request knows how long it has to last). Therefore the price becomes $u(t, T) = g(n(t), T)$, where T reflects the length of the incoming call and g is a function from $\Omega \times R$ to R.

The question again is: what is the relationship between such a dynamic pricing scheme and a static pricing scheme? The static pricing scheme can now either take T into consideration (i.e., price is $u(T)$), or not (i.e., price is u, a constant). In this section we will show that the performance of the static pricing scheme using price *independent* of T will approach that of the optimal dynamic pricing scheme when the system is large. Therefore there is really no incentive to price according to individual holding time in a large system.

We consider the special case when the range of service time can be partitioned into slots $[T_k, T_k + \Delta T_k)$, $k = 1, 2, ...$, and at each time instant the price is constant within each slot, i.e., $u(t, T) = u_k(t)$, $u(T) = u_k$, for $T \in [T_k, T_k + \Delta T_k)$.

Let $\tilde{T}_k = E\{T | T \in [T_k, T_k + \Delta T_k)\}$, and let $p_k = P\{T \in [T_k, T_k + \Delta T_k)\}$. Our idea is to decompose the original arrivals into a spectrum of substreams. Substream k has service time $[T_k, T_k + \Delta T_k)$. Its arrival rate is thus $\lambda(u)p_k$. Here, we assume that the price-elasticity of calls is independent of T. Therefore the arrival of each substream k is still Poisson. The expected dynamic revenue is then given by

$$J^* = \max_g E\left[\sum_{k=1}^{\infty} \lambda(u_k(t))u_k(t)\tilde{T}_k p_k\right].$$

The expected static revenue is

$$J_s = \max_{u_k} \sum_{k=1}^{\infty} \lambda(u_k)u_k\tilde{T}_k p_k (1 - P_{loss}).$$

where P_{loss} is the probability that a call is blocked given the static prices u_k. Note that this probability is independent of k.

Let $\lambda_k = \lambda(u_k)$. Again we write u_k as a function of λ_k. Also let $F(\lambda_k) = \lambda_k u(\lambda_k)$. We have,

Proposition 4. *If F is concave, then J* is upper bounded by J_{ub}, which is the solution of the following optimization problem:*

$$J_{ub} = \max_{\lambda_k} \sum_{k=1}^{\infty} F(\lambda_k)\tilde{T}_k p_k \tag{10}$$

$$subject\ to\ \sum_{k=1}^{\infty} \lambda_k \tilde{T}_k p_k \leq R. \tag{11}$$

Also when the system is scaled by c, i.e., $\lambda^c(u) = c\lambda(u)$ and $R^c = cR$, we have $\lim_{c\to\infty} J_s^c/c = \lim_{c\to\infty} J^{*,c}/c = \lim_{c\to\infty} J_{ub}^c/c$. Hence, the static price of the form $u(T) = u_k, T \in [T_k, T_k + \Delta T_k)$ suffices.

Proof. If we view each substream as a class, the only difference from the system we studied in last section is that it has countably infinite number of classes. Since we can still use Little's Law for each substream, the first part of this proposition can be shown by using the same idea as in the proof of Proposition 2 and by invoking the Monotone Convergence Theorem in (3).

In order to show the second part, it suffices to show that when we use the price induced by the upper bound as our static price, $P_{loss} \to 0$, as $c \to \infty$. Note that as far as P_{loss} is concerned, the system can be viewed as having only a single class. Therefore, we can reuse the result in the last section. Compared with a system with no pricing control, the arrivals in this system are "weighted" by the static price u_k according to their holding times. Hence, the arrival process is still Poisson, with arrival rate $\lambda' = \sum_{k=1}^{\infty} \lambda(u_k)p_k$, and the holding time distribution is i.i.d. with mean $T' = \left(\sum_{k=0}^{\infty} \tilde{T}_k \lambda(u_k)p_k\right)/\lambda'$. Because the constraint (11) is satisfied by such an induced static price, we have $\lambda'T' \leq R$. Using the same techniques as in the proof of Proposition 3, we can show that $P_{loss} \to 0$ as $c \to \infty$. Then, the result follows. □

Next we study the form of the optimal price u_k which maximizes the upper bound (10). We construct another bound J' as the solution of the following optimization problem:

$$J' = \max_{\lambda}\ F(\lambda) \sum_{k=1}^{\infty} \tilde{T}_k p_k \tag{12}$$

$$subject\ to\ \lambda \sum_{k=1}^{\infty} \tilde{T}_k p_k \leq R.$$

If $J' = J_{ub}$, we will be able to conclude that the optimal λ_k in (10) should be *independent* of k, i.e., it is independent of the duration T! Therefore the price $u(T)$ should also be independent of T. We now show that this is indeed the case.

From the definition of J_{ub}, $J' \leq J_{ub}$. It remains to show that $J' \geq J_{ub}$. For any set of λ_k that satisfies the constraint of (10), let

$$\lambda = \frac{\sum_{k=1}^{\infty} \lambda_k \tilde{T}_k p_k}{\sum_{k=1}^{\infty} \tilde{T}_k p_k}.$$

Then λ will satisfy the constraint of (12). Since F is concave, we have

$$F(\lambda) \geq \frac{\sum_{k=1}^{\infty} F(\lambda_k)\tilde{T}_k p_k}{\sum_{k=1}^{\infty} \tilde{T}_k p_k}.$$

Therefore $J' \geq J_{ub}$.

Now λ_k is independent of k, and hence, so is u_k. We conclude that, in the asymptotic regime, not only is there no incentive to price according to instantaneous utilization, but there is also no incentive to price based on the duration of the individual calls! Again, we recall the previous result, that the speed of convergence for this invariance result to hold is at least $1/\sqrt{c}$.

Our above result suggests that in a very broad sense, the optimal price needs only to take care of one parameter, that is, the traffic class (in a sense the total cost also depends on duration, however the relationship is linear). Note the above result is obtained under the assumption that the arrival elasticity is independent of T. If this is not true, even though the performance of our static price in the form of $u(T)$ will still approach the performance of the optimal dynamic scheme, $u(T)$ will not be a constant. In this case, we may need to treat calls of different duration as different classes.

4   Conclusion and Future Work

In this work we study pricing as a mechanism to control large networks. Our model is based on revenue maximization in a general loss network with Poisson arrivals and arbitrary holding time distributions. In dynamic pricing schemes, the network provider can charge different prices to the user according to the varying level of utilization in the network. We prove that this type of a system with feedback is asymptotically stationary and ergodic. We then analyze the performance of the static pricing scheme compared with that of the optimal dynamic pricing scheme. We prove that there is an upper bound on the performance of dynamic pricing schemes, and that the performance of an appropriately chosen static pricing scheme will approach that of the optimal dynamic pricing scheme, when the system is large. Under certain condition, this static price will only depend on the price-elasticity of each class, while being independent of the route of the call.

We then extend the result to the case when a network provider can charge a different price according to an additional factor, i.e., the individual holding time. We develop appropriate static pricing schemes for this case and show that the static scheme that is independent of the individual holding time performs as well as the optimal dynamic schemes, when the system is large.

The above results have important implication in the networks of today and in the future. Compared with dynamic pricing schemes, static pricing schemes have some desirable properties. They are less computationally intensive, and consume less network bandwidth. Their performance will not degrade as the network delay grows. Our results show that when the system is large, as in broadband networks, the difference between static pricing schemes and dynamic pricing schemes is minimal. We are currently investigating how to use these results to develop efficient and realistic algorithms to control large networks.

We also note that when the holding time distribution is heavy tailed, the sample path convergence to the mean becomes slow. We think this is caused by the fact that with the Pareto distribution, some flows will have very large duration. This fact has two implications on the sample path convergence. One is that the average of the Pareto distributed random variables converges very slowly to their mean, because of a small number of very large samples. The other is that the queue dynamics tend to have correlation over a very large timescale, which leads to very slow convergence of statistics based on queue average.

This indicates that the long term average may not be practically meaningful in such cases. The transient behavior might be more important, and deserves to be handled more carefully.

5   Appendix

Proof (of Lemma 1). We follow [1]. To show $P\{A_0\} > 0$, we note that for some $a > 0, m \geq 1$,

$$P\{A_0\} \geq P\left\{ \{\tau_0^e \geq a\} \bigcap_{k=1}^{k=m} \{\tau_{-k}^s \leq a\} \bigcap_{k=m+1}^{\infty} \left\{ \tau_{-k}^s \leq \sum_{j=-k+1}^{-1} \tau_j^e \right\} \right\}$$

$$= P\{\tau_0^e \geq a\} \prod_{k=1}^{m} P\{\tau_{-k}^s \leq a\} P\left\{ \bigcap_{k=m+1}^{\infty} \left\{ \tau_{-k}^s \leq \sum_{j=-k+1}^{-1} \tau_j^e \right\} \right\}.$$

This can be interpreted as the following: A_0 is the event that $\{Q_0 = 0\}$, i.e., all potential arrivals before time 0 has left the system. The event on the right of the inequality above says that, of all the potential arrivals before time 0, the last arrival arrives before a time interval of a, $(\tau_0^e \geq a)$; the last m arrivals all have service time less than a; and finally, the rest of the arrivals leave the system before time -1. Obviously this is a smaller event than A_0. From now on we will focus on this event only.

Now choose a such that $P\{\tau_{-k}^s \leq a\} = q > 0$, we also have $P\{\tau_n^e \geq a\} = p > 0$, since the interarrival times are exponential. Then

$$P\{A_0\} \geq pq^m P\left\{ \bigcap_{k=m+1}^{\infty} \left\{ \tau_{-k}^s \leq \sum_{j=-k+1}^{-1} \tau_j^e \right\} \right\}.$$

Now let B be the event inside the bracket on the right hand side. We only need to show $P\{B\} > 0$ for some m.

Choose $b < E\{\tau_n^e\}$. Then

$$P\{B^c\} = P\left\{ \bigcup_{k=m+1}^{\infty} \left\{ \tau_{-k}^s > \sum_{j=-k+1}^{-1} \tau_j^e \right\} \right\}$$

$$\leq P\left\{\bigcup_{k=m+1}^{\infty}\left\{\sum_{j=-k+1}^{-1}\tau_j^e < b(k-1)\right\}\bigcup_{k=m+1}^{\infty}\left\{\tau_{-k}^s \geq b(k-1)\right\}\right\}$$

$$\leq P\left\{\bigcup_{k=m+1}^{\infty}\left\{\sum_{j=-k+1}^{-1}\tau_j^e < b(k-1)\right\}\right\} + P\left\{\bigcup_{k=m+1}^{\infty}\left\{\tau_{-k}^s \geq b(k-1)\right\}\right\}.$$

Now as $m \to \infty$, the first term goes to

$$P\left\{\bigcap_m\bigcup_{k=m+1}^{\infty}\left\{\sum_{j=-k+1}^{-1}\tau_n^e < b(k-1)\right\}\right\} = P\left\{\sum_{j=-k+1}^{-1}\tau_n^e < b(k-1)\text{ i.o.}\right\} = 0$$

by Strong Law of Large Numbers (since $b < E\{\tau_n^e\}$).

On the other hand, as $m \to \infty$, the second term goes to

$$P\left\{\bigcup_{k=m+1}^{\infty}\left\{\tau_{-k}^s \geq b(k-1)\right\}\right\} \leq \sum_{k=m+1}^{\infty} P\{\tau_{-k}^s \geq b(k-1)\} \to 0$$

since $E\{\tau_n^s\} < \infty$.

Therefore we can choose m large enough such that $P\{B^c\} < 1/2$. And

$$P\{A_0\} \geq pq^m P\{B\} \geq pq^m 1/2 > 0.$$

$\square$

Proof (of Lemma 2). First note that

$$P\{X_n \in A_n \text{ at least once}\} = P\left\{\bigcup_1^{\infty} A_n\right\}.$$

Again let T be the shift operator, Let $B = \bigcup_1^{\infty} A_n$, then $TB \subset B$, and $P\{TB\} = P\{B\}$, because B is also a stationary event. Therefore TB and B differ by a set of measure zero, B is an invariant set. Since the arrivals are ergodic, $P\{B\} = 0$ or 1. However, since $P\{B\} \geq P\{A_0\} > 0$, therefore $P\{B\} = 1$, i.e., $P\{X_n \in A_n \text{ at least once}\} = 1$.

By [2], Prop 6.38.

$$E\{T_1|X_0 \in A_0\} = \frac{1}{P\{A_0\}} < \infty.$$

$\square$

References

1. A. A. Borovkov, *Stochastic Processes in Queueing Theory, translated by K. Wickwire.* New York: Springer-Verlag, 1976.
2. L. Breiman, *Probability.* Reading, Mass.: Addison-Wesley, 1968.
3. D. Y. Burman, J. P. Lehoczky, and Y. Lim, "Insensitivity of Blocking Probabilities in a Circuit-Switching Network," *Journal of Applied Probability,* vol. 21, pp. 850–859, 1984.
4. F. Kelly, A. Maulloo, and D. Tan, "Rate Control in Communication Networks: Shadow Prices, Proportional Fairness and Stability," *Journal of Operational Research Society,* vol. 49, pp. 237–252, 1998.
5. P. B. Key, "Optimal Control and Trunk Reservation in Loss Networks," *Probability in the Engineering and Informational Sciences,* vol. 4, pp. 203–242, 1990.
6. S. Kunniyur and R. Srikant, "End-to-End Congestion Control: Utility Functions, Random Losses and ECN Marks," in *Proceedings of IEEE INFOCOM,* (Tel Aviv, Israel), 2000.
7. A. A. Lazar and N. Semret, "The Progressive Second Price Auction Mechanism for Network Resource Sharing," in *8th International Symposium on Dynamic Games and Applications, Maastricht, the Netherlands,* pp. 359–365, July 1998.
8. S. H. Low and D. E. Lapsley, "Optimization Flow Control–I: Basic Algorithm and Convergence," *IEEE/ACM Transactions on Networking,* vol. 7, no. 6, pp. 861–874, 1999.
9. J. Mackie-Mason and H. Varian. "Pricing the Internet,". in *Public Access to the Internet,* pp. 269–314. The MIT Press, Cambridge, MA, B. Kahin and J. Keller ed., 1995.
10. R. McEliece and K. Sivarajan, "Maximizing Marginal Revenue In Generalized Blocking Service Networks," in *Proc.1992 Allerton Conference on Communication, Control, and Computing,* pp. 455–464, 1992.
11. I. C. Paschlidis and J. N. Tsitsiklis, "Congestion-Dependent Pricing of Network Services," *IEEE/ACM Transactions on Networking,* vol. 8, pp. 171–184, April 2000.
12. S. Shenker, D. Clark, D. Estrin, and S. Herzog, "Pricing in Computer Networks: Reshaping the Research Agenda," *ACM Computer Communication Review,* vol. 26, no. 2, pp. 19–43, 1996.
13. K. Sigman, *Stationary Marked Point Processes, An Intuitive Approach.* New York: Chapman & Hall, 1995.
14. H. Yaiche, R. Mazumdar, and C. Rosenberg, "A Game Theoretic Framework for Rate Allocation and Charging of Elastic Connections in Broadband Networks," *IEEE/ACM Transactions on Networking,* vol. 8, pp. 667–678, Oct. 2000.

Endpoint Admission Control over Assured Forwarding PHBs and Its Performance over RED Implementations

Giuseppe Bianchi[1], Nicola Blefari-Melazzi[2], and Vincenzo Mancuso[1]

[1] University of Palermo, Italy, bianchi@elet.polimi.it
[2] University of Perugia, Italy, blefari@diei.unipg.it

Abstract. The Assured Forwarding Per Hop Behavior (AF PHB) has been devised by the IETF Differentiated Services (DiffServ) working group to provide drop level differentiation. The intent of AF is to support services with different loss requirements, but with no strict delay and jitter guarantees. Another suggested use of AF is to provide differentiated support for traffic conforming to an edge conditioning/policing scheme with respect to non-conforming traffic.
Scope of this paper is twofold. First, we show that, quite surprisingly, a standard AF PHB class is semantically capable of supporting per flow admission control. This is obtained by adopting the AF PHB as core routers forwarding mechanism in conjunction with an End Point Admission Control mechanism running at the network edge nodes. The performance achieved by our proposed approach depend on the specific AF PHB implementation running in the core routers.
In the second part of the paper, we prove that changes in the customary AF PHB implementations are indeed required to achieve strict QoS performance. To prove this point, we have evaluated the performance of our admission control scheme over a simple AF implementation based on RED queues. Our results show that, regardless of the selected RED thresholds configuration, such an implementation is never capable of guaranteeing tight QoS support, but is limited to provide better than best effort performance.

1 Introduction

Two QoS architectures are being discussed in the Internet arena: Integrated Services and Differentiated Services. Nevertheless, quoting the recent RFC [R2990], *"both the Integrated Services architecture and the Differentiated Services architecture have some critical elements in terms of their current definition, which appear to be acting as deterrents to widespread deployment... There appears to be no single comprehensive service environment that possesses both service accuracy and scaling propertie*s". In fact:

1. the IntServ/RSVP paradigm [R2205, R2210] is devised to establish reservations at each router along a new connection path, and provide "hard" QoS guarantees. In this sense, it is far to be a novel reservation paradigm, as it inherits its basic ideas from ATM and the complexity of the traffic control scheme is comparable. In the heart of large-scale networks, the cost of RSVP soft state maintenance and

S. Palazzo (Ed.): IWDC 2001, LNCS 2170, pp. 232–250, 2001.

of processing and signaling overhead in the routers is significant and thus there are scalability problems. In addition to complexity, we feel that the lack of a total and ultimate appreciation in the Internet market of the IntServ approach is also related to the fact that RSVP needs to be deployed in all the involved routers, to provide end-to-end QoS guarantees; hence this approach is not easily and smoothly compatible with existing infrastructures. What we are trying to say is that complexity and scalability are really important issues, but that backward compatibility and smooth Internet upgrade in a multi-vendor Internet market scenario is probably even more important.

2. Following this line of reasoning, we argue that the success of the DiffServ framework [R2474, R2475] does not uniquely stays in the fact that it is an approach devised to overcome the scalability limits of IntServ. As in the legacy Internet, the DiffServ network is oblivious of individual flows. Each router merely implements a suite of scheduling and buffering mechanisms, to provide different aggregate service assurances to different traffic classes whose packets are accordingly marked with a different value of the Differentiated Services Code Point (DSCP) field in the IP packet header. By leaving untouched the basic Internet principles, DiffServ provides supplementary tools to further move the problem of Internet traffic control up to the definition of suitable pricing/service level agreements (SLAs) between peers. However, DiffServ lacks a standardized admission control scheme, and does not intrinsically solve the problem of controlling congestion in the Internet. Upon overload in a given service class, all flows in that class suffer a potentially harsh degradation of service. RFC [R2998] recognizes this problem and points out that *"further refinement of the QoS architecture is required to integrate DiffServ network services into an end-to-end service delivery model with the associated task of resource reservation"*. It is thus suggested [R2990] to define an *"admission control function which can determine whether to admit a service differentiated flow along the nominated network path"*.

Scope of this paper is to show that such an admission control function can be defined on top of a standard DiffServ framework, by simply making smart usage of the semantic at the basis of the Assured Forwarding Per Hop Behavior (AF PHB [R2597]). It is obvious that this function must not imply a management of per flow states, which are alien to DiffServ and which would re-introduce scalability problems. The scope of our admission control function is Internet-wise (i.e., not limited to a single domain). It is deployed by pure endpoint operation: edge nodes involved in a communication are in charge of taking an explicit decision whether to admit a new flow or reject it. These edge nodes rely upon the successful delivery of probe packets, i.e., packets tagged with a suitable DSCP label, independently generated by the end-points at flow setup. The internal differentiated management of probes and packets originated by already admitted flows is performed in conformance with the AF PHB definition. Also, following the spirit of DiffServ, the degree of QoS provided is delegated to each individual DiffServ domain, and depends on the AF PHB specific implementation and tuning done at each core router of the domain.

For convenience, we will use the term GRIP (Gauge&Gate Reservation with Independent Probing) to name the overall described operation. GRIP was originally proposed in [BB01a], although its mapping over a standard DiffServ framework was not recognized until [BB01b]. GRIP combines the packet differentiation capabilities of the AF PHB with both the distributed and scalable logic of Endpoint Admission Control [BRE00], and the performance advantages of measurement based admission control schemes [BJS00]. However, we remark that GRIP is not a new reservation protocol for the Internet (in this, differing from the SRP protocol [ALM98], from which GRIP inherits some strategic ideas). Instead, GRIP is a novel reservation paradigm that allows independent end point software developers and core router producers to inter-operate within the DiffServ framework, without explicit protocol agreements.

The organization of this paper is the following. Section 2 provides an understanding of rationales and limits of Endpoint Admission Control. Section 3 describes the GRIP operation and its support over AF PHB classes. Section 4 first qualitatively discusses the issue of performance achievable by specifically designed AF implementations; then presents numerical results that prove that GRIP's performance over AF PHB routers, implemented with RED queues ([FVJ93]), are just limited to better than best effort support. Finally, conclusions are drawn in Section 5.

2   Unfolding Endpoint Admission Control

Endpoint Admission Control (EAC) is a recent research trend in QoS provisioning over IP [BRE00]. EAC builds upon the idea that admission control can be managed by pure end-to-end operation, involving only the source and destination host. At connection set-up, each sender-receiver pair starts a Probing phase whose goal is to determine whether the considered connection can be admitted to the network. In some EAC proposals [BOR99, ELE00, BRE00], during the Probing phase, the source node sends packets that reproduce the characteristics (or a subset of them) of the traffic that the source wants to emit through the network. Upon reception of the first probing packet, the destination host starts monitoring probing packets statistics (e.g., loss ratio, probes interarrival times) for a given period of time. At the end of the measurement period and on the basis of suitable criteria, the receiver takes the decision whether to admit or reject the connection and notifies back this decision to the source node.

Although the described scheme looks elegant and promising (it is scalable, it does not involve inner routers), a number of subtle issues come out when we look for QoS performance. A scheme purely based on endpoint measurements suffers of performance drawbacks mostly related to the necessarily limited (few hundreds of ms, for reasonably bounded call setup times) measurement time spent at the destination. Measurements taken over such a short time cannot capture stationary network states, and thus the decision whether to admit or reject a call is taken over a snapshot of the network status, which can be quite an unrealistic picture of the network congestion level.

The simplest solution to the above issue (other solutions are being explored, but their complete discussion and understanding is way out of the aims of the present paper) is to attempt to convey more reliable network state information to the edge of the network. Several solutions have been proposed in the literature. [CKN00] proposes to drive EAC decisions from measurements performed on a longer time scale among each ingress/egress pair of nodes within a domain. [GKE99, SZH99,

KEL00] use packet marking to convey explicit congestion information to the relevant network nodes in charge of taking admission control decisions. [MOR00] performs admission control at layers above IP (i.e., TCP), by imposing each core router to parse and capture TCP SYN and SYN/ACK segments, and forward such packets only if local congestion conditions allow admission of a new TCP flow.

To summarize the above discussion, and to proceed further, we can state that an EAC is, ultimately, the combination of three logically distinct components (although, in some specific solutions – e.g. [BOR99, ELE00] – the following issues are not clearly distinct, this does not mean at all that these three specific issues are not simultaneously present):

1. edge nodes in charge of taking explicit per flow accept/reject decisions;

2. physical principles and measures on which decisions are based (e.g., congestion status of an internal link or an ingress/egress path, and particular measurement technique - if any - adopted to detect such status);

3. the specific mechanisms adopted to convey internal network information to edge nodes (e.g., received probing bandwidth measurement, IP packet marking, exploitation of layers above IP with a well-defined notion of connection or even explicit signaling).

In such a view, EAC can be re-interpreted as a Measurement Based Admission Control (MBAC) that runs internally to the network (i.e., in a whole domain or, much simpler, in each internal router). This MBAC scheme locally determines, according to some specific criteria (which can be as simple as non performing any measure at all, and taking a snapshot of the link state, or as complex as some of the techniques proposed in [BJS00, GRO99]), whether a new call can be *locally* admitted (i.e. as far as the local router is concerned). This set of information (one per each distinct network router) is implicitly (or explicitly) collected and aggregated at the edge nodes of the network; these nodes are ultimately in charge of performing the Y/N decision.

Put in these terms, EAC was sketched as early as in the SRP protocol specification [ALM98]. Unfortunately (see e.g., what stated in [BRE00]), SRP appeared much more like a lightweight signaling protocol, with explicit reservation messages, rather than an EAC technique with increased intelligence within the core routers. Moreover, SRP requires network routers to actively manage packets (via remarking of signaling packets when congestion occurs), and thus it does not fit within a DiffServ framework, where the core routers duty is strictly limited to forwarding packets at the greatest possible speed.

Of the three components outlined above, we argue that, in a DiffServ framework, the least critical issue is how to estimate the congestion status of a router without resorting to per flow operation. In fact, recent literature [GRO99,BJS00] has shown that *aggregate* load measurements are extremely robust and efficient. These schemes do not exploit per-flow state information and related traffic specifications. Instead, they operate on the basis of per-node aggregate traffic measurements carried out at the packet level. The robustness of these schemes stays in the fact that, in suitable conditions (e.g. flow peak rates small with respect to link capacities), they are barely sensitive to uncertainties on traffic profile parameters. As a consequence, it seems that scalable estimations can be independently carried out by the routers.

The real admission control problem is how to convey the status of core routers (evaluated by means of aggregate measurements) to the end points so that the latter devices can take *learned* admission control decisions, without *violating the DiffServ paradigm*. For obvious reasons, we cannot use explicit per flow signaling. Similarly, we do not want to modify the basic router operation, by introducing packet marking schemes or forcing routers to parse and interpret higher layer information. What we want to do is to *implicitly* convey the status of core routers to the end points, by means of scalable, DiffServ compliant procedures.

3   Grip: Endpoint Admission Control over AF per Hop Behavior

We name GRIP (Gauge&Gate Reservation with Independent Probing) a reservation framework where Endpoint Admission Control decisions are driven by probing packet losses occurring in the internal network routers. The Gauge&Gate acronym stems from the assumption that network routers are able to drive probing packet discarding (Gate) on the basis of accepted traffic measurements (Gauge), and thus implicitly convey congestion status information to the edge of the network by means of reception/lack of reception of probes independently generated by edge nodes. GRIP concepts were originally introduced in [BB01a], but its mapping over DiffServ was not fully recognized until [BB01b] (as a matter of fact, in [BB01a] we erroneously required the introduction of a new PHB to implement our scheme). The present paper moves further, and shows that the generic GRIP router operation is indeed intrinsically accounted in the Assured Forwarding (AF) PHB. In other words, the AF PHB class, with no further modifications to the specifications described in [R2597], is semantically capable of seamlessly supporting EAC.

AF PHBs have been devised to provide different levels of forwarding assurances within the Differentiated Services Framework [R2474, R2475]. Four AF PHB classes have been standardized, each composed of three drop levels. In what follows, we will use the notation $AFxj$ to indicate packet marks belonging to the AF class x, with drop level j. Conforming to [R2597], within a class x, if i<j, the dropping probability of packets labeled $AFxi$ is lower than that of packets labeled $AFxj$.

The example services presented in the appendix of [R2597] show that the primary intent of AF is to promote performance differentiation (in terms of packet drop), either among different traffic classes, e.g., marked with different drop levels, as well as within the same traffic class, e.g., marking traffic conforming to a policy specification with a lower drop level than non conforming traffic. However, low loss and low latency traffic support appears to be out of the targets of the AF model. To a larger extent, as discussed in the introduction, QoS guarantees appear not only unfeasible over AF, but also out of reach of the basic DiffServ architectural model, due to the lack of an explicit resource reservation mechanism.

Based on the discussion carried out in Section 2, it is now possible to argument that the AF PHB definition contains all the necessary semantic to support per flow admission control. Quoting [R2597], *"an AF implementation MUST detect and respond to long-term congestion within each class by dropping packets, while handling short term congestion (packet bursts) by queueing packets. This implies the presence of a smoothing or filtering function that monitors the instantaneous congestion level and computes a smoothed congestion level. The dropping algorithm uses this smoothed congestion level to determine when packets should be discarded*".

This sentence explicitly states that an AF router is capable of supporting an eventually sophisticated measurement criterion that can drive packet discarding. To run EAC over an AF PHB class it is simply necessary to clarify issue (3) presented in Section 2 (i.e., the specific mechanism adopted to convey internal network information to edge nodes). This is done by assigning to a specific AF dropping level the task of notifying internal network congestion to the end nodes *by means of packet dropping* (which, for an AF-compliant router, is the only capability we can rely on).

3.1 AF Router Operation

A particular implementation of the DiffServ router output port operation supporting the AF PHB is depicted in Fig. 1. Packets routed to the relevant output are classified on the basis of their DSCP tag and dispatched to the relevant PHB handler.

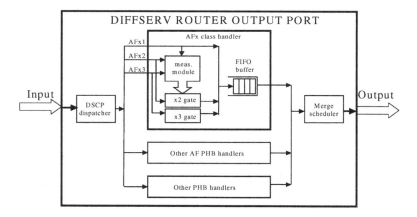

Fig. 1. Router output port operation

Let us now focus our attention to a specific module in charge of handling AF traffic belonging *to a given class x.*

A measurement module is devised to run-time measure the *aggregate* AF class x traffic (or AFx traffic). The measurement module depicted in the figure does not interact with the AFx1 packets forwarding, i.e., these packets are forwarded to the FIFO buffer placed at the output regardless of the measurements taken. On the basis of such measurements, this module triggers a suitable dropping algorithm on the AFx2 traffic. With respect to the general AF PHB operation, our AFx2 dropping algorithm depends on AFx traffic measurements. Note also that for simplicity of presentation, the drop level AFx3 is neglected until Section 3.3.

The simplest dropping algorithm is represented by a "gate" (smoother dropping algorithms for AFx2 packets - e.g. RED-like algorithms - may be considered to improve stability). When the measurement module does not detect congestion on the AFx traffic, being the notion of congestion implementation-dependent, it keeps the gate opened (we call this "ACCEPT" state). When the gate is open no AFx2 packet is dropped. Conversely, the measurement module keeps the gate closed ("REJECT" state) when congestion is detected, i.e., it enforces a 100% drop probability over AFx2 packets. Note that this operation does not violate the AF drop level relationship, as AFx1 dropping probability is lower than the AFx2 one.

While the above description is simply a particular implementation of an AF class, we now show its interpretation in terms of implicit signaling, which has important consequences for the definition of our overlay admission control function. In fact, let us assume that: i) the considered AF class x, is devoted to the support of QoS aware flows, requiring an admission control procedure; ii) traffic labeled AFx1 is generated by flows which have already passed an admission control test, iii) AFx2 packets are "signaling" packets injected in the network by flows during the setup phase (in principle, one AFx2 packet per flow).

According to the described operation, an AFx2 packet is delivered to its destination ONLY IF it encounters all the routers along the path in the ACCEPT state. This operation provides an implicit binary signaling pipe, semantically equivalent to a one-bit explicit congestion notification scheme, without requiring explicit packet marking, or, worse, explicit signaling messages, contrary to the DiffServ spirit.

The described router output port operation, combined with an endpoint admission control logic, allows overlaying an implicit signaling pipe over a signaling-unaware DiffServ framework. In fact, when an AFx2 packet reaches the destination, it implicitly conveys the information that all routers encountered across the path have been locally declared themselves in the ACCEPT state, i.e., capable of admitting new connections (see next Section).

Finally, with reference to Fig. 1, the AFx PHB class handler stores packets in a FIFO buffer, to ensure that packets are forwarded in the order of their receipt, as required by the AF PHB specification [R2597]. Packets transmission over the output link is finally managed by a scheduler, which has the task of merging the traffic coming from the different PHB handlers implemented within the router output port.

3.2 End Point Operation

For clarity of presentation, in what follows, we identify the source and destination user terminals as the network end nodes[1] . Consider a scenario where an application running on a source node within a DS domain wants to setup a one way (e.g., UDP) flow with a destination node, generally in a different DS domain. As shown in Fig. 2,

---

[1] Although, logically, user terminals are the natural nodes where the endpoint admission control should operate, this is clearly not realistic, for the obvious reason that the user may bypass the admission control test and directly send AFx1 packets. Identity authentication and integrity protection are therefore needed in order to mitigate this potential for theft of resources [R2990]. Administrators are then expected to protect network resources by configuring secure policers at interfaces (e.g. access routers) with untrusted customers.

the source node, triggered by the application via proprietary signaling, starts a connection setup attempt by sending in principle *just one single packet* (more discussion about this at the end of this Section), labeled AFx2 through the network. In the same time, a probing phase timeout is started.

The role of the Destination Node simply consists in monitoring the incoming IP packets and detecting those labeled AFx2. Upon reception of an AFx2 packet, the destination node performs a receiver capability negotiation function, eventually based on proprietary signaling, and aimed at verifying whether the destination application is able and willing to accept the incoming flow. We stress that such receiver capability negotiation is recognized as an important functionality for QoS enabled applications [R2990], and it is in an important by-product of our solution. If the destination node is willing to accept the call request, it simply relays, for each incoming probe packet, with the transmission of a feedback packet. For highest probability of delivery, the feedback packet is marked AFx1 (e.g., as an information packet).

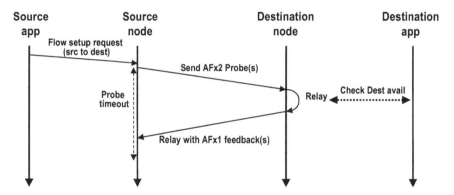

Fig. 2. End point GRIP operation

The decision whether to admit or reject the call request is driven by the eventual reception of the feedback by the source. When a feedback packet is received in response, the setting up flow is elected at the state of "accepted", and the source node can starts transmitting information packets, labeled as AFx1. Conversely, by not receiving a feedback packet within the probing phase timeout, the source node is made able to implicitly determine that at least one router along the path has declared itself not capable of accommodating additional flows, and thus the source node can abort the flow setup attempt (or reiterate the setup attempt according to some suitable backoff mechanism).

In EAC terms, packets labeled AFx2 have the meaning of probes, while AFx1 packets are meant to support already accepted traffic. The role of the drop level AFx3 is addressed in Section 3.3. We have adopted AFx1 as the label assigned to the feedback packet since the goal of the feedback is to report back to the source the information that the probe has been correctly received. In the case bidirectional flow setup is aimed at, the feedback packet has the additional task of testing the reverse path, and consequently it will be transmitted with an AFx2 label.

Note that the basic GRIP operation, i.e., a single probe packet and a single feedback packet, is compatible with the H.323 call setup scheme using UDP, which encapsulates a H.225.0v2 call setup PDU into a UDP packet. Our solution seems then perfectly compatible with existing applications. In addition, this scheme leaves the service provider free to provide optional implementation details, including:

- Addition of proprietary signaling information in the probing packet payload or in the feedback packet payload, to be parsed, respectively, at the destination node or at the source node.
- Definition of more complex probing phase operation, e.g., by including reattempt procedures after a setup failure, multiple timers and probes during the probing phase, etc.

As a last consideration, it is quite interesting to remark that this idea is extremely close to what TCP congestion control technique does, but it is used in the novel context of admission control: end points interpret failed receptions of probes as congestion in the network and reject the relevant admission requests.

3.3 Possible Roles of the AFx3 Level

In the above description, the AFx3 drop level appears in principle unnecessary. However, it may be convenient to use this level. A first possibility is that the AFx3 level be used to mark non-conforming packets, which can be eventually delivered if network resources are available. Second, AFx3 packet marking can be enforced over flows that have not successfully passed the described admission control test. This allows deploying a service model where high QoS is provided to flows that pass the admission control test, while best effort delivery is provided to initially-rejected flows. These latter flows may occasionally retry the setup procedure, by simply marking occasional packets as AFx2 (e.g. by adhering to a suitable backoff procedure), and may eventually receive the upgraded AFx1 marking when network resources become available (as testified by the eventual reception of an AFx1 feedback).

The usage of the AFx3 level as described above is targeted to increase the link utilization. However, [R2597] requires the drop probability for AFx3 to be greater (or at most equal) than AFx2. This implies that the link utilization is bounded by the possibly strict mechanism that triggers AFx2 packets dropping: when AFx2 packets receive a 100% dropping probability, all AFx3 packets must also be dropped to conform to the [R2597] specification. A more effective mechanism would consist in implementing a dropping algorithm for the AFx3 traffic not directly related to the AFx2 drop algorithm. However, this usage of the AFx3 level does not conform to the AF specification, since the AFx3 dropping probability may be eventually lower than AFx2.

A more interesting possible usage of the AFx3 level consists in providing a second control (probing) channel, in addition to AFx2. According to this solution, AFx1 traffic measurements trigger a dropping algorithm on the AFx3 traffic too, with stricter dropping conditions than the AFx2 dropping algorithm (i.e. AFx3 packets are assumed to detect congestion, and notify it via packet drop, before AFx2 packets). This AFx3 probing class could request admission for flows with e.g., higher peak rate and bandwidth requirements than flows supported via the AFx2 probing class (i.e., we are adding a second implicit signaling pipe). Also, being the AFx3 channel more reactive to congestion conditions, its usage can be envisioned to provide lower access priority to network resources. This would improve fairness and avoid some kinds of sources to "steal" a large part of network resources.

3.4 Arguments for New PHB Definitions

Although this paper leaves untouched the basic AF PHB semantic, we feel that our suggested usage of AF is different (and quite unexpected) from what intended in RFC 2597. The services that are expected to make use of admission control are RTP/UDP streams with delay and loss performance requirements, whose support is currently envisioned by means of the EF PHB. On the contrary, AF appears designed to provide better than best effort support for generic TCP/UDP traffic. Thus, our study raises the case for the transformation of the (single) EF PHB into a PHB class (i.e. by adding an associated, "paired", probing pipe with a different DSCP). An alternative is defining new "paired" PHBs.

On a different prospective, paired PHBs can be envisioned to support more general control functions than admission control. For example, the TCP fast retransmission and recovery algorithm might take advantage of isolated data packets labeled as "control", and thus expected to encounter loss if (controlled) congestion is encountered in the network.

4   Performance Issues

The described admission control semantic provides a reference framework compatible with "current" AF implementations. Scope of this section is to provide, in section 4.1, some qualitative insights about the performance achievable by the GRIP operation. Then, in section 4.2 we show that poor performance are provided over RED-like mechanisms customarily used to implement AF PHBs. These results reported in this section allow us to conclude that new explicit traffic measurement module implementations appears necessary, if tight QoS support is aimed at.

4.1 Degree of QoS Support in GRIP

Quantitative and tunable performance may be independently provided and specified by each administrative entity. Uniform implementation across a specific domain allows defining a quantitative view (e.g., a PDB [PDB01]) of the performance achievable within a considered DS domain. In this way, the refinements deemed necessary in [R2990] to provide service accuracy in the DiffServ architectural model could be considered as accomplished.

In fact, the performance achievable by the described end point admission control operation depends on the notion of congestion as the triggering mechanism for AFx2 packet discarding, which is left to each specific implementation. Each administrative entity may arbitrarily tune the optimal throughput-delay/loss operational point supported by its routers, by simply determining the aggregate AFx1 traffic target supported in each router. The mapping of AFx1 throughput onto loss/delay performance in turns depends on the link capacities and on the traffic flow characteristics offered on the AF class x.

With this approach, it is possible to construct PDBs offering quantitative guarantees. A building block of such PDBs is the definition of specific measurement modules and AFx2 dropping algorithms. A generic dropping algorithm is based on suitable rules (or decision criteria). An example of a trivial decision criterion is to accept all AFx2 packets when the measured throughput is lower than a given threshold and reject all AFx2 packets when the AFx1 measurements overflow this threshold. The resulting performance depends upon the link capacity and the traffic model.

It is well recognized that target QoS performance can be obtained by simply controlling throughput (i.e., by aggregate measurements taken on accepted traffic). This principle is at the basis of more sophisticated state of the art MBAC implementations described in [BJS00, GRO99]. As a simple quantitative example, with 32 Kbps peak rate Brady on-off voice calls (see section 4.2) offered to a 2 Mbps (20 Mbps) link, a target link utilization of 75% (92%) leads to a 99th percentile per hop delay lower than 5ms - see figure 3 (reproduced from [BCP00]).

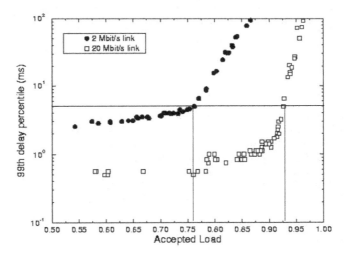

Fig. 3. 99th delay percentile versus throughput, for different EAC schemes (see [BCP00]) and related parameter settings

Tighter forms of traffic control are possible. As a second example of a decision criterion, we demonstrated that hard (loss and/or delay) QoS guarantees can be provided, under suitable assumptions on the offered traffic (i.e., traffic sources regulated by standard Dual Leaky Buckets, as in the IntServ framework) and with ad hoc defined measurement modules in the routers [BB01a]. When hard QoS guarantees are aimed at, it is furthermore necessary to solve the problem of simultaneous activation of flows. In fact, the Gauge&Gate operation implies that simultaneous setup attempts (i.e., probes arriving at the router within a very short time frame) may see the router in the ACCEPT state, and thus may lead to concurrent activation of several flows, which can in turns overload the router above the promised QoS level. Actually, this is a common and recognized problem of any MBAC scheme that does not rely on explicit signaling. Although it does not compromise the stability of described operation (overloaded routers close the gate until congestion disappears - see the mathematical formalization of such a problem and the computation of the "remedy" period in [GRO99]), this can be a critical issue if strict QoS guarantees are aimed at. In [BB01a] we have solved this problem by introducing an aggregate stack variable, which takes into account "transient" flows, i.e. flows elected at the state of "accepted" but not yet emitting information packets. This stack protection scheme avoids the concurrent activation of a number of flows, which could overload the router above the promised QoS level: the price to pay is a slight under-utilization of the available link capacity.

Another important issue is what happens when traffic flows with widely different peak rates are offered to the same link. Several approaches may be adopted. The simplest one is to differentiate traffic aggregates into classes of similar traffic characterization (e.g. similar peak rate), and associate to each class a different AF PHB class x, each with its probes AFx2 and data packets AFx1. A second possibility is to multiplex traffic onto a same AF PHB class, but differentiate probing packets by using the AFx3 drop level for traffic flows with higher peak rate (as briefly sketched in section 3.3).

Finally, we note that the AFx2 dropping algorithm must not be necessarily driven by IP-level traffic measurements. In fact, it can be driven by lower layers QoS capabilities (e.g., ATM).

4.2 Performance of GRIP over RED Implementations of the AF PHB

In this paper we have demonstrated that, quite surprisingly, admission control can be deployed over the standardized Assured Forwarding PHB. In the previous section 4.1, we have concluded that arbitrary degree of performance guarantees can be obtained by designing specific AF PHB implementations based on runtime traffic measurements. A question that comes out naturally is the following. As long as Random Early Discarding (RED) queue management is customarily considered as the "natural" AF PHB implementation[2] , what are the performance of GRIP when it is operated over RED queues?

In our simulation program, we have assumed, for convenience, only two drop levels, namely AFx1 and AFx2. We have adopted a single buffer for both AFx1 and AFx2 packets. AFx1 packets are dropped only if the buffer is completely full.

---

[2] We recall that the AF PHB specification [R2597] does not recommend, by any means, a specific implementation. Indeed, RED have emerged as the natural AF PHB implementations, since they provide improved performance when TCP traffic (i.e., the traffic traditionally envisioned for AF) is considered. We now face a very different problem, i.e. what happens to performance when widespread RED AF PHB implementations are used for a completely different purpose, i.e. to support admission controlled (UDP) traffic.

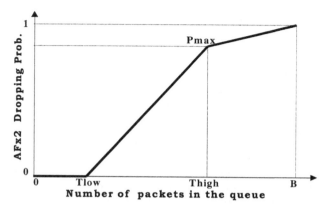

Fig. 4. Buffer management scheme

Instead, AFx2 packets are dropped according to a Random Early Discarding (RED) management, where the AFx2 dropping probability is a function of the queue occupancy, computed accounting for both AFx1 and AFx2 packets[3]. As shown in figure 4, the AFx2 dropping probability versus the queue occupancy is a piecewise linear curve: no AFx2 packets are dropped when the number of packets stored in the queue is lower than a lower threshold. As the number of packets gets greater than the lower threshold, the AFx2 dropping probability increases linearly, until a value P_{max} is reached in correspondence with an upper threshold. After this value, the dropping probability is either set at 100% (in some RED implementations), or increased linearly until the number of packets fills the buffer capacity (see figure 4). It results that a RED implementation of the AFx2 dropping algorithm is completely specified by means of 4 parameters: the buffer size, the lower and upper thresholds, and the value P_{max}.

Depending on the considered implementation, the AFx buffer occupation is either sampled at the arrival of an AFx2 packet, or suitably smoothed/filtered in order to capture the moving average of the AFx queue occupancy. However, in all implementations proposed in the literature, an AFx2 packet that finds no packets stored in the buffer is always accepted (regardless of the fact that the smoothed AFx queue occupation may give a value different from 0). We will see in what follows that, ultimately, this specific condition prevents GRIP to achieve effective performance, regardless of the RED parameter settings considered.

Performance results have been obtained via simulation of a single network link, loaded with offered calls arriving at the link according to a Poisson process. Each offered call generates a single probe packet. A sufficiently large probing phase timeout has been set to guarantee that calls are accepted when the probing packet is not dropped (in the simulation, we have attempted to simulate conditions as close to ideal as possible, in order to avoid that numerical results were affected by marginal parameters settings – e.g round trip time, probing phase timer, etc). Accepted calls

---

[3] The described operations can also be seen as a particular case of a standard WRED implementation [CIS], where the T_{low} and T_{high} thresholds for AFx1 packets both coincide with the buffer size.

have been modeled as Brady ON-OFF sources, with peak rate equal to 32 Kbps, and ON/OFF periods exponentially distributed with mean value, respectively, 1 second for the ON period, and 1.35 seconds for the OFF period (yielding an activity factor equal to 0.4255). Each call lasts for an exponentially distributed time, with mean value 120s. The link capacity has been set to 2 Mbps. Therefore, The link results temporarily overloaded when more than 146.9 active connections (2000/(0.4255x32)) are active at a given instant of time.

Figures 5 to 7 report the number of accepted calls versus the simulation time for three different load conditions: underload (normalized offered load equal to 75% of the link capacity), slight overload (110% offered load), and harsh overload (400% offered load).

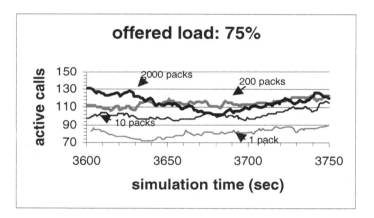

Fig. 5. Active call vs. simulation time, with 75% offered load; *"packs" indicates the AFx2 threshold.*

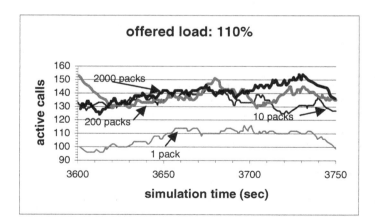

Fig. 6. Active call vs. simulation time, with 110% offered load. "packs" indicates the AFx2 threshold.

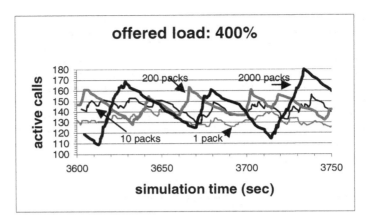

Fig. 7. Active call vs. simulation time, with 400% offered load. *"packs" indicates the AFx2 threshold.*

In the above figures, we have considered a very large AFx buffer size, such that no AFx1 packet losses occur (thus, QoS of accepted traffic is quantified in terms of AFx1 packet delay). We have tried several different RED parameters configuration, but, for simplicity of presentation, we report results related to a basic parameters settings where a single threshold is considered: all AFx2 packets are accepted whenever the number of AFx packets is lower than this threshold, while all AFx2 packets are dropped when the number of AFx packets is greater than this threshold (very similar results are obtained with more complex RED configurations, as it will be clear from the following discussion). Similarly, for simplicity, no smoothing on the buffer occupancy has been performed.

Figure 5 shows that, in low load conditions, a tight threshold setting (just 1 packet, i.e. an AFx2 packet is dropped as long as just 1 AFx packet is stored in the buffer) is overly restrictive, and exerts a high and unnecessary blocking probability on offered calls. With larger thresholds (200 and 2000 packets), we note from figure 5 that the number of accepted calls fluctuates in the range 100 to 120, meaning that just a small fraction of the offered calls are blocked by the GRIP operation (as it should ideally occur if a stateful admission control algorithm were operated).

Much more interesting and meaningful (for our purposes) are the results presented in figures 6. Here, we see that, in slight overload conditions, the only RED configuration setting that allows to keep the accepted load lower than the target 75% value (i.e. about 110 accepted calls) suggested by figure 3, is the threshold set to just 1 packet. Even with a very small threshold value, such as 10 packets, we see that the average number of accepted calls gets much greater than 110, thus resulting in unacceptable delay performance for accepted traffic.

Throughput and delay performance are quantified in table 8, for 110% and 400% offered load conditions. The table reports the AFx1 throughput, as well as the 95th and 99th delay percentiles experienced by accepted flows. Confidence intervals corresponding to a 95% confidence level are also reported in the table to quantify the accurateness of the numerical results.

AFx2 thresh	Offered load	Throughput		95th delay percentile (ms)		99th delay percentile (ms)	
1 pack	110%	68,64%±	0,28%	2,3±	0,014	3,3±	0,019
1 pack	400%	90,89%±	0,08%	80,6±	1,2	175,2±	1,3
10 packs	110%	88,71%±	0,22%	55,4±	0,7	148,444±	3.0
10 packs	400%	97,72%±	0,03%	586,3±	4,8	987,2±	10,2
200 packs	110%	93,11%±	0,28%	235,9±	4,0	433,8±	3,5
200 packs	400%	99,02%±	0,02%	1282,8±	15,9	2023,3±	75,1
2000 packs	110%	96,39%±	0,24%	1433,9±	20,0	1975,6±	26,9
2000 packs	400%	99,77%±	0,03%	4656,0±	24,9	5921,8±	58,2

Fig. 8. Throughput and delay perfomance.

From the table, we see that the only case in which we meet target QoS performance for IP telephony (i.e. 99 th delay percentile of the order of few ms) is the case of threshold set to 1 packet, and light overload. It is quite impressive to note that the smallest possible RED threshold (i.e. drop an AFx2 packet whenever the AFx queue is not strictly empty) does not succeed in guaranteeing QoS in large overload conditions. This result allows us to conclude that, regardless of the RED parameters configuration, a RED implementation is never capable of guaranteeing QoS in all load conditions. As long as a RED implementation always accepts an AFx2 packet when the AFx queue is empty[4] , performance will be at best equal to that reported in table 8, for a threshold equal to 1 packet.

As figure 7 shows, in high overload conditions, thresholds greater than 1 packet cannot even avoid that, temporarily, the number of accepted calls is greater than 146.9, i.e. that the link is overloaded. In such a case, load oscillation phenomena occur: as clearly shown in figure 7, the link alternates between periods of significant overload (in which the AFx buffer fills up), and "remedy" [GRO99] periods, where the router locks in the REJECT state, until congestion disappears. The result is that 95th and 99th delay performance are of the order of several seconds (see table 8).

To conclude this section, we observe that, although RED implementations are intrinsically uncapable of providing performance guarantees, indeed a proper parameter setting allows to achieve reasonably better than best effort performance. For example, with a threshold set to 10 packets, table 8 shows that, in very high (unrealistic) load conditions, the 99th delay is still lower than 1 second (although the link has already been congested, as proven by the link load fluctuations, of the order of 15% of the link capacity, and leading to temporary link overload, as shown in figure 7). Notably, with the same 10 packets thresholds, the 99th delay percentile drops down to less than 150ms when light overload conditions are considered. Such

---

[4] We recall that this specific rule is proper of all RED implementations. I.e., regardless of the smoothing and filtering scheme adopted on the number of AFx packets, an AFx2 packet is never dropped when an empty AFx buffer is found. Changing this rule means changing the intrinsic logic of the RED approach, i.e. moving from queue status measurements to crossing traffic measurements. What we have proven here is that such a leap is required in AF implementations if QoS guaranteed admission controlled services are to be supported.

degree of QoS support might be considered sufficient in a short-term perspective, where current AF implementations might be utilized to support admission control.

5 Conclusions

In this paper, we have shown that the standard DiffServ AF PHB is semantically capable of supporting stateless and scalable admission control. The driving idea is that accept/reject decisions are taken at the edge of the network on the basis of probing packet losses, being probes tagged with a different AF level than packets generated by accepted traffic. In turns, these losses are driven by the dropping algorithm adopted in the specific AF implementation running at each network router. It is important to understand that, following the spirit of DiffServ, the above described operation does not aim at providing a quantified level of assured QoS, but, in conformance with PHB and PDB specifications, it provides a reference framework over which quantitative performance specification may be deployed. To the purposes of this paper, it was in our opinion sufficient to show that the described operation is compliant with the specification of the AF PHB.

The key to QoS guarantees is left to each specific implementation, i.e. is left to the implementation-dependent quantification of the notion of congestion, which triggers the gate mechanism for AFx2 packet discarding. Each administrative entity is in charge of arbitrarily determine the optimal throughput/delay operational point it wants to support.

The AF PHB implementations so far considered in the literature, based on (RED) thresholds set on the queue occupancy, do not affect the described endpoint operation and thus may be considered to support admission controlled traffic. However, an addition important contribute of this paper was to prove that RED implementations are uncapable of achieving tight QoS support, regardless of their parameter settings. In fact, the "measurement" mechanism adopted is overly simple, and this translates into a poor QoS support, but still much better than best effort, since admission control is still enforced on setting up connections. Indeed, this guarantees that the described GRIP operation can be supported over already deployed AF routers to seamless improve QoS with no internal routers modification: it could be sufficient for short-term perspectives, but deeper enhancements are required in order to satisfy coming needs, in a long-term perspective.

References

[ALM98]   W.Almesberger, T.Ferrari, J. Y. Le Boudec: "SRP: a Scalable Resource Reservation Protocol for the Internet", IWQoS'98, Napa (California), May 1998.

[BB01a]   G. Bianchi, N. Blefari-Melazzi: " A Migration Path for the Internet: from Best-Effort to a QoS Capable Infrastructure by means of Localized Admission Control", Lecture Notes on Computer Science, Springer-Verlag, volume 1989, January 2001 (a more detailed technical report can be found at http://drake.diei.unipg.it/netweb/GRIP_tech_rep.pdf).

[BB01b]   G. Bianchi, N. Blefari-Melazzi: "Admission Control over AF PHB groups", internet draft, draft_bianchi_blefari_admcontr_over_af_phb.txt, March 2001, work in progress.

[BCP00]   G. Bianchi, A. Capone, C. Petrioli, "Packet Management Techniques for Measurement Based End-to-end Admission Control", KICS/IEEE Journal on Communications and Networking, June 2000.

[BJS00]   L. Breslau, S. Jamin, S. Schenker: "Comments on the performance of measurement-based admission control algorithms", Proc. of IEEE Infocom 2000, Tel-Aviv, March 2000.

[BOR99]   F. Borgonovo, A. Capone, L. Fratta, M. Marchese, C. Petrioli, "PCP: A Bandwidth Guaranteed Transport Service for IP networks", IEEE ICC'99, June 1999.

[BRE00]   L. Breslau, E. W. Knightly, S. Schenker, I. Stoica, H. Zhang: "Endpoint Admission Control: Architectural Issues and Performance", ACM SIGCOMM 2000, Stockholm, Sweden, August 2000.

[CIS]   Cisco IOS Quality of Service Solutions Configuration Guide—Online Manual.
http://www.cisco.com/univercd/cc/td/doc/product/software/ios120/12cgcr/qos_c/

[CKN00]   C. Cetinkaya, E. Knightly, "Egress Admission Control", Proc. of IEEE Infocom 2000, Tel-Aviv, March 2000.

[ELE00]   V. Elek, G. Karlsson, "Admission Control Based on End-to-End Measurements", Proc. of IEEE Infocom 2000, Tel Aviv, Israel, March 2000.

[FVJ93]   S. Floyd and V. Jacobson. "Random Early Detection Gateways for Congestion Avoidance". IEEE/ACM Transactions on Networking, vol. 1, no. 4, pp. 397-413, August 1993.

[GKE99]   R. J. Gibbens, F. P. Kelly, "Distributed Connection Acceptance Control for a Connectionless Network", 16th ITC, Edimburgh, June 1999.

[GRO99]   M. Grossglauser, D. N. C. Tse: "A Framework for Robust Measurement Based Admission Control", IEEE/ACM Transactions on Networking, Vol. 7, No. 3, June 1999.

[KEL00]   F. P. Kelly, P. B. Key, S. Zachary: " Distributed Admission Control", IEEE JSAC, Vol. 18, No. 12, December 2000.

[MOR00]   R. Mortier, I. Pratt, C. Clark, S. Crosby: "Implicit Admission Control", IEEE JSAC, Vol. 18, No. 12, December 2000.

[PDB01]   K. Nichols, B. Carpenter, "Definition of Differentiated Services Per Domain Behaviors and Rules for their Specification", draft-ietf-diffserv-pdb-def-03, January 2001, Work in progress.

[R2205]   R. Braden, L Zhang, S. Berson, S. Herzog, S. Jamin, "Resource ReSerVation Protocol (RSVP) - Version 1 Functional Specification", RFC2205, September 1997.

[R2210]   J. Wroclawsky, "The use of RSVP with IETF Integrated Services", RFC2210, September 1997.

[R2474]   K. Nichols, S. Blake, F. Baker, D. Black, "Definitions of the Differentiated Service Field (DS Field) in the Ipv4 and Ipv6 Headers", RFC2474, December 1998.

[R2475]   S. Blade, D. Black, M. Carlson, E. Davies, Z. Wang, W. Weiss, "An Architecture for Differentiated Services", RFC2475, December 1998.

[R2597]   J. Heinanen, F. Baker, W. Weiss, J. Wroclavski "Assured Forwarding PHB Group", RFC 2597, June 1999.
[R2990]   G. Huston, "Next Steps for the IP QoS Architecture", RFC2990, November 2000.
[R2998]   Bernet, Y., Yavatkar, R., Ford, P., Baker, F., Zhang, L., Speer, M., Braden, R., Davie, B., Wroclawski, J. And E. Felstaine, "A Framework for Integrated Services Operation Over DiffServ Networks", RFC 2998, November 2000.
[SZH99]   I. Stoica, H. Zhang, "Providing guaranteed services without per flow management", Proc. of ACM SIGCOMM 1999, Cambridge, MA, September 2000.

A Bandwidth Broker Assignment Scheme in DiffServ Networks

Franco Davoli and Piergiulio Maryni

Department of Communications, Computer and System Sciences,
DIST–University of Genova,
Via Opera Pia 13, 16145 Genova, Italy
{franco,pg}@dist.unige.it,
http://www.com.dist.unige.it

Abstract. Providing Quality of Service in a multimedia context in next generation Internet poses a number of open challenges, especially regarding routing and
resource management strategies. In fact, the massive amount of data that is envisaged to be transported in the near future (especially in a Differentiated Services
framework) will make it difficult to have networks that are deeply rearrangeable
in real time. A more realistic framework will most likely see a joint action of low-
level, fastly reactive, and high-level, slowly reactive, tuning operations. Within
this trend, we introduce and analyze a system in which bandwidth assignment
and routing are readjusted within a layered architecture. Low-level local tuning
procedures are fastly reactive and operate on small parts of the network; high-level
control actions act on a much longer time scale and affect the whole network: they
are able to perform global reallocations in order to cope with heavy network traffic
variations. It is shown by numerical simulation results that the use of the proposed
mechanism positively affects the overall network performance.

1 Introduction

The Internet is shifting from a mere connectionless best-effort paradigm to a network
able to guarantee (at least) some minimum set of Quality of Service (QoS) for a wider
range of traffic types(see, e.g., [1]). A notable "milestone" of such a transition has been
the introduction of the new IP protocol (i.e., IPv6 [2]), which allows to identify users'
flows while still maintaining IP's connectionless paradigm. Furthermore, a weak form
of resource reservation has been eased by the definition of reservation protocols and
frameworks such as the Integrated Services [3] (IntServ) and Differentiated Services [4]
(DiffServ) ones. Moreover, in this transition context, the Multi Protocol Label Switching
(MPLS) [5] represents a very powerful, efficient and simple solution for the actual
implementation of a large number of routing strategies.

The IntServ framework acts on a per-connection basis by running a reservation
protocol [6] at the beginning of each session in order to check for resource availability.
It has been widely accepted, however, that this solution is not scalable and, as such, is
not applicable on a global framework. Scalability is, instead, addressed by the DiffServ
solution that defines a set of "high-level pipes" satisfying a given set of QoS constraints:
the network only copes with a limited number of service classes. IntServ and DiffServ
are envisaged to cooperate in the access and in the core part of the network, respectively.

S. Palazzo (Ed.): IWDC 2001, LNCS 2170, pp. 251–265, 2001.
© Springer-Verlag Berlin Heidelberg 2001

In a DiffServ domain, the services offered by the network provider are defined in the so-called Service Level Agreement (SLA) [4]. The SLA can be either static - the most commonly used type today - or dynamic. In the latter case, the network has to be prepared to react to (a set of predefined) changes in input flow parameters, as well as in the required QoS, without human intervention. Thus it requires an automated agent and protocol (for example, a bandwidth broker [7]) to represent the differentiated service provider's domain. It is worth noting that the SLAs in DiffServ apply to aggregates of traffic and not to individual flows. Hence, while the IntServ mechanism acts on network resources on a connection-level time scale, the DiffServ one (for scalability reasons) should intervene at a slower pace in tuning its resources.

The "boundary device" between IntServ and DiffServ frameworks is the so-called DiffServ edge router. A possible role of such device is to map IntServ flows (from the access network) into pre-defined DiffServ pipes (in the core network) and to re-create the IntServ flows (to be delivered through the access network) from DiffServ aggregates. Merging and splitting flows is well supported by MPLS.

We assume a model in which, when there are too many incoming user flows, DiffServ edge routers issue a request for resource upgrade to the DiffServ bandwidth broker, along a given source-destination route. For the sake of simplicity, such requests come in the form of fixed amount of bandwidth, a sort of "bandwidth quark". This paradigm allows to re-use telephone-like (or ATM-like) resource control solutions.

In order to be fastly reactive, the bandwidth broker reserves in advance a given amount of bandwidth for each source-destination route in the core network (sources and destinations are, in this framework, edge routers). It then checks if the pre-reserved bandwidth is large enough to accommodate the request along such route. If there is not room, then an attempt is done through an alternate route. In case of success the request is accepted and buffers and scheduling controllers are re-configured properly. Otherwise, the request is rejected.

Hence, in this context, an incoming flow is to be intended as a request to add a given amount of traffic to a DiffServ pipe, i.e., from a given source to a given destination in the DiffServ network. As such, these bandwidth requests may come and go according to some arrival and departure distribution. We may envisage the edge router acting as filter: when too many IntServ user-requests are blocked, the edge router issues a "flow" request to the system. Of course, depending on resource availability on the core (DiffServ) network, also such edge router requests might be accepted or blocked.

The bandwidth broker has two main alternative approaches for supporting the DiffServ classes of service [8]: i) complete statistical multiplexing, where the resources (buffers and bandwidth) are shared among different services; ii) limited statistical multiplexing, where services with possibly widely different performance requirements and/or statistical characteristics of the traffic sources are assigned separate resources, in various combinations (*service and/or route separation*). One advantage of the first approach may be to reduce the number of blocked flows; a main limit is, to some degree, its analytical complexity. On the other hand, a service and path separation approach has the main advantage of being simple and leading to manageable and controllable models; a disadvantage may be some under-utilization of bandwidth, which, however, may be mitigated by adaptive allocation mechanisms.

In this work, we investigate an adaptive service separation scheme, the aim of which is to keep the advantage of simplicity and to enhance the bandwidth usage by introducing adaptability in DiffServ resource allocation. Furthermore, such allocation scheme, together with an appropriate joint routing strategy, allows the network to be robust against different types of failures.

The strategy adopted is along a similar line as the one initially outlined in [9]. The robustness of the proposed policy lies in its capability of responding to possible traffic changes in real-time: a part of the total capacity of any physical link is kept in a bandwidth pool and used only in case of necessity: if blocking on a DiffServ pipe exceeds some threshold, a certain amount of bandwidth can be moved from the bandwidth pool to enlarge the pipe size. On the other hand, if a pipe is under-utilized, part of its bandwidth can be given back to the pool. Reconfiguration is triggered by predefined blocking probability thresholds.

However, this mechanism only acts on a local basis and, as such, it could lead, in the long term, to a poor bandwidth distribution in the whole network. As an example, consider two pipes sharing some physical link: if the first one gets over utilized, the pool mechanism will try to enlarge it, with the possible side effect of leaving no resources for later enlargements of the second one.

Hence, periodically (but at a lower pace), a second mechanism is triggered, in order to perform a global bandwidth redistribution among all pipes on a "global information" base.

The global reallocation mechanism we have adopted, tries to distribute bandwidth evenly. A further possibility, not included in this work, would be to allow for priorities (e.g., price-based [10]) among the various DiffServ pipes.

The paper is organized as follows: in Section 2 the control system model, the routing framework, the route selection, and the bandwidth reallocation mechanism are introduced; Section 3 describes some simulation results for performance evaluation, and Section 4 contains the conclusions.

2 The Control System Model

The implementation, configuration, operation and administration of the nodes of a Diff-Serv Domain should effectively partition the resources of those nodes and the inter-node links between behavior aggregates (i.e., traffic classes) in accordance with the domain's service provisioning policy.

Hence, similarly to what happens with ATM networks [9] (with Virtual Path Connections) a DiffServ pipe may be seen as a concatenation of one or more of such partitions and identifies a direct (virtual) path between a given source–destination (SD) pair. A direct path passes through one or more (pre-configured) network routers. The definition of the direct path is left to the bandwidth broker that identifies it by optimizing resource sharing within the network, subject to QoS constraints.

Overloaded flows can be "tunneled" (in the MPLS way) to the given destination through other (less overloaded) nodes. This operation is resource consuming (i.e., it does not use the optimal route) and, as such, in our case, we restrict such alternative paths to only those composed of two hops.

A logical view of the core network is hence built on top of the physical structure; in a service separation framework, the number of such layers is proportional to the number of traffic services. DiffServ pipes can be set up by keeping into account a number of constraints that cope with network robustness and efficiency. For example, avoiding having more than one pipe, for each particular SD pair, sharing the same physical links, would increase network robustness, while having direct pipes between the most requested SD pairs would increase network efficiency. In our test-bed case, for the sake of simplicity, every SD pair is connected by a direct path and, to the extent allowed by the physical topology, by a number of alternative paths. An in-depth analysis of direct path initialization (and restoration) is out of the scope of this work: for a more detailed description of this topic see, e.g., [11,12,13,14,15].

DiffServ pipes are assigned bandwidth following the outcomes of a global high-level reallocation control policy (embedded in the bandwidth broker) that, based on the traffic matrix, computes the amount of bandwidth needed by the various pipes in order to keep their blocking probability below a given threshold (the detail of the reallocation control algorithm are given below).

After such initial set-up phase, each link may have an amount of spare bandwidth left. This remaining bandwidth is treated link by link as a "dummy" pipe, named "bandwidth pool". This bandwidth pool is useful for two purposes: the first one is to enhance the performance of the service separation policy (as we will see), and the second one is to give the network the necessary robustness in face of failures (e.g., congestion, physical failure). More in particular, the "bandwidth pool" may be used as follows. Whenever a portion of bandwidth of a given pipe is under-utilized, it will be added to the set of pools of all the physical links involved: the remaining bandwidth will then be just the necessary one to support the required QoS. Thus, if for a long period of time a pipe is only partially used, or not used at all, then its capacity will be gradually decreased and put into the pool. On the other hand, if a certain pipe is highly used and the probability of flow overload increases above a given threshold (or some other triggering event takes place), then, in real time, bandwidth can be taken from the physical link pools to enlarge the overloaded pipe. Figure 1 shows an example of this last case. The above scheme also solves the problem of the possible waste of bandwidth that may appear immediately after the initial assignment (i.e., the traffic matrix is not precisely known), since whatever initial amount of bandwidth given to a certain pipe will be adaptively adjusted over time, depending upon the actual utilization of such a path.

Another aspect regards the choice of the amount of bandwidth to be shifted between the pool and a pipe, which can be made according to various criteria [16]; in the simulations reported below, we have chosen to shift the amount corresponding to a single flow. The "optimal" choice of the bandwidth quantum is currently a matter of investigation.

As said before, this pool (low-level) reallocation mechanism (also embedded in the bandwidth broker) enlarges or shrinks single pipes whenever needed without keeping into account the global information (i.e., the traffic matrix): the first pipe that needs bandwidth can get it at the expenses of the others. Hence, a second (high-level) control mechanism is needed in order to periodically fairly reallocate resources among all pipes. Such mechanism can be triggered in a number of ways. For example, it may start when the (Euclidean) distance between the currently estimated traffic matrix and the one

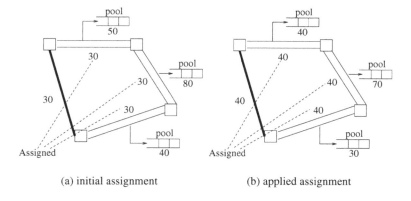

(a) initial assignment   (b) applied assignment

Fig. 1. An example of pool utilization.

used in the previous reallocation is too large; it could also be activated when the pool interventions are too many. If traffic changes very slowly with respect to flow dynamics, such mechanism could be run periodically every given amount of time.

2.1 The Routing Framework

In the proposed system, routing and bandwidth control is distributed among different levels that, therefore, can be assigned quite simple tasks. In particular, this allows to scale down the complexity of the routing mechanism to the following classical simple and fast scheme: whenever a new flow comes up try to use the direct path. If the direct path has no room, then choose the "best" alternative two-hop route (see, e.g., [17] for a number of different solutions in the choice of the alternative path); if there is not enough room over any of the alternative paths, then reject the flow.

Figure 2 shows the overall routing strategy. The scheme aims at meeting both performance and robustness requirements. In fact, the pool guarantees a higher level of control to better satisfy the required flow-level QoS by keeping the rejection rate low. Furthermore, this layered admission control mechanism (i.e., direct, alternative, pool) increases the overall robustness of the system. It is worth noting that routing a flow through an alternative path consumes twice the resources that would have been needed in normal conditions, i.e., for a direct path (in effect, even more that twice if more than two hops are allowed) and, hence, the load imposed to the network is doubled. A well-known risk in this framework is to have, in the long run, all (or the great majority of) flows set up through alternative paths, and this would dramatically decrease the network performance. A well known solution to this problem (also adopted in this work) is to reserve a minimum bandwidth in every pipe for direct flows only (trunk reservation; see, e.g., [17]).

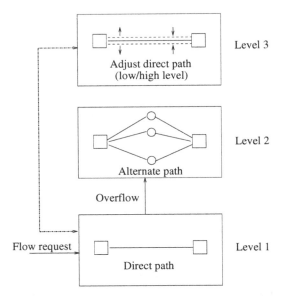

Fig. 2. Pictorial scheme of the overall multi-layer control system.

2.2 Alternate Route Selection

A route optimization mechanism may be employed in the choice of an alternative two-hop path. Generally speaking, three kinds of routing policies may be identified: static, dynamic (alternate), and adaptive routing; in all routing strategies, some technique should be used to choose the preferred route. One such technique is load sharing [17]. In the case of alternate routing, load sharing could be used not only for its simplicity, but also because it achieves some level of fairness. The static routing policy is the simplest one and also the one for which the largest collection of theoretical results is available. In its simplest form, the traffic stream between a source-destination pair is partitioned among a set of predefined paths. When an incoming flow comes up, one of the predefined paths (e.g., the one with largest residual capacity) is selected: if it is busy, then the flow is rejected. The alternate routing is quite similar to the previous one but, this time, the set of the alternative paths is ordered and the system looks for the first available one among them. The adaptive strategy does not order the list of alternative paths but, in real-time, it chooses the "best" one according to some criterion. In our case, as regards the second step of the routing mechanism (the choice of the alternative path), we adopt the third scheme with the load sharing technique, because it is tractable and meets the requirements of dividing the load over a subset of alternative paths. Since this is just a part of a more complex routing scheme and it is not adopted alone, its simplicity represents a point of strength.

However, selecting any path should not violate the QoS requirements, so that a QoS handler algorithm must anyway be applied along the traversed path. We suppose in the following that the such algorithm is based on the knowledge of the maximum number of aggregated flows $N_{r,\mathrm{max}}^{ij}$ acceptable over pipe r between nodes i and j to ensure class-h flows the appropriate QoS requirements. Moreover, the following inequality constraint is applied to determine the bandwidth assignment to a certain pipe at any moment, bearing in mind that the capacity of any pipe might be varied, over time, to comply with the traffic intensities, using the bandwidth pool strategy:

$$V_r^{h,ij} \geq V_{r,\mathrm{min}}^{h,ij} \tag{1}$$

where $V_r^{h,ij}$ is the total amount of capacity of the considered pipe r, for service class h, and $V_{r,\mathrm{min}}^{h,ij}$ is the minimum capacity that is necessary to respect the QoS requirements for the flows in progress. If the new incoming flow could not be accommodated over one of the direct paths that connect the same SD pair, then an alternative path should be selected. The process of selecting an alternative path is activated if

$$N_r^{h,ij} + 1 \geq N_{r,\mathrm{max}}^{h,ij} \tag{2}$$

where $N_r^{h,ij}$ is the number of class-h aggregated flows in progress over pipe r between nodes i and j.

In performing the load sharing mechanism, the offered load matrix to the network (i.e., external incoming traffic intensities for each SD pair) is supposed to be known. Given also a set of alternative paths, the optimization algorithm identifies the share of flow for each alternative path. For example, suppose that for a given source-destination pair and for a given service, the offered load to the alternative paths is known to be, say, the equivalent of 30 Erlangs and suppose that 3 different alternative paths (besides the direct one) are available for that service. Suppose also that the algorithm computes a value for each of them of, say 10, 5 and 15 Erlangs, respectively: hence 10/30 of the total incoming flows will be routed over the first alternative path, 5/30 will be routed over the second one and 15/30 over the third one. This could be done either with a weighted round-robin mechanism or by employing some random mechanism with an appropriate distribution. Notice that the offered load we consider here represents, in effect, the amount of traffic which exceeds the one sent through the direct paths (i.e., the "overflow" traffic), and that we use the same pipes to carry both the direct and the two-hop traffic. We model the flow interarrival and duration statistics with negative exponential distributions. The traffic matrix will contain the offered load values (in Erlangs) for each service and for each source-destination pair. For the sake of notational simplicity, let us concentrate, for the time being, on one type of service (and one pipe for each SD pair), since all that follows can be separately applied to each service class. Let us call A the external input flow matrix, A^{ij} the element which represents the external offered load from node i to node j, and let O^{ij} represent the overflow traffic on the direct path (direct pipe) ij. Let us, furthermore, suppose to have K alternative two-hop paths and let us call $O_k^{ij} = \alpha_k^{ij} O^{ij}$ (with $\alpha^{ij} \geq 0$, $\sum_{k=1}^{K} \alpha_k^{ij} = 1$) the share of the overflow traffic routed through node k, $(k = 1, \ldots, K)$. We denote with P_k^{ij} the blocking probability over the entire path through k between switching nodes i and j. By approximating the pipe blocking probabilities as independent, we will then have:

$$P_k^{ij} = 1 - (1 - B^{ik}) - (1 - B^{kj}) \tag{3}$$

where $B^{ik}, \forall i, k$ is the blocking probability over the virtual link ik. Notice that an alternative path is composed of two pipes, hence ik denotes the first and kj the second pipe involved. The B^{ik}'s are the solutions of the Erlang Fixed Point Equation [9], which, by using the Erlang B function, involves the offered traffic to pipe ik, and the maximum number of virtual circuits which can be set-up on it.

The Erlang Fixed Point Equation, used to find the link blocking probabilities B^{ik}, can be written as:

$$B^{ik} = \text{Er}(Q^{ik}, N_{\max}^{ik}) = \frac{(Q^{ik})^{N_{\max}^{ik}}/N_{\max}^{ik}!}{\sum_{l=1}^{N_{\max}^{ik}}(Q^{ik})^l/l!} \tag{4}$$

where Q^{ik} and $N_{\max}^{ik}$ represent the total offered load and the maximum number of aggregated flows that can be carried over the pipe ik (see 2) and, by using the so-called "reduced load approximation" [8,17], the total offered load to pipe ik can be expressed as

$$Q^{ik} = A^{ik} + \sum_{p\in P(i)} O_i^{pk}(1 - B^{pi}) + \sum_{s\in S(i)} O_k^{is}(1 - B^{ks}) \tag{5}$$

In 5, $P(i)$ and $S(k)$ represent the sets of predecessors of node i and successors of node k, respectively, along a two-hop path; moreover, we should keep in mind that the overflow traffic shares are represented by

$$O_k^{ij} = \alpha_k^{ij}O^{ij} = \alpha_k^{ij}A^{ij}B^{ij} \tag{6}$$

We can now write the overall blocking probability L^{ij} for the ij external traffic:

$$L^{ij} = 1 - \left[(1 - B^{ij}) + \sum_{k=1}^{K} \frac{O_k^{ij}}{A^{ij}}(1 - P_k^{ij})\right] =$$
$$= B^{ij} - \sum_{k=1}^{K} \alpha_k^{ij}B^{ij}(1 - P_k^{ij}) \tag{7}$$

The optimization task takes place by minimizing the overall average blocking probability:

$$\min_{\alpha_k^{ij}} \bar{L} = \min_{\alpha_k^{ij}} \sum_{m,n} \frac{A^{mn}}{\sum_{p,q}A^{pq}}L^{mn} \tag{8}$$

subject to:

$$\sum_{k=1}^{K} \alpha_k^{ij} = 1, \quad \forall i, j$$
$$\alpha_k^{ij} \geq 0, \quad \forall i, j, k \tag{9}$$

When A^{ij} is given, the minimization algorithm yields, for each SD pair i, j, the sharing values for each alternative path ikj ($k = 1, \ldots, K$). Since we only consider two-hop paths, the maximum number K of possible alternative paths for each SD pair will be equal to the number of nodes minus two (i.e., source and destination). This greatly decreases the computational effort of the optimization task. The latter has to be performed on-line whenever either a capacity reconfiguration takes place, or the values of the external offered load A^{ij} change significantly over time.

2.3   Bandwidth Reallocation

As said in the previous sections, the share of bandwidth held by each direct pipe is periodically updated by the two-level reallocation mechanism. In particular, the low-level (pool) mechanism looks at the blocking (estimated) probabilities computed in the previous reallocation period while, at a much lower pace, the high level control algorithm performs a global bandwidth reassignment by implementing some general policy (e.g., fairness) over all pipes.

The low-level control mechanism, having the purpose of a fast short term reallocation and working on a local network scope, does not recompute the values of the best alternate routing parameters α_k^{ij}. Such parameters are, instead, recomputed upon reallocation of bandwidth by the high-level control algorithm.

The low-level reallocation control mechanism. As regards the low-level (pool) control reallocation mechanism, the general framework is to enlarge (if possible) a direct path if its blocking probability exceeds a given threshold and shrink it (if possible) if its blocking probability is below a second threshold. The possibility to enlarge is subjected to bandwidth availability in all bandwidth pools along the DiffServ pipe, while the possibility to shrink is subjected to a lower bound on the bandwidth that is necessary to carry on the flows in progress: obviously no shrink is performed if the pipe is full.

There are (at least) five "levels of freedom" in performing the aforementioned real-location mechanism. The first is the choice of the time instants in which the reallocation procedure is to be considered; the second and the third are represented by the enlarging and shrinking threshold values; the fourth and the fifth are the amount of bandwidth that is subtracted or added to the pipe.

In this work, we have chosen to measure the average blocking probabilities over a window of L events (i.e., births and deaths). Hence, every L events a (pool) reallocation might take place (based on the new estimated values) both for enlarging or shrinking all direct pipes that are "out of threshold".

For what regards the threshold values, in this work we have kept them fixed, even if we are investigating other possibilities. More in particular, since using fixed values means acting in an "open loop" way (in enlarging or shrinking the pipe, its current capacity is not taken into account), some other mechanism able to keep track of the pipe status would be desirable. A trivial example we have tried is represented by having the thresholds proportional to the pipe capacity. However, special care has to be taken in order to have the threshold variation dynamics properly chosen, which is a non trivial task: in this field the use of fuzzy controllers might be appropriate and is the matter of current investigation.

Finally, the amount of bandwidth that is subtracted or added to the pipes has also been chosen fixed and equivalent to one flow. Other schemes are currently under investigation and, as such, they are not reported here. The general assumption, however, is that the reallocation process is able to cope with load dynamics: we suppose that the load over the pipes changes with a rate that allows the reallocation mechanism to update the bandwidth properly.

The high-level reallocation control mechanism. The high-level reallocation control mechanism tries to give enough bandwidth to the DiffServ pipes so to keep the blocking probability given in (4), introduced in Section 2.2, below a given threshold, that is, for all source destination s-d, the algorithm tries to set $B_{sd} \leq B_{\text{th}}$.

The skeleton of the algorithm works by iterating the following steps:

1. assign 0 bandwidth to all pipes.
2. pick the pipe$_{sd}$, from source node s to destination node d, with the highest blocking probability.
3. if $B_{sd} > B_{\text{th}}$ then assign a unit of bandwidth – if available – to pipe$_{sd}$
4. go to 2.

The algorithm ends either when all pipes have a blocking probability below the threshold or when it is not possible to further increase any pipes for lack of available bandwidth. At the end of the iterations, the unassigned bandwidth – if any – remains available for the pool. It might happen that the traffic matrix and the link capacities make the algorithm ending up in a situation where parts of the network have spare bandwidth, given to the pool, and other parts do not have sufficient bandwidth.

At the end of this reallocation phase, the optimal values of the best alternate routing parameters α_k^{ij} are recomputed by means of the minimization in (8) as introduced in Section 2.2.

As regards the high-level control reallocation mechanism, the choice of the appropriate triggering policy heavily depends on the type of traffic that characterizes a given network and, as such, is left to the network administrator. A simple one is to run it every given (fixed) number of incoming flows. A more efficient one waits for a given (fixed) number of low-level reallocation interventions. A further one is to trigger it whenever the "distance" (in an Euclidean way) between the latest and the present allocation matrix exceeds a given threshold. Estimated traffic matrix changes might be included in the triggering mechanism. In this work, we have chosen to set the high-level reallocation instants every R events (i.e., births and deaths).

3 Simulation Results

A number of simulation results, obtained over a simple test network, are reported and commented in this section. The aim is to investigate the performance of the above outlined global strategy.

In our simulations we considered a physical network having a topology as in Figure 3, where each link has a total capacity of 150 Mbits/s. Over this, we have created the virtual network by iterating a backtracking algorithm, which exhaustively sets up the direct pipes, by attempting to spread them evenly over the physical links. As said in Section 2, the procedure is a heuristic one, and it is not among our goals to investigate this problem, which is in itself quite challenging.

The values of the enlarging and shrinking thresholds for the activation of the pool are 0.1 and 0.01 respectively.

For robustness purposes, alternative paths for all the source-destination pairs should be chosen among all the "two-hop" alternatives that do not share physical links (if possible). However, since our aim is mainly to test the control strategy previously outlined, we

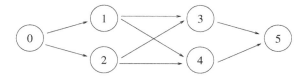

Fig. 3. The physical node topology of the testbed.

do not consider this point here, and we allow all possible two-hop paths in our example network. As regards the traffic, for the sake of simplicity, we considered a single type, in which each flow has an equivalent bandwidth of 1 Mbits/s (under service separation, the extension to multi-class traffic is rather straightforward), generated according to a Poisson model, and having negative exponential duration. The arrival parameters (in Erlangs) are given in the traffic matrix (the average flow duration time is fixed to 150 time units). The simulation runs last 300,000 events each, where an event can be either a flow arrival or completion.

The traffic matrix A in Table 1 shows the initial load (in Erlangs) for every source-destination pair.

Table 1. Initial traffic matrix (in Erlangs).

0	2.947	2.982	2.989	2.967	2.948
0	0	0	2.971	2.99	2.972
0	0	0	2.97	2.968	2.946
0	0	0	0	0	2.983
0	0	0	0	0	2.942
0	0	0	0	0	0

Given the matrix A, the high-level reallocation algorithm computes a bandwidth of approximately 6 Mbit/s for all pipes, given a blocking probability threshold $B_{th} = 0.1$.

A very first comparison has been done regarding the pool. In order to highlight its benefits, one entry of the traffic matrix – namely the one from source 0 to destination 1 – has been progressively increased (using a sine function) from its initial value of 2.947 Erlangs to a final value of about 115 Erlangs. Figure 4 show the overall average blocking probability over all SD pairs (which include both direct and alternate paths). As expected, it turns out that the presence of the pool dramatically increases the performance, regardless of the alternate choice used. The system converges toward a point in which almost all flows can be routed through the direct path. Figure 5 shows the behavior of the blocking probability of pipe 0-1 when its load increases. The pool mechanism increases the pipe size in order to keep below enlarging threshold (i.e., 0.1). Figure 6 shows the effect of the thresholds on the behavior of both pool assignments and blocking probabilities, always for pipe 0-1. In this example, the network conditions have been (appositely) set such that the blocking probability always exceeds the shrinking threshold. Because of this, the pool enlarges the pipe but never shrinks it. The system operates about 50 shrink/enlarge

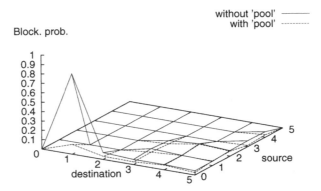

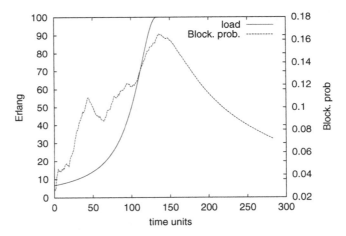

Fig. 4. Blocking probabilities: pool vs. no pool.

Fig. 5. Blocking probability vs. increasing load: the effect of the low-level (pool) mechanism.

operations on the whole network. Figure 7 shows the effect of augmenting the shrinking threshold to 0.095. In this example, always for pipe 0-1, it is evident the closer mirroring of the offered load operated by the pool with the assigned capacity. In this case, the pool reacts much more frequently: the system operates about 1600 shrink/enlarge operations on the whole network. Notice how the blocking probability much more closely sticks to its target value.

Another effect of the pool is that, being local, it works on a "first come first served" basis and, therefore, another pipe also needing part of the shared bandwidth might remain unserved. This is shown in Figure 8 where also a second entry in the traffic matrix – namely the one from source 0 to destination 3 (also traversing link 0–1) – starts increasing, from its initial value of 2.989 Erlangs to 115 Erlangs, but later in time. Since

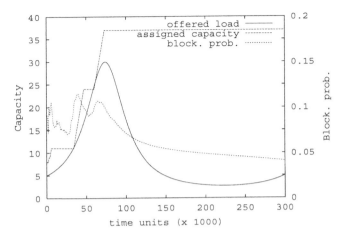

Fig. 6. The effect of the shrinking threshold (set to 0.01).

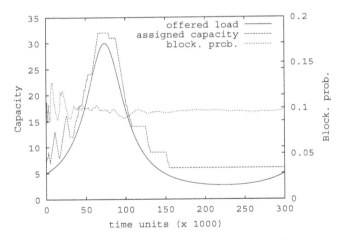

Fig. 7. The effect of the shrinking threshold (set to 0.095).

the first requiring pipe (i.e., 0–1) has already got much of the available bandwidth, the second increasing one remains unserved by the pool mechanism.

Figure 9 shows the benefits of the joint action of the two reallocation mechanisms. The plots refer to the blocking probabilities after the reallocation performed by the high-level global mechanism that redistributed evenly the bandwidth among the two pipes.

Block. prob.

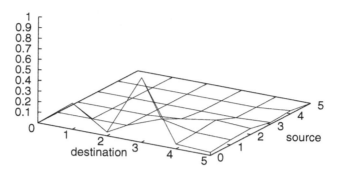

Fig. 8. Pool mechanism drawback.

Block. prob

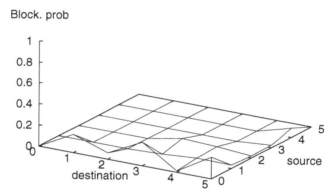

Fig. 9. The effect of the high-level control.

4 Conclusions

A measurement based real-time rearrangement policy for DiffServ pipes along with a new adaptive bandwidth reallocation scheme has been introduced and investigated. A "bandwidth pool" is proposed, to react to a high blocking probability of new incoming flows as well as to unused allocated bandwidth, while a high-level global reallocation control mechanism is adopted for adjusting distribution of shared resources over a long term basis. It has been shown by simulation that the use of the proposed mechanism positively affects the blocking rate.

References

1. Internet Traffic Engineering. IEEE Network Magazine, March/April 2000.
2. S. Deering and R. Hidden. Internet Protocol, Version 6 (IPv6) Specification. Technical report, IETF RFC 2460, December 1998.

3. J. Wroclawski. The Use of RSVP with IETF Integrated Services. Technical report, IETF RFC 2210, September 1997.
4. S. Blake, D. Black, M. Carlson, E. Davies, Z. Wang, and W. Weiss. An Architecture for Differentiated Services. Technical report, IETF RFC 2475, December 1998.
5. E. Rosen, A. Viswanathan, and R. Callon. Multiprotocol Label Switching Architecture. Technical report, IETF RFC 3031, January 2001.
6. R. Braden, L. Zhang, S. Berson, S. Herzog, and S. Jamin. Resource ReSerVation Protocol. Technical report, IETF RFC 2205, September 1997.
7. L.M. Ni X. Xiao. Internet QoS: a big picture. *IEEE Network Magazine*, 13(2):8–18, March-April 1999.
8. K.W. Ross. *Multiservice Loss Models for Broadband Telecommunication Networks*. Sprinter Verlag, London, 1995.
9. A. Dalal'ah, F. Davoli, P. Maryni, and M.S. Obaidat. Multi-layer VP Assignment in an ATM Backbone. *International Journal of Parallel and Distributed Systems and Networks*, 2(4):235–243, 1999.
10. N. Semret, R.R.-F. Liao, A.T. Campbell, and A.A. Lazar. Pricing, Provisioning and Peering: Dynamic Markets for Differentiated Internet Services and Implications for Network Interconnections. *IEEE Journal on Selected Areas in Communications*, 18(2):2499–2513, December 2000.
11. I. Chlamtac, A. Farago, and T. Zhang. Optimizing the system of virtual paths. *IEEE/ACM Transactions on Networking*, 2(6):581–587, December 1994.
12. G. Manimaran, H.S. Rahul, and S.R. Murthy. A New Distributed Route Selection Approach for Channel Establishment in Real-Time Networks. *IEEE/ACM Transactions on Networking*, 7(5):698–709, October 1999.
13. Y. Xiong and L.G. Mason. Restoration Strategies and Spare Capacithy Requirements in Self-Healing ATM Networks. *IEEE/ACM Transactions on Networking*, 7(1):98–110, February 1999.
14. M. Medard, S.G. Finn, R.A. Barry, and R.G. Gallager. Redundant Trees for Preplanned Recovery in Arbitraty Vertex-Redundant or Edge-Redundant Graphs. *IEEE/ACM Transactions on Networking*, 7(5):641–652, October 1999.
15. A. Banerjea. Fault Recovery for Guaranteed Performance Communications Conennctions. *IEEE/ACM Transactions on Networking*, 7(5):653–668, October 1999.
16. S. Otha and K. Sato. Dynamic Bandwidth Control of the Virtual path in Asynchronous Transfer Mode Network. *IEEE Transactions on Communications*, 40(7):1239–1247, July 1992.
17. A. Girard. *Routing and Dimensioning in Circuit-Switched Networks*. Addison-Wesley, Reading, MA, 1990.

A Protection-Based Approach to QoS in Packet over Fiber Networks

Patrick Thiran[1,2], Nina Taft[1], Christophe Diot[1],
Hui Zang[1], and Robert Mac Donald[1]

[1] Sprint Advanced Technology Labs, Burlingame, CA 94010, USA
[2] ICA-DSC, EPFL, CH-1015 Lausanne, Switzerland,
Patrick.Thiran@epfl.ch

Abstract. We propose a novel approach to Quality of Service, intended for IP over SONET (or IP over WDM) networks, that offers end-users the choice between two service classes defined according to their level of transmission protection. The first service class (called Fully Protected (FP)) offers end-users a guarantee of survivability: all FP traffic is protected in the case of a (single) failure. The second service class (called Best-Effort Protected (BEP)) does not offer any specific level of protection but is cheaper. When failures occur, the network does the best it can by only restoring as much BEP traffic as possible. We motivate the need for two classes of protection services based on observations about backbone network practices that include overprovisioning and an ongoing but unbalanced process of link upgrades. We use an ILP formulation of the problem for finding primary and backup paths for these two classes of service. As a proof of concept, we evaluate the gain of providing two protection services rather than one in a simple topology. These initial results demonstrate that it is possible to increase the network load (and hence revenue) without affecting users that want complete survivability guarantees.

1 Introduction

Today's internet backbone contains a large amount of unused capacity due primarily to the following three reasons: overprovisioning, duplication of equipment and unbalanced link upgrades. Overprovisioning is the current de facto solution to providing QoS.

A lot of effort is devoted to broadening the set of Internet services to a palette ranging from best-effort to real-time and streaming services. The proposed solutions for such services differ in the mechanisms they use - such as reservation or priority. However, their common goal is to provide users with a variety of service classes that differ based on their performance with respect to throughput, loss and/or delay measures. Such a differentiation is indeed useful when congestion occurs in portions of the network. But backbone networks are usually overprovisioned because it is often simpler and cheaper to buy additional hardware equipment than to run complex software for managing reservations and priorities in routers. Hence traffic rarely experiences congestion in the backbone [1],

S. Palazzo (Ed.): IWDC 2001, LNCS 2170, pp. 266–278, 2001.

making service differentiation quite useless in practice. Overprovisioning allows carriers to provide everybody with the best class of service.

Not only is the backbone overprovisoned to offer low delay and losses to all traffic, but most equipment is duplicated for protection against failures. Carriers are not willing to forgo this additional redundancy because they do not want network services to be disrupted, even rarely. (Failures are actually less rare than one might expect; [2] has recently reported failure rates of 1 per year per 300km of fiber.) Avoiding service disruption is especially critical for backbone links, where a single failure may interrupt many channels. A large fraction of the capacity in the backbone links therefore remains unused, and this situation is likely to continue as long as the bottlenecks are in the access network rather than the backbone.

On the other hand, because traffic demands grow exponentially, network operators are continuously obliged to upgrade the capacity of their backbone links. Upgrading a backbone link can be a lengthy operation, and thus in practice links are upgraded one at a time. Many months can pass between the upgrading of two links. Providing protection means that an upgrade of the capacity for primary working links, should be matched by an equivalent upgrade of the redundant protection links. However, since the network is essentially in a continual state of flux, the typical network is quite heterogeneous containing some recently upgraded high-speed links (e.g. links with a DWDM system of 80 to 160 wavelengths operating at a 10 Gbps line speed), alongside older slower-speed links (e.g. WDM fibers with only 4 to 32 wavelengths at 2.5 Gbs line speed). This situation prevents operators from making use of the capacity in recently upgraded links. To see why, consider the following scenario. Suppose all links are initially 2.5 Gbps and then exactly one of them is upgraded to 10 Gbps. The full capacity of this link cannot be used for paths spanning multiple hops for two reasons. First, other links may not be able to support the growth in traffic, and second, it is unlikely that a backup path, on the other 2.5 Gpbs links, can be found for this additional traffic.

The combination of overprovisioning, redundant capacity for failures, and partial network upgrades creates a situation in which, on a day-to-day basis, there exists a large amount of unused bandwidth in the Internet backbone. In order to leverage this unused bandwidth we propose the use of two classes of service that differ according to the protection level provided. The two service classes are intended for either IP/SONET or IP/WDM networks with IP at the logical layer and either SONET or WDM systems at the physical layer. The first one, hereafter called the *Fully Protected* (FP) class, offers users the insurance that none of their traffic will be disrupted in case of a single point of failure. The second one, hereafter called the *Best-Effort Protected* (BEP) class, does not provide any specific guarantee on service disruption. Instead, in the case of failure, it offers to restore as much of the affected traffic as possible. What BEP offers to users, as a tradeoff for a lower amount of protection, is either a larger throughput, or a cheaper price. We will discuss how having two such services allows carriers to carry the BEP traffic on the excess capacity without impacting the FP traffic.

Many proposals for service classes differentiate the classes according to their delay, loss or throughput performance. Reliability is also an important QoS performance metric and the wide variety of applications that exist today demand different levels of availability guarantees. Some applications, such as IP telephony, video-conferencing, and distance surveillance require 100% availability and hence full protection against network failures. Others, like on-line games, Web surfing, and Napster downloads are likely to be willing to tradeoff a partial and slower protection for increased throughput (or a lower price). Such tradeoffs are attractive as long as the probability of a service becoming unavailable is very small. Applications like e-mail can fall into either one of these service classes.

The reliability dimension of QoS can be quite independent of the traditional QoS parameters of delay, loss and throughput that are often correlated to one another. For example, two applications requiring similarly high levels of reliability need not have similar delay requirements. Most applications requiring full protection will be the priority traffic, but this may not always be the case. Table 1 demonstrates that categorizing applications by their protection needs can be different than categorizing them according to their traditional QoS needs. Reliability also differs from these traditional QoS measures in that delay, loss and throughput guarantees can be trivially satisfied by overprovisioning (if you are willing to pay for it), whereas reliability cannot because the amount of overprovisioning has to be carefully calculated. Overprovisioning to provide delay, loss and throughput guarantees can be done by simply inflating each link by say 20 or 30%, or by ensuring that the load on each link rarely exceeds specified thresholds (e.g., 60%). However such a per-link view of overprovisioning is insufficient for meeting reliability guarantees which requires a network-wide view of the capacity. This is because all links must be inflated proportionally if one wants to ensure that backup paths will exist for all source-destination pairs of flows. The slow and unbalanced process of link upgrades makes it very difficult to overprovision using a network-wide perspective.

The introduction of services offering different levels of protection guarantees at the WDM layer is gaining attention in the optical networking community. A classification in five classes is proposed in [3]. In our paper, we consider only two classes, but defined at the IP layer. This will result in making some SONET (or WDM) paths protected and others not, as in the work of Sridharan and Somani [4]. Despite some differences with this latter work in the problem formulation (for instance, we do not introduce different costs for each type of working or back-up paths, but we constrain the ratio between BEP and FP traffic to be less than a prescribed maximal value), we reach a similar conclusion, that the traffic load can be increased quite considerably when more then one class of protection is available in a homogeneous network, where all links have the same capacity. We show that this effect is even more accentuated in a heterogeneous network.

The rest of this paper is organized as follows. Section 2 briefly summarizes the kinds of mechanisms provided for protection at the optical and IP layers. We state our service definitions in Section 3 and describe which protection mechanisms are needed by each of the service classes. The ILP formulation of the resulting routing problem is given in Section 4. For a proof of concept demonstration, we provide an example in Section 5 that illustrates that a good deal of BEP traffic can be carried on the network without affecting the FP traffic, and thus we can substantially increase the load (and hence the revenue) the network carries. In Section 6, we extend our ILP formulation in order to secure a minimal amount of bandwdith to restore a fraction of the BEP traffic after a failure, so that this class of traffic does not suffer a complete service disruption in case of a failure, but a softer degradation. We conclude our proposal in Section 7.

Table 1. Service Categorization

Service	Fully Protected	Best Effort Protected
Low delay and losses	IP telephony, distance monitoring	cheap on-line games
Loose delay or loss requirement	professional e-mail	private e-mail, web surfing

2   Handling Failures at the IP and SONET Layers

Defining classes of service for protection requires specification of how protection is handled for each class. Before stating our proposal, we review the mechanisms that are available at each layer in the network. The optical layer provides *protection* that carries out very fast failure recovery but is often not bandwidth efficient [5, 6]. The IP layer can provide *restoration* that helps to determine more efficient routes but is typically not very fast. Most networks today rely on SONET to carry out protection.

Protection at the SONET layer. All protection techniques involve providing some redundant capacity within the network to reroute traffic in case of a failure. Protection is the mechanism by which traffic is switched to available resources when a failure occurs. It needs to be very fast; the commonly accepted standard for SONET is 50 ms. Protection routes must therefore be pre-computed, and wavelengths must be reserved in advance at the time of connection setup. Protection around the failed facility can be done at different points in the network: (i) around the two end-points of the the failed link, by *line* or *span protection* (in optical layer terminology this corresponds to protection at the line or multiplex sublayer), or (ii) by *path protection* which is between the source and destination of each connection traversing the failed link (in optical layer terminology, this corresponds to protection at the path sublayer) [7–9]. Line protection is simpler, but path protection requires less bandwidth and can better handle node failures. Here, we only consider path protection.

There are essentially two fundamental protection mechanisms. In *1+1 protection* traffic is transmitted simultaneously on two separate fibers on disjoint routes. The receiver accepts traffic from the primary fiber (also called working fiber) and only switches to accept the input from the other fiber (called protection or back-up fiber) in case the primary one fails. In *1:1 protection* traffic is transmitted only on the primary fiber. If this fiber is cut, the sender and receiver use simple signalling to jointly switch to the backup fiber. The generalization of 1:1 protection is 1:*n* protection where one back-up path protects *n* working paths. For our initial proof-of-concept analysis, we consider 1+1 and 1:1 protection schemes in this paper.

Restoration at IP layer. Since the IP layer is made up of a well meshed topology and its links are not fully loaded (due to overprovisioning), the IP layer is also capable of restoring traffic around a failed facility.

After SONET protection is done, today's routing protocols can discover routes in the new topology that are more efficient than the backup path used for failure recovery in the old topology. Within a carrier's backbone, Internal Gateway Protocols (IGP) are used for intradomain routing. IS-IS and OSPF are the most common protocols deployed today. In these protocols, routers periodically exchange hello messages to check the health of neighboring links and nodes. If a few successive messages are lost, a router deduces that a link or node is down, and begins the restoration process at the IP layer. After detection of a topology change, this process involves propagating the change information across the network and recomputing shortest paths. During the restoration process, a subset of destinations are reached through non-optimal routes (if the network supports SONET) or are briefly unreachable (otherwise). In IS-IS, the process of failure detection can take between 10-30 seconds depending upon the protocol configuration, and the rest of the recovery process can take another 10 seconds or so [10]. Although ISIS convergence today takes on the order of tens of seconds, it is believed [10] that these convergence times can be greatly reduced, potentially to the order of tens of milliseconds. The theoretical limit of link-state routing protocols to reroute is in link propagation time scales - in other words in the tens of milliseconds. Using today's technologies, restoration speed at the IP layer cannot compete with the protection and restoration speeds at SONET (or WDM) layers.

A difficulty that arises in today's networks, e.g., IP/SONET, is that each layer performs protection independently from the other layers. For example, IGP routing table updates are performed independently of SONET's line protection. This can lead to undesirable race conditions between different layers. Ideally IP and optical networks should be managed as an integrated network without overlap of functionality between layers and with sharing of information between layers. The issue of deciding exactly which aspects of protection and restoration should be carried out by which layer is still an open issue. The advantage of providing protection at the IP layer is the cost reduction that results from saving redundant equipment at the physical layer. The disadvantage is that it is slow. Providing protection at the SONET layer has the reverse tradeoff.

3 Definition and Provisioning of Service Classes

We now define our two service classes that differ in terms of their level of protection, their mechanism of protection and their cost.

Fully Protected (FP) service class.

- This service guarantees its customers that their traffic is protected against any single point of failure in the backbone within 50 msec.
- This service provides fast protection. Therefore FP traffic is protected via pre-computed, dedicated back-up paths at the SONET or WDM layer, using either by 1:1 or 1+1 protection. Failures are transparent to the IP layer for this class of traffic.
- This service is the more expensive of the two.

Best Effort Protected (BEP) service class.

- This service does not offer specific guarantees for protection against failures, but instead tries to restore as much of this traffic as possible after a the occurrence of a failure.
- For BEP traffic we offer restoration and not protection. When a failure occurs, BEP packets will be dropped at the router before the point of congestion, until IP has been able to restore this traffic by rerouting it on an alternate IP path. Actually this service can come in a variety of flavors. The simplest version of this service class is to leave BEP traffic entirely unprotected at the SONET (and/or WDM) layer. A more enhanced version of this service (and more difficult to implement) is to ensure users that in case of a single failure, they would not experience a complete service disruption but may experience a severe degradation.
- This service is cheaper than the FP service.

In order to implement two such service classes, packets would need to be marked according to their service class, and IP routers would need class-based scheduling. In normal operation, differentiation is not needed between the two types of packets. However, upon notification of a failure, FP packets continue to be served as before, while BEP packets are dropped until BEP traffic has been restored at the IP layer.

4 ILP Formulation

We formulate the problem of routing traffic flows from two service classes over a physical and logical topologies as an Integer Linear Programming (ILP) problem whose objective is to maximize the total load carried by the network, which we denote by F.

We consider here that all *physical channels* are SONET paths. They could also be WDM lightpaths, if all optical cross-connects have full wavelength conversion possibilities, or if they perform electronic conversion before switching, so that wavelength continuity constraints [5] can be ignored. Each path is assigned a unit capacity, which represents the smallest granularity level of bandwidth of a SONET path. The total capacity C_l of physical link l, with $1 \leq l \leq L$ where L is the number of physical links, is thus an integer.

The *logical topology* is the set of logical links between IP routers. Let M be the numbers of routers that are connected by a logical link. Each logical link between router s and router t has capacity d_{st}, and is a set of consecutive physical links that form a route r. Logical links are considered here as bi-directional, i.e. d_{st} is the sum of the demand from s to t and from t to s. In the following, we need thus to consider only source-destination pairs (s, t) with $s < t$. This assumption can clearly be relaxed.

Each logical link presents a demand of d_{st} capacity units at the physical layer, for which one needs to find a route r among the set of all routes $\mathcal{R}_{st}$ between the source s and the destination t, such that the capacity constaints of all links l belonging to route r are satisfied. To keep routing at the physical layer simple, we do not allow multiple working paths between a given pair of nodes. Denoting by d_{st}^r the traffic flowing on route $r \in \mathcal{R}_{st}$, we have thus that for all $1 \leq s < t \leq M$

$$d_{st} = \max_{r \in \mathcal{R}_{st}} \{d_{st}^r\}. \tag{1}$$

If multiple routes were allowed, one would have to change the maximum in this equation, by a sum.

A logical link between a given pair of nodes (s, t) carries d_{st}^{FP} traffic units of the FP class, and d_{st}^{BEP} traffic units of the BEP class:

$$d_{st} = d_{st}^{FP} + d_{st}^{BEP}. \tag{2}$$

Because of (1), both traffic classes are carried on the same route $r \in \mathcal{R}_{st}$.

In the simplest case, no precaution is taken to guarantee even a partial restoration of BEP traffic at the logical layer. This means that BEP traffic can be left unprotected, and will be restored only if resources are available after the failure has occurred. In the worst case, all BEP traffic may have to be dropped as a result of a failure.

On the other hand, FP traffic is protected on a 1+1 or 1:1 basis. This is the simplest and fastest recovery scheme, but also the most resources consuming. It requires that for each primary route $r \in \mathcal{R}_{st}$, we find a link disjoint route r' (if we have only link failures) or even a link and node disjoint route (for the general case where both link and node failures can occur) from r, that can carry d_{st}^{FP} traffic units. We consider here only the case of link failure. To state the resulting constraint, we first introduce the membership function

$$\delta_l^r = \begin{cases} 1 & \text{if } l \in r \\ 0 & \text{if } l \notin r \end{cases}$$

for any link $1 \leq l \leq L$ and any route $r \in \mathcal{R}_{st}$. We must therefore find a route $r' \in \mathcal{R}_{st}$ such that the traffic demand $d_{st}^{r',PR}$ on this protection route verifies for all $1 \leq s < t \leq M$, $r \in \mathcal{R}_{st}$

$$d_{st}^{r',PR} = d_{st}^{FP} \tag{3}$$

$$\sum_{1 \leq l \leq L} \delta_l^r \delta_l^{r'} d_{st}^r d_{st}^{r',PR} = 0. \tag{4}$$

Constraint (3) states that all traffic of the FP class, flowing on the primary route r, must be protected by a traffic allocation of the same amount on a back-up route r'. Constraint (4) ensures that it is link disjoint with r.

The finite link capacity imposes that for all $1 \leq l \leq L$

$$\sum_{s=1}^{M} \sum_{t=s+1}^{M} \sum_{r \in \mathcal{R}_{st}} \delta_l^r \left(d_{st}^r + d_{st}^{r,PR} \right) \leq C_l. \tag{5}$$

The two last constraints are provided by the actual traffic data.

The first one is the proportion of traffic belonging to both classes. We represent the amount of BEP traffic between node pairs as a given multiple ρ of the FP traffic. Clearly, FP traffic will require more resources than BEP traffic, so we need to set a maximum value to this ratio, since otherwise the optimal solution will always consist in having all traffic in the BEP class. Therefore we constrain ρ to be less than than a given maximal value $\rho_{\max}$:

$$d_{st}^{BEP} \leq \rho_{\max} \cdot d_{st}^{FP} \tag{6}$$

for all $1 \leq s < t \leq M$.

The second one is the fraction of the total load F that needs to be assigned between each pair of nodes, and which would be obtained by the IP traffic matrix data. By default, we assume here a balanced repartition of the load between each pair of IP nodes, so that the same fraction of the total load is assigned between each pair of nodes:

$$d_{st} = d_{s't'} \tag{7}$$

for all $1 \leq s < t \leq M, 1 \leq s' < t' \leq M$.

The problem amounts therefore to maximize

$$F = \sum_{s=1}^{M} \sum_{t=s+1}^{M} d_{st}$$

subject to constraints (1) to (7).

5   Example

In this section, we illustrate our ideas with a numerical example. The goal of this example is to serve as a proof of concept to demonstrate the gain that can be achieved by supporting more than one protection service class. In today's networks the only protection class is FP.

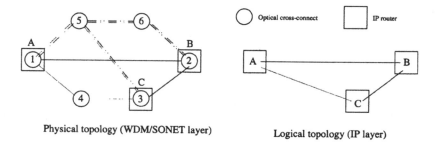

Physical topology (WDM/SONET layer)   Logical topology (IP layer)

Fig. 1. A SONET/WDM network (left) with working paths in plain and back-up paths in dashed lines. The logical topology at IP layer is represented on the right, and consists here of three logical links.

Figure 1 shows a network, consisting of $N = 6$ nodes and $L = 8$ links at the physical layer (SONET/WDM), and of $M = 3$ nodes and $M(M-1)/2 = 3$ links at the logical layer (IP). We consider here that all physical channels are SONET paths. Remember that the capacity unit is the smallest capacity of a SONET path, and that the capacity C_l of a link l is therefore an integer multiple of this capacity unit. In our example, the capacity of each phyiscal link is equal to 8, if the link has not been upgraded, and to 32, if the link has been upgraded. Figure 1 shows one possible mapping of the logical links (right) on the physical links (left), which is as follows:

- logical (IP) link (A, B) is mapped on working (SONET) physical path (or route) $\{(1, 2)\}$ and back-up (SONET) physical path $\{(1, 5), (5, 6), (2, 6)\}$;
- logical link (A, C) is mapped on working physical route $\{(1, 4), (3, 4)\}$ and back-up physical route $\{(1, 5), (3, 5)\}$;
- logical (IP) link (B, C) is mapped on working physical route $\{(2, 3)\}$ and back-up physical route $\{(2, 6), (5, 6), (3, 5)\}$.

Other mappings are of course possible, the mapping which will be eventually adopted is the one that solves the ILP described in the previous section.

We use ILOG optimizer [11] to find the solution of the ILP. Figure 2 displays the results, when the following number of links have been upgraded: (i) none, (ii) two links ($(1, 2)$ and $(2, 3)$), (iii) four links ($(1, 2)$, $(2, 3)$, $(3, 4)$ and $(1, 4)$) and (iv) all eight links. The x-axis denotes $\rho_{\max}$, which is defined by (6).

First observe the case when all links have the same capacity, either before an upgrade or after an upgrade of all links. If we compare the scenario without any BEP traffic ($\rho_{\max} = 0$), and a scenario with BEP traffic ($\rho_{\max} = 1$), we see that we can nearly double the load on the network. As $\rho_{\max}$ denotes the maximal ratio between BEP and FP, it is natural that after some value of $\rho_{\max}$, the curves become flat because no more additional traffic can be added in the system.

Second, consider the case of a partial upgrade and say $\rho_{\max} = 4$ for example. If only two links are upgraded no additional FP traffic can be carried on the

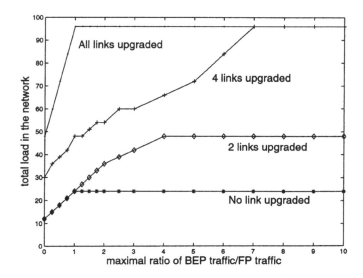

Fig. 2. Total number of demands (total load) versus maximal ratio $\rho_{\max}$ of BEP traffic over FP traffic.

network. However a good deal of BEP traffic can be added after the upgrades. In case of an upgrade of four (appropriately chosen) links, one even reaches the same capacity as with a full upgrade of all eight links, for $\rho_{\max} \geq 7$.

6   Partial Restoration of BEP Traffic at IP Layer

In the previous example, no precaution was taken to prevent BEP traffic from being dropped in case of a failure. It is however desirable that the connectivity of the IP layer be preserved after a single failure, so that BEP traffic is partly restorable (by partly restorable, we mean here that every IP node is reachable, but that queuing delays may become signifigant).

This imposes an additional constraint on the mapping of the logical topology on the physical topology, namely that a single failure leaves the logical topology connected. This problem has been shown to be NP-complete [12], and therefore requires heuristics for general logical and physical topologies. However, when the logical topology is a ring (as in our example), this constraint becomes particularly simple to state [13]: one must simply check that no physical link is shared by two logical links, since otherwise the failure of such a physical link would leave the logical topology un-connected. In other words, we now introduce the additional constraint that for all $1 \leq s < t \leq M$, $1 \leq s' < t' \leq M$, with $(s,t) \neq (s',t')$, and for any $r \in \mathcal{R}_{st}$ and $r' \in \mathcal{R}_{s't'}$

$$\sum_{1 \leq l \leq L} \delta_l^r \delta_l^{r'} d_{st}^r d_{s't'}^{r'} = 0. \tag{8}$$

In this case, the curve in Figure 2 for 2 upgraded links coincides with the curve for zero upgraded link. However, the curve for 4 upgraded links remains unchanged.

In this network, after a single failure on any link of the network, all FP traffic can therefore be rerouted on alternate routes offering the same capacity, whereas BEP traffic that used the broken link now needs to share routes with other BEP flows. As a result, congestion will occur for BEP traffic. In the example above, it is easy to check that the capacity offered to all BEP traffic after a failure is half the capacity it had before.

A better service would be provided for the BEP class, if we slightly over-provision the links taken by BEP traffic, so that it has some spare capacity from which it can benefit to absorb occassional bursts of traffic when no failure has occurred, and to offer a less severe degradation after the occurrence of failure, to rerouted BEP traffic.

Let us denote by ε the *amount of over-provisioning* we provide to the traffic class. This means that for every demand of d_{st}^{BEP} BEP traffic units between s and t, we will actually reserve $(1+\varepsilon)d_{st}^{BEP}$ capacity units. Because of the logical ring topology of our example, one can check that a single failure will then always leave a fraction $(1+\varepsilon)/2$ of the capacity needed for restoring BEP traffic at the IP layer. The value $\varepsilon = 0$ corresponds to the previous case, where BEP traffic receives half the traffic it has before a failure. A value $\varepsilon = 1$ corresponds to a fully restorable BEP traffic at the IP layer (in which case the only difference between the FP and BEP traffic is the layer, and thus the speed, at which traffic is restored).

This amounts to replacing constraint (5) by

$$\sum_{s=1}^{M} \sum_{t=s+1}^{M} \sum_{r \in \mathcal{R}_{st}} \delta_l^r \left(d_{st}^r + \varepsilon d_{st}^{r,BEP} + d_{st}^{r,PR} \right) \leq C_l$$

where $d_{st}^{r,BEP} = d_{st}^{BEP}$ because of (1) and (2). Of course, we need to keep (8) in the set of constraints.

Figure 3 shows the resulting total load when $\varepsilon = 0.5$. Because of the over-provisioning, the total load has decreased, compared to the scenario depicted in Figure 2. In this new scenario, upgrading 4 links no longer allows the network to reach the same total load level as in the case of upgrading all 8 links. This is to be expected as it illustrates the tradeoff between carrying extra load and providing (partial) restoration. However there is still a sizable gain in having two protection services. For example, in this case of partial restoration for BEP, with 4 links upgraded our approach can double the amount of new load carried as compared to a system with a single service ($\rho = 0$)

7   Conclusion

We proposed two service classes based on the level of reliability required by users. The FP class ensures fast protection at the SONET or WDM layer, and makes

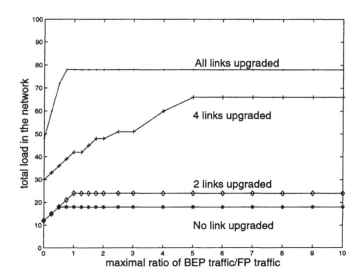

Fig. 3. Total number of demands (total load) versus maximal ratio ρ_{max} of BEP traffic over FP traffic, when $\varepsilon = 0.5$, so that a fraction of 0.75 of the BEP traffic can be restored.

failures transparent to the IP layer. The BEP class does not offer any availability guarantees after a failure, and is left unprotected at the SONET and/or
WDM layers. This proposal allows carriers to make good use of a few upgraded
backbone links that otherwise would provide limited benefit until the majority
of the backbone links have been similarly upgraded. Preliminary results show
that in heterogeneous networks resulting from partial upgrades, our approach
allows a substantial amount of additional traffic to be carried. In particular, we
showed that in the case of our simple topology, when half of the network links
are upgraded the amount of new load carried can be doubled or tripled (depending upon the amount of protection offered to BEP users) as compared to
an environment that supported only a single full protection service. Our results
demonstrate that by having a second protection class of service, carriers achieve
a new method of generating revenue without harming their existing protection
class of service.

Further research should investigate these benefits for larger physical topologies, and meshed logical topologies. This approach should also be refined more
generally (not just for ring topologies) so that the BEP traffic can secure some
level of restoration at the IP layer.

Finally, if SONET is no longer the layer handling failures, and if optical cross-
connects do not perform wavelength conversion, then MPLS may be needed to
map IP traffic directly on the lightpaths [14, 15]. The MPLS protocol indeed
offers a potential alternate mechanism for providing protection and restoration
at layers 2/3. MPLS is a general purpose tunneling mechamisn that uses a sim-

ple label-swapping forwarding technique to transport IP packets across an IP network. It creates tunnels, called Label Switched Paths (LSPs) by distributing labels along a path of MPLS-capable routers. The LSP tunnel essentially sets up a path through a network of connectionless IP routers. MPLS is suited for survivability for a few reasons. LSPs can be used as backup paths and can be computed in advance. This requires storing extra labels in a forwarding table. Also, MPLS is not dependent upon IGP convergence since backup LSPs can be established a priori. Research in the performance of MPLS restoration mechanisms is still immature. However it is hypothesized that for link failures, link protection can occur within tens of milliseconds since no signalling is required. Yet path protection is expected to take on the order of seconds because this would require some signalling to inform the head of the tunnel about the topology change.

References

1. Fraleigh, C., Moon, S. B., Diot, C., Lyles, B., and Tobagi, F.: Architecture of a Passive Monitoring System for backbone IP Networks. Sprint Technical Report TR00-ATL-101801 (2000).
2. Arijs, P. , van Caenegem, P., Demmester, P., Lagasse, P., Van Parys, W., Achten, P.: Design of Ring and Mesh Based WDM Transport Networks, Optical Networks Magazine (2000) 25–40
3. Gerstel, O., Ramswami, R.: "Optical Layer Survivability: A Services Perspective". IEEE Communication Magazine, 38(3) (2000) 104–113.
4. Sridharan, M., Somani, A. K.: "Revenue Maximization in Survivable WDM Networks." OPTICOMM 2000 (2000) Dallas.
5. Ramamurthy R. and Mukherjee B., "Survivable WDM Mesh Networks". Proc. Infocom99 (1999) 744–751 New York.
6. Doshi, B., Dravisa, S., Harshavadhana, Hauser O., Wang Y.: "Optical Network Design and Restoration." Bell Labs Technical Journal, 4(1) (1999) 58–84.
7. ITU-T G. 872, "Optical Transport Networks" (1999).
8. Bonenfant, P., Rodriguez-Moral, A.: "Optical Data Networking". IEEE Communication Magazine, 38(3) (2000) 63–70.
9. Alferness, R.C., et al: "A Pratical Vision for Optical Transport Networking". Bell Labs Technical Journal, 4(1) (1999) 3–18.
10. Alaettinoglu, C., Jacobson, V., Yu, H.: Towards Milli-Second IGP Convergence, IETF Draft (ftp://ftp.isi.edu/internet-drafts/draft-alaettinoglu-ISIS-convergence-00.txt). (2000)
11. http://www.ilog.com.
12. Crochat, O., Le Boudec, J.-Y., Gerstel O: Protection Interoperability for WDM Optical Networks. IEEE/ACM Transactions on Networking vol. 8 (2000) 384–395
13. Modiano, E., Narula-Tam, A.: Survivable Routing of Logical Topologies in WDM Networks. Proc. Infocom2001 (2001) Anchorage.
14. Awduche, D., et al: IETF Draft (ftp://ftp.isi.edu/internet-drafts/draft-awduche-mpls-te-optical-01.txt). (1999)
15. Metz, C.: IP Protection and Restoration. IEEE Internet Computing (2000), 97–102.

IP Routing and Mobility

Cristina Hristea and Fouad Tobagi

Stanford University
{christea,tobagi}@Stanford.edu

Abstract. The original design of the Internet and its underlying protocols did not anticipate users to be mobile. With the growing interest in supporting mobile users and mobile computing, a great deal of work is taking place to solve this problem. For a solution to be practical, it has to integrate easily with existing Internet infrastructure and protocols, and offer an adequate migration path toward what might represent the ultimate solution. In that respect, the solution has to be incrementally scalable to handle a large number of mobile users and wide geographical scopes, and well performing so as to support all application requirements including voice and video communications and a wide range of mobility speeds. In this paper, we present a survey of the state-of-the-art and propose a multi-layer architecture for mobility in IP networks. In particular, we propose the use of extended local area networks and protocols for efficient and scalable mobility support in the Internet.

1 Introduction

In the broadest sense, the term mobile networking refers to a system that allows users to maintain network connectivity while moving from one location to another. Mobility is often associated with wireless technologies that require mobile networks to support continuous movement, at high speeds and for long periods of time. Recently, there has been an explosive growth in wireless devices with built in access to the Internet. In the near future, large numbers of mobile users will access the Internet for a variety of high-speed multimedia services. IP packet switching has become the standard towards which many networks are converging, including those in the telecommunication sector, as less efficient, less enabling circuit-switched technologies are abandoned. Although a lot of progress has been made, supporting mobility in IP networks is still a difficult challenge.

1.1 Challenges and Early Solutions

1.1.1   Duality of IP Addresses

The IP addressing scheme was designed and optimized for a stationary environment, which makes mobility difficult. With the introduction of mobile networking, IP addresses have acquired a dual significance. On one hand, they are expected to remain fixed during the course of a connection. An important reason for this is that, while, in principle, higher layers (above IP in the protocol stack) are supposed to be

S. Palazzo (Ed.): IWDC 2001, LNCS 2170, pp. 279–294, 2001.

independent of the IP layer, in practice they make use of the IP address for basic functionality. For example, the transport layer uses this address to establish and maintain connections. If the IP address is changed during the course of a session, the connection is lost and the session is terminated. Therefore, to maintain seamless connectivity during movement, IP addresses need to be kept fixed, transparent to changes in user locations. On the other hand, IP addresses need to change dynamically as users move, since they are used for packet routing and delivery. Routers in the Internet use IP addresses in the destination field of packets to identify the subnet where the user is located, and to obtain the MAC address of the user for final delivery of the packets. Moreover, typically routers in the Internet use address based filtering to discard packets whose source IP address are from outside the subnet. Therefore, the user IP address needs to change as the user changes location in order to conform to addressing at the new location[1].

Notice that mobile networking inside a subnet is not affected by the dual significant of IP addresses. Mobile users can roam inside a subnet without having to update their IP addresses. The reason why this is possible is because LAN switches learn the users location and can route packets to them quickly using this information.

One way to resolve the duality of IP addressing is to change the transport and application layers of the protocol stack in order to handle a dynamic IP address. A mobility solution at the TCP layer is proposed in [3]. Connection migration is performed to maintain connectivity for sessions in-flight at the time of move. For this solution to work, the mobile hosts, and fixed hosts in the Internet wishing to communicate with mobile hosts would need to be upgraded to the new versions of software. While upgrading the mobile hosts may be an easier task, upgrading all the hosts in the Internet is not a possibility. Furthermore, achieving good application performance with dynamic IP addresses remains a significant challenge. Simulation results show that significant disruption is incurred during migration; moreover the solution limits movement to a single end and also may apply to TCP applications only. Another transport layer solution, proposed in [18] suggests TCP be modified to use domain names instead of IP addresses. Again, the main disadvantage of this solution is that it does not integrate easily with the existing Internet, hence it could be prohibitively expensive to deploy.

The alternative solution to the problem of IP address duality is to allow hosts to maintain a fixed IP address as they move across subnets. In turn, this would require that routers propagate host-specific routes in the Internet. However, host-specific routing requires space in the routing tables proportional to the number of hosts, slows down the routing process and consumes potentially excessive bandwidth in the Internet.

1.1.2 Mobile IP

In the 1990's, the IETF designed a solution for mobility known as Mobile IP [4], which overcomes the duality of IP addresses without requiring that routers learn host-

---

[1] Note that, to resolve this duality, in Ipv6 a host is allowed to use two addresses. One address is used as a permanent identifier while the other address is used for routing purposes. The permanent address is included in the main Ipv6 header, while the routing address is inserted in a special-purpose extension header used for routing.

specific routes. Mobile IP solves the problem by allowing a single computer to hold two addresses simultaneously. The first address is permanent and fixed. It is the address that transport and application protocols use. The second address is temporary – it changes as the computer moves, and is valid only while the computer visits a given location.

A mobile host MH is assigned a permanent home address and a home agent HA in its home subnet. DNS maps the domain name of the host to its home address. When the MH moves to a foreign subnet, it acquires a temporary care-of-address COA from an advertised foreign agent in the subnet, and it registers its new address with the HA. The HA uses gratuitous proxy ARP to capture all IP packets addressed to the MH's permanent address and uses encapsulation to forward them to the mobile's current location[2].

There are two possibilities for the packets going back from the MH to the corresponding host CH. One choice is for the MH to send out un-encapsulated IP packets with the permanent home address of the MH as the source address and the address of the CH as the destination address. However, some routers in the Internet use address based filtering and discard packets from outside the subnet. To avoid this, the MH needs to encapsulate the packet using its COA as the source address and the address of the HA as the destination address. The HA decapsulates the packet and forwards it to the CH.

One thing to notice is that packets delivered via HA typically travel further through the Internet than they would if delivered by the optimal unicast route. Apart from increasing the round-trip delay observed by the communicating parties, this also affects other users by increasing the overall load on the shared resources of the Internet. A proposed mechanism, known as route optimization, attempts to fix this, by using binding updates, containing the current COA of the MH, from the HA to the CH. A CH with enhanced networking software can learn the temporary COA and then perform the encapsulation itself, sending the packet directly to the mobile host. This avoids the overhead of indirect delivery.

1.1.3 Industrial Solutions and Mobile IP

Mobile IP, or some variant thereof, is a popular solution adopted by the majority of industrial products offering IP connectivity to mobile users.

1.1.3.1 Ricochet

The Ricochet system from Metricom [17] implements a solution for IP mobility that is similar to the Mobile IP protocol. However, it is important to point out that Ricochet was designed more than a decade ago hence it predates the Mobile IP protocol. Wireless cells are connected to IP gateways and name servers that provide security, authorization and roaming support to users. At any given point in time, a user has three addresses: one IP address, which is fixed, and two layer-2 addresses: one is fixed and unique to that user, and the other is dynamic and unique to the cell where a user is located at that point in time. When a user first connects to the network, its request is validated by the local gateway and name server. If authorized, the

---

[2] Note that for Ipv6, the extension header plays the role of the encapsulation header in Mobile IP.

gateway provides the user with an IP address that identifies a permanent virtual connection between the user and the network. All Internet traffic for the user is tunneled through the gateway to which the user was originally connected. The gateway maps the IP address of the user to the layer-2 address of the user corresponding to the cell where the user is located. As the user crosses cells, the IP address it had acquired from the gateway remains fixed. However, the mapping of this address to a cell location changes to reflect the new location of the user. In essence, this gateway performs the function of a agent in Mobile IP, by providing the user with an IP address, and tunneling the traffic for that user to its most up-to-date location.

1.1.3.2 UMTS

One example of an industrial system that uses the Mobile IP protocol is the Universal Mobile Telecommunication System (UMTS), which is proposed in [10]. UMTS aims to provide IP level services via virtual connections between mobile hosts and IP gateways connected to ISPs or corporate networks at the edges of the mobile network.

Users are assigned domain names, which are used to identify the ISP that can be accessed to provide Internet connectivity to the user. When a user logs on, it is assigned an IP address by the gateway to which that ISP is connected, also known as the home gateway. A virtual connection is established, consisting of two segments: one segment connects the mobile and some foreign gateway (via the air interface), and another, connects the foreign gateway and the home gateway (via a protocol similar to Mobile IP). The virtual connection is maintained as long as the mobile remains on and the foreign gateway can be changed as the mobile roams from the coverage area of one gateway to another. One can think of the mobile as being linked to the home gateway via an elastic global pipe. To the external world, the mobile appears to be located at the home gateway because it is this gateway that provides the IP address for the mobile. This mobility model is similar to the Mobile IP protocol.

1.1.4 Other Challenges

Another challenge of mobile networking is to support multimedia applications with stringent performance constraints, such as low packet loss and high interactivity. As users move, handoff needs to take place between the user's old point of attachment to the network and the new point of attachment to the network. Handoff may require change of state, not only at routers in the network to which a user is immediately connected, but also at routers inside the Internet that deliver packets to that user. If the number of such routers is large, or the distance from the user to these nodes is large, this change of state can take a long time. An interruption in connectivity due to a slow handoff can cause packet loss, which can significantly lower the perceived quality of these applications by the user. Packet buffering is typically used to handle packet loss. However, packet buffering may result in excessive latency overhead. Real-time applications such as voice depend on packets being delivered at a constant rate and within a certain time budget. If at times of user movement, the network cannot ensure the timely delivery of packets, they become irrelevant and would need to be discarded, to the dissatisfaction of users. Therefore, to assist moving users and maintain the continuity of multimedia traffic, the solution needs to support fast handoffs. To achieve this, handoffs should not involve propagation of information over long distances (hence should be handled in the vicinity of the user location).

Also in order to achieve smooth handoffs, it may be necessary for the mobile user to be connected to multiple points of attachments (referred to as diversity in the literature). However, this becomes difficult as speeds increase and users move continuously across space, frequently changing their points of attachment.

Unfortunately, Mobile IP does not meet the challenge of fast handoff. Rather than attempting to handle rapid network transitions such as the ones encountered in a wireless cellular system, Mobile IP focuses on the problem of long-duration moves. Let us refer to movement that requires fast handoffs as micro-mobility. The reason why Mobile IP does not perform well for micro-mobility should be clear: after it moves to a new subnet, a MH must detect that it has moved, communicate across the foreign network to obtain a COA and then communicate across the Internet to its HA to arrange forwarding. Because it requires considerable overhead after each move, mobile IP is intended for situations in which the MH crosses subnets infrequently, e.g. when the MH remains at a given location for a relatively long period of time.

2 Accelerating Micro-mobility

Many researchers have investigated ways to improve Mobile IP by accelerating micro-mobility. Subnets in the Internet are grouped into domains. Inter-domain mobility is achieved using Mobile IP, while intra-domain mobility is achieved using techniques that are particular to each research scheme. Routers or switches inside the domain keep track of users and deliver traffic to them using their learning databases. To perform this function, these devices effectively implement host-specific routing or switching. Traffic between domains is exchanged via routers typically known as gateways. The basic idea is shown in Fig. 1.

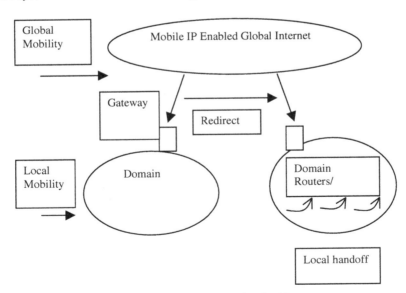

Fig. 1. Typical Micro-Mobility Architecture

2.1 HFA

In [1] hierarchical foreign agents (HFAs) are introduced to smooth out the handoff process when a mobile host MH transitions between subnets. This optimization is accomplished via hierarchical tracking of mobile hosts (MHs) by the foreign agents (FAs) and via packet buffering at FAs.

The FAs of a domain are organized into a tree structure that handles all the handoffs in that domain. The tree organization is unspecified and left up to the network administrator of that domain. One popular configuration is to have a foreign agent associated with the firewall to that domain be the root of the tree (also known as a gateway foreign agent or GFA) and all the other foreign agents provide the second level of the hierarchy.

An FA sends advertisements called Agent Advertisements in order to signal its presence to the MHs. An Agent Advertisement includes a vector of care-of addresses, which are the IP addresses of all its ancestors as well as the IP address of that FA. When an MH arrives at an FA, it registers the FA and all its ancestors with its home agent HA. The registration is seen and processed by the FA, all its ancestors and the HA.

When a packet for the MH arrives at its home network, the HA tunnels it to the GFA. The GFA re-tunnels it to the lower-level FA, which in turn re-tunnels it to the next lower level FA. Finally, the lowest-level FA delivers it to the MH. Therefore, an FA processing a registration should record the next lower-level FA as the other end of the forwarding tunnel.

Mobile IP route optimization extends the use of binding cache and binding update messages to provide smooth handoff via previous FA notification. However, tunneled packets that arrive at the previous FA before the previous FA notification are still lost. Such data loss may be aggravated if the MH loses contact with any FAs for a relatively long period of time. HFA includes an additional FA buffering mechanism. Besides decapsulating tunneled packets and delivering them directly to an MH, the FA also buffers these packets. When it receives a previous FA notification, it re-tunnels the buffered packets along with any future packets tunneled to it. Clearly, how much packet loss can be avoided depends on how quickly an MH finds a new FA, and how many packets are buffered at the previous FA. This in turn depends on how frequently FAs send out beacons or agent advertisements, and how long the MH stays out of range of any FA. To reduce duplicates, the MH buffers the identification and source address fields in the IP headers of the packets it receives and includes them in the buffer handoff request so that the previous FA does not need to retransmit those packets that the MH has already received.

While HFA helps reduce the overhead of handoff by handling handoff closer to the MH, it adds latency due to the need for packet encapsulation and decapsulation at every FA in the FA tree along the path from the CH to the MH. Moreover, scalability issues arise at the root FA and the FAs close to the root of the FA tree because of their involvement in packet tunneling for all the MHs of that domain. Finally, packet buffering results in latency overhead, while encapsulation still generates bandwidth overhead.

2.2 Cellular IP

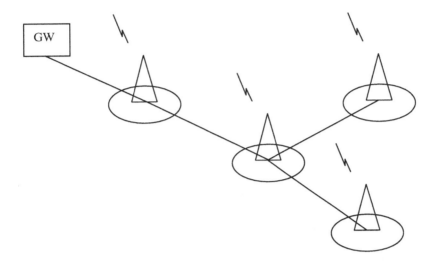

Fig. 2. A Wireless Access Network in Cellular IP. Base stations in an access network are interconnected by wired links. One gateway controls each access network.

Cellular IP access networks, depicted in Fig. 2, are connected to the Internet via gateway routers [9]. Cellular IP uses base stations for wireless access connectivity, for IP packet routing and for mobility support inside an access network. Base stations (BSs) are built on regular IP forwarding engines except that IP routing is replaced by Cellular IP routing. MHs attached to an access network use the IP address of the gateway as their Mobile IP care-of-address. The gateway de-tunnels packets and forwards them toward a BS. Inside a Cellular IP network, MHs are identified by their permanent home address and data packets routed without tunneling or address conversion. The Cellular IP routing protocol ensures that packets are delivered to the host's actual location. Vice-versa, packets sent by the MH are directed to the gateway, and from there, to the Internet.

Periodically, the gateway sends out beacons that are broadcasted across the access network. Through this procedure, BSs learn about neighbouring BSs on the path towards the gateway. They use this information when forwarding packets to the gateway. Moreover, when forwarding data packets from users to the gateway, BSs learn about the location of a user, and use that information to deliver packets sent for that user.

If a packet is received at a BS for a user that is unknown to that BS, a paging request is initiated by the BS. The paging request is broadcasted across a limited area in the access network called a paging area. The MH responds to the paging request and its route to the paging BS gets established. Each MH needs to register with a paging area when it first enters that area, regardless of whether it is engaged in communication or idle. Clearly, how fast paging occurs depends on the size of the paging area and on the efficiency of spanning tree traversal. A small paging area can help reduce the latency of paging, however it increases the number of paging area required to cover a given area, which in turn increases the signalling overhead imposed on MHs.

We observe that the paging techniques in Cellular IP are similar to those existent in the Groupe Speciale Mobile system (GSM) [13]. Mobile users are located in system-defined areas called cells that are grouped in paging areas. Every user connects with the base station in his cell through the wireless medium. Base stations in a given paging area are connected by a fixed wired network to a switching center, and exchange data to perform call setups and deliver calls between different cells. When a call arrives at the switching center for a given user, a paging request for that user is initiated across all the cells in that paging area. If the user answers, a security check on the user is performed, and if the test passes, the switching center sets up a connection for that user.

Cellular IP supports two types of handoff: hard handoff and semisoft handoff. MHs listen to beacons transmitted by BSs and initiate handoff based on signal strength measurements. To perform a handoff, the MH tunes its radio to the new BS and sends a registration message that is used to create routing entries along the path to the gateway. Packets that are received at a BS prior to the location update are lost. Just like in Mobile IP, packet loss can be reduced by notifying the old BS of the pending handoff, and requesting that the old BS forward those packets to the new BS. Another possibility is to allow for the old route to remain valid until the handoff is established. This is known as semisoft handoff and is initiated by the MH sending a semisoft handoff packet to the new BS while still listening to the old BS. After a semisoft delay, the MH sends a regular handoff packet. The purpose of the semisoft packet is to establish parts of the new route (to some uplink BS). During the semisoft delay time, the MH may be receiving packets from both BSs. The success of this scheme in minimizing packet loss depends on both the network topology and the value of the semisoft delay. While a large value can eliminate packet loss, it however adds burden on the wireless network by consuming precious bandwidth.

Cellular IP specifies an algorithm to build a single spanning tree rooted at the gateway to the access network as we described above. A spanning tree is necessary for the broadcasting of packets, to avoid packets from propagating to infinity if the topology of the access network has any loops. However, because it uses only a subset of the links inside the access network, a single spanning tree can result in link overload if traffic in the access network is high. This can be a significant drawback of Cellular IP as high-density access networks supporting many Tb/s of traffic become possible to deploy. Moreover, a single spanning tree can be prone to long periods connectivity loss. Connectivity loss would make this technology unacceptable as a replacement to wired, circuit-switched technology for telephone communications. Finally, Cellular IP specifies an interconnect between base stations that has a flat hierarchy. As access networks cover more area and exhibit higher pico-cell densities, a flat hierarchy would result in latencies of packet traversal across the access network that are unacceptable.

The description of Cellular IP assumes that originally, each wireless cell (or even pico-cell) constitutes an IP subnet. Consequently, they propose that multiple wireless cells be grouped into one subnet to improve roaming between the cells of one subnet. However, this concept is not new. For example, the 802.11 standard uses Extended Service Sets (ESS) to interconnect multiple 802.11 cells within a single subnet. Cellular IP also proposes two protocols for configuration and routing in IP subnets, however, LAN protocols already exist to accomplish these goals. For example, the algorithms for building a spanning tree and for learning as defined by the 802 standards are widely deployed and well known.

Nonetheless, it is clear that deploying wireless access networks as single subnets, like in Cellular IP is important for mobility. In this light, it becomes important to increase the size of IP subnets to the largest size possible in order to maximize their effectiveness in supporting IP mobility.

2.3 Hawaii

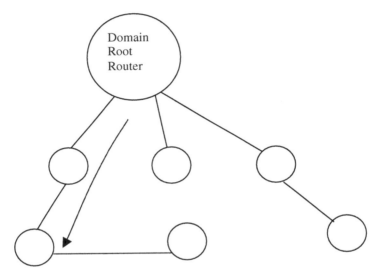

Fig. 3. Diagram of a Domain in the Hawaii Architecture. A domain root router acts as the gateway to each domain. Paths are established between the routers of a domain.

HAWAII segregates the network into a hierarchy of domains, loosely modeled on the autonomous system hierarchy used in the Internet [14]. The gateway into each domain is called the domain root router. When moving inside a foreign domain, an MH retains its COA unchanged and connectivity is made possible via dynamically established paths, as shown in Fig. 3. Path-setup update messages are used to establish and update host-based routing entries for the mobile hosts in selective routers in the domain, so that packets arriving at the domain root router can reach the mobile host. The choice of when, how and which routers are updated constitutes a particular setup scheme. HAWAII describes four such setup schemes, which trade-off efficiency of packet delivery and packet loss during handoff. The MH sends a path setup message, which establishes host specific routes for that MH at the domain root router and any intermediary routers on the path towards the mobile host. Other routers in the domain have no knowledge of that MH's IP address. Moreover, the home agent and communicating host are unaware of intra-domain mobility. The state maintained at the routers is soft: the MH infrequently sends periodic refresh messages to the local BS. In turn, the BS and intermediary routers send periodic aggregate hop-by-hop refresh messages toward the domain root router. Furthermore, reliability is achieved through maintaining soft-state forwarding entries for the mobile hosts and leveraging fault detection mechanisms built in existing intra-domain routing protocols.

HAWAII exploits host-specific routing to deliver micro-mobility. By design, routers perform prefix routing to allow for a large number of hosts to be supported in the Internet. While routing based on host-specific addresses can also be performed at a router, it is normally discouraged, because it violates the principle of prefix routing. Furthermore, host-specific routing is limited by the small number of host-specific entries that can be supported in a given router. However, this concern can be addressed by appropriate sizing of the domain and by carefully choosing the routers that are updated when a mobile is handed off. One of the problems with the implementation of HAWAII is that a single domain root router is used. This router, as well as its neighbors inside the routing tree can become bottlenecks routers for the domain for two reasons: First, they hold routing entries for all the users inside the domain. Second, they participate in the handling of all control and data packets for that domain. Another disadvantage of HAWAII comes from its use routers as a foundation for micro-mobility support. With cells becoming smaller, it is possible that a larger number of routers would be needed for user tracking and routing in a given area; however, this can become prohibitively expensive.

2.4 Multicast-Based Mobility

Numerous multicast-based mobility solutions have been proposed [2,6,7]. In [2,6], each mobile host is assigned a unique multicast address. Routers in the neighborhood of the user join this multicast address, and thus form a multicast tree for that address. Packets sent to the mobile host are destined to that multicast address and flow down the multicast distribution tree to the mobile host. In [7], packets are tunneled from the home agent using pre-arranged multicast group address, to which a set of neighboring base stations in the vicinity of the mobile host adhere. The most significant drawback of these solutions is that they require routers to be multicast capable; this capability does not exist in the Internet routers of today and would need to be added. In essence, this solution requires that routers learn multicast addresses, in the same way that routers learn unicast addresses in the other schemes for micro-mobility that we discussed. Unlike LAN switches, routers are not designed to learn host addresses, and therefore they would need to be modified for this purpose. Other drawbacks of mobility schemes based on multicast routing are that they require unique multicast addresses to be used, which creates address management complexity and limits the addressing space.

2.5 Micro-mobility and LAN Switching

In all the solutions we presented, fixed IP addresses are used to track mobile users inside a domain. This is done via learning at base stations, routers or agents. Despite the use of IP addresses, which are hierarchical, the addressing structure within a domain becomes non-hierarchical, just like in a LAN. Consequently, these addresses are tracked in the same fashion as layer-2 addresses in LANs. We make the observation that, in fact, these addresses are tracked in the same way as virtual channel identifiers in circuit-switched solutions such as ATM (employed in UMTS for the tracking of users by foreign gateways). In their original design, routers were not intended for performing tracking of individual host addresses, and consequently do not perform host-specific routing in an efficient way. It is unlikely that routers designs will be modified for this purpose. By design, layer-2 switches track host addresses, hence represent a more suitable solution for mobile tracking inside a domain.

3 A Multi-layer Infrastructure for Mobility

Our view is that an architecture to handle mobility must operate in a hierarchical fashion by providing functionality at multiple layers; namely, the MAC layer, the networking layer and DNS (or the "directory" layer). Each layer is suited for implementing mobility if specific circumstances are met. The MAC layer is ideal for delivering fine grain mobility inside homogeneous networks, by virtue of the fast, cost-effective switching technologies and the address learning schemes available at this layer. Similarly, the networking layer is best suited for implementing coarse grain mobility in cases where mobiles cross subnets and hence require new IP addresses to remain reachable, or when movement happens across heterogeneous networks where MAC layer addresses are incompatible. DNS can further support coarse grain mobility by maintaining an up-to-date directory of users and their IP addresses which can be used to simplify the operation of the networking layer. A description of this architecture as shown in Fig. 4. In the following subsections, we provide a more detailed description of the architecture. Readers who are interested in a complete description of the architecture are referred to [11].

3.1 Extended LAN

Over the past decade, we have witnessed tremendous developments in LAN technologies, such as increases in switch processing by a few orders of magnitude, and increases in link bandwidth and distances (owing to the fiber optics technology). These advances resulted in an increase in the size of LANs, and more recently, their deployment in metropolitan areas. We observed that such networks are well suited for providing mobility to portable IP devices. First, as mentioned earlier, mobile users can roam inside extended LANs without having to update their IP addresses. Secondly, the learning protocol implemented by LAN switches can be used to support diversity and adaptive routing. For these reasons, extended LANs are at the foundation of our network design for mobility.

Given the importance of LANs for mobility, it becomes paramount to answer the following questions:
1. How scalable are extended LANs in terms of number of users, user speed, application bandwidth and latency constraints?
2. What is the appropriate LAN structure and how large an area can it serve?

3. What is the protocol for tracking users in the LAN? How reliable is the LAN and what is its reconfiguration algorithm?

The answers to these questions need to take into account a variety of issues related to the wireless access networks, wired infrastructure, application traffic and requirements and user mobility. In order to minimize cost and maximize performance and reliability, the network design has to balance many parameters such as: processing power and storage capacity in the LAN switches, bandwidth across the wired links in the extended LAN, bandwidth and power consumption in the wireless cells. While a large extended LAN reduces the need for global mobility that can be inefficient, it also requires that the LAN support a larger number of users, and therefore increase the bandwidth requirements at the LAN switches and in the wireless cells in order to carry handoff control messages and user data.

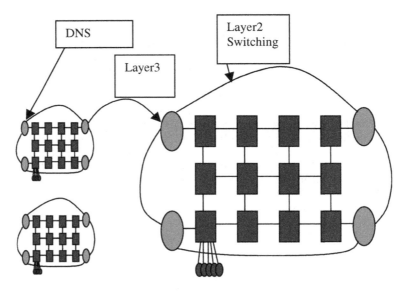

Fig. 4. A Multi-Layer Architecture for IP Mobility

3.2 Dynamic DNS

At the highest level of the protocol stack, dynamic DNS can be employed to track moving users as they change domains. The idea here is that DNS can behave like a directory that stores up-to-date, coarse-grain information on the location of mobile users. When the user enters a new domain and receives a new IP address, a DNS update is sent to ensure the most up-to-date mapping of the mobile's domain name to its IP address. Subsequently, communicating sessions that start following transition to a new domain can benefit from having the latest information and locate the user directly. However, sessions that are ongoing at the time of move cannot benefit from the directory update. This is because DNS lookups are not performed in the middle of sessions for the purpose of renewing connectivity. Instead, a network layer solution needs to be devised for this purpose. By updating the DNS database, we minimize the need for network layer mobility support and allow for efficient routing for sessions which undergo domain name resolution prior to session establishment and which start following inter-subnet crossings.

For dynamic DNS to work properly, caches that store the mapping of domain names to IP addresses at the communicating nodes need to be either:

1. Binding caches, which guarantee that the latest mapping is given
2. Disabled
3. Have a low TTL value (under a few seconds).

The first choice may give better IP address lookup performance, particularly for slow mobility, however it can be expensive to maintain fast mobility and many users. The second option removes the cost of updating caches at the expense of lookup performance. The third option is a compromise between the first two options.

It is interesting to note the parallels that exist between dynamic DNS solutions and directories techniques in cellular telephone systems such as the Groupe Speciale Mobile (GSM) or the Personal Communication System (PCS) [13,15]. In the GSM and PCS systems, users are identified by a unique phone number. One important feature of the GSM/PCS system is the automatic, worldwide localization of users. The system knows where a user currently is, and the same phone number of valid worldwide. A hierarchy of databases consisting of Home Location Registers (HLRs) and Visiting Location Registers (VLRs) is used to track users. The HLR contains information about the current VLR of a user, and the VLR knows the switching center via which the user can be reached. The HLR/VLR databases are similar to the DNS directories in our solution because they store up-to-date information on the location of every user. Notice that this feature of GSM/PCS renders the mobility management scheme a challenging problem. While this approach eliminates system-wide paging, which vastly reduces the radio link signaling, it introduces remote database lookups that may incur a large amount of wired network traffic and long call setup delay. Much effort has been focused on exploring efficient location management techniques. Extensions to standard HLR/VLR schemes, such as partial replication and caching have been developed to improve wireless call setup performance [15]. We believe that these techniques may apply to our solution in order to achieve an efficient and scalable dynamic DNS implementation.

3.3 IP Mobility

The network layer is important for providing mobility when users roam between different administrative domains, different subnets within the same domain, and possibly between heterogeneous networks. The network layer solution has two components that should be used in combination in order to deliver wide-area mobility. One component requires that a tunneling protocol be used, such as Mobile IP to redirect sessions that are ongoing at the time of a move between different domains. A second component of the solution is using host-specific entries at the routers of a domain to track groups of mobile users inside that domain, as is done in HAWAII. The first solution is important in order to eliminate the problem with DNS-based tracking that we outlined. The latter solution is important in order to improve the efficiency of roaming by giving users the ability to use a single IP address inside a large domain that extends beyond one subnet. To support a large number of users, routers that implement host-specific routing inside a domain should interconnect via a scalable fabric, and implement a scalable routing protocol.

4 A Case Study

A scalable LAN overlay is proposed to support mobile users in a metropolitan area network. A detailed description of this proposal can be found in [12]. As shown in Fig. 4, the extended LAN is implemented as a grid topology (e.g. the Manhattan Street Network). This is because the grid matches the topology of cities themselves - with the streets being rows and columns -, but also because the grid is scalable by virtue of its distributed nature. Wireless cells are connected to LAN switches in a hierarchical fashion. The hierarchy reduces the number of hops to be traversed when communicating between two access points in the grid, and therefore reduces latency. To connect to the Internet backbone, a scalable and distributed gateway router is necessary. The router needs to scale to support the aggregate traffic to and from all the cells in the LAN. For a large number of cells with many users, this bandwidth can become very large. For example, for a LAN supporting 2 million users, consuming 2 Mb/s each, the routing bandwidth is 4 Tb/s. Furthermore, the router must be physically distributed across many smaller routers to allow for load balancing at the links connecting the LAN switches to the subnet router.

A protocol is designed for the Manhattan Street Network that takes advantage of multiple links in the network and that balances the traffic load across all the links and switches in the LAN. The protocol works by partitioning switches into control and data partitions. Each control partition must have one or more switches in common with every data partition. Similarly, each data partition must have one or more switches in common with every control partition. For example, each row in the grid could be a control partition and each column, a data partition. A protocol similar to the Generic Attribute Registration Protocol (GARP) [16] is used to track users inside a given control partition according to the user location. Data packets for a given user are propagated along a given data partition (as given by the location where the data packet was first injected into the network) until the control partition for that user is reached and the packet delivered to the user.

This LAN design has a number of advantages. The LAN does not rely on a single spanning tree or root switch. This is important for scalability as the LAN extends to large geographical scopes. By exploiting control and data partitions, it minimizes the latency of user location updates without affecting the latency of packet routing inside the LAN. Finally, its operation relies on existing LAN switching techniques and protocols, which makes the solution simple, inexpensive and easy to deploy.

5 Migration Path

To transition to the mobile network of tomorrow, it may not be possible to design the supporting network infrastructure from scratch. Instead, support for mobility may need to be built on existing network structures, such as small subnets controlled by LAN switches and interconnected by IP routers with a small number of host-specific entries. Under such circumstances, one possibility is the use of Virtual Private Networks (VPN) to offer extended LAN connectivity across multiple small subnets. In order to support mobile users in the most effective fashion, the protocol to handle mobile users needs to be flexible enough to operate at different layers in the protocol stack, and versatile enough not to require changes in the implementation of the LAN switches and IP routers of that network. In particular, the protocols running on LAN switches should be based on existing 802 protocols, since they are implemented in hardware and therefore cannot be easily replaced or reprogrammed. The main challenge becomes how to use and optimize existing protocols for the purpose of efficient support for mobility.

6 Conclusions

This paper surveys the state-of-the-art in providing mobility support to mobile users in the Internet. In particular, emphasis is placed on micro-mobility techniques designed to accelerate Mobile IP. One observation is that all micro-mobility work in a similar way by requiring that network devices inside a given geographical area learn about the location of users and keep track of them as they move inside that area. The differences among these techniques are the type of device required to do the learning (it could be an IP router, Mobile IP agent or LAN switch) and the protocols for routing packets using the learning databases. This paper also presents an architecture for mobility, which exploits extended LANs, IP routing and dynamic DNS. One important feature of this architecture is its scalable and efficient LAN design, geared at optimizing IP mobility. By relying on existing technologies, and by virtue of working with Mobile IP, this architecture is also global, cost-effective, easily deployable and compatible with the Internet of today.

References

1. E. Perkins and K. Y. Wang, "Optimized Smooth Handoffs in Mobile IP". Proceedings of the IEEE Symposium on Computers and Communications, Red Sea, Egypt, June 1999.
2. J. Mysore, V. Bharghavan, "A New Multicasting-based Architecture for Internet Host Mobility". Mobicom 1997, Budapest, Hungary, September 1997
3. A. Snoeren and H. Balakrishnan, "An End-to-End Approach to Host Mobility", Mobicom 2000, Boston MA, August 2000.
4. C. Perkins, "IP Mobility Support", RFC 2002, October 1996.
5. S. Cheshire, M. Baker. "Internet Mobility 4X4". Proceedings of the ACM SIGCOMM 1996, Stanford, CA, August 1996.

6. A. Helmy, "A Multicast-based Protocol for IP Mobility Support", Second International Workshop on Networked Group Communication, Palo Alto, CA, November 2000.
7. S. Seshan, H. Balakrishnan, R. Katz, "Handoffs in Cellular Wireless Networks: The Daedalus Implementation and Experience", Kluwer Journal on Wireless Networks, 1995.
8. J. Scourias and T. Kunz, "An Activity-based Mobility Model and Location Management Simulation Framework", MSWiM, Seattle, WA, 1999
9. A. Campbell, J. Gomez, S. Kim, A. Valko, C. Wan, "Design, Implementation and Evaluation of Cellular IP", IEEE Personal Communications, June/July 2000
10. "Naming, Addressing and Identification Issues for UMTS", UMTS Forum, December 2000
11. Cristina Hristea and Fouad Tobagi "A Multi-Layer Architecture for IP Mobility", in submission
12. Cristina Hristea and Fouad Tobagi "User Tracking and Routing in Metropolitan Area Networks with a Manhattan Grid Topology", in submission
13. J. Schiller, "Mobile Communications", Pearson Education Limited 2000
14. "HAWAII: A Domain-based Approach for Supporting Mobility in Wide-Area Wireless Networks", R. Ramjee et al, Proc. IEEE International Conference on Network Protocols, 1999
15. "Efficient PCS Call Setup Protocols", Y. Cui et al, IEEE Infocom, San Francisco, CA, 1998
16. IEEE 802.1d MAC Layer Bridging Standard.
17. http://www.metricom.com
18. "An Architecture for Content Routing Support in the Internet", M. Gritter, D. Cheriton, USENIX Symposium on Internet Technologies and Systems, San Francisco, 2001

Aggregated Multicast
for Scalable QoS Multicast Provisioning

Mario Gerla[1], Aiguo Fei[1], Jun-Hong Cui[1], and Michalis Faloutsos[2]

[1] Computer Science Department, University of California, Los Angeles, CA90095
[2] Computer Science and Engineering, University of California, Riverside, CA 92521

Abstract. IP multicast suffers from scalability problem with the number of concurrently active multicast groups, while scalability of QoS multicast is even further from being solved. In this paper, we propose an approach to reduce multicast forwarding state and provision multicast with QoS guarantees. In our approach, multiple groups are forced to share a single delivery tree. We discuss the advantages and some implementation issues of our approach, and conclude that it is feasible and promising. We then describe how to use our approach to provision scalable QoS multicast. Finally, we define metrics to quantify state reduction and use simulations to show how our scheme achieves state reduction. These initial simulation results suggest that our method can reduce multicast state significantly.

1 Introduction

Multicast state scalability is the problem we address in this work. Multicast is a mechanism to efficiently support multi-point communications. IP multicast utilizes a tree delivery structure, on which data packets are duplicated only at fork nodes and are forwarded only once over each link. This approach makes IP multicast resource-efficient in delivering data to a group of members simultaneously and can scale well to support very large multicast groups. However, even after approximately 20 years of multicast research and engineering effort, IP multicast is still far from being as common-place as the Internet itself.

Multicast state scalability is among the technical difficulties that delay its deployment. A multicast distribution tree requires all tree nodes to maintain per-group(or even per-group/source) forwarding state, which grows at least linearly with the number of "passing-by" groups. As multicast gains widespread use and the number of concurrently active groups grows, more and more forwarding state entries will be needed. More forwarding entries translates into more memory requirement, and may also lead to slower forwarding process since every packet forwarding involves an address look-up. In QoS multicast, the problem becomes even worse, because not only routes but also resources(eg, bandwidth) for individual multicast group are needed to maintain. This perhaps is the main scalability problem with IP multicast and QoS multicast provisioning when the number of simultaneous on-going multicast sessions is very large.

Recently, much research effort has focused on the problem of multicast state scalability. Some schemes attempt to reduce forwarding state by tunneling[14] or by forwarding state aggregation[10,13]. Thaler and Handley analyze the aggregatability of forwarding

S. Palazzo (Ed.): IWDC 2001, LNCS 2170, pp. 295–308, 2001.
© Springer-Verlag Berlin Heidelberg 2001

state in[13] using an input/output filter model of multicast forwarding. Radoslavov et al. propose algorithms to aggregate forwarding state and study the bandwidth-memory tradeoff with simulation in [10]. Both these works attempt to aggregate routing state after this has been allocated to groups. Second, some other architectures aim to completely eliminate multicast state at routers [6,11] using network-transparent multicast, which pushes the complexity to the end-points.

Though most research papers on QoS multicast are focusing on solving a theoretical constrained multicast routing problem, there have been efforts to bring QoS into existing IP multicast architecture, such as RSVP [15], QoSMIC [2], QoS extension to CBT [7], and PIM-SM QoS extension [3]. But all these schemes are using per-flow state, keeping track of routes and resources information for each individual group, which suffers scalability problem as mentioned above.

In this paper, we propose a novel scheme to reduce multicast state and provision scalable QoS multicast, which we call aggregated multicast. Our difference with previous approaches is that we force multiple multicast groups to share one distribution tree, which we call an *aggregated tree*. This way the total number of trees in the network may be significantly reduced and thus forwarding state: core routers only need to keep state per aggregated tree instead of per group. In this paper we examine several design and implementation issues of our scheme and describe how to use aggregated multicast scheme to provision multicast with QoS guarantees. We will also present results from our initial simulation experiments in which our scheme achieves significant state reduction in the worst case scenario where group members have no spatial locality at all.

The rest of this paper is organized as follows. Section 2 introduces the concept of aggregated multicast approach and discusses some implementation related issues. Section 3 talks about QoS provisioning on the aggregated tree. Section 4 proposes metrics to quantify multicast state reduction in aggregated multicast and presents simulation results. Section 5 gives a short summary of our work.

2   Aggregated Multicast

Aggregated multicast is targeted as an intra-domain multicast provisioning mechanism in the transport network. For example, it can be used by an ISP (Internet Service Provider) to provide multi-point data delivery service for its customers and peering neighbors in its wide-area or regional backbone network (which can be just a single domain). The key idea of aggregated multicast is that, instead of constructing a tree for each individual multicast session in the core network (backbone), one can have multiple multicast sessions share a single aggregated tree to reduce multicast state and, correspondingly, tree maintenance overhead at network core.

2.1   Concept

Fig. 1 illustrates a hierarchical inter-domain network peering. Domain A is a regional or national ISP's backbone network, and domain D, X, and Y are customer networks of domain A at a certain location (say, Los Angeles). Domain B and C can be other customer networks (say, in New York) or some other ISP's networks that peer with A. A multicast session originates at domain D and has members in domain B and C. Routers D1, A1, A2, A3, B1 and C1 form the multicast tree at the inter-domain level while A1, A2, A3, Aa and Ab form an intra-domain sub-tree within domain A (there may be other routers involved in domain B and C). The sub-tree can be a PIM-SM shared tree rooted at an RP (Rendezvous Point) router (say, Aa) or a bi-directional shared CBT (Center-Based Tree) tree centered at Aa or maybe an MOSPF tree. Here we will not go into intra-domain multicast routing protocol details, and just assume that the traffic injected into router A1 by router D1 will be distributed over that intra-domain tree and reaches router A2 and A3.

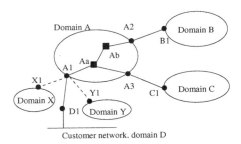

Fig. 1. Domain peering and a cross-domain multicast tree, tree nodes: D1, A1, Aa, Ab, A2, B1, A3, C1, covering group G_0(D1, B1, C1).

Consider a second multicast session that originates at domain D and also has members in domain B and C. For this session, a sub-tree with exactly the same set of nodes will be established to carry its traffic within domain A. Now if there is a third multicast session that originates at domain X and it also has members in domain B and C, then router X1 instead of D1 will be involved, but the sub-tree within domain A still involves the same set of nodes: A1, A2, A3, Aa, and Ab. To facilitate our discussions, we make some distinctions among these nodes. We call node A1 a **source node** at which external traffic is injected, and node A2 and A3 **exit nodes** which distribute multicast traffic to other networks, and node Aa and Ab **transit nodes** which transport traffic in between. In a bi-directional inter-domain multicast tree, a node can be both a source node and an exit node. Source nodes and exit nodes together are called **terminal nodes**. Using the terminologies commonly used in DiffServ[4], terminal nodes are often *edge* routers and transit nodes are often *core* routers in a network.

In conventional IP multicast, all the nodes in the above example that are involved within domain A must maintain separate state for each of the three groups individually though their multicast trees are actually of the same "shape". Alternatively, in an aggregated multicast approach, one can setup a pre-defined tree(or establish on demand) that covers nodes A1, A2 and A3 using a single multicast group address (within domain A). This tree is called an **aggregated tree** (AT) and it is shared by all multicast groups that are covered by it and are assigned to it. We say an aggregated tree T *covers* a group G if all terminal nodes for G are member nodes of T. Data from a specific group is encapsulated at the source node. It is then distributed over the aggregated tree and decapsulated at exist nodes to be further distributed to neighboring networks. This way, transit router Aa and Ab only need to maintain a single forwarding entry for the aggregated tree regardless how many groups are sharing it.

2.2   Implementation Considerations

It is not our goal to provide protocol details for aggregated multicast in this paper. However, a high-level overview of how it can be implemented in practice will provide a reality check which helps validate our work and provides some insights regarding its advantages and drawbacks.

First of all, there are various options for distributing multicast traffic of different groups over a shared aggregated tree. Regardless of the implementation, there are two basic requirements: (1)the original group address of data packets must be stored somewhere and can be recovered by exit nodes to determine how to forward these packets in the access network, and; (2)some kind of identification for the aggregated tree which the group is using must be carried in the packet header and is used by transit nodes to forward the packet. One possibility is to use IP encapsulation as said above, which, of course, adds complexity and processing overhead (at terminal nodes). A more efficient solution is MPLS (Multiprotocol Label Switching)[12] in which labels can identify different aggregated trees.

To handle aggregated tree management and matching between multicast groups and aggregated trees, a centralized management entity called tree manager is introduced. A tree manager has the knowledge of established aggregated trees in the network and is responsible for establishing new ones when necessary. It collects (inter-domain) group join messages received by border routers and assigns aggregated trees to groups. Once it determines which aggregated tree to use for a group, the tree manager can install corresponding state at the edge nodes involved, or distribute corresponding label bindings if MPLS is used. Aggregated tree construction within the domain can use an existing routing protocol such as PIM-SM, or use a centralized approach like what proposed in centralized multicast[8], or use MPLS signaling protocols extensions proposed in[9] to support the establishment of pre-calculated trees.

The set of aggregated trees to be established can be determined based on traffic pattern from long-term measurements. Let us say, for example, measurements in MCI-Worldcom's national backbone show that there are always many concurrent multicast sessions that involve three routers in Los Angeles, San Francisco and New York. Based on that knowledge, a network operator can instruct the tree manager to setup an aggregated tree covering routers in these three locations. Aggregated trees can also be established, changed (to add/remove nodes) or removed dynamically based on dynamic traffic monitoring. Knowing a set of existing aggregated trees, a tree manager can "match" a specific group, with given group membership (set of terminal nodes), to an aggregated tree that covers the group (i.e., all terminal nodes are member nodes of the tree).

2.3   Discussions

A number of benefits of aggregation are apparent. First of all, transit nodes don't need to maintain state for individual groups; instead, they only maintain forwarding state for a potentially much smaller number of aggregated trees. On a backbone network, core nodes are the busiest and often they are transit nodes for many "passing-by" multicast sessions. Relieving these core nodes from per-micro-flow multicast forwarding enables better scalability with the number of concurrent multicast sessions. In addition, an aggregated tree doesn't go away or come up as individual groups that use it, thus tree maintenance can be a much less frequent process than in conventional multicast. The benefit of control overhead reduction is also very important in helping achieve better scalability.

There are a number of concerns raised by this approach. A prime concern is membership dynamics. The problem occurs when a new edge node is added but it is not covered by the current tree, or when an edge node leaves the group and yet it still receives multicast traffic for this group (ie, bandwidth wastage). These problems can be alleviated by allowing a group to switch dynamically from one tree to another. To avoid the problems caused by membership dynamic changes, an ISP should require a customer to provide a list of group members (i.e., borders routers connecting to customer networks participating in the group) prior to the start of a multicast session and not to change group membership for the life of the multicast session - this is like providing a multi-point "VPN" (virtual private network) service. On the other hand, one may argue that, membership change on the backbone is very infrequent for many applications. For example, an Internet TV station may use an ISP's national backbone to distribute its programming to local regional networks, then to subscribers. There can be frequent membership dynamics at access networks connected to subscribers, but membership of backbone nodes is likely to be fixed or change very slowly if there is a large population of TV viewers. Another example is video-conferencing in which participants are expected to be in the group throughout the session or over a long period of time.

In group to aggregated tree matching, complication arises when there is no **perfect match** or no existing aggregated tree covers a group. A match is a **perfect** or **non-leaky match** for a group if all its leaf nodes are terminal nodes for the group thus traffic will not "leak" to any nodes that do not need to receive it. For example, the aggregated tree with nodes (A1, A2, A3, Aa, Ab) in Fig. 1 is a perfect match for our early multicast group G_0 which has members (D1, B1, C1). A match may also be a **leaky match**. For example, if the above aggregated tree is also used for group G_1 which only involves member nodes (D1, B1), then it is a leaky match since traffic for G_1 will be delivered to node A3 (and will be discarded there since A3 does not have state for that group). A disadvantage of leaky match is that certain bandwidth is wasted to deliver data to nodes that are not involved for the group. Now let's get back to the problem. When no perfect match is found, a leaky match may be used, if it satisfies certain constraint (e.g., bandwidth overhead is within a certain limit). This is often necessary since it is not possible to establish aggregated trees for all possible group combinations. The trade-off is bandwidth overhead vs. the benefit of aggregation. When no existing aggregated tree covers a group, either conventional multicast is used, or a new tree is established or an existing tree is extended (by adding new nodes) to cover that group. Of course, it is possible to enforce that aggregation is only applied to groups that are covered by a set of aggregated trees established based on long-term traffic pattern and any other group will use conventional multicast.

3 Provision Multicast with QoS Guarantees

One motivation for aggregated multicast is to provision multicast services with QoS guarantees in future QoS-enabled networks. This problem has not attracted much attention yet within IETF since the first priority so far has been to provide IP data (typically, best effort) multicast services. We note however that real time, interactive multicast applications will be in the future at least as important as(if not more than) data, or more generally, non real time multicast applications. There is an interesting reason why the support of QoS oriented multicast for interactive applications will become very important in the future Internet. Today, many non real time applications such as news, software distribution, etc, can be effectively supported by alternate techniques (to network level multicasting) such as web caching and application level multicast. In fact, these alternate techniques are often invoked to bypass the problems posed by IP level multicast. Real time (but, non interactive) applications such as video on demand can take advantage of the same alternate techniques (eg, web caching). In contrast, if we consider true interactive, real time applications such as video conferencing, distributed network games, distributed virtual collaborations (with real time visualization and remote experiment steering), distance lectures with student participation, we realize that alternate techniques such as web caching would severely affect time responsiveness. Moreover, interactive applications cannot be effectively supported by multiple unicast connections, since they typically requires many to many communications (it would be extremely costly to provision N x N connections, each with guaranteed bandwidth allocation!). As a result, it is important to address the scalability of QoS multicast, since it will be a prominent offering in the gamut of future Internet services.

As we already did for data multicast, the main technique we will be proposing in order to reduce router processing O/H and enhance scalability of QoS multicast is tree "aggregation". To this effect, we wish to note that Internet community has already embraced "flow aggregation" as the philosophy for scalable QoS provisioning. In fact, today people are backing away from the micro-flow based QoS architecture, namely the Integrated Services architecture[5], and are moving towards aggregated flow based architecture - the Differentiated Services architecture[4] - at least in the network core. The argument backing the aggregated approach is simple: the per-flow reservation and data packet handling required by Integrated Services simply do not scale to large networks.

As of now, however, the success of the Diff Serv concept as observed in the QoS unicast applications has not materialized yet in the QoS multicast word. In fact, over the past few years, several meritorious QoS multicast schemes have been proposed, all still inspired by the Int Serv model, and all dealing with individual flows.

The main criticism one can move to such schemes is poor scalability. In earlier sections we argued on the O/H required to set up and maintain routes for individual best effort multicast groups. The problem becomes much more complex if one must allocate and maintain not only routes but also resources (eg, bandwidth) for individual groups. The scalability problem, however, has not so far deterred the investigation of

Int Serv QoS multicast solutions. The main reason was the lack of incentive: given that conventional multicast routing requires per flow state, then, why should we seek non per flow state QoS multicast solutions.

The emergence of aggregated, scalable techniques like the one presented in the previous sections of this paper will clearly change the situation. If multicast routing has become scalable, then QoS provisioning to multicast applications should also be scalable. In fact, the aggregated multicast solutions presented in the previous sections will be the starting point for scalable, real time QoS multicast services in the Internet.

3.1   QoS Provisioning on the Aggregated Tree

To understand how several multicast groups can be aggregated and managed with QoS support, one may go back to the MPLS (Multi Protocol Label Switching) concept mentioned in the previous section. In essence, the aggregated multicast scheme can be viewed as the extension of MPLS from the path to the tree. MPLS plays a key role in unicast QoS support. It "pins down" the path, allowing the use of arbitrary alternate paths not available from the common routing tables. It tunnels several sessions on the same path, by encapsulating the IP packets in an MPLS envelope. It enables QoS provisioning and grooming on a per MPLS path basis (as opposed to a per flow basis). As a consequence, CAC on individual sessions is carried out at the edge node (or Border Gateway) only, with minimal latency and without engaging the intermediate nodes along the MPLS path. It enables "measurement based" resource tracking and Call Acceptance Control. Namely, the edge node need not keep track of the exact number of IP telephony calls (say) currently multiplexed on the MPLS path; it simply monitors MPLS utilization (using a proper window average) and determines available bandwidth and acceptance/rejection policy. Finally, the MPLS mechanism allows very flexible sharing of bandwidth across all flows multiplexed on the same path. The MPLS approach is consistent with the Diff Serv principles of flow aggregation. No per flow resource allocation or signaling is required at the intermediate core routers.

The proposed aggregated multicast approach will extend all the above described MPLS features to the Aggregate Tree. In particular, QoS provisioning will be done in the background, and may be coordinated between the bandwidth broker of the domain in question and the shared tree managers (assuming one resource manager per shared tree).

3.2   QoS Aggregate Tree Implementation and Operation

In the following we outline a straw-man implementation of the QoS Aggregated Multicast Tree. This implementation relies on and expands upon the basic AT implementation described in the previous section.

(a) When an AT (Aggregated Tree) is initialized, and is earmarked for the support of a particular QoS application (eg, video conference), it receives a bandwidth allocation commensurate to the traffic predictions for that application among that particular set of destinations. The AT is a permanent tree in that it is long-lived and expected to carry a large number of sessions simultaneously.

(b) A measurement based bandwidth management scheme assures that the ratio (average traffic load)/(allocated bandwidth) is adequate for the application (accounting for both traffic statistical characteristics and application QoS requirements) and for the current load. Note that different applications may require different safety margins depending on their statistical characteristics and their delay and packet loss constraints. If necessary, traffic statistics of an application can be "learned" by observing the flows at the edge nodes. The AT bandwidth manager can be centralized or distributed. In a distributed implementation, each edge node has a bandwidth agent (BA). The BA continuously monitors the above mentioned ratio and acquires/releases bandwidth as appropriate. For example, if more bandwidth is needed, the BA uses the existing intra domain tools (eg, Q-OSPF) to determine if more bandwidth is available on the path from an edge node to the Core or Rendezvous Point router (assuming a CBT approach). The peripheral BAs exchange information about bandwidth available and come up with a consistent bandwidth allocation decision (note that it would not help if one BA allocated 10 Mbps and another allocated only 5Mbps!). The BA allocation decisions may be supervised by the domain bandwidth broker, to ensure fair resource allocation across ATs supporting different applications.

(c) Call Acceptance Control is decided at the edge node, based on "measured" available bandwidth on the AT. This eliminates the very high latency typically experienced in conventional "per flow" QoS multicast approaches.

(d) Users dynamically join/leave an existing multicast AT in a totally transparent way and with zero latency - no bandwidth needs to be allocated or released. This is a dramatic improvement with respect to the node processing O/H and latency required by per flow QoS schemes.

3.3 Statistical Allocation Advantage

Typically, the use of the AT implies a "wastage" of resources since the streams are delivered to more destinations than strictly necessary. This wastage is traded off with the reduction in processing O/H and the ease of path and resource maintenance. There are situations, however, when the Aggregation approach can in fact lead to bandwidth allocation savings. We outline one such example below.

Consider a video conference with a maximum of 100 simultaneous participants placed at different locations. The multicast tree is for simplicity a star with direct, point to point links from user to RP router (see Fig. 2). Assume that 1 Mbps "equivalent" bandwidth is required by each session. Typically, at any given time only the video and audio of the person currently speaking is multicast to the group. In the traditional IntServ, "per flow" reservation approach, a full duplex 1 Mbps allocation is required (on the link from user to RV router) when a user joins the group. Thus, the "total" bandwidth allocation (counting the unidirectional bandwidth in each direction of the link) is 2S Mbps, where S is the number of current participants. If we use the AT scheme, the total allocated bandwidth is 101 Mbps, regardless of the number of simultaneous participants. This is because the AT bandwidth agent BA measures bandwidth usage, that is 1Mbps on each link in the direction RP to edge node, and 1 Mbps summed over all uplinks from edge node to RP (assuming that only one member is transmitting at a time). If we plot the allocated bandwidth as a function of S (see Fig. 3), we note that the Int Serv scheme

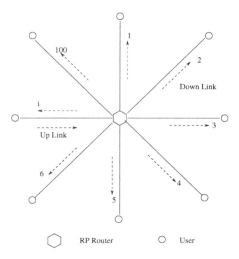

Fig. 2. Vedioconference multicast group.

requires "more" bandwidth that the AT scheme for $S > 50$. This is an unexpected results, which tells us that because of the more flexible, statistical allocation in the Aggregated Tree, we stand to save instead of waste bandwidth! The careful reader will notice that the Aggregate Tree saving leads to an allocation that is "asymmetric"; this asymmetry can be compensated across different trees, which typically have different roots and different topology layouts. Moreover, even the IntServ scheme could take advantage of the statistical sharing of uplink bandwidth among various transmitters. But, this would require checking the bandwidth allocation and state of several different multicast groups at each intermediate node!

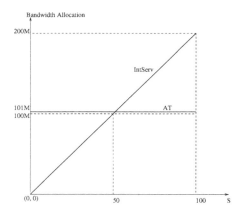

Fig. 3. Total bandwidth allocation as a function of active conference participants.

3.4 Extensions of the Basic QoS AT Scheme

Within a domain, the ISP will offer several "permanent" ATs to choose from. The multicast group manager will periodically query the AT layout database for information regarding all the installed ATs, and will dynamically select the tree that best matches its current members configuration. As the membership grows, the groups can simply switch from one tree to another. Note that by virtue of the "soft state" operation induced by the measurement scheme, this switch-over is totally transparent. It does not require any bandwidth reallocation.

In some cases, the addition of a couple of users in locations not served by the current tree may not warrant the switch-over to a new tree. The new users can then be easily accommodated by connecting them to the nearest edge node/router.

In some applications (eg, battlefield communications , distributed visualization and control, etc.) it is important to provide fault tolerant multicast. For example, consider the control of a space launch carried out from different ground stations interconnected by an Internet multicast tree. This control scenario may require the exchange of real time, interactive data and streams. One elegant way to provide fault tolerance is the use of separate, possibly node disjoint multicast trees. For added reliability and zero switch-over latency, the duplicate data could be sent on both trees simultaneously.

Another scenario in which the Aggregate Tree concept is beneficial is mobile handoff. Consider for example a video conference participant driving between Los Angeles and San Diego. Assume that both Los Angeles and San Diego are leaves of the AT tree to which the multicast group of the mobile user has subscribed. With our proposed scheme, as the user is "handed off" from the Los Angeles to the San Diego edge router, he finds a path with resources already allocated to him. There is minimal disruption of communications. This "soft handoff" is achieved with no overhead in the core network.

4   Simulation Studies for State Reduction

In this section we attempt to quantify multicast state reduction that can be achieved using aggregated multicast. It is worth pointing out that our approach of multicast "aggregation" is completely different from multicast "state aggregation" approaches in [10,13]. We aggregate multiple multicast groups into a single tree to reduce the number multicast forwarding entries, while their approach is to aggregate multiple multicast forwarding entries into a single entry to reduce the number of entries. It is possible to further reduce multicast state using their approaches in an aggregate multicast environment. Here we study state reduction achieved by "group aggregation" before any "state aggregation" is applied.

First, we introduce two state reduction metrics. Without losing generality, we assume a router needs one state entry per multicast address in its forwarding table. Here we care about the **total number** of state entries that are installed at **all** routers involved to support a multicast group in a network. In conventional multicast, the total number of entries for a group equals the number of nodes $|T|$ in its multicast tree T (or subtree within a domain, to be more specific) – i.e., each tree node needs one entry for this group. In aggregated multicast, there are two types of state entries: entries for the shared aggregated trees and group-specific entries at terminal nodes. The number of entries installed for an aggregated tree T equals the number of tree nodes $|T|$ and these state entries are considered to be **shared** by **all groups** using T. The number of group-specific entries for a group equals the number of its terminal nodes because only these nodes need group-specific state.

Thus, we come up with the concept of **irreducible state** and **reducible state**: group-specific state at terminal nodes is **irreducible**. All terminal nodes need such state information to determine how to forward multicast packets received, no matter in conventional multicast or in aggregated multicast. For example, in our early example illustrated by Fig. 1, node A1 always needs to maintain state for group G_0 so it knows it should forward packets for that group received from D1 to the interface connecting to Aa and forward packets for that group received from Aa to the interface connecting to node D1 (and not X1 or Y1), assuming a bi-directional inter-domain tree.

Let N_a be the total number of state entries to carry n multicast groups using aggregated multicast, N_0 be the total number of state entries to carry the same n multicast groups using conventional multicast. We introduce the term **overall state reduction ratio** – i.e., total state reduction achieved at all routers involved in multicast, intuitively defined as

$$r_{as} = 1 - \frac{N_a}{N_0}. \tag{1}$$

Let N_i be the total number of irreducible state entries all these group need (i.e., sum of the number of terminal nodes in all groups), **reducible state reduction ratio** is defined as

$$r_{rs} = 1 - \frac{N_a - N_i}{N_0 - N_i}, \tag{2}$$

which reflects state reduction achieved at transit or core routers.

Further more, we define another metric about aggregation overhead. Assume an aggregated tree T is used by groups $G_i, 1 \le i \le n$, each of which has a "native" tree $T_0(G_i)$, the **average aggregation overhead** for T is defined as:

$$\delta_A(T) = \frac{n \times C(T) - \sum_{i=1}^n C(T_0(G_i))}{\sum_{i=1}^n C(T_0(G_i))}$$
$$= \frac{n \times C(T)}{\sum_{i=1}^n C(T_0(G_i))} - 1, \tag{3}$$

where $C(T)$ is the cost of tree T (total cost of all T's links). Intuitively, $\delta_A(T)$ reflects the amount of extra bandwidth wasted to carry multicast traffic using the shared aggregated tree T, in percentage. Let N_g be the total number of multicast groups and N_t be the total number of aggregated trees used to support these groups, **average aggregation degree** – i.e., the average number of groups an aggregated tree "matches", is defined as

$$AD = \frac{N_g}{N_t}. \tag{4}$$

The larger this number, the larger the number of groups that are aggregated into an aggregated tree, and correspondingly the more the state reduction. This number also reflects control overhead reduction: more groups an aggregated tree supports, fewer number of trees are needed and thus less control overhead to manage these trees (fewer refresh messages, etc.).

Next we will present simulation results from a dynamic matching experiment allowing leaky matches. In this experiment, we use the Abilene[1] network core topology as our simulation network, which has eleven nodes located in eleven metropolitan areas. Distance between two locations is used as the routing metric (cost), which could result in different routes than the real ones; however, routes from UCLA to a number of universities (known to be connected to Internet 2) discovered by traceroute are consistent with what we expect from the Abilene core topology using distance as routing metric.

We randomly generate multicast groups and use the following strategy to match them with aggregated trees and establish more aggregated trees when necessary. In generating groups, every node can be a terminal node (i.e., we don't single out any node to be core node that is not directly accessible to neighboring networks); in simulation results to be presented, group size is uniformly distributed from 2 to 10. When a group G is generated, first a source-based "native" multicast tree T_0 (with a member randomly picked as the source) is computed. An aggregated tree T (from a set of existing ones, initially empty) is selected for G if the following two conditions are met: (1)T covers G; and (2)after adding G, $\delta_A(T) \leq b_{th}$; where b_{th} is a fixed threshold to control $\delta_A(T)$. When multiple trees satisfy these conditions, a min-cost one is chosen. If no existing tree satisfies these conditions, either (1)an existing tree T is extended (by adding necessary nodes) to cover G if the extended tree T' can satisfy the following condition: after adding G, $\delta_A(T') \leq b_{th}$; or (2)the native tree for G is added as a new aggregated tree. Constraints above guarantee that bandwidth overhead is under a certain threshold.

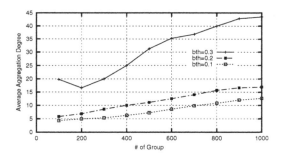

Fig. 4. Average aggregation degree vs. number of groups.

Fig. 4 plots the simulation result of average aggregation degree vs. number of groups added for different bandwidth overhead thresholds. As the result shows, as more groups are added (i.e., more concurrently active groups), the average aggregation degree increases: we can "squeeze" more groups into an aggregated tree, in average. Bandwidth overhead threshold affects aggregation degree in a "positive" way: as we lift the control threshold, more aggregation can be achieved – as we are willing to "sacrifice" more bandwidth for aggregation, we are getting more aggregation. Fig. 5 and Fig. 6 plot the results for overall state reduction ratio and reducible state reduction ratio defined in Eq. 1

and 2, and demonstrate the same trend regarding the number of groups and bandwidth overhead threshold as aggregation degree. The results show that, though overall state reduction has its limit, reducible state is significantly reduced (e.g., over 80% for a 20% bandwidth overhead threshold). This also confirms our early analysis.

In interpreting the implications of the above simulation results, we should be aware of their limitations: the network topology is fairly small and it is adopted from a logic topology and not really a backbone network with all routers at presence. Nevertheless, it should give us some feelings about the "trend". Another fine point is that, this simulation represents a worst-case scenario since all groups are randomly generated and has no correlation or pattern. In practice, certain multicast group membership pattern (locality, etc.) may be discovered from measurements and can help to realize more efficient aggregation.

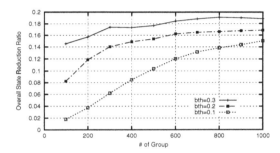

Fig. 5. Overall state reduction ratio vs. number of groups.

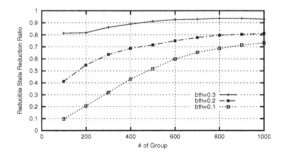

Fig. 6. Reducible state reduction ratio vs. number of groups.

5   Conclusions

In this paper, we proposed a novel approach, aggregated multicast, to provision QoS multicast within intra-domain. The key idea of aggregated multicast is to force groups into sharing a single delivery tree. This way per-flow state is is eliminated from network core and is only required at edge routers.

Our work could be summarized in the following points:

- Aggregated multicast is an unconventional yet feasible and promising approach.
- We discussed how this approach can be used to provision multicast with QoS guarantees.
- We proposed metrics to quantify multicast state reduction in aggregated multicast and our initial simulation shows promising results.

References

1. Abilene network topology. *http://www.ucaid.edu/abilene/*.
2. A. Banerjea, M. Faloutous, and E. Crawley. Qosmic: a quality of service sensitive multicast internet protocol. *Internet draft: draft-banerjea-qosmic-00.txt*, October 1998.
3. S. Biswas, R. Izmailov, and B. Rajagopalan. A qos-aware routing framework for pim-sm based ip-multicast. *Internet draft: draft-biswas-pim-sm-qos-00.txt*, June 1999.
4. S. Blake, D. Black, and et al. An architecture for differentiated services. *IETF RFC 2475*, 1998.
5. R. Braden, D. Clark, and S. Shenker. Integrated services in the internet architecture: an overview. *IETF RFC 1633*, 1994.
6. P. Francis. Yoid: extending the internet multicast architecture. *http://www.aciri.org/yoid/docs/index.html*.
7. J. Hou, H.-Y. Tyan, B. Wang, and Y.-M. Chen. Qos extension to cbt. *Internet draft: draft-hou-cbt-qos-00.txt*, February 1999.
8. S. Keshav and S. Paul. Centralized multicast. *Proceedings of IEEE ICNP*, 1999.
9. D. Ooms, R. Hoebeke, P. Cheval, and L. Wu. MPLS multicast traffic engineering. *Internet draft: draft-ooms-mpls-multicast-te-00.txt*, 2001.
10. P. I. Radoslavov, D. Estrin, and R. Govindan. Exploiting the bandwidth-memory tradeoff in multicast state aggregation. Technical report, USC Dept. of CS Technical Report 99-697 (Second Revision), July 1999.
11. Y. Chu S. Rao and H. Zhang. A case for end system multicast. *Proceedings of ACM Sigmetrics*, June 2000.
12. E. Rosen, A. Viswanathan, and R. Callon. Multiprotocol label switching architecture. *IETF RFC 3031*, 2001.
13. D. Thaler and M. Handley. On the aggregatability of multicast forwarding state. *Proceedings of IEEE INFOCOM*, March 2000.
14. J. Tian and G. Neufeld. Forwarding state reduction for sparese mode multicast communications. *Proceedings of IEEE INFOCOM*, March 1998.
15. L. Zhang, S. Deering, D. Estrin, S. Shenker, and D. Zappala. Rsvp: a new resource reservation protocol. *IEEE Network*, September 1993.

Fairness of a Single-Rate Multicast Congestion Control Scheme[*]

Gianluca Iannaccone[1,2] and Luigi Rizzo[1]

[1] Dipartimento di Ingegneria della Informazione
Università di Pisa, Italy
{gianluca.iannaccone, luigi}@iet.unipi.it
[2] Sprint ATL, Burlingame, CA, USA

Abstract. Recently, a TCP-friendly, single-rate multicast congestion control scheme called `pgmcc` was introduced by one of the authors. In this paper, we study the fairness of pgmcc in a variety of scenarios in which a multicast transfer session competes with long-lived TCP flows and web-like traffic. We evaluate fairness of the `pgmcc` scheme at different timescales and compare it with the fairness of the TCP congestion control algorithm.

Our results show that pgmcc is capable of sharing fairly the available bandwidth with competing connections. In particular, the use of a closed control loop between the sender and a group's representative – which closely mimics the TCP congestion control – guarantees that pgmcc is fair to TCP sessions and that it is capable of reacting quickly to changes of network conditions without compromising fairness.

1 Introduction

It is generally accepted that in order to be successfully deployed in today's Internet, IP Multicast needs a set of multicast congestion control mechanisms that are easily deployable, accurately tested and can co-exist with TCP. The IETF [6] has defined a set of procedures and criteria for evaluating reliable multicast protocols: the success of a proposed multicast protocol relies on its ability to compete with TCP traffic without threatening network stability.

In [9], a multicast congestion control scheme called `pgmcc` has been proposed as a viable congestion control scheme for single-rate multicast sessions. That work explains the basic `pgmcc` design choices and shows how the scheme achieves scalability, stability and fast response to changes in network conditions in a wide variety of experimental scenarios.

In this paper, we complement that work by studying in depth the fairness of `pgmcc`. Our goal is to better comprehend to what extent `pgmcc` is capable of fairly sharing the available bandwidth with competing TCP traffic.

---

[*] Work partly supported by Cisco Systems, Microsoft Research, IAT-CNR and Università di Pisa.

S. Palazzo (Ed.): IWDC 2001, LNCS 2170, pp. 309–325, 2001.

Although multicast fairness has been studied for many years [1,4,5,10,12], there is still no general consensus on what should be the relative fairness between multicast and unicast traffic. Different bandwidth allocation policies are possible [5]. For example, a multicast session could deserve more bandwidth than a TCP connection because it is intended to serve more receivers. On the other hand, it is also reasonable that a multicast session should not be given more bandwidth than TCP connections, in order not to penalize TCP connections that share portion of the path with a multicast session with a large number of receivers.

In this context, the pgmcc scheme has been designed to operate in a multicast environment where neither unicast connections nor multicast sessions are to be given any kind of preferential treatment. We consider that bandwidth is allocated following the widely popular max-min fairness model: a multicast session will be allocated bandwidth according to the most congested path in its tree. Thus, if the bottleneck link on the most constrained path of the tree has a capacity C, the pgmcc session will be allocated a share C/n where n is the number of competing sessions on that link.

We will evaluate pgmcc fairness comparing pgmcc and TCP average sending rates. Thus, for the scope of this paper, pgmcc is considered to be *fair* if its sending rate is comparable to the sending rate of an "equivalent" TCP connection (i.e. a TCP connections experiencing the same network conditions).

Note that we do not address the problem of inter-receiver fairness [4], that is fairness among receivers of the same multicast group. Indeed, pgmcc adapts its transmission rate to the bandwidth available to the "worst" receiver of the group. Thus, by definition, pgmcc does not assure any kind of inter-receiver fairness.

The rest of the paper is organized as follows. Section 2 provides a brief overview of the proposed congestion control scheme, with a description of the mechanisms that have an active role in guaranteeing the fairness of the scheme, namely: i) the election of a group's representative (*acker*), ii) the loss rate estimation performed by receivers, and, iii) the window-based flow control run between the sender and the acker that mimics TCP congestion control. In Section 3, we describe the fairness metric we use, while Section 4 describes a basic set of network scenarios that need to be analyzed in detail. Then, in Section 5 we presents the results of our simulations and Section 6 concludes the paper.

2   Overview of pgmcc

The pgmcc scheme is based on two separate but complementary mechanism: i) a window-based control loop which closely emulates TCP congestion control, and ii) a procedure to select a group's representative (*acker*).

The window based control loop is simply an adaptation of the TCP congestion control scheme to a protocol where lost data packets are not necessarily retransmitted, and so the congestion control scheme cannot rely on cumulative acknowledgements. In pgmcc, the "window" is simulated using a token-based scheme which permits to decouple congestion control from retransmission state.

One of the receivers in the group is elected by the sender as the *acker*, i.e. the node in charge of sending positive acknowledgements back to the source and thus controlling the transfer.

The procedure to elect the group's representative makes sure that, in presence of multiple receivers, the acker is dynamically selected to be the receiver which would have the lowest throughput if a separate TCP session were run between the sender and each receiver. For the acker selection mechanism, pgmcc uses a throughput equation to determine the expected throughput for a given receiver as a function of the loss rate and round-trip time. Unlike other schemes [7], the TCP throughput equation is not used to determine the actual sending rate, which is completely controlled by the window-based control loop.

In principle, pgmcc's congestion control mechanism works as follows:

1. Receivers measure the loss rate and feed this information back to the sender, either in positive acknowledgements (ACK) or negative acknowledgements (NAK).
2. The sender also uses these feedback messages to measure the round-trip time (RTT) to the source of each feedback message.
3. The loss rate and RTT are then fed into pgmcc's throughput equation, to determine the expected throughput from the sender to that receiver.
4. The sender then selects as the acker the receiver with the lowest expected throughput, as computed by the equation.

The dynamics of the acker selection mechanism are sensitive to how the measurements are performed and applied. In the rest of this section we describe specific mechanisms to perform and apply these measurements (other mechanisms are possible as described in [9]).

pgmcc operates end to end, and requires small constant state and a minimal amount of computation at both sender and receivers. We want to emphasize that, while our scheme involves positive and negative acknowledgements, we do not make any assumption on the reliability of the data transfer. This makes our scheme applicable equally well to unreliable data transfers.

2.1   Round-Trip Time Measurement

The classical way to measure the RTT without synchronized clocks is to include a timestamp in each packet from the source, and let receivers echo back the most recently received timestamp, possibly corrected with the difference between the time of reception and the time the feedback is actually sent (such delays are part of the feedback suppression schemes). The resolution of this method (especially for the correction factor) depends on the resolution of the clock at each receiver. If the latter is too coarse, the correction factor might introduce a large variance on the RTT estimates, biasing the results of the measurements in favour or against some receivers. Because we expect to deal with a large population of heterogeneous receivers, we cannot depend on the availability of a high resolution clock at all receivers.

As a consequence, and without too much loss of precision, in pgmcc we chose to measure the RTT in terms of packets: the sender simply computes the difference between the most recent sequence number sent and the most recently sequence number seen by the receiver (echoed in each NAK and ACK packet). This way we do not need to send timestamps or rely on the timer resolution at the receiver; on the other hand, for a path with a given RTT (measured in seconds), the value in packets computed by pgmcc will vary depending on the actual data rate. However this variation applies in the same way to all receivers, so it is not a source of discrimination among receivers. Furthermore, the RTT measurement in pgmcc is only used for comparing receivers, not for the actual selection of transmit rate, so any discrepancy between the real and the measured RTT cannot influence the inter-protocol fairness.

2.2   Loss Rate Measurement

The loss measurement in pgmcc is entirely performed by receivers. Again, the measurement results do not directly influence the transmit rate, but are only used for comparison purposes. As a consequence, pgmcc is reasonably robust to different measurement techniques, as long as they are not influenced too strongly by single loss events.

The method used for loss measurement is the exponentially weighted moving average (EWMA), which is formally equivalent to a single-pole digital low pass filter applied to a binary signal s_i, where $s_i = 1$ if packet i is lost, $s_i = 0$ if packet i is successfully received. The loss rate p_i upon reception or detection of loss of packet i is computed as

$$p_i = c_p p_{i-1} + (1 - c_p) p_i$$

where the constant c_p between 0 and 1 is related to the bandpass of the filter. Experiments have shown good performance with $c_p = 500/65536$, and computations performed with fixed point arithmetic and 16 fractional digits.

2.3   Acknowledgements

For each data packet (but not for retransmissions), one of the receivers is in charge of sending positive acknowledgements (ACKs). The identity of the acker is carried in each data packet.

ACKs contain loss reports (same as in NAKs), and a couple of additional fields, namely the sequence number of the data packet which elicited this ACK, and a bitmap indicating the receive status of the most recent 32 packets. This allows the sender to recover lost ACKs, and deal properly with out-of-order ACK delivery (which can occur when the acker switches between nodes on different paths).

2.4   Window-Based Controller

pgmcc uses a window-based congestion control scheme which is run between the sender and the acker, and mimics TCP congestion control. To implement this, the sender manages two state variables: a *window*[1] W, and a *token count* T, both initialized to 1 when the session starts or restarts after a stall (i.e. ACKs stop coming in and a timeout expires). The role of W is to determine how fast the window opens, same as in TCP. Tokens are instead used to regulate the generation of data packets: one token is necessary (and consumed) to transmit one packet, and tokens are regenerated by incoming ACKs.

In detail, W and T are updated as follows:

- on session restart, $W = 1, T = 1$;
- on transmit, $T = T - 1$ (consume one token);
- on ACK, $W = W + 1/W, T = T + 1 + 1/W$;
- on loss detection, $W = W/2$, ignore next $W/2$ acks.

The behaviour on normal ACKs mimics TCP's linear increase – the window expands by one packet for each round trip time. Similar to TCP, we assume a packet loss when a given packet has not been ACKed in a number of subsequent ACKs (this *dupack threshold* is set to 3 in our tests), and reproduce TCP's multiplicative decrease by halving the window. In order to match the number of outstanding packets to the window count, we need to avoid incrementing the token count for $W/2$ acks. Also, we do not react to further congestion events for the next RTT (this is easily achieved by recording the sequence number of the most recently transmitted packet).

2.5   Acker Election and Tracking

The acker election process in pgmcc aims at locating the receiver which would have the lowest throughput if each receiver were using a separate TCP connection to transfer data. Because the steady-state throughput of a TCP connection can be characterised in a reasonably accurate way in terms of its loss rate and round trip time [8], the throughput for each receiver can be estimated by using these two parameters.

Whenever an ACK or NAK packet from any of the receivers reaches the sender, the latter is able to compute the expected throughput T_i for that receiver by using the well-known TCP throughput formula:

$$T_i = \frac{1}{R_i \sqrt{p_i}} \qquad (1)$$

where R_i and p_i are the round trip time and the loss rate measured for receiver i, respectively. At any given time, the sender stores the expected throughput

---

[1] Note that the "window" used for congestion control purposes does not correspond to the "window" used for reliability or flow control.

for the current acker, T_{acker}. This value is updated every time an ACK or NAK from the current acker is received.

The selection process does not require knowledge of the whole population of receivers, or the evaluation of T_i for all of them. When we receive a NAK from node j, we can decide whether to switch to a new acker from the current one (node i) by just comparing T_i and T_j.

We should remark that the acker selection process is unavoidably approximate. Often, we only have a few RTT and loss rate samples from each potential acker, and those samples might be affected by large uncertainties. Furthermore, the formula used to elect the acker is approximate and derived under assumptions which might not be valid during the switch.

As a consequence, it is essential that we apply some histeresys when deciding to switch to a new acker, and in all cases, we should not interpret a change of acker as to a congestion signal. Rather, we assimilate the selection of a new acker to a *move* of the node in charge of sending ACKs to a path with different features. This is possible because for each data packet there is only one acker, and we have procedures to deal with duplicate, out of order and missing ACKs. Should the new acker experience congestion, we will get a timely notification by making use of the new ACKs.

3   Fairness Metric

Throughout this paper, we will compare the average send rates of TCP and pgmcc flows experiencing similar network conditions or competing for bandwidth on the same bottleneck link. Therefore, we declare pgmcc to be "fair" if its sending rate is comparable to the sending rate of a TCP connection that *experiences the same network conditions*.

The timescale at which the sending rates are measured naturally affects the values of these measurements. The use of a too coarse timescale will hide some behaviours (e.g. sending rate burstiness) while a too fine timescale will make measures very dependent on transient phenomena (e.g. retransmit timeout). For this reason we will study fairness of pgmcc on a wide range of timescale, from approximately twice the round trip time of the connections under analysis to the entire duration of the experiments.

We define the send rate R_i^{τ} of session i using b bytes long packets at the timescale τ as:

$$R_i^{\tau}(t) = \frac{b \cdot < \text{packets sent in the interval } (t, t+\tau) >}{\tau} \qquad (2)$$

Then, we compute the fairness index F_{τ} at timescale τ as follows [3]:

$$F_{\tau}(t) = \frac{(\sum_{i=0}^{n} X_i^{\tau}(t))^2}{n \sum_{i=0}^{n} (X_i^{\tau}(t))^2} \qquad (3)$$

where $X_i^{\tau}(t) = R_i^{\tau}(t)/B_i$ is the ratio between the sending rate of session i and its allocated bandwidth B_i, and n is the number of session competing for bandwidth.

The fairness index is bounded between 0 and 1, where a value of 1 indicates that network resources are fairly shared according to the targeted allocation.

4   Fairness of the Scheme

In [9], it has been shown that under normal operating conditions pgmcc and TCP share bandwidth fairly. On the other hand, [9] also mentions that there are some scenarios which need more investigation as far as fairness is concerned. These scenarios are discussed below, and the behaviour of pgmcc in those of them requiring further investigation is studied by simulations in Section 5.

Receivers behind the same bottleneck. Consider the scenario illustrated in Figure 1, where two receivers behind the same bottleneck experience very different round trip times. In this case, given that all packet losses occur on the bottleneck link L_1, both receivers will send reports with the same loss rate value. Thus, in presence of NAK suppression, there is no guarantee that the sender will elect the receiver with the largest RTT as the acker. In fact, when the difference between RTTs is very large[2], the sender will probably never receive a NAK from the receiver with the larger RTT.

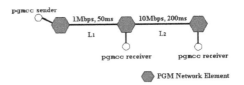

Fig. 1. In absence of other traffic, the two receivers will experience the same packet losses but very different round trip times.

This behaviour cannot be considered a source of unfairness. In fact, TCP itself favors flows with a smaller round trip time, and in presence of multiple receiver behind the same bottleneck there is no reason for the multicast session to adapt to the slowest rather than the faster of its members.

Note that, if a TCP connection competes on link L_2 with the slowest receiver, that receiver will experience more losses than the rest of the group and thus will become the acker assuring a fair allocation of bandwidth along the multicast tree. Thus, this scenario does not lead to unfair behaviour of the scheme.

Receivers with a very small round trip time. The acker election process is sensible to the timely distribution of loss reports from receivers. As soon

---

[2] at least comparable with the maximum NAK backoff time [11], the random time PGM receivers wait before sending a retransmission request.

as a receiver identifies a hole in the packets' sequence numbers, it schedules a delayed NAK transmission with an updated loss report. Reports from receivers that experience a smaller round trip time will likely reach the sender sooner than others (which will be likely suppressed, instead, along the shared path).

It is possible then for a "fast" receiver experiencing a transient period of congestion to be elected as the new acker, and to make the sender increase the sending rate (the window is increased approximately by one per round trip time of the current acker). In this case, even if "slow" receivers experience higher loss rates, the delay involved in the acker election (about 2 RTTs) will make the pgmcc session to behave unfairly toward competing TCP flows along the slow path. We will discuss this scenario in greater detail in Section 5.1.

Ambiguity in the acker election process. One of the possible caveats of using *any* TCP formula (1) for the acker election process, is the difficulty for the sender to discern between two receivers along different paths with similar estimated throughput. Variability in receivers' reports could mislead the sender, and make it choose alternatively between two or more receivers, possibly resulting overall in an unfair sharing of available bandwidth with competing TCP flows.

In fact, if acker switches are too frequent, receivers might not have a chance to send at least three acknowledgements in a row, which is the minimum amount of feedback required to signal congestion to the sender. To avoid this situation, [9] introduces some histeresys in the election process to favour the current acker. This permits to reduce the probability of an acker switch when multiple receivers have similar throughput expectations (according to the formula used). In Section 5.1, we will show a set of simulation results that show how pgmcc is capable of correctly handling such particular scenarios.

Receivers with a very high loss rate. It is known that the TCP simplified throughput formula (1) is an overestimate of the throughput for high loss rates (roughly above 5%). In the acker election process, this error can make the sender elect as the acker the wrong receiver, thus resulting in a unfair behaviour with other flows competing on the path with the actual worst receiver.

A simple fix to this problem would be the use of a more precise formula for estimating the throughput [8]. However, in case of very high loss rates, TCP congestion control and the pgmcc scheme are both dominated by timeouts, making very difficult to implement any reasonably smooth control on the sending rate.

In Section 5.2 we run a network simulation where we vary the loss rate on one path of the multicast tree to evaluate its impact on the protocol fairness.

Receivers with uncorrelated packet losses. The presence of multiple end-to-end paths in a multicast tree can make the sender assume an overall loss rate much higher than the one of each individual receiver. This could lead to an average session bandwidth far below the fair share allocated to the session [1]. To solve this problem, in pgmcc each receiver computes the loss rate and feeds it back to the sender in each NAK packet. This way, the sender can estimate the loss rate observed by each receiver. Moreover, the sender uses the loss rate only to elect the current acker, while it regulates the actual sending rate based on acknowledgement from the acker.

In Section 5.3, we present a case study where we verify the behaviour of pgmcc, computing the average bandwidth of the session in presence of uncorrelated packet losses and a very large group of receivers.

Denial of service. Another issue common to all single-rate multicast schemes is the authentication of receivers' report. Indeed, it is possible for a malicious (or malfunctioning) receiver to send fake reports that can drive the session transmit rate down to zero. However, this issue is out of the scope of this paper, because we believe that a solution to this problem cannot be found in the design of a congestion control schemes, but it requires separate mechanisms for the authentication of receivers by the sender.

5 Experimental Results

In this Section we investigate the pgmcc's behaviour in some of the scenarios presented in the previous Section. To this purpose, we have used an implementation[3] of pgmcc under the ns simulator [2]. In the scenarios described in this paper, routers are PGM-compliant Network Elements [11] and TCP sources implement the NewReno modification. Both PGM and TCP data packets are 1000 bytes long.

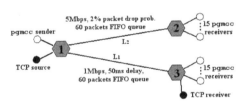

Fig. 2. A topology to exercise the acker election process in presence of different paths. Link L_2 has a variable RTT. On link L_1, three TCP flows compete for bandwidth with a pgmcc session with receivers on nodes 2 and 3.

5.1 Acker Election

An interesting network scenario to test the acker election procedures is shown in Figure 2. Here, a set of receivers lays behind a congested link, L_1, which has a high RTT – mainly due to queueing delay –, while a second set of receivers

---

[3] the source code used for the simulations described in this paper can be found at the following URL: http://www.iet.unipi.it/~luigi/pgmcc/

make use of a high bandwidth link (L_2) with a variable end-to-end delay and a non-negligible packet loss (it may model a link with a high degree of statistical multiplexing).

In this experiment, link L_1 has a capacity of 1 Mbps, a propagation delay of 50 ms and a FIFO queue with 60 slots; link L_2 has a capacity of 5 Mbps, a FIFO queue with 60 slots and a fixed 2% packet loss probability. On link L_1 the pgmcc session competes for bandwidth with 3 long-lived TCP sessions. The multicast group counts 15 receivers per node. Of course the number of receivers per node does not influence the fairness of the protocol, given that all receivers behind the same node experience the same round trip time and packet losses.

To verify the behaviour of pgmcc in different network scenarios, we vary the propagation delay of link L_2 from 10 ms to 400 ms. This way we can measure the fairness of the proposed scheme in presence of receivers with very different round trip times. On link L_1, the round trip time is dominated by the queueing delay and can be as large as 580 ms. On link L_2, no queues build up, so the round trip time is approximately twice the propagation delay.

In Figure 3[4], the curve labeled "link L2" shows the average throughput of the session for propagation delays on link L_2 varying from 10 ms to 400 ms, when no receivers on node 3 join the group. As expected, the throughput is inversely proportional to the round trip time.

When receivers on node 3 join the pgmcc session, the latter will compete for bandwidth on link L_1 with 3 TCP sessions ("tcp1".."tcp3"). The curve "pgmcc" shows the througput for a session with receivers on nodes 2 and 3.

For short propagation delay on link L_2, the pgmcc sender elects an acker behind link L_1. When the delay on L_2 increases (to more than 130 ms), the throughput on L_2 falls below the 250 Kbit/s corresponding to the fair rate on L_1 for the 4 sessions, so the acker moves to a node behind link L_2.

A few observations can be made on these results:

- as long as the control equation gives a clear indication of the worst path, the pgmcc sender behaves fairly with TCP flows (first part of the graph) or achieves the expected average throughput (second half of the graph).
- in presence of a very fast receiver (e.g. when the delay on link L_2 is 10 ms, the leftmost point in the graph) pgmcc shows a slightly unfair behaviour. This is due to the fact that any acker switch toward a receiver behind link L_2 makes the sender open the window very quickly.

In order to study in detail the fairness of pgmcc we compute the fairness index for different timescales from 1 second to 100 seconds. Moreover, to compare the fairness of pgmcc to TCP fairness, we have run a simulation where 4 TCP sessions are competing for bandwidth on link L_1.

In Figure 4 we plot the average value of the fairness index as a function of the timescale for three particular values of the propagation delay on link L_2: 10 ms,

---

[4] The graph is the result of averaging the throughput of the last 400 second of simulations over 10 runs (90% confidence intervals are shown). The simulation duration is 500 seconds. TCP flows are started at random times, uniformly distributed between 0 and 10 seconds.

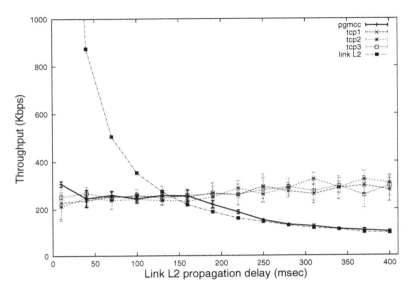

Fig. 3. Curve "link L_2" shows the throughput of a `pgmcc` session running only on link L_2. Curves "tcp1", "tcp2" and "tcp3" show the average throughput of the TCP flows competing on link L_1 with the `pgmcc` session with receivers on L_1 and L_2.

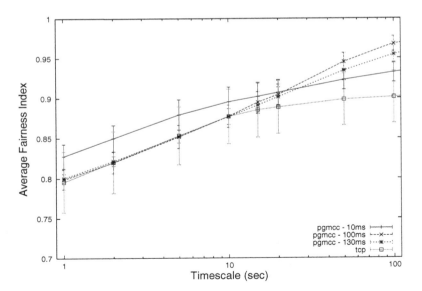

Fig. 4. Average fairness index. Curve "tcp" refers to a scenario where only 4 TCP flows compete for bandwidth, while curves "pgmcc" refers to a scenario with a `pgmcc` session and a propagation delay over link L_2 of 10, 100 and 130 ms.

100 ms and 130 ms. The curves are computed by averaging over 10 simulations runs (90% confidence intervals are shown). As we can see from the graphs, pgmcc behaves fairly at different timescales and in all the network conditions we have considered: the average fairness index is never less than 0.75 and there is no difference between the fairness of pgmcc and TCP (note that 90% confidence intervals overlap for all points in the three graphs).

These results show that pgmcc is as fair as TCP in situations where some receivers experience a round trip time which is much smaller than others, and also when two receivers have a similar throughput characterization that could lead to frequent acker switches, a potential source of unfairness.

5.2   Networks with High Packet Loss Rates

In this section we study the impact of a high loss rate path on the fairness of pgmcc. Due to the use of the simplified TCP formula, the sender may overestimate the expected throughput of a receiver behind a link with a very high loss rate, and elect as the acker a receiver along a different path resulting in an overall unfair behaviour.

To simulate such network conditions we use the simple topology shown in Figure 5. L_1 is a high capacity link (15Mbps, 50 ms delay, FIFO queue with 100 slots) that carries a high volume of bursty traffic (Web-like traffic). Link L_2, instead, has a very limited capacity (400Kbps, 50 ms delay, FIFO queue with 20 slots) but only few connections competing for bandwidth.

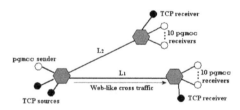

Fig. 5. Link L_1 is a high volume link with 15Mbps capacity that experience very high loss rates, while link L_2 has a relatively small capacity of 400Kbps. Both links have a propagation delay of 50 ms and a FIFO queue with a buffer size of 100 packets

On link L_1, a long-lived TCP session and a pgmcc session compete for bandwidth with a large amount of web-like cross traffic. We simulated web-like traffic with several ON/OFF UDP sources, with ON and OFF times drawn from a Pareto distribution. The mean ON time is 1 s, the mean OFF time is 2 s, and during ON time, the UDP sending rate is 500Kbps. On link L_2 instead, only one TCP session competes with the multicast session.

We vary the number of UDP sources to simulate different traffic loads on link L_1, also resulting in a loss rate ranging from 0 to about 25%. Figure 6 shows the loss rate at the bottleneck router as a function of the number of UDP sources, when only the web-like traffic is injected onto the network.

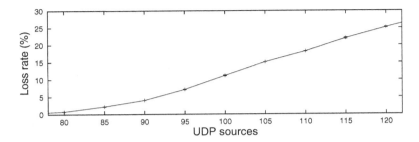

Fig. 6. Loss rate at the bottleneck in presence of ON/OFF background traffic.

At first, we are interested in measuring the fairness of pgmcc in presence of high loss rates. Therefore we run 10 simulations with a set of 10 pgmcc receivers behind link L_1 competing with a TCP connection and the UDP sources. Each simulation lasts for 2000 s.

Figure 7 shows the average throughput achieved by the multicast session and the TCP connection (90% confidence intervals are shown). As we can see from the graph, pgmcc and TCP achieve the same average throughput in presence of highly variable background traffic.

A second set of simulations has been used to verify the fairness of the protocol in presence of different paths with very different loss rates. In particular our goal is to verify whether the use of the simplified TCP formula may lead to unfairness due to errors in the estimate of the expected throughput.

In this set of simulations, 10 receivers behind link L_2 join the pgmcc session. On link L_2, pgmcc competes with a TCP session. In Figure 8 we compare the average throughput of the pgmcc session and of the two TCP sessions as a function of the number of UDP sources (90% confidence intervals are also shown).

The graph in Figure 8 shows that the presence of a path with a very high loss rate has a not negligible impact on pgmcc fairness.

Even when the *average loss rate* on link L_1 is around 5% (for about 90 UDP sources), the pgmcc's throughput is below its fair share of bandwidth. This is due to the bursty nature of the background web traffic, that causes transient but heavy congestion events on link L_1. As a direct consequence, the sender may elect as the acker one of the receivers behind L_1, and then reduce its sending rate for the entire duration of the congestion event. This explains why, when the *average loss rate* on link L_1 is around 5% (for about 90 UDP sources), the pgmcc's throughput is below its fair share of bandwidth.

On the other hand, when the loss rate increases (above 15%), pgmcc is slightly more aggressive than TCP. This behavior is mainly due to two reasons:

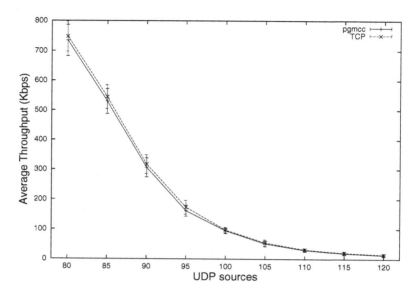

Fig. 7. TCP and `pgmcc` throughput in presence of ON/OFF background traffic.

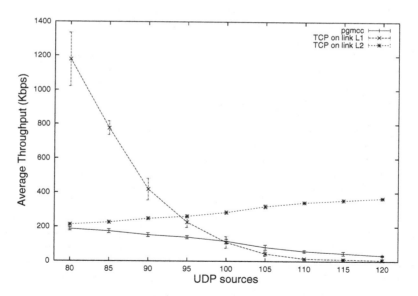

Fig. 8. Average throughput of the `pgmcc` session and TCP sessions as a function of the number of UDP sources sending on link L_1.

- pgmcc and TCP congestion control algorithms substantially differ on the slow-start mechanism and the retrasmit timeout. While TCP adapts the slow start threshold and uses an exponential backoff for the retransmit timeout, pgmcc makes use of a fixed value both for the slow start threshold and for the retransmit timeout.
- in case of two successive retransmit timeouts, the pgmcc sender may elect as acker a different receiver[5]. In this scenario, a long congestion event (that last more than twice the fixed retransmit timeout) on link L_1 may force the sender to elect as acker a receiver on link L_2 and to increase its sending rate, resulting in an unfair allocation of bandwidth on link L_1.

We believe that further investigation is needed to better understand the dynamics of the pgmcc scheme in high loss scenarios. However one should keep in mind that such experiments are highly dependent on a number of details such as network parameters (delays, buffer sizes), characteristics of competing traffic (burstiness, especially), and on the choice of protocol's parameters (such as timeouts and slow start thresholds) which also make a significant difference among different TCP flavours. As a consequence, the focus of these experiments should be on verifying the safe behaviour of the protocol (i.e. the fact that throughput decreases with increasing congestion, and that instances of different protocols do not starve each other), and not on moderate differences in the absolute throughput.

5.3   Uncorrelated Losses

An important aspect of the design of single-rate multicast congestion control schemes is the behaviour in presence of uncorrelated losses. Depending on how loss reports are handled, the source might assume an overall loss rate for the session much higher than the loss rate of each individual receiver.

To get some indications on how pgmcc works in presence of independent losses with up to 100 receivers, we ran a simulation on the topology of Figure 9, where a pgmcc source initiate a multicast session with 100 receivers behind independent lossy links with 1% packet loss rate. An additional link with the same characteristics is used for a TCP flow, in order to compare performance of the congestion control scheme.

At time 0, the TCP session and 10 pgmcc receivers are started. At time 300, 90 more pgmcc receivers join the session. Figure 10 shows the throughput of the TCP connection and of the multicast session over time. As shown from the graph, the presence of the 90 additional receivers at time 300 does not influence appreciably the throughput of the multicast session.

Much larger scale tests are certainly useful to investigate this behaviour in more detail. However, such tests cannot be run with simple retransmission-based

---

[5] In order to avoid the case where the current acker leaves the multicast session and the sender is not capable to elect a new acker and to transmit any new packet, due to the absence of acknowledgments [9].

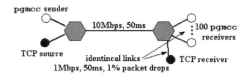

Fig. 9. A topology with 100 independent links with 1% probability of packet drop. A TCP connection is run over another link with the same characteristics.

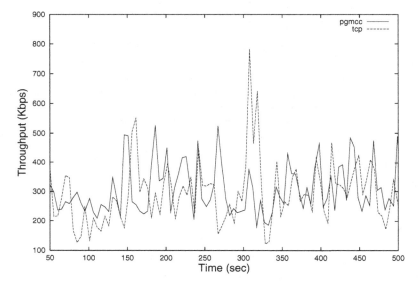

Fig. 10. Impact on throughput of multiple receivers with uncorrelated losses. Initially 10 pgmcc receivers and one TCP, then, after 300 seconds, 90 more receivers join the multicast session.

repairs, or the repair traffic would quickly dominate the actual data traffic on the link from the source.

6 Conclusion

In this paper we have discussed the performance of a recently proposed single-rate multicast congestion control scheme called pgmcc. We mainly addressed the problem of fairness and TCP friendliness of the new scheme.

We have identified a set of basic scenarios where fairness is concerned, and by simulations we have shown that pgmcc is capable of sharing available bandwidth fairly with competing TCP traffic. In particular, pgmcc behaves in a way very similar to TCP in a wide variety of network conditions, with very high loss rates or a high degree of heterogeneity among multicast receivers in terms of round trip times and packet loss rates.

We believe that more complex scenario can be built starting from the basic configurations we have analyzed. We are very confident that the performance of pgmcc will not significantly differ in such scenarios.

The strength of pgmcc relies on its simple design. Two main factors make this scheme scalable and stable: i) the fast election process that permits the sender to quickly identify the group's worst receiver, and ii) the use of a closed control loop between the sender and the group's representative that mimics TCP congestion control.

More extensive experiments certainly need to be made, but we believe that pgmcc is mature enough to be deployed on a real operational network. Therefore, future work will be devoted to specifying pgmcc features and requirements in the context of the IETF Reliable Multicast Transport Group and to implement this scheme on the most common operating systems.

References

1. S. Bhattacharyya, D. Towsely, J. Kurose "The Loss Path Multiplicity Problem in Multicast Congestion Control" *IEEE Infocom, March 1999.*
2. L. Breslau, D. Estrin, K. Fall, S. Floyd, J. Heidemann, A. Helmy, P. Huang, S. McCanne, K. Varadhan, Y. Xu, H. Yu "Advances in Network Simulation", *IEEE Computer, May 2000.*
3. R. Jain, D.M. Chiu, W. Hawe, "A quantitative measure of fairness and discrimination for resource allocation in shared systems", *TechReport DEC-TR-301, September 1984.*
4. T. Jiang, E.W. Zegura, M. Ammar, "Inter-Receiver Fair Multicast Communication Over the Internet", *ACM Sigmetrics, June 1998.*
5. A. Legout, J. Nonnenmacher, E.W. Biersack "Bandwidth Allocation Policies for Unicast and Multicast Flows" *IEEE Infocom, March 1999.*
6. A. Mankin, A. Romanow, S.Bradner, V.Paxson, "IETF Criteria for Evaluating Reliable Multicast Transport and Application Protocols" *RFC2357*
7. S. Floyd, M. Handley, J. Padhye, J. Widmer, "Equation-Based Congestion Control for Unicast Applications", *ACM SIGCOMM 2000, Stockholm, Sweden, August 2000.*
8. J. Padhye, V. Firoiu, D. Towsley, J. Kurose, "Modeling TCP throughput: a simple model and its empirical validation", *ACM SIGCOMM'98, Vancouver, CA, Sep. 1998.*
9. L. Rizzo, "pgmcc: A TCP-friendly single-rate multicast congestion control scheme", *ACM SIGCOMM 2000, Stockholm, Sweden, Aug 2000.*
10. D. Rubenstein, J. Kurose, D. Towsley "The Impact of Multicast Layering on Network Fairness" *ACM Sigcomm, August 1999*
11. T. Speakman et al., "PGM Reliable Transport Protocol Specification", *Internet Draft, draft-speakman-pgm-spec-06.txt*
12. H.A.Wang, M.Schwartz, "Achieving Bounded Fairness for Multicast and TCP Traffic in the Internet", *Proc. of SIGCOMM'98, Aug.1998, Vancouver*

Decentralized Computation of Weighted Max-Min Fair Bandwidth Allocation in Networks with Multicast Flows*

Emily E. Graves[1], R. Srikant[1], and Don Towsley[2]

[1] Department of General Engineering and Coordinated Science Laboratory,
1308 W. Main Street, Urbana, IL 61801
e-graves@comm.csl.uiuc.edu, rsrikant@uiuc.edu
[2] Department of Computer Science, University of Massachusetts,
Amherst, MA.
towsley@cs.umass.edu

Abstract. We consider the problem of designing decentralized algorithms to achieve max-min fairness in multicast networks. Starting with a convex program formulation, we show that there exists an algorithm to compute the fairshare at each link using only the total arrival rate at each link. Further this information can be conveyed to the multicast receivers using a simple marking mechanism. The mechanisms required to implement the algorithm at the routers have low complexity since they do not require any per-flow information.

1 Introduction

Multicast shows great promise as a network service for providing efficient content delivery from one sender to many receivers. However, its widespread deployment depends critically on the development of practical congestion control algorithms. One of the key challenges in developing such algorithms is to handle the heterogeneity that often characterizes multicast sessions. A single session can include many receivers with widely varying bandwidth connectivities to the sender. In cases where receivers are required to all receive data at the same rate, the sender may choose a very low data rate in order to support the smaller reception rate. On the other hand, where fully reliable reception is not required and receivers have the flexibility to receive a subset of the data it is often possible for each receiver to select a rate consistent with its network connectivity. We shall refer to the former as *single-rate* sessions and the latter as *multirate* sessions.

It was established in [13] in the context of max-min fairness that multirate reception enhances network fairness properties. Hence we focus on the problem of congestion control for multirate sessions. Moreover we focus on algorithms

---

* The work of the first two authors was supported by NSF Grants NCR-9701525 and ANI-9813710., and DARPA contract F30602-00-0542. The work of the third author was supported by DARPA contract F30602-00-0554 , and NSF grants NCR-9980552 and EIA-0121608

that produce max-min rate allocations. In particular, using ideas from [5,7,10] we formalize the problem of max-min multirate congestion control as a convex optimization problem and develop simple algorithms for its solution.

There is a tremendous amount of literature on optimization-based congestion control, e.g., [6,3,4,10,7], that formulate unicast congestion control problems as convex programs and derive congestion controllers that converge to the optimal solution of the convex programs. However, as we will show later, the multirate, multicast congestion control problem introduces certain constraints that are not easily incorporated in the framework of these earlier papers. While, in general, the convex program formulation, does not lead to implementable solutions for the multirate multicast problem, we show that there exists a particular choice of utility functions that approximate max-min fair allocation arbitrarily close and also provide implementable solutions. These algorithms are easily implemented at the receivers and require only that the network elements provide a packet marking capability. Such a capability is currently under investigation for the future Internet [12].

A number of multirate reception protocols have been proposed for the reception of streamed layered video, e.g. RLM (receiver-driven layered multicast) [11], TCP-layered multicast [17], and LVMR (layered video multicast with retransmissions) [9]. However, none of these deal with the problem of providing inter-session fairness. In addition, there have been several studies of max-min fairness in the context of multicast. Tzeng and Siu [16] first proposed its use in the context of single rate multicast sessions and presented an algorithm in the context of ATM for obtaining a max-min rate allocation in a static environment. More recently, Sarkar and Tassiulas [14] presented and analyzed an algorithm for obtaining the rate allocation in a multirate multicast network. Unlike our proposed algorithm which only requires a packet marking capability (currently under investigation for the Internet), their algorithm requires storage of per-session state information for each link in the network. Hence it appears better suited for an ATM network architecture rather than the current Internet.

The rest of the paper is organized as follows. In Section II, we first formulate the max-min fair rate allocation problem for networks with multirate, multicast flows as a convex program. We then show that there exists a simple, distributed algorithm that converges to the max-min fair rate allocation. This algorithm requires the network routers to use only the total arrival rate at each link. In Section III, we present our algorithm in detail and study its convergence properties in Section IV. Simulation results are presented in Section V, and concluding remarks are provided in Section VI.

2   Motivation for Using a Convex Program Formulation

Consider the following convex program:

$$\max \sum_{s \in S} \sum_{r \in V(s)} -\frac{w_s^n}{(n-1)x_{sr}^{n-1}} \qquad (1)$$

subject to

$$\sum_{s \in S_l} \max_{r \in V(s), l \in L_{sr}} x_{rs} \le C_l, \qquad \forall l \in L, \tag{2}$$

$$x_{sr} \ge 0, \qquad s \in S, r \in V(s). \tag{3}$$

where L is the set of links, C_l is the capacity of link l, S is the set of sessions, S_l is the set of sessions on link l, $V(s)$ is the set of virtual sessions in Session s, L_{sr} is the route (a contiguous collection of links) of Virtual Session r of Session s and x_{sr} is the transmission rate of Virtual Session r of Session s.

The above convex program corresponds to a resource allocation problem where the objective is to maximize the sum of the utilities of all the users in the network, where each user (or virtual session) has a utility function given by $-\frac{w_s^n}{(n-1)x_{sr}^{n-1}}$. The max that appears in constraint (2) reflects the property that the bandwith used by a multicast session on a link equals the maximum bandwidth required by any virtual session associated with the session that utilizes that link.

For ease of presentation, we assume that the weights w_s are the same for all virtual sessions in a multicast session although one can allow this to be more general. Note that the above utility function is a special case of the functions considered in [7]. As $n \to \infty$, this leads to a max-min fair allocation.

To simplify the notation, we consider an example network shown in Figure 1. There are three sessions, 0, 1 and 2. Session 0 has two receivers corresponding to two virtual sessions whereas Sessions 1 and 2 are unicast sessions.

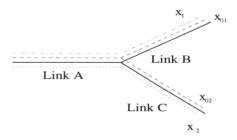

Fig. 1. A Y-network with three sessions: a multicast session with two receivers and two unicast sessions

The capacity constraints become

$$\max\{x_{01}, x_{02}\} + x_1 + x_2 \le C_A,$$

$$x_{01} + x_1 \le C_B \tag{4}$$

$$x_{02} + x_2 \le C_C. \tag{5}$$

It is more convenient to write the constraint on Link A as two linear constraints as follows:

$$x_{01} + x_1 + x_2 \le C_A, \tag{6}$$

and

$$x_{02} + x_1 + x_2 \leq C_A. \tag{7}$$

Now this problem is in the form of a standard convex program subject to linear constraints. As in [10], this can be solved using duality theory as follows:

$$p_{Ai}(k + 1) = (p_{Ai}(k) + \epsilon(x_{0i}(k) + x_1(k) + x_2(k) - C_A))^+, \quad i = 1, 2$$

$$p_B(k + 1) = (p_B(k) + \epsilon(x_{01}(k) + x_1(k) - C_B))^+$$

and

$$p_C(k + 1) = (p_C(k) + \epsilon(x_{02}(k) + x_2(k) - C_C))^+,$$

where $x_i(k)$ are calculated as

$$\frac{w_0^n}{x_{01}^n(k)} = p_{A1}(k) + p_B(k),$$

$$\frac{w_0^n}{x_{02}^n(k)} = p_{A2}(k) + p_C(k),$$

$$\frac{w_1^n}{x_1^n(k)} = p_{A1}(k) + p_{A2}(k) + p_B(k),$$

$$\frac{w_2^n}{x_2^n(k)} = p_{A1}(k) + p_{A2}(k) + p_C(k).$$

In the above equations, $p_{A1}(k)$, $p_{A2}(k)$, $p_B(k)$ and $p_C(k)$ are the estimates (at time k) of the shadow prices (or Lagrange multipliers) corresponding to (6), (7), (4) and (5), respectively.

We introduce the notation x_i^* and p_i^* to denote the optimal values of the virtual session rates and shadow prices, respectively. For the purpose of this example, we will assume that, the optimal solution to the convex program satisfies $x_{02}^* < x_{01}^*$. This implies that (7) is an inactive constraint, i.e., one that would be satisfied with strict inequality at the optimal solution.

While the above dual algorithm will converge, it is impossible to implement. Specifically, there are two issues:

- Link A needs the values of x_{01} and x_{02} to compute its shadow prices p_{A1} and p_{A2}; however, only $\max\{x_{01}, x_{02}\}$ is available. Since we have assumed that (7) is inactive, we know from standard convex optimization theory that the optimal value of p_{A2}^* is equal to zero. Therefore, it may be sufficient to only calculate a single shadow price p_A for link A using the iteration

$$p_A(k + 1) = (p_A(k) + \epsilon(\max_i\{x_{0i}(k)\} + x_1(k) + x_2(k) - C_A))^+.$$

At least locally (i.e., near the optimal solution), the iteration for p_A will be the same as that for p_{A1} and thus, $p_A(k)$ will converge to p_{A1}^* if the iteration is started in a local neighborhood of p_{A1}^*.

– Each virtual session has to know the sum of the shadow prices along its path. Suppose that p_{A2} is equal to zero in steady state; then Virtual Session 2 of Multicast Session 0 needs to know only p_C. On the other hand, Virtual Session 1 needs to know $p_A + p_B$. However, when per-flow state is not maintained, it is easiest to convey a single quantity for each link in the network. Thus, if a single p_A is computed for link A and as in [10], the sum of the shadow prices on each path is conveyed to the edges of the network, then Virtual session 1 would receive $p_A + p_B$, which is the correct information. But the sum of p_A and p_C would be conveyed to Virtual Session 2, which is incorrect. We address this problem below.

Defining $\lambda_i = p_i^{1/n}$, we obtain

$$\frac{w_0^n}{x_{01}^n} = \lambda_{A1}^n(k) + \lambda_B^n(k), \tag{8}$$

$$\frac{w_{02}^n}{x_{02}^n} = \lambda_{A2}^n(k) + \lambda_C^n(k), \tag{9}$$

$$\frac{w_1^n}{x_1^n} = \lambda_{A1}^n(k) + \lambda_{A2}^n + \lambda_B^n(k), \tag{10}$$

$$\frac{w_2^n}{x_2^n} = \lambda_{A1}^n(k) + \lambda_{A2}^n(k) + \lambda_C^n(k). \tag{11}$$

We note that, as $n \to \infty$, $(\sum_i y_i^n)^{1/n}$ approximates $\max_i y_i$ (provided that the max is achieved by only one y_i). Thus, in the limit, each source simply needs to know the maximum of the scaled shadow prices (λ's). We will now consider the implication of computing only λ_A and using it in lieu of λ_{A1} and λ_{A2} as $n \to \infty$:

– Equation (8) becomes
$$\frac{w_0}{x_{01}^*} = \max\{\lambda_A^*, \lambda_B^*\}.$$

Since we already argued that $\lambda_A^* = \lambda_{A1}^*$ in steady-state, this is consistent with the solution obtained from the dual of the convex program.
– Equation (9) yields
$$\frac{w_{02}}{x_{02}^*} = \max\{\lambda_A^*, \lambda_C^*\}.$$

This equals $\max\{\lambda_{A2}^*, \lambda_C^*\}$ only if $\lambda_A^* < \lambda_C^*$ since $\lambda_{A2}^* = 0$. This is indeed true since by our assumption $x_{02}^* < x_{01}^*$.
– Since $\lambda_{A2}^* = 0$, from (10) and (11), it is easy to see that the optimal solutions of x_1 and x_2 are unaffected by using just λ_A. For example, at the optimal solution,
$$\frac{w_1}{x_1^*} = \max\{\lambda_{A1}^*, \lambda_{A2}^*, \lambda_B^*\} = \max\{\lambda_A^*, \lambda_B^*\}.$$

Thus, the above example suggests that in the case of this particular utility function, it may be sufficient to only use the aggregate flow into a link to compute the max-min fair solution. In the next section, we present our algorithm as motivated by the above discussion. We show that the equilibrium point of the algorithm is indeed the max-min fair solution and provide conditions for the local convergence of the synchronous version of this algorithm. We also present a simple marking mechanism to implement the algorithm and finally present simulation results illustrating the global convergence of the algorithm even with asynchronous updates.

Example: We consider the network in Figure 1 to illustrate the scaling for the shadow price and show that the scaled shadow price converges to the fairshare (called the link control parameter in [14]) as $n \to \infty$. We let $w_0 = w_1 = w_2 = 1$, $C_A = 10$, $C_B = 15$ and $C_C = 5$. For this example, it is easy to calculate the max-min fair rates as

$$x_{01}^* = 3.75, \quad x_{02}^* = 2.5, \quad x_1^* = 3.75, \quad x_2^* = 2.5.$$

Alternately, we calculate the solution to the nonlinear program for integer values of n from 1 to 10, and plot the resulting values of the scaled shadow prices in Figures ?? and 3. Since $x_{02} + x_1 + x_2 < C_A$ and $x_{01} + x_1 < C_B$, for all values of n, the scaled shadow prices λ_{A2}^* and λ_B^* are always equal to zero and therefore are not plotted.

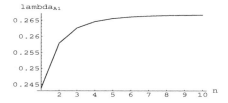

Fig. 2. Scaled shadow price λ_{A1}^* for the Y-network as a function of the utility function parameter n

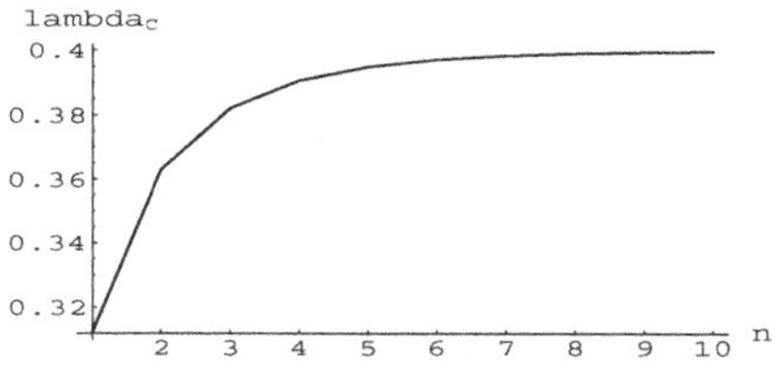

Fig. 3. Scaled shadow price λ_C^* for the Y-network as a function of the utility function parameter n.

From the figures, we see that λ_{A1}^* converges to 0.2666 and λ_C^* converges to 0.4. Using the algorithm in [14], the fairshare on Link A is 3.75 which is $1/\lambda_{A1}^*$ and the fairshare on Link C is 2.5 which is equal to $1/\lambda_C^*$. Since Link B is underutilized, the fairshare on that link can be taken to be any value greater than or equal to C_B and the optimization algorithm yields a value of ∞.   ◇

Remark 1. We note that the actual shadow prices themselves (at links A and C) go to ∞ as $n \to \infty$ and thus, do not provide insight into the max-min fair rate allocation. It is therefore important to scale the shadow prices as $\lambda_i = p_i^{1/n}$ to obtain meaningful results.   ◇

3 Weighted Max-min Fairness Algorithm

We present an algorithm motivated by the discussions of the convex program formulation of the problem discussed in the previous section. In the algorithm, we compute the scaled shadow price which, as already discussed, corresponds to the inverse of the fairshare.

The basic steps of the algorithm are summarized as follows:

1. *Compute the scaled shadow price*: As dictated by the gradient-descent iteration of the dual problem, the shadow price (λ_l for each link l) is increased or decreased depending upon whether the link is overutilized or underutilized, respectively. The fairshare for a link is the inverse of the shadow price.
2. *Compute the allowable rate for each virtual session*: Each virtual session is allowed to transmit at the minimum fairshare for the set of links that it traverses. This is the maximum allowable fairshare for that virtual session.
3. *Compute the load at each link*: The load on each link is computed as the total arrival rate at the link. When multiple virtual sessions from the same session pass through a link, the maximum of their rates is used as the arrival rate for the multicast session.

The above steps are repeated resulting in all virtual sessions being bottlenecked at max-min fairness.

Before we present the algorithm, we introduce the following terminology:

c_l: capacity of link l
$\lambda_l(k)$: scaled shadow price of link l at time k
$f_l(k)$: fairshare at link l at time k
$x_s(k)$: maximum allowable rate for session s at time k
$l_l(k)$:load on link l at time k
ϵ: scaling factor
L_v: set of links traversed by virtual session v
S_l: set of sessions traversing link l
V_{sl}: set of virtual sessions corresponding to session s on link l

The algorithm can now be described as follows:

- $\lambda_l(0)$ is chosen at random for each l in the network such that $1/c_l \leq \lambda_l(0)$.
- At each step
  1. Compute the shadow price for each link l in the network

  $$\lambda_l(k+1) = [\ \lambda_l(k) + \epsilon(l_l(k) - c_l)\ ]_{1/c_l}^{\infty},$$

  where $[y]_m^M$ denotes $\min\{\max\{y, m\}, M\}$. Depending upon the choice of the stepsize, the convergence properties of the algorithm may vary. Here we address only the situation where ϵ is a constant.
  2. Compute the fairshare of link l at time instant k, $f_l(k) = 1/\lambda_l(k)$.
  3. Compute the allowable rate for each virtual session v in the network

  $$x_v(k+1) \ = \ w_v \min_{l \in L_v} f_l(k+1) \ = \ \frac{w_v}{\max_{l \in L_v}(\lambda_l(k+1))}$$

By approximating the maximum by a sum, we can implement the computation of the maximum by a simple marking algorithm using the ideas in [10]. We will elaborate upon this later.

4. Compute the load on each link as the sum of the rates of each session across that link.

$$l_l(k+1) = \sum_{s \in S_l} \max_{v \in V_{sl}} x_v(k+1)$$

Note that the rate for Session s traversing link l is the maximum of the rates of the virtual sessions of Session s traversing the link.

3.1 Max-min Fairness of the Equilibrium Point

We now show that, if the algorithm converges to an equilibrium solution, the resulting rate allocation is max-min fair. Max-min fairness as defined in [14] is obtained iff every virtual session has a bottleneck link. By definition, for a link l to be bottlenecked with respect to a virtual session v it must satisfy the following conditions.

1. $l_l = c_l$, i.e., the bandwidth allocated to the sessions traversing link l must equal the link's capacity
2. $x_v^* \geq x_r^*$ for all $r \in S_l$.

Lemma 1. *The rate allocation at the equilibrium point of the algorithm is max-min fair.*

Proof: We claim that the route of each virtual session contains a bottleneck link. To show this, assume that a Virtual Session v does not have a bottleneck link. Suppose no link in L_v satisfies condition (1). Then, from the equation

$$\lambda_l(k+1) = [\lambda_l(k) + \epsilon(\ l_l - c_l)]_{1/c_l}^\infty ,$$

it is clear that the link will be underutilized, i.e., $l_l < c_l$. Further, the equilibrium value of the scaled shadow price λ_l will be equal to $1/c_l$. Since $x_v = \min_{l \in L_v} 1/\lambda_l$, the equilibrium value of x_v would be equal to c_l for some $l \in L_v$ which contradicts the fact that all links in the path L_v are underutilized. Thus, there exists a nonempty subset $L_v' \subset L_v$ such that all links in L_v' satisfy condition (1).

Next suppose that all links in L_v' violate condition (2). Now, consider a link $l \in L_v'$. Given $x_v < x_r$ for all $r \in S_l$, $r \neq v$, and by definition of $x_v = \min_{m \in L_v} 1/\lambda_m$, it follows that $x_v < 1/\lambda_l$. Again, $x_v = \min_{m \in L_v} 1/\lambda_m = 1/\lambda_j < 1/\lambda_l$ for some link $j \in L_v$. Suppose that $j \notin L_v'$, then it would imply that $\lambda_j = 1/c_j$ which implies that $x_v = c_j$ and thus, $l_j = c_j$, which is a contradiction. Therefore, $j \in L_v'$. This must be true for all links $l \in L_v'$. However for the case where $j = l$, then $1/\lambda_l < 1/\lambda_l$ which is contradictory, so the link cannot violate condition (2). ◇

3.2   Implementation Using One-Bit Packet Marking

Our algorithm requires the network to convey $\min_{l \in L_v} f_l(k)$ to each Virtual session v. This can be accomplished using a simple marking scheme. First, motivated by the solution to the convex program in the previous section, we use the following approximation:

$$\min_{l \in L_v} f_l(k) \approx \frac{1}{[\sum_{l \in L_v} 1/(f_l(k))^n]^{1/n}},$$

where n is some large number. Thus, we can equivalently think of the information to be conveyed to each virtual session to be $\sum_{l \in L_v} 1/(f_l(k))^n$. Now, combining this with the marking scheme presented in [1], we obtain the following algorithm to convey the fairshare to the virtual sessions:

- Each link l marks every packet that traverses it with probability $(1 - e^{-1/(f_l(k))^n})$.
- Each virtual session v keeps track of the rate at which it receives unmarked packets. From the marking scheme described above, the rate of unmarked packets for a virtual session v is

$$exp\left(-\sum_{l \in L_v} 1/(f_l(k))^n\right).$$

Thus, the virtual session can obtain an estimate of the minimum fairshare of the links in its path as $1/(-\ln u_v(k))^n$, where $u_v(k)$ is the current estimate of the rate at which it receives unmarked packets.

While the above algorithm would, in principle, provide the required information to the sources, there is a serious implementation difficulty with this approach. For large values of n, depending upon the value of $1/f_l(k)$, $1/(f_l(k))^n$ could either be very large or very close to 0. As a result , $e^{-\sum_{l \in L_v} 1/(f_l(k))^n}$ could be very close to 1 or 0. This leads to the numerically unstable condition where all packets are always marked or none of the packets are marked over long intervals of time.

To overcome this problem, we consider the following transformation:

$$\delta_l(k) = \log_b(\frac{K}{f_l(k)})$$

The idea behind the above transformation is that if K and b are chosen so that $\delta_l(k)$ remains close to one, then one can convey $\max_{l \in L_v} \delta_l(k)$ to the sources using the marking probability

$$1 - e^{(-\sum_{l \in L_v} (\delta_l(k))^n)},$$

thus avoiding numerical instability.

To find appropriate values for b and K, we assume that the fairshares at each link are constrained to lie in an interval $[f_{min}, f_{max}]$. The upperbound f_{max} could be the link capacity and the lower bound could be a minimum rate guaranteed to all sources. Now suppose that we want $\delta_l(k) \in [x, y]$ where $x < 1 < y$. For example, x could be 0.95 and y could be 1.05. Thus, we require

$$x \leq \log_b(\frac{K}{f_l(k)}) \leq y,$$

which is satisfied if

$$\log_b(K/f_{min}) = \log_b(K\lambda_{max}) = y,$$

and

$$\log_b(K/f_{max}) = \log_b(K\lambda_{min}) = x.$$

Solving these yields

$$b = (\frac{f_{max}}{f_{min}})^{\frac{1}{y-x}},$$

and

$$K = f_{max}(\frac{f_{max}}{f_{min}})^{\frac{x}{y-x}}$$

Now the marking probability algorithm is as follows:

- Packets traversing link l are marked with probability $(1 - e^{-(\delta_l(k))^n})$.
- Using this marking scheme the rate of unmarked packets received by Virtual Session v is:

$$u_v(k) = \exp\left(-\sum_{l \in L_v} (\delta_l(k))^n\right).$$

- The approximate value of $\min_{l \in L_v} f_l(k)$ is calculated as

$$\min_{l \in L_v} f_l(k) \approx \frac{K}{b^{\hat{\delta}_{vmax}(k)}},$$

where

$$\hat{\delta}_{vmax} = [-\ln u_v(k)]^{1/n}.$$

- The transmission rate for each virtual session v in the network is given by

$$x_v(k+1) = [(1 - \beta)x_v(k) + \beta w_v \min_{l \in L_v} f_l(k+1)]_{f_{min}}^{f_{max}},$$

where $\beta \in (0, 1)$ is a damping parameter.

3.3   Other Implementation Considerations

In this subsection, we discuss the amount of information required to implement our multicast congestion control algorithm. We claim that, in the context of the IP multicast architecture, no additional per-multicast flow information is required beyond what is already needed to implement multicast routing and layered multicast. It is important to note that layered multicast is typically implemented within the IP architecture by having receivers join and leave multicast groups, see the work on receiver layered multicast by McCanne, Jacobsen and Vetterli [11] and the more recent work of Rubenstein, Kurose, and Towsley [13]. Consequently, it is sufficient for the routers to propagate the marks to the receivers based on their aggregate loads. The receivers then compute the prices and determine what rates they should receive at. This in turn determines which layers they should listen to. They then may choose to add layers (by joining multicast groups) or drop layers (by leaving multicast groups). We have described how no additional per flow state is required in the context of IP. We expect this to be the case for any future multicast architectures.

As far as the congestion control algorithm goes, one may argue that the routing table contains information about the number of multicast virtual sessions passing through a router which could be used to compute the fairshare at each node. However, it is important to recognize that our algorithm is intended for networks with unicast and multicast flows and therefore, it is significant that we do not require knowledge of the number of unicast flows through the link. For example, if a router handles $100,000$ flows, out of which $10,000$ are multicast flows and the rest are unicast, it does not maintain per-flow information on the $90,000$ unicast flows. Thus, from a congestion control point of view, our algorithm does not require any additional per-flow information. In addition to aggregate flow information, we only require one-bit packet marking to convey congestion information.

4   Local Stability and Rate of Convergence

In this section, we study the conditions under which the algorithm presented is locally asymptotically stable. We make the following assumptions:

- For each Session s, there is a Virtual Session v such that $x_{sv}^* > x_{sr}^*$ for all $r \in V(s) \setminus \{v\}$, where $\{x_{sr}^*\}$ denotes the max-min fair rates. In other words, under the max-min fair rate allocation, we assume that each multicast session has a unique virtual session which determines the overall transmission rate of the multicast source.
- Under max-min fair rate allocation, each virtual session has a unique bottleneck link.
- Each link is a bottleneck for at least one virtual session.

We note that the above assumptions are required only for the convergence analysis of this section but not for the implementation of the algorithm.

Proposition 41 *The algorithm presented in the previous section is locally asymptotically stable under the following conditions on ϵ :*

$$0 < \epsilon < \min_{l \in L} 2(\lambda_l^*)^2/W_l$$

where W_l is the sum of the weights for the sessions bottlenecked at link l.

Proof: Recall that

$$
\begin{aligned}
\lambda_l(k+1) &= \lambda_l(k) + \epsilon\lambda_l(k)(c_l - l_l(k)) \\
&= \lambda_l(k) + \epsilon\lambda_l(k)(c_l - \sum_{s \in S_l} \max_{v \in V_{sl}} x_v(k)) \quad (12)
\end{aligned}
$$

$$x_v(k) = w_v \min_{m \in L_v} 1/\lambda_m(k), \quad (13)$$

where w_v, the weight of the Virtual Session v, is the weight of the corresponding multicast session to which it belongs.

Let λ_l^* be the fairshare associated with link l under weighted max-min fair rate allocation. Define

$$\hat{\lambda}_l(k) := \lambda_l(k) - \lambda_l^*$$

Let $\hat{\boldsymbol{\lambda}}(k)$ denote the vector of link fairshares. Then, (12) can be written as

$$\hat{\boldsymbol{\lambda}}(k+1) = f(\hat{\boldsymbol{\lambda}}(k)),$$

where $f(\cdot)$ is a vector whose l^{th} element is the right-hand side of (12). To prove that this system is locally stable, we linearize around $\lambda = \lambda^*$ and prove that the eigenvalues of the resulting linear system lies within the unit circle in the complex plane. Thus, the linear system is given by

$$\hat{\boldsymbol{\lambda}}(k+1) = A\hat{\boldsymbol{\lambda}}(k).$$

where

$$
A = \frac{\partial f}{\partial \hat{\boldsymbol{\lambda}}}\bigg|_{\hat{\lambda}=0}
$$

$$
= \begin{bmatrix}
1 - \dfrac{\epsilon W_{11}}{\lambda_1^{*2}} & -\dfrac{\epsilon W_{12}\lambda_1^*}{\lambda_2^{*3}} & \cdots & -\dfrac{\epsilon W_{1L}\lambda_1^*}{\lambda_L^{*3}} \\[3mm]
-\dfrac{\epsilon W_{21}\lambda_2^*}{\lambda_1^{*3}} & 1 - \dfrac{\epsilon W_{22}}{\lambda_2^{*2}} & \cdots & -\dfrac{\epsilon W_{2L}\lambda_2^*}{\lambda_L^{*3}} \\[3mm]
\vdots & \vdots & & \vdots \\[3mm]
-\dfrac{\epsilon W_{L1}\lambda_L^*}{\lambda_1^{*3}} & -\dfrac{\epsilon W_{L2}\lambda_L^*}{\lambda_2^{*3}} & \cdots & 1 - \dfrac{\epsilon W_{LL}}{\lambda_L^{*2}}
\end{bmatrix}
$$

Here W_{lm} is the sum of the weights for those sessions passing through link l that are bottlenecked at link m. L is given as the total number of links in the network.

From the definition of W_{lm}, it is easy to see that if $W_{ij} > 0$ then $W_{ji} = 0$. Note that $W_{ij} > 0$ implies that some session passes through both links i and j and is bottlenecked at link j. This means that $\lambda_j^* > \lambda_i^*$. It is not simultaneously possible to have another session passing through both links i and j resulting in a bottleneck at link i, thus $W_{ji} = 0$. In view of this, without loss of generality, let us label the link such that if a session passes through link i and is bottlenecked at link j, i is larger than j. This produces an upper triangular matrix, A, and thus, its eigenvalues are the main diagonal elements. Since the necessary and sufficient condition for the stability of a linear system is that all the eigenvalues have absolute value less than 1, the statement of the theorem follows immediately. ⋄

Remark 2. The speed of convergence is determined by the spectral radius (the maximum absolute value of the eigenvalues) of A. Since the eigenvalues of A are its diagonal elements, the spectral radius of each of the algorithms is given by $\max_l |1 - \epsilon W_l/(\lambda_l^*)^2|$ ⋄

5   Simulation Results

In this section, we study the algorithm presented in the previous sections through simulations on the simple Y-network of Figure 1, and a more complicated network, which we call the general network. We normalize time such that an update interval for the fairshare at each link is one time slot. All link capacities should then be interpreted as being measured in terms of 100 packets per time slot, i.e., if the capacity of a link is 10, then it should be interpreted as 1000 packets per time slot. All packets are assumed to be of equal size and due to the above normalization, the actual size of the packet (in bytes) is irrelevant. The parameters K and b defined in Section 3.2 are chosen such that $0.95 \leq \delta_l \leq 1.05$ for all links.

We assume that no packets are lost in the buffer. This is reasonable if we assume that active queue management (AQM) schemes (e.g., [8]) with ECN packet marking ([2]) are employed to nearly eliminate losses in the network. As in [8], this would mean that, for each link, a target utilization is chosen, say 0.98. Then, the marking algorithms would compare the arrival rate to 0.98 times the link capacity, as opposed to the full link capacity. Thus, early warning of congestion would be provided and the arrival rate at any link would almost always be less than the link capacity leading to very low levels of packet loss. These issues have been addressed extensively in [8] and therefore, are not considered here.

The following figures and simulation data are representative of the many simulations done to examine the algorithm presented. The values chosen for simulation were chosen arbitrarily within the algorithm's specifications to demonstrate max-min fair convergence.

5.1 Simple Y-Network

For each simulation the following values were used: $C_A = 30$, $C_B = 20$, $C_C = 10$, $w_0 = 1$, $w_1 = 2$, $w_2 = 3$, $n = 50$. The minimum and maximum fairshares for each flow were chosen to be 0.3 and 30, respectively. We chose the stepsize ϵ to be 0.01 and $\beta = 0.1$.

Figures 4 and 5 show the convergence of the session rates and the link fairshares, respectively, for the algorithm. For this simple network, the max-min fair rate allocation can be exactly calculated as $x_{01} = 6\frac{2}{3}$, $x_{02} = 2\frac{1}{2}$, $x_1 = 13\frac{1}{3}$, $x_2 = 7\frac{1}{2}$. Simulation of the algorithm shows convergence about the values $x_{01} = 6.6669$, $x_{02} = 2.4997$, $x_1 = 13.3337$, $x_2 = 7.4992$, with a maximum oscillation of $\pm 5\%$ due to the randomness in the marking process.

Figure 5 shows the convergence of the fairshare values for links A, B, and C. At convergence, the fairshares for links B and C oscillate around $\frac{1}{\lambda_B} = 7.9877$, $\frac{1}{\lambda_C} = 2.6650$. As link A does not contain a bottlenecked session, its fairshare is bounded by the capacity of the link. After approximately 70 iterations, $\frac{1}{\lambda_A}$ converged to 30. On the other hand, both links B and C contain a bottleneck session and therefore, their fairshare converges to a value smaller than c_l. These values are similar to the expected convergence values of $\frac{1}{\lambda_A} = 30$, $\frac{1}{\lambda_B} = 6\frac{2}{3}$, $\frac{1}{\lambda_C} = 2\frac{1}{2}$. The difference can be explained by the chosen value for n. Larger values of n show expected convergence of the fairshare values.

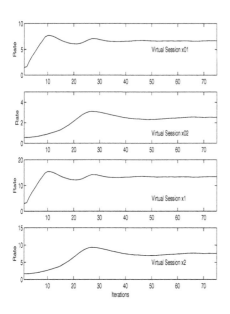

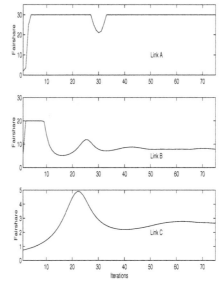

Fig. 4. Y-network: Session Rates vs Number of Iterations

Fig. 5. Y-network: Link Fairshares vs Number of Iterations

5.2   General Network

The second set of simulations were done using the network in Figure 6. The 19 links are identified by the numbers associated with them in the figure. This network carries traffic from 11 multicast and unicast sessions. The 6 multicast sessions have a total of 14 virtual sessions. Table 1 provides a list of the virtual session routes and weights, and the link capacities.

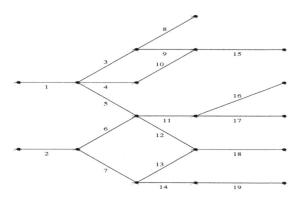

Fig. 6. General Network.

The other parameters were chosen to be $\epsilon = 0.01$, $\beta = 0.1$, and $n = 50$. The minimum and maximum fairshares for each flow were chosen to be 0.1 and 14, respectively. To simulate delays (possibly time-varying) in the network and asynchronous updates, we introduce a probability parameter p. The transmission rates of each virtual session in the network is updated at each time instant with probability p, and with probability $1 - p$, a session continues to use its old rate. In this subsection, we provide results for the case where $p = 0.5$.

Figures 7 and 8 show the convergence for four of the virtual sessions and four of the links in the network. Overall, session rates and fairshares converged within 5% of the expected values, given in Table 1. Our results show that, for sufficiently small ϵ and sufficiently large n, the algorithm converges to the max-min fair rates.

6   Conclusions

We have presented a simple, decentralized algorithm to achieve weighted max-min fairness in multirate, multicast networks. The algorithm is simple to implement at the source and receivers as well as at the routers in the network since no per-flow information is necessary. An ECN-like marking mechanism can be used to convey information from the network to the multicast receivers. The algorithm presented here can be easily generalized to the case where there are maximum and minimum rate constraints on the source transmission rates.

Table 1. Virtual Session Routes, Weights, Link Capacities, Max-min Fair Session Rates, and Link Fairshares

x_{sr}	L_{sr}	W_s	l	C_l	Session Rate	$1/\lambda_l^*$
x_{01}	1,3,8	1	1	14	0.8002	2.5613
x_{02}	1,9,15	1	2	9	1.1356	5.6035
x_1	1,4,10,15	3	3	6	3.5995	6
x_{21}	1,5,11,16	4	4	6	1.6618	5.3937
x_{22}	1,5,12,18	4	5	5	1.7136	0.5882
x_3	1,5,11,17	2	6	6	0.8134	6
x_{41}	2,6,12,18	2	7	6	3.5006	1.2832
x_{42}	2,7,14,19	2	8	4	1.4995	1.4132
x_5	2,7,13,18	3	9	4	1.5010	4
x_6	1,3,9,15	1	10	6	1.1356	5.3937
x_{71}	1,3,8	4	11	8	3.2010	8
x_{72}	1,5,11,16	4	12	7	1.6618	6.9546
x_{73}	1,5,11,17	4	13	1.5	1.6268	1.0628
x_{81}	2,6,12,18	1	14	5	1.7503	5
x_{82}	2,6,11,16	1	15	10	1.5570	10
x_{83}	2,7,14,19	1	16	5	0.7498	5
x_9	2,7,14,19	3	17	4	2.2493	4
x_{101}	1,4,10,15	2	18	9	2.3996	9
x_{102}	1,5,1,17	2	19	5	0.8134	5

There are two open issues that we plan to address in the near future. One is a proof of global convergence of the algorithm presented here. The other issue, which is of practical importance, is the impact of discrete bandwidth layers on the performance of the scheme suggested here [15].

References

1. S. Athuraliya, D. E. Lapsley, and S. H. Low. Random early marking for internet congestion control. In *Proceedings of IEEE Globecom*, 1999.
2. S. Floyd, "TCP and explicit congestion notification," *ACM Computer Communication Review*, vol. 24, pp. 10–23, October 1994.
3. R.J. Gibbens and F.P. Kelly. Distributed connection acceptance control for a connectionless network. In *Proc. of the 16th Intl. Teletraffic Congress*, Edinburgh, Scotland, June 1999.
4. R.J. Gibbens and F.P. Kelly. Resource pricing and the evolution of congestion control. *Automatica*, 1999.
5. F.P. Kelly. Charging and rate control for elastic traffic. *European Transactions on Telecommunications*, 8:33–37, 1997.
6. F. P. Kelly, A. Maulloo, and D. Tan. Rate control in communication networks: shadow prices, proportional fairness and stability. *Journal of the Operational Research Society*, 49:237–252, 1998.
7. S. Kunniyur and R. Srikant. End-to-end congestion control: utility functions, random losses and ECN marks. In *Proceedings of INFOCOM 2000*, Tel Aviv, Israel, March 2000.

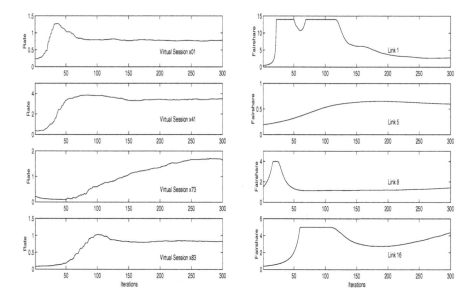

Fig. 7. General Network: Session Rates vs Number of Iterations

Fig. 8. General Network: Link Fairshares vs Number of Iterations

8. S. Kunniyur and R. Srikant, "A time-scale decomposition approach to adaptive ECN marking," to appear in the *Proceedings of INFOCOM 2001*, Alaska, Anchorage, April 2001.
9. X. Li, S. Paul, M.H. Ammar. "Layered video multicast with retransmission (LVMR): evaluation of hierarchical rate control," *Proc. INFOCOM'98*, March 1998.
10. S. H. Low and D. E. Lapsley. Optimization flow control, I: Basic algorithm and convergence. *IEEE/ACM Transactions on Networking*, 1999.
11. S. McCanne, V. Jacobson, M. Vetterli. "Receiver-driven layered multicast," *Proc. SIGCOMM'96*, Sept. 1996.
12. K.K. Ramakrishnan, S. Floyd. "A proposal to add Explicit Congestion Notification (ECN) to IP," RFC 2481, January 1999.
13. D. Rubenstein, J. Kurose, D. Towsley. "The impact of multicast layering on network fairness," *Proc. SIGCOMM'99*, Sept. 1999.
14. S. Sarkar and L. Tassiulas. Distributed Algorithms for Computation of Fair Rates in Multirate Multicast Trees. In *Proceedings of INFOCOM 2000*, Tel Aviv, Israel, March 2000.
15. S. Sarkar and L. Tassiulas. Fair Allocation of Discrete Bandwidth Layers in Multicast Networks. In *Proceedings of INFOCOM 2000*, Tel Aviv, Israel, March 2000.
16. H.Y. Tzeng, K. Siu. "On Max-Min fair congestion control for multicast ABR service in ATM," *IEEE J. on Sel. Areas in Commun.*, **15**:3, 1997.
17. L. Vicsiano, J. Crowcroft, L. Rizzo. "TCP-like congestion control for layered multicast data transfer," *Proc. INFOCOM'98*, March 1998.

Challenges of Integrating ASM and SSM IP Multicast Protocol Architectures

Kevin C. Almeroth[1], Supratik Bhattacharyya[2], and Christophe Diot[2]

[1] Department of Computer Science
University of California
Santa Barbara, CA 93106-5110
`almeroth@cs.ucsb.edu`
[2] Sprint Advanced Technology Labs
One Adrian Court
Burlingame, CA 94010
`{supratik,cdiot}@sprintlabs.com`

Abstract. The Source Specific Multicast (SSM) service model and protocol architecture have recently been proposed as an alternative to the currently deployed Any Source Multicast (ASM) service. SSM attempts to solve many of the deployment problems of ASM including protocol complexity, inter-domain scalability, and security weaknesses. However, the SSM protocol architecture is not radically different from that of ASM. This has created opportunities for integrating it into the currently deployed ASM infrastructure. In this paper, we first describe the ASM and SSM service models and associated protocol architectures, highlighting the relative merits and demerits of each. We then examine the network infrastructure needed to support both of them. Our conclusion is that integration is relatively straightforward in most cases; however there is one case—supporting ASM service over an SSM-only protocol architecture—for which it is difficult to design elegant solutions for an integrated SSM/ASM infrastructure.

1 Introduction

The original IP multicast service model was developed with the goal of creating an interface similar to that of best-effort unicast traffic[1]. For transmitters, the goal was to provide scalable transmission by allowing sources to simply transmit without having to register with any group manager or having to perform connection setup or group management functions. The application programming interface was similar to that for UDP packet transmission— an application would simply have to open a socket to a destination and begin transmitting. For receivers, the goal was to provide a way to join a group and then receive all packets sent by all transmitters to the group. In this service model, each multicast host group was identified by a class-D IP address so that an end-host could participate in a multicast session without having to know about the identities of other participating end-hosts. This eventually led to a a triumvirate of protocols to

S. Palazzo (Ed.): IWDC 2001, LNCS 2170, pp. 343–360, 2001.

build multicast trees and forward data along them: a tree construction protocol (the most widely deployed of which is) called Protocol Independent Multicast–Sparse Mode (PIM-SM), the multicast equivalent of the Border Gateway Protocol (BGP) for advertising reverse paths towards sources called the Multiprotocol Border Gateway Protocol (MBGP), and a protocol for disseminating information about sources called the Multicast Source Discovery Protocol (MSDP)[2]. In addition, the Internet Group Management Protocol (IGMP) was designed for end-hosts to dynamically join and leave multicast groups.

The wide-scale commercial deployment of this service model and protocol architecture has run into significant barriers [3]. Many of these barriers are rooted in the problem that building efficient multicast trees for dynamic groups of receivers is a non-trivial problem. As a result, the existing set of protocols is fairly complex and the learning curve is quite steep. Furthermore the "any-to-any" design philosophy of the current Any Source Multicast (ASM) service model is not suitable for commercial services. Most applications today need tighter control over who can transmit data to a set of receivers.

By trying to improve both the efficiency and reduce the complexity of current multicast protocols, the goal is to reduce the barriers to deployment. However, developing and deploying yet another set of protocols creates the additional burden of yet another round of modifications to the existing infrastructure. This may itself become an impediment to deployment efforts. Therefore, it is important to to identify the technical problems that needs to be solved before designing and deploying a new protocol architecture. For the current ASM architecture and service model, the problems include:

1. attacks against multicast groups by unauthorized transmitters
2. deployment complexity
3. problems of allocating scarce global class-D IP addresses
4. lack of inter-domain scalability
5. single point of failure problems

A new service model, Source Specific Multicast (SSM), and an associated protocol architecture have been proposed as a solution to the above problems and is beginning to be deployed[4]. In the SSM service model a receiving host explicitly specifies the address of the source it wants to receive from, in addition to specifying a class-D multicast group address.

From a deployment standpoint, the fundamental advantage of SSM is that protocol complexity can be removed from the network layer and implemented more easily, simply, and cheaply at the application layer. The tradeoff, which itself has both advantages and disadvantages, has the potential to fundamentally change how IP multicast service is provided in the Internet. Moreover there is significant overlap in the protocol architectures for ASM and SSM, thereby facilitating the rapid integration of SSM support in networks that already support ASM.

The key difference between the two service models lies in the way a receiving host joins a multicast group. As a result, there are a number of questions that arise about supporting the two. Should the two models exist simultaneously?

Should the existence of two models be made visible to the user? What part of the multicast infrastructure should be responsible for dealing with interoperability between the two service models? Answering these questions is a critical step in providing seamless interoperability between ASM and SSM.

In this paper, we describe the differences between the ASM and SSM protocol architectures and service models. We then study the challenges of deploying an integrated ASM/SSM infrastructure. We believe that SSM can solve many of the technical problems without making the existing infrastructure obsolete. But, technical challenges exist in seamlessly integrating the two without creating "black holes". Fortunately, we find that in most cases, the problems faced in integrating the two architectures are neither many nor insurmountable; they simply need to be identified and then the appropriate solutions implemented.

The remainder of the paper is organized as follows. In Section 2 we describe the multicast service models. Section 3 describes the ASM and SSM protocol architectures. Section 4 discusses the challenges in integrating ASM and SSM. The paper is concluded in Section 5.

2   IP Multicast Service Models

In order to understand the implication of offering different types of IP multicast services, we first need to make a distinction between a *protocol architecture* and a *service model*. A multicast protocol architecture refers to a set of protocols that together allow end-hosts to join/leave multicast sessions, and allows routers to communicate with each other to build and forward data along inter-domain forwarding trees. An IP multicast service model refers to the semantics of the multicast service that a network provides an end-user with. It is embodied in the set of capabilities available to an end-user at the application interface level, and is supported by a network protocol architecture. Any multicast service model is realized through:

- an application programming interface (API) used by applications to communicate with the host operating system.
- host operating system support for the API.
- protocol(s) used by the host operating system to communicate with the leaf network routers (referred to as designated routers or edge-routers).
- protocol(s) for building inter-domain multicast trees and for forwarding data along these trees.

With multiple service models and protocol architectures, the challenge therefore lies in bridging the gap between the protocol architecture deployed in the network and the service model expected by end-user applications.

Currently, there are two main IP multicast service models plus a third deriving from a combination of the two. The details of the protocol architecture supporting each of them is described in Section 3.

(a) **Any-Source Multicast (ASM)**: This is the traditional IP multicast service model defined in RFC 1112[1]. An IP datagram is transmitted to a

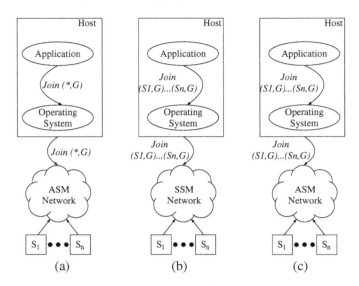

Fig. 1. Three choices for the IP multicast service model.

"host group", a set of zero or more hosts identified by a single IP destination address (224.0.0.0 through 239.255.255.255 for IPv4). This model supports one-to-many and and many-to-many multicast communication. Hosts may join and leave the group at any time. There is no restriction on the location or number of receivers, and a source need not be a member of the host group it transmits to. Host-to-network communication support for ASM is provided by the Internet Group Management Protocol (IGMP) version 2. IGMPv2 allows a receiver to specify a class-D group address for the host group it wants to join, but does not allow it to specify the sources that it wants (or does not want) to receive traffic from. This service model is shown in Figure 1(a).

(b) **Source-Specific Multicast (SSM)**: This is the multicast service model defined in [5]. An IP datagram is transmitted by a source S to an SSM address G, and receivers can receive this datagram by subscribing to channel (S,G). SSM is derived from EXPRESS[6] and supports one-to-many multicast. The address range 232/8 has been assigned by IANA[7] for SSM service in IPv4. In IPv6, an address range ($FF2x$ :: and $FF3x$ ::) already exists[8] for SSM services. IGMP version 3, which allows a receiver to specify explicitly the source address, provides host-to-network communication support for SSM. This requires upgrading most host operating systems and edge routers from IGMPv2 to IGMPv3. This also implies that the host operating system's API must now allow applications to specify a source and a group in order to receive multicast traffic. This service model is shown in Figure 1(b).

A variant of the ASM service model is known as the **Source-Filtered Multicast (SFM)** model. In this case, a source transmits IP datagrams to a host group address in the range of 224.0.0.0 to 239.255.255.255. However, each application can now request data sent to a host group G for *only* a specific set of sources, or can request data sent to host group G from *all except* a specific set of sources. In other words, applications can apply "source filtering" to the multicast data being transmitted to a to a given host group. Host-to-network support for source filtering is provided by IGMPv3 for IPv4, and version 2 of the Multicast Listener Discovery (MLD) protocol for IPv6[9].

3   IP Multicast Protocol Architectures

In this section we describe in detail the protocol architectures for supporting the ASM and SSM service models, and the relative merits and demerits of each.

3.1   ASM Protocol Architecture

The current inter-domain multicast architecture is based on the ASM service model. To become a member of a particular group, end-hosts register their membership with querier routers handling multicast group membership functionality using the IGMP version 2 (IGMPv2) protocol[10] for IPv4 or the MLD version 1 (MLDv1) protocol[11] for IPv6. With IGMPv2 and MLDv1, source-filtering capabilities are not available to receivers.

Multicast-capable routers then construct a distribution tree by exchanging messages with each other according to a routing protocol. A number of different protocols exist for building multicast forwarding trees. These protocols differ mainly in the type of delivery tree constructed[1,12,13,14,15]. Of these, the Protocol Independent Multicast Sparse-Mode (PIM-SM) protocol[14] is the most widely deployed in today's public networks. PIM-SM, by default, constructs a single spanning tree rooted at a core Rendezvous Point (RP) for all group members within a domain. Local sources then send their data to this RP which forwards the data down the shared tree to interested local receivers. A receiver joining a host group can only specify interest in the entire group and therefore will receive data from any source sending to this group. Distribution via a shared tree can be effective for certain types of traffic, e.g., where the number of sources is large since forwarding on the shared tree is performed via a single multicast forwarding entry. However, there are many cases (e.g., Internet broadcast streams) where forwarding from a source to a receiver is more efficient via the shortest path. PIM-SM also allows a designated router serving a particular subnet to switch to a source-based shortest path tree for a given source once the source's address is learned from data arriving on the shared tree. This capability provides for distribution of data from local sources to local receivers using a common RP inside a given PIM domain.

It is also possible for RP's to learn about sources in other PIM domains by using the Multicast Source Discovery Protocol (MSDP)[16]. Once an active remote source is identified, an RP can join the shortest path tree to that source and obtain data to forward down the local shared tree on behalf of interested local receivers. Designated routers for particular subnets can again switch to a source-based shortest path tree for a given remote source once the source's address is learned from data arriving on the shared tree.

The IGMPv2/PIM-SM/MSDP-based inter-domain multicast architecture supporting ASM has been deployed in IPv4 networks. It has been particularly effective for groups where sources are not known in advance; when sources come and go dynamically; or when forwarding on a common shared tree is found to be operationally beneficial.

However, there are several problems hindering the commercial deployment of these protocols. Some of these are inherent in the service model itself, while others are due to the complexity of the protocol architecture:

- **Attacks by unauthorized transmitters**: In the ASM service model, a receiver cannot specify which specific sources it would like to receive data from when it joins a given group. A receiver is forwarded data sent by *all* group sources. This lack of access control can be exploited by malicious transmitters to disrupt data transmission from authorized transmitters.
- **Deployment complexity**: The ASM protocol architecture is complex and difficult to manage and debug. Most of the complexity arises from the RP-based infrastructure needed to support shared trees, and from the MSDP protocol used to discover sources across multiple domains. These challenges often make network operators reluctant to enable IP multicast capabilities in their networks, even though most of today's routers support the IGMP/PIM-SM/MSDP protocol suite.
- **Address allocation**: This is one of the biggest challenges in deploying an inter-domain multicast infrastructure supporting ASM. The current multicast architecture does not provide an adequate solution to prevent address collisions among multiple applications. As a result two entirely different multicast sessions may pick the same class-D address for their multicast groups and interfere with each other's transmission. The problem is more serious for IPv4 than IPv6 since the total number of multicast addresses is smaller. A static address allocation scheme, GLOP[17], has been proposed as an interim solution for IPv4. GLOP addresses are allocated per registered Autonomous System (AS). However, the number of addresses per AS is inadequate when the number of sessions exceeds an AS's allocation. Proposed longer-term solutions such as the Multicast Address Allocation Architecture (MAAA)[18] are generally perceived as being too complex (with respect to the dynamic nature of multicast address allocation) for widespread deployment. Another long term solution, the unicast-prefix-based multicast architecture of IPv6[8] expands on the GLOP approach; simplifies the multicast address allocation solution; and incorporates support for source-specific multicast addresses.
- **Inter-domain scalability**: MSDP has always been something of an ugly solution. The protocol has weaknesses in terms of security and scalability. For security, it is susceptible to denial-of-service attacks by domains sending

out a flood of source announcements. For scalability, MSDP is not well designed to handle large numbers of sources. The primary reason is because the source announcements were designed to be periodically flooded throughout the topology *and* to carry data. As the number of sources in the Internet increases, MSDP will generate greater amounts of control traffic.

– **Single point of failure**: When multicast data distribution takes place over a shared tree via a core network node (RP in the case of PIM-SM), failure of the core can lead to complete breakdown of multicast communication[1]. In the ASM protocol architecture, a receiver is always grafted on to an RP-based shared tree when it first joins a multicast group. This reliance on the shared-tree infrastructure makes the ASM protocol architecture fundamentally less robust.

3.2 SSM Protocol Architecture

As mentioned before, Source Specific Multicast (SSM) defines a service model for a "channel" identified by an (S,G) pair, where S is a source address and G is an SSM address. This model can be realized by a protocol architecture, where packet forwarding is restricted to shortest path trees rooted at specific sources, and channel subscriptions are described using a group management protocol such as IGMPv3 or MLDv2.

The SSM service model alleviates all of the deployment problems described earlier:

– The distribution tree for an SSM channel (S,G) is always rooted at the source S. Thus there is no need for a shared tree infrastructure. In terms of the IGMPv2/PIM-SM/MSDP architecture, this implies that neither the RP-based shared tree infrastructure of PIM-SM nor the MSDP protocol is required. Hence the protocol architecture for SSM is significantly less complex than that for ASM, making it easy to deploy. In addition, SSM is not vulnerable to RP failures or denial-of-service attacks on RP(s).
– SSM provides an elegant solution to the access control problem. Only a single source S can transmit to a channel (S,G) where G is an SSM address. This makes it significantly more difficult to spam an SSM channel than an ASM host group. In addition, data from unrequested sources need not be forwarded by the network, which prevents unnecessary consumption of network resources[19].
– SSM defines channels on a per-source basis; hence SSM addresses are "local" to each source. This averts the problem of global allocation of SSM addresses, and makes each source independently responsible for resolving address collisions for the various channels that it creates.
– It is widely held that point-to-multipoint applications such as Internet TV will dominate the Internet multicast application space in the near future. The

---

[1] Multiple cores can certainly be used to alleviate this problem, but redundancy comes at the price of extra overhead and complexity.

SSM model is ideally suited for such applications. Thus the deployment of SSM will provide tremendous impetus to inter-domain Internet multicasting and will pave the way for a more general multipoint-to-multipoint service in the future.

A protocol architecture for SSM requires the following:

- **Source specific host membership reports**: The host-to-network protocol must allow a host to describe specific sources from which it would like to receive data.
- **Shortest path forwarding**: DR's must be capable of recognizing receiver-initiated, source-specific host reports and initiating (S,G) joins directly to the source.
- **Elimination of shared tree forwarding**: In order to achieve global effectiveness of SSM, all networks must agree to restrict data forwarding to source trees (i.e., prevent shared tree forwarding) for SSM addresses. The address range 232/8 has been allocated by IANA for deploying source-specific IPv4 multicast (SSM) services. In this range, SSM is the sole service model. For IPv6, a source-specific multicast address range has been defined[8], as a special case of unicast prefix-based multicast addresses.

We now discuss the framework elements in detail:

- **Channel discovery**: In the case of ASM, receivers need to know only the group address for a specific session. In the IGMPv2/PIM-SM/MSDP architecture, designated routers discover an active source via the RP infrastructure and MSDP, and then graft themselves to the multicast forwarding tree rooted at that source. In the case of SSM, an application on an end-host must know both the SSM address G and the source address S before subscribing to a channel. Thus the function of channel discovery becomes the responsibility of applications. This information can be made available in a number of ways, including via web pages, sessions announcement applications, etc.
- **SSM-aware applications**: The advertisement for an SSM session must include a source address as well as a group address. Also, applications subscribing to an SSM channel must be capable of specifying a source address in addition to an group address. In other words, applications must be *SSM-aware*. Specific API requirements are identified in [20].
- **Address Allocation**: For IPv4, the address range of 232/8 has been assigned by IANA for SSM. Sessions expecting SSM functionality must allocate addresses from the 232/8 range. To ensure global SSM functionality in 232/8, including in networks where edge routers run IGMPv2 (i.e., do no support source filtering), operational policies are being proposed[4] which prevent data sent to 232/8 from being delivered via shared trees.
  Note that it is possible to achieve the benefit of direct and immediate (S,G) joins in response to IGMPv3 reports in other ranges than 232/8.However, non-SSM address ranges allow for concurrent use of both the ASM and SSM service models. Therefore, while we can achieve the PIM join efficiency in

the non-SSM address range with IGMPv3, it is not possible to prevent the creation of shared trees or shared tree data delivery, and thus cannot provide for certain types of access control or assume per-source unrestricted address use as with the SSM address range.

In the case of IPv6, [8] has defined an extension to the addressing architecture to allow for unicast prefix-based multicast addresses. In this case, bytes 0-3 (starting from the least significant byte) of the IP address is used to specify a multicast group id, bytes $4 - 11$ is be used to specify a unicast address prefix (of up to 64 bits) that owns this multicast group id, and byte 12 is used to specify the length of the prefix. A source-specific multicast address can be specified by setting both the prefix length field and the prefix field to zero. Thus IPv6 allows for 2^{32} SSM addresses per scope for every source, while IPv4 allows 2^{24} addresses per source.

– **Host-to-network communication**: The currently deployed version of IGMP (IGMPv2) allows end-hosts to register their interest in a multicast group by specifying a class-D IP address for IPv4. However in order to implement the SSM service model, an end-host must specify a source's unicast address as well as an SSM address. This capability is provided by IGMP version 3 (IGMPv3). IGMPv3 supports "source filtering", i.e., the ability of an end-system to express interest in receiving data packets from only a set of *specific* sources, or from *all except* a set of specific sources. Thus IGMPv3 provides a superset of the capabilities required to realize the SSM model. Hence an upgrade from IGMPv2 to IGMPv3 is an essential change for implementing SSM.

IGMPv3 requires the API to provide the following operation (or its logical equivalent)[21]:

$$IPMulticastListen(Socket, IF, G, filter - mode, source - list)$$

As explained in the IGMPv3 specification[21], the above IPMulticastListen() operation subsumes the group-specific join and leave operations of IGMPv2. Performing (S,G)-specific joins and leaves is also trivial. A join operation is equivalent to:

$$IPMulticastListen(Socket, IF, G, INCLUDE, S)$$

and a leave operation is equivalent to

$$IPMulticastListen(Socket, IF, G, EXCLUDE, S)$$

There are a number of backward compatibility issues between IGMP versions 2 and 3 which have to be addressed. There are also some additional requirements for using IGMPv3 for the SSM address range. A detailed discussion of these issues is provided in [22].

The Multicast Listener Discovery (MLD) protocol is used by an IPv6 router to discover the presence of multicast listeners on its directly attached links, and to discover the multicast addresses that are of interest to those neighboring nodes. Version 1 of MLD[11] is derived from IGMPv2 and allows a multicast listener to specify the multicast group(s) that it is interested in. Version 2 of MLD[9] is derived from, and provides the same support for source-filtering as, IGMPv3.

– **PIM-SM modifications for SSM**: PIM-SM[14] itself supports two types of trees, a shared tree rooted at a core (RP), and a source-based shortest path tree. Thus PIM-SM already supports source-based trees; however, PIM-SM is not designed to allow a router to choose between a shared tree and a source-based tree. In fact, a receiver always joins a PIM shared tree to start with, and may later be switched to a per-source tree by its adjacent edge router.

A key to implementing SSM is to eliminate the need for starting with a shared tree and then switching to a source-specific tree. This involves several changes to PIM-SM as described in [14]. The resulting PIM functionality is referred to as PIM-SSM. The most important changes to PIM-SM with respect to SSM are as follows:

- When a DR receives an (S,G) join request with the address G it must initiate a (S,G) join and *never* a (*,G) join.
- Backbone routers (i.e. routers that do not have directly attached hosts) must be capable of receiving (S,G) joins and forwarding them based on correct RPF information. In addition, they must not propagate (*,G) joins for group addresses in the SSM address range.
- Rendezvous Points (RPs) must not accept PIM Register messages or (*,G) join messages.

In summary, the ASM service model and protocol architecture suffer from a number of serious deployment problem[3]. The SSM service model addresses many of the needs of today's commercial multicast applications. Also, the associated protocol architecture is simpler, and easy to deploy in networks that already supports ASM. The challenge then becomes integrating the two.

4   Integrating ASM and SSM

In this section, we examine interoperability issues between ASM and SSM. From our discussion so far, it is clear that there is significant overlap between the two protocol architectures. Therefore, it is possible to integrate the two. The task is to then investigate the interoperability issue from a host perspective, i.e. if a host is connected to a network, what does it have to do to properly utilize whatever multicast service is present?

Given that the two service models and two protocol architectures form a set of four combinations, the challenge is to identify any problems in providing a seamless multicast service—including both intra- and inter-domain operation. As we have discovered, most of these scenarios are trivially workable. Of those that remain, one requires minor changes to existing protocols, and one is quite challenging. Our goal is to (1) identify what the interoperability problems are, (2) identify solutions to these problems, and (3) understand the relative complexities of deploying these solutions.

In the next section we describe the four combinations of service models and protocol architectures. Following the overview, we focus specifically on solutions for the most difficult case.

4.1   Service Model and Protocol Architecture Combinations

There are four combinations of service models and protocol architectures. These are shown in Figure 2. The key challenge naturally occurs because the host does not know how the network is configured. This is actually a reasonable abstraction. The host should simply join a multicast group. If IGMPv3 is available and the application knows who the source is, this information should be passed to the network. If this information is not available or if IGMPv3 is not supported, the network should still respond in a predictable manner.

The challenge of deploying a multicast service is to provide correct operation for all kinds of multicast no matter if (1) the host is limited to only IGMPv2 and/or (2) the network is limited only to SSM support. In fact, the ability to interoperate with ASM was one of the key requirements for SSM[5]. This requirement was critical because the development of any completely new protocol architecture would mean that multicast deployment efforts would have to start completely over. Furthermore, given the near infinite lifetime of legacy architectures, ASM would in all likelihood continue to exist and need to be supported. Therefore, a new protocol architecture that did not integrate with ASM would not reduce complexity but rather increase it. Therefore, SSM was designed to interoperate with ASM. However, because SSM implements a subset of ASM functionality, there needs to be additional work to properly integrate the two.

The challenge with integrating ASM and SSM is how to handle the discontinuities between the two. If the two are not integrated properly, multicast does not work. Even more problematic is that there is no feedback from any part of the network or host that says multicast is not working. Cases when this kind of behavior occurs are called "black holes". Black holes occur when both the network and the host are operating correctly, but no multicast flows because there is a disconnect between the service model and the protocol architecture. For example, the network only allows hosts to specify an explicit list of sources but the host sends a (*,G) join. The network cannot process this kind of message and ignores it. There is no feedback to the host that the join message was not properly handled.

Before describing the specific black hole scenarios, we first describe the scenarios that are more straightforward. Figure 2 shows all four combinations. They are:

- **ASM service model and ASM network**: The upper-left scenario is the service model and protocol architecture that has been running in the Internet since 1997[2]. Theoretically, there are no black holes in this combination, though in practice, problems often occur[23,24].
- **SSM service model and ASM network**: The upper-right scenario is essentially the same protocol architecture that has been running since 1997,

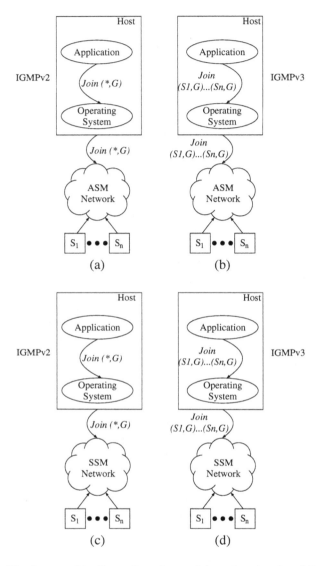

Fig. 2. The four combinations of service models and protocol architectures.

but with IGMPv3 support. Through IGMPv3, users are given the ability to specify a subset of all group receivers, thereby refining the granularity of join and leave messages to more than just one choice for all sources. In the section listing the service models, this combination is also called Source-Filtered Multicast (SFM). SSM is supported in this combination in the address range 232/8[25]. Theoretically, there are no black holes in this combination.

- **ASM service model and SSM network**: The lower-left scenario is the most problematic of the four combinations. From the network point-of-view, the service provider has opted to only provide support for SSM. But for any of a variety of reasons, a host either chooses to send or is only capable of sending (*,G) join messages. The dilemma is how to solve this problem. Several possible solutions are discussed in the next section.
- **SSM service model and SSM network**: The lower-right scenario is more straightforward than the previous case, but there is still one problem. Because the multicast address space is divided into SSM and non-SSM ranges, the straightforward behavior is when the group address is in the SSM range (232/8)[25]. Uncertainty occurs when handling (S,G) joins for the non-SSM range. There are two considerations to understand:
  - One of the dependencies here is whether the network is providing SSM support only in the 232/8 address range or whether it has been extended to cover the entire multicast address range (224/4). If the choice is only to support the 232/8 range, how should the network, host, and application handle (S,G) joins for addresses outside this range? The currently accepted practice seems to be to not allow SSM support outside of the 232/8 range and embed these semantics in the operating system.
  - If the network provider instead chooses to provide SSM support for the entire address range, a problem is created for sources. Sources transmitting on a non-SSM address will *not* have their existence announced throughout the inter-domain infrastructure. To understand why this situation occurs, consider the behavior in an ASM domain. When a source sends its first packet, the network encapsulates it and sends it to the RP. Since MSDP runs in the RP, an SA message is generated and flooded on the MSDP peering topology. Because the domain chooses to run SSM for the entire address space, there is no RP, no initial packet encapsulation, and no MSDP peer. Receivers in ASM domains will never discover the existence of this particular source. Again, the current accepted practice seems not to provide SSM-style service for addresses outside of the 232/8 range.

There are a number of solutions to solving the problem of an ASM service model running in an SSM network. These solutions are discussed in the next section.

4.2   Handling ASM Hosts in an SSM Infrastructure

For hosts that can only speak IGMPv2, operation in an SSM-only network is difficult. Even for hosts that do speak IGMPv3 there are inter-domain consistency problems if SSM behavior is enforced beyond the 232/8 address range. The first step to solving these interoperability problems is to understand more clearly what the problems are. At a minimum, applications need deterministic, predictable behavior. Ideally, applications should be able to maintain some level of abstraction from the type of multicast service. However, implementing this in the Internet looks to be quite difficult. The problem is that because there are different semantics tied to the multicast address space (232/8 vs. the rest of 224/4), different behavior is expected depending on the address used. Therefore, hosts need to have some awareness of these semantics and what the network supports. While a host does not need a complete understanding of what the network protocol architecture is, it needs to know whether its join messages are going to be processed properly.

One of the fundamental problems is that a host sending a (*,G) join into an SSM network will have the join message ignored. Therefore, some additional action must be taken by the network if IGMPv2 hosts are to be supported. The first two solutions attempt to resolve the (*,G) into a set of sources. Two choices of this type are shown in the top half of Figure 3 and described below.

– **Run an MSDP peer**: An SSM-only domain can run MSDP. Leaf routers would query the MSDP cache for source information. This solution is shown in Figure 3(a). A straightforward implementation based on existing protocols would be to run an MSDP peer at the domain boundary and then use a new protocol for communicating between the peer and the leaf routers. The obvious disadvantage of this solution is that it appears to have almost as much complexity as ASM. One savings is in not having to run an RP. A second savings is being able to run the source discovery protocol at the application layer and not embed it in the network layer. By implementing an MSDP peer as an application and then adding some basic functionality just to the leaf routers, this solution can make sense.
– **URL Rendezvous Directory (URD)**: Similar to running an MSDP peer, URD involves translating network-layer complexity into application-layer management overhead. The idea is that a host will use the web to gather information about the multicast group and will then initiate the group join based on this information. This solution is shown in Figure 3(b). When a user clicks on a link, the response is to send both the source and group information back to the client as an HTTP re-direct to the URD port (Port 659). For example, the returned link might look like the following:

`http://content-source.com:659/source-address,group-address/`

The router is intercepting traffic sent to port 659, and in combination with the IGMPv2 join message sent by the user's application, the router will be able to issue an (S,G) join to the source. The goal of URD is to be able to put all of the additional functionality necessary for SSM at the content source site and in the leaf routers. This objective is achieved because no additional modifications are required in the application or the host operating system. The application responds normally to the HTTP re-direct and the operating system issues an IGMPv2 (*,G) join. There is even support to avoid black holes in the case where a content site uses URD but the leaf router does not support it. What happens is that the router does not intercept the URL re-direct and so it reaches the content source. If this re-direct appears, the content source knows the leaf router did not intercept the re-direct and can

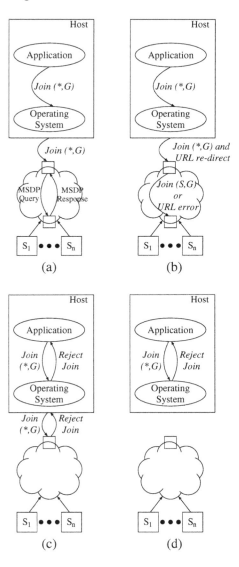

Fig. 3. Four possible solutions for ASM hosts in an SSM infrastructure.

inform the user via a web page that the join did not work. Of course, the major disadvantage is that this requires routers to snoop on port 659 and intercept traffic—not an acceptable requirement for most network providers.

Given that these two solutions require application-layer mechanisms to replace the functionality of PIM RPs and MSDP, neither solution is particularly elegant. A better solution would be to simply limit what the user can and cannot do. For example, the idea would be to prohibit joins to non-232/8 group addresses in SSM-only networks. Black holes would be avoided by letting the user/application/host know that the unsupported join actions had failed. Details on two particular solutions are shown at the bottom of Figure 3 and described below.

- **IGMPv3 with reject capability**: The idea is to modify IGMPv3 to create a more robust control path. The solution is to allow the leaf router to "reject" an IGMP join. This solution is shown in Figure 3(c). A number of possible reasons could exist for rejecting a join. The obvious case is when a (*,G) join is set for an address configured only to allow (S,G) joins. Another example might be joins sent to a group that has been listed in a site-controlled reject list. The IGMP reject message could include a return code providing a reason for the rejection. This solution would require either an addition to IGMPv3 or a new version. While this solution seems quite reasonable, it requires revising IGMPv3 which creates yet another deployment delay.
- **Host discovery of network capability**: Hosts could be given the capability to discover for themselves how the network is configured. This solution is shown in Figure 3(d). In this way, hosts could determine what group addresses require a specific set of sources and what group addresses allow the use of the "*" in join messages. This discovery provides the host with enough information to reject an application's join request. Like many of the other solutions, the problem is host behavior needs to be modified.

Either of the last two solutions is the most reasonable. But the changes require another round of deployment. The practical solution is for all routers, operating systems, and applications to assume that SSM runs only in the 232/8 range, and that some networks might only support SSM. Therefore, any (S,G) join should be expected to work, but any (*,G) join should be suspect. The simple policy should be that if an application has access to the set of one or more group sources, it should use them. Otherwise, the possibility exists that (*,G) joins will not be successful. This creates a certain amount of non-determinism but seems easy to characterize: IGMPv2 joins might not work. The incentive is to upgrade to IGMPv3 as quickly as possible.

5 Conclusions

In this paper, we have considered the integration of the two service models for IP multicast: Any Source Multicast (ASM) and Source Specific Multicast (SSM). We have described the protocol architecture for each and discussed their advantages and disadvantages. ASM is the traditional service model; however, it suffers from a number of serious problem from a commercial deployment standpoint. SSM solves most of these which makes it suitable for rapid deployment. The important advantage of SSM is the significant overlap of its protocol architecture with that of SSM. We have explored the interoperability of these two service models, and found that in most cases, the challenges are not insurmountable. In the near term, we expect these two service models to co-exist in a unified IP multicast infrastructure.

From a broader perspective, successfully integrating ASM and SSM should have a positive impact on the use and deployment of multicast. IP multicast has long suffered from the "chicken and egg" problem. The lack of popular applications has given very little incentive to ISPs to enable multicast in their networks. Furthermore, ASM has been plagued by a number of deployment problems. On the other hand, the lack of widespread deployment has resulted in very limited interest in developing new applications. At the same time, the popularity of application-layer multicast has further slowed down deployment of IP multicast. It is hoped that SSM will spur the deployment and use of IP multicast by virtue of its simplicity, ease of deployment, and its ability to be integrated into the existing infrastructure. While SSM is ideally suited for point-to-multipoint, multi-peer applications such multi-party games can easily be supported by building relays at the application level over an SSM-capable network. Such an approach represents an attractive compromise between the efficiency of network-level multicast and the ease of manageability of application-level multicast.

References

1. S. Deering, "Host extensions for IP multicasting." Internet Engineering Task Force (IETF), RFC 1112, August 1989.
2. K. Almeroth, "The evolution of multicast: From the MBone to inter-domain multicast to Internet2 deployment," *IEEE Network*, January/February 2000.
3. C. Diot, B. Lyles, B. Levine, and H. Kassem, "Requirements for the definition of new IP-multicast services," *IEEE Network*, January/February 2000.
4. S. Bhattacharyya, C. Diot, L. Giuliano, R. Rockwell, J. Meylor, D. Meyer, G. Shepherd, and B. Haberman, "An overview of source-specific ip multicast(ssm) deployment." Internet Engineering Task Force (IETF), draft-ietf-bhattach-ssm-*.txt, May 2001.
5. H. Holbrook and B. Cain, "Source-specific multicast for IP." Internet Engineering Task Force (IETF), draft-holbrook-ssm-arch-*.txt, March 2001.
6. H. Holbrook and D. Cheriton, "IP multicast channels: EXPRESS support for large-scale single-source applications," in *ACM Sigcomm*, (Cambridge, Massachusetts, USA), August 1999.
7. Z. Albanna, K. Almeroth, D. Meyer, and M. Schipper, "IANA guidelines for IPv4 multicast address assignments." Internet Engineering Task Force (IETF), draft-ietf-mboned-iana-ipv4-mcast-guidelines-*.txt, April 2001.
8. B. Haberman and D. Thaler, "Unicast-prefix-based IPv6 multicast addresses." Internet Engineering Task Force (IETF), draft-ietf-ipngwg-uni-based-mcast-*.txt, January 2001.
9. R. Vida, L. Costa, S. Fdida, S. Deering, B. Fenner, I. Kouvelas, and B. Haberman, "Multicast listener discovery version 2 (MLDv2) for ipv6." Internet Engineering Task Force (IETF), draft-vida-mld-v2-*.txt, February 2001.
10. W. Fenner, "Internet group management protocol, version 2." Internet Engineering Task Force (IETF), RFC 2236, November 1997.

11. S. Deering, B. Fenner, and B. Haberman, "Multicast listener discovery (MLD) for IPv6." Internet Engineering Task Force (IETF), RFC 2710, October 1999.
12. S. Deering, D. Estrin, D. Farinacci, V. Jacobson, G. Liu, and L. Wei, "PIM architecture for wide-area multicast routing," *IEEE/ACM Transactions on Networking*, pp. 153–162, Apr 1996.
13. D. Estrin, D. Farinacci, A. Helmy, D. Thaler, S. Deering, M. Handley, V. Jacobson, C. Liu, P. Sharma, and L. Wei, "Protocol independent multicast sparse-mode (PIM-SM): Protocol specification." Internet Engineering Task Force (IETF), RFC 2362, June 1998.
14. B. Fenner, M. Handley, H. Holbrook, and I. Kouvelas, "Protocol independent multicast - sparse mode (PIM-SM): Protocol specification (revised)." Internet Engineering Task Force (IETF), draft-ietf-pim-sm-v2-new-*.txt, March 2001.
15. S. Deering, D. Estrin, D. Farinacci, V. Jacobson, A. Helmy, D. Meyer, and L. Wei, "Protocol independent multicast version 2 dense mode specification." Internet Engineering Task Force (IETF), draft-ietf-pim-v2-dm-*.txt, June 1999.
16. D. Meyer and B. Fenner, "Multicast source discovery protocol (MSDP)." Internet Engineering Task Force (IETF), draft-mboned-msdp-spec-*.txt, May 2001.
17. D. Meyer and P. Lothberg, "GLOP addressing in 233/8." Internet Engineering Task Force (IETF), RFC 2770, February 2000.
18. M. Handley, D. Thaler, and D. Estrin, "The internet multicast address allocation architecture." Internet Engineering Task Force (IETF), draft-ietf-malloc-arch-*.txt, December 1997.
19. B. Fenner, H. Holbrook, and I. Kouvelas, "Multicast source notification of interest protocol (msnip)." Internet Engineering Task Force (IETF), draft-ietf-idmr-msnip-*.txt, February 2001.
20. D. Thaler, B. Fenner, and B. Quinn, "Socket interface extensions for multicast source filters." Internet Engineering Task Force (IETF), draft-ietf-idmr-msf-api-*.txt, June 2000.
21. B. Cain, S. Deering, B. Fenner, I. Kouvelas, and A. Thyagarajan, "Internet group management protocol, version 3." Internet Engineering Task Force (IETF), draft-ietf-idmr-igmp-v3-*.txt, March 2001.
22. H. Holbrook and B. Cain, "Using IGMPv3 for source-specific multicast." Internet Engineering Task Force (IETF), draft-holbrook-idmr-igmpv3-ssm-*.txt, March 2000.
23. K. Sarac and K. Almeroth, "Monitoring reachability in the global multicast infrastructure," in *International Conference on Network Protocols (ICNP)*, (Osaka, JAPAN), November 2000.
24. P. Rajvaidya and K. Almeroth, "A router-based technique for monitoring the next-generation of internet multicast protocols," in *International Conference on Parallel Processing*, (Valencia, Spain), September 2001.
25. G. Shepherd, E. Luczycki, and R. Rockell, "Source-specific protocol independent multicast in 232/8." Internet Engineering Task Force (IETF), draft-ietf-mboned-ssm232-*.txt, April 2001.

Integrated Provision of QoS Guarantees to Unicast and Multicast Traffic in Packet Switches

Andrea Francini and Fabio M. Chiussi

Data Networking Systems Research Department
Bell Laboratories, Lucent Technologies
Holmdel, NJ 07733, U.S.A.
{francini,fabio}@bell-labs.com
http://www.bell-labs.com/org/113480/

Abstract. Multi-stage packet switches that feature a limited amount of buffers in the switching fabric and distribute most of their buffering capacity over the port cards have recently gained popularity due to their scalability properties and flexibility in supporting Quality-of-Service (QoS) guarantees. In such switches, the replication of multicast packets typically occurs at the outputs of the switching fabric. This approach minimizes the amount of resources needed to sustain the internal expansion in traffic volume due to multicasting, but also exposes multicast flows to head-of-line (HOL) blocking in the ingress port cards. Access regulation to the fabric buffers is of the utmost importance to safeguard the QoS of multicast flows against HOL blocking. We add minimal overhead to a well-known distributed scheduler for multi-stage packet switches to define the Generalized Distributed Multi-layered Scheduler (G-DMS), which achieves full support of QoS guarantees for both unicast and multicast flows. The novelty of the G-DMS is in the mechanism that regulates access to the fabric buffers, which combines selective backpressure with the capability of dropping copies of multicast packets that violate the negotiated profiles of the corresponding flows.

1 Introduction

Many envisioned applications in the Internet, such as broadcast video, video-conferencing, multi-party telephony, and work-group applications, are multicast in nature and are expected to generate a significant portion of the total traffic. In spite of the growing importance of multicast traffic in the Internet, the support of *Quality-of-Service* (QoS) guarantees for multicast flows is still far from satisfactory in current-generation packet switches.

We consider the integration of multicast traffic in existing QoS frameworks for unicast traffic in multi-stage switches [1,2,3,4], focusing in particular on the *Distributed Multilayered Scheduler* (DMS) [3,4]. The DMS meets the throughput, delay, and delay-jitter requirements of unicast flows by closely approximating an ideal scheduling hierarchy [5] in a distributed system. It applies to multi-stage switches that include a layer of ingress port cards, a switching fabric with a

S. Palazzo (Ed.): IWDC 2001, LNCS 2170, pp. 361–380, 2001.
© Springer-Verlag Berlin Heidelberg 2001

moderate amount of buffers, and a layer of egress port cards [6,7]. The fabric may be implemented as a stand-alone shared-memory module, or expanded to achieve higher aggregate capacity in a three-stage *Memory/Space/Memory* (MSM) arrangement [4,6]. For simplicity of presentation, we restrict the scope of this paper to the stand-alone implementation of the switching fabric.

The DMS associates a separate scheduling tree with each switch output. The schedulers that distribute service at the contention points of the switch (in the ingress port cards and fabric outputs, but not in the egress port cards, which we assume to be bufferless [4]) constitute the nodes of the scheduling trees. The presence of limited amounts of buffers in the fabric decouples scheduling trees that overlap at one or more nodes, and therefore enables the association of a distinct scheduling hierarchy with each switch output. In the fabric, buffers are statically partitioned in multiple queues, called *QoS channels*. Within a traffic class, the DMS establishes a QoS channel for each input-output pair. *Selective backpressure* regulates the admission of packets to the QoS channels [4,8]: when the amount of packets in a channel exceeds a given threshold, the assertion of backpressure forces further traffic destined for that channel to remain in the corresponding ingress port card.

The generalization of the DMS for the integrated provision of QoS guarantees to unicast and multicast flows requires the adaptation of the QoS framework to the multicasting scheme of the underlying switch architecture. Our reference switch always replicates multicast packets according to a *minimum multicast tree*, that is, as far downstream as possible (and therefore never in the ingress port cards) [6]. A single instance of a multicast packet is stored in the fabric memory, and replicated (if necessary) only when the packet is transmitted to one of its outputs. After receiving a multicast packet, the fabric generates a pointer for each output in its multicast distribution (or *fanout*), and links it to the corresponding output queue (a linked pointer virtually identifies a *copy* of the packet).

The adoption of minimum multicast trees in a distributed switch with multiple layers of contention points, albeit optimal in the utilization of bandwidth and buffering resources, may compromise the provision of QoS guarantees for both unicast and multicast flows. In fact, *head-of-line* (HOL) blocking may occur if packets with different multicast distributions share portions of their *forwarding paths* before they are replicated (the forwarding path of a packet is identified by the sequence of queues that the packet traverses in the switch). In our multistage switch, the problem arises in the ingress port cards if flows with different multicast distributions are subject to the same backpressure indication.

The complete isolation of the forwarding paths of packets with different fanouts does avoid HOL blocking, but is impractical for typical switch sizes, because it requires a separate queue for each multicast distribution that can be defined at each contention point in the switch (as an example, each ingress port card of an $N \times N$ switch should contain $2^N - 1$ queues [9]).

The *overlay approach* [10] isolates unicast flows from multicast flows by assigning separate forwarding paths to the two types of traffic (all multicast flows share the same forwarding path). When applied to the DMS, the overlay approach succeeds in preserving the QoS guarantees of unicast flows, but obviously fails with multicast flows, because of the persistence of HOL blocking within their dedicated path.

The *Generalized DMS* (G-DMS) that we present in this paper aggregates unicast and multicast traffic along common forwarding paths, factually handling unicast flows as single-legged multicast flows. HOL blocking still occurs for packets with different fanouts sharing the same queues, but no longer compromises the QoS guarantees of individual flows. One of the key elements of novelty of the G-DMS is in the mechanism that triggers the assertion of backpressure: when a packet of a multicast flow is replicated in the fabric, only one of its copies is counted for backpressure purposes. The QoS channel that is charged for the presence of the packet in the fabric is the *primary QoS channel* of the multicast flow (we refer to the channels that accommodate the other copies of the packet as the *secondary QoS channels* of the flow). The primary channel is the same for all packets of a given flow. Also, the primary channel of a flow can act as a secondary channel for a different flow. In the ingress port cards, multicast flows are subject to the backpressure indication coming from their respective primary channels. In particular, the same backpressure indication applies to multicast flows having different fanouts but identical primary channel. Per-flow traffic policers in the ingress port cards and the fabric capability of selectively dropping copies of multicast packets prevent secondary QoS channels from overflowing. In the port cards, the policers mark incoming packets that violate the negotiated profiles of the corresponding flows; then, the fabric drops all the copies of marked multicast packets that are destined for congested secondary channels.

The G-DMS extends to multicast traffic all the QoS features of the DMS: throughput and delay guarantees for real-time flows that are compliant with their negotiated traffic profiles, throughput and fairness guarantees for flows with long-term bandwidth requirements, and fairness in the treatment of best-effort flows. The overhead induced by the generalization of the DMS is minimal (two counters for each queue in the fabric).

The paper is organized as follows. In Section 2, we delineate the context of application of the G-DMS by identifying the target QoS classes and switch architecture, and describe the DMS in detail. In Section 3, we discuss the available options for counting multicast packets in the switching fabric. We then overview flow-control schemes based on pointer counters (Section 4) and on cell counters (Section 5). We finalize the specification of the G-DMS in Section 6, and provide concluding remarks in Section 7.

2 Background

In this section we delineate the target QoS classes of the G-DMS and provide an overview of the underlying DMS. We also describe the algorithm that we use to enforce the coexistence of heterogeneous traffic components in the limited buffers of the switching fabric.

2.1 QoS Classes

The system that we address in this paper only handles fixed-size packets, which we call *cells*. The G-DMS applies to the case of variable-sized packets as well, but the lack of space forces us to defer the related discussion to an upcoming paper. We refer to network-wide end-to-end cell streams as *flows*, and focus on the provision of differentiated QoS guarantees to three distinct classes of flows: *Guaranteed-Delay* (GD) class, *Guaranteed-Bandwidth* (GB) class, and *Best-Effort* (BE) class.

GD flows have individually specified requirements in terms of throughput and transmission delay. In the ATM context [11], both the *Constant Bit Rate* (CBR) and the *real-time Variable Bit Rate* (rt-VBR) service categories map onto the GD class.

GB flows have individual bandwidth requirements but no specified delay requirements (or their delay requirements are very loose). In ATM, *non-real-time Variable Bit Rate* (nrt-VBR) *virtual circuits* (VC's) map onto the GB class, together with *Available Bit Rate* (ABR) and *Unspecified Bit Rate* (UBR) VC's with guaranteed *Minimum Cell Rate* (MCR).

Finally, BE flows have no negotiated bandwidth and delay guarantees. In ATM, the BE class includes ABR and UBR VC's with no specified MCR. In the absence of explicitly specified QoS parameters, we aim at the achievement of *fairness* in the relative treatment of BE flows: the switch should always deliver the same amount of traffic for BE flows that remain continuously backlogged over a common forwarding path.

2.2 The Distributed Multilayered Scheduler

The Distributed Multilayered Scheduler applies to the three-stage switch architecture of Fig. 1, which consists of a layer of ingress port cards with large buffers, a switching fabric with small buffers concentrated in a single shared-memory module, and a layer of egress port cards with no buffers.

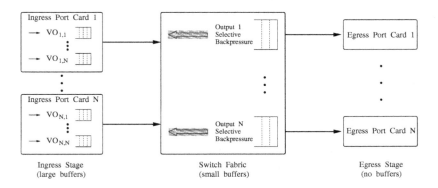

Fig. 1. Reference switch model.

The interfaces between port cards and incoming/outgoing links and between port cards and fabric are synchronized to a common timing reference: the time axis is divided into *timeslots* of fixed duration, equal to the time needed to transfer a payload unit through any of the interfaces. According to the slotted nature of the time reference, we measure transmission rates in *cells per timeslot* (*cpt*) and transmission delays in *timeslots*. During a single timeslot, each ingress port card delivers no more than one cell to the fabric, and each fabric output transfers no more than one cell to the corresponding egress port card.

The switching fabric has a relatively small amount of buffers and applies per-output selective backpressure [8] to the ingress port cards to prevent buffer overflow when congestion is detected at one or more of its outputs. Most of the buffers are located in the ingress port cards. Selective backpressure allows to achieve nonblocking behavior and buffer utilization comparable with a centralized shared-memory switch by making the buffers in the ingress port cards act as an extension of the buffers in the fabric.

Conforming to the DMS, the fabric supplies two distinct QoS channels per input-output pair (i, j), for a total of $2N^2$ QoS channels in the shared-memory module. The *Guaranteed-Delay* (GD) channel $C_{i,j}^{GD}$ conveys exclusively GD traffic, whereas the *Non-Guaranteed-Delay* (NGD) channel $C_{i,j}^{NGD}$ aggregates GB and BE traffic.

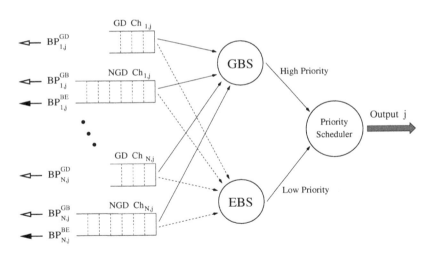

Fig. 2. The fabric-output scheduler.

At each of the N outputs of the fabric, a modular scheduler, consisting of two prioritized components, distributes service to the $2N$ channels carrying traffic from the N switch inputs (Fig. 2). The high-priority component, referred to as the *Guaranteed-Bandwidth Scheduler* (GBS), instantiates the Shaped Virtual Clock (Sh-VC) algorithm [13], a non-work-conserving worst-case-fair GPS-related scheduler [14,15]. The GBS satisfies *exactly* the minimum bandwidth requirements of the QoS channels which have non-null guaranteed service rate (the guaranteed service rate of a channel is equal to the sum of the guaranteed service rates of its associated flows in the ingress port card). The low-priority component of the fabric-output scheduler, called the *Excess Bandwidth Scheduler* (EBS), is in charge of distributing the unused GBS bandwidth to all backlogged QoS channels, thus making the whole output scheduler work conserving. An instance of the Self-Clocked Fair Queueing (SCFQ) algorithm [16] implements the EBS. The service-rate allocations for the QoS channels in the EBS are completely independent of the corresponding allocations in the GBS. A wide set of policies for the distribution of excess bandwidth can be emulated by properly assigning the EBS service rates [4]. Each QoS channel in the fabric asserts backpressure when its number of queued cells exceeds an associated static threshold.

In the ingress port card, a two-level hierarchical scheduler [5] regulates access to the interface with the fabric (Fig. 3). At the higher level of the hierarchy, a GBS-EBS pair arbitrates the distribution of service among the GD, GB, and BE classes (no service rate is allocated to the BE class in the GBS). Distinct schedulers are used at the lower level of the hierarchy for servicing the flows of the three classes. In order to properly react to selective backpressure and avoid HOL blocking, each per-flow scheduler provisions separate queues for each switch output (*virtual output queueing* (VOQ) [12]).

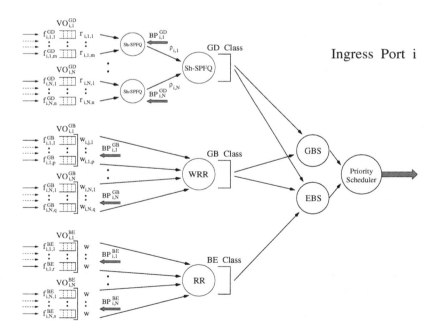

Fig. 3. The ingress-port-card scheduler.

A work-conserving worst-case-fair GPS-related scheduler such as Shaped Starting Potential Fair Queueing (Sh-SPFQ) [13] distributes bandwidth to each of the N aggregates of GD traffic associated with the switch outputs (*GD virtual outputs*). Within each GD virtual output, the individual flows get access to the available bandwidth through a local instance of Sh-SPFQ.

A Weighted Round Robin (WRR) scheduler [18] with a common *frame* reference for the whole ingress port card distributes service to the GB flows according to their respective bandwidth allocations. Proper reaction to the backpressure indication arriving from the fabric for the GB class is obtained by scheduling the backlogged GB flows through separate queues of flow pointers, each queue being associated with a distinct switch output.

The per-flow scheduler for the BE class also instantiates the WRR algorithm, with all weights set to the same value in order to enforce fairness port-wide (for simplicity, we refer to this particular instance of the WRR paradigm as a Round Robin (RR) scheduler). GB and BE flows directed to the same switch output are subject to distinct backpressure indications.

A common feature of a feasible implementation of the three per-flow schedulers that we have just summarized is the adoption of FIFO queues of flow pointers (or *flow queues*) in association with the respective virtual outputs. Several such queues are needed for the provision of delay guarantees to the GD flows even in the simplest implementations of the Sh-SPFQ algorithm [19], and at least two queues per virtual output must be deployed for implementing a WRR scheduler that applies a single frame reference to the whole ingress port card [20].

We refer the reader to [4] for a complete overview of the criteria for setting the design parameters of the DMS (such as backpressure thresholds and scheduling rates) and a performance evaluation of the QoS framework.

2.3   The Excess-Bandwidth-Detection Algorithm

Since GB and BE cells in the same input-output pair (i, j) share the same buffer space in channel $C_{i,j}^{NGD}$, a non-trivial mechanism is required to differentiate the assertion of backpressure for the two NGD traffic components. Ideally, BE traffic should be admitted to $C_{i,j}^{NGD}$ only when the channel has excess bandwidth available, since the GBS rate allocation is totally devoted to satisfying the bandwidth requirements of the GB flows. The *Excess-Bandwidth-Detection* (EBD) algorithm [3,21] monitors the availability of excess bandwidth in the NGD channel to discriminate the admission of GB and BE traffic when the channel is not fully congested (when the channel is fully congested, meaning that its occupation exceeds the associated backpressure threshold, backpressure is jointly asserted for *both* GB and BE traffic).

The EBD algorithm takes advantage of the modular nature of the fabric-output scheduler. It counts the EBS services granted to the NGD channel as additional excess bandwidth, and the arrival of a BE cell to the channel as consumed excess bandwidth. The algorithm associates an *NGD credit counter* $\alpha_{i,j}^{NGD}$ with each NGD channel $C_{i,j}^{NGD}$. Whenever the NGD channel is empty, the counter $\alpha_{i,j}^{NGD}$ is set equal to the backpressure threshold $T_{i,j}^{NGD}$. The channel increases the NGD credit counter at every EBS service that it receives, and decreases the counter at every arrival of a BE cell. When the counter becomes null, no more excess bandwidth is available, and the NGD channel selectively asserts backpressure for its BE component. In order to avoid an infinite accumulation of credits that could later penalize the GB component of the traffic aggregate, the counter is never allowed to grow above the threshold $T_{i,j}^{NGD}$.

As we will show in the presentation of the G-DMS framework, the application of the EBD algorithm is not limited to flow-control schemes, but can also be extended to cell-dropping policies.

3   Counting Multicast Cells in the Switching Fabric

Our reference switch architecture replicates multicast cells exclusively in the fabric. The fabric stores a single copy of every incoming cell in its buffer memory. Then, it generates one pointer to the memory location containing the cell for each output in its fanout, and finally links the pointers to the corresponding QoS channels. A *multicast counter* μ_h is associated with each location h in the buffer memory. The buffer-memory controller increments the value of μ_h, initially null, every time a pointer to memory location h gets queued to a QoS channel, and decrements the counter every time a copy of the cell gets transmitted to an egress port card. When μ_h becomes null again, the buffer-memory controller removes the cell from the memory location.

The cell-replication scheme introduces a mismatch between the total number of cells stored in the buffer memory of the fabric and the total number of pointers that are queued in the QoS channels. In order to charge a single QoS channel for the presence of a multicast cell in the buffer memory, we designate a *primary output* for each configured flow. The primary output of a flow is arbitrarily selected among the outputs in its multicast distribution. All the remaining outputs in the fanout are referred to as *secondary outputs*. Consistently, for each cell that enters the fabric we have a primary QoS channel and a set of secondary QoS channels (this set is obviously empty for unicast flows). The presence of a cell in the buffer memory is charged to its primary QoS channel. We refer to a flow (cell) having channel $C_{i,j}$ as its primary (secondary) channel as a primary (secondary) flow (cell) of channel $C_{i,j}$.

Each QoS channel in the fabric can simultaneously act as the primary channel for some of the cells stored in the buffer memory, and as a secondary channel for other cells. We associate two distinct counters with each QoS channel $C_{i,j}$. The *pointer counter* $\beta_{i,j}$ keeps track of the number of cell pointers that are queued in $C_{i,j}$. The *cell counter* $\gamma_{i,j}$ measures instead the number of cells in the buffer memory which have $C_{i,j}$ as their primary channel (*i.e.*, the number of primary cells of channel $C_{i,j}$ that are currently stored in the buffer memory). The counter $\gamma_{i,j}$ is incremented every time a primary cell of channel $C_{i,j}$ is stored in the buffer memory, and decremented when the same cell is removed from the buffer memory upon transmission of its last replica to any of the egress port cards.

Under heavy presence of multicast traffic, the two counters typically have different values. Considering all the QoS channels in the switching module, the sum of the pointer counters is *never* smaller than the sum of the cell counters, because each buffered multicast cell may contribute multiple times to the former, and never more than once to the latter.

The pointer counter is relevant to the activity of the fabric-output scheduler: the scheduler can select a QoS channel for service only if the associated pointer counter is greater than zero, independently of the corresponding cell counter. For the sake of backpressure assertion, on the contrary, both counters may be relevant. The backpressure threshold $T_{i,j}$ of QoS channel $C_{i,j}$ can be compared to either $\beta_{i,j}$ or $\gamma_{i,j}$ to detect the occurrence of congestion in the channel. The choice of the relevant counter has major impact on the efficiency of the scheme that integrates unicast and multicast traffic, as we argue in the following sections.

4   Flow Control with Pointer Counters

In this section, we overview three flow-control schemes that rely on pointer counters to identify congested QoS channels, and show how they all fail in supporting the QoS requirements of multicast flows. We obtain clear indication that cell counters are more promising than pointer counters as a basis for flow control in the switch fabric. We will start the investigation of flow-control schemes based on cell counters in Section 5 below.

4.1   Congestion Detection at a Single Output

The ingress port card maintains a permanent association between a multicast flow and its primary virtual output (*i.e.*, the virtual output of its primary QoS channel), while in the fabric the QoS channels assert backpressure based on the comparison of the pointer counters with the associated backpressure thresholds.

This approach allows to control the contribution of a flow to the occurrence of congestion at its primary output, but has no means to prevent its secondary outputs from overflowing. In fact, if a primary output is not congested, the fabric never applies backpressure to the corresponding virtual output in the port card, so that the flow can keep sending a continuous stream of cells to the fabric, disregarding the heavy congestion possibly induced at its secondary outputs.

4.2   Congestion Detection at Multiple Outputs

For each multicast flow, the ingress port card applies basic logical operators (AND, OR) to the portion of backpressure bitmap that overlaps with the fanout (we assume that a backpressure bit is set to 1 when the pointer counter of the corresponding channel exceeds the backpressure threshold). The idea is to control the activity of a multicast flow using some sort of aggregation of the information that is available on the congestion occurring over its fanout.

Both logical operators fall short of a balanced control of multicast traffic. The logical OR of the relevant backpressure bits is too conservative, because the occurrence of congestion at a single secondary output of flow f_k is sufficient to stop its activity. This behavior also induces heavy HOL blocking on the flows that are queued behind f_k in the ingress-port-card scheduler and have no connection with the congested secondary output of f_k. The logical AND, on the other hand, makes the flow control too loose: the presence of a single uncongested output in the fanout allows the multicast flow to keep sending traffic to all other outputs, thus overflowing the fabric.

4.3   Virtual-Output Rotation

The flow-control mechanism for multicast traffic presented in [6] inspires this solution. Instead of maintaining a permanent association with a specific output in the fanout, the multicast flow cyclically travels within the ingress port card over the virtual outputs of its multicast distribution, switching from one flow queue to the following one in the fanout at every service it receives from the port-card scheduler. If the flow ever contributes to clogging one of its outputs, it ends up being restricted by backpressure when it visits the queue of the congested virtual output.

The virtual-output rotation avoids buffer overflow, but lacks accuracy in the provision of QoS guarantees. A flow with dense multicast distribution and only a few congested outputs can unfairly subtract considerable amounts of bandwidth from other flows insisting on the same congested outputs, possibly inducing heavy violations of their throughput and delay guarantees.

We illustrate the problem with a simple example and with the support of Fig. 4. Flows f_1 and f_2 have the same bandwidth allocation, and fanouts with the same cardinality but only output j in common. All the outputs in the fanout of flow f_1 are congested, whereas output j is the only congested output in the fanout of flow f_2. The channel scheduler at the fabric output and the virtual-output scheduler in the ingress port card guarantee the transfer of traffic from virtual output j to the fabric at a rate that is not smaller than the sum of the bandwidth allocations of f_1 and f_2. The presence of congestion at output j, on the other hand, forces the aggregate rate received by the two flows to be not greater than the sum of their bandwidth allocations. Under these constraints, the bandwidth requirements at output j are satisfied for both flows only if they cycle through their respective fanouts at exactly the same frequency. However, f_2 cycles much faster than f_1 because of the immediate services it receives at all outputs different than j. As a consequence, flow f_2 receives more services than flow f_1 at virtual output j. Since f_1 has no way to recover the loss of bandwidth at virtual output j, its throughput and delay guarantees are irremediably compromised.

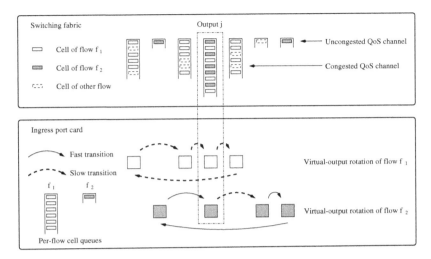

Fig. 4. Violation of QoS guarantees with virtual-output rotation.

5 Aggregation of Congestion Information over the Multicast Fanout

In the flow-control schemes that we have addressed in Section 4, the assertion of backpressure from a given QoS channel depends exclusively on *local* information, *i.e.*, on the amount of cell pointers currently queued in the channel. When this information becomes available at the ingress port card, it is already too late to extract additional elements that could be useful in the regulation of multicast traffic. The aggregation of such elements should rather occur in the fabric, upon generation of the backpressure bitmap.

In this section, we define a flow-control scheme that uses cell counters to aggregate information on the state of congestion induced by multicast flows on their output distributions. The scheme clearly outperforms the algorithms that are based on pointer counters, but still fails in supporting robust QoS guarantees, as we show with a practical example. The same example also provides directions for refining the flow-control scheme based on cell counters. We will adopt the appropriate refinements in the next section, where we complete the specification of the G-DMS.

5.1 Flow Control with Cell Counters

A flow-control scheme that is based on cell counters instead of pointer counters does aggregate information on the congestion level of multiple outputs. In the fabric, a cell contributes to congesting its primary channel until its transmission is completed over all the outputs of its fanout. If some of the secondary outputs are congested, the cell takes longer to leave the fabric, and thus increases the probability of backpressure assertion from its primary channel. In the ingress port card, a consistent use of the backpressure indication derived from the cell counters of the primary channels requires that each flow be always scheduled through its primary virtual output.

In every scheduler of the switch, the allocation of bandwidth to the traffic aggregates that include a multicast flow conforms to the following guidelines: (i) for a GD flow in the ingress port card, the guaranteed bandwidth of the flow adds to the aggregate service rate of its primary virtual output in the GD virtual-output scheduler (the GB class has no virtual-output scheduler, as shown in Fig. 3); (ii) for a GD (GB) flow in the fabric, the guaranteed service rate of the flow contributes to the GBS service rates of all the GD (NGD) channels in its fanout.

The flow-control scheme based on cell counters does not expose the fabric to buffer overflow, but still leaves some open issues in the provision of robust QoS guarantees to multicast flows. It may happen, for example, that the cell counter of a channel $C_{i,j}$ exceeds the corresponding backpressure threshold when the pointer counter of the same channel is null. In this case, traffic destined for output j can be forced to wait in ingress port card i even if output j is lightly loaded or not loaded at all. Two distinct conditions can lead to this manifestation of HOL blocking, which determines loss of aggregate throughput in the whole switch: (i) the actual load at channel $C_{i,j}$ is below its nominal level, as expressed by the GBS service rate of the channel; or (ii) the secondary outputs of some of the primary flows of $C_{i,j}$ are overloaded by flows whose primary outputs are different than j. In the former case, the individual guarantees of the primary flows of $C_{i,j}$ can still be enforced, even though the aggregate switch throughput is no longer maximized. In the latter case, the throughput and delay guarantees of the primary flows of the channel can be seriously compromised. This second condition is particularly critical for GD traffic, which typically has stringent real-time requirements. The following subsection illustrates the critical case with an example.

5.2   Violation of QoS Guarantees with Cell Counters

We consider a 4×4 switch with the flow setup summarized in Table 1. All flows in the switch belong to the GD class and have an allocated service rate that is equal to 0.01 cpt. The number of flows that are configured for each class determines the nominal load at the switch inputs and outputs, and the bandwidth allocations at each stage of the distributed scheduler. A *regulated* source is a traffic source that complies with an associated traffic profile (*e.g.*, a token-bucket regulator). An *unregulated* source does not comply with its traffic profile, and sends traffic to the switch at the highest rate allowed by the capacity of the input link, compatibly with the presence of other flows at the same input. An *idle* source sends no traffic to the switch, even if the corresponding flow has a bandwidth allocation in the ingress-port-card and fabric-output schedulers. A square $\boxed{x}$ around an output number x identifies the primary output of the flow.

Given the traffic setup, and in particular the idling behavior of the sources of class c_4, the nominal and actual loads at the switch inputs and outputs reflect the

Table 1. Critical traffic setup.

Flow Class	Number of Flows	Source Behavior	Input	Output(s)		
c_0	74	Unregulated	1	0		
c_1	49	Unregulated	2	2		
c_2	25	Unregulated	0	1	2	3
c_3	25	Regulated	0	0	2	
c_4	49	Idle	0	1	3	

distribution reported in Table 2. The idling status of the flows of class c_4 makes all the bandwidth allocated to virtual output 1 at ingress port card 0 available to the flows of class c_2. The allocation of bandwidth for the flows of class c_4 is also available to the flows of class c_2 at QoS channels $C_{0,1}^{GD}$ (the primary channel of class c_2) and $C_{0,3}^{GD}$ (one of the secondary channels of class c_2). As a consequence, the flows of class c_2 have access to the fabric at a higher rate than their nominal allocation (through virtual output 1 at input 0), and experience no contention to access outputs 1 and 3 in the fabric. The pointer counters $\beta_{0,1}^{GD}$ and $\beta_{0,3}^{GD}$ are therefore null for most of the time. Output 2, on the contrary, is heavily loaded. In particular, channel $C_{0,2}^{GD}$ is serviced at exactly its nominal GBS allocation, but is offered secondary cells from classes c_2 and c_3 at a rate that exceeds by far its GBS allocation. Whenever the secondary cells of class c_2 have access to channel $C_{0,2}^{GD}$ at a rate higher than the nominal allocation for the corresponding flows, the flows of class c_3 experience a loss of throughput. We have punctually observed this behavior in the simulation of the traffic scenario.

Table 2. Load distribution over the switch inputs and outputs.

	Inputs				Outputs			
	0	1	2	3	0	1	2	3
Nominal Load	0.99	0.74	0.49	0.00	0.99	0.74	0.99	0.74
Actual Load	1.00	1.00	1.00	0.00	1.25	0.75	2.00	0.75

Out of the bandwidth that is totally available at channel $C_{0,2}$ ($0.505\,cpt$), only $0.205\,cpt$ go to class c_3 (about 80% of the nominal allocation of the class), and $0.300\,cpt$ are taken by class c_2. The evident loss of throughput also implies the violation of the theoretical delay bounds for the flows of class c_3. The aggregate switch throughput, computed as the ratio between the number of cells delivered to the egress port cards and the total number of deliverable cells, is only equal to 0.743, because of the under-utilization of outputs 1 and 3 ($0.300\,cpt$, in spite of the availability of $0.750\,cpt$).

The observation of the simulation results leads to interesting conclusions about the possible refinements of the flow-control scheme for the generalization of the DMS framework. First, the loss of throughput for class c_3 is caused by the excessive amount of traffic admitted to the fabric for class c_2. At the same time, the aggregate switch throughput suffers from the exact opposite reason, which is the under-utilization of two of the outputs in the fanout of class c_2. The only way to admit less traffic of class c_2 to output 2 and more traffic of the same class to outputs 1 and 3 is differentiating the admission to the fabric for copies of the same cells of class c_2. In other words, we should be able to admit all copies aiming for outputs 1 and 3, and deny access to some of the copies destined for output 2. In a multicasting scheme that replicates cells exclusively in the switching module, denying access to a replica is equivalent to dropping the replica in the fabric.

6   The Generalized Distributed Multilayered Scheduler

The Generalized Distributed Multilayered Scheduler (G-DMS) extends the DMS framework to packet switches that simultaneously handle both unicast and multicast traffic. The G-DMS rejects the overlay approach and uniforms the treatment of unicast and multicast traffic, so that the former can indeed be considered as a particular case of the latter. The generalization of the DMS framework must cope with the interaction of multicast traffic with selective backpressure, subject to the architectural constraint that unicast and multicast flows are scheduled through the same flow queues in the ingress port cards.

6.1   Dropping Multicast Copies in the Fabric

The capability of selectively dropping copies of multicast cells in the switch fabric is critical to the definition of the G-DMS. In this subsection, we ponder the transport-layer implications of this option.

For a GD flow, whose transport over the packet network typically relies on connectionless datagram protocols (such as UDP), it is desirable to deliver packets to all the leaves of the multicast tree with maximum speed and minimum difference between the delivery times at distinct destinations. When the multicast distribution tree has branches that are more congested than others, it may be acceptable to drop packets on the congested branches, especially if this action is a necessary condition to preserve the desired quality of service on the uncongested branches and maximize the aggregate switch throughput.

For an adaptive flow, which runs end-to-end flow control at the transport layer (as provided in the unicast case by TCP) and may or may not have explicit allocation of bandwidth resources at the network nodes (as is the case for our GB and BE classes, respectively), dropping copies of a multicast packet in the switch fabric is not different than dropping the same copies downstream in the network. The same end-to-end mechanism that ensures reliable delivery of multicast packets to all expected destinations can therefore be invoked to recover from losses in the switch fabric.

We conclude that dropping part of the fanout of a multicast cell in the switch fabric does not collide with the requirements of both real-time and adaptive multicast applications, and is compatible with the mechanisms that support such applications throughout the packet network.

6.2   Flow Control with Cell Counters and Selective Cell Dropping

The G-DMS relies on backpressure and selective discard of multicast copies to regulate access to the fabric buffers. Per-flow policers in the port cards contribute to the fairness of the access regulation.

Per-Flow Policers.   Referring again to the traffic scenario specified in Table 1, it is clear that the source of violation for the QoS guarantees of class c_3 is in the excess of secondary cells of class c_2 that are admitted to channel $C_{0,2}^{GD}$. Some of these cells should be denied access to the fabric, but the basic flow-control scheme with cell counters has no means to explicitly regulate access to the secondary channels. We therefore need to upgrade the scheme, introducing means to control the accumulation of secondary cells in the channels. In particular, we would like to restrict admission to the fabric only for class c_2 (whose behavior is not consistent with the bandwidth allocation), and maintain the channel fully open to the flows of class c_3.

The per-flow policers that are typically available at the ingress port cards constitute an excellent instrument for detecting any discrepancy between the actual and the expected behavior of the configured flows. A policer is a device that monitors the profile of the incoming traffic on a per-flow basis, and compares it with the traffic profiles specified upon configuration of the flows. If an incoming packet falls out of profile, the policer marks it, so that it has higher probability of being dropped than an unmarked packet when they both arrive to a congested node in the network. The *Generic Cell Rate Algorithm* (GCRA) [11] provides a standard solution for checking conformance of a flow with a token-bucket profile in ATM networks. Similar policing devices can be easily defined for networks with variable-sized packets [22].

The G-DMS exploits the action of the policers at the ingress port cards to mark incompliant cells and expose them to access denial to the congested channels of the switching fabric.

Static Admission Thresholds for Secondary Channels.   We make a first attempt to complete the definition of the G-DMS with a scheme that associates a static admission threshold with each secondary channel. The channel drops the secondary copies of incompliant cells that arrive when its pointer counter exceeds the admission threshold. After implementing the algorithm, we observe substantial but not yet satisfactory improvements compared to the scheme with no cell dropping. Under the usual traffic scenario, the throughput of class c_3 increases from $0.205\,cpt$ to $0.222\,cpt$ (ideally it should be $0.250\,cpt$), and the aggregate switch throughput grows from 0.753 to 0.912 (the target is obviously 1.000).

The threshold-based admission policy fails in setting a clear discrimination between compliant and incompliant cells. Having the pointer counter below the admission threshold does not necessarily imply that the channel has bandwidth available for the transmission of incompliant secondary cells. As a consequence, such cells can still be admitted to the channel in excess of the desirable amount and thus compromise the QoS guarantees of compliant flows.

Multicast Credits for Secondary Channels. The EBD algorithm that we have reviewed in Section 2.3 allows the secondary channels to accurately detect the availability of bandwidth in excess of the amount that strictly satisfies the requirements of compliant flows. Conforming to the EBD algorithm, we equip each QoS channel $C_{i,j}$ with a *multicast credit counter* $\chi_{i,j}$ for incompliant cells. When the pointer counter of the channel is null, the credit counter $\chi_{i,j}$ is set equal to a *multicast credit threshold* ξ, which provides a common reference for all the channels in the switching module. The multicast credit counter $\chi_{i,j}$ is decremented every time the channel admits an incompliant secondary cell, and incremented every time the channel receives an EBS service. If the counter hits the threshold ξ, it is not allowed to grow any further. If the counter becomes null, the channel discards all incoming secondary cells that are marked as incompliant. It should be noticed that the admission policy has no effect on incompliant primary cells, which are taken care by the backpressure mechanism based on the cell counters. When one of such cells arrives to the channel, it is always accepted.

We have integrated the algorithm in the G-DMS, and applied it to the traffic scenario of Table 1. The throughput observed for class c_3 at channel $C_{0,2}^{GD}$ matches the expected $0.250\,cpt$, with consequent satisfaction of the delay bound expectations for all the flows in the class. The aggregate throughput measured in the switch is 0.996. The key for the achievement of all the QoS goals is in the reduction to the expected value of $0.250\,cpt$ for the throughput of class c_2 at output 2, and in the full utilization of outputs 1 and 3 (the measured throughput is $0.743\,cpt$ at both outputs).

Multicast Credits in Secondary NGD Channels. In order to complete the specification of the G-DMS, we must define the way we handle the GB and BE classes in the NGD channels. In particular, we must determine whether the EBD algorithm that discriminates the assertion of backpressure for primary GB and BE flows should use the cell counter or the pointer counter to track the credits of the NGD channel.

The EBD algorithm detects the availability of excess bandwidth in the fabric-output scheduler, and the activity of the scheduler is controlled by the pointer counters of the channels (for the scheduler, an idle QoS channel is a channel whose pointer counter has null value). For this reason, we decide to increment the credit counter of the NGD channel every time the channel receives an EBS service, and decrement the counter every time the channel receives a BE cell pointer. Notice that we still base the detection of congestion in the NGD channel, which triggers the joint assertion of backpressure for GB and BE flows, on the value of the cell counter.

By running extensive simulations with GB and BE flows in the critical traffic scenario of Table 1, we have observed the systematic achievement of all QoS expectations.

6.3   Cell Counters versus Pointer Counters in the G-DMS

With the introduction of the mechanism that selectively drops secondary cells in QoS channels with lack of excess bandwidth, the use of cell counters instead of pointer counters in the flow-control scheme for primary flows may appear to be harder to motivate. We argue that there are at least three clear reasons to maintain the role of the cell counters in the flow-control scheme for primary flows.

First, the amount of cells that can be globally admitted to the switching module is much higher when admission is regulated by the cell counters, because each cell contributes to the cell count only at the primary channel, with no contribution to the cell count at the secondary channels.

Second, for real-time traffic, the preservation of tight delay guarantees requires that the transfer of cells from the virtual outputs in the ingress port cards to the QoS channels in the fabric be as smooth as possible [4]. In order to maintain the distributed scheduler in close approximation of its ideal hierarchical reference, the virtual outputs should never experience long periods of time characterized by continuous assertion of backpressure from the corresponding QoS channels. Even if all flows configured in the system are compliant with their traffic profiles, the pointer counters can still undergo rapid fluctuations determined by the arrival of secondary cells whose flows are controlled by different primary channels. The cell counters, on the contrary, filter out these fluctuations, making the virtual-output perception of the state of the QoS channels much more stable than in the case of use of pointer counters.

Third, in the presence of selective discard of multicast cell copies, the enforcement of robust QoS guarantees relies on the accuracy of the mechanism that allocates bandwidth resources and sets the policing parameters for the configured flows. With pointer counters, any over-allocation of resources may induce buffer overflow in the fabric. With cell counters, on the contrary, the overall occupation of the buffer memory is always under control, with no risk of buffer overflow. The simulation results presented below substantiate our arguments.

Cell Counters versus Pointer Counters: Simulation Results.   The simulation in a 3×3 switch of the traffic scenario summarized in Table 3 illustrates the benefits of using cell counters instead of pointer counters in the assertion of backpressure for primary flows in the G-DMS (all flows in the table belong to the GD class). Given the traffic setup, and in particular the unregulated nature of the sources of classes c_1 and c_2, the nominal and actual loads at the switch inputs and outputs reflect the distribution reported in Table 4. Output 1 of the switch is overloaded by the presence of unregulated traffic coming from input 1 (class c_2). Similarly, the capacity of input 0 is saturated by the presence of unregulated sources sending traffic to output 2 (class c_1). The traffic conditions at

Table 3. Traffic setup showing the benefits of cell counters versus pointer counters.

Flow Class	Allocated Rate (cpt)	Number of Flows	Source Behavior	Bucket Size (cells)	Input	Output(s)
c_0	0.01	25	Regulated	2	0	1
c_1	0.01	50	Unregulated	100	0	2
c_2	0.01	50	Unregulated	100	1	1
c_3	0.0001	2400	Regulated	2	0	0 1

input 0 and output 1 critically restrict access to QoS channel $C_{0,1}^{GD}$, which aggregates unicast cells of class c_0 and secondary multicast cells of class c_3. Flows of class c_3 are strictly regulated (the leaky-bucket size is 2 cells), but their number is large, and they experience no contention at their primary output (output 0 is lightly loaded). As a consequence, the queue length of channel $C_{0,1}^{GD}$ can undergo wide fluctuations determined by the bursty arrival of compliant secondary cells of class c_3. The unicast flows of class c_0 are heavily penalized by these fluctuations when the assertion of backpressure depends on the pointer counters. The use of cell counters, on the contrary, makes the admission of secondary cells to channel $C_{0,1}^{GD}$ much smoother.

In the simulation experiment, we observe that the overall throughput for class c_0 is 99.99% of the offered load when backpressure depends on cell counters, and 93.71% of the offered load when backpressure depends on pointer counters. The worst-case transmission delay observed for class c_0 is equal to 198 timeslots with cell counters, and is instead unstable when the pointer counters drive the assertion of backpressure (as an example, at the end of a 10,000,000-timeslot simulation run we have recorded a worst-case delay of 650,736 timeslots).

Table 4. Load distribution over the switch inputs and outputs.

	Inputs			Outputs		
	0	1	2	0	1	2
Nominal Load	0.99	0.50	0.00	0.24	0.99	0.50
Actual Load	1.00	1.00	0.00	0.24	1.49	0.51

6.4 G-DMS Overhead

The G-DMS upgrades the flow-control functionality of the basic DMS while keeping its scheduling features unchanged. Table 5 lists the per-channel counters involved in the implementation of the G-DMS. The only overhead induced by the generalization of the basic DMS is given by the introduction of the cell counter $\gamma_{i,j}$ and the multicast credit counter $\chi_{i,j}$ at each QoS channel. No changes affect the number of QoS channels, the size of the backpressure bitmap, and the fabric-output and ingress-port-card schedulers.

Table 5. Per-channel counters in the G-DMS.

Channel	Pointer Counter	Cell Counter	NGD Credit Counter	Multicast Credit Counter
GD	$\beta_{i,j}^{GD}$	$\gamma_{i,j}^{GD}$	N/A	$\chi_{i,j}^{GD}$
NGD	$\beta_{i,j}^{NGD}$	$\gamma_{i,j}^{NGD}$	$\alpha_{i,j}^{NGD}$	$\chi_{i,j}^{NGD}$

7   Concluding Remarks

We presented the Generalized Distributed Multilayered Scheduler (G-DMS), a framework for the integrated provision of differentiated QoS guarantees to unicast and multicast flows in multi-stage packet switches. Two distinguishing features characterize the G-DMS: (i) a flow-control scheme that regulates access to the switching fabric based on the actual occupation of the buffer memory and not on the number of queued cell pointers, and (ii) the capability of selectively dropping copies of multicast cells that violate the traffic profiles of the corresponding flows. The algorithm that triggers the selective dropping of cell copies in the fabric exploits the modular nature of the schedulers at the fabric outputs and the presence of traffic policers in the ingress port cards.

The G-DMS meets the QoS expectations of both unicast and multicast flows while adding only minimal overhead to the implementation complexity of the underlying DMS (two counters per QoS channel).

References

1. D. C. Stephens and H. Zhang, "Implementing Distributed Packet Fair Queueing in a Scalable Switch Architecture," *Proceedings of IEEE INFOCOM '98*, March 1998.
2. D. C. Stephens, "Implementing Distributed Packet Fair Queueing in a Scalable Switch Architecture," *M.S. Thesis*, Carnegie Mellon University, Pittsburgh, PA, May 1998.
3. F. M. Chiussi and A. Francini, "Providing QoS Guarantees in Packet Switches," *Proceedings of IEEE GLOBECOM '99, High-Speed Networks Symposium*, Rio de Janeiro, Brazil, December 1999.
4. F. M. Chiussi and A. Francini, "A Distributed Scheduling Architecture for Scalable Packet Switches," *IEEE Journal on Selected Areas in Communications*, Vol. 18, No. 12, pp. 2665–2683, December 2000.
5. J. C. R. Bennett and H. Zhang, "Hierarchical Packet Fair Queueing Algorithms," *Proceedings of ACM SIGCOMM '96*, pp. 143–156, August 1996.
6. F. M. Chiussi, J. G. Kneuer, and V. P. Kumar, "Low-Cost Scalable Switching Solutions for Broadband Networking: The ATLANTA Architecture and Chipset," *IEEE Communications Magazine*, Vol. 35, No. 12, pp. 44–53, December 1997.

7. U. Briem, E. Wallmeier, C. Beck, and F. Matthiesen, "Traffic Management for an ATM Switch with Per-VC Queueing: Concept and Implementation," *IEEE Communications Magazine*, Vol. 36, No. 1, pp. 88–93, January 1998.
8. F. M. Chiussi, Y. Xia, and V. P. Kumar, "Backpressure in Shared-Memory-Based ATM Switches under Multiplexed Bursty Sources," *Proceedings of IEEE INFOCOM '96*, pp. 830–843, March 1996.
9. B. Prabhakar, N. McKeown, and R. Ahuja, "Multicast Scheduling for Input-Queued Switches," *IEEE Journal on Selected Areas in Communications*, Vol. 15, No. 5, pp. 855–866, June 1997.
10. N. McKeown,, "A Fast Switched Backplane for a Gigabit Switched Router," *Business Communications Review*, Vol. 27, No. 12, December 1997.
11. ATM Forum Traffic Management Specification, Version 4.0, April 1996.
12. Y. Tamir and G. Frazier, "High Performance Multiqueue Buffers for VLSI Communication Switches," *Proceedings of 15th Annual Symposium on Computer Architectures*, pp. 343–354, June 1988.
13. D. Stiliadis and A. Varma, "A General Methodology for Designing Efficient Traffic Scheduling and Shaping Algorithms," *Proceedings of IEEE INFOCOM '97*, April 1997.
14. J. C. R. Bennett and H. Zhang, "WF^{2}Q: Worst-case Fair Weighted Fair Queueing," *Proceedings of IEEE INFOCOM '96*, pp. 120–128, March 1996.
15. A. K. Parekh and R. G. Gallager, "A Generalized Processor Sharing Approach to Flow Control in Integrated Services Networks: The Single-Node Case," *IEEE/ACM Transactions on Networking*, pp. 344–357, June 1993.
16. S. J. Golestani, "A Self-Clocked Fair Queueing Scheme for Broadband Applications," *Proceedings of IEEE INFOCOM '94*, pp. 636–646, April 1994.
17. D. Stiliadis and A. Varma, "Latency-rate Servers: A General Model for Analysis of Traffic Scheduling Algorithms," *Proceedings of IEEE INFOCOM '96*, pp. 111–119, March 1996.
18. M. Katevenis, S. Sidiropoulos, and C. Courcoubetis, "Weighted Round-Robin Cell Multiplexing in a General-Purpose ATM Switch Chip," *IEEE Journal on Selected Areas in Communications*, vol. 9, pp. 1265–1279, October 1991.
19. F. M. Chiussi and A. Francini, "Advances in Implementing Fair Queueing Schedulers in Broadband Networks," *Proceedings of IEEE ICC '99*, June 1999 (Invited paper).
20. A. Francini, F. M. Chiussi, R. T. Clancy, K. D. Drucker, and N. E. Idirene, "Enhanced Weighted Round Robin Schedulers for Bandwidth Guarantees in Packet Networks," *Proceedings of QoS-IP 2001*, pp. 205–221, Rome, Italy, January 2001.
21. F. M. Chiussi, A. Francini, D. A. Khotimsky, and S. Krishnan, "Feedback Control in a Distributed Scheduling Architecture," *Proceedings of IEEE GLOBECOM 2000, High-Speed Networks Symposium*, San Francisco, CA, November 2000.
22. A. Francini and F. M. Chiussi, "Minimum-Latency Dual-Leaky-Bucket Shapers for Packet Multiplexers: Theory and Implementation," *Proceedings of IWQoS 2000*, pp. 19–28, Pittsburgh, PA, June 2000.

Service Differentiation in the Internet to Support Multimedia Traffic

Fouad Tobagi, Waël Noureddine, Benjamin Chen, Athina Markopoulou,
Chuck Fraleigh, Mansour Karam, Jose-Miguel Pulido, and Jun-ichi Kimura

Department of Electrical Engineering
Stanford University, Stanford CA 94305
Contact author: tobagi@stanford.edu

Abstract. The current best-effort infrastructure in the Internet lacks
key characteristics in terms of delay, jitter, and loss, which are required
for multimedia applications (voice, video, and data). Recently, significant
progress has been made toward specifying the service differentiation to
be provided in the Internet for supporting multimedia applications. In
this paper, we identify the main traffic types, discuss their characteristics
and requirements, and give recommendations on the treatment of the different types in network queues. Simulation and measurement results are
used to assess the benefits of service differentiation on the performance
of applications.

1 Introduction

The Internet is seeing the gradual deployment of new multimedia applications,
such as voice over IP, video conferencing, and video-on-demand. These applications generate traffic with characteristics that differ significantly from traffic
generated by data applications, and they have more stringent delay and loss
requirements. Voice quality, for example, is highly sensitive to loss, jitter, and
queueing delay in network node (i.e., switch or router) buffers, and the quality of
video traffic is significantly degraded during network congestion episodes. The
current best-effort infrastructure of the Internet is ill-suited to the quality of
service requirements of these applications.

In addition to emerging streaming applications, the Internet must also support interactive data applications. Good user-perceived performance of these
applications, such as telnet, video gaming, and web browsing, requires short
response times and predictability. However, these requirements are often not
met, due to the interaction of TCP with packet loss during network congestion
periods.

In this paper, we identify the main traffic types, discuss their characteristics
and requirements (Sec. 2), and examine the degree of separation of traffic types
necessary to provide adequate user-perceived performance (Sec. 3). In addition,
we describe a conceptual model for a network node port that provides service
differentiation and discuss the different required functionalities (Sec. 4). Within
this model, we demonstrate that the provision of multiple packet drop priorities

S. Palazzo (Ed.): IWDC 2001, LNCS 2170, pp. 381–400, 2001.

within a queue, in association with appropriate packet marking, can further enhance performance (Sec. 5). Throughout, simulation and measurement results support our proposals.

2   Multimedia Traffic Characteristics and Requirements

In this section we present the characteristics and requirements of voice, video, and interactive data applications. For each application, we discuss its characteristics in terms of data rate and variability, and we describe its requirements in terms of delay, jitter, and packet loss. These have to be well understood in order to determine the appropriate treatment to give each application in the network.

2.1   Voice

Voice connections generate a stream of small packets of similar size (a few tens of bytes) at relatively low bit rates. Typical stream rates range from 5 Kbps to 64 Kbps, depending on the encoding scheme, to which header overhead adds a few tens of Kbps. Therefore, voice stream rates remain on the order of tens of Kbps, regardless of the encoding scheme. For example, G.711, a simple pulse code modulation encoder, generates evenly spaced 8-bit samples of the voice signal at 125 msec intervals, resulting in a 64 Kbps stream. It is possible to reduce the rate through voice compression schemes and silence suppression, at the expense of increased variability. The suppression of samples corresponding to silence periods (which account for 45% of total time in typical conversations [6]), leads to substantial average rate reduction. For example, G.729A generates an 8 Kbps stream, while G.723 generates a 5.33 Kbps stream.

For the Internet to provide toll quality voice service, packet delay and loss must meet stringent requirements. Interactivity imposes a maximum round trip time of 200-300 msec. That is, the one-way delay incurred in voice encoding, packetization, network transit time, de-jittering, and decoding must be kept below 100-150 msec. Jitter must be limited (e.g., less than 50 msec) to ensure smooth playback at the receiver. Subjective tests have shown that periods of lost speech (clips) larger than 60 msec affect the intelligibility of the received speech [9]. Since packet loss in the Internet is bursty [2,22], the probability that consecutive voice packets are lost, resulting in long clips, is significant. Therefore, loss rates have to be kept at very low levels unless packet loss concealment is used.

2.2   Video

Video traffic is stream-oriented and spans a wide range of data rates, from tens of Kbps to tens of Mbps. The characteristics of encoded video (data rates and variability in time) vary tremendously according to the content, the video compression scheme, and the video encoding scheme.

In terms of content, more complex scenes and more frequent scene changes require more data to maintain a certain level of quality. For example, video streams of talking heads are lower-rate and less variable than those of motion pictures and commercials.

Different video compression schemes, such as H.261, H.263, MPEG-1, and MPEG-2, are designed to meet different objectives and therefore have different bit rates and stream characteristics. For instance, the applications of H.261 include video conferencing; consequently, the rates are multiples of 64 Kbps, up to 2 Mbps, and the coding is designed to achieve a fairly uniform bit rate across frames. On the other hand, prerecorded movies using MPEG-2 may have several times the picture resolution and typically require several Mbps.

The video characteristics are also affected by the video encoding control scheme used. For a given content and a given compression scheme, constant bit rate (CBR) video maintains a streaming rate that varies little over time. By contrast, variable bit rate (VBR) video traffic has been shown to be self-similar and may have a peak rate which is many times the average. Typically, VBR aims to achieve a more consistent quality for the same average bandwidth, and it is more commonly employed in practice.

The latency requirements of video depend on the application. Like voice, interactive video communication requires low delay (200-300 msec round-trip); however, one-way broadcast and video-on-demand may tolerate several seconds of delay. As is the case for voice, a packet delayed beyond the time when it needs to be decoded and displayed is considered lost. Furthermore, it has been observed that packet loss rates as low as 3% can affect up to 30% of the frames, due to dependencies in the encoded video bit stream [4]. Therefore, the packet loss rate and delay in the network should be kept small [7].

2.3   Interactive Data Applications

Data applications still account for the large majority of Internet traffic [25] and have significantly different characteristics and requirements. We focus here on interactive data applications, namely telnet (remote login) and the web.

Typical telnet sessions consist of characters being typed by a user at a terminal, transmitted over the network to another machine (server), which echoes them back to the user's terminal. The packet stream being generated consists of small datagrams (typically less than 50 bytes). Occasionally, the results of commands typed by the user are sent back by the server. This results in asymmetric traffic, with server to user terminal traffic on average 20 times the user to server traffic [21]. Packet interarrival times have been found to follow a heavy-tailed (Pareto) distribution, resulting in somewhat bursty traffic [23]. However, the inter-packet time is normally limited by the typing speed of humans, which is rarely faster than 5 characters per second [13], giving a minimum 200 msec average inter-packet time. Consequently, the traffic generated by telnet is of relatively low bandwidth and low burstiness.

Telnet is a highly interactive application and, similarly to voice over IP, has strict delay requirements on individual packets. Echo delays start to be noticeable when they exceed 100 msec, and in general, a delay of 200 msec is the limit beyond which the user-perceived quality of the interactivity suffers [13]. Furthermore, telnet traffic is highly sensitive to packet loss, since the retransmit timeout required for recovery has a minimum value that exceeds the maximum acceptable echo delay. Therefore, telnet packet loss needs to be kept at a minimum for best user-perceived performance.

Web traffic, carried by HTTP over TCP, is closely tied to the contents of web pages and to the dynamics of TCP. Trace studies of web traffic have shown that the majority of HTTP requests for web pages are smaller than 500 bytes. HTTP responses are typically smaller than 50 KB, but may also be very large when HTTP is used to download large files off web pages [18]. Indeed, HTTP responses have been found to follow a heavy tailed distribution, corresponding to that of web files in the Internet. Moreover, the aggregate traffic generated by many users of the WWW has been shown to exhibit self-similarity [5,17].

In general, short page download times (less than 5 seconds) are required for good user-perceived performance. In addition, users highly value the predictability of web response times. In other words, not only does the average download time need to be small, but so does the variance of download times. In this context, it is important to distinguish between the interactive use of HTTP, i.e. for downloading actual web pages (html file and images) which tend to be short transfers (and therefore have low rate), and the non-interactive FTP-like use, where HTTP is employed to download large files.

3   Mixing vs. Separating Traffic Types

As discussed above, voice, video, and data applications differ significantly in their traffic characteristics and requirements. Naturally, we are interested in understanding how we can support the various traffic types in a single network, such that the user-perceived performance is maximized. It appears reasonable that identifying different classes of traffic in order to separate them at the queues in the network and to treat them appropriately would achieve this goal. The questions that arise are: Where in the network would differential handling be necessary, if at all? Which types need to be separated, and which types can be safely mixed? What is the appropriate treatment of each class in its queue? In this section we attempt to answer the first two questions, and we address the third in Sec. 4 and 5.

3.1   Mixing Voice and Data

In [15], we investigate the effect of mixing and separating traffic types, i.e., which types can be mixed together in the same queue without incurring a significant loss in throughput, and which types need to be separated to meet performance objectives. We first consider the impact of data traffic on voice traffic by mixing 1 Mbps of voice traffic (11 streams) with TCP data traffic. We determine the maximum number of data sources that can be mixed with voice traffic if voice packets are to satisfy a 10 ms delay budget allocated to the section of the path being studied. We consider paths composed of either T1, 10Base-T, or 100Base-T links. Simulations reveal that mixing voice and data traffic is impossible for T1 links (1.5 Mbps). In the case of 10Base-T links, it is only possible for less bursty data flows, at the expense of a significant throughput reduction. Mixing FTP traffic with voice is possible only on 100Base-T links, in which case the link utilization must be kept below 20%. This result shows that data traffic is incompatible with voice traffic and should be separated from it.

To corroborate these simulation results, we examine [19] the transmission of voice on a VPN path between two Internet POPs in the U.S., coast-to-coast during business hours. Using delay measurements from this path, we simulate the quality of G.711 VoIP calls. In Fig. 1, we show the voice quality experienced, quantified by the MOS (Mean Opinion Score) scale. There are several series, each corresponding to different echo loss (echo cancellation) capabilities. EL=inf corresponds to perfect echo cancellation, while EL=31 corresponds to poor echo cancellation. Given that toll quality voice has a MOS of 4–5, we observe that even for good echo loss (EL=41, 51), there are many times in the course of the hour when quality is poor, due either to packet loss or excessive delay. We also note the importance of echo cancellation capabilities in contributing to voice quality.

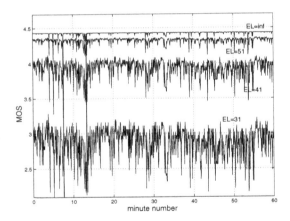

Fig. 1. MOS, averaged over 5-second intervals, for different echo cancellation capabilities.

We now simulate 1000 G.711 calls of exponential duration in the same environment. The playout deadline is optimized every 15 sec, and the echo cancellation is very good (EL=51). Even with this favorable setup over the path, we observe in Fig. 2 that the quality experienced is much worse than toll quality: around 10% of the calls experience at least one minute of poor quality (the MOS drops below 4), and 4% of all calls experience poor quality in at least 10% of their minutes. Note that these results account only for voice transit between two POPs; end-to-end, the quality suffers yet further. Clearly, voice and data traffic should be separated into different traffic classes.

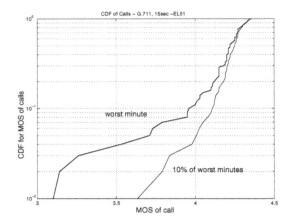

Fig. 2. Percentage of calls with MOS less than a particular value.

3.2   Mixing Voice and Video

Now that we have established the need for at least two queues, we consider whether voice and video can be mixed. First, we note that CBR and VBR video have very different characteristics, so we consider them separately.

When CBR video is mixed with voice, the increase in voice delay is contained because the CBR video streams are well behaved. Our work [15] shows that since voice and CBR video have similar characteristics and real-time delay requirements, we may allow them to be handled together. Consider a scenario in which CBR video streams are added to a 7-hop 10Base-T network carrying a 450 Kbps aggregate voice load. Voice delay exceeds 10 ms only when the link approaches full utilization. However, on a T1 link (1.5 Mbps), to achieve the target performance, only one video stream and two voice streams can be admitted, resulting in utilization as low as 55%. Therefore, unless the rate of the video stream is large compared to the link bandwidth, CBR video does not hurt voice traffic.

Consider the same scenario, with CBR video replaced by VBR video. As expected, given the characteristics of VBR video, the results of its mixing with voice traffic differ greatly from those of CBR video. Figure 3 shows the distribution of delay of voice and video packets when they are mixed and when they are separated. As we increase the number of multiplexed VBR video streams, both voice and video delays increase rapidly because of the long bursts that get injected into the queue. This implies that if latency constraints are to be met, only a small number of video streams may be mixed with voice, resulting in low network utilization. In general, the achievable throughput depends on the voice/video mix, the burstiness of the video streams, and the link speeds; however, it is clear that VBR video cannot be mixed successfully with voice.

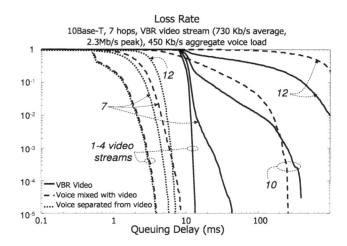

Fig. 3. Delay distributions for voice and VBR video. Video streams are added to the network.

3.3 Mixing Video and Data

We have shown that VBR video and voice are incompatible. The next question we attempt to answer is: can data and VBR video be mixed? It turns out that the answer depends on what delay can be tolerated by the video and what aggregate throughput is considered acceptable. In general, when the delay bounds are tighter, the interfering traffic load must be kept lighter. In addition, there is another important factor, which is the buffer size. Indeed, there is a tradeoff between a large buffer size, with a correspondingly large queueing delay, and a small buffer size, with the increased possibility of packet loss due to buffer overflow and the resulting decrease in throughput.

   In Table 1, we show the results of a scenario in which a constant video load and FTP streams are mixed, and a video packet loss rate of 10^{-3} is tolerated. Considering the large dependence of the results on the buffer size, we experiment with a range of buffer sizes which scale according to the link speeds considered. We determine, for network delay requirements of 100 msec and 500 msec for the video stream, the total achievable aggregate throughput. We find that the buffer size which maximizes the throughput increases proportionally to the delay bound, and the total achievable throughput can be increased if the delay bound is relaxed. Lower speed links are also more sensitive to proper buffer sizing, so it is not advisable to mix video with data even when interactivity is not required. To summarize, video should be mixed with data only on high bandwidth links.

Table 1. Video mixed with FTP traffic: maximum achievable throughput for 100 msec and 500 msec end-to-end delay bounds for video (tolerable loss rate: 10^{-3}). $D(Q_{max})$ is the maximum buffer delay, Q_{max} is the buffer size.

$D(Q_{max})$	10Base-T			T3			100Base-T		
	Q_{max} (KB)	throughput 100 ms	500 ms	Q_{max} (KB)	throughput 100 ms	500 ms	Q_{max} (KB)	throughput 100 ms	500 ms
40.1 ms	50	0%	0%	225	49%	49%	500	61%	61%
81.9 ms	100	29%	29%	450	75%	75%	1000	82%	82%
123.9 ms	150	20%	37%	675	56%	90%	1500	82%	82%
491.5 ms	600	0%	96%	2700	0%	96%	6000	0%	90%

3.4   The Case for Three Classes of Service

From the results above, we conclude that separating multimedia traffic into a minimum of three queues—voice, video, and data—is necessary for good performance for a range of network conditions. Interactive voice, with its high expectations, requires a queue of its own, for protection from the bursty traffic of VBR video and TCP data applications. Video may be mixed with data under particular conditions, but other considerations also compel us to give it its own queue. The real-time constraints of video call for higher priority in scheduling compared to data. Furthermore, the reliability levels required by video are close to 100%, whereas TCP is designed to recover from loss. Finally, in contrast to data traffic, video streams should be subjected to admission control due to their large and predictable rates. The nature of CBR video allows it to be combined with voice or with VBR video.

Yet we must be sensitive to the two extremes. High speed links tolerate better the mixing of disparate traffic types, such that on very high speed backbone routes, differential handling may be unnecessary. Likewise, very low speed links may require finer grain distinction of packets and/or packet preemption mechanisms.

4   Network Node Structure

From the discussion in the previous section, we can consider a model for the internal structure of a network node (output) port, which is shown in Fig. 4. This structure is to be used at places where packet buffering is performed within the switching node, which could be at the input ports, the output ports, or both.

We assume here that the node has an output queued implementation to avoid having to go into the details of the different possible switching architectures. We describe the components of the port (classifier, traffic conditioner, buffer management, and scheduler) in more detail below.

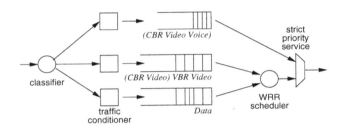

Fig. 4. The structure of a network node port containing three queues.

4.1 Packet Classification

The first step in providing differentiated services is enabling network nodes to identify the class of service of each packet they receive, possibly through a special marking carried by the packet. For example, the DiffServ architecture [10] uses the byte in the IP header previously allocated to TOS and renamed DS Field, as a priority code. In the IEEE 802.1 LAN realm, packet identification is done through a field in the recently adopted VLAN tag (added to the MAC frame header) that indicates the class of service of each packet. The priority code is checked by the classifier upon reception of a packet to determine the queue in which the packet is to be placed. Packets carrying the same marking expect to receive the same treatment in the network. The discussion in the previous section suggests that 3 queues are necessary: one for voice, another for video, and the third for data. However, allowing for more queues (e.g., 8 as in IEEE 802.1D) increases flexibility in assigning traffic to appropriate classes.

Before a packet is enqueued, it goes through traffic conditioners, which perform functions such as metering and policing. The policer ensures that the traffic entering a queue does not exceed a certain limit, determined by the queue's allocation of the link resources. This functionality is particularly needed for queues that are serviced with high priority in order to avoid starvation of lower priority traffic.

4.2 Scheduling

With traffic separated in multiple queues, a scheduler is required to service them. The scheduler's service discipline needs to be carefully designed in order to provide the appropriate delay through the node for each traffic type. In this section, we discuss the appropriate service discipline for each queue and the supporting mechanisms needed, such as admission control.

Voice. In Sec. 3, we argued for separating voice traffic in a queue of its own. The next step is to determine the appropriate scheduling discipline for providing voice with the required quality of service. In [14], we show that a strict high priority service is appropriate for handling voice traffic. Through modelling and simulation, we find that, considering the conservative 99.999^{th} percentile of packet delays, priority queueing does limit the delay and jitter of voice packets for typical link speeds. Refer to Fig. 5, where we plot the complementary cumulative distribution function (ccdf) of voice packet delays for a voice load of 1.1 Mbps over five 45 Mbps hops. We compare different link scheduling schemes, namely, priority queueing with preemption of low priority transmissions, priority queueing (PQ), weighted round robin (WRR) with a weight of 1.5 Mbps, and WRR with a weight of 10 Mbps for the voice queue. We also plot the ccdf for the case where voice packets are given a separate 10 Mbps circuit. The graphs show that, as would be expected, priority queueing with preemption achieves negligible queueing delays over the 5 hops. In addition, non-preemptive priority queueing still results in low queueing delays (the 99.999^{th} percentile is smaller than 2 msec, ignoring switching time through the node). WRR scheduling requires a large weight for the voice traffic (10 Mbps, more than 9 times the actual load) for it to compare to PQ. Note that the round robin scheduler insures that a large weight for the voice queue does not translate into wasted resources, since low priority traffic can utilize any unused bandwidth. In contrast, providing a 10 Mbps dedicated circuit for voice results in delays that are better than those of non-preemptive priority queueing and WRR, at the cost of wasted resources.

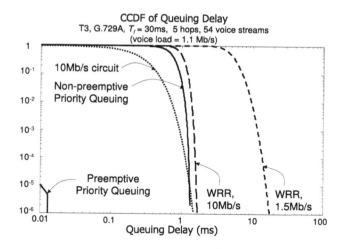

Fig. 5. Voice delay distributions for different link scheduling schemes.

While it may appear that both priority queueing and WRR can be used, the results in this graph consider only one low priority queue. Therefore, one would expect worse results if the round robin scheduler services more queues, where a voice packet may have to wait for more than one low priority packet transmission. While WRR has the well known advantage of preventing starvation of low priority queues, we believe that in the context of the Internet, no special precaution has to be taken to prevent voice traffic from starving others. Indeed, voice traffic volume is limited, and its growth rate is significantly smaller than that of data and other applications. Therefore, its share of the total traffic is decreasing. This means that not only would it not starve other traffic, but also that voice traffic may not require per-flow admission control. Rather, appropriate provisioning of the network would allow it to be serviced at highest priority.

Video. As discussed in Sec. 3, CBR video traffic is well behaved and can be mixed with voice traffic. If CBR video is mixed with voice traffic in the same queue, admission control would be needed, given the higher rates of video streams. On the other hand, VBR video streams need to be mapped to a queue separate from voice. This queue has to be serviced using round robin scheduling; otherwise, it may starve lower priority classes due to the burstiness of its traffic. Similarly to CBR video, VBR video streams may need to be subjected to admission control.

Data. Data applications could be all mapped to the same queue or, given the wide range of characteristics and requirements among them, may benefit from being mapped to multiple queues. Thus, if more than one data queue is available, low rate interactive data applications can be shielded from other data applications by separating them. In particular, telnet would benefit from having its own queue, serviced with high weight, especially on low speed links.

If the buffer sizes are chosen small enough to restrict the queueing delay of interactive packets to acceptable levels, all data applications may share the same queue. Then, differentiation among the different applications can be provided by assigning the packets to different drop priority levels within that queue.

Given that most data flows are of the short lived type [25], it is impractical to perform per-flow admission control. In addition, TCP generates bursty traffic, and therefore it is not possible to guarantee congestion-free service for data traffic without significant over-provisioning. Thus, the data queue should be serviced with a round robin scheduler to avoid starving lower-priority queues, if any.

4.3   Buffer Management

To enable high speed processing, the buffers would most likely have to be implemented as FIFO queues. Therefore, any action to be taken has to be performed before the received packet is enqueued. Possible actions are dropping or marking the packets, e.g. if explicit congestion notification (ECN) is implemented [8].

Early dropping/marking schemes such as Random Early Detection (RED) and its derivatives, aim at providing early notification of congestion before the buffer gets full, and bursty packet loss becomes necessary. Such schemes assume that an end-to-end congestion control mechanism, such as the one implemented in TCP, will react to the congestion signals. Therefore, they may not be effective or appropriate for applications which do not use such mechanisms. Indeed, different buffer management schemes should be used for different classes. Moreover, the size of the buffers should be tailored to suit the application types. Thus, small buffers would be used for packet-delay sensitive traffic to limit queueing delays, while large buffers can be used for applications that are insensitive to per-packet delay, but are sensitive to packet loss.

The priority code of each packet may indicate, in addition to the queue where it is to be placed, one of several drop priorities within that queue. This approach is used in the DiffServ AF class. Providing several drop priorities within each queue allows further differentiation among packets and may be used to achieve significant improvements in quality degradation during congestion episodes, as discussed in Sec. 5.

5   Improving Resilience to Congestion for Video and Data Applications

With traffic types appropriately mapped to different queues in the network, the dynamics of each queue depend on the particular traffic it is serving. Adequately serviced voice traffic would see little queueing delay and loss in the network, and its treatment within the queue does not require further refinement. This is not the case for video and data traffic. Since it is not possible to guarantee congestion-free delivery to unshaped bursty traffic, performance degradation may occur for such traffic. In this section, we show how providing several packet drop priorities within one queue can be used to significantly improve the user-perceived quality of applications during congestion episodes.

First, for video applications, we show that layered video in association with priority dropping is a simple, yet effective technique for providing graceful degradation in the event of packet loss. Then, we address data applications, which themselves span a wide range of requirements and characteristics. We show how identifying and prioritizing different applications can improve user experience.

5.1   Addressing Packet Loss with Layered Video

Let us consider the transmission of digital video over packet networks. One particular characteristic of video is its high sensitivity to packet loss. Video quality is greatly eroded when there is loss of data which contribute heavily to quality, such as low frequency DCT coefficients, motion vector information, or start codes needed for synchronization. In Fig. 6 we illustrate the drastic quality degradation of a video sequence resulting from random packet loss. The line segment on the left corresponds to a video sequence encoded with P frames and B frames; for the line segment on the right, only I frames were used in the encoding of the same sequence. When the sequence contains P and B frames, 1% packet loss can lead to poor quality because there is interdependence among pictures—the loss of a single packet can have an effect on multiple subsequent P and B pictures. We observe in the plot that when this interdependence is removed, such that all frames are encoded as I frames, the quality is less affected by packet loss.

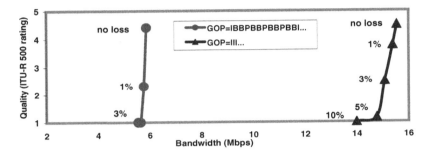

Fig. 6. Random packet loss is applied to an MPEG-2 video stream packetized using RTP. Video quality degrades sharply.

There are several possible ways to deal with packet loss. Adaptive encoding could be used, in which feedback from the receiver or a network node provides the source with information to adapt the transmission rate by modifying encoding parameters. However, this is fairly complex to implement and is limited by feedback delay. It is not suitable for networks with high variability, for multipoint communications, or for stored video. The technique of smoothing or shaping can limit the variability of the stream's traffic, and the use of admission control can limit the variability in the aggregate traffic. This, too, introduces complexity in the nodes, and it curtails statistical multiplexing gain, decreasing overall throughput. Furthermore, smoothing or shaping introduces delay, clearly undesirable when latency is of concern.

Because loss is inevitable, we must limit its effect by dealing with it intelligently, protecting important data and dealing with loss where and when it occurs. This leads us to consider layered video and priority dropping [16,24]. Simply put, layered video prioritizes information according to its importance in contribution to quality. In conjunction with priority dropping, layered video is a powerful technique for maintaining quality in the presence of loss. We show that it offers graceful quality degradation rather than the sharp drop we saw in Fig. 6, and we show how to divide a video stream into layers to maximize the perceived video quality for a particular range of network conditions.

Video layering using data partitioning. Layering mechanisms define a base layer and one or more enhancement layers that contribute progressively to the quality of the base layer. A base layer can stand alone as a valid encoded stream, while enhancement layers cannot. The key observation is that some bits are more "important" than others; we identify their importance by placing them into different layers, thus allowing a node to drop packets with discrimination.

The MPEG standards [11,12] specify four scalable coding techniques for the prioritization of video data: temporal scalability, data partitioning (DP), SNR scalability, and spatial scalability. We consider layering based on data partitioning, treating temporal scalability as a special case of it. One advantage of data partitioning is that the overhead incurred by layering is negligible. Another advantage is that it is performed after the encoding of the stream, allowing it to be easily used with pre-encoded video.

Data partitioning divides the encoded video bit stream into two or more layers by allocating the encoded data to the various layers. Naturally, the data with the most impact on perceived quality should be placed in the base layer. To indicate the portion of the data which is included in the base layer, we define a drop code for each picture type, I, P, and B. The drop code takes on a value from 0 to 7, where 0 indicates that all of the data are included in the base layer, and 7 indicates that only the header information is placed in the base layer. The partitioning of the stream data into the base and enhancement layers is completely specified by a drop code triplet, e.g., (036). There is a correspondence between the drop code value and a header field defined by the MPEG standards that can be used for data partitioning [16].

Temporal scalability is a special case of data partitioning, where the drop code is 007. In temporal scalability, entire B pictures are dropped prior to dropping any information in I or P pictures.

We illustrate in Fig. 7 the advantage of using 2-layer data partitioning in a network which supports priority dropping. In this figure, we show the quality of video using temporal scalability, data partitioning, and no layering. With temporal scalability, the dropping of B picture data degrades quality. If DC coefficients and motion vector information from B frames were not dropped, the decoder could have reconstructed enough of these frames to significantly improve the perceived quality. Therefore, not all B frame data should be assigned to the enhancement layer. Thus, data partitioning using a well-chosen drop code triplet (036) allows for graceful quality degradation as enhancement packets are randomly dropped. In the region where no base layer packets need to be dropped (above 3.8 Mbps), the quality degradation incurred varies almost linearly with the rate of packet loss. However, once we start losing base layer packets, the quality falls sharply. This establishes the need to protect the base layer from network loss with an appropriate nodal structure.

We have shown the graph for drop code triplet (036). Other layering structures (i.e., other triplets) will place the knee at different points while exhibiting the same two-piece behavior. We have, through simulation, identified the layering structures which achieve the highest quality for a given bandwidth. These

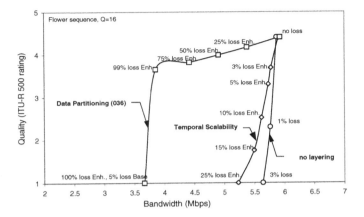

Fig. 7. Bandwidth-quality tradeoff curves for 2-layer DP, temporal scalability, and no layering.

dominant structures are the same for all sequences studied, and they correspond to the following triplets: 003, 014, 005, 015, 016, 036, 136, 046, 146, and 156 [16]. In practice, it is desirable to choose from these triplets the proper layering structure such that the linear portion of the graph is large enough to just cover the expected bandwidth range delivered by the network.

Quality degradation can be further improved with 3-layer data partitioning. In the example shown in Fig. 8, we create three layers by keeping the enhancement layer the same as in 2-layer DP with drop triplet (036). We then make the break between the middle layer and the base layer at the point where 2-layer

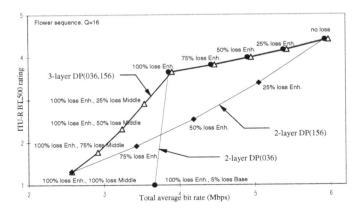

Fig. 8. Bandwidth-quality tradeoff curves for 3-layer DP compared to 2-layer DP schemes.

DP(156) does. Thus, in the 3.8–6.0 Mbps range (only the enhancement layer is dropped), the behavior parallels that of the 2-layer DP(036). The only difference is a slight increase in overhead. In the region where the middle layer is being dropped (2.5–3.8 Mbps), the quality is superior to the same region for DP(156), because the least important data have been identified and dropped first. As one would suppose, additional layers do continue to improve video quality, though with limited incremental improvement beyond 4 layers.

Multiplexing layered streams. So far, we have examined a single video stream with prioritized random loss in the enhancement layers. We now consider the case where several layered streams share a limited resource. We have observed graceful quality degradation as the number of streams is increased, as shown in Fig. 9. We have also studied SNR scalability [16] but do not make further comment here, except to remark that it performs similarly to DP, but the latter is preferred because of its negligible overhead and ease of implementation.

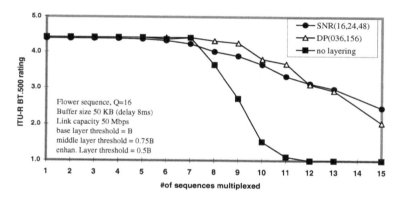

Fig. 9. Multiple video streams sharing a 50 KB buffer, for three schemes: SNR scalability, DP, and no layering.

We conclude that the combination of a simple layering of video data and a simple priority dropping mechanism at network nodes, appropriately employed, can have a significant effect on sustaining video quality in the face of packet loss.

5.2 TCP Applications

Until now, all data applications have used best-effort service for lack of an alternative. However, most Internet users have experienced times where severe quality loss is suffered. Such degradation is most distinctly perceived when associated with interactive applications. Hence, telnet interactivity is severely hindered, and web page download times become excessive during congested hours. This is particularly unacceptable for business applications which require predictable service. While large page download times can be attributed in part to server overload, we focus here on the delays caused by the interaction of TCP's congestion control mechanisms with packet loss in the network.

We observe that, similar to layered video, some data packets contribute more to user perceived quality than others. With large window sizes, the TCP fast retransmit mechanism can be used to minimize the impact of packet loss. However, when the congestion window is small, there is a much longer delay to recover from a lost packet.

Here we illustrate, using simulation results, the benefits that can be achieved for interactive applications when their packets are appropriately prioritized in the network. We consider that all data applications share one queue. Packets are marked at the source with one of 3 drop priorities. For FTP and HTTP, the SYN packet and the packets sent when the connection is operating with a small congestion window are marked as high priority because of the penalty in recovering from their loss. For telnet, all packets are marked with high priority. The aggregate high and medium priority traffic generated by each station is shaped to conform to two token bucket profiles, the goal of which is to limit the amount of high and medium priority packets injected by each user. The access to high priority tokens at the source is prioritized based on the application, with telnet receiving the highest access priority, and FTP the lowest.

The topology used for the simulations consists of 800 user stations, organized into 400 source-destination pairs of different round trip times (ranging from 20 msec to 200 msec), and connected by a symmetric tree with hierarchical link speeds (starting at 1.5 Mbps for user links, with a bottleneck of 100 Mbps). The router buffers are appropriately sized to provide low delay for telnet, while giving good performance to HTTP and FTP traffic. A randomized dropping function similar to RED is used for dropping packets for each of the three priorities. As the queue size increases, low priority packets are dropped first, followed by medium priority packets. High priority are only dropped when the queue size gets close to the maximum buffer size. Traffic consists of 1 telnet and 1 web connection per source-destination pair, with background traffic of repetitive short FTP file transfers (200 KB) in both directions.

In the following, we illustrate how the performance of interactive applications can be improved by appropriate service differentiation, at a modest cost to non-interactive applications. In Fig. 10, we plot the complementary cumulative distribution function of web download times[1] for different service differentiation scenarios. The curves marked DT and RED correspond to scenarios without service differentiation (best effort), with queues managed using Drop Tail and RED, respectively. As can be seen from the plot, 10% of the page downloads for both Drop Tail and RED suffer a delay in excess of 19 seconds.

---

[1] We show results for HTTP/1.0 traffic here, for a web page with eight 10 KB imbedded images. Up to 4 connections are opened in parallel to download the page components. Similar results were obtained for HTTP/1.1.

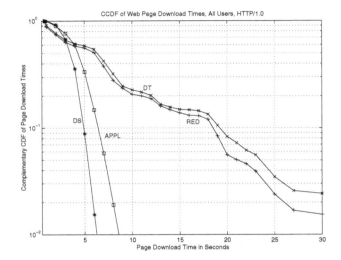

Fig. 10. The complementary cumulative distribution function of web page download times with (DS and APPL) and without (DT and RED) service differentiation.

In contrast, we show the results with service differentiation, where packets are marked at the source based on the application and TCP connection state. The corresponding curve, marked DS, clearly shows a significant improvement in terms of download times, where all downloads take between 3 and 6 seconds.

In Fig. 11, we plot the ccdf of short file transfer times, which shows that the improvement in page download times was obtained at little cost to the background traffic. A simpler form of differentiation would be to base the packet drop priority marking solely on the generating application type. Thus, packets belonging to telnet and similar low-bandwidth and delay sensitive applications such as Internet gaming would be marked high priority, and those belonging to short web page downloads would be marked medium priority. Packets generated by other, non-interactive applications such as FTP, would be marked low priority. The aggregate traffic of each priority is again shaped to limit its rate. The curves marked APPL in Figures 10 and 11 show that this technique can improve the performance of web page downloads, but rather less effectively than the more intricate method (DS) and at a higher cost in performance loss to the background traffic. Similar results can be shown for telnet in this scenario, where appropriate differentiation is provided through marking all of its packets at high priority. This allows the elimination of excessive delay of character echoes (1 sec), which result from retransmits due to packet loss. Without service differentiation, these delays occur for about one out of ten typed characters (more details can be found in [20]). In conclusion, it is possible to use multiple drop priorities to the advantage of interactive applications, in order to improve the user-perceived performance of such applications.

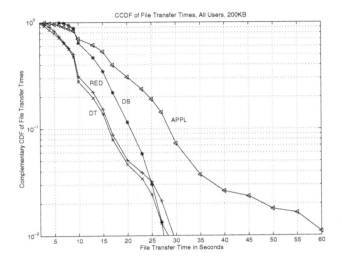

Fig. 11. The complementary cumulative distribution function of file transfer times with (DS and APPL) and without (DT and RED) service differentiation.

6   Conclusion

For the Internet to support multimedia applications, service differentiation is needed. In this paper, we describe the characteristics and requirements of voice, video, and interactive data applications, and we demonstrate the peformance improvements achieved by providing different treatments for each of these three types of traffic. We propose a three queue model for network nodes with one queue for voice traffic, one queue for video, traffic, and one queue for data traffic. The voice queue is served with strict priority, and the video and data queues share the remaining capacity using weighted round robin scheduling. We also show that the performance of the video and data queues may be further improved by using multiple levels of drop precedence within each queue and by marking packets according to their importance.

References

1. Bhatti, N., Bouch, A., and Kuchinsky, A.J. Integrating User-Perceived Quality into Web Server Design. Presented at *WWW'00*, Amsterdam, 2000.
2. Bolot, J.-C. End-to-End Packet Delay and Loss Behavior in the Internet. in *Proceedings of SIGCOMM'93*.
3. Bouch, A., Sasse, M., and DeMeer, H.G. Of Packets and People: A User-Centered Approach to Quality of Service. Submitted to *IWQoS'00*.
4. Boyce, J.M. and Gaglianello, R.D. "Packet loss effects on MPEG video sent over the public Internet. In *Proceedings of ACM MULTIMEDIA '98*, pages 181–190, Bristol, England, September 1998.

5. Crovella, M. and Bestavros A. Self-Similarity in World Wide Web Traffic: Evidence and Possible Causes. *IEEE/ACM Transactions on Networking*, December 1997.

6. Daigle, J. and Langford, J. Models for Analysis of Packet Voice Communications Systems. in *IEEE Journal on Selected Areas in Communications*, Volume 4, Number 6, September 1986.

7. Dalgic, I. and Tobagi, F. "Glitches as a Measure of Video Quality Degradation Caused by Packet Loss," In *Packet Video Workshop '96*, Brisbane, Australia, March 1996.

8. Floyd, S. TCP and Explicit Congestion Notification. In *ACM Computer Communication Review*, Volume 24, Number 5, October 1994.

9. Gruber, J. and Strawczynski, L. Subjective Effects of Variable Delay in Speech Clipping in Dynamically Managed Voice Systems. In *IEEE Transactions on Communications*, Volume 33, Number 8, August 1985.

10. IETF, DiffServ Working Group, http://www.ietf.org/html.charters/diffserv-charter.html.

11. ISO/IEC, "Generic coding of moving pictures and associated audio information" (MPEG-2), ISO/IEC 13818-2, 1995.

12. ISO/IEC, "Generic coding of audio-visual objects" (MPEG-4), ISO/IEC 14496-2, 1999.

13. Jacobson, V. Compressing TCP/IP Headers for Low-Speed Serial Links, RFC 1144, February 1990.

14. Karam, M. and Tobagi, F. Analysis of the Delay and Jitter of Voice Traffic Over the Internet. In *Proceedings of INFOCOM'01*.

15. Karam, M. and Tobagi, F. On Traffic Types and Service Classes in the Internet. In *Proceedings of GLOBECOM'00*.

16. Kimura, J., Tobagi, F., Pulido, J-M., and Emstad, P. "Perceived Quality and Bandwidth Characterization of Layered MPEG-2 Video Encoding" in *Proceedings of the SPIE International Symposium on Voice, Video and Data Communications.* Boston, Mass, September 1999.

17. Leland, W.E., Taqqu, M.S., Willinger, W., and Wilson, D.V. On the self-similar nature of Ethernet traffic, *IEEE Transactions on Networking*, Vol. 2, No. 1, Feb. 1994.

18. Mah, B. An Empirical Model of HTTP Traffic. In *Proceedings of INFOCOM'97*.

19. Markopoulou, A. and Tobagi F. Assessment of Perceived VoIP Quality Over Today's Internet. TR-CSL work in progress, Stanford University.

20. Noureddine, W. and Tobagi, F. Improving the User-Perceived Performance of TCP Applications with Service Differentiation. TR-CSL work in progress, Stanford University.

21. Paxson, V. Empirically-Derived Analytic Models of Wide-Area TCP Connections. In *IEEE Transactions on Networking*, 2(4), August 1994.

22. Paxson, V. End-to-End Internet Packet Dynamics. *IEEE/ACM Transactions on Networking*, Vol.7, No.3, pp. 277-292, June 1999.

23. Paxson, V. and Floyd, S. Wide-Area Traffic, the Failure of Poisson Modeling. In *ACM Computer Communication Review*, October 1994.

24. Pulido, J.-M. *A Simple Admission Control Algorithm for Layered VBR MPEG-2 Streams*, Engineer's thesis, July 2000. Available as Stanford University's Computer Systems Laboratory Technical report CSL-TR-00-806.

25. Thompson, K., Miller, G.J., and Wilder R. Wide-Area Internet Traffic Patterns and Characteristics. In *IEEE Network*, November/December 1997.

A Transport Infrastructure Supporting Real Time Interactive MPEG-4 Client-Server Applications over IP Networks

Haining Liu, Xiaoping Wei, and Magda El Zarki

Department of Information and Computer Science
University of California, Irvine
Irvine, CA 92697
{haining, wei, magda}@ics.uci.edu

Abstract. Nearly all of the multimedia streaming applications running on the Internet today are basically configured or designed for 2D video broadcast or multicast purposes. With its tremendous flexibility, MPEG-4 interactive client-server applications are expected to play an important role in online multimedia services in the future. This paper presents the initial design and implementation of a transport infrastructure for an IP based network that will support a client-server system which enables end users to: 1) author their own MPEG-4 presentations, 2) control the delivery of the presentation, and 3) interact with the system to make changes to the presentation in real time. A specification for the overall system structure is outlined. Some initial thoughts on the server and client system designs, the data transport component, QoS provisioning, and the control plane necessary to support an interactive application are described.

1 Introduction

Today, most of the multimedia services consist of a single audio or natural (as opposed to synthetic) 2D video stream. MPEG-4, a newly released ISO/IEC standard, provides a broad framework for the joint description, compression, storage, and transmission of natural and synthetic audio-visual data. It defines improved compression algorithms for audio and video signals, and efficient object-based representation of audio-video scenes [1]. It is foreseen that MPEG-4 will be an important component of multimedia applications on IP-based networks in the near future [4].

In MPEG-4, audio-visual objects are encoded separately into their own Elementary Streams (ES). In addition, the Scene Description (SD), also referred to as the Binary Format for Scene (BIFS), defines the spatio-temporal features of these objects in the final scene to be presented to the end user. Based upon VRML (Virtual Reality Modeling Language), the SD uses a tree-based graph, and can be dynamically updated. The SD is conveyed between the source and the destination through one or more ESs and is transmitted separately. Object Descriptors (ODs) are used to associate scene description components to the actual elementary streams that contain the corresponding coded media data. Each OD groups all descriptive components that are related to a single media object, e.g., an audio or video object, or even an

S. Palazzo (Ed.): IWDC 2001, LNCS 2170, pp. 401–412, 2001.
© Springer-Verlag Berlin Heidelberg 2001

animated face. ODs carry information on the hierarchical relationships, locations and properties of ESs. ODs themselves are also transported in ESs. The separate transport of media objects, SD and ODs enables flexible user interactivity and content management.

The MPEG-4 standard defines a three-layer structure for an MPEG-4 terminal: the Compression Layer, the Sync Layer and the Delivery layer. The Compression Layer processes individual audio-visual media streams and organizes them in Access Units (AU), the smallest elements that can be attributed individual timestamps. The compression layer can be made to react to the characteristics of a particular delivery layer such as the path-MTU or loss characteristics. The Sync Layer (SL) primarily provides the synchronization between streams. AUs are here encapsulated in SL packets. In case that the AU is larger than the SL packet, it will be fragmented across multiple SL packets. The SL produces an SL-packetized stream i.e. sequences of SL-packets. The SL-packet headers contain timing, sequencing and other information necessary to provide synchronization at the remote end. The packetized streams are then sent to the Delivery Layer.

In the MPEG-4 standard, a delivery framework referred to as the Delivery Multimedia Integration Framework (DMIF) is specified at the interface between the MPEG-4 synchronization layer and the network layer. DMIF provides an abstraction between the core MPEG-4 system components and the retrieval methods [2]. Two levels of primitives are defined in DMIF. One is for communication, between the application and the delivery layer to handle all the data and control flows. The other one is used to handle all the message flows in the control plane between DMIF peers. Mapping these primitives to the available protocol stacks in an IP-based network is still an on-going research issue. Moreover, designing an interactive client-server system for deployment in the Internet world using the recommended primitives is even more challenging [3].

The object-oriented nature of MPEG-4 makes it possible for an end user to manipulate the media objects and create a multimedia presentation tailored to his or her specific needs, end device and connection limitations. The multimedia content resides on a server (or bank of servers) and the end user has either local or remote access to the service. This service model differs from the traditional streaming applications because of its emphasis on end user interactivity. To date, nearly all the streaming multimedia applications that are running on the Internet are basically designed for simple remote retrieval or for broadcasting/multicasting services. Interactive services are useful in settings such as distance learning, gaming, etc. The end user can pick the desired language, the quality of the video, the format of the text, etc.

There are only a few MPEG-4 interactive client-server systems discussed in the literature. H. Kalva *et al.* describe an implementation of a streaming client-server system based on an IP QoS Model called XRM [6]. As such it cannot be extended for use over a generic IP network (it requires a specific broadband kernel – xbind). Y. Pourmohammadi *et al.* propose a DMIF based system design for IP-based networks. However, their system is mainly geared toward remote retrieval and very little is mentioned in the paper regarding client interactivity with respect to the actual presentation playback [7]. In [6], the authors present the Command Descriptor Framework (CDF), which provides a means to associate commands with media objects in the SD. The CDF has been adopted by the MPEG-4 Systems Group, and is part of the version 2 specification. It consists of all the features to support interactivity in MPEG-4 systems [8-10]. Our transport infrastructure subsumes that command descriptors (CDs) are used for objects in the SD. This will enable end users to interact with individual objects or group of objects and control them.

In order to support an MPEG-4 system that enables end user interactivity as described above, many issues must be considered. To list a few: 1) transmission of end user interactivity commands to the server, 2) transport of media content with QoS provisioning, 3) real time session control based upon end user interactivity, 4) mapping of data streams onto Internet transport protocols, 5) extension of existing IETF signaling and control protocols to support multiple media streams in an interactive environment, etc.

In this paper, we present our initial ideas on the design of a client-server transport architecture which will enable end users to create their own MPEG-4 presentation, control the delivery of the content over an IP-based network, and interact with the system to make changes to the presentation in real time. Section 2 presents the structure of the overall system. Server and client designs are described in section 3. In section 4, data transport, QoS provisioning and control plane message exchange are discussed. In section 5, we conclude and discuss the future work.

2 Overall System Architecture

The system we are proposing to develop is depicted in Figure 1 and consists of 1) an MPEG-4 server, which stores encoded multimedia objects and produces MPEG-4 content streams, 2) an MPEG-4 client, which serves as the platform for the composition of an MPEG-4 presentation as requested by the end user, and 3) an IP network that will transport all the data between the server and the client.

The essence of MPEG-4 lies in its object-oriented structure. As such, each object forms an independent entity that may or may not be linked to other objects, spatially and temporally. The SD, the ODs, the media objects, and the CDs are transmitted to the client through separate streams. This approach gives the end user at the client side tremendous flexibility to interact with the multimedia presentation and manipulate the different media objects. End users can change the spatio-temporal relationships among media objects, turn on or shut down media objects, or even specify different perceptual quality requirements for different media objects dependent upon the associated command descriptors for each object or group of objects. This results in a much more difficult and complicated session management and control architecture [6]. Our design starts out with the premise that end user interactivity is crucial to the service and therefore it targets a flexible session management scheme with efficient and adaptive encapsulation of data for QoS provisioning.

User interactivity can be defined to consist of three degrees or levels of interactivity that correspond to what type of control is desired:

1. Presentation level interactivity: in which a user actually makes changes to the scene by controlling an individual object or group of objects. This also includes presentation creation.
2. Session level interactivity: in which a user controls the playback process of the presentation (i.e., VCR like functionality for the whole session).

3. Local level interactivity: in which a user only makes changes that can be taken care of locally, e.g., changing the position of an object on the screen, volume control, etc.

Throughout our discussion below we will be making references to these three levels as each results in a different type of control message exchange in our system.

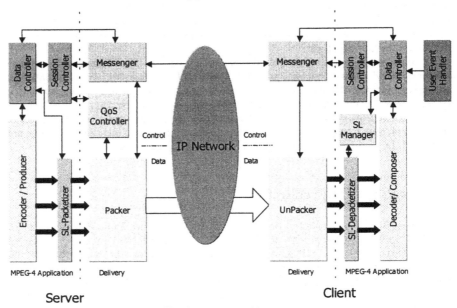

Fig. 1. System Architecture

We assume that the server maintains a database or a list of available MPEG-4 content and provides WWW access to it. An end user at a remote client side retrieves information regarding the media objects that he/she is interested in, and composes a presentation based upon what is available and desired. The system operation, after the end user has completed the composition of the presentation, can be summarized as follows:

1. The client requests a service by submitting the description of the presentation to the Data Controller (DC) at the server side.
2. The DC on the server side, controls the Encoder/Producer module to generate the corresponding SD, ODs, CDs and other media streams based upon the presentation description information submitted by the end user at the client side. The DC then triggers the Session Controller (SC) on the server side to initiate a session.
3. The SC on the server side is responsible for session initiation, control and termination. It passes along the stream information that it obtained from the DC to the QoS Controller (QC) that manages in conjunction with the Packer, the creation of the corresponding transport channels with the appropriate QoS provisions.

4. The Messenger Module (MM) on the server side, which handles the communication of control and signaling data, then signals to the client the initiation of the session and network resource allocation. The encapsulation formats and other information generated by the Packer when processing the "packing" of the SL-packetized streams are also signaled to the client to enable it to unpack the data.
5. The actual stream delivery commences after the client indicates that it is ready to receive, and streams flow from the server to the client. After the decoding and composition procedures, the MPEG-4 presentation authored by the end user is rendered on his or her display.

In the next sections, we describe in some detail the functionality of the different modules and how they interact.

3 Client-Server Model

3.1 The MPEG-4 Server

Upon receiving a new service request from a client, the MPEG-4 server starts a thread for the client, and walks through the steps described in the previous section to setup a session with the client. The server maintains a list of sessions established with clients and a list of associated transport channels and their QoS characteristics.

Figure 2 shows the components of the MPEG-4 Server. The Encoder/Producer compresses raw video sources in real-time or reads out MPEG-4 content stored in MP4 files. The elementary streams produced by the Encoder/Producer are packetized by the SL-Packetizer. The SL-Packetizer adds SL-Packet headers to the AUs in the elementary streams to achieve intra-object stream synchronization. The headers contain information such as decoding and composition time stamps, clock references, padding indication, etc. The whole process is scheduled and controlled by the DC.

The DC is responsible for several functions:

1. It responds to control messages that it gets from the client side DC. These messages include the description of the presentation composed by the user at the client side and the presentation level control commands issued by the remote client DC resulting from user interactions.
2. It communicates with the SC to initiate a session. It also sends SC the session update information as it receives user interactivity commands and makes the appropriate SD and OD changes.
3. It controls the Encoder/Producer and SL-Packetizer to generate and packetize the content as requested by the client.
4. It schedules audio-visual objects under resource constraints. With reference to the System Decoding Model, the AUs must arrive at the client terminal before their decoding time [1]. Efficient scheduling must be applied to meet this timing requirement and also satisfy the delay tolerances and delivery priorities of the different objects.

The SC likewise is responsible for several functions:

1. When triggered by the DC for session initiation, it will coordinate with the QC to set-up and maintain the numerous transport channels associated with the SL packetized streams.
2. It maintains session state information and updates this whenever it receives changes from the DC resulting from user interactivity.
3. It responds to control messages sent to it by the client side SC. These messages include the VCR type commands that the user can use to control the session.

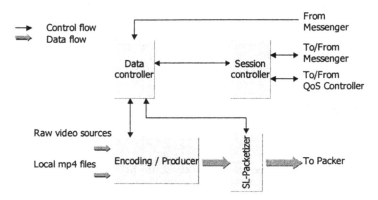

Fig. 2. Structure of the MPEG-4 Server

3.2 The MPEG-4 Client

The architectural design of the MPEG-4 client is based upon the MPEG-4 System Decoder Model (SDM), which is defined to achieve media synchronization, buffer management, and timing, when reconstructing the compressed media data [1]. Figure 3 illustrates the components of the MPEG-4 client.

The SL Manager is responsible for binding the received ESs to decoding buffers. The SL-Depacketizer extracts the ESs received from the Unpacker and passes them to the associated decoding buffers. The corresponding decoders then decode the data in the decoding buffers and produce Composition Units (CUs), which are then put into composition memories to be processed by the compositor. The User Event Handler module handles the user interactivity. It filters the user interactivity commands and passes the messages along to the DC and the SC for processing.

The DC at the client side has the following responsibilities:

1. It controls the decoding and composition process. It collects all the necessary information, e.g., the size of the decoding buffers which is specified in decoder configuration descriptors and signaled to the client via the OD, the appropriate decoding time and composition time which is indicated in the SL packet header, etc., for the decoding process.

2. It manages the flow of control and data information, controls the creation of buffers and associates them with the corresponding decoders.
3. It relays user presentation level interactivity to the server side DC and processes both session level and local level interactivity to manage the data flows on the client terminal.

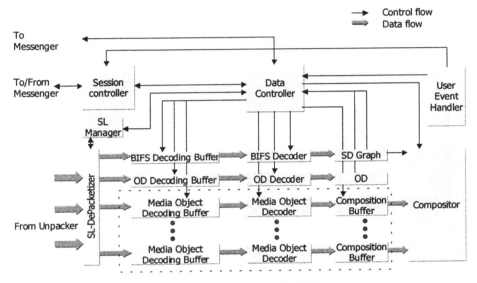

Fig. 3. Structure of the MPEG-4 Client

The SC at the client side communicates with the SC at the server side exchanging session status information and session control data. The User Event Handler will trigger the SC when session level interactivity is detected. The SC then translates the user action into the appropriate session control command.

4 Transport Architecture

The efficient and adaptive encapsulation of MPEG-4 content with regard to the MPEG-4 system specification is still an open issue. There is a lot of ongoing research on how to deliver MPEG-4 content over IP-based networks. Figure 4 shows our proposed design for the delivery layer. The following sections detail the components in this design.

4.1 Transport of MPEG-4 SL-Packetized Streams

Considering that the MPEG-4 SL defines some transport related functions such as timing and sequence numbering, encapsulating SL packets directly into UDP packets seems to be the most straightforward choice for delivering MPEG-4 data over IP-based networks. Y. Pourmohammadi *et al* presented a system adopting this approach [7]. However, some problems arise with such a solution. Firstly, it is hard to synchronize MPEG-4 streams from different servers in the variable delay environment which is common in the Internet. Secondly, no other multimedia data stream can be synchronized with MPEG-4 data carried directly over UDP in one application. Finally, such a system lacks a reverse channel for carrying feedback information from the client to the server with respect to the quality of the session. This is a critical point if QoS monitoring is desired.

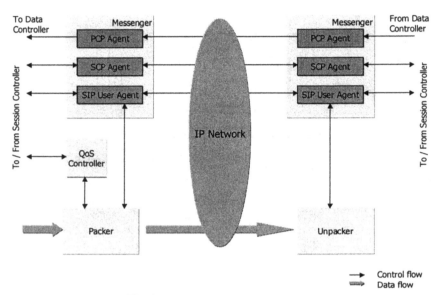

Fig. 4. Structure of the Delivery Layer

Another option is to deliver the SL packets over RTP, a standard protocol providing end-to-end transport functions for real-time Internet applications [11]. RTP has associated with it a control protocol, RTCP, which provides feedback channel for quality monitoring. In addition, the synchronization problems incurred when using UDP directly can be solved by exploiting the timing information contained in the RTCP reports. The problem arising when using RTP is the need to remove the resulting redundancy, as the RTP header duplicates some of the information provided in SL packet header. This adds to the complexity of the system and increases the transport overhead [12].

There are a number of Internet Drafts describing RTP packetization schemes for MPEG-4 data [12-14]. An approach proposed by Avaro *et al*, is attractive due to its solution regarding the duplication problem of the SL packet header and its independence of the MPEG-4 system, i.e., is not DMIF based [12]. The redundant information in the SL packet header is mapped into RTP header and the remaining part, which is called "reduced SL header", is placed in the RTP payload along with the SL packet payload. Detailed format information is signaled to the receiver via SDP. The MPEG-4 system also defines Flexible Multiplexing (FlexMux) to bundle several ESs with similar QoS requirements. The FlexMux scheme can be optionally used to simplify session management by reducing the number of RTP sessions needed for an MPEG-4 multimedia session. Van der Meer *et al*, proposed a scheme that provides an RTP payload format for MPEG-4 FlexMux streams [13].

Figure 5 shows the detailed layering structure inside the Packer and Unpacker. In the Packer, the SL-packetized streams are optionally multiplexed into FlexMux streams at the FlexMux layer, or directly passed to the transport protocol stacks composed of RTP, UDP and IP. The resulting IP packets are transported over the Internet. In the Unpacker, the data streams are processed in the reverse manner before they are passed to SL-Depacketizer.

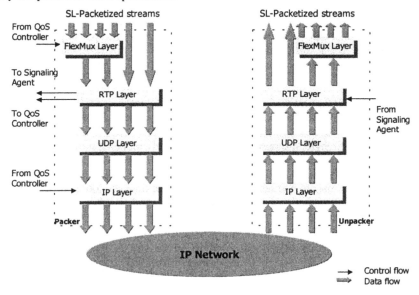

Fig. 5. Layering structure of Packer and Unpacker

In the Packer, the multiplexing of the SL-packetized streams is processed according to the QoS requirements of the streams and as such managed by the QoS controller. At the RTP Layer, the format of the encapsulation is passed to the signaling agent in SDP format to notify the client. The Packer is responsible for the allocation of transport channels, which are defined by port numbers, with the information from the QoS controller. At the IP Layer, certain actions managed by the QoS controller will be passed onto the network layer to meet the IP QoS requirements.

4.2 QoS Provisioning

According to the MPEG-4 system standard, each stream has an associated "transport" QoS description for its delivery [1]. This QoS description consists of a number of QoS metrics such as delay, loss rate, stream priority, etc. It is then up to the implementation of the delivery layer to fulfill these requirements.

In our system, the QoS Controller at the server side takes all the QoS issues into consideration as depicted in Figure 4. It is mainly responsible for managing the transport channel setup according to the required QoS parameters and controlling the traffic conditioning of IP packets. We model the IP network that our system is built on as a Differentiated Services (DiffServ) network, which is also the model for Internet 2. The QoS Controller maps the values of the QoS parameters to the values of the DiffServ Code Point (DSCP), which in turn determines the per-hop forwarding behavior of the IP packets. It then rewrites the DS byte in the IP header. It also controls the policing and rate shaping that takes place at the IP layer of the Packer.

4.3 Exchange of Signaling, Session Control and Presentation Control Messages

There are mainly three kinds of message flows exchanged between the server and the client in our system. Signaling messages are used mainly to locate the client, establish a network session, specify media information, modify, and tear-down an existing network session. Session Control messages contain commands issued by both the server and client to manage the session and provide real time session control to reflect user interactivity. Presentation Control messages are used to relay the presentation composed by the user and the presentation level user interactivity. As shown in Figure 4, three channels are maintained to carry these three distinct message flows between the server and the client.

SIP is a signaling protocol for creating, modifying and terminating sessions with one or more participants [4]. Because of its versatility, SIP fits in well with our system model. We will use SDP to deliver information such as media stream encapsulation format, network resource allocation indication, etc. The SDP messages will be embedded in the SIP message payload. SIP User Agents are placed in the MM in our system for session initiation, termination and transport of SDP messages.

RTSP is a control protocol used for real-time streaming applications [16]. Some papers have proposed schemes for adopting RTSP to provide some basic control and simple signaling for MPEG-4 applications [5]. However, as RTSP was primarily designed for media-on-demand scenarios, it cannot support the sophisticated interactivity required by our system. To reduce the overall system complexity, we have separated the signaling and control functions, and will design a Session Control Protocol (SCP) solely for exchanging control messages to manage the session (e.g., stop, resume, pause, etc.) in real time. Like SIP, the SCP agent will be placed in the MM to handle the message flow.

In our design, presentation control messages are exchanged between the client and the server via the Presentation Control Protocol (PCP). It will be designed specifically to support the presentation level user interactivity. During the presentation playback, the end user is able to interact with, and control what is being displayed at the object level. For example, VCR like functionality, such as stop, pause, resume, fast forward, can be associated with each object or group of objects. These controls will be implemented initially. More complex controls, such as the texture, dimensions, quality, etc., of an object, will be implemented as the design of the system matures and more detailed CDs are created for the objects. Similar to the other two agents, we will incorporate the PCP agent in the MM to communicate.

5   Conclusions

We presented in this paper a design for a transport infrastructure to support interactive multimedia presentations which enable end users to choose available MPEG-4 media content to compose their own presentation, control the delivery of such media data and interact with the server to modify the presentation in real-time. Detailed structures of the server and client were described. We discussed the issues associated with the delivery of media data and information exchange using a dedicated control plane that supports the exchange of signaling, session control and presentation control messages.

In summary, our work is focused on the following issues:

1. Translation of user interactivity into appropriate control commands
2. Development of a presentation control protocol to transmit user interactivity commands and a session control protocol to enable real time session management and information exchange to support user interactions
3. Processing of the presentation control messages by the server to update the MPEG-4 session
4. Transport of multiple inter-related media streams with IP based QoS provisioning
5. Client side buffer management for decoding the multiple media streams and scene composition

References

1. ISO/IEC JTC 1/SC 29/WG 11, "Information technology-Coding of audio-visual objects, Part1: Systems (ISO/IEC 14496-1)," Dec. 1998.
2. ISO/IEC JTC 1/SC 29/WG 11, "Information technology-Coding of audio-visual objects, Part6: Delivery Mulitmedia Integration Framework (ISO/IEC 14496-6)," Dec. 1998.
3. ISO/IEC JTC 1/SC 29/WG 11, "Information technology-Coding of audio-visual objects, Part8: Carriage of MPEG-4 contents over IP networks (ISO/IEC 14496-8), " Jan. 2001.
4. D. Wu, Y. T. Hou, W. Zhu,H. Lee, T. Chiang, Y. Zhang, H. J. Chao, " On End-to-End Architeture for Transporting MPEG-4 Video Over the Internet", IEEE Transactions on Circuits and Systems for Video Technology, Vol. 10, No. 6, Sep. 2000, pp. 923-941.
5. A. Basso and S. Varakliotis, " Transport of MPEG-4 over IP/RTP," Proceedings of ICME 2000, Capri, Italy, June 2000, pp. 1067-1070.

6. H. Kalva, L. Tang, J. Huard, G. Tselikis, J. Zamora, L. Cheok, A.Eleftheriadis, " Implementing Multiplexing, Streaming, and Server Interaction for MPEG-4," IEEE Transactions on Circuits and Systems for Video Technology, Vol. 9, No. 8, Dec. 1999, pp. 1299 –1311.
7. Y. Pourmohammadi, K. A. Haghighi, A. Mohamed, H. M. Alnuweiri, "Streaming MPEG-4 over IP and Broadcast Networks: DMIF Based Architectures," Proceedings of the Packet Video Workshop '01, Seoul, Korea, May 2001.
8. A. Akhtar, H. Kalva, and A. Eleftheriadis, " Command Node Implementation", Contribution ISO-IEC JTC1/SC29/WG11 MGE99/4905, July 1999, Vancouver, Canada.
9. ISO/IEC/SC29/WG11, " Text of ISO/IEC 14496-1/FPDAM 1(MPEG-4 System Version-2)," International Standards Organization, May 1999
10. Akhtar, H. Kalva, and A. Eleftheriadis, " Command Node Implementation in the IM1 Framework", Contribution ISO-IEC JT1/SC29/WG11 MGE00/5687, March 2000, Noordwijkerhout, Netherlands
11. Schulzrinne, Casner, Frederick and Jacobson, " RTP: A Transport Protocol for Real-Time Applications," draft-ietf-avt-rtp-new-08.ps, July 2000.
12. Avaro, Basso, Casner, Civanlar, Gentric, Herpel, Lifshitz, Lim, Perkins, van der Meer, "RTP Payload Format for MPEG-4 Streams," draft-gentric-avt-mpeg4-multiSL-03.txt, April 2001.
13. C.Roux et al, " RTP Payload Format for MPEG-4 Flexmultiplexed Streams," draft-curet-avt-rtp-mpeg4-flexmux-00.txt, Feb. 2001.
14. Y. Kikuchi, T. Nomura, S. Fukunaga, Y. Matsui, H. Kimata, "RTP Payload Format for MPEG-4 Audio/Visual Streams," RFC 3016, Nov. 2000.
15. Handley, Schulzrinne, Schooler and Rosenberg, " SIP: Session Initiation Protocol," ietf-sip-rfc2543bis-02.ps, Nov. 2000.
16. H. Schulzrinne, A. Rao, and R. Lanphier, "Real Time Streaming Protocol (RTSP)", RFC 2326, April, 1998.
17. M.Handley and V. Jacobson, "SDP: Session Description Protocol", RFC 2327, April 1998.

An Analytical Framework to Evaluate the Effects of a TCP-Friendly Congestion Control over Adaptive-Rate MPEG Video Sources

A. Cernuto, A. Lombardo, and G. Schembra

Dipartimento di Ingegneria Informatica e delle Telecomunicazioni
Università di Catania
V.le A. Doria 6 - 95125 Catania - ITALY
phone: +39 095 7382375 - fax: +39 095 7382397
(acernuto, lombardo, schembra)@iit.unict.it

Abstract. In the last few years researchers have made great effort to design TCP-friendly UDP-based congestion control procedures to be applied to real-time applications; however, very little effort has been devoted to investigating how real-time sources can be designed under TCP-friendly bandwidth constraints. In this perspective, this paper investigates the effects of the bandwidth variation rate during bandwidth profile variations on both the error of the rate controller in fitting the bandwidth profile, and the distortion introduced by the quantization mechanism of the MPEG video encoder. To this end we introduce an *SBBP/SBBP/1/K* queueing system modeling an MPEG video source where a feedback law is used to provide rate adaptation. The proposed paradigm addresses any feedback law, provided that the parameter to be varied is the quantizer scale.

1 Introduction

Distributed multimedia applications usually employ the UDP protocol to transmit video streams, because the delays added by relying on TCP retransmission mechanisms are unacceptable in real-time video transmission. As these applications are spreading, it is becoming increasingly important to ensure that they are able to co-exist with current TCP-based applications. In particular, as UDP does not embed any congestion control mechanism, video sessions are unresponsive sessions, that is, they do not back off their rates in time of congestion as TCP does. This behavior is unacceptable in two ways: first, coarsely fair sharing of network resources is no longer maintained as TCP sessions obviously suffer from competing with video sessions; secondly, as more and more "greedy" connections are set up across the Internet, the goodput of the network will decrease because unresponsive sessions typically send data packets at full rate even if their packets are later dropped inside the network.

It is therefore envisioned to enhance these UDP-based video communications with some kind of congestion control, in order to make them behave like "good network citizens" at times of bandwidth scarcity [1]. Achievement of the above scenario involves two challenging tasks. The first concerns the design of TCP-friendly

S. Palazzo (Ed.): IWDC 2001, LNCS 2170, pp. 413–432, 2001.

congestion control procedures for UDP-based sources: they have to be designed to provide a video connection with the same share of bandwidth as a TCP connection when they share a bottleneck over the same network path. The second concerns the definition of rate-adaptation mechanisms in the video sources which are able to shape the offered throughput according to the bandwidth profile allowed by the congestion control procedure. With respect to this task, the MPEG encoding standard is one of the most promising techniques in video encoding, thanks to its high compression ratio and flexibility. Let us note that although achievement of the first task alone, saves the network from unfairness or monopolization risks, it would not result in meaningful reception for the video application. Unresponsive video sources, in fact, send data packets at full rate even if their packets are delayed by the underlying TCP-friendly congestion control procedures; so the increasing delay experienced by the video source will result in loss of synchronization at the receiving side, which needs to skip a potentially high number of packets to recover synchronization requirements. This occurrence results in a breakdown of the QoS perceived by the end user.

In the last few years researchers have made great efforts to design TCP-friendly UDP-based congestion control procedures to be applied to real-time applications; however, very little effort has been devoted to investigating how real-time sources can be designed under TCP-friendly bandwidth constraints. In this paper we investigate the effects of the bandwidth variation rate during bandwidth profile variations on both the error of the rate controller in fitting the bandwidth profile, and the distortion introduced by the quantization mechanism of the MPEG video encoder. To this end we introduce an analytical framework modeling an MPEG video source where a feedback law is used to provide rate adaptation, as is usual in MPEG encoders [2-3]. The proposed paradigm addresses any feedback law which takes into account both the state of a counter, used to keep track of the previous encoding history, and the activity level of the next frame to be encoded; by so doing the model can be applied whatever the feedback law is, provided that the parameter to be varied is the quantizer scale.

The paper is organized as follows. Section 2 describes the scenario being addressed. In Section 3, in order to derive the analytical model, we introduce a statistical analysis of MPEG traces aimed at characterizing both the activity process and the activity/emission relationships of the output flow of an MPEG encoder. Then, Section 4 defines the analytical framework modeling the adaptive-rate MPEG video source; to this end, first the model of the non-controlled MPEG source, defined as a switched batch Bernoulli process (SBBP) [4], is introduced, following an approach similar to the one proposed by the authors in [5-6]; then the adaptive-rate MPEG source is modeled as an *SBBP/SBBP/1/K* queueing system. Section 5 describes how the analytical framework can be used to evaluate the performance of an adaptive-rate MPEG source in terms of both the output-rate statistics, and the distortion introduced by the quantization mechanism. In Section 6 the above paradigm is used to investigate the effects of the bandwidth variation frequency on the above performance. Finally, the authors' conclusions are drawn in Section 7.

2   System Description

The system we will refer to in this paper is shown in Fig. 1. It is constituted by an adaptive-rate MPEG video source over an UDP/IP protocol suite. The adaptive-rate

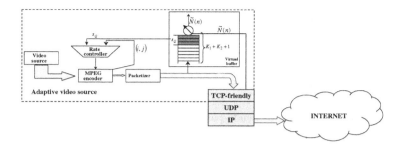

Fig. 1.: MPEG video encoder system

MPEG video source can be, for example, a TM5 MPEG video source [3]. In order to be "friendly" with TCP sources present in the network, a TCP-friendly layer is put between the UDP layer and the encoder. The TCP-friendly layer measures network delays and losses and provides the bandwidth indication to the adaptive-rate source. Many TCP-friendly protocols have been defined in literature (see for example [7-8]); they differ from each other for the method of measuring losses and delays, the measurement frequency, and the technique to choose the bandwidth to be indicated to the source.

Let us express the bandwidth indicated by the TCP-friendly layer to the adaptive-rate source in terms of packets/slot; given that this bandwidth changes in time due to network loss and delay variations, it is a stochastic process which we will indicate as $\tilde{N}(n)$. Therefore $\tilde{N}(n)$ indicates the number of packets the network accept for transmission at the slot n.

The adaptive-rate source is based on a video source whose output is MPEG encoded with a quantizer scale parameter (qsp) that varies according to the feedback provided by a rate controller, and then packetized according to the packet format imposed by the network. The rate controller works according to a feedback law with a given target. A possible target can be, for example, to encode the generic frame n with a number of packets equal to $\tilde{N}(n)$. In order to achieve its target, the rate controller monitors both the activity of the frame which is being encoded, its encoding mode, and the number of packets used to encode the previous frames. To obtain this last information, the rate controller uses a counter unit.

The *counter unit* in the rate controller is incremented at each slot by the number of packets emitted by the MPEG encoder at the UDP layer output, $\tilde{Y}(n)$, and decreased by the number of packets indicated by the TCP-friendly layer, $[-K_1, K_2]$. Thus the value of the counter unit at each slot represents the credit/debt the encoding system has with respect to the desired target; it therefore assumes positive values, that is, it registers a debit, when the encoder emitted too many packets with respect to the desired target, or negative values, that is, it registers a credit, when the encoder emitted too few packets with respect to the desired target. Too high positive or negative counter values allow very large windows to store the encoder emission history, but they may cause high output-rate burstiness; we therefore bound counter values in a range $\left[-K_1, K_2\right]$: when the counter takes values over K_2 or below $-K_1$ packets, it is truncated to K_2 or $-K_1$, respectively.

3   MPEG Statistical Properties

An MPEG encoder generates a pseudo-periodic emission process depending on the activity, the frame encoding mode (*I*, *P*, or *B*), and the *qsp* used to encode the current frame. This emission process can be described by the activity process and the activity/emission relationships. The activity process only depends on the peculiarities of the scene being encoded; the activity/emission relationships represent the number of bits resulting from the encoding of a picture characterized by a given activity, and therefore depend on both the frame encoding mode and the *qsp* used to encode each frame. Of course, the *qsp* determines the distortion introduced by the encoder; for this reason, in this section we also address the emission/distortion relationships, that is, the relationships binding the number of bits emitted and the distortion introduced for each encoding mode and for each *qsp*.

We analyzed the statistical characteristics of one hour of MPEG video sequences of the movie "The Silence of the Lambs" with the tool [9]. To encode this movie we used a frame rate of $F = 24$ frames/sec, and a frame size of $M = 180$ macroblocks. The GoP structure IBBPBB was used, selecting a ratio of total frames to intraframes of $G_I = 6$, and the distance between two successive *P*-frames or between the last *P*-frame in the GoP and the *I*-frame in the next GoP as $G_P = 3$.

Referring to this video sequence, in the next section we will first study its activity process (Section 3.1), then the activity/emission relationships (Section 3.2), and finally the emission/distortion relationships (Section 3.3).

3.1   Characterization of the Activity Process

According to the MPEG standard three elements are considered to encode a movie sequence: luminance, *Y*, chrominance *Cb*, and chrominance *Cr*. However, as luminance is the most relevant component characterizing the perceived quality, the activity of a video sequence is usually characterized using the luminance of each frame only [10]. The activity of the macroblock *p* in the frame *n* is defined as the variance of the luminance within this macroblock. So we can define the frame activity process, indicated here as the *activity process*, as the discrete-time process $L(n)$ obtained averaging the activities, $L_p(n)$, in the macroblocks within the frame *n*. In order to represent the activity process statistically, we measured its first- and second-order statistics in terms of the probability density function (pdf), $f_L(a)$, $\forall a \in \left[0, a_{MAX}^{(L)}\right]$, and normalized autocovariance function, $C_{LL}(m)$, respectively, where $a_{MAX}^{(L)}$ is the maximum value of the activity process As regards the first-order statistics, in [11] it was demonstrated that the Gamma function, defined as:

$$GAMMA_{\overline{m},\overline{\ell}}(a) = \frac{a^{\overline{m}-1}}{\Gamma(\overline{m})\,\overline{\ell}^{\overline{m}}} \cdot e^{-\frac{a}{\overline{\ell}}} \tag{1}$$

is a good approximation of the pdf, that is:

$$f_L(a) \cong GAMMA_{\overline{m},\overline{\ell}}(a) \tag{2}$$

The terms $\overline{m}$ and $\overline{\ell}$ in (1) and (2) are the so-called "shape parameter" and "scale parameter". If we indicate with μ_L and σ_L^2 the mean value and the variance of the activity process, $\overline{m}$ and $\overline{\ell}$ are defined as follows:

$$\overline{m} = \mu_L^2/\sigma_L^2 \qquad\qquad \overline{\ell} = \sigma_L^2/\mu_L \tag{3}$$

Therefore the mean value, μ_L, and the variance, σ_L^2, of the activity process are sufficient to describe the first-order statistics of this process. For the movie considered here we measured $\mu_L = 89.63$ bits/frame and $\sigma_L^2 = 2.3041 \cdot 10^4$ (bits/frame)^2.

As far as the second-order statistics are concerned, we have observed that the decreasing trend of the normalized autocovariance function curve can be approximated by a linear combination of exponential functions [4], that is:

$$C_{LL}(m) \cong \xi_{(\underline{\psi},\underline{\lambda})}(m) \equiv \sum_{w=1}^{W} \psi_w \cdot \lambda_w^m \qquad\qquad \forall m \in [1, m_{MAX}] \tag{4}$$

where m_{MAX} is the width of the interval in which we want to fit the measured normalized autocovariance function and W is the number of exponential terms needed to minimize the approximation error. From a great number of measures we have observed that two exponential terms are sufficient to achieve an acceptable error, and the use of more exponential terms produces no tangible improvement.

3.2   Characterization of the Activity/Emission Relationships

The MPEG emission process is modulated by the activity process described in the previous section. Moreover, while the activity process does not depend on the encoding mode and the *qsp* value used, the emission process does. In this section we will characterize the dependence of the emission process on the activity process when a fixed *qsp* value q is used. As an example, we used $q = 5$. Let us indicate the overall emission process which results from the encoding with a fixed *qsp* q as $X_q(n)$, and let us decompose it into G_I different emission processes, one for each frame in the GoP, $X_q^{(j)}(h)$, where $j \in J \equiv [1, G_I]$ indicates the frame position in the GoP, and h a generic GoP. Of course, we have $X_q^{(j)}(h) = X_q(h \cdot G_I + j)$. For example, in the case of $G_I = 6$ and $G_P = 3$, $X_q^{(j)}(h)$ will refer to an *I*-frame if $j \in J_I \equiv \{1\}$, a *P*-frame if $j \in J_P \equiv \{4\}$ and a *B*-frame if $j \in J_B \equiv \{2,3,5,6\}$. Given that frames of the same kind in the same GoP in any MPEG sequence have very close statistical parameters [5][12], in what follows, unless specified otherwise, we will only consider three emission processes, one for each kind of frame, $\breve{X}_q^{(\varsigma)}(m)$, for each $\varsigma \in \{I, P, B\}$. Likewise, we will consider three different activity processes, one for each frame process, and we will indicate them as $\breve{L}_q^{(\varsigma)}(m)$, for each $\varsigma \in \{I, P, B\}$. Moreover, in the following sections we will indicate the mean value of the I-, P- and B-frame emission processes as μ_I, μ_P and μ_B, and the variance of the same processes as σ_I^2, σ_P^2 and σ_B^2.

Now we can define the *activity/emission relationships* when a fixed *qsp* value q is used as the distribution of the sizes of I-, P- and B-frames once the activity a of the same frame is given:

$$y_q^{(\varsigma)}(r|a) = \lim_{m \to \infty} \mathrm{Prob}\left\{\breve{X}_q^{(\varsigma)}(m) = r, \forall j \in J_\varsigma \mid \breve{L}^{(\varsigma)}(m) = a\right\} \quad \begin{array}{l} \forall a \in \left[0, a_{MAX}^{(L)}\right] \\[4pt] \forall \varsigma \in \{I, P, B\} \end{array} \tag{5}$$

As demonstrated in [5][12], these functions can also be well approximated by Gamma distributions:

$$y_q^{(\varsigma)}(r|a) \cong GAMMA_{m_{\varsigma,a,q}, l_{\varsigma,a,q}}(r) \qquad\qquad \forall \varsigma \in \{I, P, B\} \tag{6}$$

where $m_{\varsigma,a,q}$ and $l_{\varsigma,a,q}$ can be derived as in (3) for each frame encoding mode ς.

So we can say that, for each activity a and for a given *qsp* value q, the activity/emission relationships can be exhaustively described by the mean value and variance of the I-, P- and B-frame emission processes, indicated here as $\mu_{(I|a),q}$ and $\sigma_{(I|a),q}^2$, $\mu_{(P|a),q}$ and $\sigma_{(P|a),q}^2$, and $\mu_{(B|a),q}$ and $\sigma_{(B|a),q}^2$, respectively.

Fig. 2 shows them as functions of the activity values. In this figure we can observe that the mean values present a quite linear and increasing law, while the variances follow a parabolic law. So, $\forall a \in \left[0, a_{MAX}^{(L)}\right]$ and $\forall \varsigma \in \{I, P, B\}$, we can approximate these functions by means of the following functions:

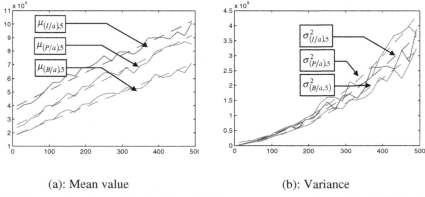

(a): Mean value   (b): Variance

Fig. 2: Mean and variance of the I-, P- and B-frame processes vs. the activity (solid line), compared with the best-fitting curves (dashed line).

$$\mu_{(\varsigma|a),q} \equiv c'_{(\varsigma|a),q} \cdot a + d'_{(\varsigma|a),q} \ ; \ \ \sigma^2_{(\varsigma|a),q} \equiv c''_{(\varsigma|a),q} \cdot a^2 + d''_{(\varsigma|a),q} \cdot a + e''_{(\varsigma|a),q} \tag{7}$$

Finally, we can state that the coefficients of these functions completely characterize the activity/emission relationships for a given qsp value q.

3.3 Characterization of the Emission/Distortion Relationships

The distortion introduced by quantization is one of the most important aspects of video encoding. When the qsp is varied during the encoding process to shape the encoder output rate, the quantization distortion is not constant: the higher the qsp values, the greater the quantization distortion. The distortion can be represented by the Peak Signal-to-Noise ratio (PSNR) process, defined as follows:

$$PSNR(n) = 10 \cdot \log_{10}\left[\frac{2^d - 1}{MSE(n)}\right] \tag{8}$$

where d is the number of bits assigned to a pixel, and $MSE(n)$ is the mean square error caused by quantization of the frame n. Fig. 3 shows the *rate-distortion curves*, $R^{(\varsigma)}(q)$, $F^{(\varsigma)}(q)$ and $V^{(\varsigma)}(\varphi)$ for each frame encoding mode $\varsigma \in \{I, P, B\}$, measured from the movie here considered. More specifically, Fig. 3a shows the rate curves, $R^{(\varsigma)}(q)$, defined as the average value of the emission process $X_q^{(\varsigma)}(n)$ versus the qsp value, q. Fig. 3b shows the distortion curves, $F^{(\varsigma)}(q)$, defined as the average value of the process $PSNR(n)$ versus the qsp value, q. Finally, Fig. 3c links the rate and distortion curves shown in Figs. 3a and 3b, by plotting the function $V^{(\varsigma)}(\varphi)$ representing the average number of bits emitted for each frame $\varsigma \in \{I, P, B\}$ in order to obtain a given average value, φ, of the process $PSNR(n)$. Any pair of these curves completely characterizes the distortion introduced by quantization.

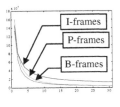
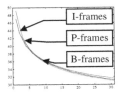
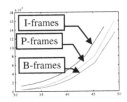

(a): Rate curves   (b): Distortion curves   (c): Rate vs. distortion curves

Fig. 3. Rate-distortion curves

4   System Model

The target of this section is to derive a discrete-time analytical model of an adaptive-rate MPEG video encoding system when its output is controlled with feedback from a TCP-friendly protocol. We will use $\Delta = 1/F$ as the slot duration, taken as being equal to the frame interval duration.

As a first step, in Section 4.2 we model the non-controlled MPEG encoder output as a switched batch Bernoulli process (SBBP) [4]. Then, the adaptive-rate MPEG video source is modeled as an *SBBP/SBBP/1/K* queueing system in Section 4.3, where $K = K_1 + K_2 + 1$ is the range of the counter state. For the sake of completeness, Section 4.1 provides a brief outline of SBBP processes, which will constitute the model of both the input and the output of the queueing system.

4.1   Switched Batch Bernoulli Process (SBBP)

An SBBP $Y(n)$ is a discrete-time emission process modulated by an underlying Markov chain. Each state of the Markov chain is characterized by an emission pdf: the SBBP emits data units according to the pdf of the current state of the underlying Markov chain. Therefore an SBBP $Y(n)$ is fully described by the state space $\Im^{(Y)}$ of the underlying Markov chain, the maximum number of data units the SBBP can emit in one slot, $r_{MAX}^{(Y)}$, and the set $(Q^{(Y)}, B^{(Y)})$, where $Q^{(Y)}$ is the transition probability matrix of the underlying Markov chain, while $B^{(Y)}$ is the emission probability matrix whose rows contain the emission pdf's for each state of the underlying Markov chain. If we indicate the state of the underlying Markov chain in the generic slot n as $S^{(Y)}(n)$, the generic elements of the matrices $Q^{(Y)}$ and $B^{(Y)}$ are defined as follows:

$$Q_{[s'_Y, s''_Y]}^{(Y)} = \lim_{n \to \infty} \left\{ S^{(Y)}(n+1) = s''_Y \middle| S^{(Y)}(n) = s'_Y \right\} \qquad \forall s'_Y, s''_Y \in \Im^{(Y)} \tag{9}$$

$$B_{[s_Y, r]}^{(Y)} = \lim_{n \to \infty} \left\{ Y(n) = r \middle| S^{(Y)}(n) = s_Y \right\} \qquad \forall s_Y \in \Im^{(Y)}, \ \forall s_Y \in \Im^{(Y)} \tag{10}$$

Below we will introduce an extension of the meaning of the SBBP to model not only a source emission process, but also a video sequence activity process, and a network bandwidth process. In the latter cases we will indicate them as an *activity SBBP* and a *bandwidth SBBP*, respectively, and their matrices $B^{(Y)}$ as the activity probability matrix and the bandwidth probability matrix, respectively.

4.2   Non-controlled Source Model

In this section we derive the SBBP process $\tilde{Y}_q(n)$, modeling the emission process of the non-controlled MPEG video source at the UDP layer output for a given *qsp*, $y_q^{(\varsigma)}(r|a)$.

The model has to capture two different components: the activity process behavior and the activity/emission relationships. More specifically, the transitions between the activity states, and between one frame and the successive one within a GoP, have to be modeled simultaneously. We will obtain the model of the non-controlled MPEG video source in three steps:

1. derivation of an activity SBBP, $G(n)$, modeling the activity process, $L(n)$;

2. derivation of the SBBP $Y_q(n)$, representing the whole measured MPEG emission process $X_q(n)$, first calculating its underlying Markov chain from the underlying Markov chain of the activity SBBP $G(n)$, and then the emission process from the activity/emission relationships defined in (5);

3. derivation of the SBBP $\tilde{Y}_q(n)$ at the packetizer output.

The desired activity SBBP, $G(n)$, has to fit the first- and second-order statistics, $f_L(r)$ and $C_{LL}(m)$, of the activity process $L(n)$. Let us indicate the state space of the underlying Markov chain of the activity process $G(n)$ as $\Im^{(G)}$. It represents the set of activity levels to be captured. For example, according to [13], we have $\Im^{(G)} = \{\text{Very low}, \text{Low}, \text{High}, \text{Very high}\}$. The activity SBBP $G(n)$ is defined by the parameter set $(Q^{(G)}, B^{(G)})$, where $Q^{(G)}$ is the transition matrix among the states in $\Im^{(G)}$, as it is customary, whereas $B^{(G)}$ does not represent an "emission probability matrix", but an "activity probability matrix". The matrix $B^{(G)}$ is defined as follows: when the underlying Markov chain of $G(n)$ is in the state representing the activity level i, the activity process takes values according to the probabilities in the ith row of the activity probability matrix $B^{(G)}$. The activity SBBP $G(n)$, defined by the parameter set $(Q^{(G)}, B^{(G)})$, can be obtained by solving the so-called *inverse eigenvalue problem* as in [5][14].

From the activity SBBP $G(n)$ modeling the activity process, we can derive the SBBP $Y_q(n)$ modeling the non-controlled MPEG source when the *qsp* q is used. To this end let us define the state of the underlying Markov chain of $Y_q(n)$ as a double variable, $S^{(Y)}(n) = (S^{(G)}(n), S^{(F)}(n))$, where $S^{(G)}(n) \in \Im^{(G)}$ is the state of the underlying Markov chain of $G(n)$, and $S^{(F)}(n) \in J$ is the frame position in the GoP at the slot n. Let us note that we have used $U = 548 \cdot 8$ instead of $S^{(Y_q)}(n)$ because the underlying Markov chain of $Y_q(n)$ is independent of q. For the same reason, below we will indicate its transition probability matrix as $Q^{(Y)}$ instead of $Q^{(Y_q)}$.

To calculate the above matrix, let us note that the admissible transitions are only between two states such that the frame at the time slot $n+1$, $S^{(F)}(n+1)$, is the one in the GoP following the frame $S^{(F)}(n)$. Thus the transition probability from the state $S^{(Y)}(n) = (i', j')$ to the state $S^{(Y)}(n+1) = (i'', j'')$, with i', $i'' \in \Im^{(G)}$, and j', $j'' \in J$, is:

$$Q^{(Y)}_{[(i',j'),(i'',j'')]} = \begin{cases} Q^{(G)}_{[i',i'']} & \text{if } (j'' = j'+1) \text{ or } (j' = G_I \text{ and } j'' = 1) \\ 0 & \text{otherwise} \end{cases} \tag{11}$$

As far as the emission probability matrix, $B^{(Y_q)}$, is concerned, its generic element depends on the frame encoding mode, the frame activity, and the *qsp* used. Once the *qsp* has been fixed, this matrix can be obtained from the activity/emission relationships. In fact, the probability of using r bits to encode the frame j when the state of the underlying Markov chain of the activity process is $i \in \mathfrak{T}^{(G)}$ and the *qsp* q is used, is:

$$B^{(Y_q)}_{[(i,j),r]} = \begin{cases} \sum\limits_{a=0}^{a^{(L)}_{MAX}} y^{(I)}_q(r|a) \cdot B^{(G)}_{[i,a]} & \text{if } j \in J_I \\ \sum\limits_{a=0}^{a^{(L)}_{MAX}} y^{(P)}_q(r|a) \cdot B^{(G)}_{[i,a]} & \text{if } j \in J_P \\ \sum\limits_{a=0}^{a^{(L)}_{MAX}} y^{(B)}_q(r|a) \cdot B^{(G)}_{[i,a]} & \text{if } j \in J_B \end{cases} \tag{12}$$

where $y^{(\varsigma)}_q(r|a)$, for each $\varsigma \in \{I, P, B\}$, is the function introduced in (6) to characterize the activity/emission relationships, while $B^{(G)}_{[i,a]}$ is the probability that the activity is $a \in [0, a^{(L)}_{MAX}]$ when the activity level is $i \in \mathfrak{T}^{(G)}$. The set $(Q^{(Y)}, B^{(Y_q)})$ defines the SBBP emission process modeling the output flow of the non-controlled MPEG encoder, when this uses a fixed *qsp* value q.

Finally, the last step is to obtain the MPEG source model at the UDP layer output. According to the UDP/IP protocol suite, let us denote the packet payload size available to the source to transmit information as $U = 548 \cdot 8$ bits. The emission process $Y_q(n)$, expressed in bits emitted per slot, can easily be transformed into an emission process $\tilde{Y}_q(n)$, expressed in UDP packets emitted per slot. In fact, its transition probability matrix and emission probability matrix can be derived as follows:

$$Q^{(\tilde{Y})}_{[s'_Y, s''_Y]} = Q^{(Y)}_{[s'_Y, s''_Y]} \qquad\qquad \forall s'_Y, s''_Y \in \mathfrak{T}^{(Y)} \tag{13}$$

$$B^{(\tilde{Y}_q)}_{[s_Y, \tilde{r}]} = \sum_{r = (\tilde{r}-1)U+1}^{\tilde{r} \cdot U} B^{(Y_q)}_{[s_Y, r]} \qquad\qquad \forall s_Y \in \mathfrak{T}^{(Y)}, \forall \tilde{r} \in [0, r^{(\tilde{Y})}_{MAX}] \tag{14}$$

In (14), $r^{(\tilde{Y})}_{MAX}$ is the maximum number of packets needed to transmit one frame and is given by $r^{(\tilde{Y})}_{MAX} = \lceil r^{(Y)}_{MAX}/U \rceil$ where $\lceil x \rceil$ is the smallest integer not less than x. In the following, for the sake of simplicity, we will indicate the number of emitted packets as r instead of $\tilde{r}$.

4.3   Adaptive Source Model

The adaptive source pursues a given output rate target by implementing a feedback law, $q = \phi(i, j, s_Q)$, in the rate controller, which calculates the qsp q to be used by the MPEG encoder for each frame. Here we will indicate the output emission process of the adaptive source, expressed in packets/slot, as $\tilde{Y}(n)$.

To model the adaptive source we have to consider the system shown in Fig. 1 as a whole, indicated here as Σ, in which the counter is incremented by the source emission process $\tilde{Y}(n)$, and decremented by the network bandwidth process $\tilde{N}(n)$. We model the counter unit with a discrete-time queueing system model with a dimension of $(K_1 + K_2 + 1)$, where $[-K_1, K_2]$ is the range of variation of the counter, as discussed in Section 2. Both the input and the output processes can be characterized as two SBBP processes, and the slot duration is the frame duration, $\Delta = 1/F$.

The *server capacity* of this queueing system, that is, the number of packets which leave the queue at each time slot, is a stochastic process which coincides with the network bandwidth process $\tilde{N}(n)$. Let us model this process with an SBBP process, called *bandwidth SBBP* in Section 3.1. Let it be characterized by the parameter set $(Q^{(\tilde{N})}, B^{(\tilde{N})})$, let $\mathfrak{I}^{(\tilde{N})}$ be the state space of its underlying Markov chain, and $d_{MAX}^{(\tilde{N})}$ be the maximum number of packets that can leave the buffer in one slot.

The bandwidth SBBP $\tilde{N}(n)$ can be equivalently characterized through the set of transition probability matrices, $C^{(\tilde{N})}(d)$, including the probability that the server capacity is of d packets. These matrices can be obtained by the parameter set $(Q^{(\tilde{N})}, B^{(\tilde{N})})$ as follows:

$$\left[C^{(\tilde{N})}(d) \right]_{[s'_{\tilde{N}}, s''_{\tilde{N}}]} \equiv \lim_{n \to \infty} \text{Prob} \left\{ \begin{matrix} \tilde{N}(n+1) = d, \\ S^{(\tilde{N})}(n+1) = s''_{\tilde{N}} \end{matrix} \middle| S^{(\tilde{N})}(n) = s'_{\tilde{N}} \right\} = Q^{(\tilde{N})}_{[s'_{\tilde{N}}, s''_{\tilde{N}}]} \cdot B^{(\tilde{N})}_{[s'_{\tilde{N}}, d]} \quad \forall d \in \left[0, d_{MAX}^{(\tilde{N})} \right] \tag{15}$$

To model the queueing system, we assume a *late arrival system with immediate access* time diagram [4]: packets arrive in batches, and a batch of packets can enter the service facility if it is free, with the possibility of them being ejected almost instantaneously. Note that in this model a packet service time is counted as the number of slot boundaries from the point of entry to the service facility up to the packet departure point. Therefore, even though we allow the arriving packet to be ejected almost instantaneously, its service time is counted as 1, not 0.

A complete description of Σ at the nth slot requires a three-dimensional state, $S^{(\Sigma)}(n) = \left(S^{(Q)}(n), S^{(\tilde{N})}(n), S^{(Y)}(n) \right)$, where:

- $S^{(Q)}(n) \in \left[-K_1, K_2 \right]$ is the virtual-buffer state in the nth slot, i.e. the number of packets in the queue and in the service facility at the observation instant;
- $S^{(\tilde{N})}(n)$ is the underlying Markov chain of the bandwidth SBBP $\tilde{N}(n)$;

- $S^{(Y)}(n)$ is the state of the underlying Markov chain of $\tilde{Y}(n)$, which coincides with that of $\tilde{Y}_q(n)$, $\forall q$, given that, as said in Section 4.2, it is independent of the qsp used.

According to the *late arrival system with immediate access* time diagram, let us note that, if we indicate the virtual-buffer state, the number of arrivals and the server capacity in the generic slot n as s'_Q, r and d, respectively, the virtual-buffer state in the generic slot $(n+1)$, s''_Q, can be obtained through the Lindley equation:

$$s''_Q = \max\left(\min\left(s'_Q + r, K_2\right) - d, -K_1\right) \tag{16}$$

Now, in order to derive the adaptive-source model, first let us characterize the queueing system input process, which is an SBBP whose emission probability matrix depends on the virtual-buffer state. To this end, we apply the algorithm introduced in Section 4.2 to obtain an SBBP model of the MPEG video source, $\tilde{Y}_q(n)$, for each qsp $q \in [1,31]$. So we have a parameter set $\left(Q^{(\tilde{Y})}, B^{(\tilde{Y}_1)}, B^{(\tilde{Y}_2)}, ..., B^{(\tilde{Y}_{31})}\right)$, which represents an SBBP whose transition matrix is $Q^{(\tilde{Y})}$, and whose emission process is characterized by a set of emission matrices, $\left\{B^{(\tilde{Y}_q)}\right\}_{q=1..31}$, consisting of one matrix for each qsp value q: at each time slot the emission of the source is therefore characterized by an emission probability matrix chosen according to the qsp value defined by the feedback law $q = \phi(i,j,s_Q)$.

More concisely, as we did in (15) for the bandwidth SBBP, we characterize the emission process of the adaptive-rate MPEG video source through the set of matrices $\left\{C^{(\tilde{Y}_q)}_{s'_Q}(r)\right\}_{q=1..31}$, $\forall r \in \left[0, r^{(\tilde{Y})}_{MAX}\right]$, each matrix representing the transition probability matrix including the probability of r packets being emitted when the buffer state is s'_Q and the qsp is q. So the generic element of the matrix $C^{(\tilde{Y}_{q'})}_{s'_Q}(r)$ can be obtained from the above parameter set, $\left(Q^{(\tilde{Y})}, B^{(\tilde{Y}_1)}, B^{(\tilde{Y}_2)}, ..., B^{(\tilde{Y}_{31})}\right)$, as a function of the qsp value q'':

$$\left[C^{(\tilde{Y}_{q'})}_{s'_Q}(r)\right]_{[(i',j'),(i'',j'')]} = Q^{(\tilde{Y})}_{[(i',j'),(i'',j'')]} \cdot B^{(\tilde{Y}_{q'})}_{[(i',j'),r]} \tag{17}$$

where $q'' = \phi(i'', j'', s'_Q)$ is the qsp chosen when the frame to be encoded is the j''-th in the GoP, its activity level is i'', and the virtual-buffer state before encoding this frame is s'_Q. Finally, if we indicate two generic states of the system as $s'_\Sigma = \left(s'_Q, s'_{\tilde{N}}, s'_Y\right)$ and $s''_\Sigma = \left(s''_Q, s''_{\tilde{N}}, s''_Y\right)$, the generic element of the transition matrix of the MPEG encoder system as a whole, $Q^{(\Sigma)}$, can be calculated, thanks to (16), as follows:

$$Q^{(\Sigma)}_{[(s'_Q,s'_{\tilde{N}},s'_Y),(s''_Q,s''_{\tilde{N}},s''_Y)]} = \sum_{d=0}^{d^{(\tilde{N})}_{MAX}} \sum_{r=0}^{r^{(\tilde{Y})}_{MAX}} \left(C^{(\tilde{N})}_{s'_Q}(d)\right)_{[(s'_Y,s'_{\tilde{N}}),(s''_Y,s''_{\tilde{N}})]} \cdot \left(C^{(\tilde{Y}_{q'})}_{s'_Q}(r)\right)_{[(s'_Y,s'_{\tilde{N}}),(s''_Y,s''_{\tilde{N}})]} \tag{18}$$

$$\underbrace{}_{\max\left(\min\left(s'_Q+r,K_2\right)-d,-K_1\right)=s''_Q}$$

Once the matrix $Q^{(\Sigma)}$ is known, we can calculate the steady-state probability array of the system Σ, $\underline{\pi}^{(\Sigma)}$, as the solution of the steady-state system.

5   Performance Evaluation

As said in Section 2, the adaptive source has the target of keeping the output traffic stream $\tilde{Y}(n)$ compliant with the bandwidth provided by the network. Therefore the main performance parameters are the output process statistics, which will be analytically derived in Section 5.1. Moreover, given the variability of the network bandwidth, other important parameters are the statistics of the distortion process, represented by the PSNR. These statistics will be obtained in Section 5.2.

5.1   Output process statistics

In this section we calculate the pdf, $f_{\tilde{Y}}(r)$, $\forall r \in \left[0, r^{(\tilde{Y})}_{MAX}\right]$, of the encoding system output process. It can be obtained as follows:

$$f_{\tilde{Y}}(r) \equiv \lim_{n\to\infty} \text{Prob}\left\{\tilde{Y}(n)=r\right\} = \lim_{n\to\infty} \sum_{s_Q=-K_1}^{K_2} \sum_{s_{\tilde{N}}\in\mathfrak{I}^{(\tilde{N})}} \sum_{(i,j)\in\mathfrak{I}^{(Y)}} \text{Prob}\left\{ \begin{matrix} \tilde{Y}(n)=r, S^{(Q)}(n)=s_Q, \\ S^{(\tilde{N})}(n)=s_{\tilde{N}}, \\ S^{(Y)}(n)=(i,j) \end{matrix} \right\} \tag{19}$$

Now, applying the theorem of total probability, and taking into account that the emission process in one slot does not depend either on the virtual-buffer state or on the bandwidth process in the same slot, we have:

$$f_{\tilde{Y}}(r) = \lim_{n\to\infty} \sum_{s_Q=-K_1}^{K_2} \sum_{s_{\tilde{N}}\in\mathfrak{I}^{(\tilde{N})}} \sum_{(i,j)\in\mathfrak{I}^{(Y)}} \text{Prob}\left\{\tilde{Y}(n)=r \left| \begin{matrix} S^{(Q)}(n)=s_Q, \\ S^{(\tilde{N})}(n)=s_{\tilde{N}}, \\ S^{(Y)}(n)=(i,j) \end{matrix} \right.\right\} \cdot \pi^{(\Sigma)}_{[(s_Q,s_{\tilde{N}},(i,j))]} = \tag{20}$$

$$= \sum_{s_Q=-K_1}^{K_2} \sum_{s_{\tilde{N}}\in\mathfrak{I}^{(\tilde{N})}} \sum_{(i,j)\in\mathfrak{I}^{(Y)}} B^{(\tilde{Y}_{\bar{q}})}_{[(i,j),r]} \cdot \pi^{(\Sigma)}_{[(s_Q,s_N,(i,j))]}$$

where $\bar{q} = \phi(i,j,s_Q)$. Finally, from the pdf $f_{\tilde{Y}}(r)$, the mean value and variance of the output process can also be easily derived.

5.2   Quantization Distortion

The target of this section is to evaluate both the static and time-variant statistics of the quantization distortion, represented here by the process $PSNR(n)$ defined in (8). To this end, let us quantize the distortion curves with a set of L different levels of distortion, $\{\varphi_1, \varphi_2, \ldots, \varphi_L\}$, each representing an interval of distortion values where the quality perceived by users can be considered as constant. As an example, for the movie "The Silence of the Lambs", from a subjective analysis obtained with 300 test subjects, the following $L = 5$ levels of distortion were envisaged: $\varphi_1 = [20.6, 34.2]$ dB, $\varphi_2 =]34.2, 35.0]$ dB, $\varphi_3 =]35.0, 36.2]$ dB, $\varphi_4 =]36.2, 38.4]$ dB, and $\varphi_5 =]38.4, 52.1]$ dB. From the distortion curves $F^{(\varsigma)}(q)$ introduced in Section 3.3, we can define the array $\gamma^{(\varsigma)}$ whose generic element, $\gamma_l^{(\varsigma)} = \{\forall q \text{ such that } F^{(\varsigma)}(q) \in \varphi_l\}$, for each $l \in [1, L]$, is the qsp range providing a distortion belonging to the lth level for a frame $\varsigma \in \{I, P, B\}$. Of course, by so doing we are assuming that a variation of q within the generic interval $\gamma_l^{(\varsigma)}$ does not cause any appreciable distortion.

From the distortion curves in Fig. 3b, we can calculate the following qsp ranges corresponding to the above distortion levels φ_l, for each $l \in [1, 5]$:

- for the I-frames: $\gamma^{(I)} = [[16, 31], [13, 15], [10, 12], [6, 9], [1, 5]]$;

- for the P-frames: $\gamma^{(P)} = [[15, 31], [13, 14], [10, 12], [6, 9], [1, 5]]$;

- for the B-frames: $\gamma^{(B)} = [[17, 31], [14, 16], [11, 13], [7, 10], [1, 6]]$.

As in the previous sections, let $q = \phi(i, j, s_Q)$ be the feedback law, linking the virtual-buffer state, $s_Q \in [-K_1, K_2]$, the activity level, $i \in \mathfrak{J}^{(G)}$, and the position of the frame in the GoP to be encoded, $j \in J \equiv [1, G_I]$, to the qsp to be used in order to fit the time-varying bandwidth. Moreover, let $\Theta_l^{(i,j)} = \{\forall s_Q \text{ such that } q = \phi(i, j, s_Q) \in \gamma_l^{(\varsigma)}\}$, for each i and j, be the range of values of the virtual-buffer state for which the rate controller chooses qsp values belonging to the level φ_l. It follows, by definition, that a variation of the virtual-buffer state within $\Theta_l^{(i,j)}$ does not cause any appreciable distortion variation.

So, we can now calculate the probability that the value of the process $PSNR(n)$ is in the generic interval φ_l, $\pi_{[l]}^{(PSNR)}$, and the pdf $f_{\delta_l}(m)$ of the stochastic variable δ_l, representing the duration of the time the process $PSNR(n)$ remains in the generic interval φ_l without interruption. They are defined as follows:

$$\pi_{[l]}^{(PSNR)} = \frac{1}{G_I} \cdot \sum_{j \in J} \pi_{[l]}^{(PSNR, j)} \qquad (21)$$

$$f_{\delta_i}(m) = \lim_{n \to \infty} \mathrm{Prob} \left\{ \begin{array}{l} PSNR(n+1) \in \varphi_i, \dots, \\ PSNR(n+m-1) \in \varphi_i, \\ PSNR(n+m) \notin \varphi_i \end{array} \middle| \begin{array}{l} PSNR(n-1) \notin \varphi_i, \\ PSNR(n) \in \varphi_i \end{array} \right\} \tag{22}$$

The term $\pi_{[l]}^{(PSNR,j)}$ in (21) is the probability that the value of the process $PSNR(n)$ is in the generic interval φ_i for the jth frame in the GoP. It can be calculated from the system steady-state probability array $\pi^{(\Sigma)}$ as follows:

$$\pi_{[l]}^{(PSNR,j)} = G_i \cdot \sum_{i \in \mathfrak{I}^{(G)}} \sum_{s_{\tilde{N}} \in \mathfrak{I}^{(\tilde{N})}} \sum_{s_Q \in \Theta_l^{(i,j)}} \pi_{[(s_Q, s_{\tilde{N}}, (i,j))]}^{(\Sigma)} \tag{23}$$

In order to calculate the pdf $f_{\delta_i}(m)$ in (22), let us indicate the matrix containing the one-slot state transition probabilities towards system states in which the distortion level is φ_i as $Q_{\to \varphi_i}^{(\Sigma)}$. It can be obtained from the transition probability matrix of the system, $Q^{(\Sigma)}$, as follows:

$$\left[Q_{\to \varphi_i}^{(\Sigma)} \right]_{[(s_Q', s_{\tilde{N}}', (i',j')),(s_Q'', s_{\tilde{N}}'', (i'',j''))]} = \begin{cases} Q_{[(s_Q', s_{\tilde{N}}', (i',j')),(s_Q'', s_{\tilde{N}}'', (i'',j''))]}^{(\Sigma)} & \text{if } s_Q'' \in \Theta_l^{(i',j')} \\ 0 & \text{otherwise} \end{cases} \tag{24}$$

Therefore, the pdf $f_{\delta_i}(m)$ can be calculated as the probability that the system Σ, starting from a distortion level φ_i, remains at the same level for $m-1$ consecutive slots, and leaves this level at the mth slot, that is [15]:

$$f_{\delta_i}(m) = \underline{\pi}^{(\Sigma 1,\varphi_i)} \cdot \left(Q_{\to \varphi_i}^{(\Sigma)} \right)^{m-1} \cdot Q_{\to (\neq \varphi_i)}^{(\Sigma)} \cdot \underline{1}^T \quad \text{where} \quad \underline{\pi}^{(\Sigma 1,\varphi_i)} = \frac{\underline{\pi}^{(\Sigma,\neq\varphi_i)} \cdot Q_{\to(\neq\varphi_i)}^{(\Sigma)}}{\underline{\pi}^{(\Sigma,\neq\varphi_i)} \cdot Q_{\to(\neq\varphi_i)}^{(\Sigma)} \cdot \underline{1}^T} \tag{25}$$

The array $\underline{\pi}^{(\Sigma 1,\varphi_i)}$ in (25) is the steady-state probability array in the first slot of a period in which the distortion level is φ_i. Instead, the array $\underline{\pi}^{(\Sigma,\neq\varphi_i)}$ is the steady-state probability array in a generic slot in which the distortion level is other than φ_i:

$$\underline{\pi}^{(\Sigma,\neq\varphi_i)} = \frac{\underline{\pi}^{(\Sigma)} \cdot Q_{\to\varphi_i}^{(\Sigma)}}{\underline{\pi}^{(\Sigma)} \cdot Q_{\to\varphi_i}^{(\Sigma)} \cdot \underline{1}^T} \tag{26}$$

At this point we have calculated the pdf of the random variable δ_i. Now its mean value can be obtained as follows:

$$E\{\delta_i\} = \sum_{m=0}^{\infty} m \cdot f_{\delta_i}(m) = \underline{\pi}^{(\Sigma 1,\varphi_i)} \cdot \left[I - Q_{\to\varphi_i}^{(\Sigma)} \right]^{-2} \cdot Q_{\to(\neq\varphi_i)}^{(\Sigma)} \cdot \underline{1}^T \tag{27}$$

where I is the identity matrix.

6   Numerical Results

Let us now apply the analytical framework proposed in the previous sections to numerically evaluate the performance of the MPEG adaptive source when the output rate is controlled by the TCP-friendly layer. To this end we address the encoding of the movie "The Silence of the Lambs", which was studied in Section 3, and we use the same feedback law used by the TM-5 standard, that is, with the target of obtaining a constant number of packets per GoP, leaving the number of packets for encoding each frame in the GoP free in order to pursue a constant distortion level within the GoP. At the first slot of the generic GoP h, for each h, the TCP-friendly algorithm calculates the number of packets to be used in that GoP, $T_G(h)$, and holds the bandwidth process value constant for all the frames of that GoP to the value $\tilde{N}(n) = T_G(h)/G_I$, for each $n \in [h \cdot G_I + 1, (h+1) \cdot G_I]$.

The qsp is chosen as follows:

$$q = \phi(i, j, s_Q) \equiv \left[\begin{array}{l} \text{minumum } \overline{q} \text{ such that :} \\ \exists [q_{j+1}, \ldots, q_{G_I}] \text{such that :} \left\{ \begin{array}{l} PSNR_k = PSNR_j \quad \forall k \in [j+1, G_I] \\ R_{i,j}(\overline{q}) + \sum_{k=j+1}^{G_I} R_{i,k}(q_k) + s_Q \le \Omega_j \end{array} \right. \end{array} \right] \tag{28}$$

where:

- $PSNR_j$ is the PSNR of the jth frame in the GoP if it is encoded using the qsp $\overline{q}$;

- $R_{i,j}(\overline{q})$ is the expected number of packets used to encode the jth frame in the GoP, if the qsp used is $\overline{q}$ and the activity state is i. It can be calculated as follows:

$$R_{i,j}(q) = \sum_{a=0}^{a_{MAX}^{(L)}} \mu_{(\varsigma|a),q} \cdot B_{[i,a]}^{(G)} \text{ where } \varsigma = \{I \text{ if } j \in J_I; I \text{ if } j \in J_I; I \text{ if } j \in J_I\} \tag{29}$$

- Ω_j is the number of packets still available to encode the jth and the remaining $G_I - j$ frames of the GoP. These packets will be distributed to maintain the same distortion level in the GoP. Given that, by this law, the counter at each slot n is decreased by $\tilde{N}(n)$, and incremented by the number of packets actually used by the encoder in the same interval, if we indicate the mean value of the bandwidth process $\tilde{N}(n)$ as $E\{\tilde{N}(n)\}$, we have:

$$E\{\tilde{N}(n)\} = \sum_{s_{\tilde{N}} \in \mathfrak{I}^{(\tilde{N})}} \sum_{d=0}^{d_{MAX}^{(\tilde{N})}} d \cdot B_{[s_{\tilde{N}}, d]}^{(\tilde{N})} \tag{30}$$

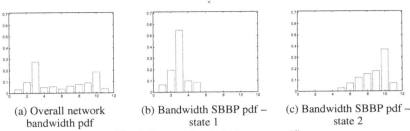

(a) Overall network bandwidth pdf

(b) Bandwidth SBBP pdf – state 1

(c) Bandwidth SBBP pdf – state 2

Fig. 4. Network bandwidth process pdf's

$$\Omega_j = G_I \cdot E\{\widetilde{N}(n)\} - s_Q - (j-1) \cdot E\{\widetilde{N}(n)\} \tag{31}$$

Thanks to the counter, the above feedback law considers the packet debit/credit resulting from both the previous GoP and the previous frames in the same GoP.

For the TCP-friendly protocol we used the TFRC protocol [7]. It is an equation-based protocol which calculates the sending rate, expressed in bytes/sec, through the TCP response function. Implementing the TFRC protocol in the TCP-friendly layer and applying it to a "greedy" source (i.e. a source which always has something to transmit) in the intranet of the Catania University Campus, the resulting output process pdf is shown in Fig. 4a. In order to apply the framework proposed in the paper, we need an SBBP model of the output bandwidth. Given that it is beyond the scope of this paper, we assume a two-state model with the same steady-state probability for both states, for the bandwidth process at the beginning of each GoP. By varying the mean state durations, this will allow us to analyze how the bandwidth variation frequency influences adaptive-rate source performance. The SBBP model we use for the bandwidth process at the beginning of each GoP is the following:

$$Q^{(GoP)} = \begin{bmatrix} (1-\Delta/T) & \Delta/T \\ \Delta/T & (1-\Delta/T) \end{bmatrix} \tag{32}$$

$$B^{(GoP)} = \begin{bmatrix} 0.067 & 0.195 & 0.55 & 0.1 & 0.878 & 0 & 0 & 0 & 0 & 0 & 0 \\ 0 & 0 & 0 & 0 & 0.03 & 0.07 & 0.123 & 0.154 & 0.18 & 0.37 & 0.073 \end{bmatrix} \tag{33}$$

where the value of Δ/T in (32) is the bandwidth changing frequency. We will analyze 8 cases, each featuring a different frequency, $T = [1,2,3,4,5,6,7,8]$ GoP's. The rows of $B^{(\widetilde{N})}$ are the pdf's of each state of the bandwidth SBBP at the beginning of each slot. They are shown in Figs. 4b and 4c. Of course, their combination gives the bandwidth pdf shown in Fig. 4a. The mean values of the pdf of the two states are 1.95 and 7.79 packets/slot, while the overall mean value of the available network bandwidth is 4.87 packets/slot.

Finally, the parameter set of the bandwidth SBBP can be found as follows:

$$Q^{(\tilde{N})} = Q^{(GoP)} \otimes Q^{(CIRC,G_I)} \quad \text{and} \quad B^{(\tilde{N})}_{[\tilde{s}_N,d]} = \begin{bmatrix} B^{(GoP)}_{[1,:]} \cdot \underline{1}^T_{G_I} \\ B^{(GoP)}_{[2,:]} \cdot \underline{1}^T_{G_I} \end{bmatrix} \quad \text{where:} \tag{34}$$

- $Q^{(CIRC,G_I)}$ is the transition probability matrix for a circulant Markov chain [14], and has been introduced to take into account the fact that the bandwidth process remains constant for the whole duration of the GoP; its generic element is defined as follows:

$$Q^{(CIRC,G_I)}_{[j',j'']} = \begin{cases} 1 & \text{if } (j'' = j' + 1) \text{ or } (j' = G_I \text{ and } j'' = 1) \\ 0 & \text{otherwise} \end{cases} \tag{35}$$

- $\underline{1}^T_{G_I}$ is the column array with G_I elements are all equal to 1.

The choice of the matrices $Q^{(\tilde{N})}$ and $B^{(\tilde{N})}$ implies that the mean network bandwidth alternates between the values 1.95 and 7.79 packets/slot, and the time interval with a constant mean network bandwidth has a mean duration of Δ/T.

In Fig. 5 we have shown the mean value of both the output rate and the PSNR. We can observe that neither the mean value of the output process nor the mean PSNR are influenced by bandwidth variations. More specifically, the mean value of the output process is equal to the mean value of the available bandwidth, i.e. 4.87 packets/slot. This means that the feedback law we are using is able to follow bandwidth variations, thanks to the memory given by the counter.

Instead, the statistics which are influenced by a change in bandwidth variation frequency regard the PSNR duration. In order to analyze this, Fig. 6 shows the mean PSNR duration of each PSNR level φ_l, for each $l \in [1,5]$, introduced in Section 5.2, and any PSNR level. From this figure we can observe that, as expected, the mean duration of all the levels increases when the network bandwidth becomes more stable. Nevertheless, the best and the worst levels, φ_5 and φ_1 respectively, present the highest mean duration values, while the intermediate levels, φ_2 and φ_3, are transient levels and are less influenced by bandwidth variation frequency changes. Finally, from Fig. 6b we can obtain the maximum frequency beyond which the image quality become unstable, due to frequent changes in the PSNR level.

7   Conclusions

The co-existence of UDP-based applications with TCP-based applications constitutes a key issue for the Internet of the near future. The problem is to enhance UDP-based multimedia applications with some kind of congestion control, in order to make them behave like "good network citizen" at times of bandwidth scarcity. In this paper we have investigated the effects of the bandwidth variation rate during bandwidth profile variations on both the error of the rate controller in fitting the bandwidth profile, and the distortion introduced by the quantization mechanism of the MPEG video encoder.

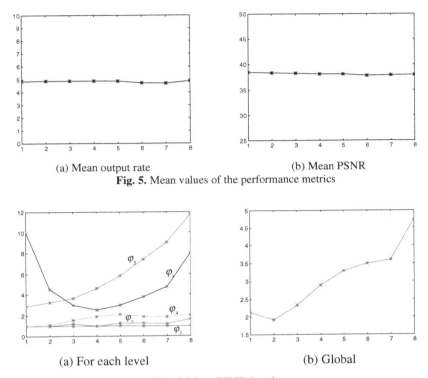

(a) Mean output rate   (b) Mean PSNR

Fig. 5. Mean values of the performance metrics

(a) For each level   (b) Global

Fig. 6. Mean PSNR duration

To this end we have introduced an *SBBP/SBBP/1/K* queueing system modeling an MPEG video source where a feedback law is used to provide rate adaptation. The proposed paradigm addresses any feedback law which takes into account both the state of a counter, used to keep track of the previous encoding history, and the activity level of the next frame to be encoded; by so doing the model can be applied whatever the feedback law is, provided that the parameter to be varied is the quantizer scale. Finally we have applied the proposed analytical framework to a real case, i.e. transmission of the movie "The Silence of the Lambs" on a TCP-friendly UDP connection. We have demonstrated that, thanks to the memory provided by the counter unit to the rate-controlled source, the output flow presents the same mean output rate as the network bandwidth. However, when the network bandwidth variation frequency is high, the mean duration of each PSNR level is so short that perceived quality is unacceptable due to instability.

References

[1] S. Floyd, K. Fall. "Promoting the Use of End-to-end Congestion Control in the Internet". IEEE/ACM Transactions on Networking, August 1999.
[2] W. Luo, M. El Zarki, "Quality Control for VBR Video over ATM Networks," *IEEE Journal on Selected Areas in Communications*, vol. 15, no. 6, August 1997.
[3] *Test Model 5*, ISO-IEC/JTC1/SC29/WG11, April 1993.
[4] A. La Corte, A. Lombardo, G. Schembra, "An Analytical Paradigm to Calculate Multiplexer Performance in an ATM Multimedia Environment," *Computer Networks and ISDN Systems*, vol. 29, no. 16, December 1997.
[5] A. Lombardo, G. Morabito, G. Schembra, "An Accurate and Treatable Markov Model of MPEG-Video Traffic," *Proc. IEEE Infocom '98*, San Francisco, USA, April 1998.
[6] A. Lombardo, G. Morabito, G. Schembra, "Traffic specifications for the Transmission of Stored MPEG Video on the Internet," *IEEE Transactions on Multimedia*, Vol. 3, No. 1, March 2001.
[7] S. Floyd, M. Handley, J. Padhye, and J. Widmer, "Equation-Based Congestion Control for Unicast Applications," February 2000. Available at: http://www.aciri.org/tfrc/.
[8] R. Rejaie, M. Handley, D. Estrin, "RAP: An End-to-end Rate-based Congestion Control Mechanism for Realtime Streams in the Internet," *Proc. IEEE INFOCOM'99*, 21-25 March 1999, New York, NY, USA.
[9] A. Cernuto, F. Cocimano, A. Lombardo, G. Schembra, MPEGViA v.01, available at "www.diit.unict.it/users/schembra/Software/MPEGViA/MPEGViA_v01.zip".
[10] J. L. Mitchell, W.*B.* Pennebaker, "MPEG video compression standard," Chapman & Hall, International Thomson Publishing, 1996.
[11] O. Rose, "Statistical Properties of MPEG Video Traffic and their Impact on Traffic Modeling in ATM Systems," University of Würzburg, Institute of Computer Science, Technical Report No. 101, Feb. 1995.
[12] A. Lombardo, G. Morabito, S. Palazzo, G. Schembra, "A Markov-Based Algorithm for the Generation of MPEG Sequences Matching Intra- and Inter-GoP Correlation," *European Transactions on Telecommunications journal*, Vol. 12, No. 2, March/April 2001.
[13] A. Puri, R. Aravind, "Motion-compensated video coding with adaptive perceptual quantization," *IEEE Transactions on Circuits and Systems for Video Technology*, vol. 1, no. 4, December 1991.
[14] S. -qi Li, S. Chong, C. Huang, "Link capacity allocation and Network Control by Filtered Input Rate in High Speed Networks," *IEEE/ACM Transactions on Networking*, vol. 3, no. 1, February 1995.
[15] M. F. Neutz, "Matrix-Geometric Solutions in Stochastic Models: an Algorithmic Approach," Johns Hopkins University Press, Baltimore, MD, USA, 1981.

Design Issues for Layered Quality-Adaptive Internet Video Playback

Reza Rejaie[1] and Amy Reibman[2]

[1] AT&T Labs- Research, Menlo Park, CA.
`reza@research.att.com`
[2] AT&T Labs- Research, Florham Park, NJ.
`amy@research.att.com`

Abstract. The design of efficient unicast Internet video playback applications requires proper integration of encoding techniques with transport mechanisms. Because of the mutual dependency between the encoding technique and the transport mechanism, design of such applications has proven to be a challenging problem. This paper presents an architecture which allows the joint design of a *transport-aware* video encoder with an *encoding-aware* transport. We argue that layered encoding provides maximum flexibility for efficient transport of video streams over the Internet. We describe how off-line layered encoding techniques can achieve robustness against imprecise knowledge about channel behavior (*i.e.*, bandwidth and loss rate) while maximizing efficiency for a given transport mechanism. Then, we present our prototyped client-server architecture, and describe key components of the transport mechanism and their design issues. Finally, we describe how encoding-specific information is utilized by transport mechanisms for efficient delivery of stored layered video despite variations in channel behavior.

1 Introduction

The design of efficient unicast Internet video playback applications requires proper integration of encoding techniques with transport mechanisms. Because of the mutual dependency between the encoding technique and the transport mechanism, design of such applications has proven to be a challenging problem.

Encoding techniques typically assume specific channel behavior (*i.e.*, loss rate and bandwidth) and then the encoder is designed to maximize compression efficiency for expected channel bandwidth while including a sufficient amount of redundancy to cope with the expected loss rate. If channel behavior diverges from expected behavior, quality of delivered stream would be lower than expected. The shared nature of Internet resources implies that behavior of Internet connections could substantially vary with unpredictable changes in co-existing traffic during the course of a session. This requires all Internet transport mechanisms to incorporate some type of *congestion control* mechanism (*e.g.*, [1], [2]). Thus to pipeline a pre-encoded stream through a congestion controlled connection, video playback applications should be able to efficiently operate over the

S. Palazzo (Ed.): IWDC 2001, LNCS 2170, pp. 433–451, 2001.
© Springer-Verlag Berlin Heidelberg 2001

expected range of connection behaviors. This means that video playback applications should be both *quality adaptive* to cope with long-term variations in bandwidth and *loss resilient* to be robust against range of potential loss rates.

In this paper, we argue that layered encoding is the most promising approach to Internet video playback. Layered encoders structure video data into layers based on importance, such that lower layers are more important than higher layers. This structured representation allows transport mechanisms to accommodate variations in both available bandwidth and anticipated loss rate, thus enabling simpler joint designs of encoder and transport mechanisms. First, transport mechanisms can easily match the compressed rate with the average available bandwidth by adjusting delivered quality. Second, transport mechanisms can repair missing pieces of different layers in a prioritized fashion over different time scales. Over short timescales (*e.g.*, a few round trip times (RTT)), lost packets from one layer can be recovered before any lost packet of higher layers. This allows transport mechanisms to control the observed loss rate for each layer. This is crucial because neither total channel loss rate nor its distribution across layers are known a priori and can be expected to change during transmission. Over long timescales (*e.g.*, minutes), the server can *patch* quality of the delivered stream by transmitting pieces of a layer that were not delivered during the initial transmission. Over even longer timescales, the server can send extra layers, to improve quality of a previously-transmitted stream without being constrained by the available bandwidth between client and server. This allows adjustment of quality for a cached stream at a proxy[3].

We have prototyped a client-server architecture (Figure 1) for playback of layered video over the Internet. The congestion control (CC) module determines available bandwidth (BW) based on the network feedback (*e.g.*, client's acknowledgment). The Loss recovery (LR) module utilizes a portion of available bandwidth (BW_{lr}) to repair some recent packet losses such that the observed loss rate remains below the tolerable rate for the given encoding scheme. The server only observes the remaining losses that have not been repaired, and uses the remaining portion of bandwidth (BW_{qa}) to perform Quality adaptation (QA)[4] by adjusting delivered quality (*i.e.*, number of transmitting layers). The QA and LR mechanisms each depend on parameters of the specific encoding and are tightly coupled. The collective performance of the QA and LR mechanisms determine the perceived quality of the video playback. A key component of the architecture is a Bandwidth Allocator (BA) that divides total available bandwidth between the QA and LR modules using information that depends on the specific encoding and on client status.

This paper describes our ongoing work to integrate transport-aware encoding with encoding-aware transport for Internet video playback. We consider a coupled design, in which the encoder and transport are each designed given knowledge of the expected behavior of the other. For transport-aware encoding, we present the main design choices and trade-offs for layered encoding, and describe how the encoding schemes can be customized based on available knowledge regarding employed transport mechanism and regarding expected channel

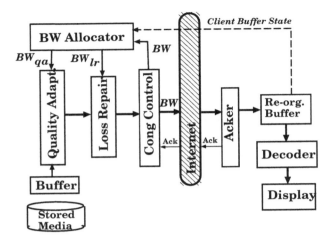

Fig. 1. Client-server Architecture

rate and loss rate. For the design of encoding-aware transport mechanisms, we focus on design strategies for the BA, QA and LR modules that are directly affected by details of the deployed layered encoding scheme.

The rest of this paper is organized as follows. In section 2, we review some of the related work. Because of the natural ordering between encoding and transport, we consider transport-aware encoding using a layered video encoder in section 3. Then in section 4, we address various components of an encoding-aware transport mechanism and examine their design tradeoff. This includes Bandwidth allocation (section 4.1), Loss Recovery (section 4.2) and Quality adaptation (section 4.3). Section 5 concludes the paper and addresses some of our future plans.

2   Related Work

Traditionally, video encoders have been designed for transport over fixed-rate channels (with fixed and known channel bandwidth) with few if any losses. This creates a bit stream that will have poor quality if transported over a network with different bandwidth or loss rate. Either a higher loss rate or a lower bandwidth would produce potentially significant visual artifacts that propagate with time. For transport over networks, the encoder design should change to be cognizant of the fact that the bit stream may have to deal with varying channel bandwidth and non-zero loss rate. Over the years, several classes of solutions have been proposed for encoding and transmitting video streams over the network as follows:

- *One-layer Encoding*: In one-layer video encodings (*e.g.*, MPEG-1 and MPEG-2 Main Profile), the trade-off between compression and resilience to errors is achieved by judiciously including Intra-blocks, which do not use temporal prediction. The choice of which blocks to code as I-blocks for a given sequence can be optimized if channel loss rate is known a priori [5].

Transport mechanisms can not gracefully adjust quality of a one-layer pre-encoded stream to available channel bandwidth. A common solution is to deploy an *Encoding-specific packet dropping algorithm* [6]. In these algorithms, the server discards packets that contain lower-priority information (*e.g.*, drops B frames of MPEG streams) to match transmission rate with channel bandwidth. The range of channel rates over which these algorithms are useful is usually limited, and the delivered quality may be noticeably degraded. Both these effects are content and encoding specific.

– *Multiple description coders*: Another approach is to use a multiple description (MD) video encoder [7]. MD coders are typically more robust to uncertainties in the channel loss rate at the time of encoding. However, they still require similar network support to adapt to varying channel bandwidth.

– *Multiple Encodings*: One alternative to cope with variations in channel behavior is to maintain a few versions of each stream, each encoded for different network conditions. Then the server can switch between different encodings in response to changes in network conditions. Limitations of this approach are the inability to improve quality of an already-transmitted portion of the stream, and the inability of such a system to quickly respond to sharp decreases in available bandwidth.

– *Layered Encodings*: Hierarchical encoding organizes compressed media in a layered fashion based on its importance *i.e.*, layer i is more important than all higher layers and less important than all lower layers. If the lower more important layers are received, a base quality video can be displayed, and if the higher less important layers are also received, better quality video can be displayed. The layered structure of encoded stream allows the server to effectively cope with uncertain channel behavior.

In summary, layered encoding has three advantages: First, layered video allows easy and *effective rate-matching*. The server can match the bandwidth of delivered stream with the available network bandwidth by changing the number of layers that are transmitted. This relaxes the need for knowing exact channel bandwidth at the time of encoding, thus it helps to decouple transport and encoding design.

Second, layered video allows *unequal error protection* to be applied during transport, with stronger error correction applied to the more important layers. This can be used effectively even if the expected loss rate is not known at the time of compression. Thus, even though one-layer video and the most important layer of layered video are equally susceptible to losses, discrepancies between the actual loss rate and the one assumed at the time of encoding can be accommodated by the transport mechanism.

Third, layered encoding allows the server to improve quality of an already-transmitted portion of the stream. We call this *quality patching*. In essence, layered structure allows the server to deliver different portions of various layers in any arbitrary order, *i.e.*, reshape the stream for delivery through uncertain channel. Thus quality of the delivered stream can be smoothed out over various timescales. Figure 2 illustrates the flexibilities of layered encoding to reshape the stream for transmission through the network. The server adjusts the number of layers when there are long-term changes in available bandwidth. Total loss rate is randomly distributed across the active layers. However, the server can prioritize loss recovery by retransmitting losses of layer i before layer j for any $i < j$. When extra bandwidth becomes available at time t_3, the server can either add the fourth layer or alternatively transmit five missing segments of the layer 2 (between t_1 and t_2). This shows how a layered encoded stream can be reshaped for delivery through the network.

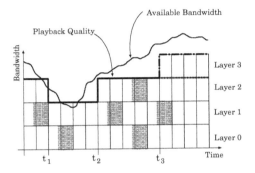

Fig. 2. Flexibility of Layered Encoding

3   Transport-Aware Layered Encoding

In this section, we consider how best to create a video bit stream to be stored at the media server in Figure 1. We argued earlier that a layered (or in this paper, equivalently a scalable) encoder will produce a better bit stream than a one-layer encoder if the values of channel bandwidth and loss rate are not known at the time of encoding. Despite these advantages inherent in layered video, good system performance requires careful design. We begin with some background on layered video encoders, and then show how incorporating at the time of encoding as much information as possible about the channel bandwidth and loss rate can generate the best bit stream for storage. We also discuss the information that needs to be shared between encoder and transport to improve quality of delivered video.

3.1   Layered Coding Background

There are two basic methods to create a layered video bit stream: 3-dimensional wavelets [8,9] and adding layering to a traditional DCT-based video coder with inter-frame temporal prediction [10,11]. The former has drawbacks of poor compression performance in sequences with complex motions, and poor video quality when temporal sub-bands are lost due to motion blurring. For these reasons, we focus on the family of layered DCT-based temporally-predictive video coders in this paper.

Among this family of coders, there has been much research and several standardized coders. In these coders, P-frames are predicted from previous frames and the DCT coefficients are coded in a layered fashion. Several important aspects of these coders are the following:

- how the DCT coefficients are partitioned between each of the layers (*i.e.*, is partitioning done in the frequency domain, the spatial domain, or the quality or SNR domain),
- whether enhancement-layer information uses temporal prediction at all, and if so whether lower-layer base information is used to predict the higher-layer enhancement information,
- how much bandwidth is allocated for each layer,
- how much of that bandwidth is redundancy in each layer,
- and how many layers are created.

These decisions all affect the compression efficiency of the layered coder, and also affect the robustness against packet loss. Thus, knowledge of the expected bandwidth and loss rates at the time of transport can have a large impact on the best choice of codec design.

While most standardized layered encoders produce two or at most a few layers [11,10], recent interest in having many layers has lead to a standard for MPEG-4 Finely Granular Scalability (FGS) [12]. The bit stream structure of MPEG-4 FGS makes it highly robust and flexible to changes in available bandwidth. However, such robustness comes with a significant penalty to the efficiency of the compression and hence to the video quality as the rate increases. More recent, not yet standardized approaches to layered encoding markedly improve the compression efficiency without too much sacrifice to the robustness [13].

The design of a layered encoder is based on underlying assumptions regarding the nature of the transport and channel. Specifically, a layered coder assumes that the transport mechanism will initially attempt to send the more important information in the available bandwidth, and that the loss recovery will be applied to the more important parts before the less important parts. This implicitly requires buffering at the client and in the transport. However, the exact bandwidth and loss rate experienced at the time of transport is typically not known at the time of encoding, and further assumptions must be made. In general, these assumptions have been implicit in the design of the layered video encoder. Here we make them explicit. We consider first the bandwidth, then consider the loss rate.

3.2 Incorporating Bandwidth Knowledge

The more knowledge available at the time of encoding regarding the expected range of operating bandwidths, the better. We focus on how knowledge of the available bandwidth impacts the prediction strategy of the encoder. If the available bandwidth is known to vary between R_{min} and R_{max}, three different prediction strategies are useful depending on the expected behavior of the bandwidth within this range.

First, consider the case where the bandwidth is nearly always close to R_{max}. Then the best design would be to rely heavily on prediction for the enhancement layers so as to improve compression efficiency. (This is essentially the strategy of a one-layer encoder.) Such a design will suffer small degradation when bandwidth dips slightly below R_{max}, but in such an environment the probability of larger degradations is small.

Second, consider the case where the bandwidth is usually near R_{min} although we'd like better quality when the rate is higher. Then the best design would be to use temporal prediction only for the base layer with rate R_{min}, and to use no temporal prediction for the higher layers. (This is the strategy used by MPEG-4 FGS.) Such a design is very robust to variations in bandwidth in $[R_{min}, R_{max}]$, but is also not very efficient at rates near R_{max}.

Third, suppose we have little knowledge of the bandwidth other than it lies in the range $[R_{min}, R_{max}]$. In this case, the best algorithm is to judiciously choose the prediction strategy for each macro block in the video, so as to balance both compression efficiency (at rates near R_{max}) and robustness (at rates near R_{min}). Figure 3 illustrates these concepts for the sequence *Hall monitor* using the scalable video coder in [13]. This coder has the flexibility of using three different methods of prediction for the different layers, which allows it to mimic both a one-layer video coder and the MPEG-4 FGS video coder through an appropriate restriction of the prediction methods. The results in Figure 3 are obtained by creating a single bit stream for each illustrated curve, and successively discarding enhancement-layer bit-planes and decoding the remainder. The x-axis shows the decoded bit-rate, and the y-axis shows the PSNR of the resulting decoder reconstruction as bit-planes are discarded. Also shown is the performance of the one-loop encoder with no loss (top dotted line). This provides an upper bound on the performance of the scalable coders.

In this figure, the curve labeled "FGS" uses the prediction strategy of the MPEG-4 FGS coder which is optimal if the available bandwidth is usually near R_{min}. The curve labeled "one-layer with loss" uses a one-layer prediction strategy, which is optimal if the available bandwidth is usually near R_{max}. The curve labeled "drift-controlled" is our coder [13] optimized to provide good performance across the range of rates.

The FGS coder performs poorly at the higher rates. The one-layer decoder with drift suffers a 2.6-4.3 dB degradation at the lowest bit-rate, compared to the drift-free FGS decoder. Relative to the FGS coder, our proposed coder suffers about 1.3-1.4 dB performance degradation at the lowest bit-rate, but significantly outperforms it elsewhere. Our coder loses some efficiency at the highest rates compared to the one-layer coder, but has noticeably less drift as bit-planes are discarded.

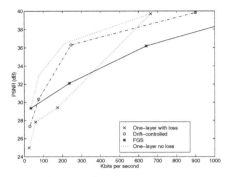

Fig. 3. Effect of Bandwidth on PSNR

Table 1 shows the PSNR averaged across different channel rates, assuming a uniform distribution of rates between the smallest and the largest rate of the one-loop encoder. The coder optimized for the range of channel rates outperforms the other coders by 0.8-2.1 dB when there is only one I-frame.

Table 1. PSNR assuming uniform distribution.

seq.	One-loop	Proposed	FGS	upper
Hall	33.14	35.25	33.11	36.77
Fore	33.33	34.13	33.09	35.03

3.3 Incorporating Loss Rate Knowledge

Next, we consider the effect of incorporating knowledge of the expected loss rate into the video encoder design. We distinguish loss rate from lowered bandwidth because losses will randomly affect the entire frame for a given layer, while lowered bandwidth can be targeted specifically toward less important parts of the bit stream within a frame for a given layer.

In the Figure 3, we showed how the choice of prediction structure used by the layered encoder is influenced by the expected bandwidth. Similarly, the expected loss rate affects the best prediction structure. Figure 4 shows the performance of a two-layer H.263 encoder under random losses in each layer. Two different prediction strategies are used for the enhancement layer. The base layer is compressed with 64 Kbps, and each enhancement layer is compressed with 128 Kbps. The curve labeled "Enh 128" corresponds to an enhancement layer which is predicted only from the base layer, while the curve labeled "Enh 128p" corresponds to an enhancement layer that also uses prediction from previous enhancement-layer pictures. Performance for "Enh 128" and "Enh 128p" assume the base layer is completely received.

For the base layer, performance degrades gradually as the loss rate increases from 0.1% to 1%, but as the loss rate continues to increase, the performance degrades significantly. Thus it will be important for the transport mechanism to keep the residual losses (after loss recovery) below 1% for the base layer.

The more aggressive prediction strategy of "Enh 128p" has visually better performance for low loss rates, but for loss rates greater than about 5%, performs significantly worse than the less efficient prediction strategy of "Enh 128". In both cases, performance with 100% loss of enhancement layer is identical to the base-only performance. If the more aggressive prediction strategy is used, then it will be important for the transport to keep the loss rate for the enhancement layer below 10%, while if the less efficient prediction strategy is used, the loss rate for the enhancement layer is less critical. However, performance will be significantly degraded in the case of few losses, or if additional layers are added beyond this first enhancement layer.

3.4   Information Provided to Transport

To enable the best design of an encoding-aware transport, the encoder should provide some meta-information to the transport mechanism. This meta-information includes two pieces of information:

1. The bandwidth-quality trade-off (e.g., Figure 3). More specifically, the encoder conveys the bandwidth (or consumption rate) of each layer (i.e., C_0, C_1, ..., C_N), and the improvement in quality caused by each layer (i.e., Q_1, Q_2, ..., Q_N).

2. The loss rate and quality trade-off (e.g., Figure 4). In some situation (such as for the base-layer or the more aggressive enhancement-layer prediction strategy in Figure 4) the loss rate vs. quality meta-information can be reduced to simple thresholds regarding the maximum tolerable loss rate for each layer (i.e., L_{max0}, L_{max1}, ..., L_{maxN}).

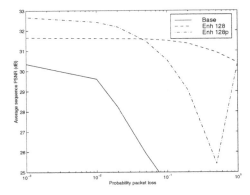

Fig. 4. Effect of Loss on PSNR

In general, this meta-information will be encoding and even content specific. Ideally, it should also include temporal variations, indicating the trade-offs for different scenes within a sequence, or for each frame in a sequence. However, in practice, it might be necessary to use static information (for the entire sequence) or generic information (for a class of sequences, like "head-and-shoulders").

The interface between the transport and the encoder is completely characterized by the loss rate and the bandwidth dedicated to video information. Therefore, the above information will be sufficient to design effective encoding-aware transport mechanisms.

4 Encoding-Aware Transport

In this section, we illustrate how a transport mechanism for layered encoded streams can leverage encoding-specific information to improve quality of delivered stream. First, we present our client-server architecture to identify the main components of the architecture and their associated design issues. Then, we explore the design space of the main components to show how encoding-specific information can be used to customized the design. Our goal is to clarify key tradeoffs in the design of a transport mechanism for layered video and demonstrate how they can use information about encoded streams to improve delivered quality over the best-effort Internet.

Figure 5 depicts our client-server architecture for delivery of Internet video playback (Figure 1) with more details. As we mentioned earlier, the architecture has four key components: 1) Congestion Control (CC) is a network-specific mechanism that determines available bandwidth (BW) and loss rate (L) of the network connection. Available bandwidth and loss rate are periodically reported to the Bandwidth Allocation (BA) module. The BA module uses encoding-specific information to properly allocate the available bandwidth between the Loss Recovery (LR) and the Quality Adaptation (QA) modules. The LR module utilizes allocated portion of available bandwidth (BW_{lr}) to retransmit the required ratio of recent losses. The remaining portion of available bandwidth (BW_{qa}) is used by the QA module to properly determine the quality of delivered stream (*i.e.*, number of transmitting layers).

All the streams are layered encoded and stored. Thus, different layers can be sent with different rates (bw_0, bw_1, ..., bw_n). The server multiplexes all active layers along with retransmitted packets into a single congestion controlled flow. The client demultiplexes different layers and rebuilds individual layers in virtually separate buffers. Each layer's buffer is drained by the decoder with a rate equal to its consumption rate (*i.e.*, C_0, C_1, ..., C_n). The client reports its playout time in each ACK packet. This allows the server to estimate the client's buffer state, *i.e.*, the amount of buffered data for each layer. Client buffering is used to absorb short-term variations in bandwidth without changing the delivered quality.

The main goal of the transport mechanism is to map the actual connection bandwidth and loss rate into the range of acceptable channel behavior expected

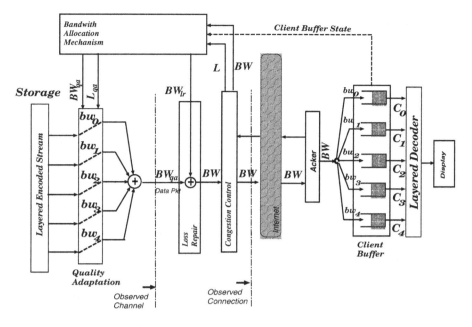

Fig. 5. Client-Server Architecture for Streaming of Layered Video

by the encoding mechanism. Layered encoding provides two levels of freedom for the transport mechanism to achieve this goal: 1) by changing number of layers, the transport mechanism can adjust the required channel bandwidth for delivery of the stream, and 2) by allocating a portion of available channel bandwidth to loss repair, the transport mechanism can reduce the observed loss rate. Therefore, there are three key issues in design of a transport mechanism for layered encoded streams that can be tailored for a given encoded stream to improve delivered quality:

- *Bandwidth Allocation strategy*: How should the transport mechanism allocate total connection bandwidth between LR and QA modules?
- *Loss Repair strategy*: How should the loss repair bandwidth (BW_{lr}) be shared among transmitting layers?
- *Quality Adaptation strategy*: How should the server adjust the quality of delivered stream (*i.e.*, the number of layers) as available channel bandwidth (BW_{qa}) changes?

Since congestion control is a network-specific mechanism, its design should not be substantially affected by application requirements. Therefore, we do not discuss design issues of the CC mechanism. The main challenge is that the behavior of a network connection (BW and L) is not known a priori, and even worse it could substantially change during the course of a connection. Thus, the server should adaptively change its behavior as the connection behavior varies.

For the rest of this section, we provide insight in each one of the above three strategies in design of transport mechanism for layered encoded stream and demonstrate how the transport mechanism can benefit from encoding-specific information. We assume that the encoding-specific meta-data (described in section 3.4) are available for each layered encoded stream.

4.1   Bandwidth Allocation

The BA module shifts the connection loss rate (L) into the range of acceptable loss rate by allocating the required amount of connection bandwidth for loss repair. Therefore, the application will observe a channel with a lower loss rate (L_{qa}) at the cost of lower channel bandwidth. We need to drive a function that presents the tradeoff between the channel bandwidth (BW_{qa}) and the channel loss rate (L_{qa}). Given the connection bandwidth (BW) and the connection loss rate (L), the total rate of delivered bits is equal to $BW(1 - L)$. Therefore, the ratio of delivered bits for the channel is $\frac{BW(1-L)}{BW_{qa}}$. Thus, we can calculate the channel loss rate as follows:

$$L_{qa} = 1 - \frac{BW(1-L)}{BW_{qa}} \quad (1), \quad \text{where } BW_{qa} \leq BW \text{ and } BW_{qa} \geq BW(1 - L)$$

Equation (1) presents L_{qa} as a function of BW_{qa} for a given connection (*i.e.*, BW, L). Figure 6 depicts this function for different set of BW and L values. Each line in Figure 6 represents possible channel behaviors for a given network connection as the BA module trades BW_{qa} with L_{qa}. For example, point A represents a connection with 1000 Kbps bandwidth and 40% loss rate. To reduce the channel loss rate down to 33% (*i.e.*, shifting point A to point B), the BA module should allocate 100 Kbps of connection bandwidth for loss repair, whereas reducing the loss rate down to 14% (*i.e.*, shifting point A to point C) requires the BA module to allocate 300 Kbps of the connection bandwidth for loss repair.

Figure 6 clearly demonstrates how the BA strategy can be customized for a given encoded streams using the information provided by the encoder. Given the bandwidth of various layers (*i.e.*, C_0, C_1, ...) and the per-layer maximum tolerable loss rates (*i.e.*, L_{max0}, L_{max1}, ...), to find the maximum number of layers (n) that can be delivered through a network connection (BW, L), the following two conditions should be satisfied:

$$\frac{\sum_{i=0}^{n} L_{maxi}}{n} \geq L_{qa} \quad (2), \qquad \sum_{i=0}^{n} C_i \leq BW_{qa} \quad (3)$$

The first condition ensures that the average loss rate for n active layers is less than the channel loss rate, whereas the second condition ensures that channel bandwidth is sufficient for delivery of n layers. Given the values of BW, L and $n = N$ (where N is maximum number of layers), the BA module should use equation (1) to search for a channel loss rate that satisfies equation (2). If such a channel loss rate can be accommodated while the corresponding channel bandwidth satisfies equation (3), then n layers can be delivered and total required bandwidth for loss repair is ($BW_{lr} = BW$ - BW_{qa}). Otherwise, the BA module decreases n by one and repeats this process.

The BA module continuously monitors the connection behavior to determine the required bandwidth for loss repair such that the channel behavior always satisfies the conditions (equation 2 and 3) for the number of active layers. If the connection loss rate increases or the connection bandwidth decreases such that these conditions can no longer be satisfied, the BA module signals the QA module to drop the top layer. This decreases n which in turn presents a new set of conditions to the transport mechanism.

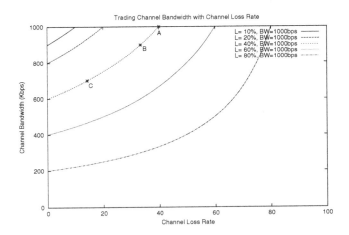

Fig. 6. Trading the channel bandwidth with the channel loss rate

In summary, the BA module determines the bandwidth share for the QA and LR modules. This allows us to separate the design of the Loss recovery mechanism from the Quality adaptation mechanism in spite the fact that their collective performance determines delivered quality.

4.2   Loss Recovery

The loss repair module should micro-manage the total allocated bandwidth for loss repair (BW_{lr}) among the active layers such that the loss rate observed by each layer remains below its maximum tolerable threshold (*i.e.*, L_{max0}, L_{max1}, ..., L_{maxN}). Since all layers are multiplexed into a single unicast session at the server, the distribution of total loss rate across active layers is seemingly random and could change in time. Thus, the bandwidth requirement for loss recovery of various layers can randomly change in time even if the total loss rate remains constant. We assume a retransmission-based loss recovery since 1) retransmission

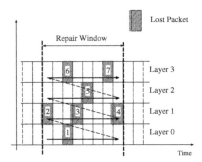

Fig. 7. Sliding Window approach to prioritized Retransmission

is feasible for playback applications with sufficient client buffering, 2) retransmission is more efficient (*i.e.*, requires less bandwidth) than other repair schemes such as FEC, and 3) retransmission allows fine-grained bandwidth allocation among active layers [1].

Since the importance of a layer monotonically decreases with its layer number, loss repair should be performed in a prioritized fashion, *i.e.*, losses of layer j should be repaired before losses of higher layers and after losses of lower layers. However, a prioritized approach to loss repair should ensure that the total repair bandwidth is properly shared among active layers and that each retransmitted packet is delivered before its playout time. To achieve this, we deploy a sliding-window approach to loss repair as shown in Figure 7. At any point of time, the server examines a recent window of transmitted packets across all layers. Losses of active layers are retransmitted in the order of their importance such that the loss rate observed by each layer remains below its maximum tolerable loss rate (*i.e.*, L_{maxi}). Figure 7 shows the order of retransmission within a window for each layer and across all layers.

The repair window should always be a few round-trip-times (RTT) ahead of the playout time to provide sufficient time for retransmission. Therefore, the repair window slides with playout time. If the BA module properly estimates the required bandwidth for loss repair, all losses can be repaired. However, if the allocated bandwidth for loss repair is not sufficient to recover all the losses within a window, this approach repairs the maximum number of more important losses. The length of the repair window should be chosen properly. A short window cannot cope with a sudden decrease in bandwidth, whereas a long window could result in the late arrival of retransmitted packets for higher layers. In summary, the sliding window approach to prioritized loss repair 1) uses maximum per-layer tolerable loss rates for an encoded stream to improve its performance, and 2) adaptively changes the distribution of total repair bandwidth among active layers.

---

[1] Although we only discuss retransmission-based loss repair, the basic idea can be applied to other post-encoding loss repair mechanisms such as unequal FEC.

4.3   Quality Adaptation

The QA mechanism is a strategy that adds and drops layers to match the quality of the delivered stream (*i.e.*, number of transmitting layers) to the channel bandwidth (BW_{qa}). When the channel bandwidth is higher than the consumption rate for active layers, the server can use the extra bandwidth and send active layers with a higher rate (*i.e.*, $bw_i > C_i$) to fill up the client buffer. The buffered data at the client can be used to absorb a short-term decrease in bandwidth without dropping any layers. Figure 8 illustrates the filling and draining phases of the client buffers where three layers are delivered. If the total amount of buffered data during a draining phase is not sufficient to absorb the decrease in bandwidth, the QA module is forced to drop a layer. Consequently, the more data that is buffered at the client during a filling phase, the bigger the reductions that can be absorbed. The amount of buffered data at the client side is determined by the strategy of adding layers.

Figure 9 compares two adding strategies. During a filling phase, the QA strategy adds a new layer after a specific amount of data is buffered. If the required amount of buffered data is small, the QA mechanism aggressively adds a new layer whenever the channel bandwidth slightly increases. In this case, any small decrease in bandwidth could result in dropping the top layer because of the small amount of buffering. Alternatively, the QA mechanism can conservatively add a new layer only after a large amount of data is buffered.

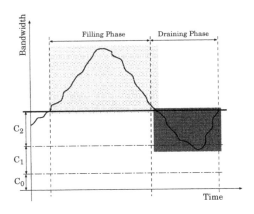

Fig. 8. Filling and Draining phases for Quality adaptation

More buffered data allows the server to maintain the newly added layer for a longer period of time despite major drops in bandwidth. Figure 10 shows an aggressive and a conservative adding strategies in action[4]. The congestion-controlled bandwidth is shown with a saw tooth line in both graphs. This experiment clearly illustrates the coupling between the adding and dropping strategies.

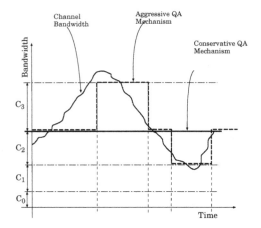

Fig. 9. Effect of adding strategy on quality changes

The more conservative the adding strategy, the longer a new layer can be kept, resulting in fewer quality changes, and vice versa.

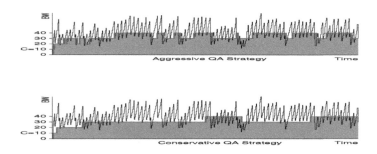

Fig. 10. Aggressive vs Conservative QA Strategies

The QA mechanism should be customized for the particular layered stream being transmitted. To explain this, we need to examine two basic tradeoffs in the design of an add and drop strategy.

– *How much data should be buffered before adding a new layer?*
  This should be chosen such that normal oscillation in channel bandwidth in the steady state does not trigger either adding or dropping a layer. Figure 8 clearly shows that the required amount of buffered data to survive a drop in

bandwidth directly depends on the total consumption rate of active n layers (*i.e.*, $\sum_{i=0}^{n} C_i$).
 − *How should the buffered data be distributed among active layers?*
 Since the streams are layered encoded, buffered data should be properly distributed across all active layers in order to effectively absorb variations in bandwidth. During a draining phase, buffered data for layer i cannot be drained faster than its consumption rate (C_i). Therefore, buffered data for layer i can not compensate more than C_i bps. Figure 11 illustrates this restriction. To avoid dropping a layer, the total draining rate of buffering layers (*e.g.*, $C_2 + C_1$) should always be higher than the deficit in channel bandwidth (BW_{def}). More specifically, during a draining phase the following conditions must be satisfied:

$$BW_{def} = \sum_{i=0}^{n} C_i - BW_{qa}, \qquad BW_{def} \leq \sum_{i \in BufLayers} C_i$$

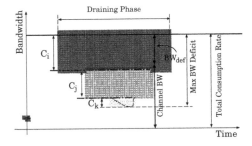

Fig. 11. Sample distribution of buffered data among buffering layers

5   Conclusion and Future Work

In this paper, we presented a joint design of encoder and transport mechanisms for playback of layered quality-adaptive video over the Internet. The main challenge is that the Internet does not support QoS. At the time of encoding, the channel bandwidth and loss rate are not known and they could significantly change during the course of a session. Therefore, traditional encoding approaches that assume static channel behavior will result in poor quality.

We argued that layered video is the most flexible solution because 1) it can be efficiently delivered over a range of channel behavior, and 2) it provides sufficient flexibility for the transport mechanism to effectively reshape the stream for delivery through the variable channel. However, to maximize quality of delivered video, the encoding should become transport-aware and the transport mechanism should become encoding-aware. Toward this end, we described several issues in the design of layered encoding mechanisms and explained how the expected range of channel bandwidth and loss rate information can be incorporated into the encoding mechanism. Furthermore, the encoding mechanism provides encoding-specific meta-information. The transport mechanism uses this information to bridge the gap between the expected range of channel behavior by the encoder and the actual connection behavior. More specifically, we provided insight on how the main components of the transport mechanism, particularly Bandwidth Allocation, Loss Repair and Quality Adaptation, can leverage encoding-specific meta-information to improve the delivered quality of layered video despite unpredictable changes in channel behavior.

Finally, we plan to conduct extensive experiments over the Internet to evaluate the overall performance of our client-server architecture. This will allow us to identify those scenarios in which our architecture can not properly cope with changes in connection behavior. Some of these problems can be addressed by the encoding mechanism through appropriate provisioning, whereas others require further tuning or modification of the transport mechanism. Our experiments should provide deeper insight about channel behavior that may suggest refinement of the layered encoding mechanism. We also plan to examine interactions among three key components of the transport mechanism, as well as implications of congestion control algorithm on other components of transport mechanism.

References

1. R. Rejaie, M. Handley, and D. Estrin, "RAP: An end-to-end rate-based congestion control mechanism for realtime streams in the internet," *Proc. IEEE Infocom*, Mar. 1999.
2. S. Floyd, M. Handley, J. Padhye, and J. Widmer, "Equation-based congestion control for unicast applications," *Proc. ACM SIGCOMM*, Sept. 2000.
3. R. Rejaie, H. Yu, M. Handley, and D. Estrin, "Multimedia proxy caching mechanism for quality adaptive streaming applications in the internet," in *Proc. IEEE Infocom*, Tel-Aviv, Israel, Mar. 2000.
4. R. Rejaie, M. Handley, and D. Estrin, "Quality adaptation for congestion controlled playback video over the Internet," *Proc. ACM SIGCOMM*, Sept. 1999.
5. R. Zhang, S. L. Regunathan, and K. Rose, "Video coding with optimal inter/intra-mode switching for packet loss resilience," *IEEE Journal on Selected Areas in Communications*, vol. 18, no. 6, pp. 966–976, June 2000.
6. Kang-Won Lee, Rohit Puri, Tae eun Kim, Kannan Ramchandran, and Vaduvur Bharghavan, "An integrated source coding and congestion control framework for video streaming in the internet," in *Proc. IEEE Infocom*, Tel-Aviv, Israel, Mar. 2000.
7. A. R. Reibman, H. Jafarkhani, Y. Wang, and M. Orchard, "Multiple description coding for video using motion compensated prediction," in *International Conference on Image Processing*, Oct. 1999.
8. D. Taubman and A. Zakhor, "Multirate 3-D subband coding of video," *IEEE Trans. Image Processing*, vol. 3, no. 5, pp. 572–588, Sept. 1994.

9. S. McCanne, V. Jacobson, and M. Vetterli, "Receiver-driven layered multicast," *Proc. ACM SIGCOMM*, Aug. 1996.
10. R. Aravind, M. R. Civanlar, and A. R. Reibman, "Packet loss resilience of mpeg-2 scalable video coding algorithms," *IEEE Transactions on Circuits and Systems for Video Technology*, vol. 6, no. 5, pp. 426–435, Oct. 1996.
11. G. Cote, B. Erol, M. Gallant, and F. Kossentini, "H.263+: Video coding at low bit rates," *IEEE Transaction on Circuit and Systems for Video Technology*, vol. 8, no. 7, Nov. 1998.
12. H. Radha et. al., "Fine-granular-scalable video for packet networks," in *Packet Video Workshop*, New York NY, Apr. 1999.
13. A. R. Reibman and L. Bottou, "Managing drift in a DCT-based scalable video encoder," in *IEEE Data Compression Conf.*, Mar. 2001.

A Multi-state Congestion Control Scheme
for Rate Adaptable Unicast Continuous Media Flows[1]

Pantelis Balaouras and Ioannis Stavrakakis

Communication Networks Laboratory (CNL)
Department of Informatics and Telecommunications
University of Athens (UoA)
15784, Athens, Greece
{P.Balaouras, istavrak} @di.uoa.gr

Abstract. In this paper a multi-state rate adaptation scheme for rate adaptable continuous media (CM) sources is presented; the network bandwidth availability information is assumed to be communicated to the source only indirectly, through a notification of the packet losses. The developed scheme is capable of detecting and responding well to static (due to access bandwidth limitations) and long-term (due to initiation/termination of CM flows) bandwidth availability, as well as locking the rate of a CM flow to an appropriate value. Its performance is compared with that under a classic Additive Increase Multiplicative Decrease algorithm and its shown to perform better in terms of a number of relevant metrics, such as packet losses, oscillatory behavior, and fairness.

1. Introduction and Background

The network layer of Internet currently provides no inherent congestion control. In current IP environments, there is no central entity to dictate an explicit rate for the flows or explicitly indicate the current network state like, for example, in an ATM environment; ECN (Explicit Congestion Notification) based schemes are still far from being adopted and widely deployed [1,2]. Thus, congestion control may be provided through mechanisms implemented either by the transport or the application layers. A flow source would use its own application or transport level information concerning packet losses to infer the state of the network, resulting in a distributed rate control scheme. Different users may infer a different network state. The Transport Control Protocol (TCP) protocol provides congestion control, whereas the User Datagram Protocol (UDP) does not.

One of the emerging services appearing in the Internet is the continuous media (CM) streaming service, such as video streaming, conferencing and "on demand" services. CM applications require more bandwidth, compared to the typical data

---

[1] This work has been supported in part by the IST Program "Videogateway" of the EU (IST-1999-10160).

S. Palazzo (Ed.): IWDC 2001, LNCS 2170, pp. 452–469, 2001.

block transfer services (e.g., Web access over TCP, file transfer via FTP, e-mail), and are sensitive to the network induced delay, delay jitter and packet loss.

TCP is not suitable for multimedia transmission, because of its window-based congestion scheme [3,4,5]. For CM services UDP is used as the base transport protocol, in conjunction with the Real Time Protocol (RTP) and Real Time Control Protocol (RTCP) [6]. RTP provides end-to-end delivery services for real-time data, such as interactive audio and video. Those services include payload type identification, sequence numbering, time-stamping and delivery monitoring. RTP itself does not provide any mechanism to ensure the timely delivery of media units nor provides for other quality-of-service guarantees, but rather relies on lower-layer services to do so (i.e., integrated or differentiated services applied at the network layer). RTP carries data with real-time properties, whereas RTCP monitors the quality of service of each participant and conveys useful information to all participating entities. Applications that use the UDP transport protocol have to implement *end-to-end congestion control* [7] otherwise (a) the TCP traffic suffers and (b) the network collapses. RTCP may be used as a feedback channel to provide for congestion notification to CM applications that implement an end-to-end congestion control scheme [8,9]. Examples of RTP/RTCP based rate control algorithms, that are TCP-friendly, are [10-14]. Other rate control protocols, such as RAP [15] and TFRCP [16], that are TCP-friendly and do not rely on RTP/RTCP, have been proposed in the literature as well.

Rate control protocols and adaptation mechanisms require sources capable of adjusting their rate on the fly (such as MPEG-4 [17]) or transcoding filters and video gateways [18,19,20] capable of transcoding a high bit-rate input stream (such as MPEG-2 [21]) to a lower bit-rate adjustable output stream. An alternative to rate control and transcoding mechanisms is the multi-layered coding with multicast transfer scheme [22]. Such a scheme allows the receiver to connect to the appropriate multicast channel(s) according to its connection bandwidth availability.

A rate adaptation scheme needs to be communicated the network state through a feedback mechanism. Most rate adaptation schemes use the current packet loss rate to infer whether the network is congested or not, and the rate is either increased or decreased, respectively. The increase function is either additive increase [23] or multiplicative increase [16], whereas the decrease function is multiplicative decrease [8,9,15] or the rate may be set to a value calculated by an equation-based formula [24] which estimates the equivalent TCP throughput [10-14,16]. The various feedback mechanisms are either RTCP-based [8-14] or ACK-based [15,16]. In the first case the feedback delay is at least 5 sec ($\pm$ 1.5 randomly determined), whereas in the second case it is equal to one round trip time. A congestion control scheme that uses the history of the packet losses has been proposed as well [25].

In this paper a rate adaptation scheme is introduced for an environment supporting CM flows. As discussed briefly below - as well as throughout this paper – the scheme will attempt to take into consideration certain peculiarities associated with a CM-supporting networking environment.

Users may be connected to the network using links of limited and fixed bandwidth (e.g., dial-up users, ADSL users), or practically unlimited bandwidth. The bit-rate of a CM flow is assumed to take any value within a specific range, where a minimum rate is considered in order to ensure a minimum perceptual quality, whereas a maximum rate represents the upper limit of the source encoder capability or is imposed to prevent unnecessarily high consumption of bandwidth and /or prevent the suppression of other non CM flows, e.g., TCP flows. The (limited) access bandwidth and the minimum and maximum source encoding rates all impose restrictions on the behavior of the source rate adaptation scheme and should be taken into consideration. For instance, probing for more bandwidth availability when the (static) access bandwidth limit is reached should be avoided by an effective rate adaptation scheme.

CM flow initiations and terminations could result in large and lasting bandwidth availability changes. Such type of changes should be distinguished from (small and short-term) typical bandwidth of fluctuations and should trigger a rapid response from the rate adaptable CM flows to avoid excessive losses or bandwidth under-utilization. Because of the nature of the CM applications, decreasing the rate upon congestion by a factor of 2 (as in TCP) should be avoided, as the perceptual QoS would suffer substantially. Fast convergence to a fair bandwidth allocation is desirable after a flow initiation or termination occurs.

The rate adaptation scheme proposed in this paper is designed for a networking environment supporting CM flows and is capable of detecting and responding effectively to static and long-term bandwidth availability. The CM flows are supposed not to be aware of any access bandwidth limitations or the number of co-existing CM flows or the overall network load. The developed scheme detects any access bandwidth limitations and locks the rate of the flow to an appropriate rate, as well as detects lasting load changes (due to the initiation or termination of CM flows) and locks the rate of the CM flows to an appropriate rate. Short-term bandwidth changes are not expected to be present in a (purely) CM supporting networking environment due to the nature of the Adaptable Constant Bit Rate (A-CBR) CM applications considered here. While an extension of this work will consider an environment where A-CBR CM flows co-exist with other traffic (such as TCP) which introduces short-term bandwidth changes, the effort there is expected to be focused on the study of the impact of such bursty traffic on the proposed rate adaptation algorithm and the introduction of the adjustments to improve its effectiveness, rather than the redesign of the algorithm to become responsive to such short term bandwidth fluctuations. To follow the latter fluctuations may be impossible due to the granularity and the hysterisis of the encoding process and the rate adaptation time lag, as well as be of negligible impact on the perceived QoS.

In this paper, a multi-state congestion control scheme for rate adaptable unicast CM flows is presented. This scheme is suitable for an environment in which the network bandwidth availability information is communicated to the source only indirectly through a notification of the packet losses that the specific source suffers over a time interval. This environment is present, for example, when CM flows are transmitted by using RTP/UDP/IP protocols. The proposed scheme aims to provide a finer rate adaptation decision space for the source by introducing a number of flow states and then basing the rate adaptation decision on the current source state in addition to the current packet loss feedback information. The introduction of states allows for an effective "summarization" of the recent bandwidth availability and adaptation history. The resulting multi-state rate adaptation algorithm has a larger decision space compared to traditional schemes, and responds well to a diversity of network situations, such as those outlined above, while exhibiting good fairness characteristics. The behavior of the proposed scheme is examined in an environment where only A-CBR CM (RTP/UDP/IP) flows are transmitted (i.e., free of TCP flows), as it may be the case under the forthcoming differentiated services.

2. Description of the Algorithm

A networking environment supporting N RTP/UDP/IP CM flows is considered. The RTCP protocol is used as a feedback notification mechanism as specified by RFC 1889 [7]. The source and receiver(s) of each flow exchange RTCP Sender and Receiver Reports (SR and RR, respectively). The sources are assumed to be capable of adapting their encoding and transmission rates on the fly, as it is the case. The source of a flow is not aware of the topology, the links, the available bandwidth and the current number of CM flows transmitted over the network. The only available information is the one carried by the RTCP RR reports. One RR report is sent every T seconds. In this paper T is set to be equal to 5 sec, as specified by the RFC 1889.

A decision module - located at the source - is responsible for deciding on the appropriate transmission rate to which the source should adapt. Rate adaptation decisions are taken at the *adaptation points*. An adaptation point occurs when a RR arrives at the source of the flow or a time-out timer expires indicating that at least one RR report is excessively delayed or lost. The time-out timer is initially set to the value of `timeout_period(i)` – for flow i – which is set to a value $T_0(i)$ greater than T to absorb (in part) the network introduced delay jitter. This counter is reset to $T_0(i)$ each time a RR is received and to $T_1(i)$, $T < T_1(i) < T_0(i)$, each time a timeout occurs. This subsequent reduction in the time-out period will lead to a reduction of the time to the next adaptation point if this is due to a time-out and, thus, will lead to a faster and more effective rate adaptation.

Let $f^k(i)$ denote the network state feedback associated with the kth adaptation point of flow i. $f^k(i) \equiv \varnothing$ (i.e., no value) if the kth adaptation point is due to a timer expiration and $f^k(i) \equiv p^k(i)$ when it is due to the arrival of a RR; $p^k(i)$ denotes the packet loss rate derived from the RR received at the kth adaptation point. Let $S^k(i)$ and $r^k(i)$ denote the state – to be defined later – and rate of flow i, just before k^{th} adaptation point. At the kth adaptation point a state transition (adaptation) from $S^k(i)$ to state $S^{k+1}(i)$ and rate transition from $r^k(i)$ to $r^{k+1}(i)$ will occur, effective just after the kth adaptation point and remaining effective until just before the $(k+1)^{st}$ (Fig.1). In addition to the received-feedback $f^k(i)$, the next state $S^{k+1}(i)$ depends on $S^k(i)$ as well as a number of other quantities defined at each adaptation point providing useful history information. Such quantities are the recent average packet loss rate $p^k_h(i)$ – defined precisely later -, the number of successive, $I^k_{succ}(i)$, and total, $I^k_{total}(i)$, visits to the current state $S^k(i)$, for certain states, and the binary lock decision function $B_{Lock}(i)$ that returns 1, if rate locking is decided. The next rate $r^{k+1}(i)$ depends on the next state $S^{k+1}(i)$ decided to be visited, the current rate $r^k(i)$, and the feedback $f^k(i)$.

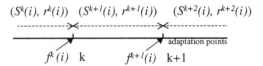

Fig. 1. State and rate transitions.

Let $\mathcal{9}$ denote the state space of the process $\{S^{k+l}(i)\}_{k\geq0}$; $\mathcal{9}=$ {just_intro, decr_due_nofeedback, increase, decrease, lock, rmax, rmin, fast_increase, fast_decrease, check_for_bw, low_increase, low_decrease, low_lock, low_check_for_bw}. A brief description of the states and the motivation behind their introduction are presented next.

Upon initiation, a flow enters state just_intro, allowing the new source to follow slightly different increase/decrease policy for a given feedback compared to existing flows. This is necessary since the initial rate of a flow is randomly selected within a range and thus different level of up or down adjustments are needed compared to older flows which have already adjusted their rate and need to respond to bandwidth availability changes.

If a rate adaptation point is due to the time-out timer expiration it is considered that at least a feedback is lost or excessively delayed and the decr_due_nofeedback state is visited while the rate is properly decreased.

States rmax and rmin are defined to capture the state of a source that attempted to select a rate beyond the corresponding rate limits. When a flow is in one of these states the source will apply such a control that the flow rate remain in the range $[r_{min}(i), r_{max}(i)]$. Also, visiting the state rmin once is an indication of a highly congested network, whereas visiting this state more times is an indication of persisting congestion, probably due to access network limitation.

States increase and decrease are defined for the normal case in which the flows compete with each other in order to share the available bandwidth. The flow enters state increase in order to *increase* its rate, whereas it enters state decrease to *decrease* the rate in response to detecting some minor congestion.

State fast_increase is for the case that the algorithm estimates that the network is rather seriously under-utilised, possibly due to bandwidth release caused by a CM flow termination(s). The transition to this state is accompanied with a rate *increase* that is considerably greater than that of state increase. State fast_decrease is for the case that the algorithm detects rather serious network congestion due to, for instance, the initiation of some new CM flow(s) and results in a rate dec*rease* that is more drastic than that of state decrease.

The algorithm is capable of detecting situations where the flows oscillate around the fairness point or around its access bandwidth limit and could optionally lock the rate. The state lock is defined for this purpose. As long as the flow is in state lock the rate is *maintained*. After spending some time in state lock, the state check_for_bw is visited and bandwidth probing rate increases take place, every time this state is (re)visited.

States   low_increase,   low_decrease,   low_lock,   and
low_check_for_bw are defined to allow for a milder reaction, which is more
appropriate when it is highly possible that the congestion is due to access network
limitations, near the $r_{min}(i)$ value.

Under no feedback $(f^k(i) = \phi)$ - that is the time-out timer expires - the state
decr_due_nofeedback is visited from any state and the rate is decreased as described in section 3. If the number of consecutive visits to this state exceeds a threshold ($max\_cons\_times\_in\_decrnofeed(i)$) then the state rmin is visited.

2.1 State Transitions under Zero Packet Loss Rate Feedback $(f^k(i)=p^k(i) = 0)$

Under zero-Packet Loss Rate (zero-PLR) feedback the state transitions shown in
Table 1 take place under the conditions specified. Such transitions are typically followed by a rate increase, as determined by the state visited and described in section 3.
Some of the state transitions are self-justifiable in view of the comments provided for
the various states in the previous section.

When a transition to a state is accompanied by a rate increase that results in a rate
exceeding $r_{max}(i)$, then the rate is set to $r_{max}(i)$ and an instantaneous transition to
state rmax occurs. State rmax may be visited from any state and transitions to rmax
are not shown on Table 1 for simplicity.

Since the initial rate of a new flow (entering state just_intro) is randomly selected, a zero-PLR feedback implies that that selection has most likely been very
conservative and, thus a transition to state fast_increase (as opposed to increase or low_increase) occurs.

Since state decr_due_no_feedback is visited upon time-out timer expiration
(indicating delayed or lost feedbacks) and the rate is decreased, a zero-PLR feedback
indicates that the previous rate decreases that occurred (upon visiting this state) have
been rather excessive and a transition to state fast_increase occurs to allow for a
fast rate "correction" (increase).

Upon zero-PLR feedback the algorithm (process) remains in (makes a transition
to) state increase, unless this type of transition has occurred for a number (equal
to go_for_fast_increase(i)) of consecutive times. When the latter occurs it is inferred
that a faster increase rate is needed and a transition to state fast_increase occurs. go_for_fast_increase(i) is a function that depends on the current rate $r^k(i)$; the
larger $r^k(i)$ the larger its value, implying that a lower rate flow enters state
fast_increase sooner than a higher rate one.

In order to maintain some smoothness in the process and avoid frequent rate oscillations the state increase (as opposed to fast_increase) is visited from states
fast_decrease upon zero-PLR feedback.

Upon zero-PLR feedback the process moves from state decrease to increase
(self-explanatory) unless a decision to lock the rate is taken (see section 4) in which
case the state lock is visited.

Upon zero-PLR feedback the process remains in state lock, unless this type of transition has occurred for a number (equal to go_for_check_bw(i)) of consecutive times, in which case the state check_for_bw is visited.

Table 1. State transitions under the zero-PLR feedback (except rmax).

Current state	Next state	Condition
just_intro	fast_increase	always
decr_due_nofeed.	fast_increase	always
rmax	rmax	always
rmin	increase	$C_1: I^k_{total}(i) < times\_in\_rmin(i)$, e.g., 3
	low_increase	$C_2: I^k_{total}(i) = times\_in\_rmin(i)$
increase	increase	$C_3: I^k_{succ}(i) < go\_for\_fast\_increase(i)$
	fast_increase	$C_3: I^k_{succ}(i) = go\_for\_fast\_increase(i)$
fast_increase	fast_increase	always
fast_decrease	increase	always
decrease	increase	C_1:not $B_{Lock}(i)$
	lock	not $C_1 : B_{Lock}(i)$
lock	check_for_bw	$C_8: I^k_{succ}(i) > go\_for\_check\_bw(i)$, e.g., 8
	lock	not C_8
check_for_bw	check_for_bw	$C_9: I^k_{succ}(i) < go\_to\_unlock(i)$, e.g., 4
	increase	not C_9
low_decrease	low_increase	always
low_increase	increase	$C_6: I^k_{succ}(i) > escape\_low\_states(i)$, e.g.,20
	low_increase	not C_6
low_lock	low_check_for_bw	$C_{10}: I^k_{succ}(i) > go\_for\_low\_check\_bw(i)$
	low_lock	not C_{10}
low_check_for_bw	low_increase	$C_{11}: I^k_{succ}(i) > go\_to\_low\_unlock(i)$, e.g., 6
	low_check_for_bw	not C_{11}

Upon zero-PLR feedback the process remains in state check_for_bw, unless this type of transition has occurred for a number (equal to go_to_unlock(i)) of consecutive times, in which case the state increase is visited and the flow is unlocked.

Upon zero-PLR feedback the process moves to state low_increase or increase depending on the total number of visits to rmin since the initiation of the flow. If this number is equal to or greater (less) than times_in_rmin(i) then state low_increase (increase) is visited. Visiting rmin for a number of times implies that it is likely that the available bandwidth is only slightly greater than $r_{min}(i)$ and that thus a lower rate increase that is associated with state low_increase would be more appropriate.

The collection of states with the prefix low form a sub-space (referred to as low-state subspace) which is entered by the process only from state rmin under zero-PLR feedback and provided that times_in_rmin(i) threshold is exceeded. While in the low-space subspace, the process evolves in a similar manner under zero-PLR feedback as described earlier, with minor decision threshold (parameter) adjustments as captured by the introduced parameters ("history" counters). The only exit point from this subspace is from state `low_increase` under zero-PLR feedback and provided that this state has been visited for a number (equal to `es-cape_low_states(i)`) of times, in which case there appears to be room for higher rate increase and the state `increase` is visited.

2.2 State Transition under Nonzero Packet Loss Rate Feedback ($f^k(i)=p^k(i) > 0$).

Under nonzero Packet Loss Rate (nonzero-PLR) feedback the state transitions shown on Table 2 take place under the specified conditions. Such transitions are followed by a rate decrease, as determined by the state visited and described in section 3. Some of the state transitions are self-justifiable in view of the comments provided for the various states in section 1. When a transition to a state is accompanied by a rate decrease that results in a rate below $r_{min}(i)$, then the rate is set to $r_{min}(i)$ and an instantaneous transition to state `rmin` occurs. State `rmin` may be visited from any state and transitions to `rmin` are not shown on Table 2 for simplicity.

It is highly desirable to rapidly detect the introduction of a new CM flow in the network and adapt the flows' rate in order to avoid excessive packet losses and provide bandwidth to the new flow. The initial rate of the new flow may be too high for a loaded network. In this case the flows will experience packets losses due to the congestion introduced by the new flow. In order to distinguish whether the congestion is due to the initiation of a new flow or not, the reported packet loss rate $p^k(i)$ is compared to a recent average packet loss rate. Let $p^k_h(i)$ denote this average, and let $diff^k(i)$ denote the difference between $p^k(i)$ and $p^{k-1}_h(i)$:

$$diff^k(i) = [p^k(i) - p^{k-1}_h(i)]^+$$

If $diff^k(i)$ is greater than $p_{new\_flow}(i)$, then it is considered that the current packet losses ($p^k(i)$) are significantly higher than the recent average losses ($p^{k-1}_h(i)$), they are attributed to the introduction of a new flow and state `fast_decrease` is visited. Otherwise, it is considered that packet losses occur due to the increases of the CM flows and the state `decrease` is visited. The value used for $p_{new\_flow}(i)$ is 0.04. When packet losses occur ($p^k(i) > 0$), the recent average packet loss rate $p^k_h(i)$ is updated as follows:

$$p^k_h(i) = c_h*p^{k-1}_h(i)+(1-c_h)*p^k(i), \ (c_h = 0.875)$$

Upon nonzero-PLR feedback the process moves from states `just_intro`, `decr_due_nofeedback`, `increase`, `decrease`, `fast_increase`, `fast_decrease` and `rmax` to state `decrease` if $diff^k(i)) \leq p_{new\_flow}(i)$ and to state `fast_decrease` otherwise. When in states `lock` or `check_for_bw` the process maintains its state under nonzero-PLR unless $diff^k(i)) > p_{new\_flow}(i)$ in which case the state `fast_decrease` is visited.

Under nonzero-PLR the process moves from state `low_increase` to state `low_increase`, unless a decision to lock the rate is taken ($B^L_{Lock}(i)$ function returns

1) and then the state `low_lock` is visited. If $\mathit{diff}^{\,k}(i)) \leq p_{new\_flow}(i)$ the process remains in state `low_lock` (no rate change) under nonzero-PLR feedback unless this type of transition has occurred for a number (equal to `cancel_low_locking(i)`) of consecutive times, in which was a wrong locking rate inferred and the state `low_decrease` is visited (rate unlocking). If $\mathit{diff}^{\,k}(i)) > p_{new\_flow}(i)$ then the process moves from state `low_lock` to state `low_decrease`. Finally, the process moves under nonzero-PLR feedback from state `low_check_for_bw` to either `low_lock` if $\mathit{diff}^{k}(i)) \leq p_{new\_flow}(i)$ or the state `low_decrease` otherwise.

Table 2. State transitions under the nonzero-PLR feedback (except `rmin`).

Current state	Next state	Condition
`just_intro`, `decr_due_nofeed.`,	`decrease`	C_{12}: $\mathit{diff}(i) \leq p_{new\_flow}(i)$
`increase`, `rmax`, `decrease`, `fast_increase`, `fast_decrease`	`fast_decrease`	not C_{12}: $\mathit{diff}(i) > p_{new\_flow}(i)$
`lock`, `check_for_bw`	`lock`	C_{12}
	`fast_decrease`	not C_{12}
`low_decrease`	`low_decrease`	always
`low_increase`	`low_lock`	C_{13}:$B^L_{lock}(i)$
	`low_decrease`	not C_{13}
`low_lock`	`low_lock`	C_{12} AND C_{14}; C_{14}: $I^k_{succ}(i) > cancel\_low\_locking(i))$
	`low_decrease`	not (C_{12} AND C_{14})= (not C_{12}) OR (not C_{14})
`low_check_for_bw`	`low_lock`	C_{12}: $\mathit{diff}(i) \leq p_{new\_flow}(i)$
	`low_decrease`	not C_{12}
`rmin`	`rmin`	always

3. The Distance Weighted Additive Increase, Loss Rate Based Multiplicative Decrease Scheme

The proposed multi-state rate adaptation algorithm belongs in the class of Additive Increase Multiplicative Decrease (AI-MD) scheme [23]. In accordance with the AI-MD characteristic of the proposed scheme, the new rate $r^{k+1}(i)$ is given in terms of the previous rate $r^k(i)$ by:

$$r^{k+1}(i) = r^k(i) + \alpha(i), \qquad \text{for the increase case}$$
$$r^{k+1}(i) = r^k(i) \ast \delta(i), \qquad \text{for the decrease case}$$

It should be noted that the increase step $\alpha(i)$ and the decrease factor $\delta(i)$ are not fixed – as in the classical AI-MD case – but the former depends on the state visited $S^{k+1}(i)$, $r^k(i)$, $r_{max}(i)$ and $r_{min}(i)$ and the latter depends on the state visited $S^{k+1}(i)$, $r_{min}(i)$ and $p^k(i)$. To capture the key fact that the increase step depends on the distance between, $r^k(i)$ and $r_{max}(i)$ and that the decrease factor is shaped by the received packet losses, the proposed scheme is referred to as the *Distance-weighted*

Additive Increase, Loss rate-based Multiplicative Decrease (D.AI-L.MD) rate adaptation scheme.

(i). In the case that the visited state is state `increase` or `check_for_bw`, the increase step $\alpha(i)$ has two attributes: the base increase rate `Incr(i)` in kbps and the rate distance factor $dist\_weight(r^k(i))$. That is,

$$\alpha(i)= dist\_weight(r^k(i))*\texttt{Incr(i)},$$
$$dist\_weight(r^k(i)) =(r_{max}(i)-r^k(i))/(r_{max}(i)- r_{min}(i))$$

Note that $dist\_weight(r(i))$ expresses the distance of the current rate from $r_{max}(i)$. If $r^k(i) \rightarrow r_{min}(i)$ then $dist\_weight(r^k(i)) \rightarrow 1$, whereas if $r^k(i) \rightarrow r_{max}(i)$, then $dist\_weight(r^k(i)) \rightarrow 0$. If flows i and j are such that $r_{max}(i)=r_{max}(j)$, $r_{min}(i)=r_{min}(j)$ and $r^k(i) > r^k(j)$, then $dist\_weight(r^k(i)) < dist\_weight(r^k(j))$. This means that flows with lower rate are increasing at a faster pace than flows with higher rate, therefore the convergence time to fairness is expected to improved.

(ii). If the visited state is state `fast_increase`, the increase step $\alpha(i)$ has also a third attribute denoted by $d(i)$. That is,

$$\alpha(i)=d(i)*dist\_weight(r^k(i))*\texttt{Incr(i)}$$

When state `fast_decrease` is visited from another state, $d(i) = d_{init}(i)=5$; $d(i)$ is decreased by one at each adaptation point at which the flow remains in state `fast_increase`. The reason for this decrease is that each time a fast increase occurs, the probability of having packet losses in the next time interval is expected to increase also. Thus, the next fast increase would be more conservative. When $d(i)$ reaches zero it is reset to the value $d_{init}(i)$ plus the number of successive visits to state `fast_increase`. The reason is that successive visits to state `fast_increase` indicates that most likely there is available bandwidth for the flow. Since, $dist\_weight()$ reduces the increase step $\alpha(i)$ as the rate increases, $d(i)$ is increased in order to compensate for this reduction, after a large number of successive visits to state `fast_increase`.

(iii). If the visited state is state `low_increase` or `low_check_for_bw` then the rate is increased by a constant step. That is,

$$a(i) = \texttt{incr\_for\_low(i)}$$

A typical value of `incr_for_low(i)` is 1kbps. Table 3 summarizes the different increase function steps $\alpha(i)$.

Table 3. Increase step function $\alpha(i)$.

Visited State	*Increase step $\alpha(i)$*
`increase, check_for_bw`	$\alpha(i)=dist\_weight(r^k(i))*\texttt{Incr(i)}$
`fast_increase`	$\alpha(i)=d(i)*dist\_weight(r^k(i))*\texttt{Incr(i)}$
`low_increase,low_check_for_bw`	$\alpha(i)=\texttt{incr\_for\_low(i)}$

The decrease factor $\alpha(i)$ has two attributes. The first attribute is the value $(1-p^k(i))$, which is a common attribute in the decrease factors when the visited states are `decrease`, `fast_decrease`, and `low_decrease`. The rate $r^k(i) \cdot (1-p^k(i))$ is roughly the rate under which no packet loss would occur, provided that packet losses occur at a constant rate $p^k(i)$ over the interval between two consecutive adaptation points. In part, because this in not likely in most cases, packet losses will continue to occur even if the new rate is set to $(1-p^k(i))*r^k(i)$. Therefore, a second attribute denoted by `decr_parameter(i)` is introduced to intensify the rate decrease. The value of `decr_parameter(i)` depends on the state visited, as shown below.

(i). If the visited state is `decrease`, `decr_parameter(i)` is less than 1; a typical value used is 0.98 but other values may be more appropriate depending on the environment.

(ii). If the visited state is `low_decrease`, `decr_parameter(i)` is set to 1. The reason is that since the rate in states `low_decrease` and `low_increase` is near $r_{min}(i)$, further rate decrease is likely to reduce the rate to $r_{min}(i)$ which is not desirable.

(iii). If the visited state is `fast_decrease`, then it is likely that a new flow is initiated, as indicated earlier. In this case, the parameter $threshold_{decr}(i)$ with a value less than one and a typical value 0.9 is used:

(a) if $p^k(i) < 1-threshold_{decr}(i)$, then `decr_parameter(i)` is set to $threshold_{decr}(i)$. Multiplicative reduction by $threshold_{decr}(i)*(1-p^k(i))$ is a considerable reduction. The reason for this drastic decrease is that first, it reduces the probability of packet losses during the next interval and second, enables a new flow (which is likely to exist) to compete for more bandwidth. Otherwise, the new flow's rate will be kept around its initial rate.

(b) if $p^k(i) \geq 1- threshold_{decr}(i)$, the factor $\beta(i)$ is set to $(1-p^k(i))$.

In the case the visited state is state `decr_due_nofeedback`, the factor used is $param_{nofeed}(i)$, with typical value of 0.75. Table 4 summarizes the different values of the multiplicative decrease factor.

Table 4. Multiplicative factor $\beta(i)$.

Visited State	Multiplicative factor $\beta(i)$
`decrease`	$\beta(i) = decr\_param(i)*(1-p^k(i))$; $decr\_param(i) = 0.98$
`low_decrease`	$\beta(i) = (1-p^k(i))$
`fast_decrease`	if $(1-p^k(i)) > threshold_{decr}(i) \{\beta(i) = threshold_{decr}(i)*(1-p^k(i))\}$ else $\{\beta(i) = (1-p^k(i))\}$
`decr_due_nofeed.`	$\beta(i) = param_{nofeed}(i)$

4. Rate Locking/Unlocking Procedure

The D.AI-L.MD scheme ensures that each flow converges to the fairness point, since it is an AI-MD scheme [23]. When a fair rate or the access bandwidth capacity (ceiling) is reached, an effective rate adaptation scheme should try to (temporarily) lock the flow's rate. This locking should not be permanent but should be reconsidered periodically by employing a proper bandwidth probing mechanism.

To lock the rate, a good estimation mechanism of the current fair rate (or the effective access bandwidth ceiling) should be employed. Since the flow's rate follows a pattern of increases and decreases, a good estimate of the locking rate should be based on recent rates (as opposed to the current rate only), which should be properly weighted. Let *local_min(i)* denote the rate just before the most recent transition from state decrease to state increase and let

$$sm\_local\_min^{m+1}(i) = f_2 * sm\_local\_min^m(i) + (1-f_2) * local\_min(i), \text{ where } f_2 = 0.9$$

denote the smoothed running average of the *local_min(i)* rates. If the current rate is above the fairness point it will be decreasing (non-monotonically in general) to reach the fairness point, as the AI-MD characteristic of the algorithm guarantees; during this period both *local_min(i)* and *sm_local_min(i)* will be monotonically non-increasing. Similarly, if the current rate is below the fairness point, it will be increasing (non-monotonically in general) and during this period *local_min(i)* and *sm_local_min(i)* will be monotonically non-decreasing. When the rate reaches the fair rate level (fairness point), both the current rate and *local_min(i)* will oscillate around it. After some time period – which depends on the value of f_2 – *sm_local_min(i)* will reach the current *local_min(i)* and this marks the time instant when the rate is locked ($B_{Lock}(i)$ returns 1). The greater the value of f_2 the more accurate the locking rate is expected to be (that is, closer to the fairness point) and the longer time it would take to lock. The locking rate is not the one in effect at the locking time instant but rather it is a smoothed rate based on recent actual rates (and not the *local_min(i)* rates), given by

$$aver\_rate\_recent^{k+1}(i) = f_r * aver\_rate\_recent^k(i) + (1-f_r) * r^k(i) \quad (1)$$

with a selected value for $f_r = 0.67$. It should be noted that locking the rate near the fairness point is possible for flows with unlimited access bandwidth or flows whose access bandwidth is higher than the fairness point. Flows whose access bandwidth is below the fairness point are locked near their ceiling rate.

As mentioned earlier, it may take a relatively long time to identify the locking time instant if the procedure mentioned above is followed. To reach a rate locking decision faster for flows whose ceiling is close to $r_{min}(i)$ (such as for dial-up users) and avoid harming rate oscillations and packet losses, a different rate locking strategy is followed for such flows. The rate is locked ($B^L_{Lock}(i)$ returns 1) to the *aver_rate_recent(i)* rate (as in (1)) as soon as more than a small number (equal to go_for_low_locking(i) and typically three) transitions occur from state low_increase to low_decrease.

While in state lock the algorithm moves to state fast_decrease and, thus, unlocks under a high nonzero-PLR feedback ($diff^k(i)) > p_{new\_flow}(i)$). After spending some time in state lock, state check_for_bw is visited, after zero-PLR feedback, from which the algorithm may unlock the rate or eventually return to state lock.

Since the procedure for visiting state low_lock is not as reliable as the one mentioned earlier and the locking rate may be incorrect, a procedure to unlock the rate is introduced in an earlier section 2.2. It is reminded that from state low_lock the state low_decrease is visited (unlocking) either if packet losses occur during the first cancel_low_locking(i)) intervals (typically two) after the locking decision or high nonzero-PLR feedback is received.

5. Simulations Results

The network simulator (ns) [26] – with some developed RTP/RTCP support enhancements – and the topology shown in Figure 2 are employed in the simulations.
Nine sources and receivers are defined. Each source s_i transmits an A-CBR flow i
over RTP/RTCP to its associated receiver r_i. Receivers r_1 and r_2 are connected at 64
kbps - simulating dialup users connected at this rate; receivers r_3 and r_4 are connected
at 0.5 Mbps - simulating ADSL users connected at this rate; receivers r_5 to r_9 are connected at 1.5 Mbps, simulating LAN or ADSL users connected at this rate. Flows 1-4
are access network limited, whereas flows 5-9 are only backbone network limited.
The backbone links are full duplex, and their bandwidth capacity is 4 Mbps. The
available memory sizes is equal to 25 packets for the interfaces that connect the
routers and 2 packets size for the interfaces that connect sources and receivers to
routers (1 and 4). The routers are drop-tail routers. The RTCP Sender and Receiver
Reports (SR, RR) are sent every five (5) seconds. No TCP flows are assumed to be
present to demonstrate the behavior of the proposed multi-state rate adaptation algorithm in an environment it is primarily designed for.

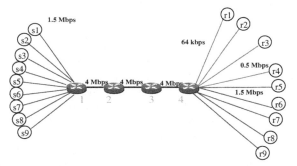

Fig. 2. Network topology and access configuration.

In all simulation configurations, the flows are initiated and terminated according to
Table 5. Flow 6 is terminated at the 1200th second, whereas flow 9 is initiated at the
1800th second to demonstrate the response of the simulated schemes to flow initia-
tion/termination. The RTP packet size is common for all configurations and flows and
set equal to 1000 bytes. The simulation duration is 2000 seconds. The following parameters are used: $r_{min}(i)$=56 kbps, $r_{max}(i)$=1.2 Mbps, $Incr(i)$ = 30 kbps for
all flows i, assuming that the sources are not aware of any access capacity limitation.

Table 5. Start/stop times of flows and rates.

flow	1	2	3	4	5	6	7	8	9
start time in sec.	0	1.33	2.17	3.22	4.5	5.17	6.3	7.1	1800
end time in sec.	2000	2000	2000	2000	2000	1200	2000	2000	2000
initial rate (kbps)	56	128	440	550	400	600	1,180	1.200	600

Two different schemes are simulated and compared in order to demonstrate the performance of the proposed D.AI-L.MD scheme. The first scheme is the classic AI-MD scheme with constant increase step of 30 kbps and decrease factor of 0.97. This factor is close to the corresponding factor for an average packet loss rate of 1% of the D.AI-L.MD scheme (0.98*(1-0.01)). A decrease factor less than this value, e.g., 0.85 or 0.5, leads to a more aggressive decrease, which may be appropriate for data but not for CM flows. Figure 3a illustrates the adaptation behavior of the AI-MD scheme.

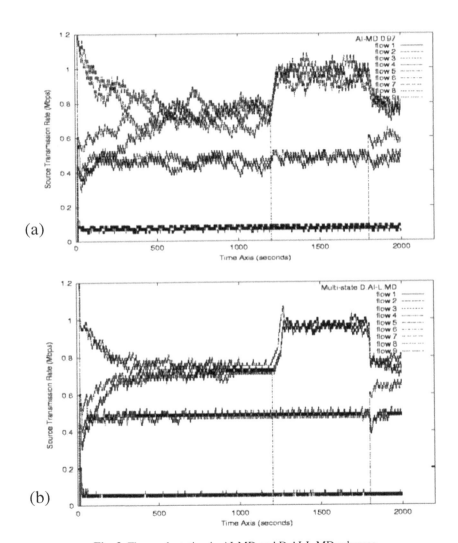

Fig. 3. Flows adaptation in AI-MD and D.AI-L.MD schemes.

The second scheme is the proposed multi-state rate adaptation algorithm (D.AI-L.MD). The following parameters of the D.AI-L.MD scheme are used: $d_{init}(i)=5$, $param_{nofeed}(i)=0.75$, $decr\_param(i)=0.98$, $threshold_{decr}(i)=0.9$, $max\_cons\_times\_in\_decrnofeed(i)=3$, $incr\_for\_low(i)=1\ kbps$, $p_{new\_flow}(i)=0.06$ for flows 1-2, $p_{new\_flow}(i)=0.04$ for flows 3-9, $times\_in\_rmin(i)=3$, $go\_for\_check\_bw(i)=8$, $go\_to\_unclock(i)=4$, $go\_for\_low\_check\_bw(i)=10$, $cancel\_low\_locking(i)=2$, $go\_to\_low\_unlo$ $ck(i)=6$, $escape\_low\_states(i)=20$. For the function $go\_for\_fast\_increase(i)$ $= fi\_min(i)+(1-dist\_weight(r^k(i))*(fi\_max(i)-fi\_min(i))$, the values of 2 and 12 are used for parameters $fi\_min(i)$ and $fi\_max(i)$, respectively. Figure 3b illustrates the adaptation behavior of the D.AI-L.MD scheme.

In view of the perceptual QoS requirement of CM flows it is highly desirable that the rate adaptations are smooth and as infrequent as possible. Oscillatory rate adaptation behavior leads to highly fluctuating perceived quality that may not be tolerated by the end users. The source also benefits from the fewer and smoother adaptations since it is less resource demanding. The congestion scheme that utilises the AI-MD scheme presents the largest deviation of the current rates from their running long-term rate average (Figure 3a), compared to proposed D.AI-L.MD scheme (Figure 3b). The latter scheme presents smoothed long-term oscillatory behavior. The AI-MD scheme induces a greater number of adaptations than that induced by the D.AI-L.MD scheme (Table 6).

Table 6. Occurred adaptations in AI-MD and D.AI-L.MD schemes.

Flows	1	2	3	4	5	6	7	8	9	Total
AI-MD	397	396	397	396	396	236	396	396	39	3049
D.AI-L.MD	218	197	180	229	371	202	361	356	30	2144

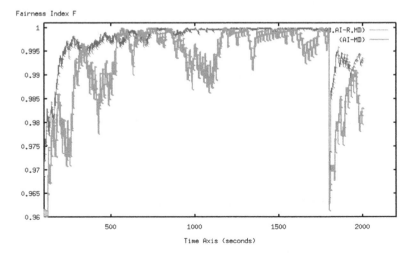

Fig. 4. Fairness index F in AI-MD and D.AI-L.MD schemes.

Fairness and convergence to fairness are requirements set when multiple flows (users) share multiple resources [23]. The fairness index F, presented in [23], is used to measure the fairness among the A-CBR CM flows. This metric is not meaningful for the limited-access bandwidth flows, i.e., the flows 1-4, and thus, the fairness index F calculation is based on flows 5-9: $F^* = (r^k(5)+ r^k(6)+ r^k(7)+ r^k(8)+ r^k(9))^2 / 5*(r^k(5)^2 + r^k(6)^2 + r^k(7)^2 + r^k(8)^{2+} r^k(9)^2)$.

In Figure 4, the line y=1 indicates the optimal fairness. The time required to reach this line and the achieved level are measures of the convergence to fairness and fairness. As illustrated in Figure 4, the D.AI-L.MD scheme requires less time to converge to the fairness point, as well as presents smaller deviation from it, which enables a more accurate locking.

Table 7. Mean long-term packet loss rate (%).

Table 8. Mean conditional packet loss rate (%).

Table 9. Total received Kbytes.

Flows	AI-MD	D.AI-L.MD
1	16.522	1.657
2	16.857	1.935
3	1.156	0.483
4	1.215	0.562
5	0.711	0.275
6	0.730	0.282
7	0.726	0.272
8	0.757	0.301
9	2.00	0.485

Flows	AI-MD	D.AI-L.MD
1	16.609	4.570
2	16.389	3.877
3	1.674	0.933
4	1.587	0,922
5	1.288	0.894
6	1.266	0.687
7	1.215	0.690
8	1.256	0.758
9	2.005	0.936

Flows	AI-MD	D.AI-L.MD
1	15,700	15,250
2	15,698	15,197
3	115,710	119,679
4	114,442	118,792
5	189,403	192,803
6	97,801	102,036
7	205,214	200,336
8	213,642	204,078
9	14,405	15,767
Total	982,015	983,938

Table 7 shows the *mean long-term packet loss rate*, that is the total number of lost packets over the total number of transmitted packets during the simulation time. Table 8 shows the *mean conditional packet loss rate*, that is the mean value of the reported packet loss rate averaged over the intervals during which nonzero losses occurred. The mean conditional packet loss rate shows the average level of losses when they occur. The D.AI-L.MD scheme presents significantly lower loss rates than the AI-MD scheme, especially for flows 1 and 2, which transmit at a low rate due to the access network limitations.

Statistics concerning the bytes sent and received are gathered. Table 9 presents the total received Kbytes during the simulation period. Figure 5 illustrates the throughput, that is the received RTP packets, over the time window of 1000-1500 seconds. The throughput during this time interval is typical for the overall simulation time. As observed from Table 9 and Figure 5, the D.AI-L.MD scheme is slightly better than the AI-MD scheme in terms of throughput.

As far as the time to recover the bandwidth released by a terminated flow (responsiveness) is concerned, the AI-MD scheme reaches the bandwidth capacity faster- due to the constant increase step – than the D.AI-L.MD scheme (Figure 6). The D.AI-L.MD scheme seizes the bandwidth at a slower pace than the AI-MD scheme, because of the distance-weight additive increase feature of the scheme. The responsiveness in the D.AI-L.MD scheme depends on the distance of the flow's rate from the $r_{max}(i)$. The larger the distance (relatively low rate flows), the faster the response. The less the distance (relatively high rate flow, as in the simulation case), the slower the response. Despite the slower response to the bandwidth release for relatively high rate flows, the D.AI-L.MD scheme is better than the AI-MD in terms of fairness, convergence to fairness, packet losses, throughput and oscillatory behavior.

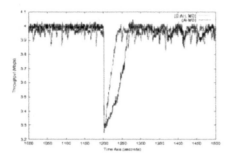 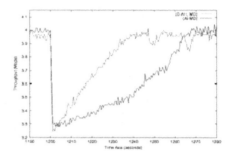

Fig. 5. Throughput of the schemes.   **Fig. 6.** Response to bandwidth availability.

6. Conclusions

In this paper, a multi-state congestion control algorithm for rate adaptable unicast CM flows in a TCP-free environment has been presented. This scheme (a) provides a finer rate adaptation decision space compared to traditional schemes, (b) enables rate locking, and (c) responds well to a diversity of network situations, such as limited access bandwidth and initiation/termination of CM flows. The proposed D.AI-L.MD scheme presents significantly less packet loss rates and oscillatory behavior compared to the AI-MD scheme. It also presents faster and closer convergence to the fairness point. The AI-MD scheme presents better responsiveness to bandwidth availability in comparison to the D.AI-L.MD scheme. Further work is in progress to further investigate the performance of the proposed scheme and extend it for environments in which both TCP/IP and A-CBR CM flows co-exist in the access and/or core network.

References

[1] Floyd, S., "TCP and Explicit Congestion Notification". ACM Computer Communication Review, V. 24 N. 5, October 1994, p. 10-23.
[2] RFC 2481. Ramakrishnan, K.K., and Floyd, S., "A Proposal to add Explicit Congestion Notification (ECN) to IP". January 1999.

[3] V. Jacobson, "Congestion avoidance control". In Proceedings of the SIGCOMM '88 Conference on Communications Architectures and Protocols (1988).

[4] RFC-2001. "TCP Slow Start, Congestion Avoidance, Fast Retransmit, and Fast Recovery Algorithms".

[5] RTP Frequently Asked Questions: http://www.cs.columbia.edu/~hgs/rtp/faq.html .

[6] RFC 1889.RTP: A Transport Protocol for Real-Time Applications.

[7] Floyd, S., and Fall, K., "Promoting the Use of End-to-End Congestion Control in the Internet". IEEE/ACM Transactions on Networking. August 1999.

[8] I. Busse, B. Defner, H. Schulzrinne, "Dynamic QoS Control of Multimedia Application based on RTP". May 1995.

[9] J. Bolot,T. Turletti, "Experience with Rate Control Mechanisms for Packet Video in the Internet'". ACM SIGCOMM Computer Communication Review, Vol. 28, No 1, pp. 4-15, Jan. 1998.

[10] D. Sisalem, F. Emanuel, H. Schulzrinne, "The Loss-Delay Based Adjustment Algorithm: A TCP-Friendly Adaptation Scheme". 1998.

[11] D. Sisalem and A. Wolisz, "LDA+: Comparison with RAP, TFRCP". IEEE International Conference on Multimedia (ICME 2000), July 30 - August 2, 2000, New York.

[12] D. Sisalem and A. Wolisz, "MLDA: A TCP-friendly congestion control framework for heterogenous multicast environments". Eighth International Workshop on Quality of Service (IWQoS 2000), 5-7 June 2000, Pittsburgh.

[13] D. Sisalem and A. Wolisz, "Constrained TCP-friendly congestion control for multimedia communication" tech. rep., GMD Fokus, Berlin Germany, Feb. 2000.

[14] D. Sisalem, A. Wolisz, "Towards TCP-Friendly Adaptive Multimedia Applications Based on RTP". Fourth IEEE Symposium on Computers and Communications (ISCC'1999), (Red Sea, Egypt), July 1999.

[15] R. Rejaie, M. Handley, D. Estrin, "An End-to-end Rate-based Congestion Control Mechanism for Realtime Streams in the Internet". Proc. INFOCOMM 99, 1999.

[16]J. Padhye, J. Kurose, D. Towsley, R. Koodli, "A Model Based TCP-Friendly Rate Control Protocol". Proc. IEEE NOSSDAV'99 (Basking Ridge, NJ, June 1999).

[17] ISO/IEC 14496. MPEG-4.

[18] J. Pasquale, G. Polyzos, E. Anderson, and V. Kompella, "Filter propagation in dissemination trees: trading off bandwidth and processing in continuous media networks". Proc. of NOSSDAV'93.

[19] E. Amirm, S.McCanne, H.Zhang, "An application-level video gateway". Proc. ACM Multimedia '95, San Francisco, Nov. 1995.

[20] IST VideoGateway project: http:/www.optibase.com.

[21] ISO/IEC 13818. "MPEG-2 - Generic coding of moving pictures and associated audio information".

[22] L. Vicisano, J. Crowcroft and L. Rizzo, "TCP-like Congestion Control for Layered Multicast Data Transfer". Proc. InfoCom'98, San Francisco, March/April 1998.

[23] R. Jain, K. Ramakrishnan, and D.-M. Chiu. "Congestion avoidance in computer networks with a connectionless network layer." Tech. Rep. DEC-TR-506, Digital Equipment Corporation, Aug. 1987.

[24] J. Padhye, V. Firoio, D. Towsley, J. Kurose, "Modelling TCP Throughput: a Simple Model and its Empirical Validation". Proceedings of SIGCOMM 98.

[25] T. Kim, S. Lu and V. Bharghavan, "Improving Congestion Control Performance Through Loss Differentiation". International Conference on Computers and Communications Networks '99, Boston, MA. October 1999.

[26] The network simulator (ns): http://www.isi.edu/nsnam/ns/ns-build.html

PCP-DV: An End-to-End Admission Control Mechanism for IP Telephony

G. Bianchi[1], F. Borgonovo[2], A. Capone[2], L. Fratta[2], and C. Petrioli[3]

[1] Universitá di Palermo
[2] Politecnico di Milano
[3] Universitá di Roma "La Sapienza"

Abstract. In this paper we describe a novel endpoint admission control mechanism for IP telephony: the PCP-DV which is characterized by two fundamental features. First, it does not rely on any additional procedure in internal network routers other than the capability to apply different service priority to probing and data packets. Second, the triggering mechanism for the connection admission decision is based on the analysis of the delay variation statistics over the probing flow. Numerical results for an IP telephony traffic scenario prove that 99th delay percentiles not greater than few ms per router are guaranteed even in overload conditions.

1 Introduction

It is widely accepted that the today best effort Internet is not able to satisfactorily support emerging services and market demands, such as IP Telephony. Real-time services, in general, and IP telephony, in particular, require very stringent delay and loss requirements (less than 150 ms mouth-to-ear delay for toll quality voice), that have to be maintained for all the call holding time. The analysis of the delay component in the path from source to destination shows that up to 100-150 ms can be spared for compression, packetization, jitter compensation, propagation delay, etc [1], leaving no more than few tens of ms for queueing delay within the many routers on the path.

Many different proposals aimed at achieving such a tight QoS control on the Internet have been discussed in IETF. IntServ/RSVP (Resource reSerVation Protocol) [2,3] provide end-to-end per-flow QoS by means of hop-by-hop resource reservation within the IP network. Such approach imposes a significant burden on the core routers, which are required to handle per flow signaling and forwarding information on the control path.

A completely different approach is provided by Differentiated Services (DiffServ) [5,6]. In DiffServ, core routers are stateless and unaware of any signalling. They merely implement a suite of buffering and scheduling techniques and apply them to a limited number of traffic classes, whose packets are identified on the basis of the DS field in the IP packet header. The result is that a variety of services can be constructed by a combination of: (i) setting packets DS bits

S. Palazzo (Ed.): IWDC 2001, LNCS 2170, pp. 470–480, 2001.

at network boundaries, (ii) using those bits to determine how packets are for-warded by the core routers, and (iii) conditioning the marked packets at network boundaries in accordance with the requirements or rules of each service.

While DiffServ easily enable resource provisioning performed on a manage-ment plane for permanent connections, their widely recognized limit is the lack of support for per-flow resource management and admission control, resulting in the lack of strict per flow QoS guarantees. Recently, a number of propos-als have shown that per flow Distributed Admission Control schemes can be deployed over a DiffServ architecture [7,8,9,10,11,12,13,14,15,16,17]. Although significantly differing in implementation details, these proposals, referred here-after with the descriptive name Endpoint Admission Control (EAC) (according to the overview paper [17]) share the common idea that accept/reject decisions are taken by the network end points and are based on the processing of "prob-ing" packets, injected in the network at set-up to verify the network congestion status.

A detailed analysis of the above EAC solutions shows that most of them rely on some advanced form of cooperation from internal network routers, e.g. prob-ing packet marking [16,17] and ad hoc probing packet management techniques [15]. More radical EAC solutions have been proposed in [8,10]. These schemes require the routers to be only capable of distinguishing between probing and data packets (e.g. via TOS precedence bits, or DSCP field of DiffServ), and to configure elementary buffering and scheduling schemes, already available in cur-rent routers. In particular, the capability to apply to probing and data packets a priority-based forwarding scheme is the only router requirement in [9], while in [8] a strict limit on the probing buffer size is also enforced. This simplicity makes these radical schemes, hereafter referred to as "pure" EAC, suited to be introduced in the Internet in a very short time frame.

In this paper, we propose an improved version of the scheme presented in [10] called PCP-DV (Phantom Circuit Protocol - Delay Variation) where the ac-ceptance test is based on delay variation analysis. We have thoroughly evaluated the performance of PCP-DV, for a wide range of parameter settings, and proved that PCP-DV is indeed capable of providing 99th delay percentiles not greater than few ms per router even in heavy overload conditions.

The paper is organized as follows. In section 2, the PCP-DV operation is described, and the rationale at the basis of the PCP-DV decision criterion is provided. Section 3 describes the simulation model, and presents the VoIP vari-able bit rate traffic scenario adopted. Section 4 is dedicated to the performance evalation and parameter tuning of PCP-DV. Conclusions are drawn in section 5.

2   PCP-DV Basic Operation

PCP-DV is an acronym for "Phantom Circuit Protocol with acceptance test based on Delay Variation analysis". The PCP-DV connection setup scheme is shown in Figure 1. A user that wants to setup a connection starts a preliminary *Probing Phase*. Scope of this phase is to verify, by means of probing packets

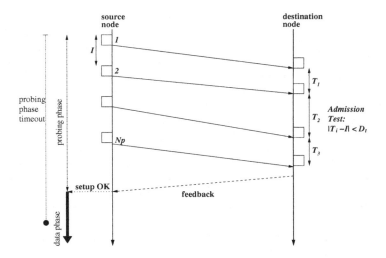

Fig. 1. PCP-DV probing and data phases

injected in the network by the source node, whether there are enough resources in the network to accept a new connection. The decision whether to accept or reject the connection is taken at the destination node, based on measures of the probing flow arrival statistics.

The probing phase lasts until an explicit accept/reject *feedback packet* is received from the destination node, or a suitable timeout expires. In case of acceptance, the sender enters the *Data phase* in which data packets (i.e. information packets) are transmitted.

The PCP-DV probing phase, graphically shown in figure 1, consists in the transmission of a fixed number N_p of packets with a fixed interdeparture time I. Once all the N_p packets are transmitted, the source waits until a feedback arrives, or the probing phase timeout expires.

The only requirement PCP-DV imposes to the core network is the capability to distinguish probe packets from data packets. Probes and data packets are tagged with a different label in the IP header (TOS or DSCP field). Core routers have the only task of forwarding packets according to a head of the line priority scheme: high priority to data packets, low priority to probing packets. This forwarding mechanism serves a probing packet only when the data packets queue is empty. Therefore, it guarantees that the probing traffic load, which is not admission-controlled, does not contend the use of bandwidth against established traffic. Furthermore, since probing packets use only resources not used by accepted calls, their flow received at the destination contains indirect information on the links congestion status that can be used to perform the accept/reject test.

In PCP-DV, the decision on whether to admit or reject a new connection is taken at the destination, and notified back to the traffic source by means of one (or more) *feedback packets*. The PCP-DV decision criterion operates as follows. Upon reception of the first probing packet, the destination node starts a timer, and measure the next probing packet interarrival time T. If the condition:

$$I - D_t \leq T \leq I + D_t \tag{1}$$

where D_t is a parameter called "acceptance delay threshold" and represents the maximum tolerance on the received probing packets jitter, is met, the timer is restarted and the above procedure is iterated until all N_p packets are received. Conversely, if the condition fails for one received probing packet or the timer exceeds the upper limit $I + D_t$ without receiving any probing packet, the connection is rejected, and a feedback rejection packet is immediately sent back to the source. The parameters, D_t and N_p allow to tune the PCP-DV effectiveness in controlling the accepted traffic load.

The rationale of the proposed scheme comes from the observation that toll quality delay performance requires a strict control of the load accepted in a link.

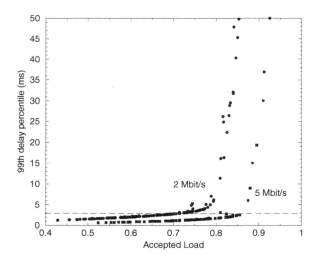

Fig. 2. Throughput/delay tradeoffs

Figure 2 shows the 99-th delay percentile of the accepted traffic, versus the accepted traffic load. These results have been obtained by simulating various EAC schemes (PCP-DV plus the approaches presented in [14,13]), with several parameter settings in a 2 and 5 Mbit/s channels. We have observed that the delay behavior is independent from the acceptance scheme adopted. The sharp knee shape curves allow to determine a threshold on the accepted load e.g. 75%

to 79% for a 2 Mbit/s link, and 86% to 88% for a 5 Mbit/s link corresponding
to a few $(3 \div 5 \text{ ms})$ of the 99-th delay percentile. The solution of the delay QoS
is then reduced to the load control on the links.

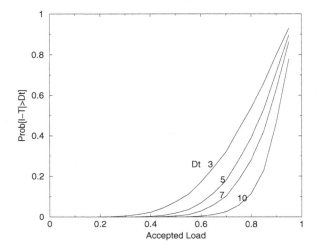

Fig. 3. Probability that the probing packets jitter exceeds a given delay threshold (3,
5, 7, and 10 ms), versus link load - 2 mbit/s link capacity

By simulation we have measured the probing packets delay variation and in
Figure 3 we show the probability that delay jitter $|I - T|$ exceeds the threshold
delay versus the accepted load. Four acceptance delay thresholds, equal to 3, 5, 7
and 10 ms, have been considered. From the figure, we see that the curves become
sharper as the D_t increases. However, none of the thresholds is able to provide
a sufficiently high rejection probability at the critical load of 0.75. To reach
the required efficiency of the test we need to consider multiple samples of the
jitter. The power of the test performed in PCP-DV is measured by the rejection
probability over $N_p - 1$ measures shown in Figure 4. The goal to limit the traffic
to 0.75 can be reached either with $D_t = 3$ ms and $N_p = 11$ or with $D_t = 10$
ms and $N_p = 77$. Even if the two cases achieve the goal to limit the traffic,
the curve with a small number of samples has an higher rejection probability
at loads below the threshold. Therefore, there is a tradeoff between the number
of probing packets (i.e. the probing phase duration) and the effectiveness of the
test at low loads.

For the correct operation of PCP-DV, it is requested that the links load actually reflects the resources seized by the accepted calls. Therefore, in case of VBR
traffic, conditioning mechanisms may be required to send dummy packets when
sources are inactive or under-utilize the assigned resources. Such shaping procedures, common to resource reservation techniques based on traffic measurements

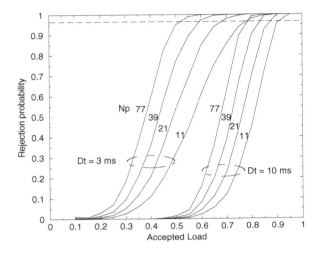

Fig. 4. call rejection probability versus load, for 3 ms and 10 ms delay thresholds, and for different number of probing pairs

(see for example [4]), constitute the price to pay for the reduced complexity of the call admission procedure.

PCP-DV, differently from stateful centralized solutions, can provide only a single QoS requirement. In fact, as long as only two priority levels (probing/data) are used within the network routers, heterogeneous real-time connections, with different loss/delay requirements are forced to share the same queue, and thus, regardless of how sophisticated the end-to-end measurement scheme is, they ultimately encounter the same performance. To overcome this limitation PCP-DV can be used to perform call admission within a DiffServ class. Isolation between DiffServ classes can then be achieved by adopting a WFQ-like mechanism assuring a given rate to the PCP-admission-controlled class. However, for the protocol to correctly operate, this mechanism must prevent admission control traffic from borrowing bandwidth from the other classes. For a thorough investigation of the architectural issues related to EAC schemes deployment, including way to implement the classes isolation described above and a discussion of mechanisms to provide multiple levels of service, see [17].

An important feature of PCP is its intrinsic stability and robustness. In fact, when an increase in the accepted traffic above the QoS limits occurs (e.g. because of rerouting of accepted connections, or because of misacceptance due to the unreliability of the measurement process), thanks to the forwarding mechanism employed, the probing traffic is throttled. This will prevent acceptance of new connections and the congestion disappears as soon as some active calls end.

3   Simulation Model

To evaluate the throughut and delay performance of PCP-PD, we have used a simulator written in C++. We have considered an IP telephony variable bit rate traffic scenario. Voice sources with silence suppression have been modeled according to the two states (ON/OFF) Brady model [18]. In particular, each voice call alternates between ON and OFF states. During the ON state, the voice source emits vocal talkspurts at a fixed peak rate $B_p = 32$ kb/s, while in the OFF state it is silent. Both ON and OFF periods are exponentially distributed with mean values equal to 1 s and 1.35 s, respectively. The actvity factor B_u defined as the fraction of time a voice source is in the ON state.

We have considered homogeneous voice sources that generate fixed-sized packets 1000 bits long. This corresponds to an interdeparture packet time equal to 31.25 ms. Probing packets are generated at a constant rate with interdeparture time $I = 26$ ms.

We have considered a dynamic link load scenario where calls are generated according to a Poisson process and have an exponentially distributed duration. For convenience, we define the normalized offered load, ρ, as:

$$\rho = \frac{\lambda}{\mu} \cdot \frac{B_p B_u}{C} \tag{2}$$

where λ (calls/s) is the call generation rate, $1/\mu$ (s) is the average call duration and C (kbit/s) is the channel rate. For a given normalized offered load, the probing traffic load depends on $1/\mu$: the greater the average call duration, the lower the probing load.

Further assumptions in our simulation model are: zero lenk propagation delay, no loss of feedback packets, and instantaneous feedback reception.

4   Performance Evaluation

An extensive simulation campaign has been performed, using the program described in the previous section, to investigate the PCP-DV performance in several scenarios. In this section we summarize the results concentrating on the accepted load and the 99-th packet delay percentile achievable by PCP-DV under different system operation conditions.

Figure 5 shows the normalized accepted load versus the offered load, for two link capacity values ($C = 2, 5$ Mb/s). All curves, which correspond to different parameter settings, show the ability of PCP-DV to guarantee a few ms 99-th delay percentile (the maximum values are indicated in the figure) and the traffic limitation below the thresholds discussed in Figure 2, i.e. 0.75 and 0.85 for a 2 and 5 Mb/s channel, respectively. This ability has been proved even in very high overload conditions, i.e. with an offered load up to 4 times the channel capacity.

However, different parameter settings show different behaviors for offered traffic loads ranging between 0.5 and 1.5. In these practical operational conditions, better performance is obtained, as anticipated in Figure 4, by adopting a

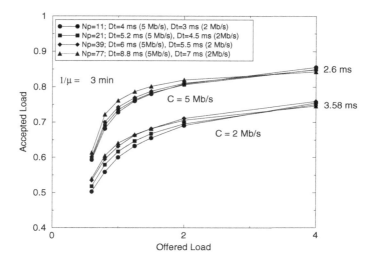

Fig. 5. PCP-DV: Accepted vs. Offered load for varying interval size and measurement period length; 2 and 5 Mb/s link capacity. 99th delay percentiles are also reported for selected samples.

longer probing phase ($N_p = 77$) and larger D_t (7 ms). With these parameter values, the acceptance test is less likely to reject calls in underload traffic conditions and results in an improvement of 20% of the accepted load for offered load equal to 1. To optimize the performance it is therefore suggested to choose the probing phase as long as possible compatibly with other performance requirements such call setup time lenght.

The robustness of the PCP-DV parameter setting with respect to the probing traffic load is proved by the results in Table 1. Increasing the call duration from 3 minutes to 10 minutes, the probing load reduces to 1/3, but the optimal delay thresholds remain practically unchanged.

Table 1. Optimal delay threshold, D_t, for call duration of 3 and 10 min

N_p	$1/\mu = 3$ min	$1/\mu = 10$ min.
11	3.0 ms	2.7 ms
21	4.5 ms	4.2 ms
39	5.5 ms	5.3 ms
77	7.0 ms	6.8 ms

To extend the PCP-DV performance evaluation from the single link case, so far considered, to a multi-link network scenario we have considered the network

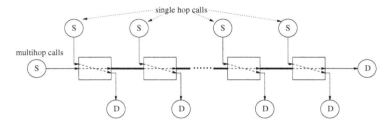

Fig. 6. Multi-link scenario.

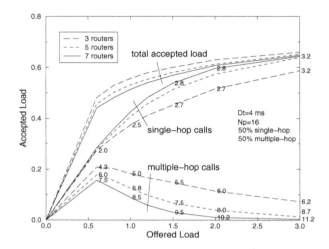

Fig. 7. Multi-link scenario: Accepted vs. Offered load for single hop and multi-hop calls and 99th delay percentiles for selected samples; 2 Mb/s links capacity.

in Figure 6, loaded by multi-hops calls, that cross all the routers, and single hop calls each one loading one link only.

We have simulated an homogeneous scenario in which all links have the same capacity, equal to 2 Mbit/s. Traffic is generated as in the single link case, with average connection duration equal to 10 minutes. Figure 7 shows the accepted versus the offered load and the 99th percentiles of the delay distribution for the two different types of calls and several network sizes (i.e. number of routers).

The number l of crossed routers has a negligible effect on the total accepted load, but, as expected, as l increases, we observe a higher percentage of admitted short calls. In fact, longer calls are more likely to detect "peak" load overflow in one of the many crossed links. Even if this is an expected behavior of any acceptance control scheme, one should note that PCP-DV tends to throttle the multi-hop traffic even in low load conditions. The 99-th delay percentiles, also reported in Figure 7, confirm that, though the multi-hop delay increases with l, the target of a few tens of ms for a backbone can be met.

5   Conclusions

In this paper we have described the "Phantom Circuit Protocol Delay Variation" (PCP-DV), a fully distributed end-to-end measurement based connection admission control mechanism able to support per flow QoS guarantees in IP networks. This scheme determines whether a new connection request can be accepted based on delay variations measurements taken on the probing packet at the edge nodes. The only capability requested to core routers is to implement a 2-priority classes forwarding procedure.

The performance evaluation presented in this paper shows that tight QoS requirements can be supported by suitably engineering the protocol parameters. We have considered the extremely challenging IP telephony scenario and measured that QoS requirements as tight as just a few ms 99-percentile delay can be guaranteed.

The PCP-DV approach conforms to a stateless Internet architecture, and it is fully compatible with the Internet architecture promoted by the Differentiated Services framework.

References

1. P. Goyal, A.Greenberg, C. Kalmanek, W. Marshall, P. Mishra, D. Nortz, K. Ramakrishnan, "Integration of Call Signaling and Resource Management for IP Telephony", IEEE Networks 13(3), pp. 24-32, June 1999.
2. R. Braden, L. Zhang, S. Berson, S. Herzog, S. Jamin, "Resource Re SerVation Protocol (RSVP)–Version 1 Functional Specification", RFC2205, September 1997.
3. J. Wroclawsky, "The use of RSVP with IETF Integrated Services", RFC2210, September 1997.
4. W. Almesberger, T. Ferrari, J. Le Boudec, "Scalable Resource Reservation for the Internet", IEEE IWQOS'98, Napa, CA, USA.
5. K.Nichols, S. Blake, F. Baker, D. Black, "Definition of the Differentiated Services Field (DS Field) in the IPv4 and Ipv6 Headers", RFC2474, December 1998.
6. S. Blake, D. Black, M. Carlson, E. Davies, Z. Wang, W. Weiss, "An Architecture for Differentiated Services", RFC2475, December 1998.
7. F. Borgonovo, A. Capone, L. Fratta, M. Marchese, C. Petrioli ,"End-to-end QoS provisioning mechanism for Differentiated Services", Internet Draft, July 1998.
8. G. Karlsson, "Providing Quality for Internet Video Services", CNIT/IEEE 10th International Tyrrhenian Workshop on Digital Communications, Ischia, Italy, September 1998.
9. F. Borgonovo, A. Capone, L. Fratta, M. Marchese, C. Petrioli, "PCP: A Bandwidth Guaranteed Transport Service for IP networks ", IEEE ICC'99, Vancouver, Canada, June 1999.

10. F. Borgonovo, A. Capone, L. Fratta, C. Petrioli, "VBR bandwidth guaranteed services over DiffServ Networks", IEEE RTAS Workshop, Vancouver, Canada, June 1999.

11. R. J. Gibbens, F. P. Kelly, "Distributed Connection Acceptance Control for a Connectionless Network", 16th International Teletraffic Conference, Edimburgh, June 1999.

12. C. Cetinkaya, E. Knightly, "Egress Admission Control" Proc. of IEEE INFOCOM 2000, Israel, March 2000.

13. G. Bianchi, A. Capone, C. Petrioli, "Throughput Analysis of End-to-End Measurement-Based Admission Control in IP", Proc. of IEEE INFOCOM 2000, Israel, March 2000.

14. V. Elek, G. Karlsson, R. Ronngren "Admission Control Based on End-to-end Measurements", Proc. of IEEE INFOCOM 2000, Israel, March 2000.

15. G. Bianchi, A. Capone, C. Petrioli, "Packet management techniques for measurement based end-to-end admission control in IP networks", KICS/IEEE Journal of Communications and Networks (special issue on QoS in Internet), July 2000.

16. F. Kelly, P. Key, S. Zachary, "Distributed Admission Control", Journal of Selected Areas in Communications, December 2000.

17. L. Breslau, E. Knightly, S. Shenker, I. Stoica, H. Zhang, "Endpoint Admission Control: Architectural Issues and Performance," in Proceedings of ACM SIGCOMM 2000, Stockholm, Sweden, August 2000

18. Brady, P.T., "A Model for Generating On-Off Speach Patterns in Two-Way Conversation", The Bell System Technical Journal, September 1969, pp.2445-2471.

DiffServ Aggregation Strategies of Real Time Services in a WF²Q+ Schedulers Network

Rosario Giuseppe Garroppo, Stefano Giordano, Saverio Niccolini,
and Franco Russo

Department of Information Engineering University of Pisa Via Diotisalvi 2 56126 Pisa – Italy
Tel. +39 050 568511, Fax +39 050 568522
{r.garroppo, s.giordano, s.niccolini, f.russo}@iet.unipi.it

Abstract. The paper presents an analysis in a DiffServ network scenario of the achievable Quality of Service performance when different aggregation strategies between video and voice real time services are considered. Each network node of the analyzed DiffServ scenario is represented by a WF²Q+ scheduler. The parameters setting of the WF²Q+ scheduler is also discussed. In particular, the necessary network resources, estimated by the WF²Q+ parameters setting obtained considering the aggregation of traffic sources belonging to the same service class, are compared with those estimated on a per flow basis. The higher gain achievable using the first approach with respect to the second one, is also qualitatively highlighted. The simulation results evidence the possible problems that can be raised when voice traffic is merged with video service traffic. As a consequence, the paper results suggest to consider in different service class queues the two kinds of traffic.

1   Introduction

A key challenge of the current telecommunication age is represented by the developing of new architecture models for IP networks in order to satisfy the recent QoS (Quality of Service) requirements of innovative IP-based services (e.g. IP Telephony and videoconferencing).

At present, the ISPs (Internet Service Provider) often provide the same service level independently from the traffic generated by their clients. Taking into account the transformation of Internet to a commercial infrastructure, it is possible to understand the need to provide differentiated services to users with widely different service requirements.

In this framework, the DiffServ approach is the most promising for implementing scalable service differentiation in IP networks. The scalability is achieved by considering the aggregate traffic flows and conditioning the ingoing traffic at the edge of the network. Aggregation obviously decreases the complexity of traffic control in the core network, but it produces some unwelcome effects, such as "lock-out" or "full-queues" phenomena, which contribute to increase end-to-end delay and jitter of the traffic flow of a single service. These two phenomena take place respectively when few flows monopolize queue space preventing other connections from getting in the queue and when it is not possible to maintain the queues non-full. Hence, the

S. Palazzo (Ed.): IWDC 2001, LNCS 2170, pp. 481–491, 2001.

effects of the aggregation mechanisms on the QoS parameters of the different aggregated flows need to be further analyzed. The first works in this field have highlighted relevant concepts to support traffic aggregation [1], however a still open issue is what kind of aggregation strategies is better to carry out. To this aim we investigate traffic aggregation strategies because there is still no clear position on what is the better configuration (standardization organisms say anything regarding this matter). On the other hand recent publication [2] suggests to divide network traffic in only two service classes (e.g. real-time and non real-time) but it seems, from our point of view, a little bit restrictive with respect to different traffic features. In the paper, we analyze the impact of the aggregation of real-time video and voice traffic in the same service class (hence, in the same queue in a per-service queueing system) on the experimented QoS parameters of the different flows. The QoS concept used in this work is to be identified with the whole set of properties which characterize network traffic (e.g. in terms of resource availability, end-to-end delay, delay jitter, throughput and loss probability). The results obtained in this scenario are then compared with those obtained considering real time video and voice as separate traffic flows.

The DiffServ architecture [3] is a good starting point but it is useless if there is no teletraffic engineering background able to provide the necessary differentiation. Therefore we should take into account also scheduling disciplines and their dimensioning, in order to understand if they may affect results (changing scheduling discipline change the way the flows are treated). Hence, we have firstly chosen to use one of the best work-conserving scheduling algorithms (WF^2Q+) instead of a non work-conserving one used in other analysis [4]. The choice derives from the assumption that the best the scheduling algorithm is (keeping acceptable its complexity) the better treatment a flow receives in terms of low end-to-end delay and jitter delay. Moreover, the parameters setting of the considered scheduling discipline has been analyzed as described in Section 4.

In the analysis, we use as traffic characterization approach, the LBAP (Linear Bounded Arrival Processes) theory [5]. Furthermore, we investigate the effects on LBAP characterization of the multiplexing of the video traffic, instead of taking into account the simple sum of the traffic descriptors obtained with the single source. By means of simulation analysis we evaluate if the multiplexing gain derived from characterization of aggregated traffic does not affect the QoS parameters.

The rest of the paper is organized as follows. In Section 2 we present the simulation scenario while in Section 3 we describe the voice source model, the video and data traffic taken into account in the simulation analysis. In Section 5, the results are discussed while Section 6 summarizes the main results presented in the paper.

2   Simulation Scenario

Our simulation scenario mainly reflects the topology of a DiffServ domain of an IP network. The simulation scenario is implemented using the OPNET Modeler vers. 6.0.L, a powerful CAMAD (Computer Aided Modeling and Design) tool used in modeling communication systems and in analyzing network performance. The considered scenario is shown in Fig. 1.

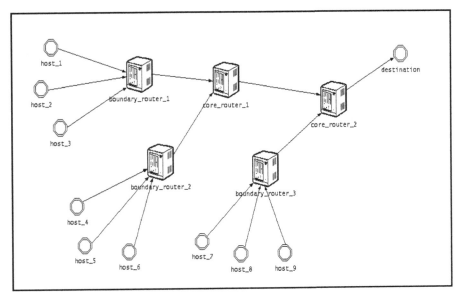

Fig. 1. Simulation Scenario

The network model is represented by edge and core routers, each one having a work conserving scheduler that permit to realize the isolation of the entering flows, based on performance guarantees. The scheduling discipline is a Worst-Case Fair Weighted Fair Queuing with the addition of a shaper, obtained using a Shaped Starting Potential Fair Queuing (SSPFQ), denoted as WF²Q+ [6]. The WF²Q+ is a GPS (Generalized Processor Sharing) approximating service discipline with high fairness properties and relatively low implementation complexity. Moreover, in order to simulate a single *DiffServ* domain, we implement at the edge router the classifier and the marker necessary to associate each packet to the selected PHB (Per Hop Behavior). Based on this classification and marking, each packet receive the suitable forwarding treatment by the core routers. The traffic sources taken into account in the simulation scenario, are the most heterogeneous possible because we want to analyze the performance of a real network; it must integrate the carrying of video, voice and data traffic. Hence, describing the scenario shown in Fig. 1 in more details, every block named as host 1, 4 and 7 contains 15 voice sources, while every block named as host 3, 6 and 9 contains a video source. The remaining hosts contain "data" module, which simulate best-effort traffic.

The statistics we have collected concern the most significant QoS parameters of real-time services, i.e. end-to-end delay and jitter delay, which are evaluated considering the connection among the different sources and the destination node shown in Fig. 1.

3   Source Models

In the simulations, we adopt a model only for the voice sources, while for the other kinds of traffic we consider actual traffic data.

The model used for the voice sources consists in an On-Off process, suggested by the typical behavior of a voice source with VAD (Voice Activity Detection): it is active or inactive depending on the talker is speaking or silent. Assuming that no compression is applied to voice signal, during active periods the source transmits at the constant bit rate of $v=64$ Kbps (this corresponds to a standard PCM codec with VAD). In-depth analyses of this traffic source, shown in literature, have emphasized that the distribution of active and inactive periods lengths can be approximated by an exponential function [7], with mean values respectively equal to $T_{on}=350$ msec and $T_{off}=650$ msec. The packet size is 64 bytes, and considering the bit rate and the header overhead (40 bytes taking into account the RTP/UDP/IP header) the source generates one packet every 3 msec.

Table 1. Statistical parameters of considered video sources

Video flow	Mean_rate (Mbps)	Peak_rate (Mbps)
GOLDFINGER	0.584	5.87
ASTERIX	0.537	3.54
SIMPSONS	0.446	5.77

The traffic data used for the video sources are described in [8], where also their statistical analysis is presented. They have been obtained collecting the output of an MPEG-1 encoder loaded by different sequences of movies half an hour long. Some relevant statistical parameters of the considered traffic data, named Goldfinger, Asterix and Simpsons, are summarized in Table 1.

The video packets are produced at application level, dividing the number of bytes produced by the encoder in the frame period, T=1/24 sec, in consecutive packets of size equal to 1500 bytes (in this case an MTU, Maximum Transfer Unit, of 1500 is supposed). Moreover, in the simulation we consider the 40 byte of overhead, related to the RTP/UDP/IP header, assuming that every packet transports 1460 byte of the traffic data produced by the encoder.

The time interval between the generation of consecutive packets in each frame period, has been considered deterministic and equal to T/N, where N is the number of packets necessary to transport all the bytes produced by the encoder during a frame period. As an example, Fig. 2 presents a case where 3200 bytes are necessary for the encoding of a frame; at application level we divide the frame in three packets, which are sent with time interval equal to T/3.

The Best Effort sources have been obtained considering the traffic data acquired at the Faculty of Engineering of the University of Pisa. In particular, we consider the traffic exchanges by the Faculty of Engineering with the external world (essentially other University sites and Internet) by means of an ATM network at 155 Mbps [9]. The peak rate of the considered traffic is equal to 11 Mbps, while a mean rate of only 400 Kbps has been observed; the high peak-to-mean ratio is a clear evidence of the high burstiness of the data traffic. During the data acquisition both the arrival time and the size of each packet have been registered. Hence, in this case the packet generation process to use in the simulations is directly obtained from the traffic data.

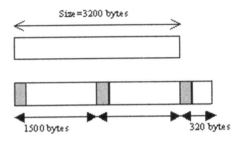

Fig. 2. An example of packet fragmentation at application level

4   Parameters Settings

In order to set the scheduler parameters, we first characterize the traffic sources by mean of the LBAP approach. The traffic characterization of the single source is obtained by the parameters (b, ρ) where b is indicated as bucket size and ρ as token rate. The physical interpretation of the LBAP parameters can be understood considering that the number of bytes produced by a single source in a time interval of $(0,\tau)$, $A(\tau)$, is upper bounded by (1).

$$A(\tau) \le b + \rho\tau \qquad \forall\ \tau{>}0 \tag{1}$$

The results presented in [10] permits to have an upper bound for the end-to-end delay experimented by the traffic when it traverses through a Latency Rate schedulers network, as that considered in our simulation scenario (i.e. a WF²Q+ schedulers network); estimation of WF²Q+ latency term is described in [11]. In particular, the delay introduced by a single node to a packet belonging to the *i-th* flow, characterized by the LBAP parameters (b_i, ρ_i) is upper bounded by (2) (in the following we suppose that for the *i-th* flow a service rate equal to ρ_i is allocated).

$$D \le \frac{b_i}{\rho_i} + \Theta_i \tag{2}$$

Where Θ_i represents the latency of the scheduler, defined as

$\Theta_i = \dfrac{L_{i,max}}{\rho_i} + \dfrac{L_{max}}{C}$. $L_{i,max}$ and L_{max} respectively represent the maximum packet

size of the *i-th* flow and of the global traffic arriving to the scheduler, while ρ_i and C, are the service rate allocated to the *i-th* flow and the global output service rate respectively. Extending the analysis to a network of K WF²Q+ schedulers, the end-to-end delay experimented by a single packet belonging to the *i-th* flow is upper bounded by (3).

$$D_i \leq \frac{b_i}{\rho_i} + \sum_{j=1}^{K} \Theta_i^j \qquad (3)$$

Where Θ_i^j indicates the latency of the *j-th* node evaluated for the *i-th* flow.

The end-to-end delay bound has been obtained considering the worst-case analysis, which is more conservative with respect to the experimented end-to-end delay. Furthermore, as shown in [12], also the LBAP traffic characterization is conservative with respect to the statistical modeling approaches. These two considerations leads us to assume that the maximum end-to-end delay of the *i-th* flow can be upper bounded simply by (4).

$$D_i \leq \frac{b_i}{\rho_i} \qquad (4)$$

This hypothesis permits to establish the buffer size and the guaranteed rate to set in the scheduler, simply evaluating the LBAP curve of the *i-th* flow.

The simulation results that will be presented in Section 5 point out the goodness of our assumption, showing the very conservative nature of the worst-case analysis.

Considering the above hypothesis, the procedure used to set the scheduler parameters consists in evaluating the LBAP curve and in finding the point where this curve intersect the straight line $b_i = \rho_i D_i$, where D_i represents the maximum delay fixed for the considered source.

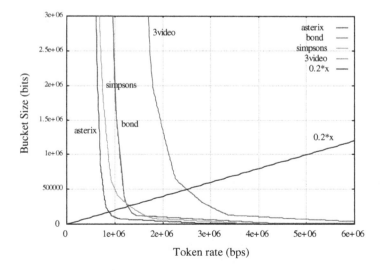

Fig. 3. Video characterization

Fig. 3 shows the presented approach in the case of the video sources. In particular, in the figure we can observe the LBAP curves for three different video sources, and the curve related to the traffic obtained aggregating these sources. Moreover, assuming a maximum end-to-end delay of 200 msec, we can observe the relate straight line and the intersection points with each LBAP curves that, as described above, give the couple (ρ_i, b_i) to consider in the setting of scheduler parameters.

The characterization results obtained with the considered three video flows (named Goldfinger, Asterix and Simpson), the aggregate of 15 voice sources (corresponding to a host in the simulation scenario) and the data traffic are reported in Table 2. In the table, the column Dmax indicated the maximum end-to-end delay analytically obtained from the estimation of scheduler parameters.

Table 2. Traffic characterization of considered sources

Traffic flow	Rate (ρ) (Mbps)	Buffer (b) (Kbit)	Dmax (ρ/b) (msec)
GOLDFINGER	1.25	250	200
ASTERIX	0.83	160	193
SIMPSONS	1.27	260	205
15 VOICE sources	1.10	30	27
Data	0.40	400	1000

In the setting of the scheduler parameters, we can choice to set a service rate equal to the sum of ρ_t and a buffer size equal to the sum of b_i, where ρ_t and b_i are the parameters referring to the *i-th* video source. Considering this approach and the results obtained with the procedure presented above, the total service rate to allocate for the video services and the related buffer size are equal to 3.35 Mbps and 670 Kbits respectively.

Table 3. Parameters of the routers

	Output Service Rate (Mbps)	Buffer size (Kbit)
Boundary routers	3.06	$B_{voice}=30$ $B_{data}=500$ $B_{video}=250$
Core router 1	6.12	$B_{voice}=60$ $B_{data}=1000$ $B_{video}=500$
Core router 2	9.18	$B_{voice}=90$ $B_{data}=1500$ $B_{video}=750$

The other approach consists in the estimation of the parameters directly from the LBAP curve of the multiplexed traffic. In this case, it is expected to obtain a multiplexing gain in the setting of the resources to guarantee to the video service. In particular, considering the same upper bound of the end-to-end delay, we need to allocate a service rate of 2.5 Mbps and a buffer size of 500 Kbits. Then, in terms of buffer size we observe a gain of 170 Kbits (corresponding to a reduction of about 25%), while in terms of service rate a gain of 850 Kbps (25%) is achieved. Considering the service rate evaluated for each kind of traffic source, it is possible to set the scheduler parameter assuming a utilization factor of the link equal to 0.9. In more details, we evaluate the sum of ρ_i, ρ_{tot}, and fix the rate of the output link equal to $\rho_{tot}/0.9$. Hence, for boundary and core routers, the parameters are set as summarized in Table 3.

The buffer size is given for each traffic class, i.e. voice, video and data. When considering the scenario related to the aggregation of voice and video flows, the buffer size for this aggregated class of traffic has been set equal to $B_{voice}+B_{video}$.

5   Simulation Results

The simulation analysis is mainly focused in the evaluation of the impact of different aggregation strategies on the QoS parameters. The low scalabity of the IntServ network architecture, suggests for the IP Telephony scenario to study a DiffServ core network architecture. Hence, it is expected that in each node of an IP Telephony core network, a per-class queueing is implemented and an appropriate scheduling algorithm guarantees the QoS requested by the real time sources.

Two relevant problems arise in this framework. The first concerns the choice of an appropriate scheduling algorithm and of a procedure for the setting of their parameters. The second issue is related to the aggregation strategies to adopt. Hence, in this Section we present the results that give insights on these problems.

We consider two different aggregation strategies: in a first one we have carried video and voice together in a premium class and data in a best effort class (in the figures the related curves are indicated with label containing the word "1 link"); in a second one we have supposed to carry video traffic in a separate queue from voice traffic (curves labeled with the noun containing the word "2 link").

Fig 4 presents the complementary probability of the end-to-end delay experimented by the voice packets when the two different strategies are considered and only one video source is activated (in this case the setting of simulation parameters have been changed according to the dimensioning procedure presented in the previous paragraph, in order to take into account the inactivity of the others video sources). Fig. 5 presents the same curves obtained when all three video sources are active.

As first analysis of the simulation results, it is possible to note that the adopted scheduling algorithm, i.e. WF^2Q+, and the procedure for its parameters setting permit to guarantee the target QoS for the voice sources if the video and voice traffic flows aren't merged in a single queue. In particular, the curves labeled as "1 Link" either in Fig. 4 and in Fig. 5 clearly show that the maximum delay observed during the simulation is under 10 msec, which is lower than the fixed delay of 27 msec, considered in the setting of the scheduler parameters.

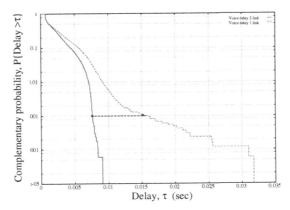

Fig. 4. Complementary probability of voice delay (One active video source)

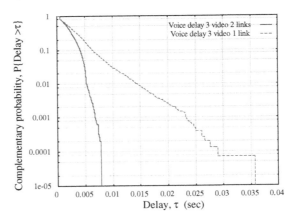

Fig. 5. Complementary probability of voice delay (Three active video source)

Both Fig. 4 and Fig. 5 show the degradation in terms of end-to-end delays of voice traffic when the video flows is merged in the same queue with the voice traffic. Indeed, in the Fig. 4, we can note that in correspondence of a probability P=0.001 a delay of 7 msec is observed in the first case (curve "1 Link"), which is lower than the 15 msec registered in the second case.

Furthermore, this degradation is amplified when the number of video sources is increased. Indeed, in Fig. 5 the maximum end-to-end delay registered for voice traffic is unvaried with respect to the previous case, i.e. about 7 msec, while it is increased to 23 msec, when all real-time flows are aggregated in the same queue.

Hence, in this second case the worsening of the delay parameter of voice service is due to the increase of the number of bursty traffic multiplexed with the voice sources. Hence, we can suppose that if there is more requested bandwidth the need for network resources increase in a non-linear way when considering a wrong strategy of aggregation, in this case the real time application may be damaged seriously.

On the other hand, the video performance take benefits from the aggregation, showing a little delay improvement that can be related to the multiplexing with the voice (the related figures are not reported for sake of simplicity).

The different performance observed with the two considered aggregation strategies, can be related to "lock-out" phenomenon, which plays a decisive role in deteriorating voice performance. Indeed, when the video sources are merged with the voice traffic, the first monopolize the queue space obstructing the second from receiving the desired service level. The "lock-out" phenomenon can be avoided using different queues for video and voice traffic, while, at the same time, the choice of appropriate scheduling disciplines can guarantees an adequate multiplexing gain.

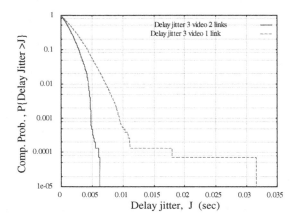

Fig. 6. Complementary probability of voice jitter delay (Three active video source)

Finally, we observed that the best effort traffic (used as background traffic) is not affected by the fusion of the two service classes because the total bandwidth share (video + voice) is unchanged.

The same worsening of performance can be observed when we consider the jitter delay parameter, as shown in Fig. 6, which plots the complementary probability estimated for this statistic (the results are related to the multiplexing of three video sources).

6   Conclusion

The main goal of the paper is the evaluation of different traffic aggregation strategies for real time voice and video services in a *DiffServ* environment. In this framework, the simulation analysis presented in the paper highlights that the wrong aggregation of traffic flows with different statistical features, such as video and voice traffic, may lead to performance worsening, which should be avoided especially in providing IP-based business services, such as IP Telephony.

On the other hand, the simulation results emphasize that with an adequate isolation between video and voice traffic flows and an appropriate dimensioning of network resources, it is possible to provide real-time services. In particular, the considered WF²Q+ schedulers network and the proposed procedure for setting the related parameters, permit to achieve the target QoS. Furthermore, analyzing the proposed procedure for the setting of schedulers parameters, based on the LBAP traffic characterization, it has been possible to highlight the multiplexing gain obtainable considering the LBAP characterization of aggregated traffic.

Finally, the simulation results have evidenced that although we have neglect the latency terms in the expression of the end-to-end delay reported in literature, the maximum delay experimented is lower than that analytically estimated (see the fourth column in Table 2). This result is a further evidence of the very conservative nature of the worst-case analysis.

Acknowledgment

The authors wish to thank D. Riso for his help in the simulation study. This work was partially supported by the project "NEBULA" of the Italian Ministry of University and Tech. Research.

References

1. K. Dolzer, W. Payer, M. Eberspacher "A Simulation study on Traffic Aggregation in Multi-Service Networks", Conference on High Performance Switching & Routing (joint IEEE ATM Workshop 2000), Heidelberg, Germany, 26-29 June, 2000
2. K. Dolzer, W. Payer "On aggregation strategies for multimedia traffic" Proceeding of the first Polish-German Teletraffic Symposium, Dresden, September 2000
3. S. Blake, D. Blake, M. Carlson, E. Davies, Z. Wang, W. Weiss "An architecture for Differentiated Services", Internet RFC 2475, December, 1998
4. H. Naser, A. Garcia, O. Aboul-Magd "Voice over Differentiated Services", Internet Draft, Diffserv Working Group, December, 1998
5. S. Keshav "An Engineering Approach to Computer Networking", Addison-Wesley, January, 1998
6. J. Bennet, H. Zhang "WF2Q: Worst-case Fair Weighted Fair Queueing", Proc. Of IEEE Infocom '96, March, 1996
7. J. N. Daigle, J. D. Langford "Models for Analysis of packet Voice Communications Systems", IEEE JSAC, Vol. 6, pp 847-855, 1986
8. O.Rose "Statistical properties of MPEG video traffic and their impact on traffic modeling in ATM systems", Un.of Wuerzburg - Inst. Of Computer Science Research Report Series. Report N. 101. February 1995
9. R.G. Garroppo, S. Giordano, M. Pagano, G. Procissi, "On the Relevance of Correlation Dependencies in On/Off Characterization of Broadband Traffic", Proc. of IEEE ICC 2000, New Orleans, Louisiana, USA, 18-22 June, 2000
10. D. Stiliadis, A. Varma, "Latency-Rate Servers: A general model for analysis of traffic scheduling algorithms", Tech. Rep. UCSC-CRL-95-38, July 1995
11. A. Charny, F. Baker et al. "EF PHB Redefined" Internet Draft, http://search.ietf.org/internet-drafts/draft-charny-ef-definition-01.txt, November, 2000
12. R. Bruno, R.G. Garroppo, S. Giordano "Estimation of token bucket parameters of VoIP traffic", Proc. of IEEE ATM Workshop 2000, Heidelberg, Germany, 26-29 June, 2000

Towards a New Generation
of Generic Transport Protocols

Patrick Sénac[1,2], Ernesto Exposito[1,2], and Michel Diaz[2]

[1] ENSICA, DMI 1 Place Emile Blouin 31056, Toulouse Cedex, France
{Patrick.Senac, Ernesto.Exposito}@ensica.fr
[2] LAAS du CNRS, 7 Avenue du Colonel Roche, 31077, cedex 4, Toulouse, France
Michel.Diaz@laas.fr

Abstract. Considering that current end to end communication services are not adapted for supporting efficiently distributed multimedia application, this paper introduces a new family of generic transport protocols directly instantiated from application layer quality of service requirements. This Generic Transport Protocol (GTP) has been successfully tested for video on demand systems and is one of the major building block of the currently under development GCAP European project. GTP allows one to apply in the transport layer powerful adaptation mechanisms to the network behavior while preserving application requirements and alleviating network bandwidth and buffering needs

1 Introduction

There is a clear need for a new generation transport protocol layer that could apply an efficient adaptation between application layer needs and network behaviour, capabilities and resources. The management of transport connections could be greatly enhanced by informing the transport layer of the reliability, ordering, synchronisation and temporal constraints associated with Application Data Units. Indeed multimedia applications (i.e. video on demand, web browsing, access to MPEG4 or SMIL documents...) do not need the full reliability and total ordering enforced by TCP. Indeed, these applications have partial order, partial reliability and specific synchronization constraints. Therefore, the use of TCP for multimedia applications induces a service that is not only unneeded by the transport service user but above all that can potentially seriously disrupt the semantics of media streams. This reason promoted UDP as the privileged transport layer for accessing to multimedia streams. However, such a solution oblige to introduce into each application complex mechanisms for enforcing application specific data ordering, synchronisation constraints and loss control. Dedicated application layer protocols such as RTP and RTCP do not greatly alleviate the load and complexity of these network aware applications which have to directly adapt their behaviour to the QoS delivered by the network. Therefore, neither UDP nor TCP are able to offer an efficient service in conformance with the great diversity of application needs. For insuring an efficient mapping between application needs and network

S. Palazzo (Ed.): IWDC 2001, LNCS 2170, pp. 492–506, 2001.
© Springer-Verlag Berlin Heidelberg 2001

behaviour and services, the transport protocol must be aware of the specific ordering, loss and synchronisation constraints related to application data units. Such a generic transport protocol entails an application layer framing approach, which consist in defining, at the application layer, self dependant data units which are also considered by the underlying communication layers as transport data units and network data units [1,2]. If the size of these data units is lower than the size of the maximum transfer unit, such an approach avoids costly fragmentation of data units while allowing the transport protocol to take advantage of data units independency.

TCP and UDP should be two specific instantiations of the considered generic transport protocol that should be able to deliver a continuum of transport services between these two extremities. At the difference of traditional application layer framing approaches which puts all the burden on the application layer and oblige to reinvent the wheel for each application, a generic transport layer has only to be designed once and can be dynamically instantiated to be adapted to specific application needs. In this paper we introduce such a generic transport protocol coupled with a simple and direct derivation technique of a transport layer service from application layer QoS requirements.

In the first part of this paper we briefly introduce a formal technique for modelling multimedia components. Then we demonstrate that this formal approach offers a simple and efficient solution for mapping application layer QoS parameters down to transport layer QoS parameters. Then this formal approach leads us to introduce a new family of Generic Transport Protocol (GTP) that can directly instantiated from the formal expression of application layer QoS requirements. In the last part of this paper we describe a platform independent Java implementation of GTP designed in the framework of the GCAP European Project. Finally two elementary experiments show that, when accessing discrete or continuous media this new generation of transport protocols delivers a service more compliant to the application needs and more efficient than the one offered by UDP or TCP.

2 From Application Layer QoS Requirements to Transport Layer Service.

Ideally, a transport protocol should realize efficient adaptations between the greet diversity of application needs and the network behavior. Current transport protocols either, like UDP, deliberately ignore the application needs and the network behavior or like TCP adapt their behavior to network conditions but deliver always a predefined service that ignores specific application needs. For realizing an effective and efficient adaptation between the network and the application, the transport layer, must be informed by the application of its needs. This customization of the transport service can be done when the application creates a transport service access point (e.g. at socket creation).

We propose a three steps approach for using such a generic transport protocol that can apply efficient adaptation decisions between application layer QoS needs and network behavior and services:

1. Definition of application layer QoS needs based on a formal model that allows the consistency of the application requirements to be checked.
2. Formal derivation of a transport service from the application layer requirements
3. Instantiation of the generic transport protocol with the previously obtained transport service.

```
<smil>
 <body>
 <seq>
 <video src="v1.mpg" region="region_1" dur=15 min=12s max=20s />
  <par begin="0" >
   <seq>
      <img src="i.jpg" region="region_1" dur=12s min=10s max=15s />
   </seq>
   <audio src="moliendo.mp3" dur=12s min=11s max=13s syncMaster=true/>
  </par>
  <video src="v2.mpg"  region="region_1" dur=30 min=25s max=35s />
 </seq>
 </body>
 </smil>                              (b)
```

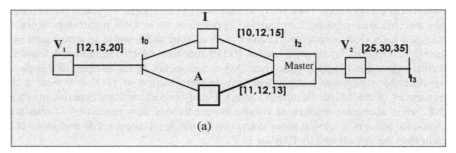

(a)

Fig. 1. Formal modeling of multimedia components. (a) A SMIL document. (b) The translation of the SMIL document into a HTSPN specification

We have previously shown that the design of complex and large scale distributed hypermedia applications can be greatly enhanced with the help of a formal model that allows the fundamental features of these applications to be specified and their properties to be checked [3,4]. Our approach is based on a temporal extension of Petri Nets, called Hierarchical Time Stream Petri Nets (HTSPN), that allows one to express simultaneously with the same formal techniques the reliability, ordering and temporal constraints associated to a multimedia or hypermedia application. Hence, this formal model allows the most fundamental QoS requirements of multimedia applications to be abstractly expressed. Moreover, taking into account the asynchronous behavior of current network services, this model allows one to specify the admissible temporal variability of multimedia components. The specification of the admissible temporal variability of multimedia components is done with the help of a 3-uple (x,n,y) called Temporal Validity Interval (TVI), where x, n and y specify respectively the minimal nominal and maximal admissible durations of the component. This model introduces a complete set of synchronization operators which define a formal semantics of synchronization bfor asynchronous or weakly-synchronous systems [5]. This formal semantics suppresses synchronization non-determinism while offering scheduling flexibility for information access, delivery and presentation.

The modeling power of HTSPNs allows the fundamental QoS requirements of advanced hypermedia components, as defined by the soon available SMIL 2.0 standard, to be abstracted and formally expressed (Figure 1).

In summary the HTSPN model allows not only the ordering requirements of the application to be specified through recursive sequential and parallel composition of media elements, but also the reliability and temporal requirements to be expressed with the help of powerful temporal synchronization rules. For instance the specification given in Figure 1-b states that the audio is the master stream of the inter-stream synchronization scheme between the audio stream and the image stream (this is graphically expressed with the help of a bold arrow). This specification is done with the help of a master synchronization transition which states that:

- The audio stream has to be fully presented to the user
- The image stream may be partially presented

3 Deriving a Transport Service from Application Requirements

3.1 The Order and Reliability Dimensions

In the previous section we have seen that the HTSPN model allows one to express three fundamental QoS features of multimedia applications:

1. Ordering constraints between the various application data units or components
2. Reliability constraints
3. Time constraints

Because order and reliability constraints are intrinsic to a wide variety of distributed applications this consideration have lead to the definition of two specific transport protocols widely used in the internet, TCP and UDP, which deliver respectively a fully ordered fully reliable service and an not ordered not reliable service. Moreover none of these two protocols take into account application layer temporal constraints. However, as exemplified by the SMIL component in Figure 1, multimedia documents or components neither needs a fully reliable and ordered service nor an unordered and unreliable service. For instance the component modeled in Figure 1 can support the partial or total loss of image I, and the audio and image I can be delivered in any order (i.e. as soon as the transport layer receive the image OR audio component it can be delivered to the service user). This statement, coupled with the gain obtained from insuring the management of partial order and reliability constraints at the transport level induce a new family of transport protocols which make the most of the application requirements for delivering an optimal service in terms of end to end delay and buffering and bandwidth needs. This new family of "application aware transport protocols" delivers a connection oriented transport service defined from the ordering, reliability and time constraints given by the application when opening a multimedia connection.

Such a generic transport protocol raise the question of a method for deriving the transport layer service from application layer QoS requirements. This mapping between the application layer specification and the transport service can be immediately obtained for the reliability and ordering constraints. Indeed the HTSPN specification defines intrinsically a partial order that is defined as follow.

Definition 1. Let A={$a_1,.....,a_n$}, with I=(1,...n), a set of ADUs associated to an application layer QoS specification specified by a HTSPN H. The partial order specification of the transport service related to H is given by the set O={ $(a_i)_{i \in p1(I)}$,....., $(a_i)_{i \in pk(I)}$ } where P={p1,...pk} is a set of permutations on I that defines the partial order on the elements of A directly derived from H.

In other words, a partially ordered transport service can deliver any sequence of ADUs that conforms with the ordering presentation requirements of the application. For example, the partial order constraints of the transport service derived from the HTSPN specification in Figure 1 is given by the set O={$(V_1,I,A,V_2),(V_1,A,I,V_2)$}.

Definition 2. A transport service s that delivers its Transport Service Data Units A={$a_1,.....,a_n$} following the order defined by o=$(a_i)_{i \in p(I)}$ is ordering conformant with the transport service specification S defined by P if and only if p∈ P.

For instance a transport service that delivers o==(V_1,I,A,V_2) is ordering conformant with the service specification S defined by O={$(V_1,I,A,V_2),(V_1,A,I,V_2)$}.

As seen previously, the synchronization operators introduced by the HTSPN model allow one to distinguish between mandatory application data units an the ones of which the presentation can be partially or totally skipped. More generally the HTSPN model allows a deterministic or probabilistic specification of admissible losses to be expressed. The following definition will consider only the deterministic point of view. The partially reliable transport service associated to an application layer QoS requirement given by a HTSPN H is defined as follows:

Definition 3. Let A={$a_1,.....,a_n$}, with I=(1,...n), a set of ADUs associated to an application layer QoS specification specified by a HTSPN H. This HTSPN defines the set R⊂A of ADUs which must be processed by the application. The specification, S, of the partially reliable transport service related to this application layer QoS is also given by the set R which, in this case, defines the set of ADUs that must be delivered to the application by the transport service.

For instance, the partially reliable transport service adapted to the multimedia component modeled in Figure 1 is defined by the set R={V1,A,V2} of the ADUs that must be delivered to the transport service user.

Definition 4. A transport service, that delivers a set r of ADUs, offers a reliability in conformance with a partially reliable transport service specification defined by the set R of mandatory ADUs if and only if R⊂r.

By combining definitions 1 and 3 we get the notion of partially ordered and reliable transport service. Such a service which delivers its TSDUs in conformance with both the ordering and reliability constraints defined for the processing of ADUs (i.e. the application layer processing schedule as defined by the formal specification is also considered as a logical access schedule to the transport service) offers the following advantages:

- The application doesn't need to buffer its ADU for reordering purpose
- The application has not to manage reliability constraints
- The receiving transport entity delivers its TSDU as soon as possible in conformance with the application needs.
- Compared to a fully reliable and ordered service, the knowledge of partial ordering and reliability requirements for ADU delivery allows the buffering needs of the receiving and sending transport entities to be reduced.
- Unnecessary retransmission of non mandatory ADUs can be avoided.
- Reliability and ordering constraints introduce some flexibility for ADU transmission schedule. Therefore, in function of the monitored network QoS, the sending transport entities can apply wise filtering or reordering decisions.

Of course, such an application aware transport service has a sound impact on the fundamental transport layer mechanisms such error control, rate control and congestion control.

3.2   The Time Dimension

Before introducing time in transport services and protocols we need first to understand the meaning of "timed transport service". A "timed transport service" can be defined as a service which delivers on time its service data units to the service user. For a multimedia component, composed of several discrete or continuous media, an application data unit must be immediately available when the previous ones have been displayed. For instance, the audio-visual sequence of the multimedia component in Figure 1 (i.e. the inter-stream synchronization scheme between A and I) can finish at any time within the relative interval (i.e. from the beginning of the audio-visual sequence) [11,13].

This time interval is obtained from the synchronization semantics of t_2 transition and from the temporal validity interval of the audio stream which is the master of this inter-stream synchronization point. Considering that this audiovisual sequence can be displayed between 11 and 13 time units the transport protocol deduces that video "V_2" can be delivered to the service user at any time during this relative time interval. Such a definition of a "timed transport service" entails that the service user (i.e. the player of the multimedia component) can adapt the rate of its presentation to the transport service delivery. This limited and accepted adaptive behavior of the application helps the transport protocol to "hide" the variations of the quality of service delivered by the network. Moreover, on time delivery of data to the transport service user spares the application of scheduling access to remote data, of using time-stamping protocols such RTP and of buffering techniques for insuring network jitter and skew reduction. When used on top of a best effort network service this temporal transport service strengthen the isolation between the application and the network and allows wise adaptation and control decision to be applied in consistency with application needs.

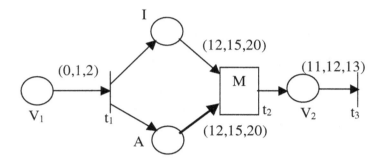

Fig. 2. Formal specification of the transport service adapted to the multimedia component modeled in Figure 1. This transport service specification supposes that the service user accepts a 2 time units control time before playing the component.

Once again such an approach raise the question of defining a temporal transport service adapted to the needs of a given application. The derivation between the application layer time constraints as expressed by a HTSPN model and the related timed transport service is less direct than for the reliability or ordering constraints. This derivation is done by using the following procedure of derivation.

Definition 5. Let us consider an application layer QoS specification modelled by a HTSPN, H. For each ADU, a, in H let us note $TVI(a)$ the Temporal Validity Interval associated to a, and $pred(a)$ the abstract place (as defined in [6]) of H that represents the immediate predecessor of a. The specification of the temporal transport service adapted to H is given by the HTSPN specification H' derived from H as follows:

- $\forall\ a \in A/\ pred(a)\#\varnothing,\ TVI(a)=TVI(pred(a))$
- $\forall\ a \in A\ /\ pred(a)=\varnothing,\ TVI(a)=(x,n,y)$, where x , n and y define respectively the minimum, nominal and maximum admissible waiting time for the delivery of the first ADU to the transport service user (i.e. the control time that the user accepts for the streaming its multimedia component)

Definition 6. A transport service that delivers a set $A=\{a_1,.....,a_n\}$ of ADUs by following the delivery schedule $T=(t(a_1),.....,t(a_n))$, where $t(a_i)$ is for the delivery time of a_i, is time conformant with a transport service specification modeled by a HTSPN H' derived from H, if and only if : $\forall\ a_i \in A/\ TVI(a_i)=(x_i,n_i,z_i)$ in H', $t(a_i) \in\ [t(pred(a_i))+x,$

t(pred(a_i))+y]. That is, every ADU is delivered in a time window that takes into account the admissible temporal variability of the previous ADUs.

Figure 2 gives the formal modeling of the timed transport service specification associated with the multimedia component modeled in Figure 1. This modeling introduces a control time (a parameter given by the application when opening the transport connection) which specifies the maximum duration accepted by the transport service user before beginning to play the component. Note that the transport service specification can be automatically and simply derived by the transport protocol from the HTSPN that models the application requirements.

3.3   The Space of Temporal Partially Reliable and Ordered Protocols

A multimedia component, as defined in section 2, induces a timed partially ordered and reliable transport service which, in turn, defines a sub-space within the space of the whole family a timed partially ordered and reliable protocols that could be used for transporting the set of application data units which compose this multimedia component (Figure 3) [7].
In this space, a Timed Partially Ordered and reliable transport servic and connection (TPOC) associated to a set of application data units A={a_1,.....a_n} is uniquely defined by a 3-uple S=(O,R,T) where:

*   O is the set of admissible sequences extracted from A,
*   R is the subset of A composed of the elements which must be delivered to the service user,
*   TVI is the set of temporal validity intervals associated to the transport service data units delivered by the service S.

For instance for the multimedia component in Figure 1, we have A={V_1,A,I,V_2}, and S=(O,R,T)   with   O={(V_1,A,I,V_2),(V_1,I,A,V_2)},   R={V_1,A,V_2}   and TVI={(0,1,2),(12,15,20), (12,15,20), (11,12,13)}.

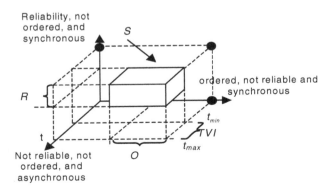

Fig. 3. The space of timed partially reliable and ordered transport services

4   Time in Transport Protocols

In the framework of best-effort networks, the transport layer has a fundamental role to play for adapting the network service to application needs. Considering that there is a gap between multimedia applications' temporal requirements and the asynchronous behavior of current networks such as the Internet (i.e. the IP network best effort service), the transport layer is a privileged place where time related QoS parameters can be controlled and enforced. This approach aims to alleviate multimedia applications of the implementation burden of sophisticated buffering and adaptive techniques. Therefore, the design effort of multimedia applications can be greatly reduced by the use of a weakly synchronous transport service (i.e. a TPOC service) which delivers multimedia information units according to time related QoS parameters derived from application level requirements. Such a new generation of transport protocols not only reduces the complexity of distributed multimedia applications, but also entails a dramatic improvement on the use of network and communication resources. Indeed, by taking into consideration at the transport level the temporal semantics of information units, this new approach allows more efficient congestion control, rate control, error control and buffer management techniques to be applied. Indeed, it is well known that the TCP congestion control technique is not adapted to the transport of multimedia streams. The slow start and congestion avoidance mechanisms take uniquely into account the QoS delivered by the network without considering the semantics of the transport service data units. Therefore, with such congestion control mechanisms variations in network QoS impact directly and blindly on the QoS delivered to the transport service user and are instantaneously perceived by the user. Such a behavior is not admissible for continuous media delivery of which the semantic is greatly dependant of time constrains. Our approach allows the transport protocol to react to congestion situations while taking into account the application requirements. This can be done by using the partial reliability and temporal flexibility offered by the concept of TPOC. Indeed, in function of the QoS delivered by the network, the TPOC can reduce it delivery rate and partially or totally suppress the sending of application data units which can be lost (i.e. the complementary of the set R in U).

Equally, window based rate control mechanisms can be efficiently replaced by rate control mechanisms which take into account the temporal semantics of the transported application data units. Indeed the sending rate can be adjusted dynamically and consistently with the admissible rate variations supported by the transport service, in function of the state of the receiving entity buffer. Note that partial reliability and order can be also used for rate control purpose.

In summary the knowledge by the transport layer of the temporal semantics of the TSDUs offers potentially the following advantages:

- Retransmissions can be more efficiently managed.
- Flow control techniques compatible with application's constraints can be applied.
- Congestion control techniques can be used by combining the flexibility offered by the weakly synchronous time constraints on ADUs.
- The temporal flexibility of ADUs offers multiplexing capabilities and allows network resources to be more efficiently used.

- Associating a temporal duration to ADUs entails a reduction of buffering needs at the receiver side (i.e. the data is received when needed) as well as at the sender side (i.e. by avoiding the buffering of out of date data).
- Avoids the retransmission of out of date data.
- Allows access schedules to ADUs to be managed by the transport layer instead of the application layer.
- Ultimately this approach induces very simple applications which have only to react to transport layer events and unload the management of time and synchronization constraints onto the transport service.

Weakly synchronous transport protocols are also useful in the framework of networks that deliver an integrated or differentiated service. By supporting some temporal admissible variability, these protocols offer some flexibility for network resources management, and offer indirectly to the user and to the network provider, tradeoff facilities between quality and price. A TPOC specification defines an envelope of services, from the worst admissible to best for which the user is ready to pay; these services have to be dynamically adapted and mapped to the network layer differentiated or integrated services. This approach allows the network service provider to satisfy its clients while optimizing its resources usage.

5   GTP Implementation

A first version of a Generic Transport Protocol based on the notion of TPOC has been developed in the framework of the 5th IST European project GCAP (Global Communication Architecture and Protocols for new QoS services over Ipv6 networks).
GCAP aims at developing for the future Internet a new generation of end-to-end multicast and multimedia transport protocols that provide a guaranteed QoS to advanced Multimedia Multipeer applications on top of heterogeneous networks
The Java language has been used for designing GTP because the Java environment offers a multi-platform implementation that delivers the performance required for a transport protocol. Indeed, we have experimented that although C performs better than Java, Java performances are acceptable enough for designing a transport protocol in the user space and for offering an efficient support to multimedia applications [8]. In its current implemented version GTP offers a generic support to the partial order and reliability constraints required by the transport service user. Temporal services are not yet offered but should be available in the next version.
GTP uses a pull approach, where the receiver side initiates the connection establishment and termination. At the sender side, a server waits for a connection request from the receiver. When the connection request arrives a sender socket is instantiated and connected to the receiver socket. The receiver sends an object request to the sender side in order to get a multimedia component. This object request includes the identification of the multimedia component, and QoS parameters (i.e. a compact representation the TPOC specification of service). Therefore the sending and receiving entities share a common specification of the transport service which allows then to apply effi-

cient adaptation mechanisms between application needs and network behaviour. The GTP API is similar to the TCP standard java API as defined by the socket class of the java.net package (Table 1).

Table 1. The GTP API

Class GTPServerSocket
This server waits for a connection request from a receiver side
Constructor GTPServerSocket (local Address, local Port, max-Conn)
This constructor creates a socket server using a local address, a local port and specifies the maximum number of connections
GTPSenderSocket accept()
This method waits for a connection and accepts it, instantiating a GTPSenderSocket connected to the receiver side

Class GTPSenderSocket
The sender socket is instantiated with the specification of the transport service sent by the peer entity.
Request accept Request()
The accept Request waits for a request from the receiver side
void ackRequet(GTP.Request request)
The ackRequest method acknowledges the requests
void send(GTP.GTPPacket fp)
This method allows one to send GTPpackets to the receiver side

Class GTPReceiverSocket
The receiver socket is able to send the object requests. The "receive" method allows one to read the GTPpackets ready to be delivered to the user
Constructor GTPReceiverSocket(remoteAddress, remotePort, localAddress, localPort)
This constructor creates a GTPReceiverSocket using the local and remote addresses and the ports specified
void closeRequest()
This method allows one to request the termination of the connection
ObjectRequest mediaObjectRequest(GTPMediaObject pmo)
This method instantiates the transport connexion with a specification of service and send a request for a media object to the server side
GTPPacket receive()
The receive method allows one to get a GTPpacket

5.1   First Experiment

This first experiment aims to evaluate the gain obtained by using a partial order proto-col on top of a non-reliable network environment. The experiment consists in testing the contribution of the GTP protocol for transferring simple JPEG still images be-tween an image server and a remote images player. The considered client server ap-plication is able to respectively receive and send independent segments composed of group of macro-block (i.e. these segments can be decoded and displayed in any order). For comparison purpose, this client-server application has been tested successively on top of UDP, GTP, and a fully reliable and ordered instantiation of GTP, hereinafter referred to as TCP*, which aims to simulate the TCP behaviour without its congestion control mechanisms (in its current version GTP does not apply any congestion control technique). Figure 3 illustrates the dummynet based emulation platform used for the experiment. In this experiment, the sender and receiver side are two Windows 2000 systems located in two different subnets. A third computer running the FreeBSD sys-tem insures the routing services between the two subnets and supports the network layer emulation environment. Dummynet emulation capabilities have been used for they allow to tune very easily the main network QoS parameter such as bandwidth, losses and delays.

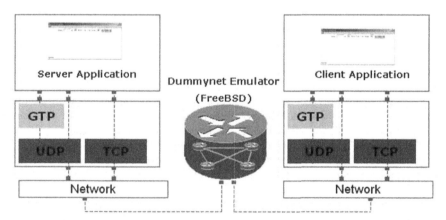

Fig. 4. The emulation platform

Figure 5 graphically illustrates the end to end transmission delay required for image 218 Kbytes image. UDP (i.e. no order no loss) , GTP with full reliability and no order, and TCP* (i.e. full reliability and order) protocols are respectively used on top of a network service that induces 0, 5, 10, 15 and 20 percent of losses. The results show that GTP, though instantiated for delivering a fully reliable service, requires almost the same duration than UDP to transmit the data. In contrast, from 10 % of losses, TCP exhibits a dramatic increase of its transmission duration.

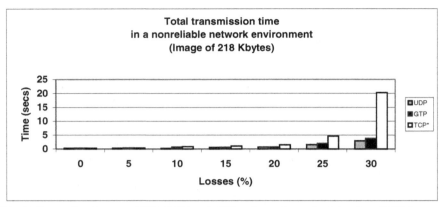

Fig. 5. Comparison between UDP, TCP* and GTP end to end transfer delay for a JPEG image

5.2   Second Experiment

The main goal of the second experiment consists in comparing the partially reliable service offered by GTP with the fully reliable service of TCP* (once again TCP is emulated by a fully reliable and ordered GTP connection). The integration capabilities of GTP with existent multimedia applications represent another feature evaluated in this experiment. For this purpose, we have used the Java Media Framework player (JMF Player) for its capabilities to integrate dynamically new transport services and protocols on top of UDP. The same platform that the one described for the experiment 1 has been used for the test bed and for network emulation purpose. The media object to transmit consists in a 1.6 Mbytes MJPEG video stream composed of 411 frames. The partial reliability service of GTP has been instantiated for supporting 30 % of losses. It is important to understand that, in this elementary experiment, because of the fragmentation of TPDU by the network protocol (i.e. video frames are considered as monolithic transport segments), 5% of network losses entails around 35% of losses for the transport layer segments. Note that for avoiding such a segmentation each video frame could be fragmented in independent ADUs like in experiment 1. Such an approach has been described in [9] and [10] where we have proposed a technique that allows MPEG video frames to be segmented into independent ADUs that can benefit not only from a partially reliable transport service but also from a partially ordered one.

In presence of 5 % of network losses and with an admissible partial reliability of 30 %, for the GTP service, GTP is able to deliver, within 39 seconds, 70% of the video frames to the transport service user. In contrast a fully reliable service requires 57 seconds for transmitting the full video stream and entail at the application layer long blocking periods that are incompatible with the continuity constraints that must be satisfied for displaying correctly a video stream. Indeed, in this case, the use of a fully ordered transport service results in 87% of video frames missing their presentation deadlines, to be compared to 47 % with the partially reliable one considered in the experiment.

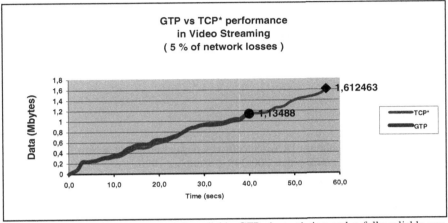

Fig. 6. Comparison between a partially reliable GTP   instantiation and a fully reliable one (denoted TCP*) for the transport of MPEG video in presence of 5% of network losses.

More generally a partially reliable service offers a trade-off capabilities between the controlled percentage of losses (uncontrolled with an unreliable service such as UDP) and the transmission delay (systematically longer with a reliable transport service such as TCP).

6   Conclusion

On top of highly performing networks there is a clear need of new transport services adapted simultaneously to the new types of services delivered by these networks and to the QOS needs of distributed multimedia applications which are pervading the Internet. We have introduced in this paper a new family of transport services and protocols which satisfy this double requirement. The contribution of this advanced transport services have been successfully experimented for videoconference and video on demand applications [9] and is one of the major building blocks of the currently designed GCAP 5th IST European Program. This new generic transport protocol delivers a service and can exploit rate control, congestion control and error control mechanisms which are directly derived from a formal specification of the application QoS needs. When used on top of best effort networks such an approach enhance dramatically the QoS perceived by the user. This approach is also very promising when used on top of differentiated or integrated services because she offers flexibility for the management of network and end-systems resources, hence GTP allows the user as well the network provider to trade off between performances and cost .

This new field of timed partially reliable and ordered transport protocols offers several open issues. Particularly, the compatibility of the currently studied congestion control mechanisms with TCP connections and TCP receiver or sender entities is a critical research topics which must be solved for insuring the successful dissemination of this new family of protocols. Moreover, extending the point to point approach introduced in this paper for delivering a multicast timed transport service partially reliable and ordered on top of heterogeneous networks is currently experimented in the framework of the GCAP European project.

References

1. Clark D, Tennenhouse D, Architectural Considerations for a New Generation of Protocols, *SIGCOMM* pages 200-208, September 1990.
2. Chrisment I, Huitema C, Evaluating the Impact of ALF on Communication Subsystems, Design and Performance, *Journal of High Speed Networks*, vol. 5, num. 2, Netherlands 1996.
3. Diaz M, Senac P, "Time Stream Petri Nets, a Model for Timed Multimedia Information", Published in Application and Theory of Petri Nets, Lecture Notes in Computer Science, Springer Verlag Editor, 1994.
4. Senac P., M. Diaz, P. de Saqui-Sannes, A. Leger, "Modeling Logical and Temporal Synchronization in Hypermedia Systems", *IEEE Journal on Selected Areas in Communication, special issue on multimedia synchronization*, vol. 14, n° 1, pp. 84 -103, Jan. 1996
5. Courtiat J.P., M. Diaz, R. Cruz, P. Sénac, Formal Models for the Description of Time Behaviors of Multimedia and Hypermedia Distributed Systems, *Computer Communications*, 1996.
6. P. Sénac, P. de Saqui-Sannes, R. Willrich, M. Diaz, "Hierarchical Time Stream Petri Net: A Model for Hypermedia Systems", Published in Application and Theory of Petri Nets (IEEE), Lecture Notes in Computer Science, Springer Verlag Editor, 1995
7. Amer P., Chassot C., Connolly T., Diaz M., and Conrad P., "Partial order transport service for multimedia applications", *IEEE/ACM Transactions on Networking*, Vol.2, N°5, pp.440-456, Octobre 1994
8. Apvrille L, Dairaine L, Rojas-cardebas L, Sénac P, Diaz M, "Implementing a User Level Multimedia Transport Protocol in Java", The Fifth IEEE Symposium on Computers and Communication (ISCC'2000), Antibes-Juan les Pins, France, July 2000
9. L. Rojas, E. Chaput, L. Dairaine, P. Sénac, Michel Diaz, Transport of Video on Partial Order Connections, published in the Journal of Computer Networks and ISDN Systems, 1998.
10. L. Rojas, L. Dairaine, P. Sénac, M. Diaz, Error Recovery Mechanisms Based on Retransmissions for Video Coded with Motion Compensation Techniques, Packet Video 99, New-York, April 99.bb

Activating and Deactivating Repair Servers in Active Multicast Trees*

Ping Ji, Jim Kurose, and Don Towsley

Department of Computer Science
University of Massachusetts/Amherst
{jiping,kurose,towsley}@cs.umass.edu

Abstract. For time-constrained applications, repair-server-based active local recovery approaches can be valuable in providing low-latency reliable multicast service. However, an active multicast repair service consumes resources at the repair servers in the multicast tree. A scheme was thus presented in [10] to *dynamically* activate/deactivate repair servers with the goal of using as few system resources (repair servers) as possible, while at the same time improving application-level performance. In this paper, we develop stochastic models to study the distribution of repair delay both with and without a repair server in a simple multicast tree. From these models, we observe that the application deadline, downstream link loss rates, the number of receivers, and the upstream round trip time of a repair server all influence the overall value of activating an active repair server. Based on these observations, we propose a modified dynamic repair server activation algorithm that considers the packet loss rate, the number of downstream receivers, and the round trip time to the nearest upstream active repair server when activating/deactivating a repair server. From simulation, we observe that our modified dynamic repair server activation algorithm provides a significant reduction in the latency of successful packet delivery (over the original algorithm) while using the same amount of system resources. We also find that much of the performance gains achievable by having active repair servers can be obtained by having only a relatively small fraction of repair servers actually being active.

1 Introduction

Delay-sensitive applications such as teleconferencing, distributed simulation, multiplayer games, and Internet telephony all have timing constraints on the successful delivery of data from source to destination(s). In such applications, data that do not arrive at receivers prior to some application-determined deadline, are considered lost and can result in impairments in application-level performance. For a real-time video stream, for example, a missed deadline can result in playout jitter or breaks in the playout stream. For a conversation with interaction among multiple speakers, the delay from when a user speaks or moves until the action is manifested at the receiving hosts should be less than a few hundred milliseconds [6].

---

* This work is supported by the Defense Advanced Research Projects Agency (DARPA) under contract N66001-9117-411V

Many reliable multicast protocols exploit local recovery [2] [13] [7] [14] [11] to reduce the delay in successful data delivery, making them attractive for supporting delay-sensitive applications. Similarly, by using active repair servers (RS) in a multicast tree to provide retransmission (i.e., error recovery) service, repair-server-based local recovery schemes can successfully reduce the repair latency, as well as suppress NAK implosion, and provide retransmission scoping [4]. On the other hand, active repair servers inside the network require additional resources (e.g., buffering and processing). Thus it is desirable to activate as few repair servers as possible, while at the same time providing enough repair servers to improve repair latency.

In this paper, we study the tradeoff between repair delay (equivalently, the time needed to successfully deliver data to the receiver(s)) and system resource consumption by focusing on the benefit gained by using server-based active error recovery (AER) [4]. We begin by developing stochastic models to study the distribution of repair delay both in the presence, and in the absence, of repair servers. Based on these models we observe that the application deadline, downstream link loss rates, the number of receivers, and the upstream round trip time of a repair server are all important criteria influencing the decision of whether to activate/deactivate an RS.

We then consider specific algorithms for dynamic RS activation/deactivation. We begin with the algorithm from [10], which introduced a protocol to dynamically activate/deactivate repair servers on the basis of the packet loss rate within the RS's subtree. For delay-sensitive applications, it is also appropriate to consider time-based measures such as the upstream RTT in deciding whether or not to activate an RS. We thus modify the algorithm in [10] to consider not only the packet loss rate and number of downstream receivers, but also the round trip time to the nearest upstream active repair server (or the sender) when making the activation/deactivation decision. We study via simulation the tradeoff that exists between the fraction of RSs that are activated and the repair delay. We show that our modified dynamic RS activation algorithm provides a significant reduction in repair delay over the original algorithm [10], while using no more system resources than the original algorithm. We also find that much of the performance gains achievable by having active repair servers can be obtained by having only a relatively small fraction of repair servers actually being active.

The remainder of this paper is organized as follows. In Section 2, we describe a multicast loss recovery architecture and a reliable multicast protocol that uses active repair services. Section 3 presents the analytic model that we use to evaluate the decrease in repair delay when using an active repair server for time-constrained applications in a simple multicast tree. Section 4 proposes a modified algorithm for dynamically activating/deactivating repair servers, and presents simulation results comparing the performance of the modified algorithm with that of the original algorithm. Section 5 concludes this paper.

2   Real-Time Reliable Active Multicast: Motivation and Algorithms

Several recent efforts have focused on providing *better-than-best-effort* service for delay-sensitive applications. Maxemchuk et al. [9], Lucas et al.[8] and other researchers have proposed various distributed *receiver*-based local recovery schemes. The Active Error Recovery (AER) protocol [4] uses a *repair-server*-based local recovery scheme, in which a repair server (RS) is attached to a router to perform active local error recovery (i.e., buffering and retransmission) for downsteam receivers and RSs. AER was shown to achieve low latency error recovery, NAK suppression, retransmission scoping, and superior bandwidth utilization. However, the use of additional active nodes inside the network comes at a price - additional system resources, such as buffers and computation, are needed at the active nodes. Several questions immediately arise - how many active RSs are needed and where should they best be placed; what is the tradeoff between the number of activated RSs and the repair latency; must a repair server always be active, or can a repair server dynamically monitor performance and then self-activate/deactivate according to observed performance. These are some of the questions we address in this paper.

Rubenstein et al.[12] have proposed a *static* centralized tree-based protocol to construct a *repair graph* that uses RSs to retransmit lost packets to receivers before their deadline. However, when the multicast tree structure and/or link loss rates change over time, such a static approach may not be appropriate. In [10], Osland et al. presented an algorithm to *dynamically* activate/deactivate RSs in response to changing network conditions. However, their approach uses only the packet loss rate to trigger RS activation/deactivation. We will see shortly that for delay-sensitive applications, delay-based criteria can be effectively used in making dynamic activation/deactivation decisions at an RS.

Let us now describe a reliable multicast protocol known as AER (Active Error Recovery [4] [5] [10] [1]) that implements active server-based repair services; we will subsequently modify this protocol to implement dynamic RS activation/deactivation. In AER, *subcast* is defined to be a multicast transmission from an RS to the downstream multicast subtree rooted at the RS. We describe the algorithm in the context of a sender that periodically multicasts data to a multicast address that is subscribed to by all receivers and participating Repair Servers; other scenarios are also possible. The algorithm operates as follows:

- **Packet forwarding, buffering.** When a new packet arrives at a router associated with an RS, it is multicast downstream by the router and stored in the buffer of the repair server.
- **NAK suppression.** On detecting a loss, a receiver (or repair server), after waiting for a random period of time (the NAK suppression time), sends out a NAK to its nearest upstream active server (a repair server or the sender). At the same time, it starts a NAK retransmission timer. If the receiver (or repair server) receives a NAK for the same packet from its upstream repair server prior to sending its own NAK, it suppresses its own NAK transmission.
- **NAK timeout.** The expiration of the NAK retransmission timer at a repair server (or a receiver), without prior reception of the corresponding repair, serves as the detection of a lost packet for the repair server (or the receiver) and a NAK is retransmitted.
- **Repair packet transmission, NAK aggregation and propagation.** When a repair server receives a NAK from a downstream node, it subcasts the packet if it has the packet in its buffer. Otherwise, if there is already a pending NAK for that lost packet, the new incoming NAK is suppressed. If there is neither a pending NAK

for this packet, nor a buffered repair packet, the RS *immediately* subcasts the NAK downstream and sends a NAK upstream after waiting for a random period of time, as well as keeping a pending NAK until the repair packet arrives.
- **Repair packet retransmission.** On receiving a NAK from a downstream repair server, the sender re-subcasts the requested packet to all the receivers and repair servers. As mentioned above, these repairs are received by intermediate repair servers and forwarded down the tree only if there is a pending NAK for that repair.
- **Repair packet reception.** If a repair packet is received, the repair server first checks whether there is a pending NAK for that packet. If so, that repair packet will be buffered and subcast downstream. If there is no pending NAK for this repair packet, the RS discards this packet.

The AER protocol successfully reduces NAK implosion by using randomized timer-based NAK suppression and by using the repair server hierarchy for NAK aggregation. We next present a simple model of the delay performance of this protocol in a simple multicast tree.

3   A Simple Model for Understanding the Benefits of Using a Repair Server in Time-Constrained Applications

In this section we develop an analytic model to study the distribution of repair delay both with and without a repair server for time-constrained applications. Our goal is to gain insight into the effect of parameters such as application deadline, downstream link loss rates, the number of receivers, and the upstream round trip time of a repair server.

Figure 1 shows a generic model for a single RS co-located at a router. Node A is the sender. Node B represents an intermediate router, to which a repair server (that is denoted as a square node in this graph) can be attached. On receiving a packet from sender A, router B multicasts that packet on its subtree. $R = \{R_1, R_2, ..., R_N\}$ denotes the set of receivers on the tree. The solid line between node A and node B represents the transmission path between the sender and the router (or its corresponding repair server). Similarly, the lines between router B and receivers R_1 to R_N represent the paths between the router and receivers.

The notation we will use is as follows:

X_i: Time required to successfully deliver a packet from the sender to receiver R_i.

X_{ij}:   Time to transmit a packet from i to j in the absence of packet loss, where $i \in \{A, B\}$ and $j \in \{\{B\} \cup R\}$. We model X_{ij} as a fixed value.

X_{ji}:   Time to transmit a NAK from j to i, where $j \in \{\{B\} \cup R\}$ and $i \in \{A, B\}$. We model X_{ji} as a fixed value.

τ_i: A joint timer of i, that integrates the NAK suppression timer, which is a function of the RTT between i and its nearest upstream active node (the sender or the *active* RS), and the NAK retransmission timer of i, which is a function of the RTT from i to the sender. Here $i \in \{\{B\} \cup R\}$.

D: The application-dependent deadline

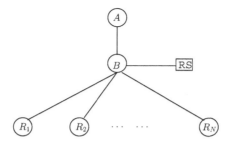

Fig. 1. A Simple Multicast Topology

p_{ij}:   Loss probability of path ij, where $i \in \{A, B\}$ and $j \in \{\{B\} \cup R\}$.
λ: Recall that we are modeling a constant rate sender. Data packets arrive
            at A for first-tie transmission at constant rate λ.

We denote the probability that a packet is delivered after the application deadline at
receiver R_i as $P\{X_i > D\}$.

Based on this single RS configuration, we now describe two simple models that
characterize the distribution of repair delay both with and without RSs.

3.1   Evaluation of $\sum_{i \in R} P\{X_i > D\}$ when an RS Is Used

Let us first consider the transmission delay, X_i, in the presence of a repair server (Figure
1). We introduce the following random variables. Let K denote the number of losses
prior to the first successful transmission of a packet on path AB. Let K_i denote the
number of losses prior to the first successful transmission of a packet on path BR_i. We
are interested in the value of X_i conditioned on $K = k$ and $K_i = k_i$. We denote this as
$X_i(k, k_i)$ and compute its value as follows:

$$X_i(k, k_i) = \begin{cases} X_{AB} + X_{Bi} & k = 0, k_i = 0 \\ X_{AB} + \Delta_i + (k_i - 1)\tau_i + X_{iB} + X_{Bi} & k = 0, k_i \geq 1 \\ \Delta_{RS} + (k - 1)\tau_B + X_{BA} + X_{AB} + X_{Bi} & k \geq 1, k_i = 0 \\ \Delta_{RS} + (k - 1)\tau_B + X_{BA} + X_{AB} + \Delta + \\ (k_i - 1)\tau_i + X_{iB} + X_{Bi} & k \geq 1, k_i \geq 1 \end{cases} \quad (1)$$

A timeline of X_i, when $k \geq 1$ and $k_i \geq 1$, is shown in Fig. 2. The loss detection duration,
Δ_{RS}, is a function of the constant transmission rate λ and the delay of path AB. At the
repair server, the time until the next successful receipt of a packet follows a geometric
distribution with mean $\frac{1}{1-p_{AB}} \cdot \frac{1}{\lambda}$. Thus we model $\Delta_{RS} = X_{AB} + \frac{1}{1-p_{AB}} \cdot \frac{1}{\lambda} + \tau_B'$,
where τ_B' is the NAK suppression timer of repair server B. Similarly we model $\Delta_i = X_{Bi} + \frac{1}{1-p_{Bi}} \cdot \frac{1}{\lambda} + \tau_i'$, where τ_i' is the NAK suppression timer of receiver R_i. Here, Δ
denotes the time between the arrival of a retransmitted packet at the repair server and
the receiver's transmission of the next NAK. From Figure 2, we observe that Δ follows
a uniform distribution in $[0, \tau_i]$.

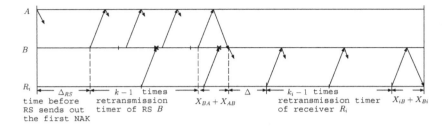

Fig. 2. Timeline of $X_i(k, k_i)$, Repair Service is Used

For a given k and a given k_i, from equation (1), we can easily evaluate the value of $X_i(k, k_i)$. Consequently, we can determine the probability that $X_i(k, k_i)$ is greater than the application deadline D. The probability $P\{X_i > D\}$ can be evaluated by removing the conditioning on the values of k and k_i:

$$\sum_{i \in R} P\{X_i > D\} = \sum_{k=0}^{\infty} (\sum_{i \in R} \sum_{k_i=0}^{\infty} \mathbf{1}(X_i(k, k_i) > D) \cdot P\{K_i = k_i\}) \cdot P\{K = k\} \quad (2)$$

where, $P\{K_i = k_i\} = p_{Bi}^{k_i} \cdot (1 - p_{Bi})$, $P\{K = k\} = p_{AB}^k \cdot (1 - p_{AB})$ and $\mathbf{1}(P)$ is one when the predicate P is true and zero otherwise.

3.2   Evaluation of $\sum_{i \in R} P\{X_i > D\}$ when RS Is Not Used

The analysis is more complicated when there is no repair server attached at router B of Figure 1, because the end-to-end link losses are correlated. Thus, the probability that a packet is delivered after the application deadline cannot be solved independently at each receiver. In this case, we can use the following alternative approach to evaluate $\sum_{i \in R} P\{X_i > D\}$. Again, we focus on a randomly selected packet.

$$\sum_{i \in R} P\{X_i > D\} = E[\text{number of receivers that receive the packet after the deadline}]$$

$$= N - E[\text{number of receivers that receive the packet by the deadline}] \quad (3)$$

where N denotes the number of receivers. To evaluate the expected number of receivers that successfully receive the data packet before its deadline, we must define some additional notation.

Let $T^{(l)}$ be the number of receivers that successfully receive that data packet before or at the l-th retransmission of that packet. Thus, from formula (3), we can derive

$$\sum_{i \in R} P\{X_i > D\} = N - E[T^{(d)}] \quad (4)$$

where d is the maximum number of retransmissions allowed to meet the deadline. Here d is a function of the application-dependent deadline, D, and link transmission delays. Yamamoto et al. [15] provide a general model for evaluating the delay performance of receiver-initiated reliable multicast protocols. In our study, we assume that a NAK-receipt suppression timer is used by the sender. Within a NAK-receipt timeout interval, the sender considers multiple NAKs it receives for a given packet as being redundant and only retransmits (multicasts) the packet once. This timer is a function of the longest RTT to receivers. We simplify our problem to the case that all receivers have the same RTT to the sender. Therefore, at receiver R_i, during each timer τ_i, there can be at most one retransmission for a specific lost packet. Thus we have $d = \lfloor \frac{D - \Delta_i - X_{iA} - X_{Ai}}{\tau_i} + 1 \rfloor$, where Δ_i denotes the loss detection duration of receiver R_i and is a function of p_{Ai}, X_{Ai} and λ. Consequently, from equation (4), we have

$$\sum_{i \in R} P\{X_i > D\} = N - \sum_{i=1}^{N} i \cdot P\{T^{(d)} = i\} \tag{5}$$

Given the application deadline and the corresponding maximum number of retransmissions d, we can calculate the probability $P\{T^{(d)} = i\}$ as follows:

$$P\{T^{(l)} = i\} = \sum_{j=0}^{i} P\{T^{(l)} = i | T^{(l-1)} = j\} \cdot P\{T^{(l-1)} = j\} \qquad l = 1, 2, ..., d \tag{6}$$

where the conditional probability of $P\{T^{(l)} = i | T^{(l-1)} = j\}$ is shown in equation (7).

$$P\{T^{(l)} = i | T^{(l-1)} = j\} = \begin{cases} p_{AB} + (1 - p_{AB}) p_{Bi}^{N-j}, & i = j; \\ (1 - p_{AB}) \binom{N-j}{i-j} p_{Bi}^{N-i} (1 - p_{Bi})^{i-j}, & N > i > j; \end{cases} \tag{7}$$

with the initial condition:
$$P\{T^{(-1)} = 0\} = 1$$

Thus $P\{T^{(d)} = i\}$ can be recursively computed and the corresponding $\sum_{i \in R} P\{X_i > D\}$ can be evaluated.

3.3 Numerical Case Study

Using the analysis techniques described above, we can evaluate the cumulative distribution function of the X_i, the time needed to successfully deliver a packet to a receiver in the presence/absence of a RS for various application-level deadlines, D. A numerical case study for the CDF of X_i is shown in Figure 3. Here, we assume the number of receivers is 5, path AB experiences a loss rate as $p_{AB} = 0.05$. For each downstream path, we assume a loss rate of $p_{Bi} = 0.10$. The one-way delay for path AB is 30ms, and the one-way delay between router B and each receiver R_i is 10ms. Figure 3 shows that when a repair server is active, there is a higher probability that a packet is successfully delivered within a given deadline than the case when the repair server is not active. For

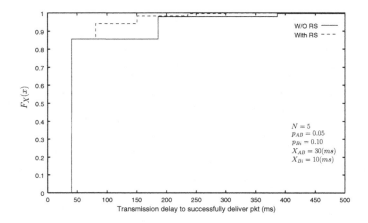

Fig. 3. Compare the CDF of Transmission Delay for With/Without RS Cases

this set of parameters, we see the largest gain when the application deadline D is between $80ms$ to $190ms$.

As the values of the model parameters change, so too does the distribution of X_i. Due to space considerations, we only briefly explore the parameter space to identify general trends; in [3] we provide a more complete study. In [3] we define a general cost function that weights the benefits of decreased repair latency and the cost of resources (e.g., buffering and computation) required by active repair servers. The cost difference between the case of having active repair servers and not having active repair servers, is denoted by ΔC, and is defined as:

$$\Delta C = c_a P\{X_i^{noRS} > D\} - (c_a P\{X_i^{withRS} > D\} + c_b)$$

where c_a and c_b are weights reflecting the unit performance degradation penalty cost for an overly delayed packet and the resource cost for recovering a packet, respectively.

We illustrate the effect of an application's deadline on the cost difference ΔC in Figure 4. We considered four downstream loss rates and observed that as the downstream link loss probability p_{Bi} increases, so too does the relative benefit of having active repair servers. For the parameters considered in Figure 4, the benefit is largest when the application deadline lies in the range of 2 to 4.6 times the one-way delay (about $80ms$ to $190ms$). As the application deadline increases beyond this value, the relative performance benefits decrease and actually become negative when $D \geq 380ms$. This is because as the application deadline increases the probability of successfully recovering a packet within the deadline increases (approaching 1) both with and without repair servers, while the with repair server case incurs a resource cost, c_b. In [3], we also consider the effects of downstream loss rates (p_{Bi}), upstream round trip time (X_{AB}) and number of receivers (N) on cost difference ΔC.

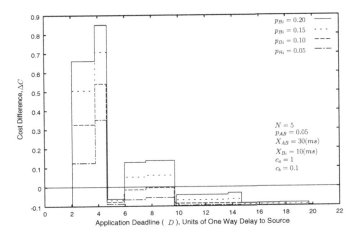

Fig. 4. The Effect of Application Deadline on the Saving of Transmission Delay by Setting a RS

4   An Algorithm for Dynamic Activation/Deactivation of Repair Servers

In [10], Osland et al. modified the basic AER protocol to include a two-threshold algorithm for dynamically activating and deactivating repair servers. In their approach, an RS estimates the loss rate to receivers in its subtree over intervals of time. At the end of each interval, if the measured loss rate is greater than an upper threshold value, the repair server is activated; if the measured loss rate is less than a lower threshold value, the corresponding repair server is deactivated.

Our goal is to develop a new algorithm that dynamically activates/deactivates a RS for the purposes of supporting time-constrained applications. In Section 3, we studied how system parameters such as the upstream RTT, number of receivers, application deadline, and link loss rate determine the benefits possible with the use of active RSs. Given that our primary application-level performance metric of interest - the probability that a packet is successfully delivered within its deadline - is time-related, it is natural to include time-based considerations into the activation/deactivation decision.

Informally, our time-sensitive RS activation/deactivation algorithm operates as follows. As in [10], an RS estimates (observes) the loss rate to receivers in its subtree over intervals of time of length τ. We introduce variable N_{loss} to denote the number of lost packets during a time interval, and N_{pkt} to denote the number of packets sent by the sender during the same time interval. N_{loss} and N_{pkt} are measured at the RS. Let p_{loss}^k denote the loss probability of an RS's subtree during time interval k. The formula for computing p_{loss}^k is:

$$p_{loss}^k = \frac{N_{loss}}{N_{pkt}} \qquad (8)$$

Rather than use the measurement of lost packets to define p_{loss}^k above from [10], we can derive the following expression for p_{loss}^k,

$$p_{loss}^k = 1 - \prod_{i=1}^{n}(1 - p_i) \qquad (9)$$

where n is the number of receivers suffering loss and p_i is the link loss rate between the RS and receiver R_i. Our earlier analysis showed that the number of receivers and the individual link loss rates were important factors in determining performance. We note that both of these factors appear explicitly in equation (9).

In addition to using the number of receivers and the individual link loss rates in the activation/deactivation decision, we will want to include time-based measures as well. In our analysis in Section 3, we saw that the upstream RTT (the round trip time from an RS to the closest upstream RS or the sender itself - X_{AB} in Figure 1) was an important factor affecting the delay distribution. Thus, we would like to incorporate the upstream RTT into our dynamic activation/deactivation algorithm. It is worth mentioning here that a crucial parameter affecting performance is the *application deadline D*. However, D is completely application-dependent and can easily be determined only at the receivers. Thus, we choose not to include D in our dynamic repair server activation/deactivation algorithm.

4.1   Algorithm Description

We now modify the RS activation/deactivation algorithm presented in [10] to account for the *upstream RTT* of a RS with p_{loss}^k. Intuitively, if an RS is very close to its upstream RS (or the sender), activating the downstream RS will not significantly reduce the repair delay. On the other hand, if the RTT between an RS and its upstream active repair node is large, activating this RS can result in *local* repair service to downstream nodes (i.e., repairs can be supplied by the RS itself), providing the possibility for a significant reduction in repair delays. This intuition tells us that the larger the upstream RTT of a RS, the more important it is to activate that RS. Therefore, we use the product of the upstream RTT of an RS and its packet loss rate during time interval k as a metric to control the activation/deactivation decision at the end of a time interval.

In the original dynamic RS activation/deactivation algorithm [10] (which we will refer to as AL1), during a time interval (of length $\tau = 4sec$), the number of lost packets and the number of packets sent by the sender are measured by an RS. The packet loss probability of a RS's subtree for time interval k, p_{loss}^k, is then calculated using equation (8). The exact mechanism for estimating the packet loss probability, $\hat{p}_{loss}^k$ is

$$\hat{p}_{loss}^k = (1 - \alpha) \cdot \hat{p}_{loss}^{k-1} + \alpha \cdot p_{loss}^k$$

where α is a smoothing parameter. At the end of each time interval, AL1 compares $\hat{p}_{loss}^k$ with two thresholds. If $\hat{p}_{loss}^k$ is greater than the upper threshold, then the corresponding RS will be active during the next time interval, otherwise if $\hat{p}_{loss}^k$ is less than the lower threshold, the corresponding RS will be inactive during time interval $k + 1$.

AL1 only considers the packet loss probability of a RS's subtree, $\hat{p}_{loss}^k$, as the single metric to control a RS's activation/deactivation decision. In our modified algorithm (which we will refer to as AL2), during each time interval the packet loss rate of a RS's subtree, as well as the RS's upstream RTT, is measured. The product of the upstream RTT of an RS and $\hat{p}_{loss}^k$ of its subtree is then used as the metric for RS activation/deactivation. We denote the product of the upstream RTT and $\hat{p}_{loss}^k$ as ϕ:

$$\phi = R_u \times \hat{p}_{loss}^k \tag{10}$$

where R_u denotes the round trip time from the current repair server to its nearest upstream repair server (or the sender). Dynamic RS activation/deactivation is controlled by a two-threshold mechanism based on the value of ϕ. If ϕ is greater than the modified upper limit, then the corresponding RS will be activated during the subsequent time interval; if ϕ is smaller than the modified lower limit, then the corresponding RS will be deactivated.

4.2 Comparison of AL1 and AL2 via Simulation

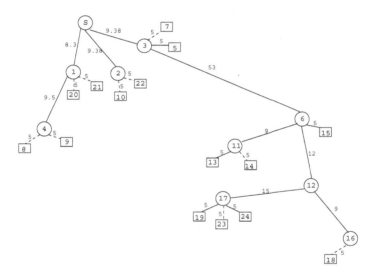

Fig. 5. Simulation Topology 1

We evaluate the new two-threshold policy on three topologies[1]. Figure 5 illustrates the first topology. Focusing on the topology in Figure 5, nodes 1 2 3 4 6 11 12 16 17 are routers attached to repair servers. S represents the sender (source). The remaining nodes are receivers. Link propagation delays are marked in the graph in units of milliseconds. Lossy links are represented by dashed lines. We assume that other links are lossless.

---

[1] The three topologies were originally generated by Diane Kiwior of The Analytic Sciences Corporation, TASC

In order to characterize the system-wide extent to which the repair servers are activated, we introduce the active fraction, ρ as follows:

$$\rho = \frac{\text{Total active duration of all RSs}}{\text{Total running time} \times \text{Number of RSs}} \tag{11}$$

ρ is taken as a measure of operational cost. The other measure of interest to us is the average repair latency. We focus on the tradeoff between ρ and the average repair latency provided by algorithms AL1 and AL2. This is done through simulation. Here by *average repair latency* we mean the average latency of recovering lost packets at all of the receivers.

Note that by varying the thresholds, the fraction of repair servers that are active will also vary. Figure 6 shows the relationship between active fraction ρ and average repair latency, the curves being generated by changing the value of the thresholds. In our figures, AL1 represents the original algorithm, in which loss probability $\hat{p}_{loss}^{k}$ was the only metric considered, while AL2 represents the modified algorithm that combines the parameters of upstream round trip time and $\hat{p}_{loss}^{k}$. We observe from Figure 6 that AL2 can produce a lower average repair latency for the same value of ρ. That is, while consuming the same amount system resources as AL1, AL2 results in the successful delivery of packets with lower average delay.

As a second comparison of interest, Figure 7 plots the *worst average repair latency* - the largest average repair latency experienced over all receivers - for AL1 and AL2. Consistent with the results from Figure 6, the modified algorithm also reduces the worst average repair latency (for the same value of ρ).

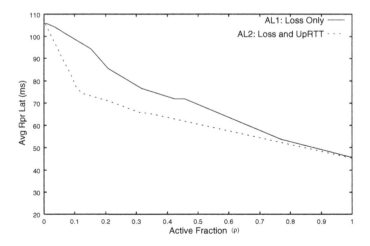

Fig. 6. Average repair latency versus active fraction, topology 1

Another interesting result illustrated by Figures 6 and 7 is that the average repair latency and worst average repair latency decrease rapidly as the fraction of active repair servers increases from zero (i.e., no active repair servers are ever active) to an active fraction of approximately 10 percent. This indicates that much of the performance gain by having active repair servers can be obtained by having only a relatively small fraction of repair servers being active.

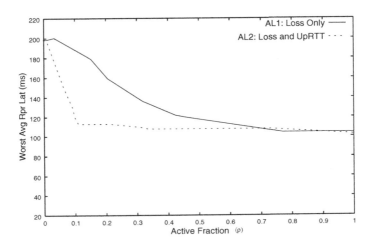

Fig. 7. Worst Average Repair Latency versus active fraction, topology 1

We next introduce a new metric to compare the throughput of AL1 and AL2. The *system repair throughput overhead* counts to total number of link traversals by all re-transmitted packets during the simulation. A smaller system repair throughput overhead indicates that less link bandwidth is used in the network to recover from lost packets. We observe, for topology 1, that AL2 results in a smaller repair throughput overhead than AL1, as shown in Figure 8.

Recall that AL2 uses the upstream RTT in determining the activation/deactivation status of an RS. That means, among several subtrees suffering a similar loss rate, the RS with a long-delay link to its parent will be activated first. For a given amount of repair server resource usage, this could possibly result in higher repair traffic for the system as a whole. Figure 9 shows a second topology, for which AL2 reduces (over AL1) both the average repair latency over all receivers and the largest average repair latency, as shown in Figures 10 and 11. However, for this topology, from Figure 12, we notice increased retransmission throughput overhead incurred by using AL2 in comparison to AL1 - the opposite of what we observed in topology 1. This indicates that the retransmission throughput overhead gains in AL1 versus AL2 are topology dependent. A third topology and its corresponding simulation results can be found in [3].

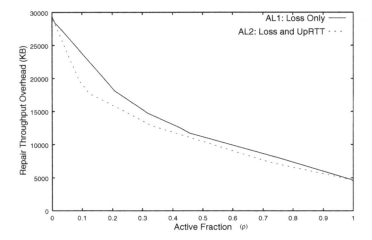

Fig. 8. Repair Throughput Overhead versus active fraction, topology 1

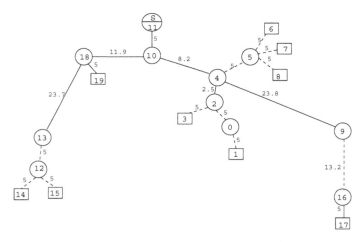

Fig. 9. Topology 2

5   Conclusion

In this paper, we have studied the tradeoff between the time needed to successfully deliver data to the receiver(s) and system resource consumption in server-based active error recovery multicast networks. We began by developing stochastic models to study the distribution of repair delay both in the presence, and in the absence, of repair servers. Based on these models we observed that the application deadline, downstream link loss

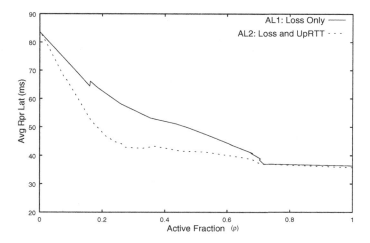

Fig. 10. Average repair latency versus active fraction, topology 2

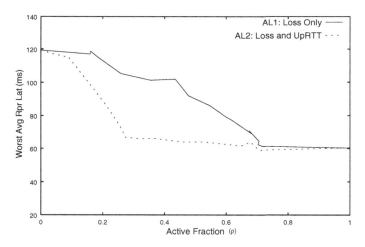

Fig. 11. Worst average repair latency versus active fraction, topology 2

rates, the number of receivers, and the upstream round trip time of a repair server were all important criteria influencing the decision of whether to activate/deactivate an RS.

Recognizing the value of explicitly consider time-sensitive parameters in determining when an RS should be active, and when it should not, we modified the algorithm in [10] to consider not only the packet loss rate and number of downstream receivers, but also the round trip time to the nearest upstream active repair server (or the sender) when making the activate/deactivate decision. We studied the tradeoff that exists between the fraction of RSs that are activated, the amount of overhead traffic, and the latency of suc-

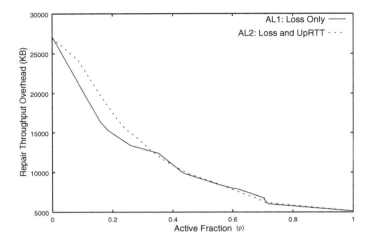

Fig. 12. Repair throughput overhead versus active fraction, topology 2

cessful packet delivery. We found that our modified dynamic RS activation algorithm provides a significant reduction in repair delay over the original algorithm [10], while using no more system resources than the original algorithm. We also found that much of the performance gains achievable by having active repair servers can be obtained by having only a relatively small fraction of repair servers actually being active.

Our future research in this area will investigate performance under bursty link loss rates. We expect the advantages here (over the case of statically configured repair servers that are always active) to be even more significant, as repair servers can be adaptively activated whenever burst loss occurs. An additional interesting area for study is to develop theoretical models that characterize the performance gains possible with active within-the-network servers, as a function of the density of such servers.

Acknowledgement

The authors would like to thank Zihui Ge for his useful suggestions and Diane Kiwior for her help in simulation.

References

1. http://www.tascnets.com/panama/aer/index.html active error recovery (aer) website.
2. S. Floyd, V. Jacobson, S. McCanne, C. Liu, and Zhang L. A reliable multicast framework for light-weight sessions and application level framing. *IEEE/ACM Transactions on Networking*, 5(6):784–803, December 1997.
3. Ping Ji, James Kurose, and Don Towsley. Activating and deactivating repair servers in active multicast tree. Technical Report 01-18, University of Massachusetts at Amherst, June 2001.

4. Sneha K. Kasera. *Scalable Reliable Multicast in Wide Area Networks*. PhD thesis, University of Massachusetts at Amherst, 1999.
5. Sneha K. Kasera, Supratik Bhattacharyya, Mark Keaton, Diane Kiwior, Jim Kurose, Dowsley, and Steve Zabele. Scalable fair reliable multicast using active services. IEEE Networks Magazine, 2000.
6. James F. Kurose and Keith W. Ross. *Computer Networking: A Top-Down Approach Featuring the Internet*, chapter 6. ADDISON WESLEY, 2000.
7. L. Lehman, S. Garland, and D. Tennenhouse. Active reliable multicast. In *IEEE Infocom*, March 1998.
8. M.T.Lucas, B.J.Dempsey, and A.C.Weaver. Mesh: Distributed error recovery for multimedia streams in wide-area multicast. In *IC3N'97*, 1997.
9. N.F.Maxmchuk, K.Padmanabhan, and S.Lo. A cooperative packet recovery protocol for multicast video. In *ICNP'97*, October 1997.
10. Per-Oddvar Osland, Sneha K. Kasera, Jim Kurose, and Don Towsley. Dynamic activation and deactivation of repair servers in a multicast tree. Technical report of computer science department, University of Massachusetts at Amherst, 1999.
11. C. Papadopoulos, G. Parulkar, and G. Varghese. An error control scheme for large-scale multicast applications. In *IEEE Infocom*, March 1998.
12. Dan Rubenstein, Nicholas F. Maxemchuk, and David Shur. A centralized, tree-based approach to network repair server for multicast streaming media. In *NOSSAV*, 2000.
13. S.Paul, K.K.Sabnani, J.C.Lin, and S.Bhattacharyya. Reliable multicast transport protocol (rmtp). *IEEE Journal on Special Areas in communications*, April 1997.
14. Tony Speakman, Dino Farinacci, , Steven Lin, and Alex Tweedly. Pragmatic general multicast internet draft. August 1998.
15. Miki Yamamoto, Jim Kurose, Don Towsley, and Hiromasa Ikeda. A delay analysis of sender-initiated and receiver-initiated reliable multicast protocols. In *IEEE Infocom*, 1997.

Looking for Science
in the Art of Network Measurement

Matthias Grossglauser and Balachander Krishnamurthy

AT&T Labs - Research, 180 Park Ave, Florham Park NJ 07932, USA
{mgross,bala}@research.att.com

Abstract. Network measurements are crucial both to drive research and in network operations. We introduce a taxonomy and survey the state of the art of network measurement. We compare measurements available at the network layer with those available at higher layers, specifically the DNS and the Web. Both the DNS and the Web can be viewed as logical networks; this allows the direct comparison of measurement methods available at these layers with those available at the network layer. We argue that measurement support within the DNS and the Web is insufficient in light of the fact that they affect end-user performance as much as the network layer. We derive some recommendations for the reuse of network layer measurement methods in the DNS and the Web.

1 Introduction

The Internet has evolved into a system of astonishing scale and complexity, fraught with conflicting economic interests, comprised of subsystems from an unprecedented range of vendors, and burdened by many short-sighted fixes to fundamental problems (such as the deployment of NATs in response to a shortage in IP addresses). The research community has no hope of completely modeling the Internet [36]; only crude abstractions that focus on very specific subproblems are within our reach. Therefore, we are called upon to *discover* the behavior of the Internet, in addition to *modeling* aspects of it. The collection, analysis, and interpretation of measurements is key to this aspect of Internet research. It parallels other fields of study that are concerned with systems that escape exhaustive modeling, such as econometrics and biometrics. For example, measurements have revealed surprising statistical features in network traffic that were not predicted by any models before their discovery [27]; clearly, exploratory measurement studies deserve an important place in the research agenda.

Measurements are also crucial in the operation of the Internet. Traffic control and engineering, i.e., routing and resource allocation to make the best use of network resources while maximizing performance, directly depends on traffic measurements [9]. Other examples include accounting and billing, intrusion and attack detection, verification of service level agreements (SLAs), measurement-based admission control [20,14,18], etc.

At higher layers, measuring traffic at popular Web sites as well as examining problems like flash crowds [23, Chapter 11] (sudden surge in traffic aimed at a

S. Palazzo (Ed.): IWDC 2001, LNCS 2170, pp. 524–535, 2001.

site) require measurements where typically administrative access is not available. Identifying the location of clients, mirroring popular sites and resources [28], improving the performance of individual servers, and examining ways to move popular services to edge of the network require large scale measurements.

In this paper, we give a brief survey of the state of the art of network measurement. For this purpose, it is not useful to think of the Internet is terms of the traditional layered model. Rather, we should think of the Internet as being composed of multiple subsystems that exhibit network structure themselves. Chief among these overlay networks are the domain name system (DNS) and the World Wide Web (consisting of clients, origin servers, proxies, caches). Other examples include content distribution networks (CDNs) and peer-to-peer networks, such as Napster [29] and Gnutella [15].

We argue that measurement efforts within these overlay networks face similar challenges and pitfalls as measurement efforts at the network layer, which are arguably more mature. To structure the comparison, we examine three classes of measurements for the network layer, for the DNS, and for the Web: topology, state, and traffic. The *topology* is the static, underlying structure of each network (e.g., physical links between routers, or the static configuration of a proxy in a browser). The *state* refers to dynamic changes in the active topology and other variables *not directly related to traffic* (for example, the utilization of a link would not be considered a state variable, while the operational state of a link is). The *traffic* refers to the flow of work through the network (e.g., packets through the IP network, name resolution requests and responses through the DNS network).

Our systematic comparison illustrates the fact that logical networks, which affect end user performance, are severely under-instrumented today. The visibility into the DNS and the Web is much more limited than at the network layer, with the result that troubleshooting, control and engineering is significantly more challenging for these networks. The purpose of this paper is to point out some of these shortcomings, and to suggest possible remedies inspired by measurement support at the network layer.

The paper is structured as follows. Section 2 briefly summarizes the state of the art in measurement at the network layer. Section 3.1 and section 3.2 give an analogous assessment for the DNS and the Web. We summarize our findings and proposed remedies in Section 4.

2   IP Network Measurements

In this section, we give an overview of measurements available at the IP network layer. For each type of measurement - topology, state, and traffic - we describe what measurements are available externally (i.e., without having administrative control over the network) and internally (with administrative access).

2.1   Topology

The topology of the Internet can thought of as having two levels of hierarchy: the autonomous system (AS) level and the network domain level. We mainly focus on the domain level.

External measurements. Several classes of methods have been proposed to infer the topology of the Internet from external measurements. The first class of methods, commonly referred to as *topology discovery*, relies on a combination of probing methods such as ping and traceroute, and of heuristics to sample the IP address space in an intelligent way to find new nodes and links [17,5].

The second class of methods relies on correlation in the packet loss process of a multicast session. Specifically, a packet loss in a multicast session is experienced by all the receivers downstream from the link where the loss occurred. Thus, by observing a large number of packets at these receivers, the structure of the multicast tree can be approximately inferred [3,31]. This corresponds to finding a subgraph of the network topology.

The third class of methods focuses on inferring other static attributes of the network, such as the capacity of links through active probing [2], or the scheduling discipline [26]. On the Web, identifying and characterizing intermediaries (such as HTTP/1.0 and HTTP/1.1 proxies) is attempted via probing techniques.

Internal measurements. There are several additional sources of information about network topology and configuration when one has administrative control over a network domain. Chief among these are the router configuration files, which provide a router's local view of the topology, including its neighbors, links to and from these neighbors and their capacities. From this, it is conceptually easy to completely determine the physical network topology of the domain [11]. Complications can arise because direct manipulation of router configuration files by operations personnel can lead to inconsistent configurations.

Inferring the topology at the AS level (a graph with currently approximately 10000 nodes) is impossible today. While it is easy to obtain a list of all the ASs, their connectivity depends on local public and private peering arrangements and the routing policies put in place by ISPs. Some heuristic methods to infer subgraphs of the full AS topology are described in [16,8,13].

2.2   State

Next, we compare inferring the state of the network, assuming that the underlying topology is known. Network state includes the operational state of links and routers, the routing and forwarding tables in effect, and other variables that do not directly depend on traffic (e.g., temperature of the CPU).

External measurements. Essentially the same tools used for external topology discovery can be relied upon to discover the operational state of links and routers in a domain. The obvious drawbacks, as in any polling scheme, is that there is a potential delay between a state transition and the time it is discovered; this delay depends on the poll cycle. Also, the absence of a response to a ping packet can imply that the target router or interface is down, or that the path to that router/interface is down. Therefore, the results of multiple pings must be carefully combined to infer link and router state correctly.

To obtain a snapshot of the state of routing, traceroute has been successfully used [30]. This basically amounts to temporally sampling a small subset of routing table entries. Pathchar [19] is an extension of traceroute that is able to obtain rough estimates of additional path characteristics (loss and delay). Beyond this, it is virtually impossible to measure other state variables from the outside.

Internal measurements. Observing the state of network elements is the realm of network management protocols such as SNMP [34]. SNMP enables a network management station to query remote state variables through an agent. The remote variables are standardized as a MIB (management information base) tree. In addition to this polling mode, SNMP allows the definition of events that alert the management system synchronously to state changes. In practice, SNMP tends to incur a relatively high overhead in routers, and its usefulness for fine-grained tracking of network state is limited.

Another method consists in intercepting link state advertisement messages exchanged by the intra-domain routing protocol (e.g., OSPF). This approach has the advantage of being authoritative, in the sense that the observed state is exactly the one that computation of routing tables is based upon [33].

2.3   Traffic

We next examine methods to measure the domain-wide traffic flow. This includes both the load (or demand) imposed on the network domain, the routes followed by the incoming traffic, and its loss and delay characteristics.

External measurements. It is virtually impossible to estimate traffic as a whole through a large measurement domain through external probing of that domain. The only method that falls into this category are recent proposals for the inference of sink trees in distributed denial-of-service (DDoS) attacks, such as IP traceback [32]. In this method, routers randomly encode their address into the identification field of the IP header. With enough samples, a target site of a DDoS attack can reconstruct the sink tree of attack traffic through these encoded addresses.

Internal measurements. There are several methods proposed in the literature that infer domain- wide traffic statistics from different types of measurement.

The first class of methods called *network tomography* relies only on measurements of link utilizations over time. The goal of these methods is to infer the traffic matrix, i.e., the traffic intensity between every ingress and every egress point [37,4,35].

The second class of methods relies on *aggregate flow measurements* at network ingress and/or egress points [10]. A flow is an artificial abstraction of a set of IP packets with identical source-destination addresses (or address prefixes) that are observed close together in time [6]. This method has some drawbacks in terms of overhead, implementation cost, and delay; nevertheless, careful post-processing can yield satisfactory estimates of the domain-wide traffic flow [10].

The third class of methods uses packet sampling. Packet sampling can either be used at all ingress points (in analogy to flow aggregation), relying on measured or simulated routing tables to infer the flow through the domain. Another method called *trajectory sampling* relies on pseudo-random sampling based on hash functions computed over packet content to directly observe these paths [7].

3   DNS and Web Measurements

The various problems we have described in the network layer are reflected largely in other layers as well. In general, having administrative access over all aspects in applications that cross the network layer is harder. As examples, we examine two application areas: Domain Name System (DNS) and the Web.

The most popular application on the Internet currently is the World Wide Web. In terms of traffic on the Internet, the Web is currently responsible for 75% of the packets on the Internet. The rate of growth of traffic between the millions of Web users and Web sites has grown steadily for a decade. Presently there is significant growth at the Intranet level as well. Traffic in peer to peer networks due to the popularity of Napster and Gnutella is growing but they are a much smaller part of the overall traffic and remain largely concentrated in college campuses.

Increase in user-perceived latency, redundant transfer of popular content across the network led to deployment of caches between the clients and the origin servers where resources reside or are generated. Once the usefulness of caching began to crest, offloading of content delivery became popular and led to the advent of content distribution networks (CDNs). Most CDNs use DNS-based redirection which has caused a significant increase in DNS traffic on the Internet.

We begin with some background information and then examine why inferring topology, state, and traffic is difficult at each of these areas.

3.1   DNS Measurements

Topology. The topology of the DNS network consists of a collection of top-level domains (such as .com, .edu, .it etc.) that are just below the root of an hierarchy. These are then organized into separately administered zones (e.g., att.com). The individual zones are responsible only for registering the names and IP addresses of a set of authoritative DNS servers with the root servers. Client requests for translation of names to IP addresses and vice-versa are typically sent by a resolver library that contacts a local DNS server. The local DNS server will check its cache for the request and if it does not have any pertinent information it will forward the request to a root server. The root server will return the names and addresses of the authoritative DNS server that can help answer the query. The queries may proceed iteratively with each query resulting in a pointer to the next server to be queried or recursively, whereby the queried server will do the necessary work and return the result. Positive caching (for hits) and negative caching (for failures) with a specific time to live value is routinely employed at the DNS servers. Most client sites have more than one local DNS server—one or two more serve as secondary servers for backup purposes. Figure 1 shows the various steps involved in resolving the address of the server component embedded in a URL

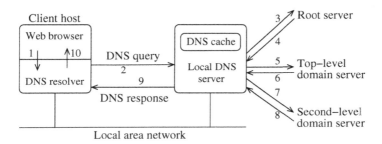

Fig. 1. DNS resolver and local DNS server

A host relying on DNS may be able to examine the static configuration file (e.g., `/etc/resolv.conf` on many UNIX systems) but more recent systems rely on DHCP (Dynamic Host Configuration Protocol) for more automated configurations. The extent of information available to a client is often limited to the names of local DNS servers. The set of authoritative DNS servers that may have to be contacted by the local DNS servers to resolve all the various addresses of interest is simply too large. The set of authoritative DNS servers change from time to time. The local DNS servers have to be kept operational at all times since in their absence applications will not be able to conduct any remote network activity.

Even when the set of DNS servers are within a single administrative domain, there is no simple mechanism to obtain a list of the configured collection since they are distributed across a large number of machines. An attempt to walk the DNS tree hierarchy by using tools like *dig* [1] starting at the zone from a server and examining the name server (NS) records would still miss servers that are not authoritative (caching-only servers) and the unofficial secondary servers used for fault tolerance purposes. Additionally, access control lists placed on zones will make this process even harder. The model of interaction with remote DNS servers is hop by hop with little or no knowledge available about anything beyond the first hop. There is no equivalent of traceroute at the DNS layer.

State. Changes occur often at the DNS layer: new domains are registered, old ones change, cache information becomes stale, etc. The only way to learn about changes is upon a request and failure is often based on a timeout model. Consider the problem of a common mis-configuration known as *lame delegation*: a set of IP addresses have been registered as authoritative for a particular domain. Suppose one of them is incorrect: it is either not an authoritative server for that domain or does not even run a name server. Queries sent to this server will time out and be resent leading to additional unnecessary traffic. Note that a lame server may be contacted directly or as a result of redirection from another authoritative server for the zone.

It has been estimated that there may be up to 25% of all DNS zones with lame delegations. There are two problems in identifying and removing lame delegations. The lame delegations are rarely discovered since often the only indication of their presence is additional latency and a redundant server will eventually answer the query. Even if they are identified, fixing the problem requires interactions with administrators (who are hard to locate). Finally, there is no guarantee that the problem will indeed be fixed. Although the rate of change of information at the DNS layer is significantly less than at the network layer, issues of scale and distributed control make it harder to determine and correct state problems.

Traffic. There are no known mechanisms to trace the DNS traffic as it percolates through the hierarchy. Although logs are maintained on several servers (primarily to detect scan attempts by hackers) they disclose at best partial information. The handing off of control to a backup server via DNS's zone transfer mechanism (a common occurrence) is not known to other servers since the mechanism is meant to be transparent to the end users. The recent significant increase in the DNS traffic due to overlay networks like Akamai (whereby URLs of content distributed resources are replaced with alternate CDN company specific ones whose address resolution can be controlled for load balancing purposes) became visible only because the rewritten URLs can be seen in the HTML text. Other CDNs do dynamic URL rewriting making such detection even harder. DNS servers of CDN companies often give out short TTLs to have more fine-grained control over the use of the mirror servers. With each short TTL expiration the CDN can balance the load on its network of servers. However, this results in additional DNS traffic with questionable performance improvements. Furthermore, obtaining a complete list of sites to which requests get redirected is hard since the CDN resolution is designed to give different answers at different times depending on the client's location.

Inside the network, one can obtain flow records via tools like *netflow* or by running packet tracing programs like *tcpdump*. In both cases, the task of extracting DNS traffic for purpose of identifying problems is pretty complex although there is currently some work in progress in this direction [24]. The problems at the internal level include obtaining a complete view of all DNS traffic entering and exiting the network. Even if the complete information is obtained at high cost, the local configurations are not known which results in at best a partial view of what is actually occurring.

3.2 Web Measurements

As mentioned earlier, Web traffic accounts for 75% of all packets on the Internet. A Web transfer requires interaction with a variety of protocols and often with multiple entities on the Internet. The suite of protocols include the Domain Name System (DNS) protocol, the transport layer protocol (often TCP) for transporting the requests and responses reliably between the Web client and the server, and HTTP (HyperText Transfer Protocol [12,23]—the protocol underlying the World Wide Web and serves as the language of Web messages.

A single user click can result in the involvement of several entities [23, Chapter 15] including the following:

- A browser (and often a client side cache).
- Several Web intermediaries such as Web proxies or gateways. The proxy may be directly configured, be invisible to users (interception proxy), or be part of a large proxy farm.
- A surrogate server in front of the actual Web server deployed to balance the load at the Web site.
- The DNS server at the client/proxy side and additional redirected lookups due to rewritten URLs (due to content distribution overlay networks such as Akamai or Digital Island) and advertisement servers (who contribute images and text to the full container document).

The number of parties involved on an end to end basis in a single Web transaction can be more than a handful.

Topology. The resources requested on the Web by clients like browsers (or quite often programs like spiders), can often be served from different locations either locally through surrogates on the server side or remotely at mirror sites. The set of entities on the Web include clients, proxies, gateways, servers, surrogates, mirrors. The choice of mirrors is dynamically decided. A user's request may traverse several of these entities. Often the user has only control over the next-hop proxy and thus identifying all the entities is hard. The recent advent of HTTP/1.1 version of the protocol [12,23,22] allows intermediaries to identify themselves through a new HTTP header (Via). But given the widespread prevalence of HTTP/1.0 proxies in the Internet for the foreseeable future, those entities may not participate in this enhancement. Furthermore, the presence of interception proxies (the ones that dip into the network layer to examine suspected HTTP traffic and possibly redirect them) exacerbates the difficulty of knowing all the entities that are involved in a transaction.

State. A resource may be cached at a proxy or at a dynamic mirror site. For load balancing and fault tolerance purposes a set of resources may be available from multiple sites in case some sites are inaccessible at any given time. Learning about the state of a specific server is hard due to possible redirection at the HTTP or DNS level.

The widespread presence of caching proxies makes it much harder to determine the actual number of requests generated for a particular resource. Downloading a container document (such as an HTML file which includes links to one or more embedded resources, such as images or animations) requires contacting several servers due to the growing use of CDNs and the presence of advertisements (often located on remote machines). Since the CDNs often dynamically decide the mapping between strings and the actual machines that serve the distributed content, it is not possible to obtain stable latency metrics. The end to end measurements of interest include user perceived latency, load on the network, and load on the servers. However, inferences regarding load on a remote server is very hard to obtain. Simple hacks like examining TCP sequence numbers can be risky in the presence of redirections at the HTTP and DNS layer. Caches introduce the well known problem of staleness: a significant fraction of HTTP requests are validation queries to ensure that a cached resource is the same as the current instance on the origin server. There are risks to a cache assigning freshness time overriding the origin server's wishes.

Traffic. The end to end traffic on the Web is both simply too large and too complex to estimate. Companies like Media Metrix and Keynote attempt to present sampled figures by examining traffic at a few interchanges and extrapolating from them. As discussed above, such studies miss the cached responses, failures, etc. Some studies have been carried out to perform end to end measurements [25] that examine improvements due to the new version of the protocol such as reduction in the number of TCP connections due to persistent connections feature, cache effectiveness, content delivery from multiple sites (CDNs, ad servers etc.), and latency reduction due to the ability of downloading partial responses. Yet, there is not a statistically reliable sampling technique for estimating end to end traffic, due to the complexity of the Web, the widespread prevalence of intermediaries, and implementations that are not compliant with the protocol specification, etc.

Even on an intra-net level, a Web server may not know what fraction of requests directed towards it reach it eventually due to the possibility of proxy cache farms in the path. The server would have to know about the configuration information of all clients in order to obtain a good estimate or indulge in cache busting [23].

Currently the best known traffic artifacts are logs maintained at the proxy and server level. The logs record several fields including the IP address of incoming request, time of request, the HTTP method, URL, protocol version, response code and content length. Even this relatively small subset of items logged has problems associated with how they are interpreted. The 'client' IP addresses recorded could be the last hop proxy and not the original client. A long response in transit may never reach the client who may have aborted the request; yet the server log might indicate that several thousands bytes of response were sent.

4   Conclusion

We have argued that network measurements are crucial both to drive fundamental discoveries and for the purpose of control and engineering. At the IP network layer, this need is fairly obvious, and instrumentation support at the network layer is reasonably mature as a result of pressure on vendors to include measurement support in their products. However, end-user performance depends as much on the performance of logical networks such as the DNS and the Web as on the network layer proper. This suggests that the granularity and scope of measurements available at these layers should match that at the network layer. We have illustrated in this paper that this is not the case through a direct comparison of the state of the art at the network layer with the DNS and the Web.

We have described network measurements for topology, state, and traffic. There is a range of methods for each category. Administrative access to a network domain is usually required to gain access to measurements of sufficient quality to perform traffic engineering. Nevertheless, a range of clever methods are also available to obtain useful snapshots of network topology, routing state, and traffic loads.

At the higher layers the problem of inference is more complicated since topology, state, and traffic have several additional entities and hidden artifacts (some of which are known, such as lame delegations at the DNS layer). Additionally, interesting questions such as user-perceived latency or end-to-end delay are inherently more complicated due to the involvement of multiple protocols and intermediaries. The problem is further compounded by implementations of Web components that are not fully compliant with the protocol specification [21]. In some cases application level protocols have attempted to mimic some of the useful ideas in the network layer. For example, HTTP/1.1 introduced the `Via` and `Max-Forwards` header to expose some of the topology information about intermediaries and to better target the requests.

We believe that the additional measurement support at the application layer could be inspired by tools and methods that have proved valuable at the network layer. For example, an equivalent of *traceroute* in DNS to track a request, *pathchar* in HTTP to derive performance properties of nodes (proxies) on the path to the server, or native support for request sampling, would be useful for troubleshooting, testing, and control. While the details of such a suite of higher-layer tools would certainly reflect the application area, we hope that the foregoing discussion has shown conceptual similarity between the network layer and other application areas to motivate the reuse of the expertise gained at the network layer.

References

1. Paul Albitz and Cricket Liu. *DNS and BIND*. O'Reilly, third edition, September 1998. ISBN 1565925122.
2. J-C. Bolot. End-to-end packet delay and loss behavior in the Internet. In *Proc. ACM SIGCOMM '93*, San Fransisco, CA, September 93.
3. R. Cáceres, N.G. Duffield, J. Horowitz, F. Lo Presti, and D. Towsley. Loss-based Inference of Multicast Network Topology. In *Proc. 1999 IEEE Conference on Decision and Control*, Phoenix, AZ, December 1999.
4. J. Cao, D. Davis, S. Vander Wiel, and B. Yu. Time-Varying Network Tomography. *Journal of the American Statistical Association*, December 2000.
5. Bill Cheswick, Hal Burch, and Steve Branigan. Mapping and Visualizing the Internet. In *Usenix 2000*, San Diego, CA, 2000.
6. Cisco Corp. Netflow Services and Applications (white paper). August 1999.
7. N. C. Duffield and M. Grossglauser. Trajectory Sampling for Direct Traffic Observation. *IEEE/ACM Transactions on Networking*, June 2001.
8. Michalis Faloutsos, Petros Faloutsos, and Christos Faloutsos. On power-law relationships of the Internet topology. In *Proc. ACM SIGCOMM '99*, pages 251–262, August/September 1999.
9. A. Feldmann, A. Greenberg, C. Lund, N. Reingold, and J. Rexford. NetScope: Traffic engineering for IP networks. *IEEE Network Magazine*, March 2000.
10. A. Feldmann, A. Greenberg, C. Lund, N. Reingold, J. Rexford, and F. True. Deriving traffic demands for operational IP networks: Methodology and experience. In *Proc. ACM SIGCOMM*, Stockholm, Sweden, August 2000.
11. Anja Feldmann and Jennifer Rexford. IP network configuration for intradomain traffic engineering. *to appear in IEEE Network Magazine*, 2001.
12. R. Fielding, J. Gettys, J. C. Mogul, H. Frystyk, L. Masinter, P. Leach, and T. Berners-Lee. Hypertext Transfer Protocol — HTTP/1.1. RFC 2616, IETF, June 1999. Draft Standard of HTTP/1.1.
    http://www.rfc-editor.org/rfc/rfc2616.txt.
13. Lixin Gao. On inferring autonomous system relationships in the Internet. In *Proc. Global Internet 2000*, November 2000.
14. R. J. Gibbens, F. P. Kelly, and P. B. Key. A Decision-theoretic Approach to Call Admission Control in ATM Networks. *IEEE Journal on Selected Areas of Communications*, pages 1101–1114, August 1995.
15. Gnutella.
    http://gnutella.wego.com.
16. Ramesh Govindan and Anoop Reddy. An analysis of Internet inter-domain topology and route stability. In *Proc. IEEE INFOCOM '97*, April 1997.
17. Ramesh Govindan and Hongsuda Tangmunarunkit. Heuristics for Internet Map Discovery. In *Proc. of IEEE Infocom 2000*, Tel Aviv, Israel, April 2000.
18. M. Grossglauser and D. N. C. Tse. A Framework for Robust Measurement-Based Admission Control. *IEEE/ACM Transactions on Networking*, 7(3), June 1999.
19. Van Jacobson. Pathchar.
    ftp://ftp.ee.lbl.gov/pathchar.
20. S. Jamin, P. B. Danzig, S. Shenker, and L. Zhang. A Measurement-Based Admission Control Algorithm for Integrated Services Packet Networks. *IEEE/ACM Transactions on Networking*, 5(1), February 1997.
21. Balachander Krishnamurthy and Martin Arlitt. PRO-COW: Protocol Compliance on the Web—A Longitudinal Study. In *Proc. USENIX Symposium on Internet*

Technologies and Systems, March 2001.
`http://www.research.att.com/~bala/papers/usits01.ps.gz`.

22. Balachander Krishnamurthy, Jeffrey C. Mogul, and David M. Kristol. Key Differences between HTTP/1.0 and HTTP/1.1. In *Proc. Eighth International World Wide Web Conference*, May 1999.
`http://www.research.att.com/~bala/papers/h0vh1.html`.

23. Balachander Krishnamurthy and Jennifer Rexford. *Web Protocols and Practice: HTTP/1.1, Networking Protocols, Caching, and Traffic Measurement.* Addison-Wesley, May 2001. ISBN 0-201-710889-0.

24. Balachander Krishnamurthy, Oliver Spatscheck, Chuck Cranor, Emden Gansner, and Carsten Lund. Characterizing DNS traffic. Document in preparation.

25. Balachander Krishnamurthy and Craig E. Wills. Analyzing Factors that influence end-to-end Web performance. In *Proc. World Wide Web Conference 2000*, pages 17–32, May 2000.
`http://www.research.att.com/~bala/papers/www9.html`.

26. A. Kuzmanovic and E. Knightly. Measuring Service in Multi-Class Networks. In *Proc. of IEEE INFOCOM 2001*, Anchorage, Alaska, April 2001.

27. Will E. Leland, Murad S. Taqqu, Walter Willinger, and Daniel V. Wilson. On the Self-Similar Nature of Ethernet Traffic (Extended Version). *IEEE/ACM Trans. on Networking*, 2(1):1–15, February 1994.

28. Andy Myers, Peter Dinda, and Hui Zhang. Performance Characteristics of Mirror Servers on the Internet. In *IEEE INFOCOM '99*, New York City, March 1999.

29. Napster.
`http://www.napster.com`.

30. V. Paxson. End-to-End Routing Behavior in the Internet. *IEEE/ACM Transactions on Networking*, 5(5):601–615, October 1997.

31. S. Ratnasamy and S. McCanne. Inference of Multicast Routing Trees and Bottleneck Bandwidths using End-to-End Measurements. In *IEEE INFOCOM '99*, New York City, April 1999.

32. Stefan Savage, David Wetherall, Anna Karlin, and Tom Anderson. Practical Network Support for IP Traceback. In *ACM SIGCOMM 2000*, September 2000.

33. Aman Shaikh, Mukul Goyal, Albert Greenberg, Raju Rajan, and KK Ramakrishnan. An OSPF Topology Server: Design and Evaluation. *submitted for publication*, 2001.

34. William Stallings. *SNMP, SNMP v2, SNMP v3, and RMON 1 and 2 (Third Edition)*. Addison-Wesley, Reading, Mass., 1999.

35. C. Tebaldi and M. West. Bayesian Inference on Network Traffic. *Journal of the American Statistical Association*, June 1998.

36. V. Paxson and S. Floyd. Why We Don't Know How To Simulate The Internet. In *Proceedings of the 1997 Winter Simulation Conference*, December 1997.

37. Y. Vardi. Network Tomography. *Journal of the American Statistical Association*, March 1996.

Design and Deployment
of a Passive Monitoring Infrastructure

Chuck Fraleigh[1], Christophe Diot[2], Bryan Lyles[2], Sue Moon[2],
Philippe Owezarski[3], Dina Papagiannaki[2], and Fouad Tobagi[1]

[1] Stanford University, Stanford, CA, USA
{cjf,tobagi}@stanford.edu
[2] Sprint Advanced Technology Lab, Burlingame, CA, USA
{cdiot,lyles,sbmoon,dina}@sprintlabs.com
[3] LAAS-CNRS, Toulouse, France
owe@laas.fr

Abstract. This paper presents the architecture of a passive monitoring system installed within the Sprint IP backbone network. This system differs from other packet monitoring systems in that it collects packet-level traces from multiple links within the network and provides the capability to correlate the data using highly accurate GPS timestamps. After a thorough description of the monitoring systems, we demonstrate the system's capabilities and the diversity of the results that can be obtained from the collected data. These results include workload characterization, packet size analysis, and packet delay incurred through a single backbone router. We conclude with lessons learned from the development of the monitoring infrastructure and present future research goals.

1 Introduction

Network traffic measurements provide essential data for networking research and operation. Collecting and analyzing such data from a tier 1 ISP backbone, however, is a challenging task. The traffic volume ranges from tens of Mb/sec on OC-3 access links to 10 Gb/sec on OC-192 backbone links. The measurement equipment must be installed in commercial network facilities where physical space and power are constrained, and which are, in some cases, not staffed by any human operators. Data analysis involves processing terabytes of data, and must account for unusual phenomena such as routing loops and malicious network users.

This paper presents our experiences developing the Sprint IP Monitoring System, a traffic measurement system for the Sprint Internet IP network. The Sprint Internet network is a tier 1 IP backbone connecting 20 Points of Presence (POPs) in the United States. The monitoring system is designed to collect synchronized traces of traffic data from multiple links for use in in-depth research projects where aggregate statistics are insufficient.

The Sprint IP Monitoring System consists of three basic components:

S. Palazzo (Ed.): IWDC 2001, LNCS 2170, pp. 556–575, 2001.

- A set of data collection systems (IPMON systems) which collect TCP/IP headers of every packet transmitted on various links in the Sprint network, along with a measurement system that collects BGP and IS-IS routing information. Currently 11 IPMON systems are deployed at one POP in the Sprint network. Two additional POPs, each with 10 systems, are scheduled to be installed.
- A data repository which archives the traces collected by the IPMON systems.
- A 16 node computing cluster used for data analysis.

The design of the system focuses on four major issues: collecting packet traces from high speed links, synchronizing the traces, manipulating the large data sets, and administering the system. Collecting traces from high speed OC-3 (155 Mb/sec) and OC-12 (622 Mb/sec) links has been addressed by several prior measurement systems [1][2]. Our system follows the same general design, and we discuss some of the challenges when extending it to OC-48 (2.48 Gb/sec) speeds. Synchronizing the clocks that generate the packet timestamps is accomplished using a stratum-1 GPS reference clock distributed to the IPMON systems. The data set for a single 24 hour trace collected on all of the IPMON system is 1.1 TB. This data is compressed and transmitted from the IPMON systems to the data repository over a dedicated OC-3 link. The data repository and data analysis systems are interconnected using a gigabit Ethernet network. All system administration functions for the IPMON systems may be performed from the lab. In cases of extreme failures, the entire operating system may be reinstalled over the network.

The remainder of the paper describes the details of how we address these design issues in the IP Monitoring System and presents some sample results that demonstrate the system's capabilities. Section 2 discusses other work on IP monitoring systems and compares them with our monitoring infrastructure. Section 3 describes the architecture of the network in which our monitoring systems are installed. Section 4 presents the design requirements and details of the monitoring system. Section 5 presents traffic measurements which demonstrate the capabilities of our system and which evaluate the system performance. Section 6 concludes and discusses areas of future research.

2   Related Work

There has been much work on active and passive network measurement systems. A measurement system is called active if it injects measurement traffic, such as probe packets, in the network. Passive measurement systems, on the other hand, do not inject any measurement traffic but rather observe the actual traffic flowing in the network.

Active measurement systems include NIMI, MINC, Surveyor, AMP, and IEPM. The NIMI (National Internet Measurement Infrastructure) project developed an architecture for deploying and managing scalable active measurement systems [3]. The NIMI system uses tools such as *ping, traceroute, mtrace*, and *treno* to perform the actual measurements [4][5]. MINC (Multicast-based Inference of Network-internal Characteristics) measurement systems transmit multicast probe messages to many destinations, and infer link loss rates, delay distributions, and topology based on the observed correlations of the received packets [6]. Surveyor uses a set of approximately 50 GPS synchronized hosts to measure one-way and round-trip delay over various Internet paths [7]. The AMP (NLANR Active Measurement Project) system consists of a set of monitoring stations which measure the performance of the vBNS backbone [8]. IEPM (Internet End-to-end Performance Monitoring) monitors network performance between high energy nuclear and particle physics research institutions [9]. In addition, companies such as Keynote and Matrix conduct commercial network performance measurements [10][11].

Passive measurement systems include Simple Network Management Protocol (SNMP)-based network traffic measurement tools, *tcpdump*, NetFlow, and CoralReef. SNMP is the most widely used network management protocol in today's Internet [12]. Agents and remote monitors update a management information base (MIB) within network routers, and management stations retrieve MIB information from the routers using UDP. Most routers support SNMP and implement public MIBs as well as vendor-specific private MIBs. Using SNMP, for example, network operators can keep track of the number of packets and bytes that have arrived on an interface, the number of packets and bytes that have been dropped on an interface, and the number of transmission errors that have occurred on a link. Another common network monitoring tool is *tcpdump*. It collects packets transmitted and received by systems running the Unix operating system [13]. NetFlow is a monitoring system available on Cisco routers which collects flow statistics observed by an interface [14]. NetFlow provides more detailed information than is available through SNMP, but it requires an external system to record the NetFlow data. The CoralReef suite, developed by CAIDA and originally based on the OC3MON developed at MCI, collects timestamped packet traces from various ATM and SONET links [15][1]. This system is very similar to our monitoring system, but it does not have GPS synchronization.

Other efforts have been made in routing and standardization of metrics. The Internet Performance Measurement and Analysis (IPMA) project investigates routing behavior and network failures [16]. The IETF IP Performance Metrics (IPPM) working group standardizes metrics for evaluating network performance based on observations realized within the projects described above [17]. To our knowledge, the only project to address network-wide traffic analysis in a comprehensive manner has been developed at AT&T [18]. This project relies on packet-level information collected by packet sniffers called PacketScopes, flow statistics collected using Cisco's NetFlow tools, and routing information. It also includes active components which collect loss, delay, and connectivity statistics. Results from active and passive components are combined to be used in network monitoring and management of the AT&T backbone.

Our project is unique in that it allows trace collection at different points in a commercial backbone network, and it provides the capability to correlate these traces through highly accurate timestamps. Many other traffic measurement systems are installed in either research or access networks [19][8][20]. Some results from commercial IP backbones are available in [21][2], but they do not have the GPS synchronization capabilities of our system.

3 Backbone Network Description

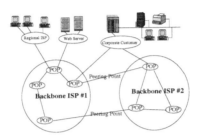

Fig. 1. Tier 1 IP backbone network

A backbone IP network provides connectivity over a geographically wide area. The network topology consists of a set of nodes known as Points-of-Presence (POPs) connected by high bandwidth backbone links. These links are typically 2.5 Gb/sec OC-48 links or 10 Gb/sec OC-192 links. Each POP also contains links to customers (e.g. large corporate networks, regional ISPs providing dial-up access, large web servers), ranging from 1.5 Mb/sec T1 links to 622 Mb/sec OC-12 links. Figure 1 shows the architecture of a backbone network.

A backbone network connects to other backbone networks at private peering points or public network access points (NAPs). Two networks can peer at multiple points to accommodate the traffic volume between them as shown in Figure 1. The peering points are intended to carry traffic that originates from a customer connected to one backbone ISP and is destined to a customer of another backbone ISP. Most peering agreements prohibit transit traffic, or traffic whose source and destination are not customers of the backbone ISP. For example, Backbone ISP #2 would not accept traffic from Backbone ISP #1 if the destination was not one of Backbone ISP #2's customers.

In case of the Sprint Internet backbone, there are 20 POPs located in the continental United States. The POPs in the Sprint Internet backbone have a two-level hierarchical structure as shown in Figure 2. At the lower level, customer links are connected to access aggregation routers. The access routers are in turn connected to the higher level backbone routers. The backbone routers provide connectivity to other POPs, and they also connect to public and private peering points. The backbone links connecting the POPs are optical fibers with a bandwidth of 2.5 Gb/sec (OC-48) or 10 Gb/sec (OC-192). All of the links

Fig. 2. POP architecture

carry IP traffic using a proprietary version of the Packet-over-SONET (POS) protocol similar to that proposed in [22].

4   System Description

The goal of the Sprint IP Monitoring System is to collect data from the Sprint Internet backbone that is needed to support a variety of research projects. The particular research projects include studying the behavior of TCP, evaluating network delay performance, investigating the nature of denial of service attacks, and developing network engineering tools. While each project could develop a customized measurement system, many of the projects require similar types of data and installing monitoring equipment in a commercial network is a complex task making a general purpose measurement system more preferable.

To meet this goal, the IP Monitoring system is designed to collect and analyze synchronized packet level traces from selected links in the Sprint Internet backbone. These packet traces consist of the first 44 bytes of every packet carried on the links along with a 64 bit timestamp. The clocks which generate the timestamps are synchronized to within 5 μs using a GPS reference clock. This provides the capability to measure one-way network delays and study correlations in traffic patterns.

Trace collection and analysis is accomplished using three separate systems. A set of data collection components (IPMON systems) collect the packet traces. The traces are transferred to a data repository which stores the traces until they are needed for analysis. Analysis is performed on a cluster of 16 Linux servers. The remainder of this section describes the requirements and architecture of each of these components.

4.1   IPMON Systems

The monitoring systems, called IPMON systems, are responsible for collecting packet traces from the network. These system consists of a Linux PC with a large disk array and a SONET network interface, known as the DAG card [23][24]. To collect the traces, an optical splitter is installed on selected OC-3, OC-12, and OC-48 links in the network, and one output of the splitter is connected to the DAG card in the IPMON system.

Table 1. Packet Record Format

bytes	field description
0	8 byte
4	timestamp
8	record size (fixed to 64 bytes)
12	POS frame length
16	HDLC header
20	First 44 bytes of IP packet
⋮	
64	

The DAG card decodes the SONET payloads and extracts the IP packets. When the beginning of a packet is identified, the DAG card generates a timestamp for the packet, extracts the first 48 bytes of the POS frame which contains 4 bytes of POS header and 44 bytes of IP data, and transfers the packet record to main memory in the PC using DMA. The format of the packet record is shown in Table 1. If the packet contains fewer than 44 bytes, the data is padded with all 0's. Once 1 MB of data has been copied to main memory, the DAG card generates an interrupt which triggers an application to copy the data from main memory to the hard disk. It would be possible to transfer the data from the DAG card directly to the hard disk, and bypass the main memory. Main memory, however, is necessary to buffer bursts of traffic as described later in the section.

The IPMON system has 5 basic design requirements:

– Support data rates ranging from 155 Mb/sec (OC-3) to 2.5 Gb/sec (OC-48)
– Provide synchronized timestamps
– Occupy a minimal amount of physical space
– Prevent unauthorized access to trace data
– Be capable of remote administration

Next we describe how each of these requirements are met in the system design.

Data Rate Requirements. The data rate requirements for OC-3, OC-12, and OC-48 links are summarized in Table 2. The first line of the table shows the data rate at which the DAG card must be able to process incoming packets. After the DAG card has received a packet and extracted the first 44 bytes, the timestamp and additional header information is added to the packet record and copied to main memory. If there is a sequence of consecutive packets whose size is less than 64 bytes, then the amount of data that is stored to main memory is actually greater than the line rate of the monitored link. The amount of internal bandwidth required to copy a sequence of records corresponding to minimum size TCP packets (40 byte packets) from the DAG card to main memory is shown on the second line of Table 2. To support this data rate, the OC-3 and OC-12 IPMON systems use a standard 32 bit, 33 MHz PCI bus which has a capacity of 1056 Mb/sec (132 MB/sec). The OC-48 system, however, requires a 64 bit, 66 MHz PCI bus with a capacity of 4224 Mb/sec (528 MB/sec). It is possible to have non-TCP packets which are smaller than 40 bytes resulting in even higher bandwidth requirements, but the system is not designed to handle extended bursts of these packets as they do not occur very frequently. It is assumed that the small buffers located on the DAG card can handle short bursts of packets less than 40 bytes in size. The impact of this design decision is evaluated in Section 5.

Table 2. Data rate requirements

	OC-3	OC-12	OC-48
link rate (Mb/sec)	155	622	2480
peak capture rate (Mb/sec)	248	992	3968
1 hour trace size (GB)	11	42	176

Once the data has been stored in main memory, the system must be able to copy the data from memory to disk. The bandwidth required for this operation, however, is significantly lower than the amount of bandwidth needed to copy the data from the DAG card to main memory as the main memory buffers bursts of small packets before storing them to disk. Only 64 bytes of information are recorded for each packet that is observed on the link. As reported in prior studies, the average packet size observed on backbone links ranges from about 300-400 bytes during the busy periods of the day [21]. For our design, we assume an average packet size of 400 bytes. The disk I/O bandwidth requirements are therefore only 16% of the actual link rate. For OC-3 this is 24.8 Mb/sec; for OC-12, 99.5 Mb/sec; and for OC-48, 396.8 Mb/sec. To support these data rates, we use a three-disk RAID array for the OC-3 and OC-12 systems which has an I/O capacity of 240 Mb/sec (30 MB/sec). The RAID array uses a software RAID controller available with Linux. To support OC-48 we use a five-disk RAID array with higher performance disks that can support 400 Mb/sec (50 MB/sec) transfers. To minimize interference with the data being transferred from the DAG card to memory, the disk controllers use a separate 32 bit 33 MHz PCI bus.

Timestamp Requirements. In order to correlate the traces, the packet timestamps generated by each IPMON system need to be synchronized to a global clock signal. This is accomplished using a dedicated clock on board the DAG card. The clock runs at a rate of 16MHz which provides a granularity of 59.6 ns between clock ticks.

Unfortunately, the oscillators used to drive the clock run just a little bit faster or a little bit slower than 16 MHz based on system temperature and the quality of the oscillator. Therefore, it is necessary to fine tune, or discipline, the clocks using an external stratum 1 GPS receiver located at the POP sites. The GPS receiver outputs a 1 pulse-per-second (PPS) signal which is distributed to all of the DAG cards located at the POP.

The clocks synchronization on board the DAG card operate in the following manner [25]. At the beginning of trace collection the clock is loaded with the absolute time from the PC's system clock (e.g. 7:00 am Aug 9, 2000 PST). The clock then begins to increment at a rate of 16 MHz. When the DAG card receives the first 1 PPS signal after initialization, it resets the lower 24 bits of the clock counter (note: 24 bits will count from 0 to 16 million. If the lower 24 bits of the clock are all 0 it represents the beginning of a second). Thereafter, each time the DAG card receives the 1 PPS signal, it compares the lower 24 bits of the clock to 0. If the value is greater than 0, the oscillator is running a little bit fast and the DAG card decreases the frequency slightly. If the value is less than 0, the oscillator is running a little bit slow and the DAG card increases the frequency slightly.

In addition to synchronizing the DAG clocks, the IPMON systems must also synchronize their own internal clocks so that the DAG clock is correctly initialized. This is accomplished using NTP. A broadcast NTP server is installed on the LAN which is connected to the monitoring systems and is capable of synchronizing the system clocks in the PC to within 200 ms. This is sufficient to synchronize the beginning of the traces, and the 1 PPS signal is used to further synchronize the DAG clock. There is an initial period where the 1 PPS is attempting to correct the initial clock skew, so we ignore the first 30 seconds of each trace to account for this.

There are several sources of error that may occur in the synchronization system. The first is clock skew between the 1 PPS signals generated by different GPS receivers located at different POPs. This error is minimal as we are using stratum 1 GPS receivers which are guaranteed to have a maximum clock skew of 500 ns. Another source of error is the difference in propagation time for the 1 PPS signal. The 1 PPS signal is distributed to the DAG cards using a daisy chain topology. The difference in cable length between the first and last systems is 8 meters, which corresponds to a propagation delay of 28 ns. Finally, the clock synchronization mechanism cannot immediately adjust to changes in the oscillator frequency, it needs to wait for the next 1 PPS signal. To test this aspect of performance, we measure the maximum clock error that is observed when the DAG card receives a 1 PPS interrupt. The maximum value we have seen in lab tests is 30 clock ticks which represents an error of 1.79 μs. The median error observed during these tests was 1 clock tick, or 59.6 ns. Adding all of these factors, the worst case skew between any two DAG clocks is less than 2 μs.

Another source of error is that packets are not timestamped immediately when they arrive at the DAG card. They must first pass through a chip which implements the SONET framing. As this chip was initially designed to operate on 53 byte ATM cells, it is possible for two packets (one 40 byte packet and the first 13 bytes of the next packet) to be placed in a cell buffer at the same time. Since this buffer is read as an entire unit, both packets will have identical timestamps. However, their actual inter-arrival time is 2 μs if we are measuring an OC-3 link. This results in an additional 2 μs of timestamp error. This is only an issue for the OC-3 and OC-12 DAG cards. The OC-48 systems use a newer SONET framing chip which was designed to support POS directly and does not use 53 byte buffers.

The total effect of these errors is a maximum of 5 μs of clock skew between DAG cards. However, we are interested in measuring the delay experienced by packets as they traverse the network. This delay is typically measured on the order of milliseconds, so the 5 μs skew is acceptable. The only case where the clock skew affects the measurements are when we are interested in measuring the delay through a single router in the network. The minimum delay we have observed is 30 μs, so a 5 μs skew represents a 16% error in the measurement.

Physical Requirements. In addition to supporting the bandwidth requirements of OC-3, OC-12, and OC-48 links, the IPMON systems must also have a large amount of hard disk storage capacity to record the traces. As the systems are installed in a commercial network facility where physical space is a scarce resource, this disk space must be contained in small form factor. Using a rack-optimized system, the OC-3 and OC-12 IPMONs are able to handle 108 GB of storage in only 4U of rack space[1]. This allows the system to record data for 9.8 hours on a fully utilized OC-3 link or 2.6 hours on a fully utilized OC-12 link. The OC-48 systems have a storage capacity of 360 GB, but in a slightly larger 7U form factor. This is sufficient to collect a 2 hour trace on a fully utilized link. Fortunately, the average link utilization on most links is less than 50%, allowing for longer trace collection.

The physical size constraint is one of the major limitations of the IPMON system. Collecting packet level traces requires significant amounts of hardware. These traces are critical for conducting research activities, but trace collection is not a scalable solution for operational monitoring of an entire network. The ideal solution is to use traces collected by the IPMON system to study the traffic and develop more efficient monitoring systems targeted towards exhaustive monitoring of all links in the network for operational purposes.

Security Requirements. The IPMON systems collect proprietary data about the traffic on the Sprint Internet backbone. Preventing unauthorized access to this trace data is an important design requirement. This includes preventing access to trace data stored on the systems as well as preventing access to the

---

[1] 1U is a standard measure of rack space and is equal to 1.75 inches or 4.45 cm.

systems in order to collect new data. To accomplish this, the systems are configured to accept network traffic only from two applications: *ssh* and NTP. *ssh* is an authenticated and encrypted communication program similar to *telnet* that provides to access a command line interface to the system. This command line interface is the only way to access trace data that has been collected by the system and to schedule new trace collections. *ssh* only accepts connections from a server in our lab and it uses an RSA key based system to authenticate users. All data that is transmitted over the *ssh* connection is encrypted.

The second type of network traffic accepted by the IPMON systems is NTP traffic. The systems only accept NTP messages which are transmitted as broadcast messages on a local network used exclusively by the IPMON systems. All broadcast messages which do not originate on this network are filtered.

Remote Administration Requirements. In addition to being secure, the IPMON systems must also be robust against failures since they are installed, in some cases, where there is no human presence. To detect failures, a server in the lab periodically sends query messages to the IPMON systems. The response indicates the status of the DAG cards and of the NTP synchronization. If the response indicates either of these components fails, the server attempts to restart the component through an *ssh* tunnel. If the server is not able to correct the problem it notifies the system administrator that manual intervention is required. In some cases, even the *ssh* connection will fail, and the systems cannot be accessed over the network. To handle this type of failure, the systems are configured with a remote administration card that provides the capability to reboot the machine remotely. The remote administration card also provides remote access to the system console during boot time. In cases of extreme failure, the system administrator can boot from a write-protected floppy installed in the systems and completely reinstall the operating system remotely.

The one event that cannot be handled remotely is hardware failure. The monitoring systems play no role in the operation of the backbone, and thus we decided not to provide hardware redundancy to handle failures.

4.2   Data Repository

The data repository is a large tape library responsible for archiving the trace data. Once a set of traces has been collected on the IPMON systems, the trace data is transferred over a dedicated OC-3 connection from the IPMONs to the data repository.

A single 24-hour-long trace from all of the monitoring systems currently installed consumes approximately 1.2 TB of disk space (this will increase to 3.3 TB when the additional 20 systems are installed). The tape library has 10 individual tape drives which are able to write data at an aggregate rate of over 100 MB/sec. The rate at which the data can be transferred from the remote systems, however, is limited to 100 Mb/sec which is the capacity of the network interface cards on the IPMON systems. At this rate the raw data would take

26.4 hours to transfer from the IPMON systems to the tape library. To improve transfer time and decrease the storage capacity requirements, the trace data is compressed before being transferred back to the lab. Using standard compression tools such as *gzip*, we are able to achieve compression ratios ranging from 2:1 to 3:1 depending on the particular trace characteristics. This reduces the transfer time to about 12 hours.

This transfer time presents another difficulty when exhaustively monitoring a network for operational purposes. An alternative solution would be to avoid the data repository and perform the analysis on the monitoring systems themselves. This is a good solution if there is a single type of analysis that is being performed on the traces. However, the data is used for many research projects, and some of the analysis performed in several projects requires multiple iterations through the trace. In addition, we would like to keep an archive of the collected data so that it may be used for future projects.

4.3   Analysis Platform

All data analysis is performed off-line by a cluster of 16 Linux PCs. Some of the data analysis, such as measuring packet size distributions, could be performed on-line but others, such as measuring network delay, cannot. Measuring network delay requires identifying a packet on multiple traces. This involves exchanging significant amounts of data between two (or more) IPMON systems for every packet whose delay is measured. This could be accomplished on a local network if both systems are located in the same POP, but it would significantly increase the network load when processing data collected at multiple POPs. Since we do not want to perturb the network during trace collection, it is necessary to perform the analysis off-line.

There are two categories of analysis that are performed by the analysis platform:

- Single trace analysis involves processing data from a single link to measure traffic characteristics. This type of analysis includes, for example, determining packet size distributions, flow size distributions, and determining what types of applications are using different links. To efficiently perform this type of analysis, traces from different links are loaded onto separate PCs in the cluster and processed in parallel.
- Multi-trace analysis involves correlating traffic measurements among different links. This includes performing delay measurements and looking at round trip TCP behavior. This type of analysis is performed by dividing each trace into several time segments and loading the different time segments onto different machines. For example, PC #1 might contain the first 30 minutes from a set of 5 traces, and PC #2 might contain the next 30 minutes of those same 5 traces.

One key requirement when performing multi-trace analysis is to be able to identify an individual packet as it travels across multiple links in the network. The only two pieces of information that should change as a packet travels through the network are the TTL and checksum fields in the IP header. By comparing the remaining 41 bytes of data we collect for each packet, we can identify a packet at multiple locations in the network. However, it is possible for two different packets to have the same 41 bytes. In theory this should happen infrequently since the ID field for each packet generated by a particular source should be unique. However, in the traces we collect we do observe duplicate packets due to systems generating incorrect IP id fields or due to link layer retransmissions, but these packets only represent .001% to 0.1% of the total traffic volume. In these cases, we typically ignore all packets which have duplicate values.

5   Measurement Results

In this section we present a sample of measurement results to demonstrate the types of data that can be collected using the IPMON measurement facilities. The presented data are not intended to make any generalizable statement on the nature of the traffic on an IP backbone. The intent is to validate our monitoring infrastructure and to demonstrate its capabilities using a few simple trace analyses.

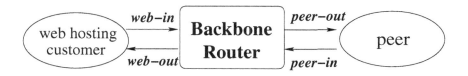

Fig. 3. Measurement Configuration

For brevity, we present data from only two of the nine bi-directional monitored links. Both links are connected to the same core router in one POP within the Sprint IP backbone network. The trace *web-out* was collected on a link from the backbone router to an access router connected to a web hosting company, and *web-in* is the link from the access router to the backbone router. The *peer-out* trace was collected on a link from the backbone router to a peering point, and *peer-in* is the link from the peering point to the backbone router. Both the peering link and the web hosting link are OC-3 links. Figure 3 shows a diagram of the monitored links. It is important to note that there are other, unmonitored, links which are also connected to the backbone router. We do not exhaustively monitor all links in the POP. Table 3 provides the trace start times, trace end times, and trace sizes.

Table 3. Trace Statistics

Link	Start Time (PST)	End Time (PST)	Number of Packets (millions)
web-out	9:56, Wed, 8/9/2000	19:57, Wed, 8/9/2000	568
web-in	9:56, Wed, 8/9/2000	23:18, Wed, 8/9/2000	852
peer-out	9:56, Wed, 8/9/2000	9:55, Thurs, 8/10/2000	816
peer-in	9:56, Wed, 8/9/2000	9:55, Thurs, 8/10/2000	794

5.1   Workload Characterization

First we present the general characteristics of the traces. Figures 4 and 5 plot link utilization averaged over one minute periods. Figure 6 shows the application mix on the *web-out* trace. From these figures we make several observations:

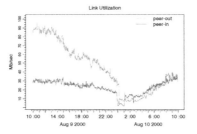
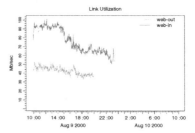

Fig. 4. Peering link utilization in Mb/sec   **Fig. 5.** Web link utilization in Mb/sec

- As reported in many other studies, the dominant source of traffic is http. We only present the results from the *web-out* link, but the results are similar on the other links.
- The link utilization on the peering link changes dramatically (from nearly 100 Mb/sec to under 10Mb/sec) over a 24 hour period.
- Link utilization is not symmetric for either the peering link or the web hosting link. While this is expected for the web hosting traffic, as the web servers should generate much more data than they receive, the traffic was expected to be more symmetric on the peering point. All of the links we monitor exhibit such asymmetric characteristics. It may be possible to take advantage of this fact when allocating disk space for the traces if a single system is used to monitor both directions of a link.
- Link utilization is typically under 50% with peaks reaching just over 60%. The data we present represents two of the most heavily utilized monitored links. The drop in link utilization around midnight corresponds to a maintenance period.

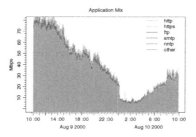

Fig. 6. *web-out* traffic breakdown by application

Another point to note is that the web link trace reflects the limited scalability of our monitoring system. The reason why the trace in Figure 5 is truncated around midnight is that there was insufficient disk space. In an effort to increase the number of monitored links, we configured one IPMON system to monitor both directions of the web link, and tried to optimize the disk allocation to take advantage of the traffic asymmetries. However, this was not successful. The disk space on a single IPMON system is not sufficient to capture a full 24 hour trace for both of these links. Later traces collected on the web link used one IPMON system for each direction.

Table 4. Trace statistics

trace	TCP packets	UDP packets	Other packets	min packet size	average packet size	max packet size	IP fragments	IP options	TCP options
web-out	530 million	33 million	4 million	20	339	1500	57,549	2068	67 million
web-in	806 million	34 million	13 million	20	540	1500	420,269	1548	66 million
peer-out	740 million	106 million	7 million	20	590	1500	292,450	828	79 million
peer-in	635 million	143 million	14 million	20	315	63,945	164,924	2915	67 million

Other traffic characteristics are summarized in Table 4. This table shows the number of TCP/UDP/other packets; the minimum, average, and maximum packet sizes; the number of packets which were IP fragments; and the number of packets with IP and TCP options. The number of packets with IP and TCP options affect the amount of information that is provided by the trace data. If a packet contains either type of option, then the size of the TCP/IP header can exceed the 44 bytes of data we collect. If this is the case, we may lose information about the TCP port numbers, sequence numbers, or flags. As can be seen in the table, the number of packets which contain IP options is less than .0004% of the total traffic volume. The number of packets with TCP options, on the other hand, can be up to 12% of the total traffic. However, the only part of the header which we do not capture on these packets is the TCP options. We are able to record the source/destination port, the TCP sequence numbers, and flags.

5.2   Packet Size Analysis

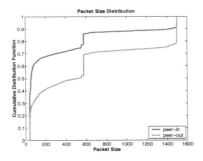

Fig. 7. packet size distribution on peering link

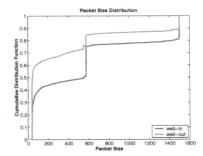

Fig. 8. packet size distribution on web link

The packet size characteristics of a link impacts two system design parameters:

– the duration of the trace that may be collected
– the rate at which the IPMON systems must process incoming packets

The IPMON systems record 64 bytes for each packet. Therefore, the duration of the trace is limited to a particular number of packets. If two links are running at the same link utilization, the system monitoring the link with the higher average packet size will be able to record a longer trace.

The packet size distribution we observe follows the same tri-modal distribution as observed in other studies [21][2][20]. The cumulative distribution function of the packet size distribution is shown in Figures 7 and 8. The packet size distribution has peaks at 40 bytes (minimum size TCP packets, 1500 bytes (maximum size Ethernet packets), and at 552 and 576 bytes (maximum size TCP packets from TCP implementations which do not perform MTU discovery). The minimum, average, and maximum packet sizes for each link are shown in table 4. The packet size distribution on the web links explains the reason for the discrepancy in trace durations. Each trace was configured to use an amount of disk space proportional to the link utilization (i.e. the *web-in* link was configured with approximately twice the capacity of the *web-out* link). The goal was to collect the same duration of trace from each link. However, as can be seen from the packet size distributions, the *web-out* link (which is the output link from the network to the web server customer) carries a large number of small packets containing web requests and acknowledgements, while the *web-in* link carries a large number of maximum size packets containing web data. Therefore, when determining the amount of disk space required to collect the traces, the traffic volume in terms of packets per second is a better measure than the traffic volume in terms of bits per second.

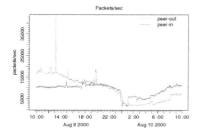 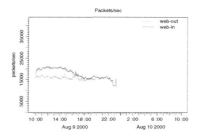

Fig. 9. Peer link traffic volume in pack-   **Fig. 10.** Web link traffic volume in
ets/sec   packets/sec

The traffic volume in terms of packets/sec are shown in Figures 9 and 10. These values are computed over one minute averages. The figures demonstrate that the data rate in packets per second on the *web-in* link is nearly the same as the data rate on the *web-out* link, rather than twice the data rate as predicted by the bits per second data rate.

The figures also indicate how the packet size characteristics affect the data rate requirements of the IPMON systems. Figure 9 shows several peaks in the traffic volume in packets/sec on the peering point links. These peaks, however, do not correspond to equivalent peaks in overall traffic volume in bytes (i.e. there is no equivalent peak observed in Figure 4). The peaks correspond to bursts of small packets that occur in the network. In this case, the peaks actually represent a large number of SYN packets which are all transmitted to one particular destination, a common denial-of-service attack.

Regardless of the source of the traffic, a sequential arrival of small packets imposes a performance burden on the IPMON system. To evaluate this burden we count the number of packets of similar sizes which arrive sequentially (e.g. the number of 40 byte packets that arrive back-to-back) and plot the distribution in Figure 11. We categorize packets into three classes: small, medium, and large. Small packets are less than 500 bytes, medium packets are between 500 and 1000 bytes, and large packets are longer than 1000 bytes. Sequential arrivals of medium and large packets are similar on both *peer-in* and *peer-out* links. For small packets, the number of sequential arrivals can be quite large, and in the case of *peer-out*, even reach 599.

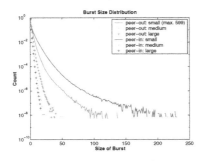

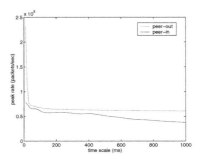

Fig. 11. Sequential packet arrivals on peer link

Fig. 12. Peak rate for peering link

The number of sequential arrivals, however, does not tell how close in time the packets arrive. To capture the temporal aspects of packet bursts, we examine the peak arrival rate in packets per second on the monitored links shown in Figure 12. This figure shows the peak traffic rate (in packets/sec) at time scales ranging from 10 ms to 1 sec. The data is generated by computing the average arrival rate over 10 ms intervals for the entire trace. Using this data we compute the peak arrival rate observed over any of the 10 ms intervals. We then evaluate the peak rate at a range of time intervals ranging from 10 ms to 1 sec. Even at the smallest time scale, 10 msec, the peak arrival rate is only 231,000 packets/sec, well within 1.94 million packets/sec supported by the IPMON systems.

The peak rate determines the data rate required to copy traffic data from the DAG cards to main memory in the system. The packet sizes also affect the data rate required to store the data to disk. The systems were originally designed to support traffic with an average packet size of 400 bytes, which does correspond to the average packet size across all the links. However, from our measurements, the average packet size on a single link can be closer to 300 bytes. The disks on the OC-3 and OC-12 monitors have enough bandwidth to support the smaller average packet size, but the OC-48 monitors can only support up to 1.9 Gb/sec of traffic if the average packet size is 300 bytes. Fortunately, the OC-48 links we plan to monitor are not run at full link utilization. While there may be bursts of traffic which increase link utilization to 100%, Figure 12 indicate that these types of bursts only occur at small time scales on OC-3 links, and similar behavior is expected on the OC-48 links. The OC-48 systems are configured with a 512 MB memory buffer which can buffer eight seconds of data, and should be able to accommodate bursts which may occur.

5.3 Delay Measurements

One of the unique aspects of the IPMON system is its capability to measure network delays for actual network traffic. Most current delay measurements are performed using a set of probe packets which are transmitted at periodic or random time intervals. While these systems are able to provide a general idea about the delay performance of the probe packets, it is difficult to determine if the performance of the probe packets represent an accurate sampling of the delays experienced by actual network traffic.

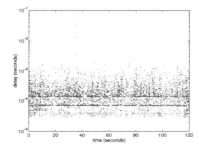

Fig. 13. Delay from *web-in* to *peer-out*

The IPMON systems allow us to measure the delays seen by every packet that is transmitted between two points in the network. This is accomplished by identifying the same packet in two different traces and computing the difference in the timestamps. Figure 13 shows a two minute sample data set. The x-axis shows the packet arrival time, and the y-axis shows the delay experienced by the packet for traffic between the *web-in* and *peer-out* links. Since both links are located on the same core router, this data represents the single hop delay experienced by packets in the backbone. We currently have systems installed only in one location in the network, but additional systems are in the process of being installed. With data from these systems, we will be able to measure delays across many hops in the backbone.

There are several points to note from this figure. First, the minimum delay across the entire interval remains almost constant, around 30 μs. This is the smallest delay interval that it is possible to measure in the network, so the 5 μs error that may be introduced by the clock synchronization mechanism is acceptable. Second, there is a rather large increase in delay at time 30 seconds. The delay experienced increases to nearly 30 ms, which is unusually large for a single hop delay. This type of delay is also observed at a small number of points elsewhere in the trace. These excessive delays are the types of data that are difficult to observe using probe traffic. The long delays are experienced only by a small number of packets, but the impact on these packets is extremely large. The source of these long delays is currently under investigation, but it is believed to be due to a pathologic behavior of the router we are observing rather than actual queuing delays.

6   Conclusion

We describe the Sprint IP Monitoring system, a passive monitoring system that collects packet-level traces from the Sprint Internet IP backbone. The systems are capable of supporting OC-3, OC-12, OC-48 data rates, and are synchronized to within 5 μs using a stratum-1 GPS reference clock. We present the system design and demonstrate the performance of the system with several sample measurements.

The advantage of our system is it provides the capability to collect traces from multiple locations in the network and correlate the traces through highly accurate timestamps. This provides the capability to study both single link characteristics (e.g. workload and packet size distributions), as well as characteristics which require data from multiple links (e.g. delay, TCP behavior, network provisioning). It is also very flexible in that the data is not targeted towards a single use. The packet traces are useful in many diverse research projects.

The disadvantage of the system is that the amount of data collected is very large. Data from a single 24 hour period exceeds 3.3 TB. This requires both a large amount of resources to be installed in network facilities for data collection purposes and a large amount of resources to perform data analysis. While our system supports monitoring 31 different links and can be extended to monitor several dozen additional links, scaling the system to exhaustively monitor the entire network is impractical.

In future work we plan to analyze in depth the traffic observed on various links on the network. These results will be used to:

- Design provisioning and dimensioning tools to better anticipate customers needs and increase customer satisfaction, eventually making it possible to provide various classes of service.
- Gain a better undestanding of the traffic characteristics on an IP backbone, and design more accurate traffic models.
- Work with router designers to design embedded measurement facilities.

Acknowledgements

In the course of this project we received help from so many people that it is impossible to acknowledge them individually. Instead we choose to acknowledge their institutions: SprintLink, Sprint ISC, Sprint ATL, the University of Waikato, and CAIDA.

References

1. Joel Apsidorf, K.C. Claffy, Kevin Thompson, and Rick Wilder, "OC3MON: Flexible, affordable, high performance statistics collection," in *Proceedings of INET*, June 1997.

2. KC Claffy, Greg Miller, and Kevin Thompson, "The nature of the beast: Recent traffic measurements from an internet backbone," in *Proc. INET'98*, Geneva, Switzerland, July 1998.

3. A. K. Adams and M. Mathis, "A system for flexible network performance measurement," in *Proceedings of INET 2000*, June 2000.

4. "Mtrace ftp site," ftp://ftp.parc.xerox.com/pub/net-research/ipmulti.

5. "Treno web page," http://www.psc.edu/networking/treno_info.html.

6. R. Cáceres, N. Duffield, D. Towsley, and J. Horowitz, "Multicast-based inference of network-internal loss characteristics," *IEEE Transactions on Information Theory*, vol. 45, no. 7, pp. 2462–2480, November 1999.

7. S. Kalidindi and M. J. Zekauskas, "Surveyor: An infrastructure for internet performance measurements," in *Proceedings of INET '99*, June 1999.

8. T. McGregor, H.-W. Braun, and J. Brown, "The NLANR network analysis infrastructure," *IEEE Communications*, vol. 38, no. 5, May 2000.

9. W. Matthews and L. Cottrel, "The PingER project: Active internet performance monitoring for the HENP community," *IEEE Communications*, vol. 38, no. 5, pp. 130–136, May 2000.

10. "MIQ ratings methodology," http://ratings.miq.net/method.html.

11. "The interaction of web content and internet backbone performance," http://www-.keynote.com/services/html/wp_compdata.html, Keynote white paper.

12. William Stallings, *SNMP, SNMPv2, SNMPv3, and RMON 1 and 2*, Addison Wesley, 3rd edition, 1999.

13. "Tcpdump web page," http://ee.lbl.gov.

14. "NetFlow services and applications," http://www.cisco.com/warp/public/cc/pd/-iosw/ioft/neflct/tech/napps_wp.htm, 2000, Cisco white paper.

15. "CoralReef website," http://www.caida.org/tools/measurement/coralreef.

16. C. Labovitz, A. Ahuja, and F. Jahanian, "Experimental study of internet instability and wide-area backbone failures," in *Proceedings of the 29th International Symposium on Fault-Tolerant Computing (FTSC-29)*, Madison, Wisconsin, June 1999.

17. V. Paxson, G. Almes, J. Mahdavi, and M. Mathis, "Framework for IP performance metrics," RFC 2330, IETF, May 1998.

18. Anja Feldmann, Albert Greenberg, Carsten Lund, Nick Reingold, and Jennifer Rexford, "Netscope: Traffic enginnering for ip networks," *IEEE Network*, March/April 2000.

19. Vern Paxson, "Empirically-derived analytic models of wide-area tcp connections," *IEEE/ACM Trans. on Networking*, vol. 2, no. 4, August 1994.

20. Sean McCreary and K.C. Claffy, "Trends in wide area ip traffic patterns," in *ITC Specialist Seminar*, Monterey, California, May 2000.

21. Kevin Thompson, Greg Miller, and Rick Wilder, "Wide area internet traffic patterns and characteristics," *IEEE Network*, Nov 1997.

22. W. Simpson, "PPP in HDLC-like framing," rfc 1662, IETF, July 1994.

23. J. Cleary, S. Donnelly, I. Graham, A. McGregor, and M. Pearson, "Design principles for accurate passive measurement," in *PAM 2000*, Hamilton, New Zealand, April 2000.

24. "Dag 4 SONET network interface," http://dag.cs.waikato.ac.nz/dag/dag4-arch-.html.

25. "Dag synchronization and timestamping," http://dag.cs.waikato.ac.nz/dag/docs-/dagduck_v2.1.pdf.

Network Delay Tomography
from End-to-End Unicast Measurements*

N.G. Duffield[1], J. Horowitz[2], F. Lo Presti[1,3], and D. Towsley[3]

[1] AT&T Labs–Research, 180 Park Avenue, Florham Park, NJ 07932, USA
{duffield,lopresti}@research.att.com
[2] Dept. Math. & Statistics, University of Massachusetts, Amherst, MA 01003, USA
joeh@math.umass.edu
[3] Dept. of Computer Science University of Massachusetts, Amherst, MA 01003, USA
towsley@cs.umass.edu

Abstract. In this paper, we explore the use of end-to-end unicast traffic measurements to estimate the delay characteristics of internal network links. Experiments consist of back-to-back packets sent from a sender to pairs of receivers. Building on recent work [11, 5, 4], we develop efficient techniques for estimating the link delay distribution. Moreover, we also provide a method to directly estimate the link delay variance, which can be extended to the estimation of higher order cumulants. Accuracy of the proposed techniques depends on strong correlation between the delay seen by the two packets along the shared path. We verify the degree of correlation in packet pairs through network measurements. We also use simulation to explore the performance of the estimator in practice and observe good accuracy of the inference techniques.

1 Introduction

Background and Motivation. As the Internet grows in size and complexity, it becomes increasingly important for users and providers to characterize and measure its performance and to detect and isolate problems. Yet, because of the sheer size of the network and the limit imposed by administrative diversity, it is not generally possible to directly access and measure but a small portion of the network. Consequently, there is a growing need for practical and efficient procedures that can take an internal snapshot of a significant portion of the network.

A promising approach to network measurements, the so called Network Tomography approach, addresses these problems by exploiting the end-to-end traffic behavior to reconstruct the network internal performance. The idea is that correlation in performance seen on intersecting end-to-end paths can be used to draw inferences about the performance characteristics of their common portion, without cooperation from the network. Multicast traffic is in particular well suited for this since a given packet only occurs once per link in the multicast

---

* This work was supported in part by DARPA and the AFL under agreement F30602-98-2-0238

S. Palazzo (Ed.): IWDC 2001, LNCS 2170, pp. 576–595, 2001.

distribution tree. Thus multicast traffic introduces a well structured correlation in the end-to-end behavior observed by the receiver that share the same multicast session. This correlation allows to infer the performance characteristics as packet loss rates, [1], packet delay distributions, [11], and packet delay variance, [6].

Fig. 1. 2-Leaf Tree.

To illustrate the idea behind multicast based delay inference, consider the simple tree in Figure 1 with the source (the root node) sending multicast packets to the two leaf nodes L and R and assume we collect the end-to-end measurements at the two receivers. If we consider the events where the delay seen by L is zero (assume for simplicity that the transmission and propagation delay are zero), the corresponding additional delays seen at R can be attributed to the link from C to R alone. We can thus form an estimate of the delay distribution for the link from C to R. The delay distribution of the other links can be derived by similar arguments.

Despite the encouraging results, multicast measurements suffer from two serious limitations. First, large portions of the Internet do not support network-level multicast. Second, the internal performance observed by multicast packets often differs significantly from that observed by unicast packets. This is especially serious given that unicast traffic constitutes the largest portion of the traffic on the Internet.

To overcome the limitation of multicast measurements, methods to extend the inference techniques to unicast measurements have been recently proposed in [3, 7] for the inference of loss rates and [4, 5] for delay distributions. The key idea is to design unicast measurement whose correlation properties closely resemble those of multicast traffic, so that it is possible to use the inference techniques developed for multicast inference; the closer the correlation properties are to that of multicast traffic, the more accurate the results. The basic approach, which has been further refined in [7] for the estimation of the loss rates, is to dispatch two back-to-back packets (a packet pair) from a probe source to a pair of distinct receivers. The premise is that, when the duration of network congestion events exceeds the temporal width of the packets, packets experience very similar behavior when they traverse common portions of their paths. Difference in the packets behavior occurs because congestion events may not affect packets uniformly: packet loss could not be uniform if lossy periods last less than the time between the arrival of the two packets; delays will differ because of the interleaving of background traffic. Still, if the packets experience very similar behavior, the error in using the multicast based estimator is very small.

As an example, consider again the tree in Figure 1 with the source now sending two packets, back-to-back, the first to L and the second to R. In correspondence of the events where the delay seen by L is zero we will still attribute the additional delays seen at R to the link from C to R. But, because the two packets will possibly experience slightly different delays along the link from 0 to C, our estimate of the delay distribution for the link from C to R will contain an error roughly equal to the difference in delay seen by the two packets along common link. The smaller this difference, the more accurate the estimates.

We observe that a more accurate approach would consist in taking into account the difference experienced by the two packets along the shared link and incorporating it in our model. Unfortunately, we found out that it is not possible to estimate its value, at least not without additional assumptions. Therefore, here we rely on small deviations from the ideal behavior and proceed as the two packets experience the same delay along the shared path.

Contributions. In this paper we describe efficient techniques for the estimation of link delay characteristics, namely, the per link delay distribution and per link delay cumulants, via end-to-end packet pairs measurements.

For the distribution analysis, our starting point is the work by Lo Presti, et al. [11] and subsequent work by Coates and Novak in [5, 4]. Following [11], we model link delay by non-parametric discrete distributions. The discrete distribution can be a regarded as binned or discretized version of the (possibly continuous) true delay distribution, where we explicitly trade-off the detail of the distribution with the cost of calculation. A potential limitation of this approach lies in the accuracy/complexity trade-off itself. Since the complexity of the analysis is function of the numbers of bins, it results that under the usual discrete model, whereby delay is discretized using a fixed bin size q, a small q to ensure a desired level of accuracy in the estimates results in too many parameters (bins) and excessive computational costs.

To overcome these limitations, here, we describe a novel approach to delay modeling. The idea is to discretize delay using variable sized bins. Smaller bins are used only in correspondence of concentrations of probability mass to ensure adequate resolution while larger bins are used otherwise. Intuitively, this allows us to reduce the number parameters (bins), and hence complexity, significantly, without losing accuracy. A complication with this approach is that a discrete model with variable bin size does not lend itself to analysis. To this end, we propose an approach to variable bin size modeling which, while restricting the possible choices of bin size to a specific format, lends itself to analysis. In particular, we can formulate the estimation problem for the proposed variable bin size model, by generalizing the Maximum Likelihood formulation of [5]. Estimation is carried out by adapting the Expectation-Maximization (EM) algorithm used in [5] to compute the MLE estimates.

Then, we also describe an efficient method to directly infer the per link delay variance. By a simple argument, we show that it is possible to express the link delay variance in terms of the covariance of the end-to-end delays. Therefore, we can estimate the variance directly from the sample covariance of the end-to-end delays. The same method can be extended for the estimation of higher order cumulants. Distribution and cumulants are closely related: knowledge of (all) the cumulants of a random variable is equivalent to know its distribution.

The rest of the paper is organized as follows. In Section 2 we specify the tree and delay model. In Section 3 we describe the estimators of the delay distribution. In Section 4 we describe the link delay variance estimator (for lack of space, we omit the extension to higher order cumulants). In Section 5 we use the National Internet Measurement Infrastructure (NIMI) [13] to gather end-to-end data from a diverse set of Internet paths, and verify the conditions for the accuracy of our methods. In Section 6 we use network level simulation to evaluate the accuracy of the estimators. We conclude in Section 7.

Related Work. There exist several tools and methodologies for characterizing link-level behavior from end-to-end unicast measurements. One of the first methodologies focuses on identifying the bottleneck bandwidth on a unicast route. The key idea is that, in an uncongested network, two packets sent back-to-back will arrive at the receiver with a spacing that is inversely proportional to the lowest link bandwidth on the path. This was noted by Jacobson [9], and analyzed by Keshav [10].

Use of end-to-end measurements of packet pairs in a tree connecting a single sender to several receivers for estimation of the link delay has been first considered in [5]. The inference of the link delay distribution is formulated as a maximum likelihood estimation problem which is solved using the Expectation Maximization (EM) algorithm. In [5, 4] the authors extend this approach to the nonstationary case and in [14] investigate unicast based inference in context of passive monitoring, whereby inference is based on observation of ongoing unicast sessions. Preliminary results on these methods reported in these papers show promise.

Our approach extend the results in [5] in that we consider a more general form of discrete model which allows us to significantly improve the accuracy/complexity trade-off. We remark that the variable bin size scheme presented in this paper can be used in other setting, *e.g.*, multicast based inference techniques.

2   The Tree and Delay Models

Tree Model. We represent the underlying physical network as a graph $G_{phys} = (V_{phys}, L_{phys})$ comprising the physical nodes V_{phys} (e.g. routers and switches) and the links L_{phys} between them. We consider a single source of probes $0 \in V_{phys}$ and a set of receivers $R \subset V_{phys}$. We assume that the set of paths from 0 to each $r \in R$ is stationary and forms a tree T_{phys} in (V_{phys}, L_{phys}); thus two such paths never intersect again once they have diverged. We form the logical source tree $\mathcal{T} = (V, L)$ whose vertices V comprise 0, R and the branch points of $\mathcal{T}_{\text{phys}}$. The link set L contains the link (j, k) if one or more of the probe paths in $\mathcal{T}_{\text{phys}}$ pass through j and then k without encountering another element of V in between. We will sometimes refer to link $(j, k) \in L$ simply as link k. For $k \neq 0$, $f(k)$ denotes the parent of k. We write $j \succ k$ if j is an ancestor of k in $\mathcal{T}$. $i \vee j$ denotes the minimal common ancestor of i and j in the $\preceq$-ordering.

Packet Pair and Delay Model. Let $\langle i, j \rangle$ denote a packet pair dispatched to destination nodes i, j in that order. The paths traverse a common set of links down to node $i \vee j$. Let $p(i, j)$ denote the set of nodes traversed by at least one member of the packet pair. For $k \in p(i, j)$ let $G(k) \subseteq \{1, 2\}$, where 1 and 2 denote the two packets sent in order to i and j, denote the set of packets that transit k. We describe the progress of the packet pair in $\mathcal{T}$ by the variable $X_k(l)$, $l \in G(k)$, which represents the accrued queueing delay of packet d along the route to k. We assume that we only observe the end-to-end delay $X_{ij} = (X_i(1), X_j(2))$ at receivers i and j.

We specify a delay model for the packet pair. We associate with each node k a pair of random variables D_k and D'_k that take values in the extended positive real line $R_+ \cup \{\infty\}$. By convention $D_0 = D'_0 = 0$. D_k (D'_k) is the delay that would be encountered by the first (second) packet attempting to traverse the link $(f(k), k) \in L$. A delay equal to ∞ indicates that the packet is lost on the link. We assume that delays are independent between different pairs, and for packets of the same pair on different links. The delay experienced by packet 1 on the path from root 0 to node k is $X_k(1) = \sum_{l \succeq k} D_l$. The delay experienced by packet 2 is $X_k(2) = \sum_{l \succeq (i \vee j) \vee k} D'_l + \sum_{(i \vee j) \vee k \succ l \succeq k} D_l$. Note that $X_k(\cdot) = \infty$ iff any delay along the path to k is infinite, i.e. if the packet is lost on some link between nodes 0 and k.

For any $k \in V$, $E_k = D'_k - D_k$ is the difference between the delays experienced by the back-to-back packets of a packet pair traversing k. Ideally, $E_k = 0$, and the packet pair behaves like a notional multicast packet sent to the two receivers. In practice, we expect the two delays to be different. This is because congestion events at intervening nodes may not affect packets uniformly if they are not back-to-back. This occurs, for example, because of the packets being spaced apart as a result of traversing a bottleneck (low available bandwidth) link, and the interleaving of background traffic in between. Observe that $E_k \neq 0$ even in the case of perfectly back-to-back packets, *e.g.*, packet 2 suffers on additional delay due to the time required to transmit packet 1.

Measurement A measurement experiment consists of sending, for each pair of distinct receivers $i, j \in R$, n packet pairs $\langle i, j \rangle$. As a result of the experiment we collect a set of measurements $X^{i,j} = (X^{i,j(m)})_{m=1,\ldots,n}$, where $X^{i,j(m)} = (X_i(1)^{(m)}, X_j(2)^{(m)})$ and $(X_i(1)^{(m)}, X_j(2)^{(m)})$ is the end-to-end delay of the m-th packet pair $\langle i, j \rangle$. Let $X = (X^{i,j})_{i \neq j \in R}$ denote the complete set of measurements.

3   Non-parametric Estimation of Delay Distribution

In this section we describe techniques for the estimation of the probability distribution of the per link variable delay D_k. We quantize the delay to a finite set of values Q. We assume that once quantized, $D_k = D'_k$. In other words, we assume E_k small enough that it can be ignored in the discrete model. We consider two cases. First, in Section 3.1 we consider the most usual form of discretization, where we discretize delay to a set $Q = \{0, q, 2q, \ldots, Bq, \infty\}$, where q is a suitable fixed bin size. Then, in Section 3.2 we consider a different approach whereby delay is discretized to a more general set Q.

For the analysis, we thus model the link delay by a nonparametric discrete distribution that we can regard as a discretized version of the (possibly continuous) actual delay distribution. We denote the distribution of D_k by $\alpha_k = (\alpha_k(d))_{d \in Q}$, where $\alpha_k(d) = P[D_k = d]$, $d \in Q$. We will denote by $\alpha = (\alpha_k)_{k \in V}$ the set of links distributions.

3.1   Delay Analysis with a Fixed Bin Size Discrete Model

Here, we consider the usual discrete model wherein D_k takes a value in $Q = \{0, q, 2q, \ldots, Bq, \infty\}$, where q is a suitable fixed bin size. The point ∞ is interpreted as "packet lost" or "encountered delay greater than Bq". We define the bin associated to $iq \in Q$ to be the interval $[iq - \frac{q}{2}, iq + \frac{q}{2})$, $i = 1, \ldots, B$, and $[Bq - \frac{q}{2}, \infty)$ the one associated to the value ∞. Because delay is non negative, we associate with 0 the bin $[0, \frac{q}{2})$. We denote this model as the (q, B) model.

Our goal is to estimate α using maximum likelihood based on the overall observed data X (also discretized to the set Q). Denote by $\Omega = Q \times Q$ the set of possible outcomes for the packet pairs delays For each outcome $x_{i,j} \in \Omega$ denote $n(x_{i,j})$ the number of pairs $\langle i, j \rangle$, $m = 1, \ldots, n$, for which $X^{i,j(m)} = x_{i,j}$. Let $p_\alpha(x_{i,j}) = P_\alpha[X_{i,j} = x_{i,j}]$ denote the probability of the outcome $x_{i,j}$. $p_\alpha(x^{i,j})$ can be expressed in terms of convolutions of the distribution α_k, $k \in p(i, j)$.

The log-likelihood of the measurement X is

$$\mathcal{L}(X; \alpha) = \log P_\alpha[X] = \sum_{i \neq j \in R} \sum_{x_{i,j} \in \Omega} n(x_{i,j}) \log p_\alpha(x_{i,j}) \qquad (1)$$

We estimate α by the maximizer of the likelihood (1), namely, $\widehat{\alpha} = \arg\max_\alpha \mathcal{L}(\alpha)$. Unfortunately, given the form of (1) we have been unable to obtain a direct expression for $\widehat{\alpha}$. Instead, we follow the approach in [4, 5], and employ the Expectation Maximization (EM) algorithm to obtain an iterative approximation $\widehat{\alpha}^{(\ell)}$, $\ell = 0, 1, \ldots$, to (local) maximizer of the likelihood (1). The basic idea behind the EM algorithm is that, rather then performing a complicated maximization, we "augment" the observed data with unobserved or latent data so that the resulting likelihood has a simpler form. Following [5], we augment the observations X with the unobserved actual delay experienced by the packet pairs along each link, namely, $D = (D_k^{i,j})_{k \in p(i,j), i \neq j \in R}$, where $D_k^{i,j} = (D_k^{i,j(m)})_{m=1,\ldots,n}$ are the delays experienced by the n packet pairs $\langle i, j \rangle$ along link k. The pair (X, D)

represents the *complete data* for our inference problem. The log-likelihood of the *complete data* (X, D) is

$$\mathcal{L}(X, D; \alpha) = \log \mathsf{P}_\alpha[X, D] = \log \mathsf{P}_\alpha[X|D] + \log \mathsf{P}_\alpha[D]. \tag{2}$$

The first term is 0 (since D uniquely determines X, we have that $\mathsf{P}_\alpha[X|D] = 1$). Expansion of the second term yields

$$\log \mathsf{P}_\alpha[D] = \sum_{i \neq j \in R} \sum_{k \in p(i,j)} \log \mathsf{P}_\alpha[D_k^{i,j}] = \sum_{k \in V} \sum_{d \in Q} n_k(d) \log \alpha_k(d) \tag{3}$$

where $n_k(d)$ is the total number of packets pairs that experienced a delay equal to d along link k. Should D be observable, the counts $n_k(d)$ would be known, and maximization of (3) would directly yield the MLE estimate of $\alpha_k(d)$,

$$\widehat{\alpha}_k(d) = \frac{n_k(d)}{\sum_{d \in Q} n_k(d)} \tag{4}$$

Since D and $n_k(d)$ are not known, the EM algorithm uses the complete data log-likelihood $\mathcal{L}(X, D; \alpha)$ to iteratively find $\widehat{\alpha}$ as follows:

1. *Initialization.* Select the initial link delay distribution $\widehat{\alpha}^{(0)}$. As shown in Appendix, we select $\widehat{\alpha}^{(0)}$ as an estimate of α we compute by adapting the approach in [11].
2. *Expectation.* Given the current estimate $\widehat{\alpha}^{(\ell)}$, compute the conditional expectation of the log-likelihood given the observed data X under the probability law induced by $\widehat{\alpha}^{(\ell)}$, $Q(\alpha'; \widehat{\alpha}^{(\ell)}) = E_{\widehat{\alpha}^{(\ell)}}[\mathcal{L}(X, D; \alpha')|X] = \sum_{k \in V} \sum_{d \in Q} \widehat{n}_k(d) \log \alpha'_k(d)$ where $\widehat{n}_k(d) = E_{\widehat{\alpha}^{(\ell)}}[n_k(d)|X]$. $Q(\alpha'; \widehat{\alpha}^{(\ell)})$ has the same expression as $\mathcal{L}(X, D; \alpha')$ but with the actual *unobserved* counts $n_k(d)$ replaced by their conditional expectations $\widehat{n}_k(d)$. To compute $\widehat{n}_k(d)$, observe that we can write the counts $n_k(d)$ as $n_k(d) = \sum_{i \neq j \in R: k \in p(i,j)} \sum_{m=1}^n \mathbf{1}_{\{D_k^{i,j(m)}=d\}}$. Then

$$\widehat{n}_k(d) = \sum_{i \neq j \in R: k \in p(i,j)} \sum_{m=1}^n \mathsf{P}_{\widehat{\alpha}^{(\ell)}}[D_k^{i,j(m)} = d|X^{i,j(m)}] \tag{5}$$

$$= \sum_{i \neq j \in R: k \in p(i,j)} \sum_{x_{ij} \in \Omega} n(x_{ij}) \mathsf{P}_{\widehat{\alpha}^{(\ell)}}[D_k = d|X_{ij} = x_{ij}] \tag{6}$$

3. *Maximization.* Find the maximizer of the conditional expectation $\alpha^{(\ell+1)} = \arg\max_{\alpha'} Q(\alpha', \widehat{\alpha}^{(\ell)})$. The maximizer is given by (4) with the conditional expectation $\widehat{n}_k(d)$ in place of $n_k(d)$.
4. *Iteration.* Iterate steps 2 and 3 until some termination criterion is satisfied. Set $\widehat{\alpha} = \widehat{\alpha}^{(\ell)}$, where ℓ is the terminal number of iterations.

Convergence. Because the complete data likelihood can be shown to derive from a standard exponential family, the EM iterates $\widehat{\alpha}^{(\ell)}$ converge to a stationary point of the likelihood α^*, *i.e.*, $\frac{\partial \mathcal{L}(X, D; \alpha)}{\partial \alpha}(\alpha^*) = 0$, (see *e.g.* [15]) . This implies that when there are multiple stationary points, *e.g.* local maxima, the EM iterates may not converge to the global maximizer. Unfortunately, we were not able to establish whether there is a unique stationary point or conditions under which unicity holds. Therefore, in general the estimates $\widehat{\alpha}^{(\ell)}$ converge to a local (but not necessary global) maximizer. Since the point of convergence depends on the initial estimate, we must carefully choose the initial estimate $\widehat{\alpha}^{(0)}$. Here we select as initial distribution the estimate of α obtained by using the approach in [11] (see the Appendix). We expect that, for large enough n, $\widehat{\alpha}^{(0)}$ (which converges to α), is close enough to the actual likelihood maximizer so to ensure, in most cases, the desired convergence.

Complexity. The complexity of the algorithm is dominated by the computation of the conditional expectation $\widehat{n}_k(d)$ which can be accomplished in time that is $O(npB^2)$, where p is the average number of links between the source and a leaf node, using the upward-downward probability propagation algorithm [8].

Choice of bin size. Since packet delay is essentially continuous in nature, the use of a discrete model introduces a quantization error, which is a function of the bin size q. The choice of q is thus primarily dictated by the trade-off between accuracy and computational complexity: a smaller q provides better accuracy but at rapidly increasing computational cost (observe that since the product qB is constant the complexity is basically $O(np/q^2)$); on the other hand, use of larger bin size, reduces the computational complexity but may not be adequate to accurately capture very small delays.

We must also consider that, in the context of unicast measurements, the delay resolution must be large enough so that, once discretized to $\mathcal{Q}$, we can use the approximation $D_k \approx D'_k$. Our network experiments in Section 5 suggest that q should not be smaller than $1msec$ to satisfy this condition.

3.2   Delay Analysis with Variable Bin Size Discrete Model

Here we consider a more general form of discrete model in which D_k takes values in a more general finite set $\mathcal{Q}$. This is motivated by the observation that the use of a fixed bin size may be too restrictive in the analysis of large networks where delay characteristics significantly vary from node to node: a value of q chosen to adequately capture the delay behavior of very fast links would result in too many parameters if slower or congested links are also present. Ideally, to overcome the limitations of the accuracy/complexity trade-off of the fixed bin size models, it is preferable to discretize delay to a suitable set $\mathcal{Q}$, which guarantees the desired resolution in the delay range of interest while keeping the overall number of bins sufficiently small. For example, smaller bins could be used only in correspondence of concentrations of probability mass to ensure adequate resolution while larger bins could be used otherwise. Intuitively, this would allow us to reduce the number parameters (bins), and hence complexity, significantly, without losing accuracy.

A complication with this approach is that a discrete model with a general set of values $\mathcal{Q}$ does not lend itself to analysis. The problem, is that a general discrete set $\mathcal{Q}$ is not closed under the sum operation. Therefore, we cannot express the observable delay (discretized to $\mathcal{Q}$) in terms of sum of link delays (also discretized to $\mathcal{Q}$).

To overcome these difficulties, we now describe a simple approach to variable bin size modeling which, while restricting the possible choices of $\mathcal{Q}$ to a specific format, lends itself to analysis. The key idea is to consider variable bin size models, the analysis of which can be reduced to that of a set of fixed bin size models. We proceed as follows: (1) we define a variable bin size model as the *composition* (in the sense described below) of fixed bin size models; and (2), we choose the constituent fixed bin size models so that the estimates of the distribution for these models can be *composed* to form the estimate of the distribution of the variable bin size model itself. By appropriate choice of the fixed bin size models, the resulting variable bin size model has a better accuracy/complexity trade-off. We detail the approach below.

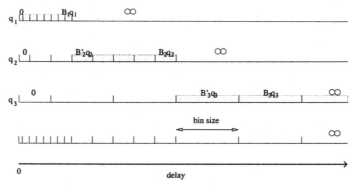

Fig. 2. Variable bin size model as composition of fixed bin size models.

We define the variable bin size discrete model as the *composition* of M uniform bin size discrete models $((q_l, B_l))_{l=1,...,M}$, with increasing bin size, $0 < q_1 < ... < q_M$ and such that $B_1 q_1 < ... < B_M q_M$ (see Figure 2). We assume that for $l = 2, ..., M$, each bin of level l either corresponds to an integer number of level $l-1$ bins (*i.e.*, the boundaries of the bin of level l correspond to boundaries of a group of adjacent bins of level $l-1$) or is contained in the ∞ bin of level $l-1$. We let $g_l(j)$ denote the set of level $l-1$ bins which corresponds to the j-th level l bin, $j = 0, ..., B'_l < B_l$ where B'_l is the first level l bin contained in level $l-1$ last bin (the one corresponding to ∞).

In the variable bin size model, D_k takes values in $\mathcal{Q} = \{0, q_1, ..., B_1 q_1, B'_2 q_2, ..., B'_M q_M, \infty\}$. We define the bin associated to $iq_l \in \mathcal{Q}$ as the interval $[iq_l - \frac{q_l}{2}, iq_l + \frac{q_l}{2})$, and $[B_M q_M - \frac{q_M}{2}, \infty)$ the one associated to ∞. With this definition, we create a correspondence between the bins of the variable bin size model and bins of the fixed bin size models (the shaded bins in Figure 2). This allows us to express the distribution α in the variable bin size model in terms of the delay distribution in the M uniform bin size models. For $k \in V$, denote $\alpha_k(d; q_l) = P[D_k = d]$, $d \in Q_l = \{0, \ldots, B_l q_l, \infty\}$ the distribution for the model with fixed bin size q_l. The distribution of D_k in the variable bin size model is then $\alpha_k = (\alpha_k(d))_{d \in Q}$, where $\alpha_k(d) = \alpha_k(i q_l; q_l)$, $d = i q_l \in Q$, and $\alpha_k(\infty) = \alpha_k(\infty; q_M)$, $k \in V$. We will take advantage of this correspondence for the estimation.

With the above definition, we are limited to variable bin size models where the bin size progressively increases. However, we do not believe this choice to be restrictive. Indeed, we expect that in most cases it is desirable to have smaller bins in correspondence with small delay values and larger bins otherwise; while the small bins guarantee enough resolution for very fast or uncongested links, the larger bins prevent the explosion of the number of parameters due to the large delays experienced by the slower and congested links.

Example. We consider the *ternary* variable bin size model defined, for a given base bin size q and number of levels M, as $((3^{(l-1)}q, 2))_{l=1,\ldots,M}$. Delay is thus discretized to the set $\{0, q, 3q, 9q, \ldots, 3^{M-1}q, \infty\}$ (see Figure 3). This can be considered as an extreme case where each level has only three bins, 0, $3^{(l-1)}q$ and ∞, and the bin size grows exponentially with the level. Observe that this model covers the delay range from 0 up to a maximum value d_{max} with only $O(\log_3 \frac{d_{max}}{q})$ bins.

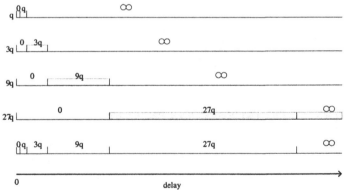

Fig. 3. Example: The Ternary Variable Bin Size Model ($M = 4$).

We estimate the distribution α of the variable bin size model indirectly by taking advantage of the relationship between the bins of the variable bin size model and those of the component fixed bin size models. Basically, we estimate the probabilities of the former by the corresponding estimates of the latter. More precisely, estimation of α proceeds by computing recursively the MLE estimate of M discrete models (q_l, B_l), starting with $l = 1$ as follows:

1. Discretize the delays to the set Q_l.

2. Estimate the probabilities $\alpha_k(d; q_l)$, $d \in Q_l$, $k \in V$. For $l = 1$, we use the EM algorithm directly. For $l > 1$, to have consistency between the estimates of the different models, we compute the estimates of the probabilities of level l bins corresponding to a group of level $l - 1$ bins, directly as the sum of the probabilities of those bins. In other words, we let $\widehat{\alpha}_k(d; q_l) = \sum_{j \in g_l(d/q_{l-1})} \widehat{\alpha}_k(jq_{l-1}; q_{l-1})$ for $d \leq q_l B'_l$. We then use the EM algorithm to estimate the remaining probabilities $\alpha_k(d; q_l)$ for $d \geq q_l B'_l$ assuming the probabilities $\alpha_k(d; q_l)$, $d \leq q_l B'_l$, as known parameters (set equal to the estimates above). This is equivalent to the EM algorithm shown in Section 3.1, where we replace (4) with

$$\widehat{\alpha}_k^{(\ell)}(d; q_l) = \left(1 - \sum_{d' \leq B'_l q_l} \widehat{\alpha}_k(d'; q_l)\right) \frac{\widehat{n}_k(d)}{\sum_{d' > B'_l q_l} \widehat{n}_k(d')} \qquad (7)$$

3. Iterate 1 and 2 for $l = 1, \ldots, M$.
4. Compose the estimates of the M models to estimate α, i.e., set $\widehat{\alpha}_k(d) = \widehat{\alpha}_k(iq_l; q_l)$, $d = iq_l \in Q$, $k \in V$.

Complexity. The computational cost equals the sum of the costs of computing the MLE estimates of each model. Assuming for simplicity that the number of iterations required by the EM algorithm does not vary, the complexity is then $O(np \sum_{l=1}^{M} B_i^2)$.

Choice of the Variable Bin Size Model. The use of the variable bin size model provides great flexibility in terms of both accuracy and computational cost. We consider two examples below. To ensure high accuracy a simple solution lies in using a variable bin size model with only two levels, i.e., $M = 2$: the first level has a small bin size, chosen according the desired level of accuracy and enough bins to include most of the probability mass, e.g., B_1 large enough that $P[D_k \leq B_1 q_1] > 0.999$; the second level has a larger bin size and covers the rest of the delay interval. We expect that capturing the tail of the distribution with a larger bin size can provide a significant reduction in the computational cost without accuracy degradation. At the other extreme, we might consider the solution which has the smallest complexity. Since the complexity is proportional to $\sum_{l=1}^{M} B_i^2$, we simply have to minimize the number of bins per level and use as many levels as necessary. We thus obtain the ternary variable bin size model. In between these extreme cases, it is possible to consider several models which provide the desired accuracy complexity trade-off. In general we expect the model to be determined either *a priori* or based on the measurements themselves.

3.3 Comparison of the Variable and Fixed Bin Size Model

We illustrate the potential benefit of the variable bin size model using model-based simulations in which link delays are independent, exponentially distributed random variables. We assume no packet loss. We conducted 1000 independent experiments over the 2-leaf tree in Figure 1. In each experiment, we sent 1000 packet-pairs down the tree. We assumed that the back-to-back packets have the same delay along the common link. Average link delays were chosen independently with a uniform distribution in the interval $[0.1, 10]msec$.

For the analysis, we consider three different discrete models: the first two models are the two bin size models (1msec, 100) and (10msec, 10); the third model is the ternary model $(3^{l-1}1\text{msec}, 2)_{l=1,...,5}$. The number of bins in each model was chosen so that the largest finite delay in each model was about $100msec$. We used the EM algorithm for the estimation. Initialization was performed as described in the Appendix. The termination criterion for the EM algorithm was that successive iterates of any probability should have an absolute distance less of 10^{-3}.

Complexity. The computational costs differ substantially. To compare the costs, observe that each iteration has a complexity proportional to the square of the number of bins, which for the three models is 10,000, 100 and 9. The average number of iterations for the different models, was respectively, 22, 13 and 31 (the last number is the sum over the 5 fixed bin size models). Thus, the fixed bin size model with bin size $1msec$ requires about two orders of magnitude more operations than what is needed by the variable bin size model, which, despite the need of executing the EM algorithm multiple times, has the smallest computational complexity.

Table 1. Median of the Absolute Relative Error of the Average Delay Estimates.

	fixed bin size		variable bin size
	$q = 1msec$	$q = 10msec$	
all links	2.6%	8.1%	17.6%
links with average delay < $1msec$	9.7%	64.4%	14.6%

Accuracy. We now compare the accuracy of the different approaches. In order to quantify the accuracy, in Table 1 we list the median of the absolute relative error of the average delay estimates. As expected the best performance is achieved by the fixed bin size model with $q = 1msec$; use of a larger bin, while greatly reducing the complexity, resulted in very poor accuracy for the smaller delays (if we consider only links with average delay smaller than 1msec, the typical error was 64.4%). By contrast, the variable bin size model achieves good accuracy across the entire delay range, while at the same time enjoying a low computational cost.

4   Non-parametric Estimation of Link Delay Variance

In this section we present a class of non-parametric estimators of the link delay variance. We assume initially that all delays are finite: $P[D_k = \infty] = 0$. We will later relax this assumption.

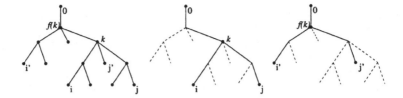

Fig. 4. Logical multicast Tree (left) and the two subtrees traversed by the pairs $\langle i, j \rangle$ (center) and $\langle i', j' \rangle$ (right).

For a node $k \in V$, consider the packet pairs $\langle i, j \rangle$ and $\langle i', j' \rangle$, dispatched to the nodes i and j and i' and j', respectively, such that $i \vee j = k$ and $i' \vee j' = f(k)$; see Figure 4. From the assumption that delays along different links are independent and the bilinearity of the covariance, for the packet pair $\langle i, j \rangle$ it follows that

$$\text{Cov}[X_i(1), X_j(2)] = \text{Cov}[X_k(1) + (X_i(1) - X_k(1)), X_k(2) + (X_j(2) - X_k(2))] \quad (8)$$
$$= \text{Cov}[X_k(1), X_k(2)] \quad (9)$$
$$= \text{Var}[X_k(1)] + \text{Cov}[X_k(1), X_k(2) - X_k(1)] \quad (10)$$
$$= \text{Var}[X_k(1)] + \sum_{l \succeq k} \text{Cov}[D_l, E_l]. \quad (11)$$

Similarly, for the the packet pair $\langle i', j' \rangle$ we have that $\text{Cov}[X_{i'}(1), X_{j'}(2)] = \text{Var}[X_{f(k)}(1)] + \sum_{l \succeq f(k)} \text{Cov}[D_l, E_l]$. Observe that $X_k(1) = X_{f(k)}(1) + D_k$, and $X_{f(k)}(1) = \sum_{l \succeq f(k)} D_l$ and D_k are independent. Therefore, $\text{Var}[D_k] = \text{Var}[X_k] - \text{Var}[X_{f(k)}]$ which we can rewrite

$$\text{Var}[D_k] = \text{Cov}[X_i(1), X_j(2)] - \text{Cov}[X_{i'}(1), X_{j'}(2)] - \text{Cov}[D_k, E_k] \quad (12)$$
$$\approx \text{Cov}[X_i(1), X_j(2)] - \text{Cov}[X_{i'}(1), X_{j'}(2)] \quad (13)$$

under the assumption that $|\text{Cov}[D_k, E_k]| \ll \text{Var}[D_k]$ (Observe that $\text{Cov}[D_k, E_k] = 0$, in particular, if E_k and D_k are independent, or if E_k is constant). (13) expresses the variance of the packet delay along link k in terms of the covariance of delays measured at receivers. We can form an estimator of $\text{Var}[D_k]$ (which is unbiased if $\text{Cov}[D_k, E_k] = 0$) from the unbiased estimators of the end-to-end covariances. More precisely, abbreviate $\text{Cov}[X_i(1), X_j(2)] = s_{ij}$ and $\text{Var}[D_k] = v_k$. We can then estimate v_k by the difference $\widehat{s}_{ij} - \widehat{s}_{i'j'}$ of the unbiased estimators of s_{ij} and $s_{i'j'}$, namely

$$\widehat{s}_{ij} = \frac{1}{n-1} \left(\sum_{m=1}^{n} X_i^{i,j}(1)^{(m)} X_j^{i,j}(2)^{(m)} - \frac{1}{n} \sum_{m,m'=1}^{n} X_i^{i,j}(1)^{(m)} X_j^{i,j}(2)^{(m')} \right) \quad (14)$$

and similarly for $\widehat{s}_{i'j'}$.

More generally, let $Q(k) = \{\{i,j\} \subset R \mid i \vee j = k,\}$ be the set of distinct pairs of receivers whose $\prec$-least common ancestor is $k \in V$. Measurements of the packet pairs $\langle i,j \rangle$, $\{i,j\} \in Q(k)$ and $\langle i',j' \rangle$, $\{i',j'\} \in Q(f(k))$ yields estimates of v_k, namely, $\widehat{s}_{ij} - \widehat{s}_{i'j'}$ as does any convex combination $\sum_{\{i,j\} \in Q(k), \{i',j'\} \in Q(f(k))}$ $\eta_{iji'j'}(\widehat{s}_{ij} - \widehat{s}_{i'j'})$ (where the $\eta_{iji'j'}$ are non negative and sum to 1), which we can rewrite as

$$V_k(\mu, \widehat{s}) := \sum_{\{i,j\} \in Q(k)} \mu_{ij}(k)\widehat{s}_{ij} - \sum_{\{i',j'\} \in Q(f(k))} \mu_{i'j'}(f(k))\widehat{s}_{i'j'} \qquad (15)$$

where $\widehat{s} = \{\widehat{s}_{ij} : \{i,j\} \in Q(k), k \in V\}$, $\mu(k) = (\mu_{ij}(k))_{\{i,j\} \in Q(k)}$, $\mu_{ij}(k) = \sum_{\{i',j'\} \in Q(f(k))} \eta_{iji'j'} \geq 0$, $\sum_{\{i,j\} \in Q(k)} \mu_{ij}(k) = 1$, and similarly for $\mu_{i'j'}(f(k))$. Finally, denote $\mu = (\mu(k), \mu(f(k)))$. An example is the uniform estimator where all $\mu(k)$ and $\mu(f(k))$ are constant. The uniform estimator has the disadvantage that a high variance of any summand may result in a high variance in the overall estimate. By proper choice of the weights we can determine the estimator $V_k(\mu, \widehat{s})$ of minimum variance.

The next theorem characterizes the asymptotic behavior of $V_k(\mu, \widehat{s})$ and gives a form for the estimator of minimum variance. The proofs follow the same lines of those for the multicast case in [6] and are omitted. Define $Z_i(l) = X_i(l) - \mathrm{E}[X_i(l)]$, $i \in R$, $l = 1, 2$, and let $w_{ij} = \mathrm{Var}[Z_i(1)Z_j(2)]$, $i \neq j \in R$ and $m_k = \mathrm{Cov}[D_k, E_k]$.

Theorem 1. *For each $k \in V$:*

(i) *the random variables $\sqrt{n} \cdot (\widehat{s}_{ij} - v_k + m_k)$, $\{i,j\} \in Q(k)$ are independent and converge in distribution to a Gaussian random variable with mean 0 and variance w_{ij};*

(ii) *for any choice of μ, $\sqrt{n}(V_k(\mu, \widehat{s}) - v_k + m_k)$ converges in distribution to a Gaussian random variable of mean zero and variance $\sum_{\{i,j\} \in Q(k)} \mu_{ij}^2(k)w_{ij} + \sum_{\{i',j'\} \in Q(f(k))} \mu_{i'j'}^2(f(k))w_{i'j'}$;*

(iii) *the minimal asymptotic variance of the estimator $V_k(\mu, \widehat{s})$ is achieved when*
$$\mu_{ij}(h) = \mu_{ij}^*(h) := \frac{w_{ij}^{-1}}{\sum_{\{i',j'\} \in Q(h)} w_{i'j'}^{-1}}, \quad h \in \{k, f(k)\}. \text{ The corresponding}$$
asymptotic variance of the estimator is $\dfrac{1}{\sum_{\{i,j\} \in Q(k)} w_{ij}^{-1}} + \dfrac{1}{\sum_{\{i',j'\} \in Q(f(k))} w_{i'j'}^{-1}}$.

Theorem 1 shows that $V_k(\mu, \widehat{s})$ is asymptotically normal. We define the estimator bias as $b_k = |E[V_k(\mu, \widehat{s}) - v_k]|$, $k \in V$. For large n, we can use the approximation $b_k \approx |\mathrm{Cov}[D_k, E_k]|$. Thus, under the assumption that $|\mathrm{Cov}[D_k, E_k]| \ll \mathrm{Var}[E_k]$, we have $\mathrm{E}[V_k(\mu, \widehat{s})] \approx v_k$.

Operationally, the weights μ need to be calculated from an *estimate* $\widehat{w}_{i,j}$ of the variances w_{ij}. These can be computed as shown in [6]. The resulting estimator $V_k(\widehat{\mu}^*, \widehat{s})$, where μ^* is obtained by using $\widehat{w}_{ij}$ in place of w_{ij}, has the same asymptotic behavior of $V_k(\mu^*, \widehat{s})$.

Impact of Loss on the Estimators. We now relax the assumption of finite delays. We associate infinite delays to packet losses. Although lost packets will not provide delay samples at receivers, clearly, the foregoing still applies to cumulant estimation based on the end-to-end delays of the received packets. For any packet pair $\langle i, j \rangle$, define $I_n(i,j) \subset \{1, \ldots, n\}$ the set of pairs for which both packets reach the leaf nodes; define $N_n(i,j) = \#I_n(i,j)$ the number of such pairs. Denote by $B(i,j) = \prod_{k \succeq i,j} \mathsf{P}[D_k < \infty]$ the probability that the two packets of the packet pair reach the leaf nodes. $N_n(i,j)/n$ converges almost surely to $B(i,j)$ as $n \to \infty$. For large n we have approximatively $N_n(i,j) \approx B(i,j)n$ delay measurements from both packets of the pair $\langle i, j \rangle$.

We adapt the approach of the foregoing theory by estimating s_{ij} using only the measurements from the pairs in $I_n(i,j)$. This corresponds to replacing n with $N_n(i,j)$ and $\sum_{m=1}^{n}$ with $\sum_{m \in I_n(i,j)}$ in (14). The effect of packet loss is to reduce the number of packet pairs available for estimation, thus increasing the variability of the estimates. The asymptotic behavior is characterized by results similar to Theorem 1 where we replace w_{ij} by $\frac{w_{ij}}{B(i,j)}$.

5   Network Experiments

The accuracy of the techniques described in Sections 3 and 4 rely on the assumptions that: (1) the back-to-back packets in the packet pair experience roughly the same delay on each link along their common path, $i.e.$, $D_k \approx D_k'$; (2) the additional delay experienced by the second packet is uncorrelated to the delay experienced by the first packet (or practically so), $i.e.$, $|\mathsf{Cov}[D_k, E_k]| \ll \mathsf{Var}[D_k]$. In this section we investigate conformance of both of measurements of packet pairs transmitted across a number of end-to-end paths in the Internet to both of these assumptions. Although these experiments did not access the transmission properties of individual links (which are very difficult to measure), they are able to detect link-wise departures from the assumptions, since these would also be reflected in the properties of end-to-end paths over non-conformant links.

Measurement Infrastructure. We conducted the experiments using the National Internet Measurement Infrastructure (NIMI) [13]. NIMI consists of a number of measurement platforms deployed across the Internet (primarily in the U.S.) that can be used to perform end-to-end measurements. We made the measurements using the zing utility, which sends UDP packets in selectable patterns, recording the time of transmission and reception. zing was extended to transmit packets pairs with minimal spacing between packets. The resulting inter-packet spacings were of about $40\mu sec$.

Measurements were performed along end-to-end paths, by sending packet pairs from a sender to a receiver host. These measurements did not allow us to directly study the delay behavior of the pair along internal links, which would have required measurement inside the network.

Here we report the results from 13 successful measurements made between 11 NIMI sites (two of which are in Europe). Each measurement recorded at both sender and receiver the transmission of 6000 back-to-back packet pairs sent at exponentially distributed intervals with a mean of $100msec$. All measurements were made at either 2PM EDT (a busy time) or 2AM EDT (a fairly unloaded time) separated by a mean of $100msec$. Since our focus is on the variable portion of the delay, in the results reported below we normalize each delay measurement by subtracting the minimum delay seen at the receiver. A delay equal to the minimum delay is thus regarded as a variable delay equal to 0. In other words, we interpret the observed minimum delay as the constant propagation and transmission delay along the path, under the assumption that at least one packet experienced no queueing delay along the path.

Delay Characteristics. In Table 2 we display the relevant delay statistics (measured in $msec$) along each path, ordered in increasing average delay.

Table 2. Summary Delay Statistics (in $msec$).

$E[D_k]$	$\sqrt{\mathrm{Var}[D_k]}$	$E[E_k]$	$\sqrt{\mathrm{Var}[E_k]}$	$\frac{\mathrm{Cov}[D_k, E_k]}{\mathrm{Var}[D_k]}$
0.58	0.40	0.06	0.14	$-2.73 \cdot 10^{-2}$
0.62	10.22	0.08	0.22	$7.81 \cdot 10^{-4}$
0.89	2.63	1.03	0.36	$-3.16 \cdot 10^{-4}$
1.27	7.50	0.18	0.88	$2.23 \cdot 10^{-3}$
1.58	8.74	0.10	0.22	$1.34 \cdot 10^{-4}$
1.95	1.62	1.01	0.18	$-2.78 \cdot 10^{-3}$
2.64	3.61	0.08	0.04	$-1.81 \cdot 10^{-3}$
4.30	23.31	0.25	0.34	$-1.10 \cdot 10^{-4}$
5.44	42.70	0.29	0.90	$2.80 \cdot 10^{-5}$
5.62	60.31	0.32	0.45	$-5.66 \cdot 10^{-5}$
37.28	47.55	0.26	0.83	$2.79 \cdot 10^{-5}$
63.72	65.67	0.61	0.82	$5.98 \cdot 10^{-5}$
65.97	62.16	0.42	1.57	$5.19 \cdot 10^{-5}$

The average delay ranged from $0.6msec$ to $66msec$, a span of two orders of magnitude. The entries in Table 2, with either a large average delay, a large standard deviation, or both, correspond to six experiments involving sites in Europe (the last rows in Table 2). Despite the delay diversity in our measurements, the difference in the average delay seen by back-to-back packets was somewhat more uniform, increasing with larger delay, but with an average and standard deviation typically below $1msec$. This suggests that in practice we can use the approximation $D_k \approx D'_k$ as long as we adopt a delay resolution larger than $1msec$, i.e., we discretize delay with bin size larger than $1msec$. At these resolutions, indeed, the delays seen by the two packets can be considered identical.

Finally, we turn our attention to the bias of the variance estimator. In Table 2, we list the relative bias $\frac{\mathrm{Cov}[D_k, E_k]}{\mathrm{Var}[D_k]}$. The results show that variability of E_k is much smaller than that of D_k. The bias is only 3% in the worst case.

6   Simulation Results

The experiments of Section 5 show that the delay properties of back-to-back packets make packet pairs suitable for delay inference. In this Section, we employ simulation to evaluate how accurate the estimators might be in practice.

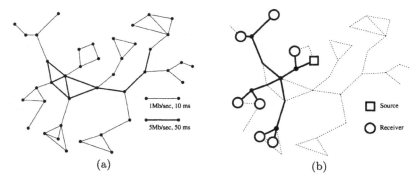

Fig. 5. Network Topology and Logical Source Tree used in the Simulations.

We used the **ns** simulation environment [12]; this enables the representation of transport-protocol details of packet transmissions. The simulations reported in this paper used the 39-node topology of Figure 5(a). The buffer on each link accommodated 20 packets. Background traffic came from 420 sessions comprised of a mixture of TCP sessions and exponential and Pareto on-off UDP sources.

We performed different sets of experiments. In each set we fixed a source and a set of receivers and conducted 100 experiments across the logical tree spanning those nodes. Measurement probes comprised packet pairs with a 1μsec interpacket time. The packet pairs were generated periodically with an inter-packet time of 16 msec by cycling through pairs $\langle i, j \rangle$ sent to distinct receivers i, j. In each experiment, for each pair of distinct receivers $i, j \in R$, $n = 1000$ packets pairs $\langle i, j \rangle$ were transmitted.

In order to evaluate the inference methods, we compare inferred delay statistics, namely, mean and variance, against the actual link delay as determined by instrumentation of the simulation. Here we will report the results for the logical 7 receiver tree in Figure 5(b) which covers part of the network.

Table 3. Simulations Summary Delay Statistics (in *msec*) .

	$E[D_k]$	$\sqrt{\mathrm{Var}[D_k]}$	$E[E_k]$	$\sqrt{\mathrm{Var}[E_k]}$	$\frac{\mathrm{Cov}[D_k, E_k]}{\mathrm{Var}[D_k]}$
min.	0.3	1	0.04	0	$2.1 \cdot 10^{-6}$
median	17.4	17.6	0.27	1.1	$2 \cdot 10^{-3}$
max	55	38.7	0.47	2	$2.03 \cdot 10^{-2}$

Link Statistical Properties. We first examine the statistical properties of the underlying link processes. Characteristics vary considerably across the different links and in the different simulations. The average delay ranged from 0.3msec to 55msec, and the delay variance from 1msec^2 to $1,500$msec^2. The link loss rates ranged from 0% to 18%. The link delay statistics are displayed in Table 3. The behavior and range is very similar to that observed in the network experiments. The important observation is that for 98% of the packet pairs the difference in the delay seen by back-to-back packets was less than $1msec$. We can thus use the approximation $D_k \approx D'_k$ as long as the delay resolution is larger than this value. The bias due to ignoring the term $\mathsf{Cov}[D_k, E_k]$ in the estimation of the variance is negligible and only about 2% in the worst case.

Accuracy of Inference. We now compare inferred and actual link delay in the simulations. Here we focus on the estimation of the summary delay statistics, namely mean and variance. Given the large delay spread across the different links (delay was as large as a few hundreds $msec$), to infer the average delay we estimated the link delay distribution using the variable bin size model. We used the ternary variable bin size model $(3^{l-1}msec, 2)_{l=1,\ldots,5}$. For the analysis, delay was thus discretized to the set $\mathcal{Q} = \{0, 1, 3, 9, 27, 81, 243, \infty\}msec$, only eight bins. The estimate of the average delay is then $\mathsf{E}[D_k|\widehat{D_k} < \infty] = \dfrac{\sum_{d \in \mathcal{Q} \setminus \{\infty\}} d\widehat{\alpha}_k(d)}{\sum_{d \in \mathcal{Q} \setminus \{\infty\}} \widehat{\alpha}_k(d)}$, $k \in V$. As shown below, even if this model can be considered too coarse to adequately capture the probabilities of large delays, it allowed us to compute the estimates of the average delay efficiently and accurately. To estimate the link delay variance, we used directly the method described in Section 4.

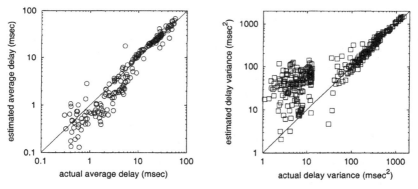

Fig. 6. Inferred vs. actual average and variance of link delay in simulations. Scatter plot for 100 experiments: (a) average link delay; (b) link delay variance.

In Figure 6, we display scatter plots of inferred vs. actual link delay mean and variance. Accuracy increases for higher values as exhibited by the clustering about the line $y = x$. In order to quantify the accuracy of the estimates, we computed the median of the absolute error of the estimates of the link delay mean and variance. The median was 22.05% for the mean and 40% for the delay variance. Estimates were more accurate for larger delays: if we consider delay means larger than $10msec$ or delay variances larger than $10msec^2$, the median of the relative absolute error fell to 10% and 11.75%, respectively.

We can attribute the larger inference errors for smaller delays only in part to the fact that $D_k \neq D'_k$ and $\mathsf{Cov}[D_k, E_k] \neq 0$. Observe, indeed, that especially for the variance estimates, the relative errors are quite large despite $|\mathsf{Cov}[D_k, E_k]| \ll \mathsf{Var}[D_k]$. We ascribe these larger errors to departure of the actual packets delay from the independence assumption of the model. We calculated the coefficient of correlation of packet delays on consecutive links. The median was 0.09, the maximum value 0.57. We believe that the higher correlations are a result of the small scale of the simulated network. In general, we expect correlations to be smaller in real networks because of the wide traffic and link diversity. The large effect that correlation has on the estimates of the variance can be explained by observing that, because of the independence assumption we ignore all the cross-correlation terms when we derive the expression for the variance estimator (equations (8)-(13)). In the presence of correlation, these terms are not negligible and can be significantly larger than the smaller variances. On the other hand, we observed that estimation of the average delay is more robust and does not significantly suffer from the violation of the independence assumption. This is not unexpected since unlike the variance, the mean of a sum is always equal to the sum of the means irrespective of the underlying correlation structure.

7   Conclusions

In this paper, we explored the use of end-to-end unicast traffic measurements to estimate the delay characteristics of internal network links. Measurement experiments consist of back-to-back packets (a packet pair) sent from a sender to pairs of receivers. We develop efficient techniques to estimate the link delay characteristics, namely, delay distribution and delay variance.

For the estimation of the delay distribution, building on previous work in [11] and the recent work of the authors of [4, 5], we proposed a novel approach for the estimation of the link delay distribution. The key idea is the use a variable bin size model, wherein smaller bins are used in correspondence of concentrations of probability mass and larger bins otherwise. We consider a variable bin size model the analysis of which can be reduced to that of fixed bin size models. Through examples, we showed that, compared to previous approaches, we are able to significantly reduce the computational complexity, without losing accuracy.

We also provided methods to directly estimates the link delay variance. We express the link delay variance in terms of the covariance of the end-to-end delays. Therefore, we can estimate the variance from the sample covariance of the end-to-end delays. The method can be extended to the estimation of higher order cumulants.

Accuracy of the proposed approaches depends on strong correlation between the delay seen by the two packets along the shared path. We verified the degree of correlation in packet pairs through network measurements. We also used simulation to explore the performance of the estimator in practice and observed good accuracy of the proposed inference techniques, although violation of some of the model assumptions, e.g., spatial correlation, introduces systematic errors. This will be object of further study.

References

1. R. Caceres, N.G. Duffield, J.Horowitz and D. Towsley, "Multicast-Based Inference of Network Internal Loss Characteristics", *IEEE Trans. on Information Theory*, November 1999.

2. R. Carter, M. Crovella, 'Measuring bottleneck link-speed in packet-switched networks,' *Performance Evaluation*, 27&28, 1996.
3. M. Coates, R. Nowak, "Network loss inference using unicast end-to-end measurement", *Proc. ITC Conf. IP Traffic, Modeling and Management*, Monterey, CA, September 2000.
4. M. Coates, R. Nowak, "Sequential Monte Carlo Inference of Internal Delays in Nonstationary Communication Networks," submitted for pubblication, Jan 2001.
5. M.J. Coates and R. Nowak, "Network Delay Distribution Inference from End-to-end Unicast Measurement," *Proc. of the IEEE International Conference on Acoustics, Speech, and Signal Processing*, May 2001.
6. N.G. Duffield and F. Lo Presti, "Multicast Inference of Packet Delay Variance at Interior Network Links", *Proc. IEEE Infocom 2000*, Tel Aviv, March 2000.
7. N.G. Duffield, F. Lo Presti, V. Paxson and D. Towsley "Inferring Link Loss Using Striped Unicast Probes", *Proc. IEEE Infocom 2001*, Anchorage, AK, April 2001.
8. B. Frey. Graphical Models for Machine Learning and Digital Communication. MIT Press, Cambridge London (1998).
9. V. Jacobson, "Congestion Avoidance and Control," *Proc. SIGCOMM '88*, pp. 314-329, August. 1988.
10. S. Keshav. " A control-theoretic approach to flow control," *Proc. SIGCOMM '91*, 3–15, September 1991.
11. F. Lo Presti, N.G. Duffield, J.Horowitz and D. Towsley, "Multicast-Based Inference of Network-Internal Delay Distributions", submitted for publication, September 1999.
12. ns – Network Simulator. See http://www-mash.cs.berkeley.edu/ns/ns.html
13. V. Paxson, J. Mahdavi, A. Adams, M. Mathis, "An Architecture for Large-Scale Internet Measurement," *IEEE Communications Magazine*, Vol. 36, No. 8, pp. 48–54, August 1998.
14. Y. Tsang, M.J. Coates and R. Nowak, "Passive Network Tomography using EM Algorithms," *Proc. of the IEEE International Conference on Acoustics, Speech, and Signal Processing*, May 2001.
15. C.F. Jeff Wu, "On the convergence properties of the EM algorithm", Annals of Statistics, vol. 11, pp. 95-103, 1982.

A   Computation of $\widehat{\alpha}^{(0)}$

We illustrate the method for the computation of $\widehat{\alpha}^{(0)}$. Let $A_k(d) = \mathsf{P}_\alpha[X_k(1) = d]$, $k \in V$ the probability that the first packet of the pair reaches k in d unit of time. For each pair $\{i, j\} \in Q(k)$, we use the approach in [11] to compute an estimate $\widehat{A}_k^{i,j}(d)$ of $A_k(d)$ from the empirical distribution of $X_{i,j}$ by solving a system of polynomial equations. Since $X_k(1) = X_{f(k)}(1) + D_k$ and $X_{f(k)}$ and D_k are independent we obtain an estimate of the distribution of D_k by deconvolution of the estimates of the distributions of $X_k(1)$ and $X_{f(k)}(1)$. We use this estimate as initial distribution. More precisely, for $k \in V$, we let $\widehat{\alpha}_k^{(0)}(d) = (\widehat{A}_k(d) - \sum_{d' \in \mathcal{Q}, d' \leq d} \widehat{A}_{f(k)}(d')\widehat{\alpha}_k^{(0)}(d - d'))/\widehat{A}_{f(k)}(0)$, $d \in \mathcal{Q} \setminus \infty$, where $\widehat{A}_k(d) = \frac{1}{\#Q(k)} \sum_{\{i,j\} \in Q(k)} \widehat{A}_k^{i,j}(d)$, and let $\widehat{\alpha}_k^{(0)}(\infty) = 1 - \sum_{d \in \mathcal{Q} \setminus \infty} \widehat{\alpha}_k^{(0)}(d)$. It is possible to show that $\widehat{\alpha}^{(0)}$ is a consistent estimator of α and, as n goes to infinity, $\sqrt{n}(\widehat{\alpha}^{(0)} - \alpha)$ converges in distribution to a multivariate Gaussian random variable with mean 0 and covariance matrix σ_α.

On the Contribution of TCP to the Self-Similarity of Network Traffic *

Biplab Sikdar and Kenneth S. Vastola

Department of ECSE, Rensselaer Polytechnic Institute
Troy, NY 12180 USA
{bsikdar,vastola}@networks.ecse.rpi.edu

Abstract. Recent research has shown the presence of self-similarity in TCP traffic which is unaffected by the application level and human factors. This suggests the presence of protocol level contributions to network traffic self-similarity, at least in certain time scales where the effect of protocol behavior is most prominent. In this paper we show how TCP's retransmission and congestion control mechanism contributes to the self-similarity of aggregate TCP flows. We develop a mathematical formulation which shows that TCP's retransmission and congestion control mechanism results in packet dynamics of a TCP flow being analogous to a number of ON/OFF sources with OFF periods taken from a heavy tailed distribution. Using well known limit theorems, we then show that this contributes to the self-similar nature of TCP traffic. Our model shows a direct correlation of the loss rates to the degree of self-similarity. Measurements on traces collected by us also exhibit this relationship predicted by our model.

1 Introduction

Research on the causes of self-similarity in network traffic have primarily focused on the application level characteristics of high-speed networks and the human factors involved. In [21], the causes of the self-similarity are investigated at the source level. In [1] the authors cite the distribution of file sizes, the effects of caching and human factors like response time and preference as possible causes for the self-similarity in WWW traffic. On the other hand, protocol level causes of self-similarity in network traffic has been investigated in [2] and [13] which showed that closed loop protocols like TCP lead to much richer scaling behavior than open loop protocols like UDP.

In this paper we show that TCP can contribute to the self-similarity of network traffic and its contribution is visible in the time scales ranging from milliseconds to tens of seconds. Thus though TCP may not be able to contribute at higher time scales, the observed self-similarity in these scales can be attributed to application and human level causes which inherently operate at time scales of

---

* Supported in part by DARPA contract F30602-00-2-0537 and in part by DoD MURI contract F49620-97-1-0382.

S. Palazzo (Ed.): IWDC 2001, LNCS 2170, pp. 596–613, 2001.

minutes and hours. Also, though in the pure mathematical sense a self-similar process should exhibit the same statistical characteristics over all possible time scales, this is not possible in real systems due to physical limitations. We show that TCP is capable of causing scaling in 3 to 4 time scales (few milliseconds to 10s of seconds) and it is in this sense that we call TCP traffic is self-similar. This range of timescales is generally sufficient for traffic modeling purposes as shown in citeGrBo99, since the range of relevant timescales is determined by the finite buffer sizes of real systems.

In [19], the authors attribute the self-similarity of TCP traffic to the chaotic nature of TCP's congestion control mechanism. The adaptive nature of TCP's congestion control is suggested as the cause for the propagation of self-similarity in the Internet in [20]. The main aim of our paper is to understand the effects of TCP's retransmission and congestion control mechanism on the observed self-similarity of TCP traffic. Our results show that the timeout and exponential backoff mechanisms in TCP play a crucial in inducing self-similarity. We also show that the degree of self-similarity has a direct relationship with the losses experienced by a flow with the traffic no longer self-similar, i.e. $H \approx 0.5$ for very low loss rates. While similar phenomena have been reported recently (after this paper was completed), their models to explain the self-similarity either require unrealistic loss rates to induce self-similarity [7] or are able to show long-range dependence over very small time scales [5]. In this paper, we present a model of TCP based on ON/OFF processes which explains the self-similarity of TCP traffic and validate it using TCP traces collected from the Internet. We also give a mathematical formulation of how TCP's congestion control mechanism leads to self-similarity in the traffic it generates and account for the effects of the network in terms of the loss probabilities and the presence of other flows.

The rest of the paper is organized as follows. In Section 2 we first present the results of tests on traffic traces generated by individual TCP transfers over the Internet showing proof of self-similarity. We then present a model which explains this self-similarity and experimentally validate our model using the same TCP traces. In Section 3 we provide a mathematical foundation for our model and investigate the mechanisms of TCP which contribute to self-similarity in greater detail. Finally, Section 4 presents the discussions and concluding remarks.

2   Self-Similarity of TCP Flows

In this section we provide experimental evidence of the self-similarity of individual TCP flows which motivates the investigation of TCP dynamics for causes of self-similarity. In [19] the authors showed that the data sent by an isolated TCP flow from the superposition of a number of TCP flows shows evidence of self-similarity and attribute it to the chaotic nature of TCP's congestion avoidance mechanism. All previous reports of self-similarity in network traffic concentrated on the self-similar characteristics of the aggregated traffic. However, the results in [19] were generated by carrying out experiments using the simulator *ns* which is not an exact reflection of the actual scenarios in the Internet. Hence to dis-

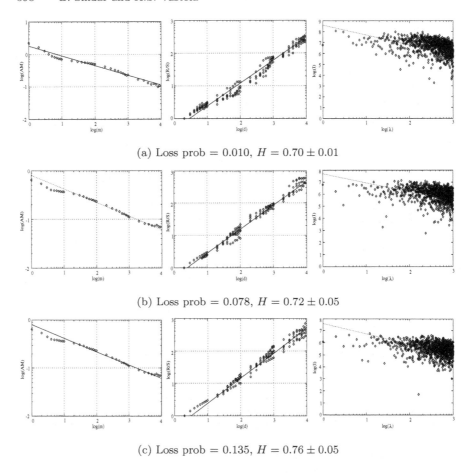

(a) Loss prob = 0.010, $H = 0.70 \pm 0.01$

(b) Loss prob = 0.078, $H = 0.72 \pm 0.05$

(c) Loss prob = 0.135, $H = 0.76 \pm 0.05$

Fig. 1. Tests for self-similarity for the various traces to Columbus, Ohio. For each trace, the figures show the results from the absolute value method (left), R/S statistics method (middle) and the Periodogram method (right).

pel any doubts about the self-similar nature of single TCP microflows, we first present the results from tests for long-range dependence on traces collected from real life TCP connections over the Internet.

We first give a brief description of the datasets. We collected traces for data transfers originating from a machine running Solaris 2.6 at RPI, Troy, NY. The destinations for the transfers were in Ohio State University, Columbus, OH (HP-UX), University of California, Los Angeles, CA (FreeBSD Cairn-2.5), Massachusetts Institute of Technology, Boston, MA (Linux 2.0.36) and University of Pisa, Pisa, Italy (FreeBSD 3.3). Due to space restrictions, we show results for only the transfers to Ohio and Italy. The results for the others are similar.

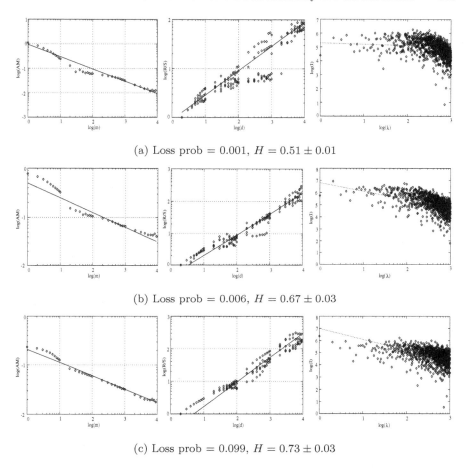

(a) Loss prob = 0.001, $H = 0.51 \pm 0.01$

(b) Loss prob = 0.006, $H = 0.67 \pm 0.03$

(c) Loss prob = 0.099, $H = 0.73 \pm 0.03$

Fig. 2. Tests for self-similarity for the various traces to Pisa, Italy. For each trace, the figures show the results from the absolute value method (left), R/S statistics method (middle) and the Periodogram method (right).

Each trace is 2000 seconds or around 33 minutes long and was collected using tcpdump which did not lose any packets. The transfers were done over periods in 1999 and 2000 at various times of the day and week. Depending on the prevalent network conditions, the loss rates experienced by each flow is different and we use this to classify transfers between a source-destination pair.

Figure 1 shows the results of the tests for long-range dependence on three traces to Ohio which had loss rates of 0.010, 0.078 and 0.135. Figure 2 shows the results of similar tests on the traces collected from transfers to Pisa which had loss rates of 0.001, 0.006 and 0.099. We tested for long-range dependence using three of the widely used methods [18]: the absolute value method, R/S

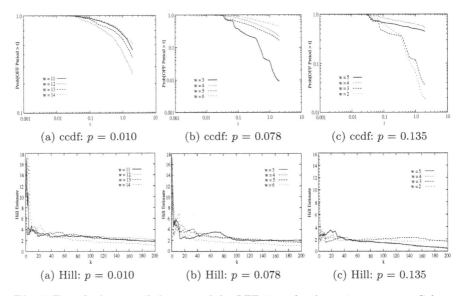

Fig. 3. Tests for heavy-tailed nature of the OFF times for the various traces to Columbus, Ohio. For each trace, the figures show the ccdf plots (top) and the corresponding Hill's estimates (bottom) for various values of w.

statistics method and the periodogram method. The results clearly show the long-range dependence in the individual TCP flows. Also the degree of long-range dependence, as indicated by the Hurst parameter, is clearly dependent on the loss rate experienced by the flow, with higher loss rates leading to larger values of H. Also note that for extremely low probabilities (less than 0.001) the traffic is no longer self-similar as indicated by the Hurst parameter of approximately 0.5 as shown in section (a) of Fig. 2. We describe this in detail in the following subsection and in Section 4.

This poses the following questions. What are the underlying mechanisms which are responsible for the direct influence of the loss probabilities on the self-similarity of TCP traffic? What role does TCP's fast-retransmit and timeout mechanisms play in all this? In this paper we address these issues and show how TCP's retransmission and congestion avoidance mechanisms contribute to the self-similar nature of network traffic.

2.1 ON/OFF Model Based Explanation and Its Validation

TCP follows a window based flow control mechanism and transmits a certain number of packets in each "round". We define a round as in [12]. A round begins with the back to back transmission of a window of packets. After these packets are transmitted, no other packet is transmitted till an ACK is received for one of these packets. The receipt of an ACK marks the end of the round.

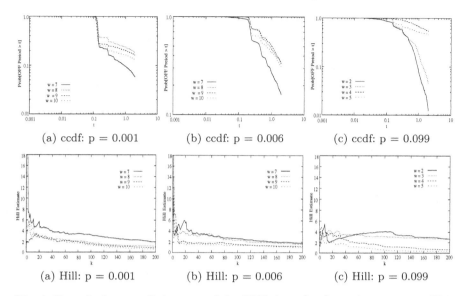

Fig. 4. Tests for heavy-tailed nature of the OFF times for the various traces to Pisa, Italy. For each trace, the figures show the ccdf plots (top) and the corresponding Hill's estimates (bottom) for various values of w.

To give an explanation for TCP's effect on the self-similarity of network traffic, we consider a TCP flow to be composed of the superposition of W_{max} ON/OFF processes. Each process corresponds to each of the possible values that the *cwnd* of the flow might have since W_{max} is the receiver's advertised maximum buffer size and is the upper limit on *cwnd*. A *cwnd* of w, corresponding to the wth ON/OFF process, $1 \leq w \leq W_{max}$, implies a deterministic ON time which is equal to the time to transmit the w packets with the packets generated at a constant rate in during this period. We note that though in practice there might be a small variation in the time between two successive packets in a round, these are generally very small and with high speed networks these variations are negligible when compared to the RTTs. Also, as described after a few paragraphs, the way we demarcate the end of ON periods ensures that the spacing between the packets in the ON period is almost constant.

The OFF period for the wth process, $1 \leq w \leq W_{max}$, corresponds to the time interval between two successive instants where *cwnd* has the value w. Now, if the distribution of these times has a heavy tail, their complementary cumulative distribution function (ccdf) $F_c(x)$ behaves like

$$F_c x \sim lx^{-\alpha}L(x) \quad \text{with } 1 < \alpha < 2 \tag{1}$$

where $l > 0$ is a constant, $L(x)$ is a slowly varying function at infinity, i.e., $\lim_{x \to \infty} L(tx)/L(x) = 1, \forall\ t > 0$ and the relation $f(x) \sim g(x)$ implies $\lim_{x \to \infty} f(x)/g(x) = 1$. We can now use the following Theorem from [17] which says that the superposition of a number of these processes converges in the limit to fractional Brownian motion (fBm) and thus exhibit self-similarity.

Consider M independent ON/OFF processes. Let $F_{1c}^{(r)}$ $(F_{2c}^{(r)})$, $\mu_1^{(r)}$ $(\mu_2^{(r)})$ and $\sigma_1^{2(r)}$ $(\sigma_2^{2(r)})$ be the ccdf, the mean duration and the variance of the ON (OFF) period of the ON/OFF process of type r. Now, if $W_M(Tt)$ represents the aggregated packet count in the interval $[0, Tt]$ due to the contribution from all the M sources then

Theorem 1. (Taqqu, Willinger and Sherman) *As $M^{(r)} \to \infty$, $r = 1, \cdots, R$ and $T \to \infty$, the aggregated cumulative packet traffic $\{W_M(Tt), t \geq 0\}$ behaves statistically like*

$$Tt \sum_{r=1}^{R} \frac{M^{(r)}\mu_1^{(r)}}{\mu_1^{(r)} + \mu_2^{(r)}} + \sum_{r=1}^{R} T^{H^{(r)}} \sqrt{L^{(r)}(T)M^{(r)}} \sigma_{\lim}^{(r)} B_{H^{(r)}}(t)$$

where the $B_{H^{(r)}}(t)$ are independent fractional Brownian motions and $H^{(r)}$ and $\sigma_{\lim}^{(r)}$ are as defined in [17].

In our case, R corresponds to the maximum window size allowed for any of the flows in the network and the limiting conditions are reached when we have a large number of flows in the network each contributing its ON/OFF processes to the superposition. Now we just need to show that the distribution of the OFF times indeed corresponds to the form of Eqn. 1. In Figs. 3 and 4 we plot the ccdf of the OFF times for various window sizes for the traces for Ohio and Italy and the heavy tailed nature of each is clearly evident. While the ccdf plots often provide solid evidence for or against the presence of heavy tails, an eyeballing method is statistically unsatisfactory and the rough estimates of α obtained from these plots may be unreliable. A statistically more rigorous method for estimating the slope of the tails and thus α is the *Hill's estimator* [8]. The presence of heavy tails is indicated by a straight line behavior of the Hill's estimate $\hat{\alpha}_n$ as the number of samples used in the calculation of the estimate increases while a steadily decreasing pattern is a strong indication of the data being not from a heavy-tailed distribution. Figs. 3 and 4 also plot the Hill's estimates for the OFF time distribution for various window sizes for the Ohio and Italy traces respectively and clearly they are consistent with the form of Eqn. 1. Thus we can conclude that the superposition of such ON/OFF process from a number of TCP flows will converge in the limit to fBm and thus exhibit self-similarity.

It is interesting to note the ccdf and Hill estimate plots for the Italy trace with $p = 0.001$. From Figure 2 the Hurst parameter for this trace can be seen to be around 0.5, i.e. the trace does not exhibit self-similarity. We note from Figure 4 that the Hill estimates for all the ON/OFF process corresponding to this trace are decaying constantly and thus do not have a heavy tailed nature. Thus the ON/OFF processes corresponding to this trace do not satisfy the conditions of Theorem 1 and as a result the trace is not self-similar. In Section 3.3 from our derivation of a lower bound of the ccdf it will be clear why low loss rates fail to give rise to heavy tails.

An important assumption here is the independence of the window sizes of different flows, which need not be the case for *all* the flows in a link. Simulation studies have indicated that the window sizes of TCP flows sharing a common bottleneck link may get synchronized though such synchronization is hard to observe in the Internet [11]. Also, most of the simulation studies focus on very heavily congested bottleneck links while link loads in practice tend to be comparatively much lower. Also, note that the independence requirements fail to be satisfied only when nearly all the flows in a link are correlated. To prove that the independence assumptions of Theorem 2 of [17] are satisfied, we analyzed some of the traces reported in [14]. The results of our statistical tests on these traces to see if the individual TCP flows are indeed independent indicate that amongst the longer flows in the traces, roughly 35-70 % of the flows are mutually independent, providing enough independent flows in the superposition.

An important part in the calculation of the OFF times is what criterion we use to define a OFF period. We define an ON period to be over whenever the distance between two successive packets in the trace exceeds a length δ dependent on the packet transmission time on the link. By keeping δ sufficiently small we can ensure that the spacing between the packets in the ON period is almost constant thus satisfying the requirement of Theorem 1. Also, as in [21], the exact numerical choice of δ does not affect the results and the heavy tailed nature of the ccdf remains an invariant independent of the choice of δ.

3   Investigating the Role of TCP

Having presented a model explaining the self-similarity of TCP traffic we now pinpoint the sources in TCP's retransmission and congestion avoidance mechanism which are responsible for this phenomena. We then derive a lower bound on the tail of the OFF time distribution and show that it decays according to a power law providing a firm mathematical foundation to our model. In this paper we concentrate on TCP Reno as it the most widely deployed variant of TCP. The effect of the other versions of TCP is discussed in Section 4. We assume that the reader is familiar with the basic concepts of TCP like the congestion window *cwnd*, slow start, delayed acknowledgments etc and refer the reader to [16] for details on TCP's algorithms.

3.1   The Impact of Timeouts

From the explanation for the observed self-similarity in TCP traffic given in Section 2 it is obvious that the central aspect of the phenomenon lies in the infinite variance or the heavy tailed nature of the OFF time distributions. Let us now consider the features of TCP which lead to such a behavior.

In the following we assume an infinite or steady state flow currently in the congestion avoidance mode to make the visualization easier. Consider a TCP flow with a current window size of w, $w < W_{max}$. In every round that follows, the window now increases linearly until it reaches W_{max} and we need a loss for the window to drop back so that we get a window of size w again. Note that if the window reaches a value greater than $2w$ before a loss indication and it results in a fast retransmit, the subsequent congestion avoidance mode will start with a window greater than w leading to even longer times before a window of w is reached. However the occurrence of heavy tails is mainly due to the loss indications which lead to timeouts. This is due to the following reasons. A timeout represents a significant duration when no packets are transmitted and acts as a boundary between ON and OFF periods of the flow as a whole leading to a bursty nature of TCP traffic. The durations of timeouts are generally an order of magnitude greater than the RTT [12] and with coarse TCP timer granularities and variations in the RTT measurements can be quite large. Again, if the retransmitted packet following a timeout is also lost, the silent period is doubled and from the traces reported in [12] the occurrence of multiple consecutive timeouts is frequent. Also, a majority of the losses experienced by TCP flows lead to timeouts which can be attributed to the fact loss that most routers in the Internet deploy droptail queues. Correlated loss models, where all the packets following the first dropped packet in a round are also dropped are an appropriate models for the losses arising from these queues [12]. This coupled with the fact that a single loss in a window less than 4, two or more losses in a window less than 8 and three or more losses for higher windows in TCP Reno will lead to a timeout contributes to the large proportion of timeouts in the observed loss indications. Before moving on to the derivation of the lower bound on the tail of the ccdf, we first derive the probability that a loss in a window of size w leads to a timeout.

3.2   Probability of Timeouts

Consider a round with window w and let the probability that a loss of any packet in this round will lead to a timeout be denoted by $Q(w)$. We assume that the receiver sends one ACK for every two packets it receives. We assume that all losses are due to packet drops at intermediate queues and that losses due to data corruption are negligible. We also assume droptail queues and the correlated loss model of the previous subsection. Packet losses in a round are assumed to be independent of losses in other rounds and the packet loss probability is denoted by p.

For window sizes less than 4, any packet loss leads to a timeout and thus $Q(w) = 1$ for $1 \leq w \leq 3$. For windows with $4 \leq w \leq 8$ (or $K+1$ to $2(K+1)$) two or more packet losses in a round leads to a timeout. If only one packet is lost in the current round, if we lose any packet in the following round, the flow will eventually timeout. In addition the retransmitted packet must also be transmitted successfully to avoid a timeout. Thus the probability that a packet loss *does not* lead to a timeout for this range of window values is given by

$$1 - Q(w) = \frac{p(1-p)^{w-1}}{1-(1-p)^w}(1-p)^{w-1}(1-p) \tag{2}$$

The first term corresponds to the probability of exactly one packet loss in a window of w. The second last two terms correspond to the probability that all the $w - 1$ packets in the following round and the retransmitted packet are received correctly. Thus

$$Q(w) = 1 - \frac{p(1-p)^{2w-1}}{1 - (1-p)^w} \qquad \text{for } 4 \leq w \leq 8 \qquad (3)$$

For window sizes greater than 8, three or more losses in a round will lead to a timeout. Also we have to ensure that the retransmitted packet is received successfully along with the fact that none of the packets in the succeeding round are lost. Neglecting the extremely few possibilities in which it is possible to recover a single loss in the succeeding round without going into a timeout, we thus have

$$Q(w) = 1 - \frac{p(2-p)(1-p)^{2w-2}}{1 - (1-p)^w} \qquad \text{for } 9 \leq w \leq W_{max}$$

3.3   A Lower Bound on the OFF Time Distribution

We now derive a lower bound on the ccdf by identifying the possible ways in which the time between two successive windows of the same size can exceed a given value. We concentrate on the most likely paths that the $cwnd$ is likely to follow while not accounting for the others as their contribution to the ccdf is negligible. In this derivation, we measure time in units of the round trip time.

Let us assume that the current window size is w and we want to find the probability that the time until the next instant where $cndw = w$ is greater than 100. The most obvious possibility is that the flow does not experience any loss for the next 100 rounds so that after some round the $cwnd$ stays at W_{max}. However, with higher loss probabilities this event is unlikely and the probability tail based on just this mechanism has an exponential decay. Another possibility could be that after i rounds (when $cwnd > 2w$) the flow experiences a loss which results in a fast retransmit. The flow then transmits the next $100 - i$ rounds without any loss. As a variation of this we could have a number of successive fast retransmits without reaching a window of w. Note that each of these possibilities are mutually independent and their individual contribution the tail of the distribution has an exponential decay, each having its own rate. Yet another line of possibilities is timeouts. Let us denote the average duration of a timeout (in terms of RTTs) by $E[TO]$. As the first possibility we could have that there are no losses in the first $100 - E[TO]$ followed by a timeout. We could also have i initial rounds without loss and then n timeouts (with n sufficiently large) before the window gets a chance to increase to w. Other possibilities include cases where we have timeout periods of length $2E[TO]$, $4E[TO]$ and so on. Again, each of these cases represent independent possibilities whose individual contribution to the tail of the OFF time distribution has an exponential decay, the rate of which depends on the corresponding probability of the loss indications and their effects.

The tail of the OFF time distribution for each window size and the corresponding ON/OFF process can thus be seen as the superposition of a large number independent exponential tails each with its own rate of decay. The mix of these independent exponentials leads to a composite distribution which has a heavy tail over the region of our interest. The following theorem by Bernstein [4] provides the link between the mixture of exponentials and a completely monotone probability density function (pdf).

Theorem 2. (Bernstein) *Every completely monotone pdf f is a mixture of exponential pdfs, i.e.,*

$$f(t) = \int_0^\infty \lambda e^{-\lambda t} dG(\lambda), \qquad t \geq 0 \qquad (4)$$

for some proper cdf G.

It can be shown that the commonly used heavy tailed distributions like Pareto and Weibull are completely monotonous. Also, in [3] it is shown that the superposition of a number of properly chosen exponentials can be used to model heavy tailed distributions in the region of primary interest. Having shown the basic construction of how the mix of exponentials lead to heavy tails in the OFF time distribution, we now obtain the probabilities corresponding to each of the possible paths that we described.

Case 1: The no loss case. Let us begin with the simplest case where there are no losses. Consider the wth ON/OFF process which corresponds to a *cwnd* of w, $1 < w < W_{max}$ excluding the special cases with *cwnd*s of 1 and W_{max}. Assume that the current round has a window of size w. The probability that the next window of size w occurs after t units of time (i.e. t RTTs) assuming there are no losses in between is given by

$$P\{T > t\} = (1-p)^{N(t)} \qquad (5)$$

where $N(i)$ represents that number of packets that are transmitted in the i rounds following the round with size w and is given by

$$N(i) = \begin{cases} iw + \lceil \frac{i}{2} \rceil \left(i - \lceil \frac{i}{2} \rceil \right) & \text{if } i \leq j \\ jw + \lceil \frac{j}{2} \rceil \left(j - \lceil \frac{j}{2} \rceil \right) + (i-j)W_{max} & \text{else} \end{cases} \qquad (6)$$

where $j = 2(W_{max} - w) - 1$ and represents the time it takes for the *cwnd* to reach W_{max}, assuming no losses.

Case 2: Fast retransmission losses. We now consider the more likely cases where a flow experiences n losses between two successive windows of the same size which are far apart in time. Consider again the wth ON/OFF process, $1 < w < W_{max}$. We can have a OFF time greater than t if we have loss indications at windows greater than $2w$ which result in fast retransmits. For simplicity, we consider only those cases where the loss occurs in a window of size W_{max}. The flow first transmits packets without loss for the first i rounds during which its window reaches W_{max}. It then experiences a loss which is recovered by a fast retransmit. Since $w < \lceil W_{max}/2 \rceil$ the desired window size is not achieved at the beginning of the congestion avoidance mode. Also, following each loss there are $2(W_{max} - m) - 1$ rounds with $W_{max}(W_{max} - 1) - m(m + 1)$ packets till $cwnd$ reaches W_{max} again with $m = \lceil W_{max}/2 \rceil$. Thus there are $t - n - n(2(W_{max} - m) - 1) - 2(W_{max} - w) + 1$ rounds with successfully transmitted windows of W_{max}. The total number of correctly transmitted packets, after algebraic simplifications, is thus

$$N_c(w, t) = W_{max}(t - (n + 1)W_{max} + 2w + 2nm - 4n)$$
$$-w(w + 1) - nm(m - 1) \qquad (7)$$

Now, since there are $M = t - 2nW_{max} + 2w + 2(n - 1)m - 2n + 3$ rounds with a $cwnd$ of W_{max} with n of them having losses, the probability that the OFF time is greater than t is given by

$$P\{T > t\} = \binom{M}{n} (1 - (1 - p)^{W_{max}})^n (1 - Q(W_{max}))^n (1 - p)^{N_c(w,t)} \qquad (8)$$

Also, since each loss is associated with $2(W_{max} - m) - 1$ rounds where the window is not W_{max}, the maximum possible losses in t rounds can be shown to be limited by

$$n_{max} = \left\lfloor \frac{t - 2(W_{max} - w) + 1}{2W_{max} - 2m - 1} \right\rfloor \qquad (9)$$

Case 3: Loss indication resulting in a timeout. Let us now consider the case when the TCP flow experiences a single loss indication which results in a timeout. Consider the case when the loss occurs after i rounds from the round with a window of w. The number of packets transmitted in these i rounds, $N(i)$ is given in Eqn. 6 and the value of the $cwnd$ in the ith round w_i is given by

$$w_i = \min\{W_{max}, w + \lceil i/2 \rceil\} \qquad (10)$$

To find the number of packets transmitted in the slow start phase which follows a timeout, we use the model of [15] which models the window increase pattern more accurately than the commonly used approximation where the window always increases 1.5 times every RTT. From [15], the number of rounds spent in the slow start phase is given by

$$t_{ss}(w_i) = \left\lfloor 2\log_2 \left(\frac{2m}{1 + \sqrt{2}} \right) \right\rfloor - 1 \qquad (11)$$

where $m = \lceil \frac{w_i}{2} \rceil$ and the number of packets transmitted in the slow start phase can be expressed as

$$N_{ss}(w_i) = \left\lfloor 2^{\frac{t_{ss}(w_i)+1}{2}} + 3 \cdot 2^{\frac{4t_{ss}(w_i)-3}{8}} - 2 - \frac{3\sqrt{2}}{2} \right\rfloor \qquad (12)$$

If $w > m$ we also have a linear phase where the window increases linearly from m to w. The total time required by the flow to reach a window of w again following the timeout is thus

$$D_{nl}(w, w_i) = \begin{cases} t_{ss}(w) + E[TO] + 1 & \text{if } w \leq m \\ t_{ss}(w_i) + E[TO] + 2(w - m) & \text{else} \end{cases} \quad (13)$$

Now, the probability that we have a loss in a round of size u following the timeout, before the window reaches w, $P_{TO}(u, w_i)$, $1 \leq u < w$, is given by

$$P_{TO}(u, w_i) = \begin{cases} (1-p)^{N_{ss}(2u)}(1 - (1-p)^u) & \text{if } u < m \\ Q(u) \\ (1-p)^{N_{ss}(w_i)}(1 - (1-p)^{2u}) & \text{else} \\ Q(u)(1-p)^{u(u-1)-m(m-1)} \end{cases} \quad (14)$$

Note that the $1 - (1-p)^{2u}$ term in the second case has an exponent $2u$ because in the linear phase we have two consecutive rounds with the same window size. Then, the probability that there is another timeout before the window reaches a window of w is given by

$$P_s(w, w_i) = \sum_{u=1}^{w-1} P_{TO}(u, w_i) \quad (15)$$

Note that in the summation above some of the values of $P_{TO}(u, w_i)$ are zero if $u < m$ and $cwnd$ skips these values of u due to the exponential increase pattern. After the ith round, on an average 2 more round of packets are sent (where the first couple of losses may be recovered) before the timeout period begins. Thus if $i \geq t - D_{nl}(w, w_i) - E[TO] - 2$, the probability that the off time is greater than t is given by

$$P\{T > t\} = \begin{cases} (1-p)^{N(i)}(1 - (p)^{w_i})Q(w_i) & \text{if } i \geq I_l \\ (1-p)^{N(i)}(1 - (p)^{w_i})Q(w_i)(1 - P_s(I_l - i)) & \text{else} \end{cases} \quad (16)$$

where $I_l = t - E[TO] - 2$. The factor $(1 - P_s(I_l - i))$ in the second case gives the probability that we do not have another loss before the window reaches w. It is absent in first case since $i + E[TO] + 2 \geq t$ and we do not have to consider whether the packets following the timeout period are transmitted correctly or not.

Case 4: When the retransmitted packet is lost. When the first retransmitted packet following a timeout is also lost, the retransmission timer backs off exponentially with a factor of 2 and can thus lead to very large silent periods. The duration of a sequence of n consecutive losses in lengths of $E[TO]$ is given by

$$L_n = \begin{cases} 2^n - 1 & \text{for } n \leq 6 \\ 63 + 64(n - 6) & \text{else} \end{cases} \quad (17)$$

Each of the losses following the initial loss indication occur with probability p. Also, the linear phase of the $cwnd$ following the second loss begins after $cwnd$

reaches 2. Now consider the case when the flow experiences n loss indications, $n-1$ of them being losses of retransmitted packets and that the first loss occurred after i rounds. Then, if $i > t - L_n E[TO] - 2(w - 2) - 1$ the probability that the off time for window w is greater than t is given by

$$P\{T > t\} = \begin{cases} (1-p)^{N(i)}(1-(1-p)^{w_i})Q(w_i)p^{n-1} & \text{if } i \geq I_l \\ (1-p)^{N(i)}(1-(1-p)^{w_i})Q(w_i)p^{n-1}(1-P_s(I_l - i)) & \text{else} \end{cases}$$
$$(18)$$

where $I_l = t - L_n E[TO] - 2$. The presence of $(1 - P_s(I_l - i))$ in the second case can be explained as before.

Case 5: n isolated timeouts. Let us now consider the case where there are n isolated timeouts each of length $E[TO]$. After the first loss after i rounds, the slow start phase lasts till $cwnd$ reaches $m = \lceil \frac{w_i}{2} \rceil$. The second loss occurs before $cwnd$ reaches a values of w. The expected duration between the first and the second loss indications is given by

$$D_l(w_i) = \begin{cases} E[TO] + 2 + \frac{1}{1-P_s(w-1)} \left(\sum_{u=2}^{w-1} u P_{TO}(u) \right) & \text{if } w < m \\ E[TO] + 2 + \frac{1}{1-P_s(w-1)} \left(\sum_{u=2}^{m-1} u P_{TO}(u) \right. & \text{else} \\ \left. + \sum_{u=m+1}^{w-1} (u + 2(u - m) - 0.5) P_{TO}(u) \right) \end{cases}$$
$$(19)$$

In the above expression, the second summation in the second case corresponds to the linear increase phase where we have two consecutive windows with the same size. After the initial loss indication, each of the succeeding loss indications can occur at a window between 1 and $w - 1$. For each of these, we model the average duration between two successive losses by $D_l(w)$. Also, the probability that there is another loss following first loss (before the window reaches w) leading to a timeout is given by $P_s(w, w_i)$. Correspondingly, we model the same probability for all losses after the second loss by $P_s(w, w)$. Also, after the last loss, it takes $t_{ss}(w - 1) + 2(w - \lceil \frac{w-1}{2} \rceil) - 1$ rounds for the window to reach a size of w. Since $t - D_l(w_i) - i$ rounds comprise the duration for the rest of the losses following the first loss indication, we need at least

$$n = \left\lceil \frac{t - D_l(w_i) - i}{D_l(w)} \right\rceil + 1$$
$$(20)$$

losses for the off time to exceed t. Then if $n > 1$ (the case $n = 1$ had already been considered) the probability that the off time is greater than t is given by

$$P\{T > t\} = \begin{cases} (1-p)^{N(i)}(1-(1-p)^{w_i})Q(w_i) & \text{if } i \geq I_l \\ P_s(w, w_i)(P_s(w, w))^{n-2} & \\ (1-p)^{N(i)}(1-(1-p)^{w_i})Q(w_i) & \text{else} \\ P_s(w, w_i)(P_s(w, w))^{n-2}(1 - P_s(I_l - i)) & \end{cases}$$
$$(21)$$

where $I_l = t - D_l(w_i) - (n-2)D_l(w) - E[TO] - 2$.

Case 6: Multiple consecutive losses. We now consider the cases where there are n losses which are successfully recovered using a single timeout and l

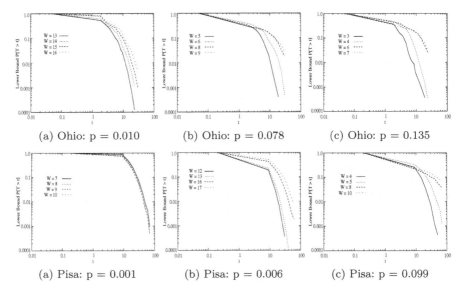

(a) Ohio: p = 0.010   (b) Ohio: p = 0.078   (c) Ohio: p = 0.135

(a) Pisa: p = 0.001   (b) Pisa: p = 0.006   (c) Pisa: p = 0.099

Fig. 5. The lower bound on the tails of the ccdf for the Ohio and Italy traces for various values of w. The time t is in seconds.

losses in which the retransmitted packet is also lost resulting in silent periods which are multiples of $E[TO]$. Let the l periods of consecutive timeouts be all due to j consecutive losses. The probability of each of these n periods is $P_s(w, w)p^{j-1}$ and the probability of the single loss indications is $P_s(w, w_i)$ and $P_s(w, w)$ for the first and the rest of the $n - 1$ losses respectively. For a given n and l we can have a sequence corresponding of $n + l$ losses in t rounds only if $t - D_l(wi) - (n + l - 1)D_l(w) - l(2^j - 2)E[TO] \leq i < t - D_l(wi) - (n + l - 2)D_l(w) - (l-1)(2^j - 2)E[TO]$. For the values of i falling in this range, the probability that the off time is greater than t is given by

$$P\{T > t\} = \begin{cases} (1 - p)^{N(i)}(1 - (1 - p)^{w_i})Q(w_i)P_s(w, w_i) \text{ if } i \geq I_l \\ (P_s(w, w))^{n+l-2}p^{l(j-1)} \\ (1 - p)^{N(i)}(1 - (1 - p)^{w_i})Q(w_i)P_s(w, w_i) \text{ else} \\ (P_s(w, w))^{n+l-2}p^{l(j-1)}(1 - P_s(I_l - i)) \end{cases} \quad (22)$$

where $I_l = t - D_l(w_i) - (n + l - 2)D_l(w) - l(2^j - 2)E[TO] - E[TO] - 2$.

3.4 Numerical Results

We now present the numerical evaluation for the lower bounds for the parameters from all the Ohio and the Pisa traces considered in Section 2. In Fig. 5 we show the ccdf for the various window sizes for both destinations. The heavy tailed nature of the tails is evident and as expected, the rate of decay reduces with increasing loss probabilities. Also, to see the impact of timeouts on the tails of

Table 1. Contribution of various losses to the ccdf. $t = 200RTTs$, $w = 10$, $W_{max} = 18$.

Type of	$p = 0.100$		$p = 0.001$	
Loss	prob	ccdf	prob	ccdf
Case 1	0.0000	0.0000	0.0304	0.0304
Case 2	0.0000	0.0000	0.0000	0.0304
Case 3	0.0000	0.0000	0.0123	0.0428
Case 4	3.99E-4	3.99E-4	3.18E-6	0.0428
Case 5	0.0116	0.0120	0.0156	0.0584
Case 6	0.1306	0.1426	3.57E-5	0.0584

the ccdf, in Table 1 we show the contribution to the tails by the various cases involving timeouts that we considered in the previous subsection. As expected, the contribution from the timeouts have a large contribution to the tails, specially higher loss probabilities. For very low loss rates, the contribution due to multiple losses is negligible and the tail is made of just 3-4 exponentials. For higher losses, the probability of multiple timeouts increases and we have a large number of exponentials with different rates the superposition of which leads to a heavy tailed distribution.

4   Conclusions and Discussions

In this paper we provided an explanation of how TCP can cause self-similarity in network traffic. Using traces of actual TCP transfers over the Internet, we showed that individual TCP flows, isolated from the aggregate flow on the link also have a self-similar nature. Our results also showed that the degree of self-similarity is dependent on the loss rates experienced by the flow and increases with increasing loss rates with the traffic no longer self-similar at very low loss rates. We then proposed a model explaining the contribution of TCP to traffic self-similarity. The model is based on considering each TCP flow as the superposition of a number of ON/OFF processes where the OFF times have a heavy tailed distribution. We verified the model empirically and then provided a firm mathematical basis to the empirical observations of heavy-tailed distributions in the OFF times by deriving a heavy tailed lower bound on the ccdf.

The loss rate experienced by a TCP flow is an important indicator of the degree of self-similarity in the network traffic. A natural construction of the extremely bursty nature of TCP traffic comes from timeouts which represent "silent" periods and separate periods of activity. Since a majority of loss indications under current Internet scenarios lead to timeouts, losses increase the burstiness and the heavy tails in the OFF times. The degree of self-similarity or H being dominated by the heaviest tail in the superposition, higher loss rates thus lead to higher values of H. In contrast when the loss rate is extremely low, TCP transmits W_{max} packets in every round and behaves like a CBR source. Thus the bursty nature is absent at low loss rates and consequently the OFF times have an exponential tail with the traffic no longer being self-similar. This explains the observations in Section 2 where flows with loss rates less than 0.001 had a Hurst parameter of approximately 0.5. Our findings show that the loss probability is a faithful indicator of the "network's effect" on TCP traffic in terms of both the effects of superposition with other flows and the degree of self-similarity of the traffic.

While TCP Reno is the most widely implemented version of TCP, other versions of TCP are currently under research, the most notable amongst them being TCP SACK. TCP SACK provides robustness against multiple packet losses in a single window and recovers them without resorting to timeouts. However, it does not completely eliminate timeouts since it requires the receipt of K (usually 3) duplicate ACKs before the retransmission mechanism kicks in. Thus timeouts are inevitable for small windows and will be present even for larger windows for correlated losses. Consequently we expect self-similarity to be present in TCP SACK traces also, though the loss rates at which $H > 0.5$ will be greater than those for TCP Reno.

References

1. Crovella, M., Bestavros, A.: Self-similarity in World Wide Web traffic: Evidence and possible causes. IEEE/ACM Trans. on Networking. **5** (1997) 835-846
2. Feldmann, A., Gilbert, A. C., Willinger, W.: Data networks as cascades: Investigating the multifractal nature of Internet WAN traffic. Computer Communications Review **28** (1998) 42-58
3. Feldmann, A., Whitt, W.: Fitting mixtures of exponentials to long-tail distributions to analyze network performance models. Proceedings of IEEE INFOCOM. (1997) 1096-1104
4. Feller, W. E.: An introduction to probability theory and its application. Wiley, New York (1971)
5. Figueiredo, D. R., Liu, B., Misra, V., Towsley, D.: On the autocorrelation structure of TCP traffic. Technical Report TR 00-55, Computer Science Department, University of Massachusetts. (2000)
6. Grossglauser, M., Bolot, J-C.: On the relevance of long-range dependence in network traffic. IEEE/ACM Trans. on Networking. **7** (1999) 629-640
7. Guo, L., Crovella, M., Matta, I.: TCP congestion control and heavy tails. Technical Report BU-CS-2000-017, Computer Science Department, Boston University. (2000)
8. Hill, B. M.: A simple general approach to inference about the tail of a distribution. Annals of Statistics. **3** (1975) 1163-1174
9. Jacobson, V.: Congestion avoidance and control. Proceedings of ACM SIGCOMM. (1988) 314-329
10. Kumar, A.: Comparative Performance Analysis of Versions of TCP in a Local Network with a Lossy Link. IEEE/ACM Trans. on Networking. **6** (1998) 485-498
11. May, M., Bonald, T., Bolot, J-C.: Analytic evaluation of RED performance. Proceedings of IEEE INFOCOM. (2000) 1415-1424
12. Padhye, J., Firoiu, V., Towsley D., Kurose, J.: Modeling TCP Reno performance: A simple model and its empirical validation. IEEE/ACM Trans. on Networking. **8** (2000) 133-145
13. Park, K., Kim, G., Crovella, M.: On the relationship between file sizes, transport protocols, and self-similar network traffic. Proceedings of International Conference on Network Protocols, (1996) 171-180

14. Paxson V., Floyd, S.: Wide area traffic: The failure of Poisson modeling. IEEE/ACM Trans. on Networking. **3** (1995) 226-244
15. Sikdar, B., Kalyanaraman, S., Vastola, K. S.: TCP Reno with random losses: Latency, throughput and sensitivity analysis. Proceedings of IEEE IPCCC, (2001) 188-195
16. Stevens, W. R.: TCP/IP illustrated. vol. 1. Addison Wesley (1994)
17. Taqqu, M. S., Willinger, W., Sherman, R.: Proof of a fundamental result in self-similar traffic modeling. Computer Communication Review. **27** (1997) 5-23
18. Taqqu, M. S., Teverovsky, V.: On estimating long-range dependence in finite and infinite variance series. In: Adler, R. J., Feldman, R. E., Taqqu, M. S., (eds.): A Practical Guide to Heavy Tails: Statistical Techniques and Applications, Birkhauser, Boston, (1998) 177-217
19. Veres, A., Boda, M.: The chaotic nature of TCP congestion control. Proceedings of IEEE INFOCOM. (2000) 1715-1723
20. Veres, A., Kenesi, Z., Molnar, S., Vattay, G.: On the Propagation of Long-Range Dependence in the Internet. Proceedings of ACM SIGCOMM. (2000) 243-254
21. Willinger, W., Taqqu, M. S., Sherman R., Wilson, D. V.: Self-similarity through high-variability: Statistical analysis of Ethernet LAN traffic at the source level. IEEE/ACM Trans. on Networking. **5** (1997) 71-86

Bandwidth Estimates
in the TCP Congestion Control Scheme

Antonio Capone and Fabio Martignon

Politecnico di Milano, Italy

Abstract. Many bandwidth estimation techniques, somehow related to the TCP world, have been proposed in the literature and adopted to solve several problems. In this paper we discuss their impact on the congestion control of TCP and we propose an algorithm which performs an explicit and effective estimate of the used bandwidth. We show by simulation that it efficiently copes with the packet clustering and ACK compression effects without leading to the biased estimate problem of existing algorithms. We present numerical results proving that TCP sources implementing the proposed scheme with an unbiased used-bandwidth estimate fairly share the bottleneck bandwidth with classical TCP Reno sources. Finally, we point out the benefits of using the proposed scheme compared to TCP Reno in networks with wireless links.

1   Introduction

The Transmission Control Protocol (TCP) is based on the assumption that the network does not provide any explicit feedback to the sources. Therefore each source must form its own estimates of the network path properties, such as round-trip time (RTT) or usable bandwidth, in order to perform efficient end-to-end congestion control.

The TCP congestion control has actually the twofold aim to prevent congestion events and achieve a fair share of bandwidth among different connections. Therefore, according to the guidelines in [1] and [2], it's worth to define the *available bandwidth* as the maximum rate at which a TCP connection, exercising correct congestion control, should ideally transmit, and the *used bandwidth* as the rate at which the source is actually sending data.

The most widely deployed TCP implementations (TCP Reno and its extensions as SACKS [3] or New Reno [4]) do not explicitly estimate the available bandwidth. Instead, the end-systems maintain two state variables to regulate the transmission rate: the congestion window (*cwnd*), which usually determines the transmission window, and the slow start threshold (*ssthresh*) that marks the *cwnd* value which discriminates between the slow-start and the congestion avoidance phases. At the beginning of the connection, as the *ssthresh* is set to a big value, the source exponentially increases the number of packets in flight (slow start) until the network drops packets, thus signaling congestion. In response to congestion TCP Reno sets the *ssthresh* to one half of the bytes in

S. Palazzo (Ed.): IWDC 2001, LNCS 2170, pp. 614–626, 2001.
© Springer-Verlag Berlin Heidelberg 2001

flight, and rapidly enters congestion avoidance phase during which the *cwnd* is linearly increased.

In [5] it has been shown that the general scheme of *additive increase and multiplicative decrease* (AIMD), on which the congestion control scheme of the TCP is based, leads to a fair share of the network bandwidth among different connections in an ideal scenario where all TCP connections take decisions in a synchronized fahion. So, ideally, the *ssthresh* gives an implicit estimate of the available bandwidth and the congestion avoidance is used to gently probe for extra bandwidth.

Unfortunately, it is well known that in real scenarios TCP Reno fails to achieve fair allocation of the bandwidth among connections sharing the same bottleneck when the connections experiment different conditions on the end-to-end path (as for example path delays). For these reasons the *ssthresh* can be considered as an implicit estimator only of the *used* rather than the *available* bandwidth.

Moreover, in TCP Reno, the implicit bandwidth estimate is strictly dependent on the congestion control events experienced by the connection. Therefore, as TCP Reno actually does an implicit estimate of the bandwidth it is using, we may ask whether it is worth performing an explicit run-time estimate of the used bandwidth and how this estimated value can be used by the congestion control scheme.

Various bandwidth estimation techniques, somehow related to the TCP world, have been proposed in the literature and adopted to solve different problems [2,6,7,8,9,10,11]. In this paper we first review these techniques pointing out their impact on the behavior of the TCP congestion control (Section 2). We then propose an algorithm which performs an explicit and effective estimate of the used bandwidth and show by simulation that it efficiently copes with the packet clustering and ACK compression effects without leading to the biased estimate problem of the algorithm proposed in [10,11] (Section 3). We show, however, that the best way to use the estimated value is to set the *ssthresh* to the byte-equivalent of the bandwidth/delay product only after congestion events as proposed in [10,11]. Moreover, we present numerical results proving that TCP sources implementing the proposed scheme with an unbiased used bandwidth estimate fairly share the bottleneck bandwidth with classical TCP Reno sources provided that the end-to-end path conditions are the same. Finally, we point out the benefits of using the explicit used bandwidth estimate compared to the implicit bandwidth estimate of TCP Reno when wireless links are on the path.

2   Estimation Techniques

In a classical IP architecture, to provide best effort service the network resources must be shared by all flows in an as fair as possible way. A centralized controller could in principle regulate the rate of all flows to ensure fairness based on the knowledge of the number of flows and the routing paths. However, a central controller is unfeasible and too far from the IP philosophy, so the network must somehow estimate the bandwidth availability in a distributed way.

In Core Stateless Fair Queing (CSFQ) scheme [6], bandwidth estimate is performed at the IP router level. The router, knowing the bandwidth B_k of its k-*th* outgoing link and the number n_k of active flows by means of packet classification, estimates by B_k/n_k the bandwidth available to each flow. Through a run-time estimate of the bandwidth actually used by each flow, the router can decide to drop packets belonging to connections using bandwidth in excess, i.e. connections sending at a rate greater than the available bandwidth B_k/n_k. This approach has been shown to solve problems of unfairness among connections having different round trip times, and can be the basis for mechanisms designed to regulate *non TCP-friendly* or *unresponsive* flows [12].

CSFQ has the great advantage of forcing flows to fairly share bandwidth even when the congestion control mechanism of the transport protocol is not accurate. However, it requires relevant modifications in the IP routers and it cannot be easily deployed over the Internet.

If only the end-systems are in charge of the rate regulation without any explicit support from the network, some kind of bandwidth estimate must be performed at the TCP level. Explicit bandwidth estimation algorithms have been proposed to be used by the TCP sources at the beginning of the connection. Their main goal is to set the first value of the *ssthresh* in order to mitigate the effect of multiple losses due to the high default value commonly used [7]. Though the *ssthresh* should be set to the *available* bandwidth, most of the proposed schemes estimate the *bottleneck* bandwidth, a quantity which can be more easily tracked by analyzing the timing structure of received acknowledgments (ACKs). The "Packet Pair" algorithm [8] is based on the assumption that if two packets are sent with closely spaced timing, the interarrival time of the ACKs strictly reflects bottleneck bandwidth. However, as shown in [1], this technique performs poorly if implemented at the sender side, mainly due to the ACK compression [13] which alters the ACK spacing. Some variants of "Packet Pair" consist in tracking "Closely Spaced ACKs" (CSAs) [2,7].

A more sophisticated bandwidth estimation scheme which runs throughout the connection has been adopted in TCP Vegas [9]. While TCP Reno relies on packet losses in order to estimate the available bandwidth of the network, TCP Vegas estimates the available bandwidth of the network based on the difference between the expected and the actual flow rate. The expected and actual rates are given by $cwnd/baseRTT$ and $cwnd/RTT$, respectively, where $baseRTT$ is the minimum RTT ever recorded by the TCP source and RTT its last value. By this mechanism, when the network is not congested, the actual flow rate is close to the expected one, while, when network is congested, the actual rate is smaller than the expected flow rate.

TCP Vegas builds over this explicit and continuous bandwidth estimate a new congestion control scheme which leads to convergence of the congestion window to an equilibrium point. It has been shown, however, that even TCP Vegas fails to obtain a fair allocation of bandwidth especially in an heterogeneous environment. Moreover, it is known that TCP Vegas is greatly penalized by the aggressive nature of TCP Reno, and so it receives very little bandwidth while Reno easily captures the rest [14]. Even the use of RED gateways [15], while bettering the situation, fails to fill the gap between Reno and Vegas. Finally, in [16] it has been pointed out that even in a homogeneous environment, TCP Vegas may fail to achieve fairness, fundamentally due to the convergence to fixed but different values of the *cwnd* parameters of competing connections.

TCP Westwood, recently proposed in [10,11], performs an estimate of the available bandwidth by measuring the returning rate of acknowledgments, and uses this estimate to set the *ssthresh* and the *cwnd* after congestion events such as the receipt of three duplicate ACKs or coarse timeout expirations. TCP Westwood uses this faster recovery mechanism to avoid the blind halving of the sending rate as in TCP Reno after packet losses. Therefore, this explicit bandwidth estimation scheme has a deep impact over the performance of TCP Westwood sources, especially in presence of random, sporadic losses typical of wireless links or with paths with high bandwidth/delay product.

The bandwidth estimation algorithm performed by TCP Westwood as reported in [11] is described by the following pseudocode:

```
if (ACK is received)
   sample_BWE[k] = (acked * pkt_size * 8)/(now - lastacktime);
   BWE[k]= beta*BWE[k-1] + (1 - beta)*(sample_BWE[k] +
                                       sample_BWE[k-1])/2;
endif
```

Here, *acked* indicates the number of segments acked by the latest ACK, *pkt_size* indicates the segment size in bytes, *now* indicates the current time, *lastacktime* indicates the time the previous ACK was received, k and k-1 indicate the current and previous value of the variables, BWE is the low-pass filtered measure of the available bandwidth, and *beta* is the pole used for the filtering (in [11] a value of $beta = 19/21$ is suggested).

The basic idea of the proposed scheme is to low-pass filter the bandwidth signal to obtain an accurate estimate of the bandwidth not affected by sporadic losses. Unfortunately, filtering directly the samples of BWE presents some drawbacks when the packet interarrivals are significantly different. The example depicted in Figure 1 shows a simple and typical situation where this scheme fails to correctly estimate the bandwidth:

Fig. 1. Packet timing structure

Here L is the packet length expressed in bits, T_p the time interval between contiguous packets, T the total observed time. The bandwidth used by the connection is $\frac{4L}{T}$, and for simplicity $T = 9 * T_p$. The algorithm above, however, estimates approximately the value $\frac{7L}{T}$, as it averages the rates. By filtering we extract the average value of the rate, which is different from the used bandwidth. To be slightly more rigorous let the random variables Y and X represent the packet length and the interarrival time respectively. The average rate is given by:

$$E[\frac{Y}{X}] = E[Y] * E[\frac{1}{X}] \tag{1}$$

being X and Y independent. This value is in general different from the used bandwidth which is given by μ_y/μ_x, where $\mu_x = E[X]$ and $\mu_y = E[Y]$. If we expand the function $1/X$ around the value μ_x, up to the third term, we obtain:

$$E[\frac{Y}{X}] \approx \mu_y * [\frac{1}{\mu_x} + \frac{\sigma_x{}^2}{\mu_x{}^3}] \tag{2}$$

where $\sigma_x{}^2 = E[(X - \mu_x)^2]$. Even if the validity of the expression (2) is limited due to the approximations, it shows that the estimate is biased and the error depends on the variance of the interarrivals.

Figure 2 shows the bandwidth estimated by one of 20 TCP Westwood connections performing the rate estimation algorithm and sharing the same 10 Mb/s bottleneck. Similar curves are observed for the other connections. The bottleneck queue was designed to hold a number of packets equal to the bandwidth/delay product. The one-way-RTT was 50 ms, the test lasted 600 simulated seconds to simulate an FTP session, and *beta* was set to 0.995. These results as all the others presented in this paper were obtained using the Network Simulator, 'ns' ver.2 [17]. To provide a comparison, the dotted line represents the bandwidth estimated by a TCP Reno source running the DFT algorithm we propose and describe in detail in the next section. Since the fair-share value is 500 kb/sec, while TCP Westwood algorithm estimates more than 8 Mb/s we conclude that variance of packet interarrivals is quite large.

The interarrivals would be almost regular if packets belonging to different connections could alternate on the channel. On the contrary, it has been shown that TCP transmissions tend to be clustered so that on a channel we usually observe many consecutive packets of the same connection [13]. Note that the bias on the bandwidth estimate does not depend on the value *beta* chosen for the pole of the IIR filter. It's easy to understand that any fixed value of the pole leads to the same problem.

Filtering directly the rate measured considering the ACK arrival times also exposes the algorithm to the phenomenon of the ACK compression. This happens when the time spacing between returning ACKs is altered due to congestion of the routers on the return path. As one or more ACKs spend some time in the queue of the congested router in the reverse path, subsequent ACKs may reach each other and their original spacing is lost. It has also been shown that ACK compression is quite relevant for real networks operation [18], and therefore it

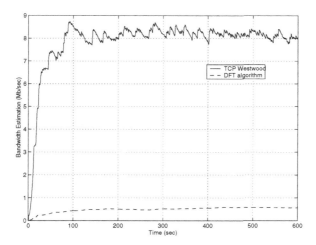

Fig. 2. Bandwidth estimated by TCP Westwood

cannot be neglected. To evaluate the impact of the ACK compression on the algorithm we have considered a scenario with two connections sharing the same 10 Mb/s bottleneck link, but transmitting data in opposite directions, as described in [13]. The end-to-end propagation delay was 100 ms, and the bottleneck queue could contain a number of packets equal to the bandwidth/delay product. The two routers at each end of the bottleneck are therefore charged with both packets from one connection and ACKs of the other, thus leading to the situation of ACK compression described before. The results show that the impact on the estimate of TCP Westwood is dramatic since the estimate is about 25 times higher than that shown in Figure 2.

Finally, we point out that an algorithm implemented in the TCP source according to the TCP Westwood approach can estimate the *used bandwidth* and not the *available bandwidth*. Therefore, the benefits of the new scheme are not due to the estimate of the available bandwidth which cannot be estimated end-to-end, but on the possibility to explicitly estimate the used bandwidth taking into account the short-medium term history of packet arrivals. Moreover, the bandwidth which the algorithm tries to explicitly estimate is the same implicitly considered by TCP Reno and reflected by the *ssthresh* in steady-state conditions. So, if the estimate is accurate enough we expect that the bandwidth used by TCP Reno and by a TCP exploiting a bandwidth explicit estimate are almost the same. This can provide a fair behavior in homogeneous (all sources using the algorithm) and heterogeneous scenarios (with also classical TCP Reno sources).

3   Double Filtering Technique

In this section we present a new technique, the Double Filtering Technique (DFT), which using the basic idea of TCP Westwood succeedes to obtain correct estimates of the bandwidth used by the TCP source.

Fig. 3. Packet timing structure

To explain the rationale of DFT let us refer to the example in Figure 3 where transmissions occurring in a period T are considered. Let n be the number of packets belonging to a connection and $L_1, L_2...L_n$ the lengths, in bits, of these packets. The average bandwidth used by the connection is simply given by $\frac{1}{T} \sum_{i=1}^{n} L_i$. If we define $\overline{L} = \frac{1}{n} \sum_{i=1}^{n} L_i$, we can express the bandwidth (Bw) occupied by the connection as:

$$Bw = \frac{n\overline{L}}{T} = \frac{\overline{L}}{\frac{T}{n}} \tag{3}$$

The basic idea is to perform a run-time sender-side estimate of the average packet length, $\overline{L}$, and the average interarrival, $\frac{T}{n}$, separately. Following the TCP Westwood approach this can be done by measuring and low-pass filtering the length of acked packets and the intervals between ACKs' arrivals. However, since we want to estimate the used bandwidth we can also low-pass filter directly the packets' length and the intervals between sending times.

Note that sending time intervals can be very very short when groups of packets are generated by TCP sources. However, this is not a problem for DFT since the estimate is performed directly on the interarrival samples. Different would be the case for algorithms that filter the bandwidth samples, such as TCP Westwood, since these samples are close to infinity.

The pseudocodes of the two bandwidth estimation schemes are the following:

1)Processing the stream of *sent packets*:

```
if (Packet is sent)
  sample_length[k] = (packet_size * 8);
  sample_interval[k] = now - last_sending_time;
  Average_packet_length[k] = alpha * Average_packet_length[k-1] +
                            (1-alpha)*sample_length[k];
  Average_interval[k] = alpha * Average_interval[k-1] +
                            (1-alpha )* sample_interval[k];
  Bwe[k] = Average_packet_length[k] / Average_interval[k]
endif
```

where *packet_size* indicates the segment size in bytes, *now* indicates the current time, *last_sending_time* the time the previous packet was sent, *k* and *k-1* indicate the current and previous values of the variables. *Average_packet_length* and *average_interval* are the low-pass filtered measures of the packet length and the interval between sending times. *Alpha* is the pole of the two low-pass filters. *Bwe* is the measure of the available bandwidth.

2)Processing the stream of *received ACKs*:

```
if (Packet is received)
  sample_length[k] = (acked * packet_size * 8);
  sample_interval[k] = now - last_ack_time;
  Average_packet_length[k] = alpha * Average_packet_length[k-1] +
                             (1-alpha)*sample_length[k];
  Average_interval[k] = alpha * Average_interval[k-1] +
                        (1-alpha )* sample_interval[k];
  Bwe[k] = Average_packet_length[k] / Average_interval[k]
endif
```

where the quantities are the same as before. Here, *acked* indicates the number of segments acked by the latest ACK. In order to compute this value, the algorithm shown in [11] must be used.

If we consider the minimum RTT measured by the TCP source (RTT_{min}) as a good estimator of the end-to-end propagation delay, then we can set:

$$Ssthresh = Bwe * RTT_{min} \qquad (4)$$

The *ssthresh* is set to the value of equation (4) only after three duplicate ACK's, or after a coarse-grained timeout expiration, following the guidelines of [11].

Simulation results show that DFT is not biased, and obtains bandwidth estimates which oscillate around the fair-share value when all TCP sources experience almost the same path conditions. In order to smooth these oscillations and ensure an estimate closer to the right value, we propose to further filter the value of *Bwe* as follows [6]:

$$Bwe[k] = (1 - e^{\frac{-T[k]}{T_0}}) * \frac{Average\_packet\_length[k]}{Average\_interval[k]} + e^{\frac{-T[k]}{T_0}} * Bwe[k-1] \quad (5)$$

where $T[k]$ is the instantaneous time interval between two estimates and T_0 is a time constant we set equal to 1 second in our simulations. By binding the value of the pole to $T[k]$, we perform an adaptive filtering which exploits the oscillations of the signal *Bwe* in order to quickly follow variations in the available bandwidth.

Figures 4a and 4b show the behavior of DFT without and with the filtering performed by Equation (5), respectively. For both the figures the scenario consists of a single TCP connection running over a 10 Mb/s link. In the interval between 200 and 300 seconds, an UDP flow, having the same priority as TCP, transmits at a rate of 4 Mb/s. Then, in the interval between 300 and 400 seconds,

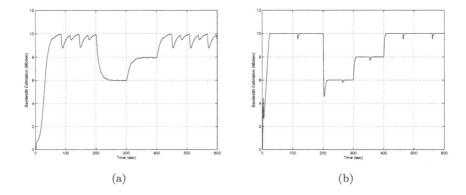

(a)   (b)

Fig. 4. DFT bandwidth estimate (a) without adaptive filtering (b) with adaptive filtering

another UDP flow starts transmitting at 2 Mb/s. The bottleneck queue, managed with a drop-tail policy, was designed to contain a number of packets equal to the bandwidth/delay product, and the simulation lasted 600 simulated seconds. In Figure 4a, the oscillations are evident, even if the estimate is unbiased. Moreover, the algorithm adapts slowly to changes in the bandwidth available to the connections. In Figure 4b, instead, the oscillations have been smoothed, and the estimate follows more quickly bandwidth variations in the underlying network.

Following the approach in [11] we set the *ssthresh* to the estimated value only after congestion events. This choice is supported by the worst performance observed with more frequent updating of the *ssthresh* value as shown in Figure 5. The scenario and the parameters considered are the same as in Figure 4, but the *ssthresh* is continuously updated. We observe that the estimate is not accurate mainly because the continuous updating of the *ssthresh* to the estimated used bandwidth value forces the source in the congestion avoidance phase and prevents to follow available bandwidth variations. Similar results have been obtained with a periodic updating with period equal to 0.5 s.

Figure 6 compares the performance of DFT algorithm (the version filtering the stream of *sent* packets), and the TCP Reno, referring to a simulation scenario that considers 10 connections sharing a single bottleneck link of 10 Mb/s with an end-to-end delay of 100 ms. The buffer contains a number of packets equal to the bandwidth/delay product, and FIFO queueing management is adopted, to test DFT even in absence of a somewhat fair queueing. Several simulations have been run and the results have been averaged in order to eliminate phase effects [19]. We have numbered the connections from 1 to 10: the first 5 used DFT algorithm, the other 5 TCP Reno. We observe that an almost fair division of the link has been obtained and both algorithms use almost the same bandwidth.

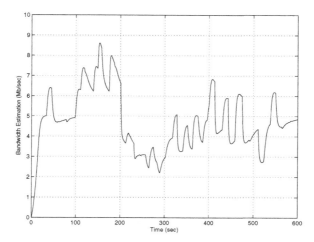

Fig. 5. Bandwidth estimate with continuous *ssthresh* updates

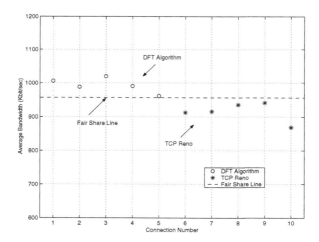

Fig. 6. DFT fairness towards TCP Reno in a 10 Mb/s bottleneck (*alpha* = 0.99)

We have run also simulations over different scenarios covering link bandwidths ranging from few kb/s to 150 Mb/s, varying the number of competing connections and using also a more complex topology with multiple congested gateways. The conclusions obtained are the same: DFT obtains a no worse level of fairness than TCP Reno. Simulation results also show that the strength of DFT lies in its scalability: as more connections share the bottleneck link, as the estimate variance reduces. The presence of constant rate flows, such as UDP flows for IP telephony or video conference, makes DFT perform better as it reduces the dimension of packet clusters.

So far we have proved the accuracy of the DFT algorithm and shown that TCP sources using this algorithm are fair to other sources. To complete the performance evaluation we need to verify the ability to achieve high throughput in presence of links affected by sporadic losses as achieved by TCP Westwood.

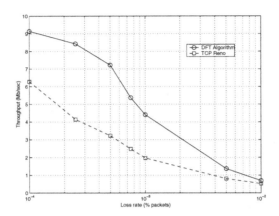

Fig. 7. DFT and RENO throughput vs link error rate

To this purpose, in Figure 7 we compare the throughput achieved by a connection running the DFT algorithm to that of a TCP Reno connection over a link with random errors. The link has a capacity of 10Mb/s, and the FIFO queue can contain a number of packets equal to the bandwidth/delay product. The one-way RTT is 50 ms, and the link drops packets according to a Poisson process with average ranging from 0.01% to 1%. We observe that DFT can sustain higher throughput than TCP Reno at all drop rates considered. This is due to the filtering process which keeps in account also the past history of the bandwidth estimates avoiding to confuse network congestion signals due to queue drops with losses due to link errors.

4   Conclusions

In this paper we proposed the DFT algorithm which performs an explicit and effective run-time estimate of the used bandwidth of a TCP source. It is based on separate filtering of both the intervals between sending times of TCP packets and the packets' lengths.

Following the approach of TCP Westwood, we used the estimate to set the *ssthresh* after congestion events. We proved by simulation that the accuracy of the estimation algorithm allows the proposed scheme to be fair with TCP Reno connections sharing the same bottleneck channel. As a result it is suitable to be gracefully adopted in the IP world with no coexistence problems. In addition the new scheme, differently from TCP Reno, is effective to cope with channel random errors as it occurs in wireless environments.

Acknowledgements

The authors wish to thank Professor Luigi Fratta for his help and suggestions.

References

1. V. Paxson. End-to-End Internet Packet Dynamics. *IEEE/ACM Transactions on Networking*, 7(3):272–292, 1997.
2. M.Allman and V.Paxson. On Estimating End-to-End Network Path Properties. In *Proceedings of ACM SIGCOMM'99*, 1999.
3. S.Floyd, M.Mathis, J.Mahdavi, and A.Romanow. TCP Selective Acknowledgement Option. *RFC 2018*, April 1996.
4. S.Floyd and T.R.Henderson. The NewReno Modifications to TCP's Fast Recovery Algorithm. *IETF RFC 2582*, 26(4), April 1999.
5. D.Chiu and R.Jain. Analysis of the Increase and Decrease Algorithms for Congestion Avoidance In Computer Networks. *Computer Networks and ISDN Systems*, 17:1–14, 1989.
6. I.Stoica, S.Shenker, and H.Zhang. Core-Stateless Fair Queueing: Achieving Approximately Fair Bandwidth Allocations in High Speed Networks. In *Proceedings of ACM SIGCOMM'98*, Vancouver, Canada, September 1998.
7. J.C.Hoe. Improving the Start-up Behavior of a Congestion Control Scheme for TCP. *ACM SIGCOMM Computer Communications Review*, 26(4):270–280, October 1996.
8. S.Keshav. A control-theoretic approach to flow control. In *Proceedings of ACM SIGCOMM'91*, pages 3–15, September 1991.
9. L.S. Brakmo and L.L. Peterson. TCP Vegas: End-to-End Congestion Avoidance on a Global Internet. *IEEE Journal on Selected Areas in Communications*, 13(8):1465–1480, October 1995.
10. C. Casetti, M. Gerla, S.S. Lee, S. Mascolo, and M. Sanadidi. TCP with Faster Recovery. In *Proceedings of Milcom 2000*.
11. S. Mascolo, C. Casetti, M. Gerla, S.S. Lee, and M. Sanadidi. TCP Westwood : congestion control with faster recovery. Technical report, UCLA CS Technical Report #200017, 2000.
12. Sally Floyd and Kevin Fall. Promoting the Use of End-to-End Congestion Control in the Internet. *IEEE/ACM Transactions on Networking*, 7(4):458–472, Aug.1999.
13. L.Zhang, S.Shenker, and D.Clark. Observations on the Dynamics of a Congestion Control Algorithm : The Effects of Two-Way Traffic. In *Proceedings of SIGCOMM'91 Symposium on Communications Architectures and Protocols*, pages pages 133–147, Zurich,September,1991.
14. J.Mo, V.Anantharam, R.J.La, and J.Walrand. Analysis and Comparison of TCP Reno and Vegas. In *Proceedings of ACM GLOBECOMM'99*, 1999.
15. S.Floyd and V.Jacobson. Random early detection gateways for congestion avoidance. *IEEE/ACM Transactions on Networking*, 1(4):397–413, August 1993.
16. Go Hasegawa, M.Murata, and H.Miyahara. Fairness and Stability of Congestion Control Mechanisms of TCP. In *Proceedings of INFOCOM'99*, 1999.

17. ns-2 network simulator (ver.2).LBL. URL: http://www.isi.edu/nsnam.
18. J.C.Mogul. Observing TCP Dynamics in Real Networks. In *Proceedings of ACM SIGCOMM'92 Symposium on Communications Architectures and Protocols*, pages 305–317.
19. S.Floyd and V.Jacobson. On Traffic Phase Effects in Packet-Switched Gateways. *Internetworking:Research and Experience*, 3(3):115–156, September 1992.

Implementation and Performance Evaluation of a Differentiated Services Test-Bed for Voice over IP Support

Romano Fantacci[1], Gianluca Vannuccini[1],
Alessandro Mannelli[2], and Luca Pancani[2]

[1] Electronics and Telecommunications Department,
via di S,Marta, 3, 50139 Florence, Italy
[2] Alcatel Network Services Division,
via Prov. Lucchese, 33, Sesto F.no, Florence, Italy

Abstract. This paper deals with a Differentiated Services test-bed, developed in Alcatel Labs in order to evaluate and compare the performance offered by three different queuing policies, namely FIFO, Priority Queuing and Custom Queuing, when supporting Voice over IP traffic. A Premium class has been assigned to voice traffic, while Best Effort class traffics have been considered as disturbing traffic and they have been generated by means of suitable TCP and UDP software tools. Performance comparisons have been developed for the considered queuing schemes. In particular, Priority Queuing has resulted to be the best solution for Premium class quality in the developed test-bed. For this scheme a suitable analytical approach has been also proposed and validated by comparing analytical predictions with experimental results.

1 Introduction

The present Internet is widely known as a best-effort network, meaning that it is generally not possible to guarantee a pre-defined quality of service over IP protocols. However, a general trend in communication networks is the attempt to provide packet services with the typical guarantees of a switched network. In this direction, IP Telephony is one of the most promising real-time technologies over IP networks [1] [2]. In order to support voice services on the Internet, many Quality of Service techniques are being extensively studied [3], and all of them are trying to differentiate the treatment that flows receive from the network depending on their priority level. This paper deals with the Differentiated Services technique, where flows are treated with different policies depending on the Differentiated Services Code Point (DSCP) written in their IPv4 Type of Service (TOS) field [4]. We have implemented a Differentiated Services test-bed in Alcatel Labs, Florence, Italy, in order to evaluate the performance of resource allocation techniques under consideration. We have developed several tests where a telephone call was established between two users, with traffic profile of the so-called Premium class. The quality provided to the Premium class, in terms of

S. Palazzo (Ed.): IWDC 2001, LNCS 2170, pp. 627–641, 2001.
© Springer-Verlag Berlin Heidelberg 2001

perceived intelligibility, has been verified through opinion score tests. Best Effort traffic has been generated in order to test the Quality of Service degradation under an increased network congestion level. Several queuing techniques, like FIFO, Priority Queuing (PQ), and Custom Queuing (CQ), have been implemented in the network nodes, and their effects on the overall system performance have been evaluated. This paper is organized as follows: a brief review of the available QoS management strategies is presented in Sec. 2, where some main issues concerning the overall characteristics of the chosen queuing strategies are discussed. In Sec. 3, the implemented test-bed is presented in detail, while describing the developed experimental tests. Then in Sec. 4 we have discussed an analytical model of the router queue when using PQ, which, as the preliminary evaluation given in Sec.2 has outlined, seems to be the best solution in the scenario of this work, in particular in case of low voice traffic amount. We have used the typical $M/G/1$ queues analysis methods with priority differentiation [5] , and an average packet delay time has been evaluated through this method for a generic IP packet crossing the network while PQ is used on the router queues. In Sec. 5, a more detailed performance comparison between the experimental tests, when using FIFO, CQ and PQ, is done, and some general observations are given concerning their behavior in the test-bed. Then, the obtained analytical predictions have been verified by comparison with experimental results. Finally, in Sec.6 our conclusions have been discussed.

2   QoS Management Schemes for IP Networks

In order to support real-time services over IP packet networks, and to let delay-sensitive and best-effort traffic coexist in the same network, some main performance parameters must be kept under control through suitable management strategies. According to typical Quality of Service studies, three parameters seem to be critical for real-time traffic support over packet networks, namely: the packet loss [6], the one-way delay time [7] and the instantaneous voice packets delay variation (IPDV) [8]. Typical values for these parameters depend on the considered application, and on the requested service level agreement. However, for medium-size IP networks, an acceptable QoS for Voice over IP services can be accomplished by having almost 10%, 150ms and 40ms as maximum values for, respectively, the three QoS parameters described above.

Towards this goal, many bandwidth reservation strategies have been studied in the last decade. The two main approaches in the IP QoS field are the Integrated Services (IntServ) and the Differentiated Services (DiffServ) architecture [4]. The IntServ model is commonly referred to as a per-flow architecture, where in each network node a particular bandwidth amount is reserved to each traffic flow between two end-points. In order to exchange bandwidth reservation information between routers, a signaling protocol is requested, which is the Resource Reservation Protocol (RSVP) [9]. When the number of nodes, and users, is growing up it is indeed difficult to implement this model on the limited memory of network routers, where for each flow a specific bandwidth reservation policy has to be recorded and periodically refreshed, and therefore the amount of per-flow information increases proportionally with the number of flows to be managed by the IntServ network. Then, the IntServ model is said to be generally non-scalable to large networks, and a possible solution has been proposed in the DiffServ architecture. In DiffServ a per-behavior treatment is adopted in each router, meaning that a particular behavior is associated to each traffic type, and, more specifically, to each QoS traffic class, by marking its IPv4 TOS field with a proper DSCP. Each router will then read the TOS field of each packet, it will classify it and then it will treat the packet according to the classified behavior, reserving a corresponding bandwidth amount to that packet. This can for example be achieved by providing several queues in the router instead of the single FIFO queue. If we associate at each queue a different priority, we can obtain a final overall differentiated treatment for each packet. Thus, it will be sufficient to record in each router a simple information on the behavior associated with each DSCP, and even though the network size grows, the same amount of memory will be requested to the network routers. This approach seems to be much more scalable than the IntServ model, even if a less dynamic bandwidth allocation is achievable with DiffServ than with IntServ. Many other further studies have been developed in the last five years, intended to solve the problems of each approach, but since the main concept of per-behavior approach seems to be the most promising among the proposed solutions, we have chosen to implement a DiffServ-like network, and to reserve bandwidth resources according to the TOS field of the transmitted IP packets. In each network node, after being classified, either via a DiffServ or an Intserv method, the IP packet meets generally a node internal queue, where a proper treatment must be reserved to it according to the classification results. Towards this goal, we have considered two queuing approaches that are currently implemented in many commercially available routers, namely the Priority Queuing - PQ - and the Custom Queuing - CQ - methods [10]. In these queuing strategies, an input flow is classified via, for example, the TOS field, and then a certain number of virtual queues is created in the router, each one with the proper priority, and the output scheduler will read packets from a queue or from another at a rate depending on their priority level. For PQ, two queues are created, one with higher and the other one with lower priority, and delay-sensitive traffic is obviously addressed to the higher priority queue, thus receiving a much lower delay in the router crossing. In CQ, many queues can be created, each one with a specific priority level, in order to let many services have their fair bandwidth share, thus avoiding congestion situations. A router internal queue is assigned to each packet class, and the output scheduler reads packets in a round-robin mode. To each packet class, and therefore to each queue, a certain fixed bandwidth amount is reserved, and the output scheduler, by reading the corresponding bytes number from each queue, will reproduce in the output port the decided bandwidth reservation scheme. It is evident that PQ is better performing than CQ when high QoS performance are required for the Premium class Voice over IP traffic flow. In fact, as it will be verified in Sec.5, with PQ it is possible to achieve much lower delay variation spreading when network congestion level is high, since in CQ a certain bandwidth amount is delivered to Best Effort traffic in any case. PQ can therefore be a good choice when a high priority Premium class traffic must be served with excellent QoS parameters, while CQ seems to be more conserving and less flexible when also Best Effort type traffics are to be maintained within acceptable QoS features. Since the aim of this work was to evaluate suitable QoS algorithms to provide VoIP traffic of modest entity with acceptable quality of service, PQ will be proven to be the best choice for our scopes.

In the sequel, the test-bed used for our experiments will be described in detail, and measurement conditions will be highlighted.

3   Test-Bed Description

In order to study and evaluate the performance of a VoIP traffic over a DiffServ-like network, we have developed a test-bed in the Alcatel Florence laboratories, and we have used standard devices in order to emulate the traffic behavior of a common backbone IP network. As shown in Fig. 1, a backbone network has been emulated through three Cisco routers, namely two 2600 with VoIP ports acting as access routers, and one 4500 acting as the backbone router. Ethernet interfaces have been used to connect the routers. Two groups of PCs and workstations have been linked through switches and hubs to the access routers, thus emulating the connection between two larger LANs via an IP network. Two analog phones have been plugged into the two corresponding Cisco 2600 routers, where users were able to communicate via a Voice over IP service. The communication between the two analog phones has been chosen to be the Premium class service, to which the suitable priority algorithm will deliver the required bandwidth amount. According to the DiffServ architecture, a specific DSCP will be written in the TOS field of the IPv4 packets generated by the two VoIP ports, thus allowing the three routers to reserve a particular treatment to packets having this DSCP. The PCs and workstations have been used to generate disturbing traffic, namely TCP traffic via the TTCP software [11], and UDP traffic via the MGEN software [12]. Disturbing traffic, either of a TCP type or of an UDP type, is considered as Best Effort class traffic, and then its packets will be properly classified by the three routers through the corresponding DSCP codes. The general step sequence of our experiments is described hereafter. First, while the telephone call is established, we have gradually increased the congestion level of the network, and we have measured the corresponding QoS parameters degradation in the end-to-end connection by using the default FIFO configuration in the three routers queues. Then, we have implemented priority managing algorithms in the routers queues, and we have measured the overall system performance in terms of the chosen QoS parameters, while making voice quality intelligibility tests. For CQ, we have chosen two modes of operation, by using two different byte counts values for the two CQ queues [10]. In particular, in CQ-I we have reserved a 10% bandwidth amount to the Premium class telephone traffic, which for our purposes was far enough for a single-connection voice call, and the re-

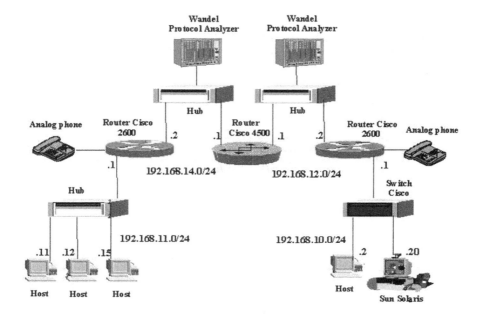

Fig. 1. Test-bed in Alcatel Labs

maining bandwidth to Best Effort traffic. Then, in CQ-II we have reserved a 1% bandwidth to Premium traffic, leaving the remaining bandwidth to Best Effort traffic. Since our goal is to provide a low amount of VoIP traffic with an acceptable QoS, while letting Best Effort class traffic be delivered without performance degradation, the more flexible features offered by PQ has proven to be the best solution in comparison with the fixed bandwidth allocation offered by CQ. For this reason, since our network supports a low amount of voice traffic, PQ seems to be the best solution, in the next section we have proceeded in the study of an analytical model of a PQ router, which will be used as a theoretical comparison for the experimental tests.

4   Analytical Model

A functional model for the internal queuing system of a router is shown in Fig. 2, where its I/O interfaces are modeled as input and output queues, with an internal queue representing the CPU processing time. In order to make a realistic modeling, we have considered both the traffic directions that the router faces. We consider the voice Premium traffic coming from the telephone #1 (input line 1), the Best Effort UDP and TCP traffic coming from the LAN #1 (input line 3), and the corresponding voice and Best Effort traffic coming from the other trunk

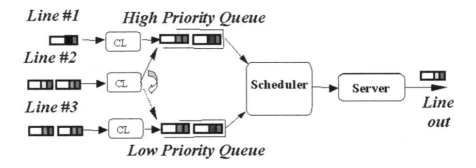

Fig. 2. Router model for analytical performance evaluation

of the network, which represents in this case the backbone network. The output line is therefore the output port of the router facing the backbone. According to Fig. 2, at each input line, a classifier is modeled, whose aim is to read TOS field in IP packets, and to re-route them in the correct queue. Two internal queues are considered, one for the high priority Premium traffic, and the other for Best Effort traffic. We can model the whole system in Fig. 2 as an only M/G/1 type queue with differentiated serving time [5], with two priority classes managed in a non-preemptive way. Since our goal is to calculate the average delay time spent by a general IP packet in the router queue, we can define the average total time spent in the router from a general IP packet as:

$$T_{router} = T_{classifier} + T_{queue} + T_{scheduler} + T_{CPU} + T_{codec+pack} \qquad (1)$$

where $T_{classifier}$ and $T_{scheduler}$ can be neglected, since they are several orders of magnitude lower than the considered delay values, $T_{codec+pack}$, due to codec and packetization processes, can be assumed as almost $6ms$ for G.711 codecs [13]. T_{CPU} is the required service time for a packet to be processed in the output queue, and therefore it corresponds to the CPU processing time for each packet. T_{queue} is the average delay time spent by each packet in the priority queuing system, and it will be addressed in the following as W. The average service time, for respectively Premium and Best Effort packets, is given by:

$$X_i = \frac{L_i}{L_{rate}}, i = 1, 2 \qquad (2)$$

where L_i is the average bit length of a Premium or Best Effort packet, and L_{rate} is the average bit rate of the input line expressed in bit/s. In the following, we will denote with the suffix "1" the Premium Class packets, and with the suffix "2" packets belonging to Best Effort traffic. In order to develop the analytical theory for our system, we must first investigate under which assumptions the considered router is a stable queuing system. To achieve this goal, we can first evaluate the total arrival rate for Premium and Best Effort type traffic, given in packets/s:

$$\lambda_{tot} = \lambda_1 + \lambda_2 \tag{3}$$

We then can calculate the arrival probabilities for Premium and Best Effort packets, given by:

$$P_1 = \frac{\lambda_1}{\lambda_{tot}} \tag{4}$$

and

$$P_2 = \frac{\lambda_2}{\lambda_{tot}}. \tag{5}$$

The average service time for both Premium and Best Effort packets is then equal to:

$$\bar{X} = P_1 X_1 + P_2 X_2. \tag{6}$$

The average total traffic utilization factor for the considered queuing system is therefore:

$$\rho = X\lambda_{tot} \tag{7}$$

In order to guarantee system stability, it must be $\rho < 1$, which happens, for the considered architecture, when the network congestion level remains below the 75%. Once we have verified the limit for the system stability, let us proceed in the mathematical evaluation of the average delay time spent by IP packets in the router queues. If we call W_1 the queuing time experienced by a premium class packet (high priority), and W_2 the queuing time of a Best Effort packet (low priority), we have that the average waiting time spent in the high priority queue is:

$$W_1 = R_m + N_1 X_1 \tag{8}$$

where R_m is the residual average time required to finish the processing of the packet that was already served in the queue (the system is a non preemptive queuing system), X_1 is the average service time of Premium class packets, given in (2), $N_1 = \lambda_1 X_1$ is the average number of Premium class requests in the queue, with λ_1 being the average arrival rate for Premium class packets. Equation (8) can be then rewritten as:

$$W_1 = \frac{R_m}{(1 - \lambda_1 X_1)} = \frac{R_m}{1 - \rho_1} \tag{9}$$

where ρ_1 is the router utilization factor due to Premium class packets. We can then proceed to calculate W_2 as:

$$W_2 = R_m + \rho_1 W_1 + \rho_2 W_2 + \rho_1 W_2 \tag{10}$$

where $\rho_1 W_1$ is the average time spent to serve the Premium class requests stored in the queue, $\rho_2 W_2$ is the average time required to serve the Best Effort class requests, and $\rho_1 W_2$ is the average time needed to satisfy the Premium class requests which arrived while the Best Effort packet was being served. Following the same method used in (9) we have:

$$W_2 = \frac{R_m}{[(1 - \rho_1)(1 - \rho_1 - \rho_2)]} \qquad (11)$$

In order to calculate W_1 and W_2 from (9) and (11), we must explicit R_m. Following a well-known approach [14], we obtain:

$$R_m = \frac{1}{2} \sum_{i=1}^{2} \lambda_i X_i^2 \qquad (12)$$

which allows us to derive a final numerical value for W_1 and W_2. Recalling eq.(1), we obtain a total average delay time both for Premium class packets and Best Effort class packets in the router. We then have:

$$T_{router\,Premium} = W_1 + X_1 + T_{codec+pack} = \frac{1}{2}\frac{\sum_{i=1}^{2}\lambda_i X_i^2}{(1 - \rho_1)} + X_1 + T_{codec+pack} \quad (13)$$

and

$$T_{router\,BestEffort} = W_2 + X_2 + T_{codec+pack} = \frac{1}{2}\frac{\sum_{i=1}^{2}\lambda_i X_i^2}{(1 - \rho_1)(1 - \rho_1 - \rho_2)} + X_2 + T_{codec+pack}$$
$$(14)$$

which are all known quantities. We then have evaluated the average delay time spent by Premium class and Best Effort class IP packets in each router. To achieve the end-to-end delay, according to the implemented network shown in Fig. 1, we need to sum up the network delays encountered from voice packets in the network and, in a first approximation, since the routers are three, and each of them is associated with similar queuing techniques, we can assume that the total end-to-end delay experienced by an IP packet is given by three times the values given in (13) and (14), respectively when the packets belong to Premium and Best Effort traffic classes. This is only a first rough approximation, but we will verify its practical applicability, and the single-router delay above-described evaluation, in the next section, where the experimental measurements will be compared to analytical predictions.

5   Experimental Results and Performance Evaluation

In order to study and evaluate the performance of the considered network, we have implemented many tests, by using the three above mentioned queuing algorithms, and by then observing the overall system performance in the various settings. We have increased the network congestion level by using MGEN and TTCP, respectively to generate UDP and TCP disturbing traffic. We have then started a telephone connection between the two analog phones, and we have observed the main QoS parameters values, with FIFO queuing management, with

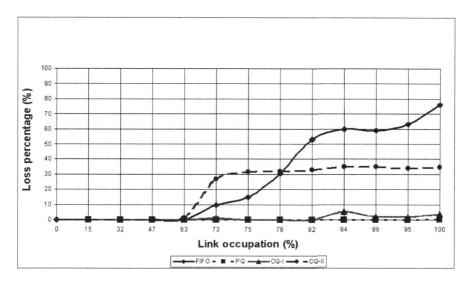

Fig. 3. Packet Loss for Premium Class traffic

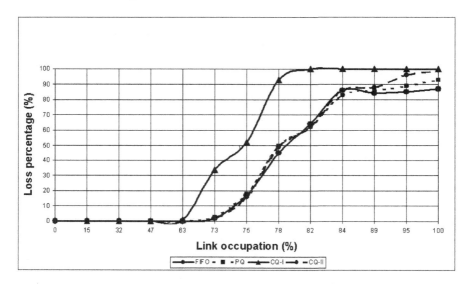

Fig. 4. Packet Loss for Best Effort Class traffic

PQ, and with the two operational modes CQ-I and CQ-II, described in Sec. 3. For each of the following measurements, a G.711 codec is used for voice coding and decoding at the Cisco 2600 routers. Fig. 3 shows the packet loss experienced by Premium class packets in the network crossing, versus the network congestion level. As it is shown, the use of the described queuing strategies fixes the voice packets loss on merely noticeable percent values, even though for FIFO and CQ-II strategies we observe higher packet losses. Fig. 4 shows the packet loss for Best Effort class packets. Here, the packet loss obviously increases with the network congestion level, and it is evident that CQ-I shows the worst performance; this difference with the other queuing policies is due to the large amount of resources that CQ-I reserves to voice packets, that obviously penalizes Best Effort class traffic flows. We have then focused on the total average end-to-end delay that Premium and Best Effort packets received in the network, due to the router internal queues. Fig. 5 shows the delay behavior versus network congestion level, depending on the implemented queuing strategy, for Premium class traffic. It is evident how, by using any of the QoS management methods, instead of FIFO, the average end-to-end delay remains under 50ms, a very acceptable value for such a network, and in general for VoIP performance standards, as described in Sec.2. Fig. 6 then shows the overall end-to-end delay received from Best Effort class traffic, which reaches unacceptable values when the network congestion level goes beyond the 75% value, which is, as described in Sec. 4, the corresponding maximum throughput level for each router. It can be noted that, in this case, the different queuing policies have given similar results, and this is due to the fact that, although CQ-I causes high packet losses as shown in Fig. 4, when Best Effort packets are accepted into the router they are given the same delay in the lower priority internal queue, for every queuing policy. We then compare the overall average end-to-end delay calculated with the analytical model described in Sec. 4 with the experimentally evaluated delay in the testbed. Fig. 7 shows the performance comparison for Premium class traffic, while Fig. 8 deals with Best Effort class traffic. It is evident how our analytical model is in both cases well suited to the experimental measurements and, due to its worst-case nature, it can be considered as a reference in feasibility analysis. As for the packet delay variation (IPDV) [8], which is the third main QoS parameter that must be controlled for VoIP services, as described in Sec.2, due to analytical complexity in deriving this term in a closed form, we have resorted to report here only experimental measurements, and in Fig. 9, Fig. 10 and Fig. 11 we show the measured performance for the delay variation received from Premium class voice packets using, respectively, FIFO, PQ and CQ-I queuing strategies. In such figures, it can be noted that the reduction of the standard deviation is an evident proof of the better performance of CQ-I, and especially of PQ, in the considered network. CQ-II performance was not reported here due to its completely unsatisfactory performance with regards to IPDV. It is evident that PQ outperforms both FIFO and CQ-I alternatives, due to the fact that, by giving always priority to Premium class traffic, it sensibly reduces the overall delay variation received by voice packets at the receiving router buffer. Thus, PQ seems to be the best solution also to reduce and control this third critical QoS parameter, thus allowing VoIP services to receive a good overall performance. Further extensions to the proposed architecture of Fig. 1, where more than one voice call flows

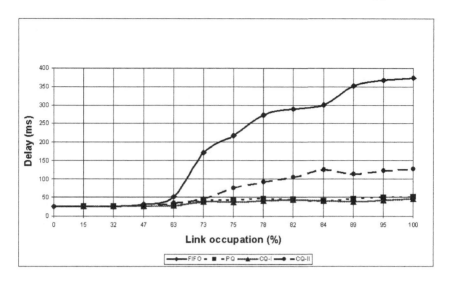

Fig. 5. Average end-to-end delay time for Premium Class traffic

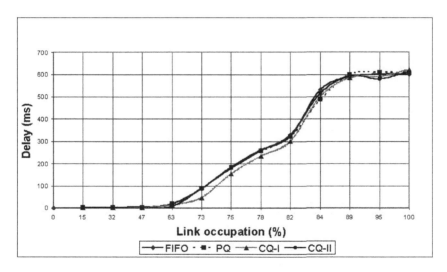

Fig. 6. Average end-to-end delay time for Best Effort Class traffic

are considered as Premium class traffic, are presently under development in the Alcatel labs in Florence.

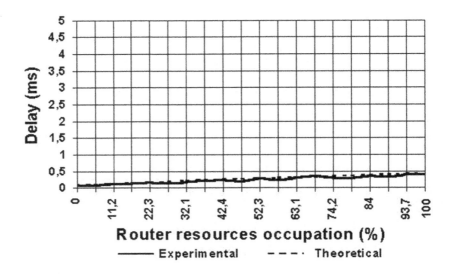

Fig. 7. Performance comparison for average end-to-end delay time for Premium Class traffic with PQ

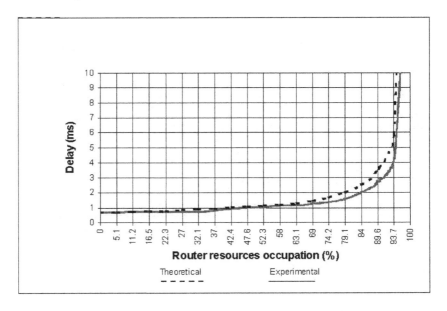

Fig. 8. Performance comparison for average end-to-end delay time for Best Effort Class traffic with PQ

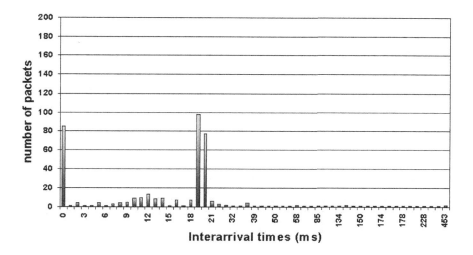

Fig. 9. Delay variation received from Premium voice packets with FIFO queuing

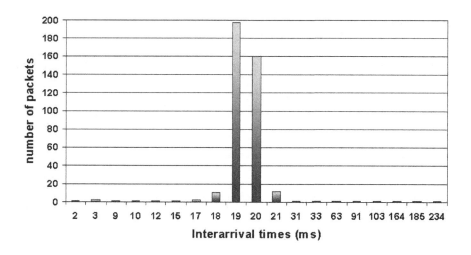

Fig. 10. Delay variation received from Premium voice packets with PQ

6   Concluding Remarks

In this paper a Differentiated Services IP network with Voice over IP traffic support have been considered, which has been implemented in Alcatel Labs, Florence, Italy. Three queuing algorithms have been used in the router queues, namely FIFO, Priority and Custom Queuing, in order to reserve a suitable band-

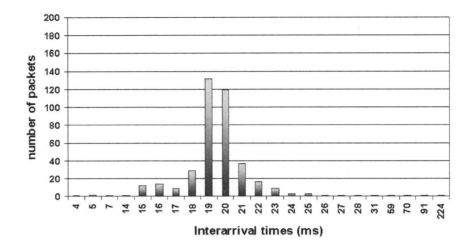

Fig. 11. Delay variation received from Premium voice packets with CQ

width amount to Premium class voice over IP traffic, while letting Best Effort class traffic operate without interrupting its service, and by reserving it the same bandwidth amount as in the FIFO queuing management case. We have developed experimental measurements of packet loss, average delay time and delay variation received from Premium and Best Effort class packets during the network crossing, while increasing the network congestion level through disturbing traffic generators. Since for the considered network Priority Queuing has resulted to be the best solution, we have then developed an analytical estimation of the average delay time for a router using PQ, and we have compared the analytical predictions with the experimental measurements results, thus verifying the theoretical analysis.

Acknowledgements

Authors wish to thank Alcatel Network Service Division, Florence, for having supported this research work.

References

1. J. Rosenberg H. Schulzrinne, "The IETF internet telephony architecture and protocols," *IEEE Network Magazine*, May, Jun 1999.
2. Atiqzzaman M. Hassan, A. Nayandoro, "Internet telephony: Services, technical challenges, and products," *IEEE Network Magazine*, Apr 2000.
3. V. Peris, R. Guerin, "Quality of Service in packet networks: basic mechanisms and directions," *NH Elsevier Computer Networks*, vol. 31, pp. 169–189, 1999.

4. C. Metz, "IP QoS: Traveling in first class on the Internet," *IEEE Internet Computing*, Mar-Apr 1999.
5. L. Kleinrock, *Queueing Systems*, Wiley, 1976.
6. IETF, "RFC 2330, http://andrew2.andrew.cmu.edu/rfc/rfc2330.html".
7. M. Zekauskas, G. Almes, S. Kalidindi, "RFC 2679: One way delay metric for IPPM".
8. P. Chimento C. Demichelis, "Instantaneous Packet Delay Variation Metric for IPPM; ippm draft, work in progress," .
9. C. Metz, "RSVP: General purpose signaling for IP," *IEEE Internet Computing*, May-Jun 1999.
10. "Cisco Technical Documentation: Configuration guides and command references," http://www.cisco.com/univercd/cc/td/doc/product/software/ios120/12cgcr.
11. ""Test TCP" Performance evaluation tool for TCP and UDP for Linux," http://www.leo.org/ elmar/nttcp/.
12. Naval Research Laboratory (NRL), "The multi generator (MGEN) toolset," Release x.y, http://manimac.itd.nrl.navy.mil/MGEN/.
13. "Cisco technical documentation," *http://www.cisco.com*.
14. R. Gallager D: Bersekas, *Data Networks*, Prentice Hall, 1992.

Dynamic SLA-Based Management
of Virtual Private Networks

M. D'Arienzo, M. Esposito, S.P. Romano, and G. Ventre

University of Napoli "Federico II", Via Claudio 21 – 80125
{maudarie, mesposit, spromano, giorgio}@unina.it

Abstract. This paper aims at shading some light on the concept of
Service Level Agreements (SLAs) and on their usefulness in the context
of the so-called Premium IP Networks. Such networks provide users with
a portfolio of services, thanks to their intrinsic capability to perform a
service creation process while relying on a QoS-enabled infrastructure.
We will introduce a definition of SLAs and we will then focus on an
actual example, namely the negotiation and management of QoS-aware
Virtual Private Networks (VPNs). We believe that VPNs are a significant
application due both to their importance in corporate scenarios and to
the high revenues they guarantee to service providers. We will discuss
the issues related to the effective SLA-based management of resources
in those cases where the need arises for an entity that is capable of
optimizing resource utilization in the presence of network infrastructures
shared by a community of users. Finally, a novel component, named *SLA
Manager*, that accomplishes these tasks, will be presented.

1 Introduction

QoS has been in the last years one of the major research topics in the networking
community. First in Academia, then in Industry, the issues related to the provision of guarantees in the performance achievable when offering communication
services have been subject to an intense debate that is still continuing in various
fora (as for example the IETF).

It should be noted, however, that most of such activity has been focused
on the technological aspects of the assurance of QoS within network elements,
nodes and terminals. Most of the work has indeed been performed in the area of
the mechanisms and architectures required for assuring that in packet switched
networks data belonging to certain flows can be differentiated from others. This
analytic approach has led to the definition of basic mechanisms and standards to
be adopted within and across network elements. We might define such results as
the basic elements for the provisioning of QoS "at a low level", or also *micro-QoS*.

In our opinion, however, this path towards technological development is missing a critical evolution factor, i.e. the definition of technologies for a system-wide
approach to the provision of QoS-aware communication services. These technologies should be the ones responsible for the provisioning of QoS "at a high level"
across complex network infrastructures with a real process and business model

S. Palazzo (Ed.): IWDC 2001, LNCS 2170, pp. 642–659, 2001.

oriented philosophy. We will define them as technologies for *macro-QoS* provisioning.

Currently, from a Network Operator point of view the task of creating a QoS-aware communication infrastructure is obliged to be simply that of assembling new, advanced network components and adding such new infrastructure to the existing ones, trying to maintain and re-use the existing business models and architectures. Manufacturers of network equipment introduce continually new solutions and products characterized by conformance to the existing or proposed standards and recommendations for what concerns the micro-QoS issues, but at the same time adopting specific and proprietary solutions for the provisioning of macro-QoS functionality.

The aim of this paper is to bring theoretical and practical contributions exactly in this area, with the goal of allowing the definition of tools, procedures and processes for the offering of advanced communication services in Premium IP networks across global infrastructures which might have a high degree of complexity in terms not only of scale, but also of the number of operators and level of technological heterogeneity.

The paper is organized in six sections. The reference framework where this work has to be positioned is presented in section 2. Section 3 discusses the main issues related to the definition, negotiation and activation of Service Level Agreements in Premium IP Networks. An actual example of the applicability of these new concepts is shown in section 4, where we will expand on the issues related to the negotiation and management of QoS-aware Virtual Private Networks (VPNs). This will allow us to introduce a model for a novel network component, named the *SLA Manager (SLAM)*, whose main objective is the dynamic, SLA-based management of corporate traffic (section 5). Finally, section 6 provides some concluding remarks, together with some information concerning our future work in this field.

2   Reference Framework

This section introduces the general architecture proposed for the dynamic creation and provisioning of QoS based communication services on top of Premium IP networks [1]. Such an architecture includes key functional blocks at the user-provider interface, within the service provider domain and between the service provider and the network provider. The combined role of these blocks is to manage user's access to the service, to present the portfolio of available services and to appropriately configure and manage the QoS-aware network elements available in the underlying network infrastructure. Their internal operations comprise activities such as authentication, aggregation and a mediation procedure that includes the mapping of user-requested QoS to the appropriate service/network resources, taking into account existing business processes.

In our view, network architectures are expected to be highly heterogeneous in terms of variety of systems and nodes, owing to the fact that they should be able to support dynamic service creation and service configuration on top of generic QoS-aware IP networks. Closely related to this activity is the management of those resources in the underlying networks that are reserved at registration/subscription, as well as those that are used – and maybe subsequently modified – when the service is invoked/configured. Associated with the reservation and usage of resources is the automated production and presentation of the corresponding SLAs to the user and the translation from the SLA to the corresponding Service Level Specification(s) (SLS) [2].

To the purpose, we have introduced three major components [8] (figure 1) that we believe are needed to supervise the dynamic service creation and service configuration process:

– *Access Mediator (AM)*;
– *Service Mediator (SM)*;
– *Resource Mediator (RM)*.

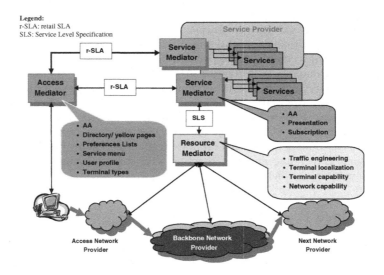

Fig. 1. The reference framework for Premium IP Networks.

The Access Mediator is the device into which users input their requests to the system. It adds value for the user, in terms of presenting a wider selection of services, ensuring the lowest cost, and offering a harmonised interface: the Access Mediator presents to the user the currently available services. The source of the services is a so-called "Service Directory" database, but the Access Mediator performs processing of the raw information. For example, it can select the cheapest offer if a movie is available from more than one service provider, and it can notify the user as soon as a new movie becomes available that matches the stored user's profile. Its main role thus consists in assisting and easing the service selection process. This functionality may be under the control of a trusted third party and appears to offer excellent novel opportunities for a value-added service provider. The usage of a service generally involves two business processes: registration to be a user of the service and invocation of the service at the moment when it is used (any modification of the service parameters during a session can be considered as a new invocation). The following sequence of events is broadly applicable to both processes:

- after authentication, the user requirements are captured, and the Access Mediator sends the information to Service Mediators (which in turn employ the Resource Mediators) to map the requested and (subsequently) selected service into the deployed physical network;
- once the service selection has been agreed with all parties, the SLA is "signed" between the user and the Service Mediator;
- records of usage and the associated SLAs are stored in the Access Mediator for future reference.

A graphical user interface associated with the Access Mediator is expected to provide a harmonised interface to the user for all the available service offers.

The Access Mediator may form associations with one or more Service Mediators to which requests are issued. Generally off-line, the Service Mediator will supervise the incorporation of new services, their presentation in the "Service Directory" and the management of the physical access to these services via the appropriate underlying network, using the Resource Mediator(s). It is the task of the Service Mediator to prepare the SLA for the user to sign, and subsequently map the SLA from the Access Mediator into the associated SLS(s) to be instantiated in cooperation with the Resource Mediator(s).

The Service Mediator has an important role, as this is the place where services are created and from where the impacts of service reconfigurations are communicated to the network resource management. The Service Mediator has to inform the Access Mediators (usually via the Service Directory) of all new service offerings, so to allow them to present the updated portfolio to their users. It also has to check that the addition of a new service, or invocation of an existing one, will not affect the services that are currently operational.

In this scenario, a policy based approach is a possible solution to ensure the correct operation of the network. Subsequent to the service creation, a policy extension could be applied to the network to ensure that all services can be managed correctly. The system would have a global view of the configuration of the devices (including an accounting system) and of the policy rules to be applied. In such a case, it would be the function of the Service Mediator to update the service level management system with new rules and configuration as required, in conjunction with the Resource Mediators.

The communication between the Service Mediator and the Resource Mediator should be generic (i.e. independent of the technology employed by the underlying network). According to our design, it is the Resource Mediator that will hold the current end-to-end view of the network QoS, by communicating with all the appropriate underlying network management systems. A network provider wishing to offer its resources should support an interface capable of handling messages defining an SLS, from its network management system to one or more Resource Mediator(s). The SLS templates we envision are in line with the descriptions in [2,3].

Since there can be more than one Resource Mediator, a Service Mediator can issue identical requests for information about network resource availability to several Resource Mediators. The Resource Mediators will either act on their own image of the network, or explicitly enquire to the individual network management systems, before returning an answer to the Service Mediator. The Service Mediator will accept the best offer, on the basis of the current policy decisions.

In order for the Resource Mediator to maintain and update its end-to-end network view of the current QoS availability, it may use a set of policy rules that are agreed with the underlying network management systems [4].

A common feature of the communication process surrounding the Access Mediator, Service Mediator and Resource Mediator components is a "one-to-many, search-and-selection" mechanism. In particular:

− the Access Mediator is responsible for selecting the appropriate Service Mediator(s), according to the user's request;
− the Service Mediator is responsible for finding – and, in some cases, building from individual elements – the service, requesting information from (and then selecting) the appropriate Resource Mediator(s);
− the Resource Mediators are responsible for selecting the appropriate network capabilities, given several available options.

The potential for developing a common protocol for all of these similar actions is described in [7].

3   Service Level Agreements

In this section we address the issues related to the definition of the Service Level Agreements [6] suitable for the services envisaged in the framework of Premium IP (PIP) Networks. As we saw in the previous section, PIP networks are capable of delivering new services to the end-users. Such services are characterized by different levels of Quality of Service (QoS). In this scenario, the definition of a service creation framework [5] plays a major role, given its aim to dynamically create application-level services by appropriately combining and configuring pre-existing service components. Based on these considerations, we will provide definitions for SLA and a number of scenarios describing their relationship with the service creation and resource management framework. This set of definitions is part of a comprehensive conceptual model (SLA modeling framework), extensible in nature and capable to include sections specific to the different technologies or service architectures and cover SLA modeling for both transport services and end-user services.

3.1   Service

A service is a concept which may be modeled from different perspectives. Whether acquiring services in a retail or whole-sale mode (see below) a service provisioning/creation life cycle starts with a service description, which then requires a proper degree of formalization. Emphasis is on the different types of information that need to be modeled and on the needed ability to model relations between them.

3.2   SLA

An SLA is a contract between the customer and the provider of a specified service. Such a contract is signed upon subscription to the service itself. An SLA is prepared from templates specifically conceived for the available services.

SLA templates are used during customer negotiation to define the required level of service quality. The production of an SLA template is an intrinsic part of service development. These SLA templates may either relate to standard product/service offerings, where they are used "as-is" to define the required level of service quality, or provide a baseline for custom negotiation (either automated or human-assisted). SLAs are defined on something perceived by the customer (i.e. explicitly subscribed to), that is the service elements composing the service product offering.

3.3   Retail vs Whole-Sale SLA

The retail SLA refers to the agreement between an end-user and a service provider. The end-user might be either a single person or a user organisation (e.g. a corporate or a public institution). Such an end-user could be induced to establish a SLA with his provider in order to support different kinds of applications. Some of the applications which we deem worth investigating are, for example:

- Adaptive Multimedia Applications (e.g. Video on Demand, Video Conference, etc.);
- Voice over IP (VoIP);
- Virtual Private Networks (VPNs).

SLAs trigger the negotiation of hierarchical agreements between different contractors. In the case of multi-domain scenarios, service providers may need to create inter-network agreements in order to support their end-user SLAs. We call whole-sale SLAs these inter-provider contracts. A whole-sale SLA takes into account traffic aggregates flowing from one domain to another. In general, there is no direct connection between r-SLAs and w-SLAs. In particular, w-SLAs might not be based on parameters related to a single service but might focus on statistical indicators related to the Grade of Service of the entire bundle provided by one provider to one of its neighbors. The focus of this paper is mainly on retail SLAs.

3.4   Static vs Dynamic SLA

As already stated, an SLA is a contract between two parties. To date, the general trend has been to consider only static SLAs: the contracts are instantiated after negotiations by human agents and their terms cannot be modified during their lifetime. We do believe that dynamic features are needed in order to better match the requirements of real-world operational scenarios. We envision at least two different flavors of dynamic behavior:

– time-varying user requirements (with different time-scales of time variability induced by specific application characteristics);
– time-varying network conditions (of which the user is made aware via feedback signals raised by the network itself).

Both flavors may induce the end-user to change over time the terms of a pre-established SLA. Typical usage scenarios are shown below.

1. No time-varying user requirements, No time-varying network conditions:

   in this situation the end-user establishes a static SLA, with no feedback. This implies that, once successfully terminated the negotiation phase, no modifications to the contract are allowed. In case of an admission control failure, the user is provided with no information concerning the reasons behind it. Thus, he has no clues on how to better re-formulate his request.

2. Time-varying user requirements, Time-varying network conditions:

   this case refers to the most complex possible scenario, which requires re-negotiable SLAs. During the negotiation and usage phases, the network may provide the user with useful information for tuning his request. The contract may be re-negotiated at any time.

3. No time-varying user requirements, Time-varying network conditions:

   in this scenario the network is capable of keeping users informed about its state, even if the users themselves cannot re-negotiate contracts on the fly. To exploit network hints they are compelled to tear down pre-existing contracts and re-formulate their requests from scratch.

4. Time-varying user requirements, No time-varying network conditions:

   here, users feel free to change their contracts, according to their new requirements. It is obvious that in this case, users' requests may incur admission control failures, due to the absence of specific data concerning the current state of network resources.

Figure 2 summarizes the aforementioned scenarios.

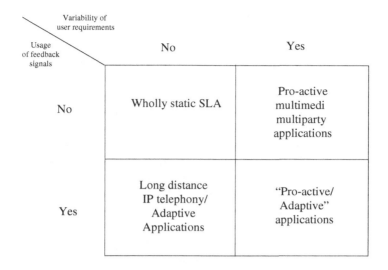

Fig. 2. Dynamic SLAs and related applications.

3.5   Content-Based SLAs

The service creation framework we depicted in the previous section envisages a scenario where users contact an Access Mediator (AM) in order to gain access to a number of value-added services, by means of negotiation of specific Service Level Agreements. The AM, in turn, needs to interact with one or more Service Mediators, each providing a certain set of services, to retrieve information about the characteristics of the services themselves. Afterwards, it organizes these information in order to let the user choose the service that most appropriately fits his needs. Once a specific service has been chosen, the involved Service Mediator(s) is (are) in charge of interacting with one or more Resource Mediators which, eventually, configure network elements so to efficiently satisfy the negotiated requests.

The process described foresees the generation of a number of documents (Service Level Agreement, Service Level Specification, policy rules), each describing the same instance of the service at a different level of abstraction and thus requiring creation/interpretation by the modules (Access Mediator, Service Mediator, Resource Mediator) belonging to the corresponding level of the overall architecture.

Digging into the details of such mechanisms, we can see in figure 3 that the Service Level Agreement is a contract between the end-user and the Service Mediator, negotiated via mediation of the Access Mediator. Once this contract has been signed, the Service Mediator is in charge of translating it into an appropriate Service Level Specification, containing a technical description of the service itself. This translation is a uni-directional process, requiring some additional information on the SM's side in order to retrieve, where necessary, service-specific data.

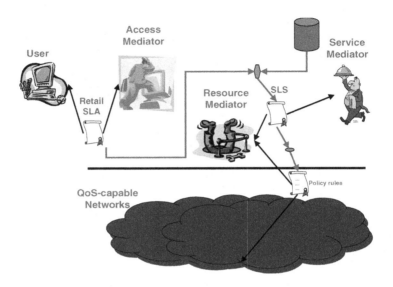

Fig. 3. SLAs, SLSs, policy rules.

The SLS is in turn given to the Resource Mediator, which translates it into a format that is the most appropriate for the QoS-capable network it manages. For example, it might build a list of policy rules, needed inside Policy Decision Points (PDP) in order to configure the underlying network elements (or Policy Enforcement Points - PEPs) via a policy protocol like COPS [4] .

3.6   Issues Related to the Definition of Service Level Agreements

As a general remark, an SLA should give the user the possibility to negotiate a certain type of service, among those offered by the network operator. We expect that most users will simply ignore the details of the service they expect from the network (especially those concerning the traffic characterization), either because such information is not available at all, or because they lack the motivation or the necessary technical skills required to understand their semantics. To ease the process of filling the contract template, a number of different SLA models might prove useful: the contract would become easier to understand, being focused on the actual needs expressed by the user. These SLAs may be considered as formed by two different parts, one containing information that does not depend on the particular application and the other containing application-specific data. The first might, for example, include:

- service level;
- user authentication module;
- information concerning availability/reliability of the service;
- encryption services;
- pricing and billing policies;
- options(enabling, for example, contract re-negotiation in case of unavailability of the required resources).

The second part of the agreement is analyzed in further detail in the next sub-section, devoted to examples on possible operational scenarios.

What we want to point out here is that from the network perspective the need arises to unambiguously specify all of the details and the characteristics of the service the user is willing to receive. SLAs should thus be translated into related Service Level Specifications (SLSs) containing all of the technicalities associated to the service itself. SLSs should be independent from both the high-level applications they stem from and the low-level network infrastructures on which they operate. Work is in full swing in the Internet community to define the main aspects related to SLS definition.

3.7   Example Operational Scenarios for SLAs

Interactive Multimedia Applications

Such applications include audio/video transmissions where a user connects to a video-server archive containing a number of movies that can be sent, in a streaming fashion, to a client host. In the same category we can also put those applications, like Video Conference and Tele-medicine, where video and audio data are generated from live sessions. For these applications, a mechanism is required to grant access to either the movie list or the session directory, in order to let the user choose the file/event he is interested in. The user is required to indicate the service he is willing to perceive, optionally specifying service lifetime. After defining these parameters, the translation module has to retrieve the traffic characterization associated to the specified files/sessions, in order to insert it inside the network SLS.

Virtual Private Networks

In this section we will consider only issues related to the creation of an SLA for a VPN service with respect to the problems linked to the provision of Quality of Service guarantees. We will therefore not cope with any aspect related to security or fault tolerance. We envision a scenario where a company, or in general an organization based on multiple facilities or sites, asks for the provisioning of a Virtual Private Network service as a way to efficiently interconnect its networking infrastructures. We expect, therefore, to see the VPN service forward traffic generated by a variety of users, network infrastructures, services, and applications. For example, we might consider a situation where two or more sites of a company are connected via a VPN service so to have both data-like connectivity (LAN-to-LAN) and voice-like connectivity (VoIP interconnections). The SLA will therefore be related to the provision of services to a mix of traffic, with different requirements in terms of bandwidth and QoS.

In the case of VPNs, the r-SLA negotiated with the Service Mediator might not be as fine-grained as needed to accommodate the internal needs of the company/institution for which the VPN is being set up. Thus, the customer may apply further traffic management on its own premises. We will focus on these issues in the next two sections: more precisely, in section 4 we will propose an SLA template for VPNs, while in section 5 we will show how effective management of the VPN resources may be achieved thanks to the introduction of a novel component, called SLA Manager (SLAM).

4   SLAs for Virtual Private Networks

While designing a template for a Service Level Agreement for VPNs, we kept in mind the fact that users are looking for simple solutions to complex problems: that is to say, when buying an enhanced VPN service, they would just like to express their needs at a high level of abstraction, with no need to bother with all the underlying technicalities. Thus, we exploited the capabilities of the aforementioned mediation components in order to fill the semantic gap between the user's and the provider's perspectives on the same service.

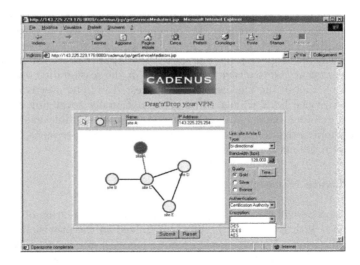

Fig. 4. The "Drag&Drop" VPN.

We provide the user with a friendly graphical interface (figure 4) enabling him to design his own private network by simply drawing a graph representing the sites he wishes to interconnect, together with the related tunnels (represented by the edges in the graph). Furthermore, for each tunnel, we envisaged the possibility to choose its specific features:

- type: mono or bi-directional;
- required bandwidth;
- desired quality;
- time schedule: permanent or time-dependent (either periodic or related to a specified time frame);
- authentication procedure;
- encryption algorithm.

This graphical representation of the VPN is then passed, in the form of an incidence-like matrix, to the Access Mediator, which in turn builds a formal description (e.g. in a language like XML) of the agreement the user is willing to negotiate (figure 5).

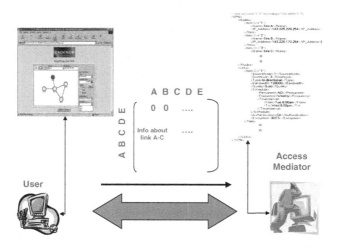

Fig. 5. User to AM communication of the VPN features.

This template SLA is the grounds on which lays the negotiation process involving all of the mediators and described in the previous sections. In particular, the Service Mediator, upon reception of this information, is in charge of performing the following tasks:

- retrieve information about links;
- for each link:
    - create an SLS, based on the Virtual Leased Line concept [2];

- compute the cost associated to this VLL, by contacting the involved Resource Mediators (this might require "SLS/policy rules" translation).
- compute overall cost (e.g. sum of all link costs);
- send cost back to the AM.

Figure 6 gives an idea of how the SLS for a single Virtual Leased Line might look like.

```
-Scope: one-to-one (143.225.229.254,143.225.170.254)

-Flow description: (143.225.229.0,143.225.170.0) DSCP=EF

-Traffic Conditioning: token bucket (b,r): r=128kbps

-Excess Treatment: dropping: only in-profile packets allowed

-Delay guarantee: qualitative (e.g. delay="low")

-Loss guarantee: p = 0 (implying a throughput guarantee R = r)

-Service Schedule: Weekly, 8am on Tuesday to 8pm on Wednesday

-Reliability: May be specified

- Options: Authentication (from CA),3DES encryption
```

Fig. 6. SLS for a Virtual Leased Line of the VPN.

After retrieving overall service costs, the AM is able to build the candidate SLAs and send them to the user: once the user has chosen a specific offer, it can store the associated SLA inside the user's repository and notify SMs about the final decision, so to let them appropriately configure network resources.

It should be noticed that in this scenario, one SLA corresponds to multiple SLSs and only at the SLA level of abstraction the concept of a VPN does exist: the only component who speaks "Dragged&Dropped VPNs" is the AM. Furthermore, the SLSs are both service-independent and network-independent.

5   A Model for an SLA Manager

One of the issues to be considered in SLA based Premium IP networks is related to the possible complexity of the interactions between the users and the network entities responsible for the presentation and negotiation of SLAs. An SLA is a complex set of data, pertaining to different aspects of the provision of a network service or application: as already mentioned, in the case of a video delivery service, an SLA can include content dependent elements. In this business scenario, the user is requested simply to accept the purchase of a service (for example, the delivery of a movie or of a multimedia document) and to ask the network (via the Access Mediator) for the provision of a communication service suited to the performance requirements of that specific content/service. Even though novel services could introduce new, more complex business scenarios, we expect that such kind of interaction between an end-user and the Access Mediator will remain quite simple, and perfectly manageable directly by the final user himself, for example via a web based interface.

However, we believe that there exist cases where such interactions could become much more complex, due to the presence of multiple technical and commercial aspects related to the nature of the offered service and to the specific needs of what we define the end-user. This is the case, for example, of the provision of a VPN service for the interconnection of multiple network infrastructures via public Premium IP networks.

Virtual Private Networks are being considered as the real killer applications for future networks. From the network provider's standpoint this is primarily due to the complexity of their setup and management. Users, on the other side, are mostly interested in the possibility of exploiting an infrastructure capable to provide services with guaranteed QoS.

In a QoS-capable network, a VPN service will be based on the idea of provisioning, on top of purely virtual links, a tailored communication service that could be differentiated from existing ones in terms of:

- performance and QoS;
- monitoring and accounting;
- security and privacy.

As far as the first issue, we expect that a VPN should offer a bouquet of services corresponding to the possible different QoS requirements that the data flowing across the VPN itself might demand, according to policy decisions that are — and should be kept — free to be defined and set by the user. Indeed, since VPNs are more and more linked to the need of creating and operating the so called "network companies", one of the major issues will be dynamic management of the virtual communication infrastructure.

We introduced two differently-grained SLA types: (retail) r-SLA and (wholesale) w-SLA. The former is negotiated between an end-user and a service provider on a relatively small time scale, involving generally small resource amounts. The latter, instead, is negotiated between two different network domains less frequently than r-SLA, in order to create a pipe which merges different flows relating to r-SLAs already instantiated.

In the case of a users' organization, for example, an organisation representative signs an SLA with the access provider, reserving a large amount of resources. Then, single organisation members can apply for a smaller part of these resources up to the bulk available. Further requests have to be declined or a new contract has to be (re)negotiated by the organization.

In such a scenario the necessity arises for an entity that manages bulk resources, assigns sub-portions of them and is capable, in case of need, to (re)negot-

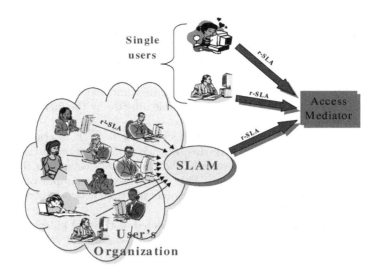

Fig. 7. The Service Level Agreement Manager.

iate larger quantities of resources. It can be considered as a mediator between a single user and a larger generic service provider (reached via an Access Mediator). This entity is in charge of negotiating with the Service Mediator a retail SLA that applies to the users' organization as a whole. We call *SLA Manager (SLAM)* this entity (figure 7). A SLAM should have the following major functions:

- towards the single users:
  - AAA (Authentication, Authorization & Accounting);
  - internal negotiation of the r-SLA, in the form of what we call "r^2-SLA" (figure 8);
  - providing a friendly GUI;
  - enabling a user preference profile to be "bookmarked" for future use.
- towards the Access Mediator:
  - (re)negotiation of the r-SLA.
- from a more global perspective:
  - policing and/or shaping of single user's flows;
  - admission-control;
  - providing a fair sharing of total resources;
  - triggering local network configuration.

The SLAM is therefore a new entity, placed between an existing Access Mediator and the end-user, but surely pertaining to the user's domain.

We are currently prototyping a SLAM entity for VPNs, based on a model where QoS is taken into account at different levels of granularity inside the VPN

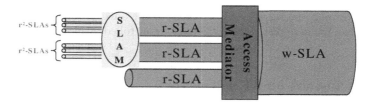

Fig. 8. The role of the SLAM.

tunnels. More precisely, a small and fixed number of traffic classes are considered in the core of the network (as delimited by the LAN gateways) and share the available bandwidth in a controlled way (exploiting, for example, a Priority Queuing algorithm [10]). Inside each class, a further level of discrimination is applied, by identifying the single micro-flows and deserving to each of them an ad hoc treatment via an additional scheduling (e.g. Weighted Fair Queuing [10]). Figure 9 gives an idea of how this scenario is provided.

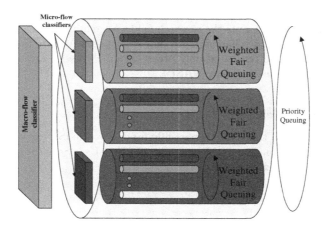

Fig. 9. Scheduling flows inside VPN tunnels

We implemented such scheduling algorithms on a programmable networks platform [9], where they may be combined in a structured fashion, thus bringing to complex configurations, ranging from simple series/parallel structures to nested ones. The situation depicted in the figure might relate, for example, to the case where all IP telephony flows are given priority over those generated by other applications (as guaranteed by the external scheduler) and are further discriminated among each other, by appropriately configuring the internal WFQ scheduler (that acts as a "gate keeper" element). In this case, the SLAM is responsible for negotiating the r-SLA related to the bulk resources assigned to the VPN tunnel, while re-distributing them in a controlled fashion in the inner local network.

6 Conclusions and Future Work

In this paper we presented a novel approach to the design and development of a global architecture for the effective deployment of value-added Internet services upon Premium IP networks. As we saw, this is an ambitious goal, requiring a comprehensive understanding of all the procedures involved, from user-to-network interaction all the way through appropriate configuration of network devices, passing by a formal description of the service. Based on these considerations, we pointed out the need for a thorough definition of the concept of Service Level Agreements and associated Service Level Specifications. We then presented a model based on a modular decomposition of tasks involved in the deployment process, exploiting at their best the concepts of "mediation" and of recursive group communication. Finally, to give an idea of how this concepts apply in a real-world operational scenario, we focused on an actual example related to dynamic, SLA-based management of Virtual Private Networks.

This definitely represents the research field we are mostly interested in further investigating, due to the challenging topics it proposes. We are firmly convinced that the approach we presented may prove extremely useful for next generation Value Added Service Providers (VASP).

Acknowledgements

Research outlined in this paper is partially funded by the IST project CADENUS IST-1999-11017 "Creation and Deployment of End-User Services in Premium IP Networks", http://www.cadenus.org/

References

1. IST project CADENUS (IST-1999-11017) "Creation and Deployment of End-User Services in Premium IP Networks", http://www.cadenus.org/
2. D. Goderis et al, "Service Level Specification Semantics and Parameters", draft-tequila-sls-00.txt, November 2000, expires March 2001.
3. Y. T'Joens et al, "Service Level Specification and Usage Framework", draft-manyfolks-sls-framework-00.txt, October 2000, expires April 2001.
4. J. Boyle, R. Cohen, D. Durham, S. Herzog, R. Rajan, A. Sastry, "The COPS (Common Open Policy Service) Protocol". IETF Proposed Standard RFC2748, January 2000.
5. M. Smirnov, "Service Creation in SLA Networks", work in progress, Internet Draft draft-cadenus-slan-screation-00.txt, November 2000, http://www.cadenus.org/papers/

6. S. P. Romano, M. Esposito, G. Ventre, G. Cortese, "Service Level Agreements for Premium IP Networks", work in progress, IETF draft `draft-cadenus-sla-00.txt`, November 2000, `http://www.cadenus.org/papers/`
7. M. Smirnov (Ed.), "SLA Networks in Premium IP", CADENUS Deliverable D1.1, March 2001.
8. G. Cortese (Ed.), "Mediation Components Release 1 — Requirements and Architecture", CADENUS Deliverable D3.1, March 2001.
9. R. Maresca, M. D'Arienzo, M. Esposito, S. P. Romano and G. Ventre, "An Active Network approach to QoS-enabled Virtual Private Networks", submitted to 2nd Workshop on Quality of Future Internet Services (QofIS2001), 24-26 September 2001, Coimbra, Portugal.
10. S. Keshav, "An Engineering Approach to Computer Networking. ATM Networks, the Internet, and the Telephone Network", *Addison Wesley Professional Computing Series*, 1996.

Distributed Service Platform
for Managing Stream Services

Rami Lehtonen[1] and Jarmo Harju[1]

[1]Tampere University of Technology, Telecommunications Laboratory,
PL 553, 33101 Tampere, Finland
{rampe, harju}@cs.tut.fi

Abstract. The number of Internet and IP-capable devices and networks will increase dramatically in the near future. In order to exploit their potential in multimedia communications, we need a service platform that can handle the management and control of the multimedia streams independently of the access networks. In this paper we introduce a distributed service platform that is based on CORBA and we discuss the user and device management in such system in more detail. In addition to that, we try to discover various problems, design issues and possible ideas linked to this kind of system, so that we can use this information later on while designing similar horizontal service platform solutions for commercial use.

1 Introduction

Traditionally each access network has its own set of services and only lately we have seen attempts to combine the service offering from various access networks; e.g. WAP (or I-mode) devices are able to use the Internet services in addition to the operator's WAP services. The ultimate goal might be that the access network doesn't limit the amount and type of available services. On the other hand, some of the services in the fixed network side are not interesting for the mobile users and vice versa. Yet, it might still be fruitful to combine things related to the user, devices, billing, etc. at the management side. Our approach assumes that all services are accessible from every access network. This doesn't mean, however, that we should be able to watch movies from a GSM phone. The idea of the distributed service system, discussed in this paper, is that a device from any access network could be used to *order* and *control* stream services and the processing of the stream can be handled on some other device like PC or future IP-TV. This approach adapts quite easily to current service platforms, because many access networks already take care of the user authentication and billing; a GSM user can be automatically authenticated and charged, while an ordinary PC user necessarily cannot.

Generally, bit rates are increasing, but they are not in the required level, where we could use the broadband (from appr. 100 kbits/s to several Mbits/s) stream services from any access network. Only certain fixed networks and WLAN connections are already able to provide this type of multimedia services. However, the bit rates are not the only thing slowing down the evolution of multimedia services; the pricing models are not up-to-date, and currently high service prices are preventing the usage of the

S. Palazzo (Ed.): IWDC 2001, LNCS 2170, pp. 660–673, 2001.

broadband multimedia services. On the other hand, the lack of services might have something to do with this also. The pricing models should be able to handle multimedia services properly, e.g. the users should be charged only from the services they use, not from the control traffic they have to generate while making the order. This kind of service specific billing means that the operators must co-operate to enable these services and agree on the service expenses, e.g. roaming, without additional payment for the user. The operator expenses could be covered in the service payment, but for the user this means a clear and practical way to order services, compared to the current model.

From the operator's point of view it is evident that the number of different access networks and also the number of users will increase by the time. This creates general management problems and some day current vertical service solutions, where each access network has its own set of services, don't scale anymore for the user's and operator's purposes. The horizontal solution, which is independent of access networks in the application layer, brings us a possibility to manage users and services efficiently. The user identity and profile information and also the view to the service system stays the same, independently of the access mechanism. All-IP solution, which is also the ultimate goal for the Universal Mobile Telecommunications System (UMTS) according to 3GPP's vision [1], enables this kind of horizontal service concept. Along with the All-IP solution we will eventually get IPv6 [2] to all mobile devices and that also forces the fixed world to move towards IPv6 networks and devices. That means that we have to address all those new requirements (i.e. how to identify the mobile devices?) coming from the All-IP world.

The distributed service platform introduced in this paper attempts to discover various problems and possible ideas that can be used later on while designing similar horizontal service platform solutions for commercial use. It is assumed that all network elements are capable of communicating with IP (either IPv4 or IPv6) and because we have chosen CORBA [3] as the control mechanism, it creates also requirements for the Media Client, which handles the actual service processing. The next section describes the distributed architecture briefly by introducing all the important entities involved in the service control process. The third section discusses about user and device management concentrating on problems and key issues on solving them. Section 4 covers additional design issues related closely to the user and device management and Section 5 concludes the paper.

2   Distributed Architecture

The distributed stream service architecture is based much on the earlier work with the stream services done in [4] and [5]. The current scenario introduces four network elements that form the basis of the stream service architecture; WWW server, Access Server, Media Server and Media Client. The WWW server is responsible mainly for providing the user interface for ordering the services while the other network elements are involved in the actual stream control and negotiation. Figure 1 will give a view of the network elements involved in the architecture.

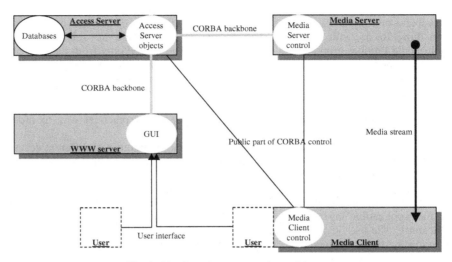

Fig. 1. Distributed stream service architecture

The WWW server, Access Server and Media Server communicate via a CORBA backbone that can be isolated from the public IP network e.g. by using a VPN, SSL or some other secure mechanism. The Media Client part of the CORBA control is not directly connected to the CORBA backbone mainly because the user should not be part of the operator's control network and secondly because the control information sent to/from the Media Client is non-critical and not too confidential. The following subsections give more detailed information on the network elements and their functions.

2.1   WWW Server Objects

The main functionality of the WWW server is to provide the interface towards the users accessing the stream services. Those users might have different capabilities in terms of access network bandwidth, display size, etc. and the WWW server must be able to fulfill those different requirements when providing the user interface. In this current architecture we have considered the WWW server to be accessed only via HTTP protocol. This comes mainly from the fact that the bit rates of the mobile networks (GPRS, UMTS, Bluetooth, WLAN) and display sizes of the mobiles (PDAs, Communicators) are gradually increasing and the WAP protocol will become unpractical over the time. However, it is possible to add a WAP gateway between the user and the WWW server if there is need for content conversion to the WAP enabled devices.

The WWW server provides a personalized view on the services available for the user and it supports different user roles and profiles for payments and browsing. It is also responsible for communicating and forwarding the selected service information to the Access Server, which is the main control element in the service system.

2.2   Access Server Objects

The Access Server is the most important element in our service system and it consists of several functional entities, which can be distributed to several computers to balance the load. The functional entities are

- Access Manager,
- Customer Manager,
- Device Manager,
- Order Manager,
- Service and Content Manager,
- Billing Manager,
- User, device, service and billing databases, and
- Naming Service.

The Access Manager is responsible for user and device authentication and granting the access rights. The WWW server forwards the authentication parameters from the user to the Access Manager and the Access Manager checks the user properties from the database. If the user is authenticated, the Access Manager creates a Customer Manager object to handle the user specific parameters and settings. Similarly, for valid devices the Access Manager creates a Device Manager to handle the device specific functions.

The Customer Manager takes care of the user roles and profiles and it is also responsible for user authorization. The user can modify his/her profile and change the role (e.g. from anonymous to identified) through the functions provided by the Customer Manager. The WWW server also uses directly some of the functions of the Customer Manager to modify the view of the web pages. The Customer Manager also creates Order Manager objects based on user selections to take care of the actual service.

The Device Manager is quite similar to the Customer Manager. The biggest difference is that it handles the roles and registrations of the user devices without any user interaction. The devices can inform the service system whether they are online or not, and what kind of network and stream parameters they are willing to accept.

The Order Manager controls the actual stream service transactions based on the information it gets from the user and the Customer Manager. When the user wants to make an order, the Order Manager takes care that the device and user parameters do not conflict. After the order parameters are selected and the user has confirmed the service, the Order Manager contacts an appropriate Media Server and forwards the service request to it. The Order Manager is also responsible for communicating with the Billing Manager and initiating the billing transaction.

The Service and Content Manager handles the media information registration to the databases. The Media Servers register their media content, when they are started at the first time or when new content is stored. Many of the parameters are only stored in the Media Servers, but the parameters, which affect the user decision or describe useful information to the user, are forwarded to the Service and Content Manager.

The Billing Manager is also part of the Access Server and it collects charging information throughout the service transaction. Both the Order Manager and the Media Server inform the billing system about the status of the stream service. Based on that information the Billing Manager can charge the user correctly. The user itself (Media Client) is not part of the billing, because the Media Server is in this current architecture aware of the status of the stream service.

The rest of the Access Server is composed of several databases and naming service functionality. The databases contain information of the service system users, identified devices, services and billing. The databases are queried through a number of database adapters. The naming service provides name to IOR (Interoperable Object Reference) mapping and all the objects within the service system, including the Media Client, must register to it. Figure 2 shows the relationships between the different Access Server objects.

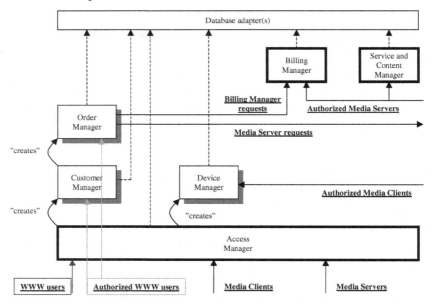

Fig. 2. Access Server objects and their relationships

2.3   Media Server Objects

The reason for separating Media Server from the Access Server comes from the idea that the media information can be achievable through a well-known place, Access Server, and the actual (sometimes high bandwidth) media data can be stored in several Media Servers. The Access Server can select the Media Server that is located near the end-user, thus keeping the bandwidth usage somehow limited. The Media Server and also the Media Client are based somewhat on the scheme that is introduced in the A/V Stream specification [6]. The A/V Stream specification gives us the building blocks that can handle stream controlling between two devices.

The intelligence is in the Stream Controller, which is responsible for setting up and configuring the stream and the network connection between two Multimedia Devices (Media Server and Media Client). Parameter negotiation is considered as part of the stream and network connection configuration. In addition to the A/V Stream part, the Media Server contains the Media Server control object, which is responsible for contacting the Media Client and requesting a permission to connect to the Multimedia Device of the Media Client. It also handles billing and forwarding of the stream and the network parameters to the Stream Controller. The control part of the Media Server is also responsible for providing the Stream Controller's IOR to such an entity, which may want to control the stream remotely (play, pause, rewind, etc.).

2.4   Media Client Objects

In our scenarios we have defined the Media Client as an entity, which is able to process streaming audio and video, and has decent network connection characteristics. It can be a future TV, an ordinary PC or some other terminal with adequate multimedia capabilities. Before the user can order anything to the Media Client device, the device must register its identity to the Access Server. This registration can include device specific information like network and stream capabilities, but it is enough to just register the IOR of the Media Client and the activity status (ON/OFF). The user can be associated with one or several Media Clients and vice versa. For that purpose the registration can include AccessID(s) for that specific Media Client.

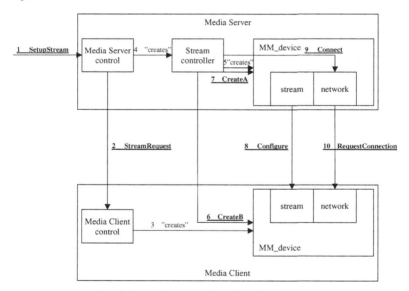

Fig. 3. Media Server and Media Client architecture

In addition to the registration and other control functionalities, the Media Client includes also A/V Stream specific objects in order to be able to negotiate and configure the network and stream parameters with the Media Server. Figure 3 depicts a more detailed version of the Media Server and Media Client objects and connections. The stream model is based on [6].

2.5   Example Scenario

Before describing the user and device management in this kind of system, Figure 4 shows a scenario, where all the defined network elements are involved in the control of the stream service ordered by the user.

A user, who wants to order something to a Media Client contacts the WWW server, authenticates himself and selects a stream service. The WWW server provides a personalized view for the user based on the user profile and role throughout the transaction. If the user has associated some devices with himself, the Access Server provides that information to the user through the WWW server and the user is able to select e.g. "my TV" as the desired Media Client. When the service details are agreed, the Access Server forwards the service information to the Media Server, which initiates the actual service after it has negotiated the stream and network parameters with the Media Client.

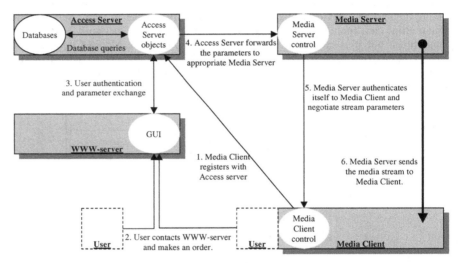

Fig. 4. Example scenario

3   User and Device Management

In order to handle the service system operation as smoothly as possible, we should be able to connect the users, devices and services in such a way that the user doesn't have to care about registrations or addressing and, moreover, the services should be somehow associated with the device capabilities. This chapter deals with the user and device management problematics and proposes solutions for those problems.

3.1   Connecting the User, Device and Service

The ease of service usage comprises of many important things related to connecting the user, device and service smoothly to each other. One of the principle questions in this kind of service system, where the purpose is to order multimedia services with device A and to deliver the service to device B, is how we take into account the properties of the destination device (B, Media Client) while we are browsing and selecting the service. The are at least two different possibilities; 1) we choose some service from a broad selection of services and then we point out the destination device where this service should be sent to, or, 2) we select the destination device first and we choose the service from the list of services that are suitable for that specific device. Table 1 describes the advantages and disadvantages of these two different ordering models from the viewpoint of both user and service provider.

Table 1. Comparison of the ordering models

Ordering Model	Advantages	Disadvantages
Service => Device	- better selection of services	- device does not necessarily support the service
	- clearer classification model	- cumbersome way to choose services possible user dissatisfaction
Device => Service	- device is capable of processing the services	- some of the available services are not visible to the user with limited capabilities
	- better customer satisfaction	

The users should be able to easily identify the destination device (Media Client), which is used to process the media service, and therefore the service platform databases should contain information of the user-associated devices, their status and capabilities. Generally, one user can be associated with several devices and, on the other hand, some device can be associated with multiple users. A device must register itself to the Access Server (device database), before we can order anything to it. This can happen automatically in a such way that the Media Client control object contacts the Access Manager and authenticates itself. After the authentication the Access Manager creates a Device Manager object to handle the actual device capability registration. Figure 5 shows the Media Client registration scenario.

Before the registration can be successfully handled, the Media Client must be connected to some user (owner), which has valid account in the Access Server system. So a device is registered to one user, but several users could have association with it. All devices in the device database have unique deviceID and associated to that deviceID there is information about the device capabilities like

- Supported media types (MIMEs),
- Min/max bandwidth for audio and video,
- Min/max video framerate,
- Min/max video resolution,
- Supported audio and video codecs,

- Control method for device (SIP, CORBA),
- Device information (name, type, model, etc.),
- Device address (SIP-URL, IPv6, IPv4, IOR), and
- Device activity.

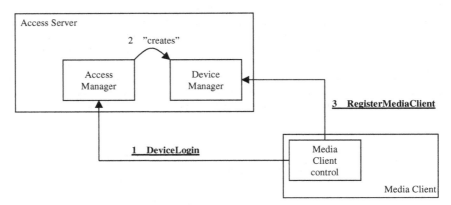

Fig. 5. Media Client registration with the Access Server system

This set of information is delivered to the Access Server while registering the Media Client. The information consists mainly of the stream and network (audio/video) parameters, but there is also information about control methods and device activity. The device activity parameter informs the Access Server, whether the Media Client is ready to take stream requests or not.

The user database contains information about user characteristics, profiles, roles and device associations. A user can be associated with several devices and the user can name these devices quite freely, however, they should be named uniquely within the specific user. The relationships between users and devices are clarified in Figure 6.

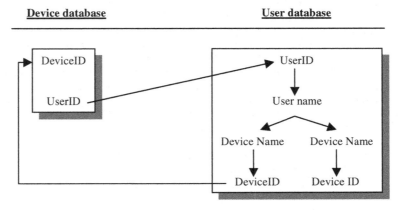

Fig. 6. Relationships between users and devices

When the required information about users and devices are in the databases, the users can see the registered and associated devices (Media Clients) while making the service order and because of that they can easily select the destination device with knowing the exact IP addresses, port numbers and so on. The other possibility is to select the destination device among the non-associated devices. Then the user has to know the address of the device.

3.2   User and Device Addressing

Before the users can identify the devices unambiguously, the user and device addressing must be unambiguous and location independent. IP addresses are complex and hard to remember, especially IPv6 addresses. Same goes for the DNS addresses, which does not always have any connection to the device identifier. Because all the service system users must be registered to the Access Server and every device is associated at least with one user (device is registered to one user), the user and device addressing is handled in the following way:

```
ACS address  = devicename "@" username

devicename   = *( unreserved | escaped | "&" | "=" | "+"
               | "$" | "," )

username     = *( unreserved | escaped | "&" | "=" | "+"
               | "$" | "," )

reserved     = ";" | "/" | "?" | ":" | "@" | "&" | "="
               | "+" | "$" | ","
```

The syntax is described using Augmented Backus-Naur Form following the style and rules shown in the RFC 2543 [7]. The username takes in a way the form of a domain name in this syntax and the devices belong to that user (domain). Here are few examples of valid ACS (Access Server) addresses:

pc.sonera@rami.lehtonen

pc106.tampere@sonera

The first address could point to the computer in the Sonera's office, which is used by Rami Lehtonen. The association is between the device (some PC) and the user. The other address could point to the same PC, however, the association is this time between the device and Sonera (as a community). By identifying just the user/community, we could get information on the associated devices for that user/community. The ACS addresses are valid only within the service system.

In any case the addressing will be mapped to IP level addressing, but with this system specific addressing, we can hide those issues from the users. What comes to the IP level addressing, we handle the Media Clients as servers, so they have to be addressable from the public IP network. The lack of public IPv4 addresses creates also requirements for the IP level infrastructure. This kind of service system requires IPv6 network to be fully functional, mainly because the Media Server must be able to connect to the Media Client and in that case the Network Address Translator (NAT) [8] systems do not help.

The change in device's IP address can be handled in the registration phase by setting the new IP address to the device database. If the device's IP address changes while the device is online, it should renew its registration to the service system. The actual ACS address remains the same even though the device's IP address may change during the time. So, this system works in a way better than the DNS systems excluding Dynamic DNS systems. This addressing model is not, contrary to the IP/DNS addressing, location dependent and hierarchical and thus enables more practical use of addressing and mobility of the devices.

4   Design Issues

This section focuses on a few special design aspects of this service system. First we compare the functionality and features of the distributed model to the service system that we have designed and implemented earlier. Then we look at the security aspects of the CORBA control connections like secure CORBA backbone and explain how to guard the Media Client against different types of misuse.

4.1   Comparison to the Earlier Model

The previous version of the service system that allowed us to send control information to the Media Client via the Access Server is described in the papers [4] and [5]. The earlier approach was based much on the Service Control Protocol (SCP) that handled the parameter forwarding and negotiation between the Access Server and the Media Client. In the distributed model the parameters are circulated through the Media Server, before they reach the Media Client. So the Access Server is not anymore in direct control connection with the Media Client (exception: Registration). The reason why we had to forward the stream and network parameters directly from the Access Server to the Media Client was that fact that we worked with commercial or public domain Media Servers, which couldn't handle our control mechanism. In the distributed system described here, we can examine the effects of the stream parameters in more detail and we can also examine the billing scenarios, because now we can get hands on the actual service delivery. By moving the parameter forwarding and negotiation to the Media Server instead of the Access Server, we can also ease the load in the Access Server and thus have a more scalable architecture. What comes to the billing, the Media Server can now inform the Billing Manager about the service properties that can be used in the billing process. If the stream delivery is interrupted, we can signal the Billing Manager and it can take proper actions for correcting the charge. The following Table 2 will summarize the main differences between the two systems.

Table 2. Comparison between two different models for stream management

Old Model	New Model
- Centralized model (Access Server)	- Distributed model (Access Server)
- Own control protocol (SCP)	- CORBA based control (A/V Streams)
- Control connection from Access Server to Media Client	- Control connection from Media Server to Media Client
- Stream connection initiated by Media Client	- Stream connection initiated by Media Server
- Commercial Media Servers	- Media Servers actively involved in control
- No billing management	- Billing management possible

4.2   Security Aspects

The Access Server has a lot of valuable information about the users, devices and services and therefore the users must trust that their information is safe within the Access Server. In order to assure of that the Access Server must be behind a firewall and that way separated from the public IP network. The Access Server must, however, communicate with many Media Servers and also the Media Clients must be able to register themselves to the Access Server. The firewalls can be quite easily restricted to allow only the IIOP (Internet Inter-ORB Protocol) protocol and so the Media Clients are able to connect to the Access Server. Media Clients also have to authenticate themselves to the Access Manager, and in case of successful authentication, the Access Manager creates a Device Manager object to handle the Media Client registration. The Media Client can access only those two objects. The control traffic between the Media Client and the Access Server must be always encrypted using e.g. SSL, IIOP level encryption or IPsec. Similar encryption can also be used between the Access Server and Media Servers, but we can also create VPN (Virtual Private Network) type of connection between those two entities, because they are likely to be quite static network elements. The VPN connection can be secured e.g. by using IPsec. So the secure CORBA backbone consists of those VPN connections and also from the connections between the Access Server and the WWW server, which are always isolated from the public IP network.

If we look the security aspects from the user's point of view, we are concerned mainly with the security of the access connection that is used to connect the user to the WWW server. That's the most critical interface from the user's perspective, because the billing is based on the information changed over that interface. Normally we can have access network authentication for the users (e.g. GSM authentication) and then above that we authenticate users in the application level. The application level connection should be encrypted and we can use e.g. SSL on the top of the TCP/IP layer to secure the information change.

4.3   Restricting Media Client Usage

The other part of the client security is the Media Client connections. Because the Media Client acts as a server and waits for the Media Server requests, we must be somehow sure that the requesting side is a valid Media Server and that the order originated from some trusted and allowed user. Therefore we have defined a parameter called AccessID, which can be considered as a password that must be included in the StreamRequest (see Figure 3) message. The user that makes the order through the service system must give the AccessID, and the Access Server forwards this parameter along with the stream and network parameters to the Media Server responsible for delivering the media content. The Media Client can accept a variable amount of AccessIDs and based on those AccessIDs it can make service and user differentiation. An AccessID can be individual, it can be common to some group or it can be a device specific identifier. This decision depends on the Media Client configuration. In some cases, where the Media Client wants to avoid unnecessary control traffic to reach it, it can inform the Access Server about the valid AccessIDs and then the unauthorized users can be blocked out at an early stage.

The control traffic between the Media Server and the Media Client does not contain any sensitive information except the AccessID. If we avoid sending the AccessID over this control connection, the control connection can be implemented without any encryption. One way of doing this can include one-way functions, so that we calculate "a checksum" from the AccessID by using some one-way function. This checksum is sent to the Media Client and it uses the same function to calculate another checksum. If the checksums match we can be sure that the AccessIDs also match. However, this does not protect against replay attacks. In that case we need to include also some other parameters (like timestamp, nonce, source IP address, etc.) to the checksum calculation. The last option is to encrypt also the control connection between the Media Server and the Media Client.

5   Conclusions

This paper describes a system that can be used for instance to order stream based personalized services for a number of terminals, wired or wireless, with different capabilities. Building this kind of system is not a trivial task and various issues like user and device management must be carefully examined. By introducing CORBA for the control purposes we can model the system in a higher abstraction level and we can combine a number of various elements like billing and service management with the core system. Future networks and terminals will be IP-based and the terminals (mobile and fixed) also move towards open platforms. Together they enable a lot of new possibilities for network operators and service providers. The key elements in those systems include a flexible control mechanism, efficient and scalable architecture and well-designed management functionality.

References

1. 3GPP's All-IP vision. http://www.3gpp.org/news/mobile_news_2000/page3.htm
   (see also http://www.nokia.com/3g/solutions_all-ip.html)
2. Steve Deering and Robert Hinden, "Internet Protocol, Version 6 (IPv6)," Internet RFC 2460, December 1998.

3. Object Management Group, Inc., "The Common Object Request Broker: Architecture and Specification. CORBA V2.3.1," October, 1999.
4. Rami Lehtonen and Jarmo Harju, "Access Network Independent Service Control System for Stream based Services," EUNICE2000 Innovative Internet Applications proceedings, pp. 23-30, September 2000.
5. Rami Lehtonen, Petteri Heinonen, Jani Peltotalo, Sami Peltotalo, Jarmo Harju and Veikko Hara, "Implementation of a Flexible Control System for Launching Stream Services Independently of the Access Network," GLOBECOM2000 Service Portability workshop, December 2000.
6. Object Management Group, Inc., "Audio/Video Stream Specification," January 2000.
7. M. Handley, H. Schulzrinne, E. Schooler, and J. Rosenberg, "SIP: Session Initiation Protocol," Request For Comments 2543, Internet Engineering Task Force, March 1999.
8. P. Srisuresh and M. Holdrege, "IP Network Address Translator (NAT) Terminology and Considerations," Request For Comments 2663, Internet Engineering Task Force, August 1999.

IP QoS at Work: Definition and Implementation of the AQUILA Architecture

B. Koch[1] and S. Salsano[2]

[1] Siemens AG, ICN WN CC EK A 19
Berthold.Koch@icn.siemens.de
[2] Dip. Ingegneria Elettronica, University of Rome "Tor Vergata"
CoRiTeL – Consorzio di Ricerca sulle Telecomunicazioni
salsano@coritel.it

Abstract. The IST project AQUILA[1] aims to develop a flexible, extendable and scalable Quality of Service architecture for the existing Internet. The core network will be an enhanced DiffServ network providing several dynamically manageable traffic classes with specific QoS parameters, per hop behaviors, and other "guidelines" that realize different network services. A new logical layer has been defined on top of the DiffServ network, the Resource Control Layer. The task of this layer is to control the underlying network in order to provide QoS features to the customers of the network. Resource Control Agent, Admission Control Agent and End-user Application Toolkit (EAT) are the logical components of this architecture. Legacy as well as new QoS-aware applications running on the hosts will use the EAT middleware to benefit from the QoS capabilities of the AQUILA architecture. The AQUILA project is split into two phases. Now it has completed the definition and implementation and it has run the first trial. The architectural definition and the trial results are discussed in this paper.

1 Introduction

In order to satisfy the huge commercial demand for Quality of Service (QoS) solutions over IP networks, the project AQUILA defines, evaluates, and implements an enhanced architecture for Quality of Service. The Differentiated Service (DiffServ) approach for QoS provisioning in IP networks is used as basis for the specification of this architecture. A goal of the project is to verify the achieved technical solutions by testbed experiments and by trials involving end-users. The trials include QoS demanding on-line services like multimedia services [1].

Hereafter, the overall objectives that were established at the beginning of the project will be presented, followed by a discussion on the main innovation expected from the project. Key objectives of the project are:

---

[1] This work is partially founded by the European Commission as project IST-1999-10077 AQUILA

S. Palazzo (Ed.): IWDC 2001, LNCS 2170, pp. 674–690, 2001.
© Springer-Verlag Berlin Heidelberg 2001

- To enable dynamic end-to-end QoS provisioning in IP networks for QoS sensitive applications e.g. Internet telephony, premium web surfing and video streaming. Static resource assignments will be considered as well as dynamic resource control.
- To design a QoS architecture including an extra layer for resource control for scalable QoS control and to facilitate migration from existing networks. The DiffServ architecture for IP networks will be enhanced introducing dynamic resource and admission control. Main features are:
  - The architecture will be usable by any relevant kind of IP application.
  - The architecture will be cost-effective, scalable and backward compatible for the provisioning of QoS in IP networks covering both the inter- and intradomain QoS.
- To implement prototypes of the QoS architecture as well as QoS based end-user services and tools in order to validate the technical approach of the solution design. This includes
  - Development of a novel resource control layer extending Bandwidth Broker functionality.
  - Provision of an End-user Application Toolkit (EAT) in order to support the establishment of QoS by end-users and applications.
  - Creation of tools for QoS provisioning, monitoring and management in order to facilitate operators to control QoS IP networks.
  - Development of a distributed measurement infrastructure for end-to-end QoS parameters.
- To validate the QoS architecture in a field trial involving a commercial online service. To prove the concepts for larger scale networks, higher network load and different kinds of end-user services within a distributed testlab and by simulations.

Under the following headlines, the specific innovations of the project are explained in more detail. The strength of the project is to address all the individual issues under a unified perspective, while achieving specific innovations in each single aspect.

- *Scalable and flexible Resource Control Layer*

  A major innovation of the project is a new layer on top of the DiffServ network, called *Resource Control Layer* (RCL), which is used for controlling and managing the network resources. This layer can be seen as a distributed Bandwidth Broker in the DiffServ architecture. A node in the Resource Control Layer is called a *Resource Control Agent (RCA)*. The Resource Control Layer will include both the intra-domain aspects (i.e. the dialogue of the RCAs with end-users and with the networking devices like Edge devices and core routers) and the Inter-domain aspects (i.e. the dialogue among RCAs belonging to different domains). In comparison to the current bandwidth broker proposals and prototypes, the project will introduce:
  - a distributed architecture as a base for scalability and reliability,
  - active mechanisms in the Resource Control Layer in order to adapt network configuration according to current traffic loads,
  - measurement based admission control in Resource Control Agents,

- a highly reliable Resource Control Layer design
- integration of defined DiffServ classes and mechanisms (Per Hop Behaviors - PHB) into the layered QoS architecture. In particular, suitable control/management interfaces between IP networking devices and the resource control layer will be defined.

- *End-user Application Toolkit*

In order to provide QoS for end-user applications, the AQUILA project comprises the development of an End-user Application Toolkit (EAT). This toolkit can be used by application developers and provides reusable and generic components for both client and server sides of applications.

The EAT enables the construction of QoS-aware applications as well as the migration of legacy applications to QoS-awareness, and ensures compatibility with various methods and protocols for communicating QoS requests between end-user applications and the Resource Control Layer. It provides also scalable mechanisms for applications, which produce a large number of short-lived sessions with unclear requirements or where dedicated reservations are inappropriate, and it offers control mechanisms for the support of bi-directional services.

- *QoS Management Tool*

In order to deploy and operate QoS mechanisms in large scale IP networks, the project develops a *QoS Management Tool* (QMTool) that
- allows high level management of user related QoS policies,
- enables the interaction among different ISPs and administrative domains,
- foresees the interworking of different mechanisms (i.e. IntServ / DiffServ / MPLS / traffic engineering).

The current lack of tools to handle these complex issues hinders deployment of QoS into operational IP based networks. The QMTool simplifies the task of a service provider in operating QoS aware networks.

- *Traffic and Admission control, Traffic engineering*

The innovative architectural concepts require novel algorithms for traffic control, admission control and traffic engineering in order to optimize the network performance and achieve efficient usage of the network resources. A goal of the project is to define a clear reference framework to operate with the different facets of traffic handling in a consistent way. Traffic handling is composed of:
- Provisioning and Traffic Engineering procedures operating at hours/days/weeks time scale
- Admission Control functions operating at seconds/minutes time scale
- Packet level traffic control that includes all the mechanisms for classifying, marking, scheduling and policing the IP packets (it operates at milliseconds time scale).

The AQUILA project has undertaken a two phased approach in the definition of the architecture and in its development and trials. The first phase focused on single domain QoS, while no focus has been put on the QMTool and the 1st trial has been a lab trial. The first phase now almost completed and most results from 1st trial are available. In sections 2, 3 and 4 the definition of AQUILA first phase architecture is given. Section 5 discusses the trial results, while section 6 gives some information on the ongoing work for the definition of the second phase architecture.

2   AQUILA QoS Architecture

The current Internet architecture is not designed to support QoS, and there exist different approaches for providing QoS over IP-based networks. Due to the different underlying mechanisms of these approaches and the complexity of end-to-end QoS, there is currently no solution suitable for global operation. Integrated Services, Differentiated Services, Multi Protocol Label Switching, QoS Routing, Bandwidth over-provisioning are among the most important techniques for QoS in IP networks. For a comprehensive discussion on the IP QoS, see [2], [3].

Most of the above-mentioned technical solutions on how to bring QoS into IP networks are still under discussion. Some of them are divergent, while some are complementary. No integrated scaleable solutions are available right now. Furthermore, management and interoperability aspects of the mentioned approaches are currently treated poorly. The project assumes the DiffServ architecture as the most promising starting point for its work. The project develops extensions of this architecture in order to avoid the statically fixed pre-allocation of resources to users. Dynamic adaptation of resource allocation to user requests is enabled in a way that keeps the overall architecture scalable to very large networks. As an example for an alternative approach, the RSVP protocol defined in the Integrated Services architecture should be present in the access network where the load is small and the scalability issue is not important.

2.1 The AQUILA Resource Control Layer

The *Resource Control Layer* (RCL) is an overlay network on top of the DiffServ core network [4]. The RCL mainly has three tasks, which are assigned to different logical entities:

- To monitor, control and distribute the resources in the network. This task is assigned to the Resource Control Agent (RCA).
- To control access to the network by performing policy control and admission control. This task is assigned to admission control agents (ACA). Each edge router or border router is controlled by an ACA. As each access request necessarily means usage of resources, the RCA may be directly or indirectly involved in handling admission requests.
- To offer an interface of this QoS infrastructure to applications. This task is assigned to the End-user Application Toolkit (EAT). From the network point of view the EAT acts as an RCL front-end. From the user point of view, the EAT provides a QoS portal.

The entities defined above are associated to network elements within the underlying domain as shown in the Fig. 1. An EAT instance can be responsible for a single host as well as for a set of hosts. The latter might be the case, when not a single host, but a whole sub-network is connected to an edge router. The resource control layer assumes an underlying DiffServ network. The DiffServ code points (DSCP) and the PHBs of this network are assumed to be predefined by management. They are not under control of the RCL for the 1st trial of the project AQUILA. For each traffic class however, there is a specific amount of bandwidth available in each link of each edge router, border router or core router. So bandwidth is the main resource, which is handled by the RCL.

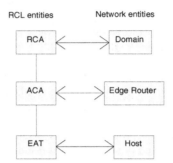

Fig. 1: Mapping of RCL entities to the underlying network entities

In the first phase of the project (1st trial), the RCL implements a dynamic admission control by distributing the pre-configured, static resources of the core network among edge routers and border routers. In the 2nd trial, dynamic reconfiguration of core network resources is also taken into consideration.

In the following, a single DiffServ domain is assumed. Inter-domain aspects are not covered as they will be handled in the second phase of the project.

2.2   Resource Control Agent

The support of QoS, as proposed above, leads to the introduction of a new logical layer, the Resource Control Layer (see Fig. 2). The Resource Control Layer provides an abstraction of the underlying layers. A node in the Resource Control Layer is called a *Resource Control Agent* (RCA) and represents a portion of the IP network, which has internally the same QoS control mechanisms. An RCA is a generalization of the concept of the Bandwidth Broker in the DiffServ architecture. RCAs are logical units that run on several physical configurations, e.g. one server per RCA or several RCAs co-located on one server. The QoS control mechanisms used in the underlying network are of varying nature, e.g. in some part the routers may not even support DiffServ (which means that there is only a trivial best-effort QoS control), while in other parts they may be DiffServ capable. Moreover, some parts of the network may allow dynamic reconfiguration of resources, e.g. by adding ATM connections, others may have a more or less fixed configuration, e.g. pure SDH or WDM sub-networks. Another reason for the introduction of separate RCAs is that sub-networks are domains managed by different operators.

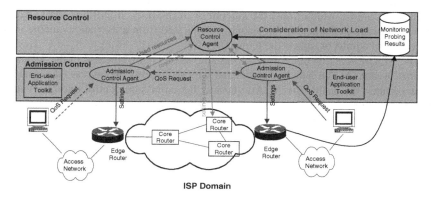

Fig. 2. AQUILA QoS Architecture: the Resource Control Layer

A Resource Control Agent is able to observe and in some sense to influence the actual configuration in the network portion it represents. Configuration parameters may describe the fraction of a network connection devoted to a specific DiffServ traffic class or the existence of a virtual connection (in ATM networks) with a specified bandwidth. For this administration purpose, a mechanism is required to access IP routers and possibly other network elements.

The RCAs employ distributed computation that adapts the network according to user requests for QoS. The RCAs always try to establish a situation where the network configuration is "slightly over-dimensioned" such that user requests can be immediately satisfied by just checking the admissibility of the user and recording the additional resource usage. As soon as some "watermarks" are reached, the RCAs start a dynamic reconfiguration process in order to avoid congestion. The adaptation algorithm of an RCA is the reason why it is called an "agent" (even an intelligent agent) since the RCA can act autonomously, in contrast to admission control which is often non-local. For the purposes of a first prototype, locally fixed agents (which communicate over CORBA) seem to be sufficient. However, it is basically feasible to make use also of mobile agents here e.g. to move the "master" agent of an Internet backbone provider to a different physical location when its current home gets overloaded.

The RCA must also support flexible accounting schemes for QoS services, including both end-user-ISP and ISP-ISP accounting. In addition, the RCA interacts with network management, firstly for configuration and secondly for partial network failures. For configuration and service creation, the RCA interacts with the QMTool, which offers integrated and flexible service control.

2.3 Admission Control Agent

A DiffServ network can only provide quality of service, if it is accompanied by an admission control, which limits the amount of traffic in each DiffServ class. Admission control checks, whether the resources requested from a user are available in the network and admits or rejects the request.

The AQUILA architecture uses a local admission control located in the *Admission Control Agent* (ACA), which is associated with the ingress and egress edge router or border router. To enable the ACA to answer the admission control question without interaction with a central instance, the RCA will locate objects representing some share of the network resources nearby the ACA. Resources are assigned to these objects proactively. For the ACA, these objects represent a "consumable Resource-Share".

Admission control can be performed either at the ingress or at the egress or at both, depending on the reservation style.

The ACA will just allocate and de-allocate resources from its associated consumable ResourceShare. The ACA is not involved in the mechanisms used by the RCA to provide this resource share, to extend and to reduce it.

Resource distribution is performed on a per DiffServ class basis. In the 1st trial, there is no dynamic reconfiguration of DiffServ classes. So, the resources of each class can be handled separately and independently of each other. This per class distribution however is not appropriate for edge devices, which are connected via small bandwidth links to the core network. In this case, additional mechanisms apply.

Resources are handled separately for incoming traffic (ingress) and for outgoing traffic (egress). The following description of resource distribution applies to both.

Resource distribution is performed by the RCA in a hierarchical manner using so-called R*esource Pools*. For this purpose it is assumed, that the DiffServ domain is structured into a backbone network, which interconnects several sub-areas. Each sub-area injects traffic only at a few points into the backbone network. This structuring may be repeated on several levels of hierarchy.

When considering the resources in the backbone network, all traffic coming from or going to one sub-area can be handled together. So it is reasonable to assign a specific amount of bandwidth (incoming and outgoing separately) to each sub-area.

2.4 End-User Application Toolkit

The *End-user Application Toolkit* (EAT) is an application that aims to provide access to end-user applications to QoS features. The EAT is a middleware between the end-user applications (for example a video conferencing tool or a video-on-demand service) and the network infrastructure (for example the AQUILA network).

The tasks of the EAT are to allow legacy applications (QoS-aware and non-QoS-aware) to benefit from QoS features (1st trial), and to allow the implementation of QoS-aware EAT-based applications by making use of an API (2nd trial).

In particular, the EAT

- enables the construction of QoS-aware applications as well as the migration of legacy applications to QoS-awareness,
- ensures compatibility with various methods and protocols for communicating QoS requests between end-user applications and the Resource Control Layer. Special schemes for integrating application level protocols (e.g. H.323, SIP) with QoS control are taken into consideration,
- provides scalable mechanisms for applications, which produce a large number of short-lived sessions with unclear requirements (e.g. WWW) and where dedicated reservations are inappropriate,
- offers control mechanisms for the support of bi-directional services. This enhances the DiffServ approach that supports only simple uni-directional flows.

The EAT is transparent for legacy applications, but will be mostly transparent for new EAT-based applications (using the API).

3 Service Level Specifications, Network Services and Traffic Classes

Two important aspects of QoS are *QoS guarantees* and *QoS differentiation*.

In order to provide QoS differentiation, a limited set of Network Services have been defined in the AQUILA project, which represent the services sold by the provider to its customers: Premium Constant Bit Rate (PCBR), Premium Variable Bit Rate (PVBR), Premium Multimedia (PMM), Premium Mission Critical (PMC) and Standard Best Effort (STD). Each Network Service is meant to support a class of applications with substantially similar *requirements* and *characteristics*. The Network Services are internally mapped by the operator into a set of Traffic Classes. The Traffic Classes use DiffServ based packet handling mechanisms and are defined in term of queuing and scheduling mechanism in the routers. More details on the set of traffic classes defined by AQUILA will be discussed in the next section.

In order to provide QoS guarantees an ISP must somehow regulate the amount of traffic entering the network regarded as a limited set of resources. In the AQUILA approach this is accomplished by the distributed Resource Control Layer, whose architecture has been defined in the previous section. The RCL embeds different mechanisms to regulate the traffic at different time-scales – Initial Provisioning, Dynamic Resource Pool and Admission Control. These mechanisms will be presented in the next section as well. As already mentioned, in the first phase of the AQUILA project the focus is on QoS in a single domain.

An important aspect of the QoS provisioning is the definition of the agreement between the user and the provider about the scope of the QoS contract: which flows should receive QoS, which are their traffic characteristics, which is the expected QoS level. The specification of this agreement is commonly referred to as "Service Level Specification" - SLS. It has been recognized that a common standardized way to express the semantic of the SLS could be useful in an open multi-provider environment. Some work has been produced in this direction ([9], [10]) and related discussion within IETF is ongoing. If a "static" QoS provisioning approach is envisaged the agreement is negotiated "off-line" between the user and the provider, may be involving human intervention. A formal SLS can be useful to have a clear and commonly understood picture of the service and of the required QoS. If a dynamic approach is used, where the user application can automatically send QoS requests to the network, the SLS should also be mapped into signaling information exchanged by QoS aware elements. This is the approach followed by the AQUILA project, where the EAT sends their reservation request messages to the ACA, specifying an SLS. The semantic definition of the AQUILA SLS is described in [11]. The reservation request messages, as all the AQUILA control messages, are transported using CORBA.

4   Traffic Handling Approach

This section focuses on the *traffic handling* mechanisms in AQUILA. Traffic handling is used here as a general term for a set of coordinated mechanisms that operate at different time scales:

- *Traffic control* refers to the mechanisms operating at milliseconds time scale like packet scheduling, policing, queue management.
- *Admission control* refers to the algorithms to decide about the acceptance of a new flow in the network, operating at the time scale of seconds to tens of minutes.
- *Resource Pools* refers to the algorithm for short-term resource redistribution, to cope with local fluctuations in offered traffic, operating at the time scale of tens of minutes to hours.
- *Provisioning* refers to the algorithm for medium/long term resource allocation and redistribution, operating at the time scale of hours to days.

The relationships among these logical components (Provisioning, Resource Pools, Admission Control and Traffic Control) are described hereafter. A very high-level view of the process that enables QoS in the AQUILA architecture is given in Fig. 3., while Fig. 4 gives a simplified pictorial view of the relationships between the different mechanisms.

The Provisioning phase is run off-line before the network operation, and gives the required input to the RCL elements as well as configuration values for setting the router parameters. The initial provisioning algorithm takes as an input global information about the topology, the routing (costs of links), the expected traffic distribution between Edge Routers for each Traffic Class (TCL), and any further constraints on the link bandwidth sharing between TCLs. It performs a sort of global computation and produces as output:

- the expected amount of traffic for each Traffic Class on each link, called *provisioned rate*. This is used for the router configuration, i.e. to chose the appropriate setting for the scheduling / queuing parameters (WFQ weights, WRED thresholds) at each router interface.
- the Admission Control Limits for each Traffic Class at each Edge Router, which are used by AC algorithms during the operation phase.
- Definition of the Resource Pools sets.

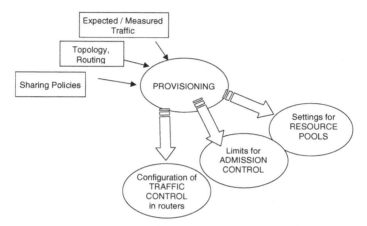

Fig. 3. Enabling QoS in the AQUILA architecture

The Traffic Control mechanisms define how the packets of the different classes are handled by the Edge and Core Routers in the AQUILA network. They includes traffic conditioning (also referred to as policing), that is enforced at ingress ERs only, and scheduling / queuing algorithms, implemented at any router interface.

The configuration of the scheduling / queuing mechanism is "static", i.e. the relevant parameters are configured in the routers at start up. An off-line procedure computes these parameters starting from the provisioned rates produced by the Initial Provisioning algorithm. Obviously, the configuration of per-flow traffic conditioning parameters at the ingress edge is done run-time according to the admitted requests.

The Admission Control procedure is intended to restrict traffic in order to avoid congestion. The AC procedure is operated on-line, but the AC reference limits, or AC Limits, are computed during the off-line initial provisioning phase and configured during the start-up phase.

The assignment of AC Limits to each Edge Router for each TCL represents a resource assignment to the relevant traffic aggregates. As the AC Limits are computed based on the expected offered traffic at each ER, some deviation can occur during the operation phase between the actual offered traffic and the resource distribution between ERs. The Resource Pools mechanism represent a way to dynamically change the AC Limit to some extent, so as to dynamically track short term fluctuations in traffic requests. Such mechanisms are based on the concept of RP, which are sets of Edge Routers that can exchange resources with each other. Such sets are defined during the Initial Provisioning phase.

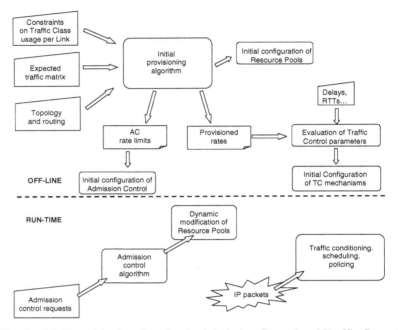

Fig. 4. Initial Provisioning, Res. Pools, Admission Control and Traffic Control

4.1 Traffic Classes and Packet Level Traffic Control

AQUILA has defined a set of five Traffic Classes. Each TCL is associated a different queue in the router output interface, and a bandwidth portion on each link. The queue dedicated to TCL-1 is served with strict priority over the others. All queues are served by a WFQ scheduler. Fig. 5 shows the inter-TCL scheduling scheme.

TCL-5 is intended to support the Standard Service (STD), i.e. the traditional best-effort traffic. The traffic accessing the STD service is not delivered any QoS and is not regulated by any AC and/or policing function inside the ER. Nevertheless, a non-null amount of bandwidth will be guaranteed to this traffic on each link.

TCL-1 and TCL-2 are intended to support non-reactive (open loop) traffic with stringent QoS requirements. In particular TCL-1 will be characterized by very high QoS performance (very low delay and very low losses), accomplished by a conservative AC scheme. In the AQUILA architecture TCL-1, which is somehow similar to the EF PHB defined in [12] will exclusively support the PCBR service. Typically, TCL-1 will be entered by flows originated by real-time streaming applications like

VoIP, etc. On the other hand, TCL-2 will deliver a lower QoS level (low delay and low losses) to those streaming application with high emission rate variability and/or large packets. TCL-2 will mainly support the PVBR network service. A "severe" purely dropping traffic conditioning at ingress point is associated to TCLs 1 and 2, i.e. all packets exceeding the declared profile are discarded. The traffic profile for TCL-1 is described in terms of a Single Token Bucket, limiting the flow peak rate. The traffic profile for TCL 2 is described in terms of a Dual Token Bucket, controlling both the peak and mean rate of the flow.

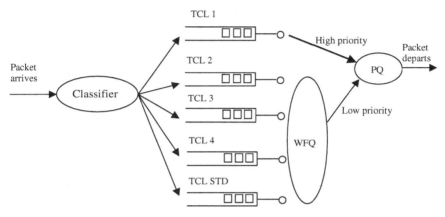

Fig. 5. Design of router output port

TCL-3 and TCL-4 are dedicated to reactive flows (TCP and TCP-like). In particular, TCL-3 will support PMM service and serve long-lived TCP connections (for long file transfers) or other adaptive application flows (audio/video download, adaptive video). A Single Token Bucket descriptor is used to declare the mean rate only. Traffic conditioning at ingress point is based on 2 colors marking: out-of-profile packets are not discarded but simply marked as such with a different DSCP value. At router interfaces, the TCL-3 queue uses a WRED management algorithm with two different sets of parameters for in-profile and out-of-profile packets. TCL-4 instead will support PMC service and will receive non-greedy elastic flows, typically short-lived TCP connections originated by some critical transaction application (e.g. finance) or interactive games. Dual Token Bucket is used descriptor is used to declare both mean and peak rate. Traffic conditioning and queue management are similar to TCL-3.

4.2 Admission Control

At each Edge Router (ER), each TCL is assigned a bandwidth value, which is used to limit the maximum amount of traffic that the ER can inject into the network for the specific TCL. This value, referred to as "AC Limit", will be used by the Admission

Control algorithms to decide about the acceptance of new flows. Different admission control algorithms have been defined for each different traffic class. The AC algorithms are partly derived from the results developed in the context of ATM traffic control. The TCL-1 class uses peak rate allocation scheme. In case of TCL-2 traffic class the REM (*Rate Envelope Multiplexing*) multiplexing scheme is assumed for guaranteeing low packet delay [13]. In case of TCL-3 each flow is characterized by parameters of single token bucket algorithm that correspond to the sustained bit rate (SBR) and the burst size (BSS). One has to consider that the traffic flows submitted to TCL-3 class are TCP-controlled and rough QoS guarantees can be acceptable. The admission control procedure will only check that the sum of sustainable rates declared by the sources is less than the AC Limit. In case of TCL-4, a flow is characterized by parameters of dual token bucket algorithm. The proposed admission control algorithm evaluates an effective bandwidth and then checks that the sum of the effective bandwidths is less than the AC limit. The goal is to provide a very low packet loss rate for in profile packets. A complete description of the Admission Control algorithms can be found in [5].

5   QoS at Work: Prototype Implementation and Trial Experiences

As mentioned earlier the project follows a two phased approach, each containing the full circle of prototypical software development with design, implementation, integration and trial. The first phase ending with the 1st trial was meant to give early results on the experimental verification of the AQUILA concepts and the operability of the proposed architecture. Therefore the 1st trial was planned to run as a lab trial in three different operator sites, each focussing on different configuration aspects and network topologies, and following different trial scenarios.

During the first year of the project the QoS IP network architecture was defined and prototypes of the ACA, RCA and EAT were implemented. In addition a Distributed Measurement Architecture (DMA) was developed in AQUILA [6] for generation of foreground traffic, active network probes and collection of QoS monitoring information from the routers. It is designed to provide the validation of the end-to-end QoS provision in AQUILA based on mappings between the measured end-to-end-QoS, the used network service and the end-to-end QoS that is required by the user.

It also offers monitoring of QoS information (e.g. packet loss) for the resource control layer.

This measurement system consists of two main parts, the measurement server and measurement clients. The measurement server is situated in the core of the network, while the measurement clients are distributed in the network leafs. For one-way delay measurements, the clients are equipped with GPS hardware for time synchronization.

After successful integration of all components that were developed at various partner sites all over Europe – the consortium consists of 12 partners out of six European countries - the 1st trial experiments were prepared and performed in three partner sites, where the main site is located in Warsaw, and two others are in Vienna and

Helsinki [7]. The following figure (Fig. 5) shows the configuration of the Warsaw testbed as an example.

These testbeds differ in network configurations, main focus of test scenarios and in the number of routers. Appropriate software developed inside the AQUILA project was integrated in these testbeds. In particular, EAT, RCA, ACA and measurement tools were integrated with router configurations.

The trial experiments are mainly focused on the evaluation of previously defined network services providing QoS: Premium CBR (PCBR), Premium VBR (PVBR), Premium Multimedia (PMM) and Premium Mission Critical (PMC). For the purpose of the AQUILA demonstrator, testing of chosen Internet applications is also included in the plans.

The following major objectives for the trials were defined:

- In Warsaw: two network services, PCBR and PMM. A mixture of network services is also being tested in order to check the ability of the AQUILA architecture for providing service differentiation. Additionally, the correctness of admission control as well as resource pool algorithms is verified. The experiments are provided under artificially generated traffic and real Internet applications.
- In Vienna: two network services, PVBR and PMC.
- In Helsinki: measurements of the performance of the RCL.

Several trials with the pre-defined network services were performed, either using a single service or a combination of services. Details of the trial results can be found in the First Trial Report [7].

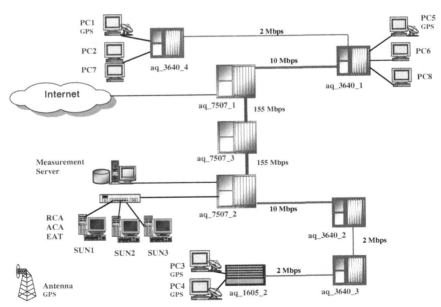

Fig. 5. Configuration of the Warsaw testbed

The - still ongoing - trial experiments allowed to verify the AQUILA architecture concept. The introduction of different network services for serving streaming and elastic traffic is justified. The matching of PCBR for streaming traffic, of PVBR for video conferencing and of PMM for elastic traffic is verified. It is under consideration the need for the second class for elastic traffic (PMC). The project is looking for real application that can benefit of this class. The need of admission control mechanisms for providing QoS is justified and the effectiveness of implemented admission control mechanism was tested. In particular it was found that the utilization of the network was less than expected, i.e. the admission control algorithms are too conservative.

Another lesson learned from implementation and trial experience is that it is difficult to provide correct Traffic Parameters for the different applications. Together with the previous observation about a possible inefficiency of admission control algorithm, this suggests to improve the overall architecture with input from measurements. This input can improve the admission control efficiency - introducing some concepts from measurement-based approach - and can relieve the application from the need to provide complex set of Traffic Descriptor parameters.

6 Current Work

For the second phase of the project the encouraging results of the 1st trial will be fed back to the specification and design of the enhanced architecture elements. This will include in particular the feedback from measurements in the Traffic handling procedures. Focus of the second phase will be also on the Inter-domain aspect and on the running of the 2nd trial involving end-users. In this section we will briefly deal with Control Loops and we will provide some comments on the Inter-domain aspects.

The logical process for QoS provisioning as defined in section 4 can be seen as an open loop from Provisioning to configuration of Resource Pools, Admission Control and Traffic Control. Fig. 6 gives a pictorial representation of the inclusion of control loops in the AQUILA architecture. The only point where a feedback from the actual network status is included in the 1st trial architecture is the dynamic adaptation of Resource Pools according to reservation request messages (feedback "A" in Fig. 6). The time scale for this control loop is in the order of minutes/hours.

Taking into account the input from "on-line" measurement two additional control loops can be envisaged. One is aimed at improving efficiency in network utilization by enhancing Admission Control functionality. The Declaration Based admission control used in the 1st trial will be enhanced with concepts of Measurement Based Admission Control (feedback "B" in Fig. 6). The time scale of operation of this control loops is in the order of few seconds.

The second additional control loop is referred to the initial provisioning phase. This phase is bases on estimates of traffic matrix. The measurement of actual traffic can obviously be used to tune the provisioning (feedback "C" in Fig. 6). The time scale of this control loop is in the order of hours/days.

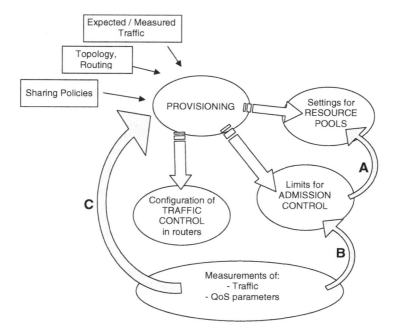

Fig. 6. Control loops in the AQUILA architecture

As for the problem of Inter-domain QoS, it will be tackled by trying to define efficient and scalable mechanisms for aggregating Inter-domain resource reservation and for conveying the needed signaling information across the domains. The basic approach is derived from the BGRP proposed in [14], which will be adapted to suit the AQUILA needs.

7   Conclusions

In this paper the results of the ongoing IST project AQUILA have been discussed. The AQUILA project has defined and implemented a QoS architecture for IP networks. It exploits DiffServ-like functionality and introduces a Resource Control Layer for requesting and managing QoS resources. A stepwise approach in the development has been undertaken. Lab trials related to the first phase have been carried out. The benefits of the proposed architecture have been investigated and considerable experience has been gained. Work is now ongoing on the specification of the second phase.

References

[1]   IST project AQUILA deliverable D002 "Project Presentation", April 2000, http://www-st.inf.tu-dresden.de/aquila (Publications)

[2]   X. Xiao, L.M. Ni "Internet QoS: A Big Picture", IEEE Networks, March 1999

[3]   W. Zhao, D. Olshefski and H. Schulzrinne, "Internet Quality of Service: an Overview" Columbia University, New York, New York, Technical Report CUCS-003-00, Feb. 2000.

[4]   IST project AQUILA deliverable D1201 "System architecture and specification for first trial", June 2000, http://www-st.inf.tu-dresden.de/aquila (Publications)

[5]   IST project AQUILA deliverable D1301 "Specification of traffic handling for the first trial", July 2000, http://www-st.inf.tu-dresden.de/aquila (Publications)

[6]   IST project AQUILA deliverable D2301 "Report on the development of measurement utilities for the first trial", September 2000, http://www-st.inf.tu-dresden.de/aquila (Publications)

[7]   IST project AQUILA deliverable D3201 "First Trial Report", June 2001, http://www-st.inf.tu-dresden.de/aquila (Publications)

[8]   B. Koch "A QoS architecture with adaptive resource control - The AQUILA approach" Interworking'2000 (Fifth International Symposium on Interworking), Bergen, Norway, October 3-6, 2000 available at http://www-st.inf.tu-dresden.de/aquila (Publications)

[9]   Proposed Chapter for SLSU WG (version 1) -- 23/02/01; http://www.ist-tequila.org/slsuwgv1.txt

[10]   Y. T'Joens et al, "Service Level Specification and Usage Framework", draft-manyfolks-sls-framework-00.txt, October 2000;

[11]   S. Salsano et al, "Definition and usage of SLSs in the AQUILA consortium", draft-salsano-aquila-sls-00.txt", November 2000; http://www-st.inf.tu-dresden.de/aquila/

[12]   B. Davie et al., "An Expedited Forwarding PHB", Internet Draft, draft-ietf-diffserv-rfc2598bis-01.txt, April 2001.

[13]   Final report COST 242, Broadband network teletraffic: Performance evaluation and design of broadband multiservice networks (J. Roberts, U. Mocci, J. Virtamo eds.), Lectures Notes in Computer Science 1155, Springer 1996.

[14]   "BGRP: A Tree-Based Aggregation Protocol for Inter-domain Reservations", P. Pan, E, Hahne, and H. Schulzrinne, Journal of Communications and Networks, Vol. 2, No. 2, June 2000, pp. 157-167

An Adaptive Periodic FEC Scheme
for Internet Video Applications

Tae-Uk Choi, Myoung-Kyoung Ji, Seong-Ho Park, and Ki-dong Chung

Department of Computer Science, Pusan National University,
Kumjeoung-Ku, Pusan, South Korea
{tuchoi,bluesky,shpark,kdchung}@melon.cs.pusan.ac.kr

Abstract. When transmitting packets compressed with a high compression standard such as MPEG or H.261, the loss of single packet has a considerable effect on the following frames because of motion estimation and compensation. There are many techniques that prevent this error propagation. We classify the video error control techniques into codec-level and network-level schemes and investigate the effect of various combinations. As the result, we propose a new FEC-based video error control scheme, Periodic FEC, which provides error resilience, adaptability to network conditions and the ability to combine with other scheme. Through experiments, it is confirmed that the Periodic FEC scheme is superior to other schemes such as FEC, RPS etc, and the performance of this scheme can be maximized when combined with other schemes.

1 Introduction

Real-time video transmission over the Internet is very challenging. In the current Internet, because network loss and delay are variable and bandwidth is limited, the QoS of video applications is not guaranteed and video transmissions require high compression efficiency. Video compression standards such as MPEG and H.261 are not designed for transmission over a lossy channel. Although they can achieve very impressive compression efficiency, even a small amount of data loss can severely degrade video quality. Namely, because the codec uses motion estimation and compensation to remove temporal redundancy in a video stream, packet loss in a frame can be propagated to the subsequent frame and get amplified.

Many video error control techniques have been proposed for the prevention of error propagation [1][2][3]. The simplest approach is to use intra-coded frames at periodic intervals. But, this clearly has large bandwidth requirements. Another approach is to intra-code and transmit only those blocks in a frame that change more than some threshold. Such a process is referred to as conditional replenishment and is used in nv [2] and vic [3].

We classify these video error control techniques into codec-level and network-level schemes. The codec-level scheme is a scheme that is implemented within an encoder and a decoder, and the network-level scheme is a scheme that can be implemented

S. Palazzo (Ed.): IWDC 2001, LNCS 2170, pp. 691–702, 2001.

only at the transmission level without codec's help. ET (Error Tracking) [4] and RPS (Reference Picture Selection) [5] are representative schemes of the codec-level. Also, retransmission and FEC (Forward Error Correction) are two major schemes at the network-level.

These schemes have their own strengths and weaknesses. To improve the performance, the schemes at one level can be combined with the techniques at another level. For example, FEC and RPS is a combination that provides good performance. When FEC cannot recover a lost frame due to long burst losses, RPS can prevent the loss from propagating to its successive frames. However, this combination still has its own weaknesses of the additional bandwidth overhead by FEC and supplementary buffer space by RPS.

After analyzing various combinations, we propose a new FEC-based error control scheme, called Periodic FEC, which provides error resilience with a smaller bandwidth than that of FEC and stops error propagation without a feedback channel. Moreover, the scheme can dynamically adjust the redundant information depending on the network loss rate, and its performance can be maximized when other feedback-based error control mechanisms are combined.

Through experiments, we show the effectiveness, adaptability and combinableness of the Periodic FEC. As the result, Periodic FEC scheme is superior to other schemes such as FEC, RPS etc and the performance can be maximized when it is combined with RPS.

The rest of this paper is organized as follows. Section 2 presents an overview of related works. Section 3 describes the Periodic FEC scheme and its combination schemes. Section 4 presents the effectiveness of Periodic FEC and its combination schemes through experiments. Section 5 presents the conclusion.

2 Related Works

Error control schemes can be classified into three types. The first type consists of codec-level error control schemes that are implemented at the coding level. The second type consists of network-level error control schemes that are implemented in at the transmission level. The third type consists of techniques that combine codec-level and network-level techniques. The following describes works related to each level.

2.1 Codec-Level Schemes

Recently, H.263+ incorporated two feedback-based error control techniques: error tracking (ET) and reference picture selection (RPS) [6]. ET requires the encoder to know the location and extent of erroneous image regions in displayed images. This scheme requires feedback messages from the receiver. The receiver sends information about missing packets and the encoder estimates the region of error propagation in the displayed images and intra-codes the blocks contained the region. This scheme is attractive since it does not require any modifications of the bit stream syntax of the motion-compensated coder.

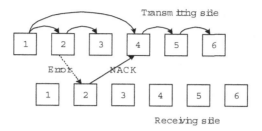

Fig. 1. Illustration of a typical reference picture selection

RPS allows the encoder to select one of several previously decoded frames as a reference picture for motion estimation. It is designed to support a coding technique called NEWPRED [5][7]. Fig. 1 illustrates the basic operation of a typical reference picture selection. When the receiver does not correctly decode a picture, it notifies the transmitter. Based on this notification, the transmitter determines the picture it will select as a reference. As shown in the figure, an error in picture 2 is propagated up to picture 3 because each picture refers to the previous picture or the last decoded picture. However, picture 4 refers to picture 1, which is correctly decoded. Thus, the receiving side eliminates error propagation without additional bandwidth overhead.

ET and NEWPRED have a limitation in that they should modify their picture coding patterns based on specific information about lost packets. Thus, continuous feedback from a receiver is essential for their performance.

2.2 Network-Level Schemes

Retransmission and FEC are the two major error control schemes at the network-level. The former can provide good error resilience without incurring much bandwidth overhead because packets are retransmitted only when there are indications of packet loss. Because retransmission always involves additional transmission delay, it is ineffective for interactive real-time video applications. However, it can be still a very effective technique for improving error resilience in interactive real-time video applications. In [8], a new retransmission-based error control technique is proposed that effectively alleviates error propagation in the transmission of interactive video.

In the latter case, redundant information is transmitted along with the original information so that the lost original data can be recovered based on the redundant information. This scheme is attractive because it provides resilience to loss without increasing latency. Many FEC-based schemes involve exclusive-OR operations [9]. These increase the send rate of the source by a factor of $1/k$, and they add latency since k packets must be received before the lost packet can be reconstructed. Bolot and Turletti [1] proposed an interesting FEC scheme for packet video where a packet contains the redundant information of some previous packets. The redundant information is created by encoding the image blocks contained in the previous packets with a large quantization step. They claimed that if the video source is not bursty and long burst losses are rare, and then their scheme would work well for video.

2.3 Combined Schemes

There are some schemes that combine codec-level and network-level schemes. In [10], packets arriving after their display times are not discarded but instead used to reduce error propagation. Although a periodic frame may be displayed with errors because of some loss of its packets, the errors will stop propagating beyond the next periodic frame because these losses can be recovered within a PTDD (Periodic Temporal Dependency Distance). When this scheme is combined with retransmission or FEC, the performance is effectively improved. However, under high motion scenes, it would be less effective, and it would need extra computation and buffer space at the receiver side.

3   Combined Error Control Schemes

3.1 Simple Combined Schemes

Simple combination schemes are considered in order to discover the effects of various combinations. When RPS or ET scheme of codec-level is combined with the retransmission scheme of the network level, the two are not complementary to each other because all of the combinations utilize a feedback-based mechanism. These schemes may require much delay time to retransmit the lost packet and wait the feedback message for RPS or ET. FEC, however, can combine with RPS or ET to improve error resilience because the two are complementary to each other. If FEC fails to recover the lost packet at the transmission level, RPS or ET will stop error propagation at the coding level. In the combination of FEC and ET, the bandwidth overhead may be large because of the redundant information of FEC and intra-coded blocks of ET. The combination of FEC and RPS is a good combination because when a lost frame is unrecoverable by FEC, RPS can eliminate error propagation with a low overhead.

To study the effectiveness of various combinations, we investigate a simple combination scheme. Fig.2 shows the combination of FEC and RPS, which RPS is used in the codec-level, and RES-based FEC scheme is used in the network-level. A Reed-Solomon erasure correcting code (RSE code) [11] is a commonly used FEC encoder where k source packets of P bits are encoded into $n(>k)$ packets of P bits (namely, k data packets plus n-k parity packets). This group of n packets is called an FEC block. The RSE decoder on the receiver side can reconstruct the source data packets using any k packets out of its FEC block. As shown in Fig. 2, the RSE decoder fails to recover frame 3 because three packets are lost in network. Thus, the receiver sends a NACK message to the sender. Based on these feedback messages, the encoder on the sender side modifies the reference of frame 4 to depend on frame 2. Thus, the error propagation due to the loss of frame 3 can be stopped. Consequently, while FEC helps reduce the occurrences of independent losses at the network level, RPS stops error propagation when losses are unrecoverable at the network level. However, since FEC needs additional bandwidth and RPS requires additional buffer space, this combination still has some weaknesses.

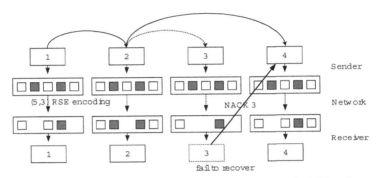

Fig. 2. Combination of RPS and FEC with a (5, 3) RSE code

3.2 Periodic FEC

Through the analysis of the simple combined schemes, we found that a new scheme should have the following features: Firstly, error propagation should be prevented and the redundant information should be minimized. Secondly, when it is combined with other schemes, the strengths should be maximized while weaknesses should be minimized. Lastly, it should adapt to the network condition to control the overhead.

Considering these features, we design a new FEC-based error control scheme: Periodic FEC. This scheme consists of the frame reference method in the encoder and the transmission method in the network. As shown in Fig. 3, every p-th frame is referred to as a periodic frame, and frames between two consecutive periodic frames are referred to as nonperiodic frames. Fig. 3 (a) shows that a periodic frame depends on its previous periodic frame for motion estimation, while every nonperiodic frame depends only on its immediately preceding periodic frame. Thus, if periodic frames are received safely, errors in nonperiodic frames do not propagate. For the safe transmission of periodic frames, the conventional FEC scheme is used. The redundant information of only periodic frames is encoded and then transmitted along with original data. Fig. 3 (b) shows how transmitted redundant and original data can be. The redundant data of a periodic frame is created based on its previous {n-i} periodic frames. We refer to the maximum value of i as the order of Periodic FEC. As shown in Fig. 3, when the order value is 1, if the periodic frame denoted as n is lost, then the receiver waits for periodic frame $n+1$, decodes the redundant information, and displays the reconstructed information. Consequently, this method stops error propagation without any feedback information.

Bolot et al. proposed an adaptive FEC scheme in which the amount of redundant information can be adjusted to the packet loss rate in the network by increasing or decreasing the order values [1]. As the order value becomes large, both the amount of redundant information and the probability of packet recovery increase. Like Bolot et al.'s scheme, Periodic FEC could adjust the amount of redundant information over time depending on the measured loss rate of periodic frames in the network. However, the packet loss process should be observed during long periods because the interval of periodic frames is long and the loss probability of periodic frame is reduced by FEC. Thus, this mechanism cannot quickly react to the network conditions. We propose a NACK-based redundancy control algorithm that can quickly react to network condition and easily be combined with other feedback-based error control algorithms.

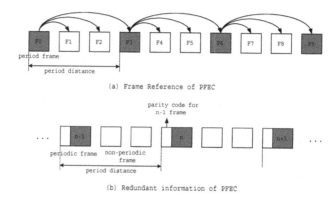

Fig. 3. Frame reference and redundant information of Periodic FEC

Fig. 4 shows the proposed algorithm based on the length of burst NACK of periodic frames. Initially, the order value is set to 0: no redundancy data is sent. When a NACK message is received, the order value is set to 1. And, when two NACK messages are received successively, the length of the burst NACK is 2, and the order value is set to 2. On the other hand, when the NACK message is not received for constant time MAX_WAITING_TIME, the order value is set to 0. Consequently, this algorithm can quickly react to NACK messages, reduce the additional redundancy of the Periodic FEC scheme and improve the video quality by dynamically adjusting the order value.

Another approach to adjusting the amount of redundant information is to increase or decrease the distance of a period. As the distance increases, the amount of redundant data decreases because only a periodic frame is transmitted along with the redundant data in a period. When a NACK message is received, this mechanism reduces the distance of a period, and when a NACK message is not observed for a constant time, it increases the distance. However, this approach cannot recover successive packet losses because the order value is fixed. In worst case, if the period distance is 1, the Periodic FEC will functions like the conventional FEC.

```
IF a NACK messages is received THEN
    BEGIN
        check the length of Burst NACK L_{Burst_NACK};
        IF L_{Burst_NACK} <> order THEN order = L_{Burst_NACK};
    END
ELSE
    BEGIN
        Check the elapse time T_{elaps} since the last NACK
was  received;
        IF T_{elaps} >= MAX_WAITING_TIME THEN order = 0;
    END
```

Fig. 4. A NACK-based redundancy control algorithm

3.3 Complementary Schemes

In Periodic FEC, a periodic frame may not be recovered from the redundant information if packet losses of periodic frames would occur consecutively in the network. Then, any frames in the period cannot be decoded, and the error may be propagated to the next periodic frame. To prevent this error propagation, Periodic FEC can be combined with a complementary scheme such as retransmission, intra-frame replenishment, or reference picture selection.

One approach is the combination with the inter-frame replenishment scheme. In RESCU[6], periodic frames cannot be recovered before the decoding of their dependant frame. This leads to error propagation and thus intra-frame replenishment was used to prevent error propagation. The receiver notifies the sender about the irrecoverable losses, and the notification triggers the sender to code the next frame as an intra-frame. Like RESCU, Periodic FEC can be combined with intra-frame replenishment.

Another approach is retransmission. When the receiver notifies the sender about the loss of a periodic frame, the sender retransmits only the redundant information of the lost periodic frame along with the next periodic frame.

The third approach is to use the RPS method in the codec. When the receiver cannot recover a periodic frame, it sends a NACK message to the sender. Using this feedback message, the encoder uses for motion prediction the periodic frame that is not reported missing. Thus, error propagation is eliminated without increasing bandwidth. Moreover, this approach can be nicely matched with the NACK-based redundancy control algorithm of Periodic FEC.

4 Experiments

The objective of our experimental work is to study the potential effectiveness of Periodic FEC and its combination schemes. To achieve this objective, it is first shown that Periodic FEC has the ability to adapt to various network conditions. And, the superior performance of the Periodic FEC scheme to other existing recovery schemes is investigated in terms of SNR and bit overhead.

We modified the telenor H.263 codec source to implement the Periodic FEC scheme and other schemes, and conducted actual video transmission tests over the Internet from Pusan to Seoul, South Korea. The transmission tests were conducted between 2 p.m. and 6 p.m. for 3 days. The average loss rate was about 10%, and the average delay was about 100ms. During the tests, the byte rate transmitted from the sender and SNR at the decoder were measured to compare their performance. The byte rate indicates bandwidth overhead, and SNR indicates video quality.

For convenience, we refer to the Periodic FEC scheme as PFEC, and the combination scheme of FEC and RPS is referred to as FEC+RPS. Also, PFEC of which the order value is fixed is called static PFEC, and PFEC of which the order value is changed depending on network loss is called dynamic PFEC, which is referred to as DPFEC

4.1 A Comparison between Static and Dynamic PFEC

To show the ability to control redundant information according to various packet losses, we compared dynamic PFEC to static PFEC that had the order value of 1. As shown in the first half of the graphs in Fig. 5, the dynamic PFEC sends no redundant data because packet loss does not occur, while its SNR value is similar to that of the static PFEC. And in the latter half of the graphs, the order value is increased up to 2, due to the consecutive losses of periodic frames. Thus, more of the redundant data of the dynamic PFEC is transmitted more, and the SNR value increase more than that of the static PFEC. Consequently, the dynamic PFEC is superior to the static PFEC because the static PFEC encodes and transmits the fixed amount of redundant data of all frames, while the dynamic PFEC encodes the redundant data only when lost frames exist. Table 1 shows the average performance of the two schemes. The dynamic PFEC can reduce the byte rate by 20% while its SNR is similar to that of the static PFEC.

Table 1. Average SNR and byte rates of the static and the dynamic PFEC

	Avg. SNR	Avg. Byte rate
Static PFEC	18.2	1348.2
Dynamic PFEC	19.1	1111.8

4.2 A Comparison between the Dynamic PFEC and Other Schemes

The dynamic PFEC is compared with conventional schemes such as FEC, RPS to show the superior performance, and combination schemes such as FEC+RPS, DPFEC+ RPS to show the effectiveness of combination. It is assumed that the order of the FEC scheme is 1. Table 2 shows the average SNR and the byte rate of each scheme.

Table 2. The average SNR and the byte rate of five types of schemes

	Avg. SNR	Avg. Byte rate
FEC	18.8	1607.5
RPS	16.8	1047.4
FEC+RPS	20.5	1640.8
Dynamic PFEC	18.7	1111.8
DPFEC+RPS	19.2	1136.9

Fig. 6 shows the SNR and the byte rate of the FEC and the dynamic PFEC. The FEC provides additional bandwidth overhead because it encodes the redundant data of all the frames. Also, its SNR value is not high because it has no mechanism for preventing error propagation. However, the dynamic PFEC can stop error propagation and minimize the amount of redundant data. As shown in Table 2, the FEC transmits 1.4 times more redundant data than the dynamic PFEC does, while their SNR values are similar.

Fig. 7 shows a comparison of the dynamic PFEC to RPS. RPS depends on the previously decoded frames for motion estimation. In consecutive packet losses, the interval between the encoding frame and the reference frame is increased and redundant information is reduced, resulting in a high frame rate. As shown in the latter part of the graphs in the Fig. 7, where the packet losses are heavy, the byte rate of RPS increases, and its SNR value becomes lower than that of the dynamic PFEC. Table 2 shows that the dynamic PFEC can improve video quality (SNR) with small additional amount of redundant information.

Fig. 8 shows a comparison of dynamic PFEC to RPS+ FEC. RPS+FEC provides a good performance in terms of the SNR value and a poor performance in terms of the byte rate. This is a typical example that indicated the tradeoff between redundancy and video quality: the greater the amount of redundancy, the higher the performance. However, as shown in table 2, RPS+FEC encodes 1.4 times more redundant data, but the improvement of SNR was at most 9%.

Fig. 9 shows a comparison of dynamic PFEC to the combination scheme, DPFEC+ RPS. As shown in the figure, the SNR and the byte rate of DPFEC+RPS are similar to those of Dynamic PFEC. However, when packet losses are heavy, the dynamic PFEC could not recover periodic frames and prevent error propagation. Thus, under situation of high packet loss, the SNR of DPFEC+RPS is higher than that of Dynamic PFEC.

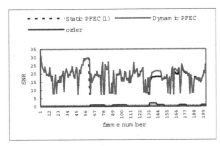
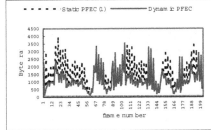

Fig. 5. Comparison between the dynamic PFEC and the static PFEC

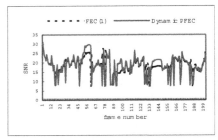
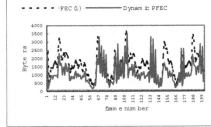

Fig. 6. Comparison between the dynamic PFEC and FEC

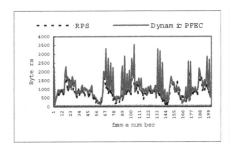
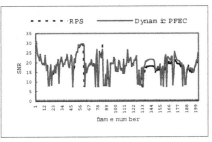

Fig. 7. Comparison between the dynamic PFEC and RPS

In summary, the dynamic PFEC is superior to other schemes because it provides a good SNR at the cost of only a small amount of redundancy. Moreover, when it is combined with RPS, this scheme provides the maximum performance by preventing error propagation perfectly.

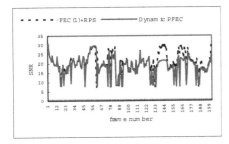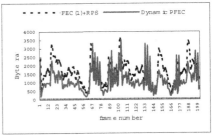

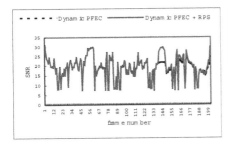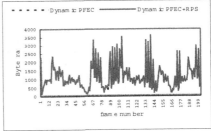

Fig. 8. Comparison between the dynamic PFEC and FEC+RPS

Fig. 9. Comparison between the dynamic PFEC and DPFEC + RPS

5 Conclusion

This paper classifies the video error control techniques into codec-level and network-level schemes, and investigates the effects of various combinations of these schemes. After considering strengths and weakness of combination schemes, we propose a new FEC-based video error control scheme, the Periodic FEC, which prevents of error propagation and provides a loss recovery mechanism with small bandwidth overhead. Also, the scheme can adjust the amount of redundant data, depending on network conditions. Through experiments, we show the effectiveness and adaptability of the dynamic PFEC and the performance of this scheme is maximized when combined with RPS. Future studies will investigate the effectiveness of combinations of other schemes not discussed in this paper.

References

1. J-C.Bolot, T.Turletti, "Adaptive error control for packet video in the Internet", in Proceedings of International Conference on Internet Protocols, Lausanne, September, 1996

2. R. Frederick, "Experiences with real-time software video compression", Proc. 6th Packet Video Workshop, Portland, OR, Sept. 1994
3. S.McCanne and V.Jacobson,"Vic: A flexible framework for packet video." In Proc. ACM SIGCOMM, Philadelphia, PA, Sept.1990
4. E.Steinbach, N.Farber, and B. Girod, "Standard compatible extension of H.263 for robust video transmission in mobile environment," IEEE Trans. Circuits Syst. Video Tech., vol.7, pp.872-881, Dec.1997
5. Kimata, H. Tomita, Y. Yamaguchi, H. Ichinose, S. "Study on adaptive reference picture selection coding scheme for the NEWPRED-receiver-oriented mobile visual communication system", GLOBECOM 1998, Volume 3.
6. ITU-T, Recommendation H.263 version 2, "Video coding for low bitrate communication", Jan. 1998
7. LBC, "An error-resilience method based on back channel signaling and FEC", Document, LBC-96-033(ITU-T SG 15/1), 1996
8. Injong Rhee, "Error control techniques for interactive loww-bit rate video transmission over the Internet", in Proc. ACM SIGCOMM'98, Vancouver, Canada, Sept. 1998.
9. N.Shacham, P.McKenney, "Packet recovery in high-speed networks using coding and buffer management", in Proceedings of IEEE Infocom'90, San Fransisco, CA, May 1996.
10. I.Rhee, S.R.Joshi, "Error recovery for interactive video transmission over the Internet", IEEE journal on selected on areas in communications, VOL.18, NO 6, June, 2000.
11. Anthony J. MaAuley, "Reliable broadband communication using a burst erasure correcting code", In Proc. ACM SIGCOMM, Philadelphia, PA, Sept. 1990.

Efficient Interaction with Virtual Worlds over IP

R. Bernardini and G.M. Cortelazzo

Dipartimento di Elettronica ed Informatica
Università di Padova
Via Gradenigo 6A, 35100 Padova
{bernardi, corte}@dei.unipd.it

Abstract. The construction of photo-realistic virtual worlds is at reach of current computer graphics. Unfortunately, the philosophy currently adopted for the diffusion of virtual worlds over the Internet has a fundamental drawback: it calls for downloading at client side the 3D virtual world description. Recently, the split-browser approach was proposed in order to solve this issue by transforming the problem of interacting with a virtual worlds in a image transmission task. This work addresses the crucial issue of compressing the image stream generated in the split-browser approach. The proposed compression scheme first decomposes the virtual world as a union of objects, successively it approximates each object as a polyhedron and finally, it compresses the image of each face of the polyhedron by predicting it with a projective transformation.

1 Introduction

Three-dimensional virtual environments are often described in VRML, a language tailored for the Internet exchange of 3D virtual worlds. Unfortunately, the current way of accessing VRML described virtual worlds (i.e. downloading the VRML file to the user's computer, then render it) is plagued by several problems such as long downloading times (a good quality VRML file can be as large as 100 MB), poor interaction (due to the huge complexity of VRML files which would require very powerful computers in order to have a smooth navigation) and loss of copyright control (producing a good quality VRML file can require several months of work, but if the user can download it, the author looses the copyright control).

In order to solve such problems, in [1] the alternative *split-browser* approach for virtual worlds browsing was proposed. The split-browser approach is based on the idea of performing the rendering at the server's side and sending the resulting images to the client, transforming the interaction with the VRML file in a problem of image transmission.

It is clear that a major problem that must be solved in order to make the split-browser approach effective, is the compression of the generated images. Although such a problem could be solved by means of off-the-shelf solutions such as JPEG or MPEG, the images transmitted with the split-browser approach have the peculiarity of being artificially generated by the server and this suggests that a specifically tailored compression technique could result in a greater efficiency.

S. Palazzo (Ed.): IWDC 2001, LNCS 2170, pp. 703–712, 2001.

This paper presents a compression scheme that we are currently developing in order to allow efficient access to virtual worlds over the Internet. In the proposed compression scheme the sequence of rendered views is compressed by splitting it in smaller images, which in turn are compressed by means of an MPEG-like approach, but using "generalized motion vectors" representing projective transformations.

This paper has 5 sections. Section 2 briefly recalls the split-browser approach proposed in [1], Section 3 describes the compression scheme that we are developing, Section 4 briefly describes other issues related to the efficient access to remote 3D virtual environments. Section 5 gives the conclusions.

2   A Split-Browser Structure

The recent development of 3D imaging technologies, e.g. the availability of range cameras and semiautomatic 3D modeling tools, makes feasible the construction of 3D models of real objects and suggests their use for interactive applications such as virtual visits.

The only currently available way to view a 3D model, is by locally downloading a file (e.g. in VRML format) and having the client rendering it. This approach, however, has several drawbacks

- Descriptions of 3D models can be quite large. As an example, the full, uncompressed model of "Madonna con bambino" by Giovanni Pisano (Fig. 1) is approximately 100 Mbytes.
- Rendering with good quality a VRML file requires a lot of memory and computational power, often not available to home users. Interacting with virtual objects and environments can be quite frustrating if the rendering is too slow.
- 3D model construction is a labor-intensive task and it is reasonable to assume that one may want to keep the copyright of the 3D model. If the electronic description of the model can be downloaded via Internet, copyright control becomes very challenging.

In order to solve these problems, in [1] the split-browser approach was proposed. In order to make this paper self-contained, we will briefly recall the main ideas behind the split-browser concept. As a first step, let us analyze how a VRML browser works. In Fig. 2a one can see the internal structure of a generic VRML browser: the end user interacts with a Graphical User Interface (GUI) which, by means of events, controls a virtual world and a graphical engine (GE) whose task is to produce 2D views of the virtual world.

The solution proposed in [1] splits the browser in two: the part with the GE is moved to the server and the part with the GUI remains at the client (see Fig. 2b). The internal events generated by the GUI in Fig. 2a are now transmitted to the server along the network. At the server's side the GE responds by sending back to the client the updated views as images. This approach, which essentially turns the interactive inspection of a 3D model into an image transmission task, has the following advantages

- The amount of data transmitted from the server to the client is much smaller than sending the full 3D model.

Fig. 1. 3D View of the complete model of "Madonna con Bambino" by Giovanni Pisano (Arena Chapel, Padova).

- Visualization at the client side does not require strong computational capabilities anymore.
- The copyright control of the 3D data is preserved, since the end user does not receive the whole model, but only 2D views of it.

Observe also that the split-browser idea allows several users to interact with the same virtual environment, making possible applications like network video game and sharing of scientific data.

Although the proposed solution is conceptually very simple, several questions must be answered in order to make it suitable for applications. A fundamental issue is about compression schemes suited to make more efficient the downloading of the updated views. Minor (but important) issues concern the transport protocol to be used for sending the compressed images and how to avoid problems due to packet loss when sending the packets over IP. In the following sections we are going to address such questions.

3   A Compression Scheme

Although off-the-shelf solutions (e.g., JPEG or MPEG) could be used in order to compress the rendered view, it is well worth developing a compression scheme tailored to this particular application in order to exploit the fact that the images sent to the client are artificially generated.

3.1   Images of Planar and Quasi-planar Objects

Consider the situation depicted in Fig. 3 where the virtual world contains just a planar object $O(\mathbf{x})$ (a picture, for example), V_1 is the position of the user's "virtual camera" and $L_1(\mathbf{x}) : \mathbb{R}^2 \to \mathbb{R}$ is the image shown on the user's screen. Suppose now the user

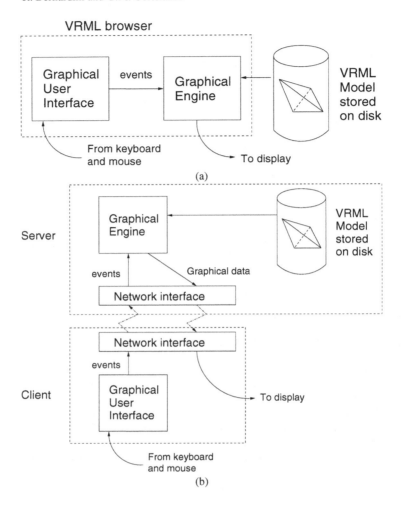

Fig. 2. The Split-browser approach: (a) The internal structure of a generic VRML browser. (b) In the split-browser approach the graphical engine remotely runs at the server and sends the computed views to the client.

moves to V_2 and let $L_2(\mathbf{x})$ be the corresponding view. Our goal is to compress $L_2(\mathbf{x})$ by (eventually) exploiting the fact that L_1 and L_2 are different views of the same object.

It is well-known that if O is planar, there is a simple relationship between L_1 and L_2, more precisely,

$$L_2(\mathbf{x}) = L_1\left(\frac{\mathbf{A}\mathbf{x} + \mathbf{b}}{\mathbf{c}^t\mathbf{x} + d}\right) \tag{1}$$

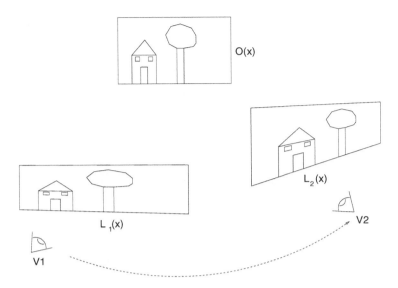

Fig. 3. Two views L_1 and L_2 of the same planar object are related by a projective transformation.

where $\mathbf{A}$ is a 2×2 real matrix, $\mathbf{b}, \mathbf{c} \in \mathbb{R}^2$ and $d \in \mathbb{R}$. Transformations of type (1) are called *projective transformations*. Parameters $\mathbf{A}, \mathbf{b}, \mathbf{c}$ ad d can be easily be computed by knowing the positions of the planar object O and the user's virtual camera.

If the object is not planar, but it can be fairly well approximated by a planar object, (i.e., it is a *quasi-planar* object)[1], it is reasonable to expect that (1) "almost" holds, in the sense that the power of error image

$$E(\mathbf{x}) \stackrel{\triangle}{=} L_2(\mathbf{x}) - L_1 \left(\frac{\mathbf{A}\mathbf{x} + \mathbf{b}}{\mathbf{c}^t\mathbf{x} + d} \right) \tag{2}$$

is small. This suggests to transmit L_2 by sending the four parameters $\mathbf{A}, \mathbf{b}, \mathbf{c}$ and d and the error image E compressed by means of a lossless or lossy technique. This approach resembles the approach used in MPEG, but with the motion vectors replaced by the "generalized motion vector" represented by $\mathbf{A}, \mathbf{b}, \mathbf{c}$ and d.

3.2 Images of General Objects

It is clear that the hypothesis of planar (or quasi-planar) object is a very strong one since most objects cannot be considered quasi-planar. In such cases, the object is first roughly approximated by means of a polyhedra whose faces correspond to quasi-planar zones of the original object. Successively, the region relative to each face of the polyhedron is compressed as a quasi-planar object. The rough approximation of the object can be easily obtained by simplifying the original model with well-known algorithms.

---

[1] Note that the condition of being quasi-planar it depends both on the object and on the viewing distance

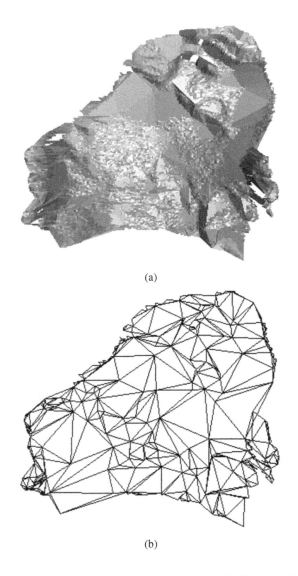

(a)

(b)

Fig. 4. (a) Close-up of the 3D model of *Madonna con Bambino*. (b) Example of a rough approximation of (a) by means of a planar faces.

More precisely, the compression algorithm we suggest is the following

1. Determine the set $\mathcal{F}$ of the faces visible in both the old and the new view
2. For each face $F \in \mathcal{F}$ found at the previous step
   (a) Determine the projection P_F of F on the old image

(b) Determine the parameters $\mathbf{A}_F$, $\mathbf{b}_F$, $\mathbf{c}_F$, d_F of the corresponding projective transformation

(c) Compute the distorted version

$$M_F(\mathbf{x}) = L_1 \left(\frac{\mathbf{A}_F\mathbf{x} + \mathbf{b}_F}{\mathbf{c}_F^t\mathbf{x} + d_F} \right) \chi_{P_F} \left(\frac{\mathbf{A}_F\mathbf{x} + \mathbf{b}_F}{\mathbf{c}_F^t\mathbf{x} + d_F} \right) \tag{3}$$

where $\chi_{P_F}(\mathbf{x}) = 1$ if $\mathbf{x} \in P_F$ and $\chi_{P_F}(\mathbf{x}) = 0$ otherwise.

3. Compute the predicted image

$$\hat{L}_2(\mathbf{x}) = \sum_{F \in \mathcal{F}} M_F(\mathbf{x}) \tag{4}$$

4. Compute the prediction error

$$E(\mathbf{x}) = L_2(\mathbf{x}) - \hat{L}_2(\mathbf{x}) \tag{5}$$

5. For each face $F \in \mathcal{F}$ transmit
   - The parameters $\mathbf{A}_F$, $\mathbf{b}_F$, $\mathbf{c}_F$, d_F
   - The vertices of P_F
6. Transmit the prediction error $E(\mathbf{x})$, suitably compressed.

Observe that the overall scheme is lossy or lossless, depending on the compression method used for E. A schematic picture of the proposed compression scheme can be found in Fig. 5.

It is worth noting in Fig. 5 the presence of a cache. The motivation for its introduction is that while browsing a virtual world is very common to return in an already visited positions. It is clear that in such a case the rendered image will be equal to the image already sent to the client. In order to exploit this, it is convenient to keep a cache where the more recently generated views are saved.

Note from Fig. 5 that the server uses the cached images also as reference images to be used in the compression scheme. This allows the server to use as reference any previously generated image and not necessarily the current one.

3.3   The Sprite Model

Until now we made the hypothesis that the virtual world contains only one object. Although this could be true in some applications (for example in a close-up of a statue during a virtual visit or in an e-commerce context), it is clear that several other applications (e.g. video games) will have virtual worlds populated by several objects.

The compression scheme presented in the previous sections can be nevertheless still used by exploiting the *sprite model* introduced in [1]. Within the sprite model, each image sent to the client is considered as a set of several independent layered images (with non-rectangular support) called *sprites*. Each sprite is the rendered view of a single object or a part of an object.

A sprite is completely described by its support and by the RGB values associated to points of its support. The support can be described as a polygon (for simple supports, like the support of the sprite relative to a check-board) or by means of a transparency component, also known as alpha-channel.

It is worth observing that the task of composing several sprites into a single image is well within the capabilities of current personal computers.

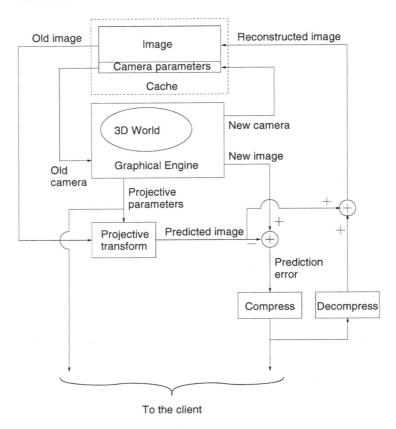

Fig. 5. Proposed compression scheme.

4   Other Issues

4.1   Progressive Transmission

The prediction error can be compressed both in a lossy or lossless way. If a lossy scheme is chosen, it can be convenient to use a wavelet-based one such as SPIHT [2] or EZW [3] which allow for progressive transmission (i.e., a coarse approximation of the compressed image can be reconstructed from any initial part of the resulting bit-stream). In order to understand why this can be convenient, consider a virtual world with a moving object. Each time the object moves the GE must generate a new view and send it to the client. If the object moves very fast, it could happen that the GE generates a new view before the complete description of the previous one is sent to the client. In applications like network games this would bring to an unacceptable loss of synchronization between the virtual world and the view shown to the users. However, since the view is changing rapidly, it is reasonable to assume that it should be possible to lower the quality of the compressed image since the user will not notice it. If the bit-stream allows for progressive transmission, this can be achieved by just stopping the transmission of the old image and beginning to transmit the new one.

4.2   Choosing the Transport Protocol and Resilience to Packet Loss

At the moment, in our experimental implementation, we are using TCP as a transport protocol because of its simplicity. However, the retransmission mechanism used in TCP could introduce too high an overhead in the interaction between the user and the virtual world. Because of this, it could be convenient to switch to an UDP-based protocol, such as RTP. This choice, however, introduces the problem of dealing with packet losses. Since the bit-stream created by the split-browser approach contains more precious data (projective transform parameters and low-resolution components) and less precious ones (high-resolution error components), we plan to use priority encoded transmission [4], [5].

5   Conclusions

The construction of articulated and photo-realistic 3D virtual worlds is at reach of current computer graphics and 3D imaging technology. The philosophy currently adopted for the diffusion of 3D virtual worlds over the Internet via the VRML and/or current related extensions has a fundamental drawback: it calls for downloading at client side the 3D virtual world description. The split-browser approach solves this issue by giving an alternative solution which transforms the problem of interacting with a virtual worlds in an image transmission task.

This work addresses the crucial issue of compressing the image stream in order to allow smooth interaction between the user and the virtual world. The proposed compression scheme first decomposes the virtual world as a union of objects, successively it approximates each object as a polyhedron, and, finally, it compresses the image of each face of the polyhedron by predicting it with a projective transformation and sending the parameters of the projective transformation together with the prediction error suitably compressed. In order to allow automatic bandwidth/image quality tradeoff, the prediction error should be compressed with a multiresolution scheme which allows for progressive transmission (e.g. SPIHT or EZW).

Future research will address the issue of the transmission protocol to be used for the split-browser structure and (eventually) how to deal with packet loss.

References

1. R. Bernardini and G. M. Cortelazzo, "An efficient network protocol for virtual worlds browsing," in *The 12th Tyrrhenian International Workshop on Digital Communications "Software Radio Technologies and Services"*, (Portoferraio, Italy), CNIT, Springer-Verlag, Berlin, Sept. 2000.
2. A. Said and W. A. Pearlman, "A new fast and efficient image codec based on set partitioning in hierarchical trees," *IEEE Transactions on Circuits and Systems for Video Technology*, vol. 6, pp. 243–250, 1996.

3. J. Shapiro, "Embedded image coding using zerotrees of wavelet coefficients," *IEEE Transaction on Signal Processing*, vol. 41, pp. 3445–3462, Dec. 1993.
4. J. Albanese, Blomer, J. Edmonds, M. Luby, and M. Sudan, "Priority encoded transmission," *IEEE Transaction on Information Theory*, vol. 42, pp. 1737–1744, Nov. 1996.
5. A. Mohr, E. Riskin, and R. Ladner, "Unequal loss protection: Graceful degradation of image quality over packet erasure channels through forward error correction," in *Proc. of DCC*, 1999.

MPEG-4 Video Data Protection for Internet Distribution

F. Bartolini[1], V. Cappellini[1], R. Caldelli[1], A. De Rosa[1], A. Piva[1], and M. Barni[2]

[1] Dipartimento di Elettronica e Telecomunicazioni, Università di Firenze,
Via S. Marta 3, 50139 Firenze, Italy
{barto, cappellini, caldelli, derosa, piva}@lci.det.unifi.it
[2] Dipartimento di Ingegneria dell'Informazione, Università di Siena
Via Roma 56, 53100 Siena, Italy
barni@dii.unisi.it

Abstract. The increasing availability of multimedia contents and particularly the rapid development of the Internet have determined new ways to distribute information documents. In this scenario a copy protection system allowing to control the distribution of multimedia data is hardly required. A new technology useful for copyright protection is watermarking: a digital code (watermark), indicating the copyright owner, is directly embedded into the video signal. In this paper specific attention has been paid to the standard MPEG-4 and a digital video watermarking system has been designed in such a way that the complexity of the standard and the diversity of its applications are considered. In particular, the possibility of the MPEG-4 standard to directly access objects within a video sequence introduces a new constraint to the watermarking process: even if a video object is transferred from a sequence to another, the copyright data of the single object has to be correctly detected. Moreover, to make the watermarking system robust against format conversions, the code has to be inserted before compression. The method proposed in this paper satisfies the previous requirements by relying on an image watermarking algorithm which embeds the code in the Discrete Wavelet Transform of each frame.

1 Introduction

The day by day increasing availability of multimedia contents and the rapid development of communication networks, in particular the Internet, have determined that a new and unusual way to distribute information is offered. The huge growth of Internet users has created the chance to easily reach many people all over the world and a new market has been starting in the last few years. Anyway, to effectively devolop, this trade needs to grant security during transactions and to avoid all the partecipating partners to be afraid of making their business on the net. Here after the specific case of video, according to the MPEG-4 standard, has been taken into account and the problems concerned with this medium has been investigated. The MPEG-4 standard [1] is very attractive for a large set of applications such as video editing, internet video distribution, wireless video communications. Each of these applications has a set of requirements regarding protection of the information it manages. Hence a copy protection system allowing to limit the duplication of multimedia data and to make broadcast monitoring possible is considered a mandatory requirement by multimedia data owners [2]. A new technology

S. Palazzo (Ed.): IWDC 2001, LNCS 2170, pp. 713–720, 2001.

that seems useful for copyright protection is watermarking: a watermarking system embeds a digital code (watermark), that can be used to indicate the copyright owner, directly into the video signal.

A digital video watermarking system has to be designed so that some basic requirements are satisfied: the embedded watermark should be perceptually invisible; the copyright information should be robust against processing which do not seriously degrade the quality of the image. The false positive rate should be extremely low. Finally, the embedded watermark should be robust against attacks like cutting one or more frames of the video. To avoid this kind of attacks it is necessary to insert the copyright information continuously in the video sequence.

In general, existing video watermarking methods have been conceived to work with MPEG-1 or MPEG-2 streams. These algorithms can be classified into two main classes according to the type of content watermarking is applied to: raw-video watermarking algorithms ([3,4,5,6,7]) which add the watermark before compression, or bit-stream watermarking systems ([8,9]) which embed the code after compression.

A digital video watermarking system for MPEG-4 video has to be designed so that the complexity of the standard and the diversity of its applications have to be considered: in particular the main difference from previous video standards such as MPEG-1 and MPEG-2 is that MPEG-4 coding is content-based, i.e. single objects are coded individually. Each frame of an input sequence is segmented into a number of arbitrarily shaped regions, Video Object Planes (VOPs), and the shape, motion and texture information of the VOPs belonging to the same Video Object (VO) are coded into a separate Video Object Layer (VOL). Hence, the possibility of the MPEG-4 standard to directly access and manipulate objects within a video sequence introduces a new constraint to the watermarking process. An object watermarking system has to be designed in such a way that, even if a video object is transferred from a sequence to another, it is still possible to correctly detect the copyright data relating to the single object.

A watermarking technique designed for MPEG-4 video streams has been proposed in [10]. This algorithm embeds a watermark in each video object of an MPEG-4 coded video bit-stream by imposing specific relationships between some predefined pairs of quantized DCT middle frequency coefficients in the luminance blocks of pseudo-randomly selected macroblocks. The quantized coefficients are recovered from the MPEG-4 bit-stream, they are modified to embed the watermark and then encoded again. The main drawback of this technique is that, since the code is directly embedded into the compressed MPEG-4 bit-stream, the copyright information is lost if the video file is converted to a different compression standard, like MPEG-2. In order to be robust against format conversions, the watermark has to be inserted before compression, i.e. frame by frame. The problem is that if a Video Object (VO) of the scene is very small, it is very difficult to embed into it a watermark robust to the MPEG-4 compression. The method proposed in this paper satisfies the previous requirements: it belongs to the category of raw-video watermarking algorithms, since it operates frame by frame by casting a different watermark in each video object of an MPEG-4 coded video bit-stream. Watermarking relies on the image watermarking algorithm presented in [11], which embeds the code in the Discrete Wavelet Transform (DWT) domain [12]. The proposed system is invariant to

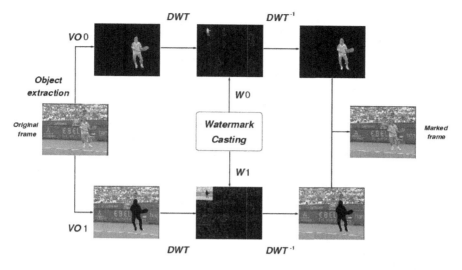

Fig. 1. Watermark casting process

the conversion to a different compression standard, as described later, and it is able to detect a watermark also in very small regions of an image [11].

2 Watermark Embedding

To embed the watermarks, the MPEG-4 coded video bit-stream is decoded obtaining a sequence of frames. The procedure described in Figure 1 is then applied frame by frame. Objects contained in the frame are extracted obtaining different images; in each image a different code is embedded by means of the system presented in [11] and resumed in the following.

The image to be watermarked is first decomposed through DWT in four levels: let us call I_j^θ the sub-band at resolution level $j = 0, 1, 2, 3$ and having orientation θ where $\theta = LL, LH, HL, HH$. The watermark, consisting of a pseudo-random binary sequence, is inserted by modifying the wavelet coefficients belonging to the three detail bands at level 0, i.e. I_0^{LH}, I_0^{HL} and I_0^{HH}. This choice is motivated by experimental tests showing that it offers the best compromise between robustness and invisibility. Before adding it to the DWT values, each binary value is multiplied by a weighting parameter which is obtained by a noise sensitivity function. In this way the maximum tolerable level of disturb (i.e. watermark coefficient) is added to each DWT coefficient. The construction of the sensitivity function is mainly based on the analysis of the degree of image activity in the neighborhood of the pixel to be modified.

In particular, given a code sequence $x_i \in \{+1, -1\}$, with $i = 0, \ldots 3MN - 1$, where $2M \times 2N$ is the image size, the three detail sub-bands are modified in this way:

$$\tilde{I}_0^{LH}(i,j) = I_0^{LH}(i,j) + \alpha w^{LH}(i,j) x_{iN+j}$$
$$\tilde{I}_0^{HL}(i,j) = I_0^{HL}(i,j) + \alpha w^{HL}(i,j) x_{MN+iN+j} \tag{1}$$
$$\tilde{I}_0^{HH}(i,j) = I_0^{HH}(i,j) + \alpha w^{HH}(i,j) x_{2MN+iN+j}$$

where α is a parameter accounting for watermark strength and $w(i,j)$ is a weighting function taking into account the local sensitivity of the image to noise.

The inverse DWT is computed. The watermarked images are then mixed together in order to obtain the frame containing the copyright information concerning the considered objects. When all the frames have been marked, the sequence is compressed, obtaining the watermarked MPEG-4 coded bit-stream.

3 Watermark Detection

In this section the process of watermark detection is analysed. The watermarked MPEG-4 coded video bit-stream is decoded obtaining a sequence of frames. Once again, the objects present in the scene are extracted frame by frame, obtaining a different image for each object. The DWT of each image is then computed; next, the code corresponding to the object is detected by means of the correlation between the watermark, and the DWT marked coefficients. The value of the correlation is compared to a threshold to decide if the watermark is present or not. An optimum threshold has been theoretically set to minimize the probability of false positive detection. In particular, the value of this threshold depends on the variance of the DWT coefficients of the watermarked image, and can thus be computed *a-posteriori*, without the need of knowing data concerning the original frame or the watermark embedding process.

Since the detection process is computed frame by frame, the watermark embedded into a video object can be revealed also if the VO is transferred from a sequence to another. However, since DWT is not invariant to translation, if the video object is placed in a different position of the new scene, the synchronization between the watermark and the VO is lost. To cope with this, the watermark detector needs to compute the correlation for all shifts of the frame. To reduce the computational complexity, the 2 dimensional Fourier Transform can be used: the peak position of the Fourier Transform in fact indicates the translation of the VO required to recover the synchronization before computing the watermark detection.

4 Experimental Results

The proposed watermarking algorithm has been tested on different video sequences. Here, the results concerning the CIF sequence *Stefan* are shown. Each frame is composed by two different objects, the background (Video Object 1) and the player (Video Object 0). We introduced a different random sequence into each object, as in Figure 1, and after watermarking the sequence was compressed obtaining a MPEG-4 coded video bit-stream, with a rate of 500 Kb/s per Video Object Layer (VOL). The video stream was next decompressed and in each frame the two objects were separated obtaining two different images, where the detection process was applied. Detection results are shown in Figure 2 (a) for VO 0, and (b) for VO 1: the response of the watermark detector to 200 marks randomly generated shows that the response to the embedded watermark sequence (i.e. no. 10 for VO 0, no. 20 for VO 1) is much larger than the response to the others and it is higher than the detection threshold. This result is interesting, since the player is a very small object that was heavily compressed after watermark embedding. Moreover, the detection process was also applied to the complete frame, without selecting the objects: as demonstrated in Figure 2 (c), the two watermarks embedded in the two objects are easily detected; let us note that the correlation peak of the tennis player (VO 0) is lower than the response of the background (VO 1), due to the fact that a low watermark energy can be embedded into the tennis player. The correct detection of the two objects indicates that the system is robust to conversion from MPEG-4 to MPEG-2, since the two watermarks are revealed in the frame, even if, clearly, in this case it is no longer possible to associate the watermark to the corresponding Video Object.

A new test has been done by introducing into the video sequence *Flowers*, previously watermarked with the code number 100 (see Figure 3 (a)), the Video Object represented by the player (Figure 3 (b)), obtaining a new sequence where Stefan seems playing in the flower garden (Figure 3 (c)). We applied the watermark detector to this new sequence,

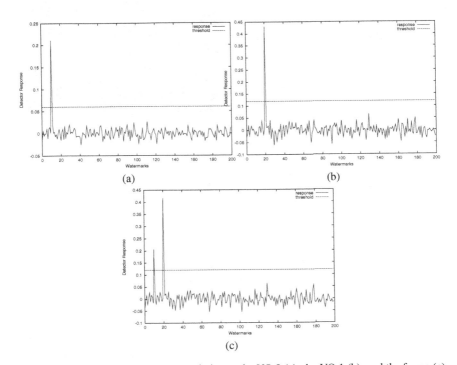

Fig. 2. Watermark detection response relating to the VO 0 (a), the VO 1 (b), and the frame (c).

obtaining the results shown in Figure 4: again, the two objects are clearly detected, even if, as already explained, the correlation peak of the player is lower than the peak of the background.

5 Conclusions

The MPEG-4 standard is revealing very attractive for a wide variety of multimedia applications, where a copy protection system allowing to limit the duplication of multimedia data and to make broadcast monitoring possible is required. A new technology that seems useful for copyright protection is watermarking. With regard to the complexity of the MPEG-4 standard and the diversity of its applications, new constraints are introduced to the watermarking process: in particular, the possibility of the MPEG-4 standard to directly access and manipulate objects within a video sequence implies that even if a video object is transferred from a sequence to another, the copyright data of the single object has to be correctly detected. Another important requirement is that, in order to be

(a)   (b)

(c)

Fig. 3. A frame of the video sequence *Flowers*, previously watermarked (a), the Video Object represented by the player (b), and the new sequence where Stefan seems playing in the flower garden (c).

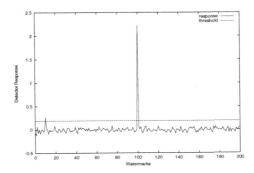

Fig. 4. Detection results of the new video sequence represented in Figure 3 (c).

robust against format conversions, the watermark has to be inserted before compression, i.e. frame by frame.

The method proposed in this paper satisfies the previous requirements: it belongs to the category of raw-video watermarking algorithms, since it operates frame by frame by casting a different watermark in each video object of an MPEG-4 coded video bitstream. Watermarking relies on the image watermarking algorithm presented in [11], which embeds the code in the Discrete Wavelet Transform (DWT) domain. Future work will be dedicated to test the system against a larger set of attacks.

References

1. T. Sikora, "The MPEG-4 video standard verification model," *IEEE Trans. on Cicruits and Systems for Video Technology*, vol. 7, no. 1, February 1997.
2. T. Kalker, "Digital video watermarking," in *Proc. IEEE Internat. Conf. Multimedia Computing and Systems ICMCS'99*, Florence, Italy, June 1999.
3. F. Hartung and B. Girod, "Digital watermarking of raw and compressed video," in *Digital Compression Technologies and Systems for Video Communications*, N. Ohta, Ed., Berlin, Germany, October 1996, Proc. SPIE 2952, pp. 205–213.
4. C. Hsu and J. Wu, "Digital watermarking for video," in *Proc. IEEE Intern. Conf. on Digital Signal Processing*, July 1997, vol. 1, pp. 217–220.
5. M. D. Swanson, B. Zhu, and A. H. Tewfik, "Multiresolution scene based video watermarking using perceptual models," *IEEE Journal of Selected Areas in Communications*, vol. 16, no. 4, May 1998.
6. T. Kalker, G. Depovere, J. Haitsma, and M. Maes, "A video watermarking system for broadcast monitoring," in *Security and Watermarking of Multimedia Contents*, P. W. Wong and E. J. Delp, Eds., San Jose, California, January 1999, Proc. SPIE 3657, pp. 103–112.
7. F. Deguillaume, G. Csurca, J. O'Ruanaidh, and T. Pun, "Robust 3D DFT video watermarking," in *Security and Watermarking of Multimedia Contents*, P. W. Wong and E. J. Delp, Eds., San Jose, California, January 1999, Proc. SPIE 3657, pp. 113–124.
8. F. Hartung and B. Girod, "Digital watermarking of MPEG-2 coded video in the bitstream domain," in *Proc. Internat. Conf. on Acoustic, Speech, & Signal Processing ICASSP'97*, Munich, Germany, April 1997, vol. 4, pp. 2621–2624.

9. G. C. Langelaar, R. L. Lagendijk, and J. Biemond, "Watermarking by DCT coefficient removal: A statistical approach to optimal parameter settings," in *Security and Watermarking of Multimedia Contents*, P. W. Wong and E. J. Delp, Eds., San Jose, California, January 1999, Proc. SPIE 3657, pp. 2–13.

10. M. Barni, F. Bartolini, V. Cappellini, and N. Checcacci, "Object watermarking for mpeg-4 video streams copyright protection," in *Security and Watermarking of Multimedia Contents II, Wong, Delp, Editors, Proceedings of SPIE Vol. 3671*, S. Jose', CA, January 2000.

11. M. Barni, F. Bartolini, V. Cappellini, A. Lippi, and A. Piva, "A dwt-based algorithm for spatio-frequency masking of digital signatures," in *Security and Watermarking of Multimedia Contents, Wong, Delp, Editors, Proceedings of SPIE Vol. 3657*, S. Jose', CA, January 1999, pp. 31–39.

12. N. J. Jayant, J. Johnston, and R. Safranek, "Signal compression based on models of the human perception," *Proc. of the IEEE*, vol. 81, pp. 1385–1422, 1993.

Author Index

Lecture Notes in Computer Science

For information about Vols. 1–2084
please contact your bookseller or Springer-Verlag

Vol. 2124: W. Skarbek (Ed.), Computer Analysis of Images and Patterns. Proceedings, 2001. XV, 743 pages. 2001.

Vol. 2125: F. Dehne, J.-R. Sack, R. Tamassia (Eds.), Algorithms and Data Structures. Proceedings, 2001. XII, 484 pages. 2001.

Vol. 2126: P. Cousot (Ed.), Static Analysis. Proceedings, 2001. XI, 439 pages. 2001.

Vol. 2127: V. Malyshkin (Ed.), Parallel Computing Technologies. Proceedings, 2001. XII, 516 pages. 2001.

Vol. 2129: M. Goemans, K. Jansen, J.D.P. Rolim, L. Trevisan (Eds.), Approximation, Randomization, and Combinatorial Optimization. Proceedings, 2001. IX, 297 pages. 2001.

Vol. 2130: G. Dorffner, H. Bischof, K. Hornik (Eds.), Artificial Neural Networks – ICANN 2001. Proceedings, 2001. XXII, 1259 pages. 2001.

Vol. 2132: S.-T. Yuan, M. Yokoo (Eds.), Intelligent Agents. Specification. Modeling, and Application. Proceedings, 2001. X, 237 pages. 2001. (Subseries LNAI).

Vol. 2136: J. Sgall, A. Pultr, P. Kolman (Eds.), Mathematical Foundations of Computer Science 2001. Proceedings, 2001. XII, 716 pages. 2001.

Vol. 2138: R. Freivalds (Ed.), Fundamentals of Computation Theory. Proceedings, 2001. XIII, 542 pages. 2001.

Vol. 2139: J. Kilian (Ed.), Advances in Cryptology – CRYPTO 2001. Proceedings, 2001. XI, 599 pages. 2001.

Vol. 2141: G.S. Brodal, D. Frigioni, A. Marchetti-Spaccamela (Eds.), Algorithm Engineering. Proceedings, 2001. X, 199 pages. 2001.

Vol. 2142: L. Fribourg (Ed.), Computer Science Logic. Proceedings, 2001. XII, 615 pages. 2001.

Vol. 2143: S. Benferhat, P. Besnard (Eds.), Symbolic and Quantitative Approaches to Reasoning with Uncertainty. Proceedings, 2001. XIV, 818 pages. 2001. (Subseries LNAI).

Vol. 2146: J.H. Silverman (Eds.), Cryptography and Lattices. Proceedings, 2001. VII, 219 pages. 2001.

Vol. 2147: G. Brebner, R. Woods (Eds.), Field-Programmable Logic and Applications. Proceedings, 2001. XV, 665 pages. 2001.

Vol. 2149: O. Gascuel, B.M.E. Moret (Eds.), Algorithms in Bioinformatics. Proceedings, 2001. X, 307 pages. 2001.

Vol. 2150: R. Sakellariou, J. Keane, J. Gurd, L. Freeman (Eds.), Euro-Par 2001 Parallel Processing. Proceedings, 2001. XXX, 943 pages. 2001.

Vol. 2151: A. Caplinskas, J. Eder (Eds.), Advances in Databases and Information Systems. Proceedings, 2001. XIII, 381 pages. 2001.

Vol. 2152: R.J. Boulton, P.B. Jackson (Eds.), Theorem Proving in Higher Order Logics. Proceedings, 2001. X, 395 pages. 2001.

Vol. 2153: A.L. Buchsbaum, J. Snoeyink (Eds.), Algorithm Engineering and Experimentation. Proceedings, 2001. VIII, 231 pages. 2001.

Vol. 2154: K.G. Larsen, M. Nielsen (Eds.), CONCUR 2001 – Concurrency Theory. Proceedings, 2001. XI, 583 pages. 2001.

Vol. 2157: C. Rouveirol, M. Sebag (Eds.), Inductive Logic Programming. Proceedings, 2001. X, 261 pages. 2001. (Subseries LNAI).

Vol. 2158: D. Shepherd, J. Finney, L. Mathy, N. Race (Eds.), Interactive Distributed Multimedia Systems. Proceedings, 2001. XIII, 258 pages. 2001.

Vol. 2159: J. Kelemen, P. Sosík (Eds.), Advances in Artificial Life. Proceedings, 2001. XIX, 724 pages. 2001. (Subseries LNAI).

Vol. 2161: F. Meyer auf der Heide (Ed.), Algorithms – ESA 2001. Proceedings, 2001. XII, 538 pages. 2001.

Vol. 2162: Ç. K. Koç, D. Naccache, C. Paar (Eds.), Cryptographic Hardware and Embedded Systems – CHES 2001. Proceedings, 2001. XIV, 411 pages. 2001.

Vol. 2164: S. Pierre, R. Glitho (Eds.), Mobile Agents for Telecommunication Applications. Proceedings, 2001. XI, 292 pages. 2001.

Vol. 2165: L. de Alfaro, S. Gilmore (Eds.), Process Algebra and Probabilistic Methods. Proceedings, 2001. XII, 217 pages. 2001.

Vol. 2166: V. Matoušek, P. Mautner, R. Mouček, K. Taušer (Eds.), Text, Speech and Dialogue. Proceedings, 2001. XIII, 452 pages. 2001. (Subseries LNAI).

Vol. 2170: S. Palazzo (Ed.), Evolutionary Trends of the Internet. Proceedings, 2001. XIII, 722 pages. 2001.

Vol. 2172: C. Batini, F. Giunchiglia, P. Giorgini, M. Mecella (Eds.), Cooperative Information Systems. Proceedings, 2001. XI, 450 pages. 2001.

Vol. 2176: K.-D. Althoff, R.L. Feldmann, W. Müller (Eds.), Advances in Learning Software Organizations. Proceedings, 2001. XI, 241 pages. 2001.

Vol. 2177: G. Butler, S. Jarzabek (Eds.), Generative and Component-Based Software Engineering. Proceedings, 2001. X, 203 pages. 2001.

Vol. 2181: C. Y. Westort (Eds.), Digital Earth Moving. Proceedings, 2001. XII, 117 pages. 2001.

Vol. 2184: M. Tucci (Ed.), Multimedia Databases and Image Communication. Proceedings, 2001. X, 225 pages. 2001.

Vol. 2186: J. Bosch (Ed.), Generative and Component-Based Software Engineering. Proceedings, 2001. VIII, 177 pages. 2001.

Vol. 2188: F. Bomarius, S. Komi-Sirviö (Eds.), Product Focused Software Process Improvement. Proceedings, 2001. XI, 382 pages. 2001.

Vol. 2189: F. Hoffmann, D.J. Hand, N. Adams, D. Fisher, G. Guimaraes (Eds.), Advances in Intelligent Data Analysis. Proceedings, 2001. XII, 384 pages. 2001.

Vol. 2190: A. de Antonio, R. Aylett, D. Ballin (Eds.), Intelligent Virtual Agents. Proceedings, 2001. VIII, 245 pages. 2001. (Subseries LNAI).

Vol. 2191: B. Radig, S. Florczyk (Eds.), Pattern Recognition. Proceedings, 2001. XVI, 452 pages. 2001.

Vol. 2193: F. Casati, D. Georgakopoulos, M.-C. Shan (Eds.), Technologies for E-Services. Proceedings, 2001. X, 213 pages. 2001.